AA002913

Metabolic Engineering 11

Awaji Island, Japan
26 - 30 June 2016

ISBN: 978-1-5108-3211-4

Printed from e-media with permission by:

Curran Associates, Inc.
57 Morehouse Lane
Red Hook, NY 12571

Printed by Curran Associates, Inc. (2017)

AIChE
120 Wall Street, FL 23
New York, NY 10005-4020

Phone: (800) 242-4363
Fax: (203) 775-5177

www.aiche.org

Additional copies of this publication are available from:

Curran Associates, Inc.
57 Morehouse Lane
Red Hook, NY 12571 USA
Phone: 845-758-0400
Fax: 845-758-2633
Email: curran@proceedings.com
Web: www.proceedings.com

TABLE OF CONTENTS

KEYNOTE PRESENTATIONS

SESSION 1: METABOLIC ENGINEERING FOR FUELS AND CHEMICALS

SESSION 2: MOLECULAR PARTS AND METABOLIC ENGINEERING

SESSION 3: SYNTHETIC BIOLOGY AND METABOLIC ENGINEERING

SESSION 4: COMPUTATIONAL TOOLS AND METHODS FOR METABOLIC ENGINEERING

SESSION 5: METABOLIC ENGINEERING FOR CHEMICALS AND MATERIALS

SESSION 6: METABOLIC ENGINEERING OF INDUSTRIAL MICROORGANISMS

SESSION 7: OMICS SCIENCE AND METABOLIC ENGINEERING

SESSION 8: METABOLIC ENGINEERING FOR CELL CULTURE AND MEDICAL APPLICATIONS

SESSION 9: DESIGN AND ASSEMBLY OF METABOLIC PATHWAYS

SESSION 10: INDUSTRIAL APPLICATIONS OF METABOLIC ENGINEERING

SESSION 11A: METABOLIC ENGINEERING: METHODS AND APPLICATION A

SESSION 11B: METABOLIC ENGINEERING: METHODS AND APPLICATION B

SESSION 11C: METABOLIC ENGINEERING: METHODS AND APPLICATION C

RAPID FIRE POSTER SESSION 1

POSTER SESSION 1

RAPID FIRE POSTER SESSION 2

POSTER SESSION 2

RAPID FIRE POSTER SESSION 3

POSTER SESSION 3

Metabolic Engineering of Yeast

Nielsen, J., Chalmers University of Technology

Originally presented on: 6/26/2016 16:00:00 - 16:55:00

Metabolic Engineering relies on the Design-Build-Test cycle. This cycle includes technologies like mathematical modeling of metabolism, genome editing and advanced tools for phenotypic characterization. In recent years there have been advances in several of these technologies, which has enabled faster development of metabolically engineered strains that can be used for production of fuels and chemicals.

The yeast *Saccharomyces cerevisiae* is widely used for production of fuels, chemicals, pharmaceuticals and materials. Through metabolic engineering of this yeast a number of novel industrial processes have been developed over the last 10 years. Besides its wide industrial use, *S. cerevisiae* also serves as an eukaryal model organism, and many systems biology tools have therefore been developed for this organism. These tools can be used for detailed phenotypic characterization as well as for metabolic design.

In this lecture it will be demonstrated how the Design-Build-Test cycle of Metabolic Engineering has allowed for development of yeast cell factories for production of a range of different fuels and chemicals. Some examples of different technologies will be presented together with examples of metabolic engineering designs, in particular for development of platform strains that can be used for production of a fatty acid derived products, e.g. fatty alcohols and alkanes. It will be argued that with advancement in genome-editing technologies and novel methods for rapid phenotypic screening, advancement in the field is hampered by our design abilities, i.e. to predict genotype-phenotype connections. For this genome-scale metabolic models is a strong technology, and in the presentation recent advancements in mathematical modeling for cell factory design will be presented. Finally, the presentation will also demonstrate how the Design-Build-Test cycle can be expanded to incorporate adaptive laboratory evolution to identify targets for engineering complex traits, such as improved tolerance to toxic metabolites like elevated temperatures or low pH.

Amino Acids Production and Metabolic Engineering

Kojima, H., Ajinomoto Co., Inc.

Originally presented on: 6/26/2016 16:55:00 - 17:50:00

Industrial amino acids production has more than 100 years of history. L-Glutamate was firstly produced for the umami taste substance from hydrolyzed wheat protein. However, this production method was not so appropriate from the view point of equipment and huge environmental burden. Production of low content amino acids in the protein or acid sensitive amino acids such as tryptophan was difficult to produce. Thus, chemical synthesis method was developed for the production of various kinds of amino acids including glutamate. However, most of the amino acids have optical isomer(s) and usually specific isomer, L-amino acid, is required. Separation of specific isomer from racemic mixture required significant production costs.

Around 60 years ago, microbial fermentation production of L-glutamate started after the discovery of L-glutamate producing bacteria in Japan. Promptly after the development of L-glutamate fermentative production method, producing strains of various kinds of amino acids, L-lysine, L-glutamine, L-threonine and so on, have been developed. The development of production technologies also enhanced the expansion of the application of amino acids in many fields such as animal nutritions, pharmaceuticals, and cosmetics. Production technology of amino acids was based on the deepening knowledge of various kinds of research areas; biochemistry, molecular biology, recombinant DNA technology and metabolic engineering. Amino acids production from the view point of metabolic engineering will be reviewed and future expectation to the metabolic engineering will be presented.

How (Not) to Force a Cell to Eat and Produce

Kinetically Accessible Yield (KAY)
Synthetic methanol auxotrophy

James C. Liao
Chemical and Biomolecular Engineering, UCLA
Chemistry and Biochemistry, UCLA
UCLA-DOE Institute for Genomics and Proteomics
BioEnergy Science Center

UCLA-DOE Institute for Genomics and Proteomics — UCLA — BESC BioEnergy Science Center

0

Difficulties in Metabolic Engineering

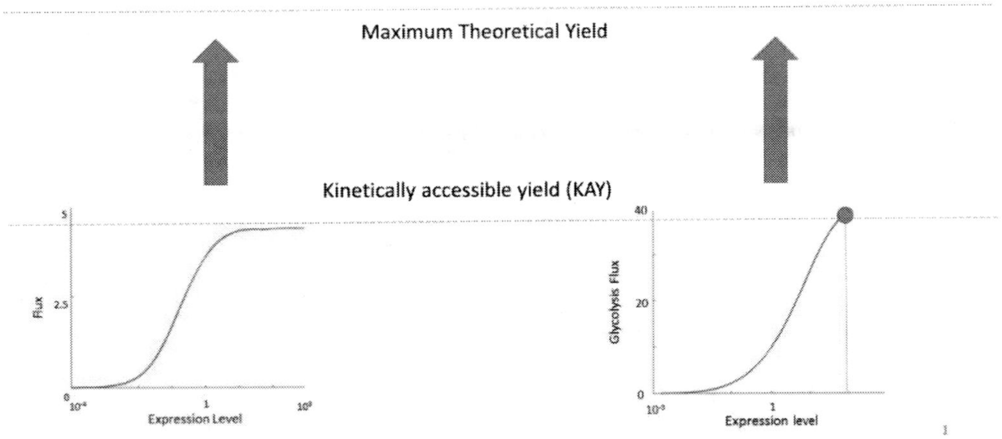

1

Stability bifurcation may be an issue

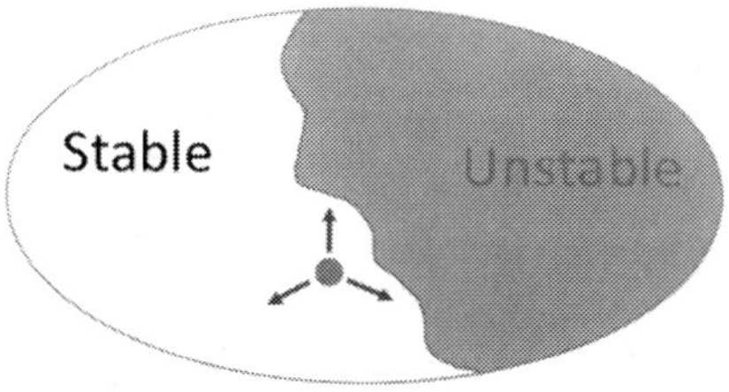

2

A toy kinetic model

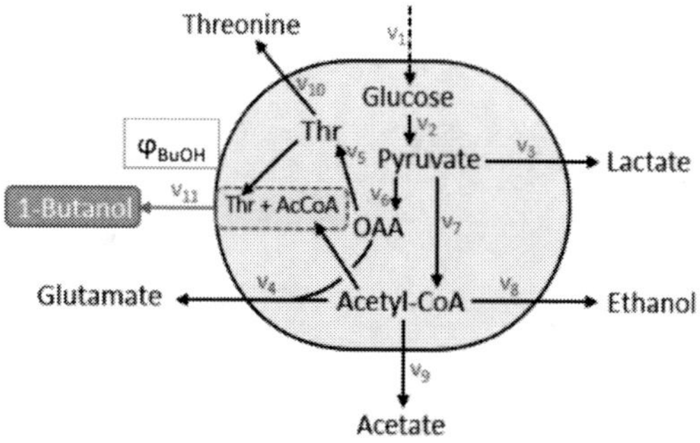

$$\frac{k_{14}}{\frac{k_{15}}{X_{Thr}} + \frac{k_{16}}{X_{AcCoA}} + \frac{k_{15}k_{16}}{X_{Thr}X_{AcCoA}} + 1}$$

Reaction Symbol or Description	Rate Law or Relationship	Reaction Stoichiometry
V1: Gluc_In	$\frac{k_{23}\left(1-\frac{X_{Gluc}}{k_{24}}\right)}{k_{25}\left(\frac{1}{k_{25}}+\frac{X_{Gluc}}{k_{26}}+1\right)}$	(Glucose out)<-> Glucose (rev, in)
V2: Glyc	$\frac{k_1}{\frac{k_2}{X_{Gluc}}+1}$	Glucose -> 2 Pyruvate
V3: Ldh	$\frac{k_3}{\frac{k_4}{X_{Pyr}}+1}$	Pyruvate -> (lactate out)
V4: GlutSyn	$\frac{k_5}{\frac{k_6}{X_{OAA}}+\frac{k_7}{X_{AcCoA}}+\frac{k_6k_7}{X_{OAA}X_{AcCoA}}+1}$	OAA + AcCoA -> (glutamate out)
V5: ThrSyn	$\frac{k_8}{\frac{k_9}{X_{OAA}}+1}$	OAA -> Thr
V6: Pyc	$\frac{k_{10}}{\frac{k_{11}}{X_{Pyr}}+1}$	Pyruvate -> OAA
V7: Pdh	$\frac{k_{12}}{\frac{k_{13}}{X_{Pyr}}+1}$	Pyruvate -> AcCoA
V8: Adh	$\frac{k_{17}}{\frac{k_{18}}{X_{AcCoA}}+1}$	AcCoA -> (ethanol out)
V9: AcetS	$\frac{k_{19}}{\frac{k_{20}}{X_{AcCoA}}+1}$	AcCoA -> (acetate out)
V10: Thr_Out	$\frac{k_{21}}{\frac{k_{22}}{X_{Thr}}+1}$	Thr -> (threonine out)
*V11: ButSyn	$\frac{k_{14}}{\frac{k_{15}}{X_{Thr}}+\frac{k_{16}}{X_{AcCoA}}+\frac{k_{15}k_{16}}{X_{Thr}X_{AcCoA}}+1}$	Thr + AcCoA -> (butanol out)
Φ_{BuOH}: BuOH_Flux_Only	$k_{27,1}$	Thr + AcCoA -> (butanol out)
Φ_{Gly}: Glyc_Flux_Only	$k_{27,2}$	Glucose -> 2 Pyruvate
ThrSyn Repression	$\frac{\Delta k_8}{k_8} = -4 \cdot 10^{-4}\left(\frac{\Delta k_{14}}{k_{14}}\right)$	Repression relationship such that as k_{14} increases 1000-fold, k_8 decreases 40%.

3

Kinetically Accessible Yield (KAY)

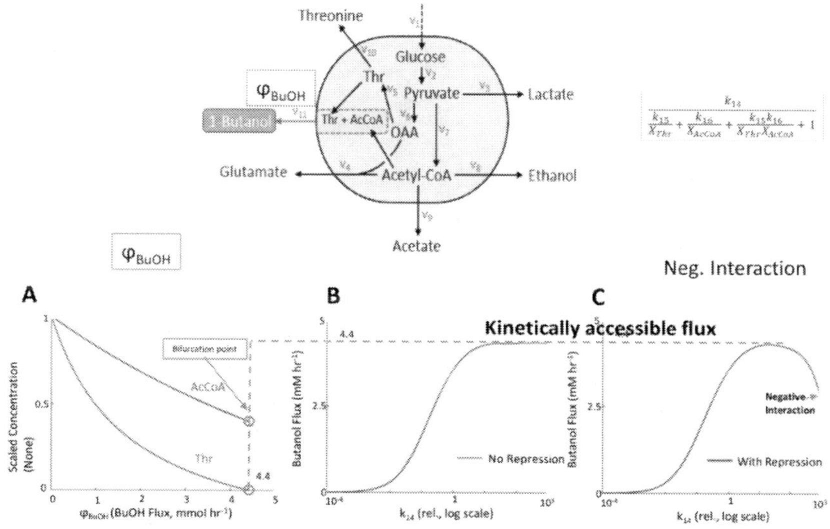

4

Kinetically Accessible Yield (KAY)

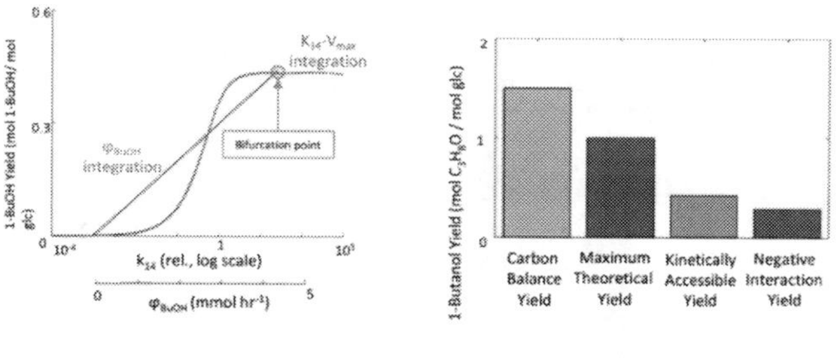

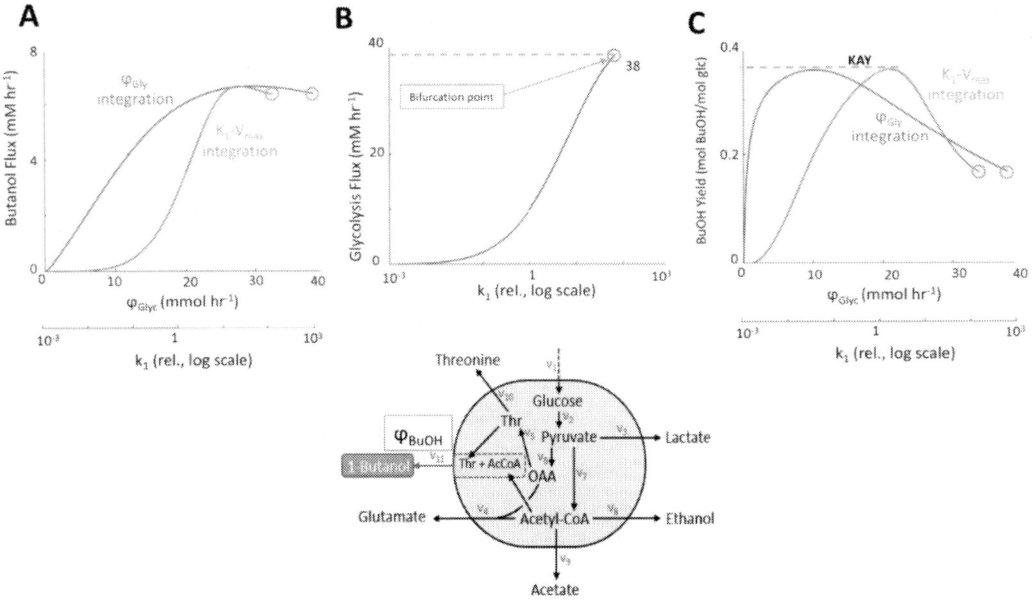
A
B
C
Butanol Flux (mM hr⁻¹)
φGly integration
K1-Vmax integration
φGlyc (mmol hr⁻¹)
k1 (rel., log scale)
Glycolysis Flux (mM hr⁻¹)
Bifurcation point
38
BuOH Yield (mol BuOH/mol glc)
KAY
Threonine
Glucose
Thr
Pyruvate
Lactate
φBuOH
1-Butanol
Thr + AcCoA
OAA
Glutamate
Acetyl-CoA
Ethanol
Acetate

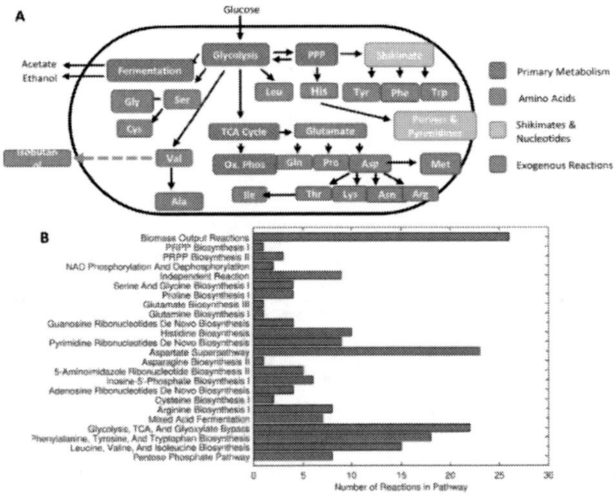
A
Glucose
Acetate
Ethanol
Fermentation
Glycolysis
PPP
TCA Cycle
Glutamate
Primary Metabolism
Amino Acids
Shikimates & Nucleotides
Exogenous Reactions
B
Number of Reactions in Pathway

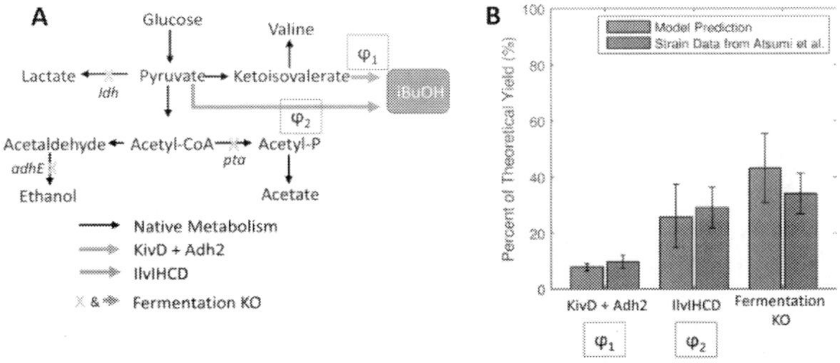

Engineer *E. coli* for methanol conversion

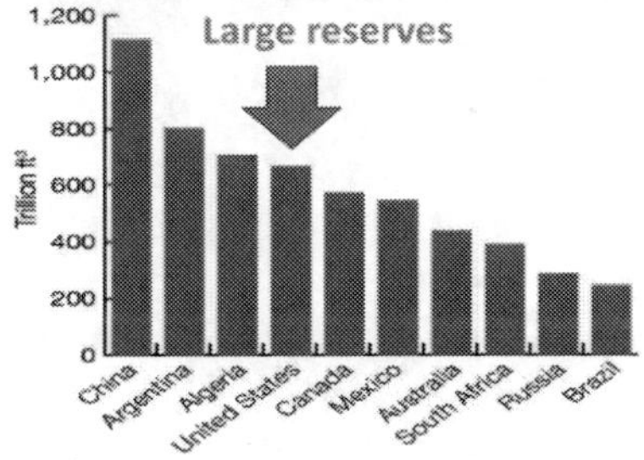

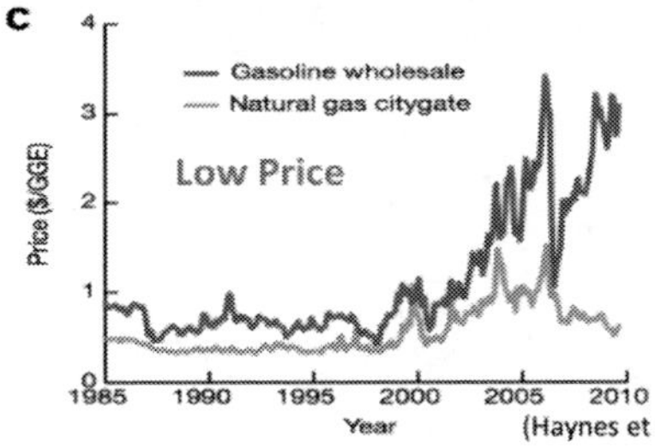

(Haynes et al., 2014)

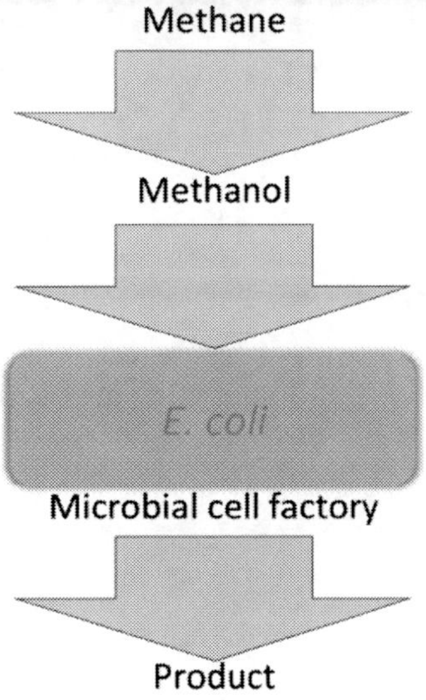

9

Natural Methylotrophy

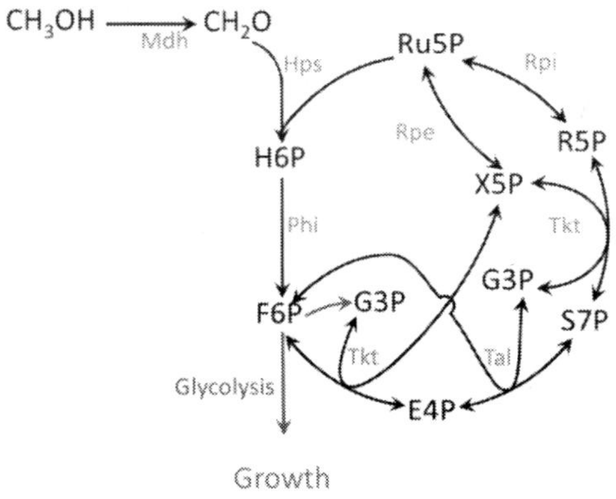

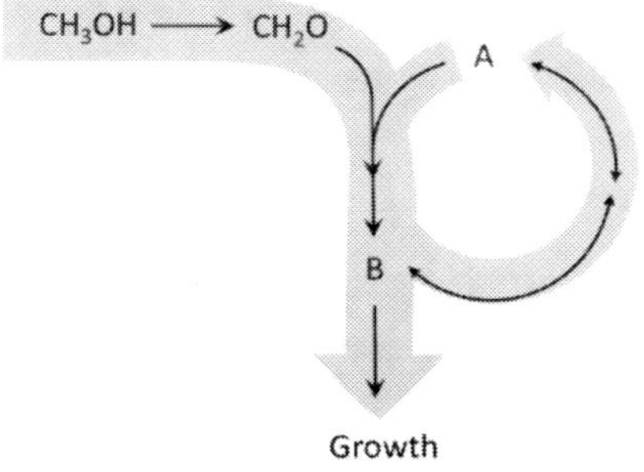

10

Contents lists available at ScienceDirect

Metabolic Engineering

journal homepage: www.elsevier.com/locate/ymben

Engineering *Escherichia coli* for methanol conversion

Jonas E.N. Müller [a], Fabian Meyer [a], Boris Litsanov [a], Patrick Kiefer [a], Eva Potthoff [a],
Stéphanie Heux [b,c,d], Wim J. Quax [e], Volker F. Wendisch [f], Trygve Brautaset [g,h],
Jean-Charles Portais [b,c,d], Julia A. Vorholt [a,*]

AEM

Journals.ASM.org

Metabolic Engineering of *Corynebacterium glutamicum* for Methanol Metabolism

Sabrina Witthoff, Katja Schmitz, Sebastian Niedenführ, Katharina Nöh, Stephan Noack, Michael Bott, Jan Marienhagen

Institute of Bio- and Geosciences, IBG-1: Biotechnology, Forschungszentrum Jülich, Jülich, Germany

11

Synthetic Methanol Auxotrophy

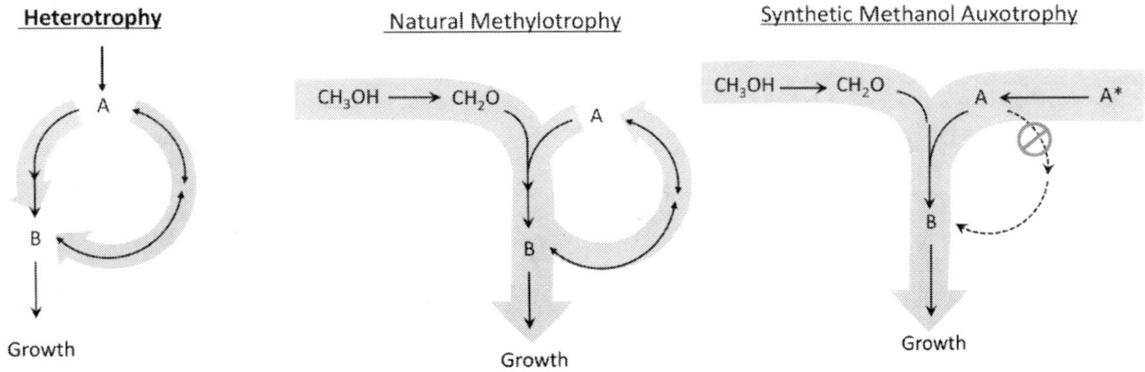

12

Design of synthetic methanol auxotrophy: Δ*rpe*

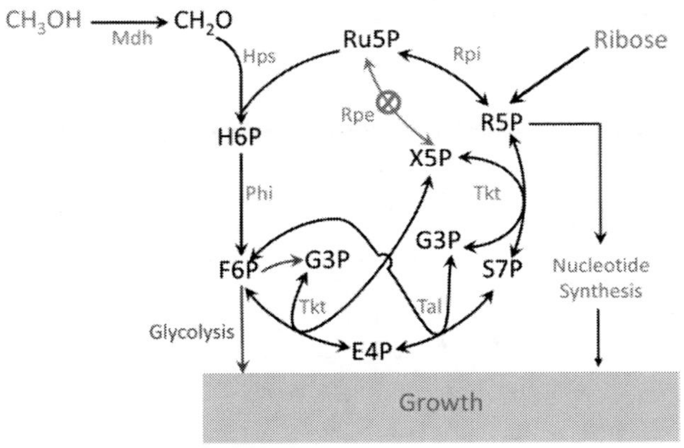

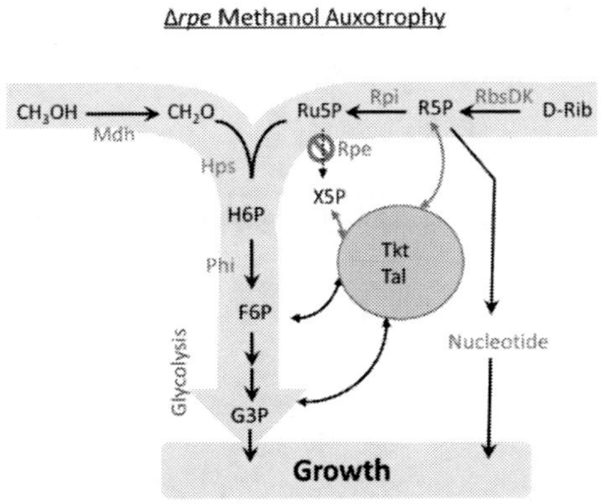

Design of synthetic methanol auxotrophy : Δ*rpiAB*

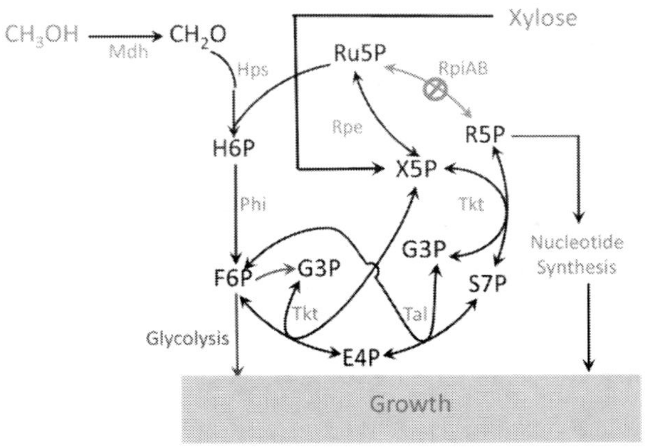

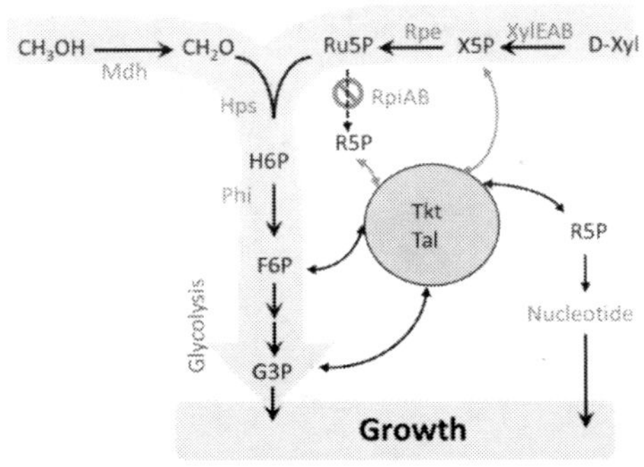

Growth Phenotype of Δ*rpe* strain (IB405)

Expected Growth Phenotype

- Ribose.........no growth
- Xylose.........no growth

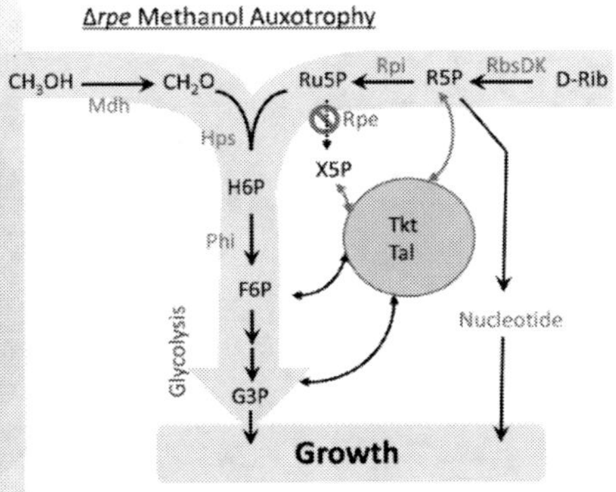

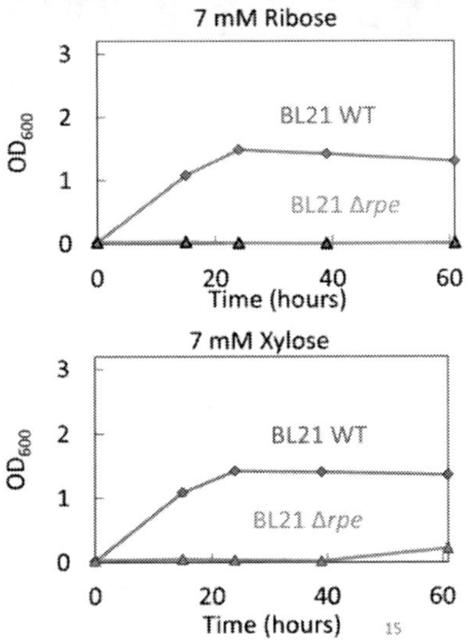

15

Growth Phenotype of Δ*rpe* strain (IB405)

Expected Growth Phenotype

- Glucose............growth

Glucose Growth

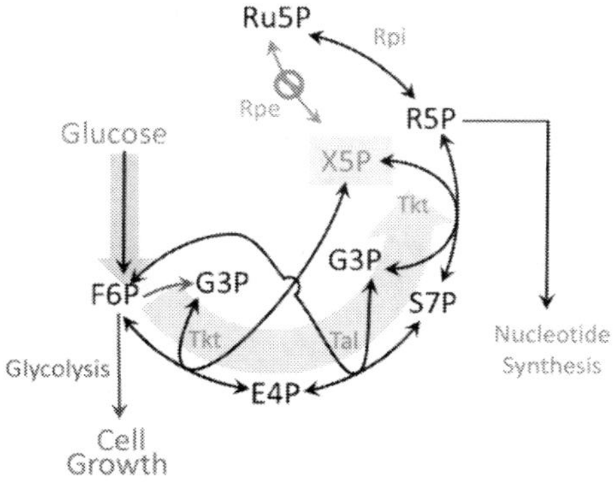

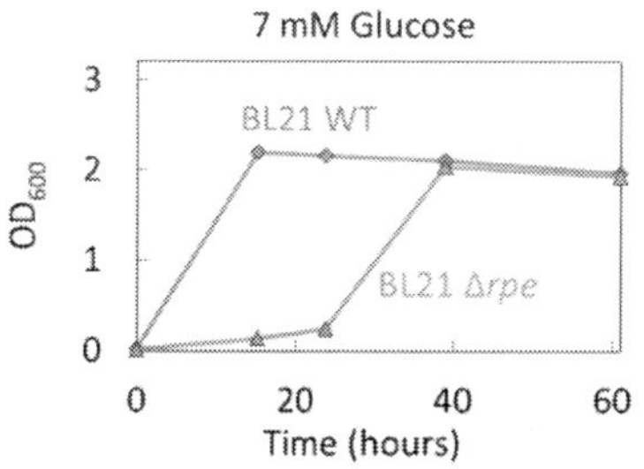

- *rpe* KO might result in accumulation of X5P
- Result in enzyme inhibition or unfavorable regulation

16

Growth Phenotype of Δ*rpe* strain (IB405)

Expected Growth Phenotype

- Rib+Xyl.............growth

Xyl+Rib Growth

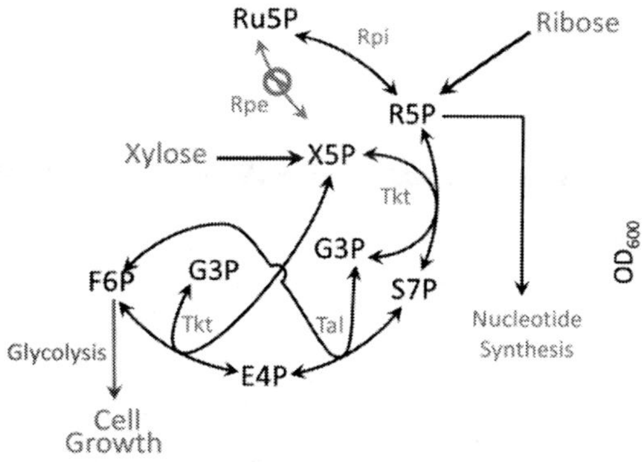

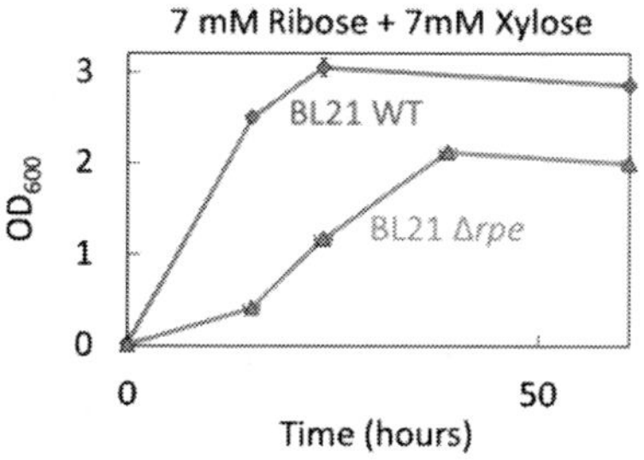

- Deficient conversion of X5P to R5P: Cells need X5P and R5P in a perfect 1:1 ratio to grow: perturbation in any one of them can cause more severe growth retardation.

17

17

Loophole of Δrpe strain

Loophole of Δrpe strain may be developed during competent cell preparation

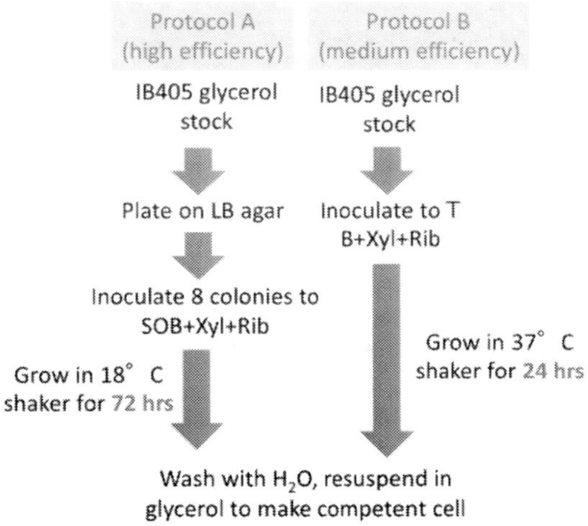

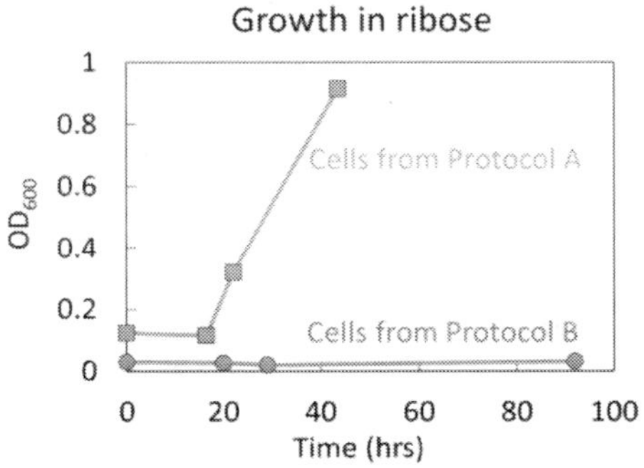

- Protocol B can be used for enrichment of a small library (~10^5 variants based on the transformation efficiency).
- Further study is required for the Δrpe loophole developed.

Study of Δ*rpe* loophole strain

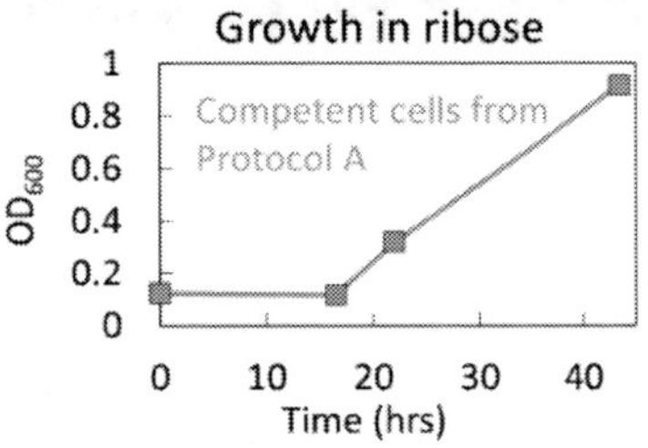

Streak out and separate 3 colonies
Repeat growth in ribose

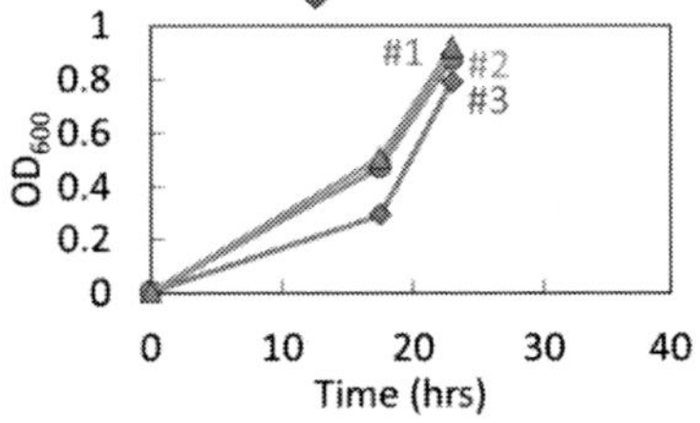

Literature indicates two possible genes encoded enzymes with rpe activity: *sgcE* (predicted epimerase) and *alsE* (allulose-6-phosphate 3-epimerase)

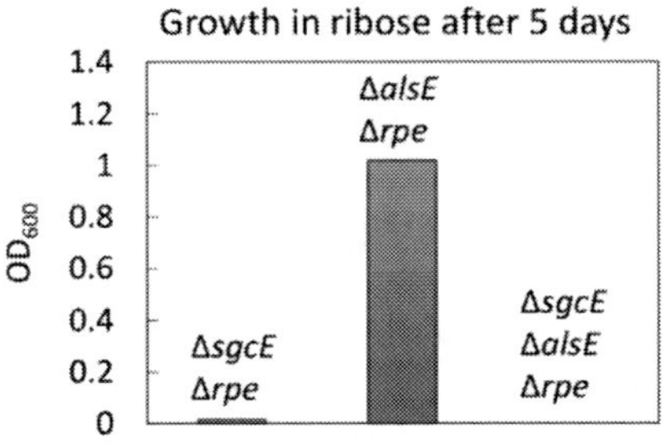

Further experiments are ongoing verifying the absence of loophole in Δ*sgcE*, Δ*rpe* strain.

19

Design of synthetic auxotrophy : Δ*rpiAB*

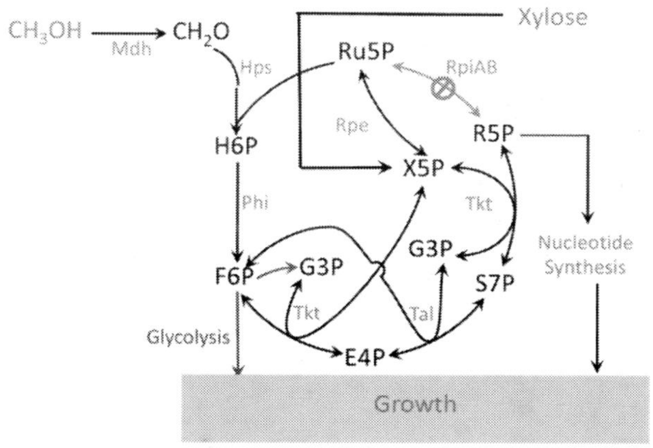

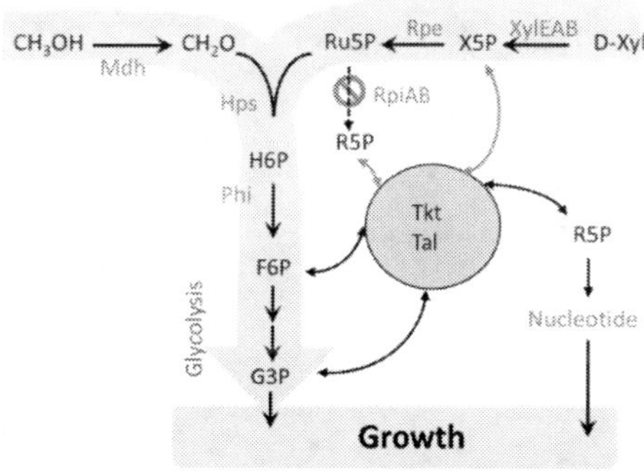

Growth Phenotype of unevolved ΔrpiAB strain

Expected Growth Phenotype

- Ribose.........no growth
- Xylose.........no growth

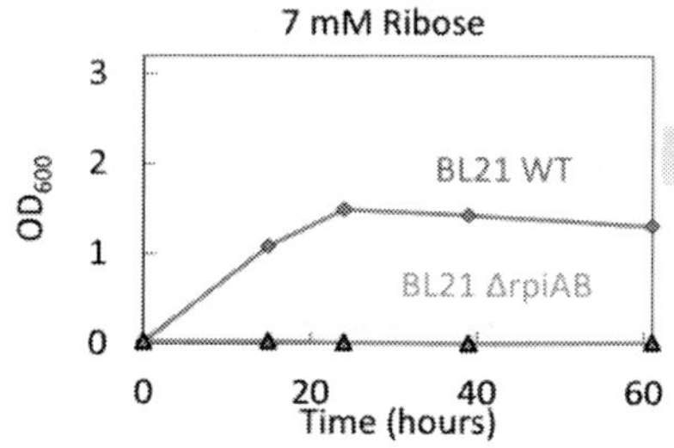

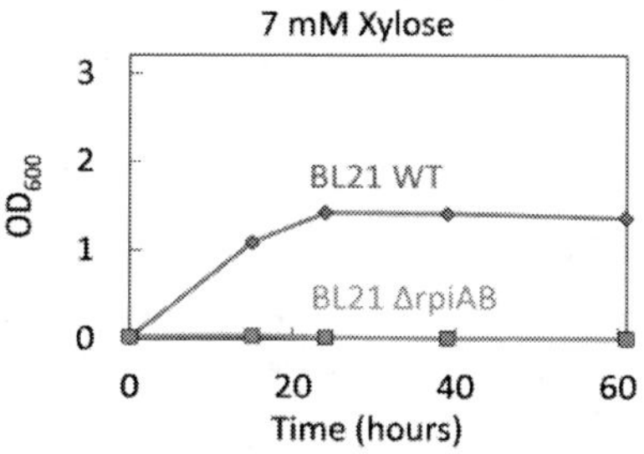

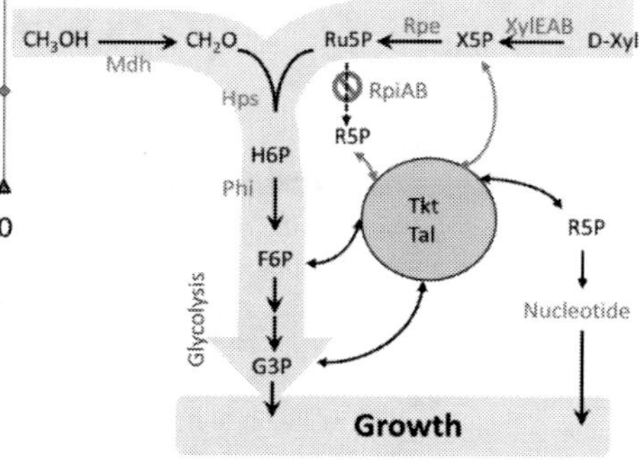

21

Growth Phenotype of unevolved ΔrpiAB strain

Expected Growth Phenotype

- Glucose............growth
- Ribose.........no growth
- Xylose.........no growth
- Rib+Xyl............growth

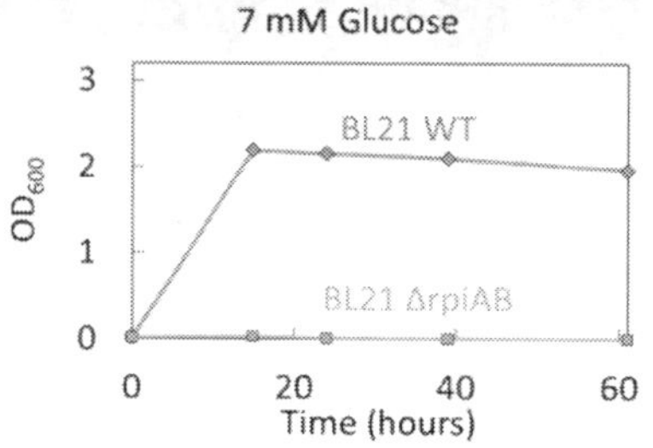

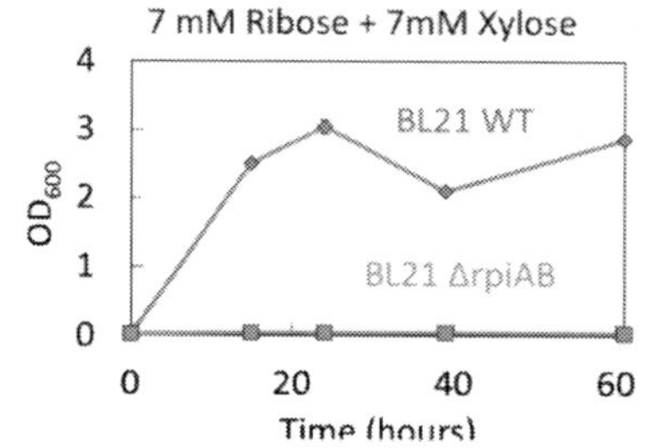

Possible Explanations of no growth in glucose and Rib+Xyl

Glucose Growth

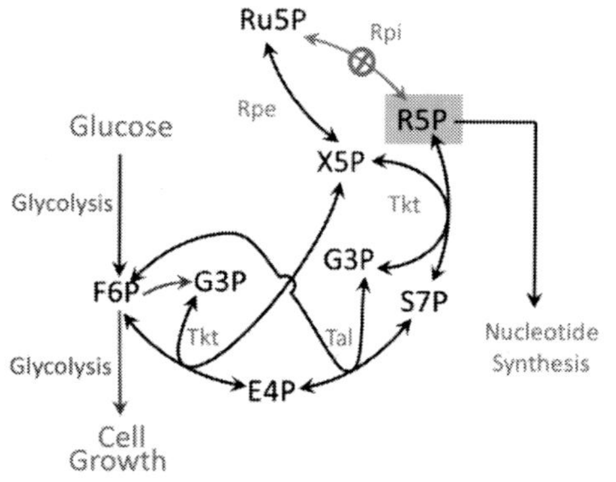

- *rpiAB* KO might result in accumulation of Ru5P
- Insufficient R5P for nucleotide synthesis (critical for cell growth as well)

Xyl+Rib Growth

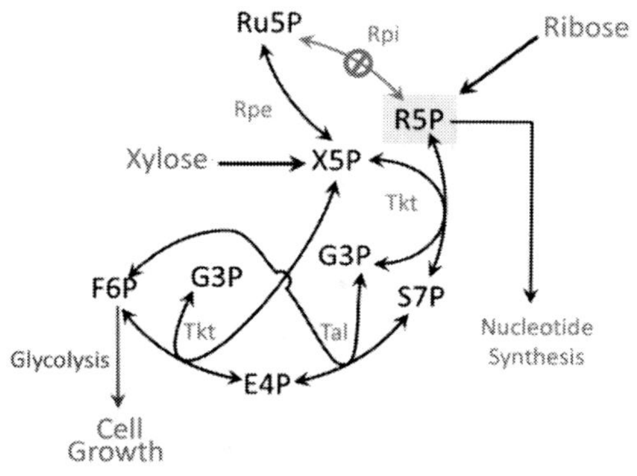

- Deficient conversion & imbalanced utilization of X5P to R5P
- Cells are not used to nucleotide synthesis via direct Ribose uptake

Construction of synthetic methanol auxotrophy

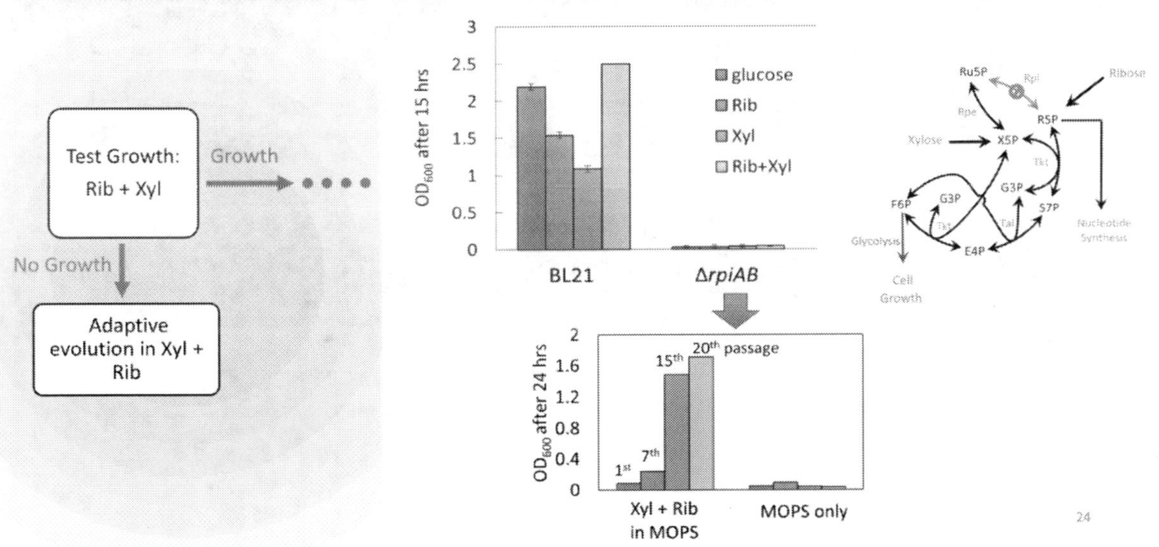

Ensemble Modeling for Robustness Analysis (EMRA)

ΔrpiAB methanol+ xylose growth

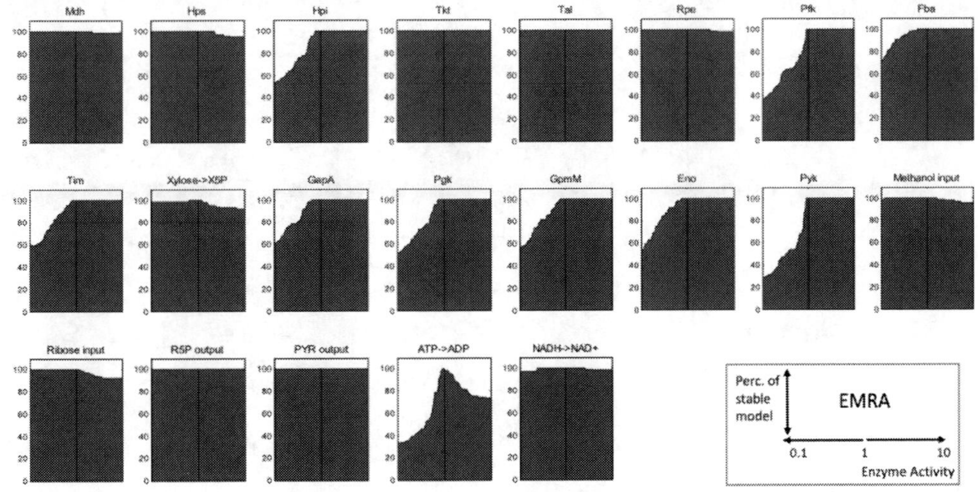

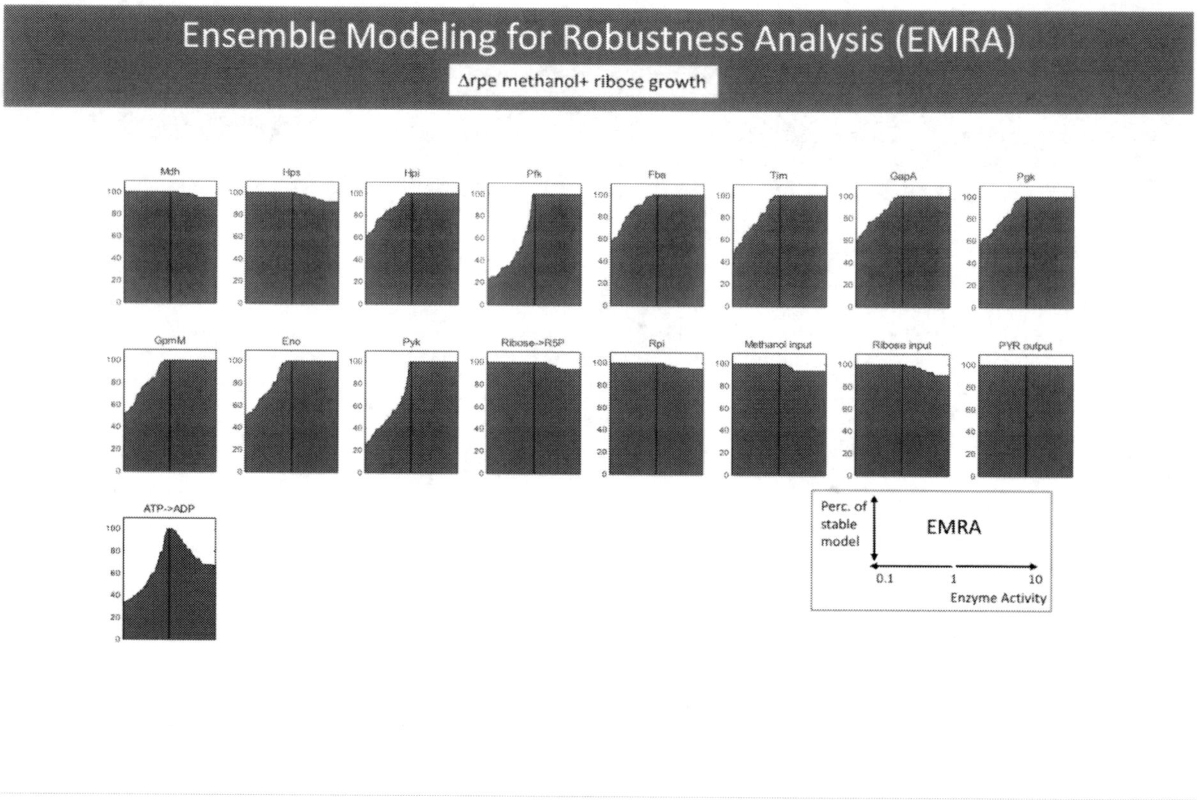
Ensemble Modeling for Robustness Analysis (EMRA)
Δrpe methanol+ ribose growth
Mdh
Hps
Hpi
Pfk
Fba
Tim
GapA
Pgk
GpmM
Eno
Pyk
Ribose->R5P
Rpi
Methanol input
Ribose input
PYR output
ATP->ADP
Perc. of stable model
EMRA
0.1
1
10
Enzyme Activity

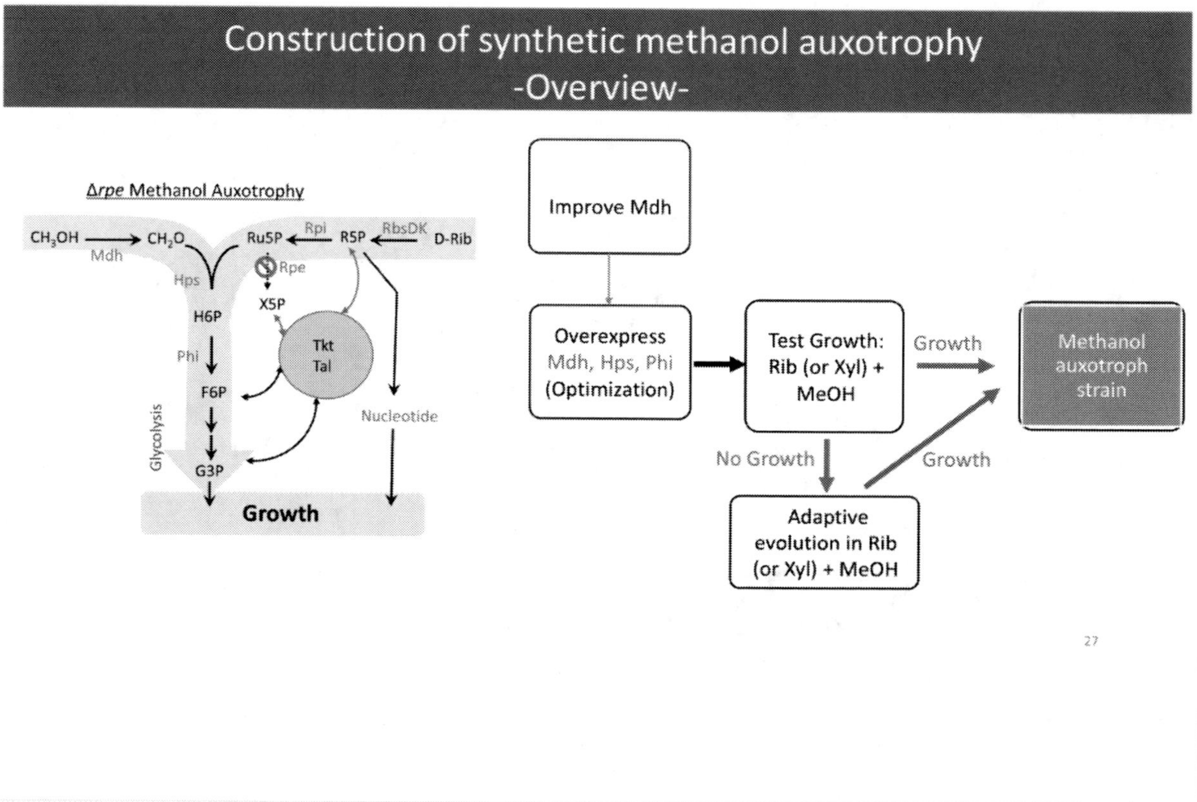
Construction of synthetic methanol auxotrophy
-Overview-
Δrpe Methanol Auxotrophy
CH3OH
Mdh
CH2O
Hps
Ru5P
Rpi
R5P
RbsDK
D-Rib
Rpe
X5P
H6P
Phi
F6P
Tkt
Tal
Nucleotide
Glycolysis
G3P
Growth
Improve Mdh
Overexpress
Mdh, Hps, Phi
(Optimization)
Test Growth:
Rib (or Xyl) +
MeOH
Growth
Methanol
auxotroph
strain
No Growth
Growth
Adaptive
evolution in Rib
(or Xyl) + MeOH
27

Activator (ACT) independent methanol dehydrogenase, Mdh2 from *C. necator* N-1

PQQ-dependent Mdhs
NAD-dependent Mdhs
(group III Adh).

Bacillus methanolicus

Corynebacterium glutamicum

Cupriavidus necator N-1

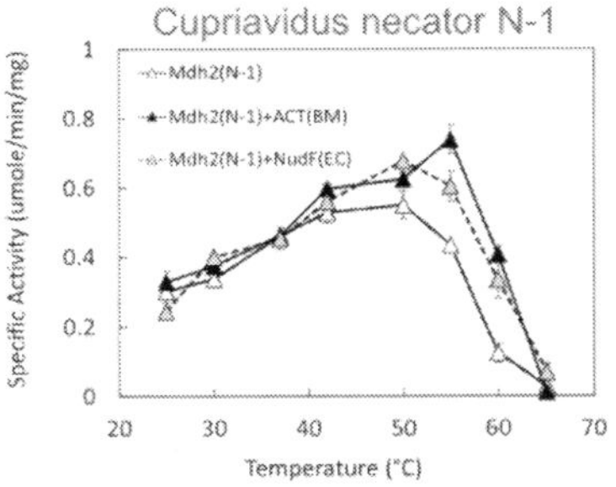

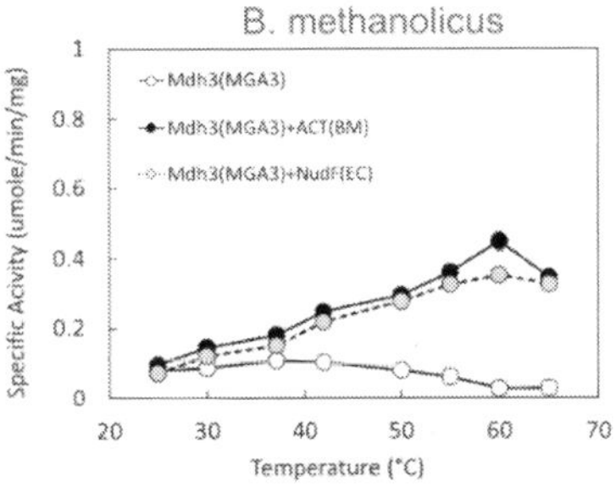

Substrate	K_m (mM)	K_{cat} (s^{-1})	K_{cat}/K_m ($M^{-1} s^{-1}$)	V_{max} (U/mg)	V_{max} (U/mg) (MGA3 Mdh3, 45 °C)
Methanol	132.1±15.4	0.22±0.01	1.6	0.32	0.07
Ethanol	0.77±0.1	11.1±0.3	14483	16	1.3
Propanol	0.54±0.1	9.6±0.2	17759	14.2	2.8
n-butanol	7.2±1	6.5±0.2	906	9.6	2.6
NAD^+	2.1±0.2	0.36±0.02	170	0.54	-
$NADP^+$	N.D.	N.D.	N.D.	N.D.	-

Mdh High-throughput screening (HTS)

Colorimetric Nash reaction for Mdh activity assay

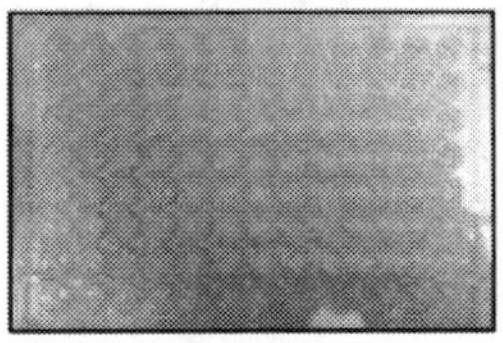

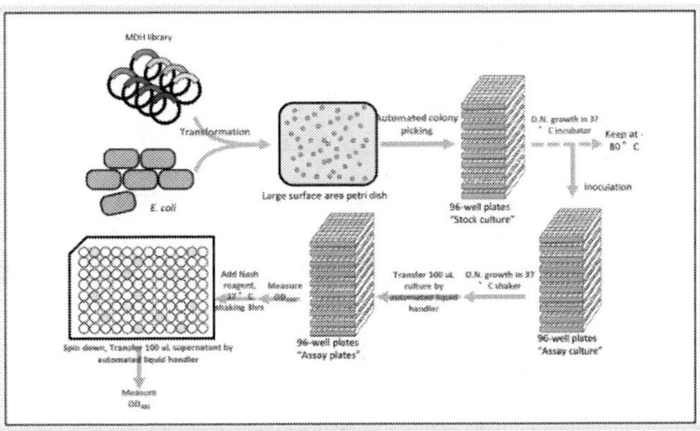

System optimization

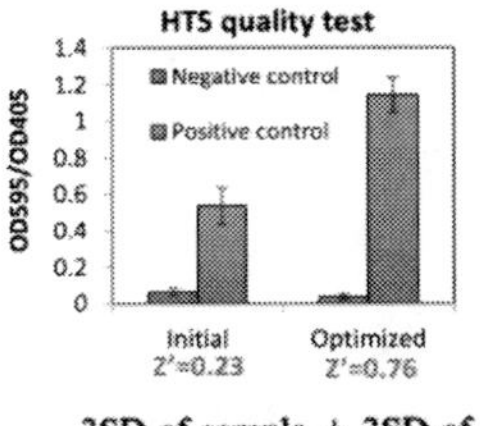

$$Z = 1 - \frac{3SD\ of\ sample + 3SD\ of\ control}{|mean\ of\ sample - mean\ of\ control|}$$

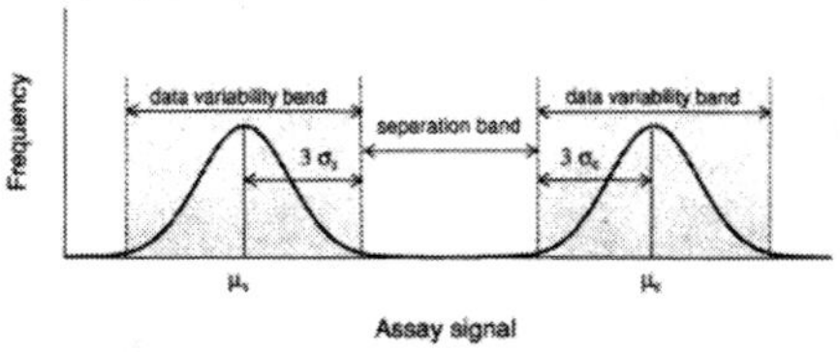

29

Kinetic Parameters of engineered MDH2 variants to methanol and n-butanol.

Mdh2 variant	Library	Mutations	Nash activity	methanol K_m (mM)	methanol K_{cat} (s^{-1})	methanol K_{cat}/K_m ($M^{-1} s^{-1}$)	*n*-butanol K_m (mM)	*n*-butanol K_{cat} (s^{-1})	*n*-butanol K_{cat}/K_m ($M^{-1} s^{-1}$)
WT	-	-	**1.00**	**132±15.4**	**0.22±0.0**	**1.6**	**7.2±2.1**	**6.5±0.1**	**903**
CT1-1	Round 1	A31V	1.45	129±11.9	0.25±0.0	1.9	11.3±1.3	7.2±0.2	637
CT1-2	Round 1	A169V	1.47	26.9±2.7	0.13±0.0	4.8	66.5±11.8	3.3±0.1	50
CT2-1	Round 2	A26V, A169V	2.96	30.7±3.6	0.21±0.0	6.8	75.9±1.1	5.0±0.2	66
CT2-2	-	A26V	ND	ND	ND	ND	ND	ND	ND
CT4-1	Recombination	A26V, A31V, A169V	3.46	21.6±1.5	0.20±0.0	9.3	120±17.9	5.7±0.4	48
CT5-1	Round 3	F279L	3.46	58.2	0.28	4.8	-	-	-

30

Alcohol specificity of MDH2 and its engineered variants

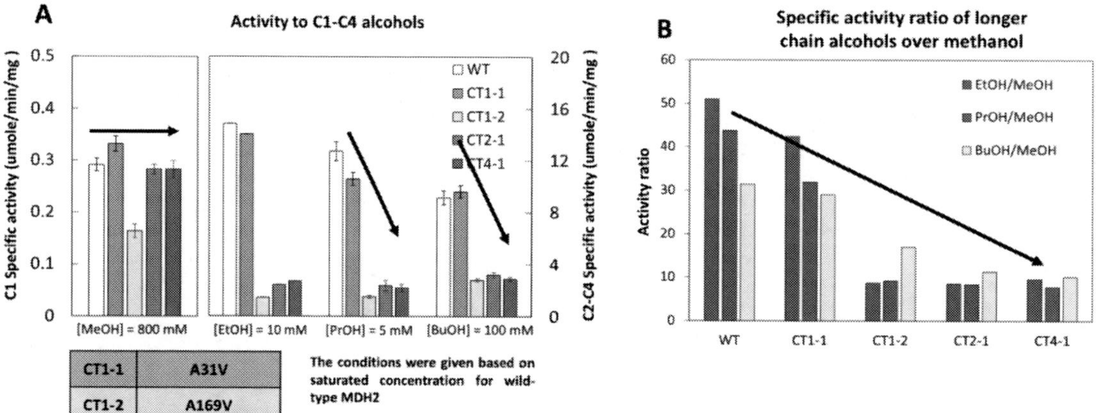

CT1-1	A31V
CT1-2	A169V
CT2-1	A26V, A169V
CT4-1	A26V, A31V, A169V

The conditions were given based on saturated concentration for wild-type MDH2

31

Construction of synthetic methanol auxotrophy -Overexpress Mdh, Hps, Phi-

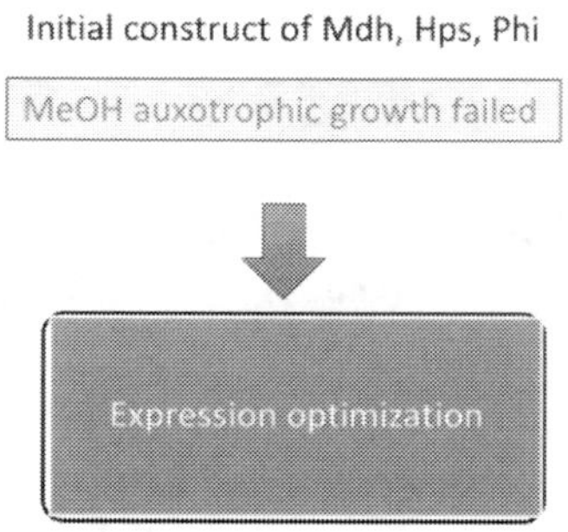

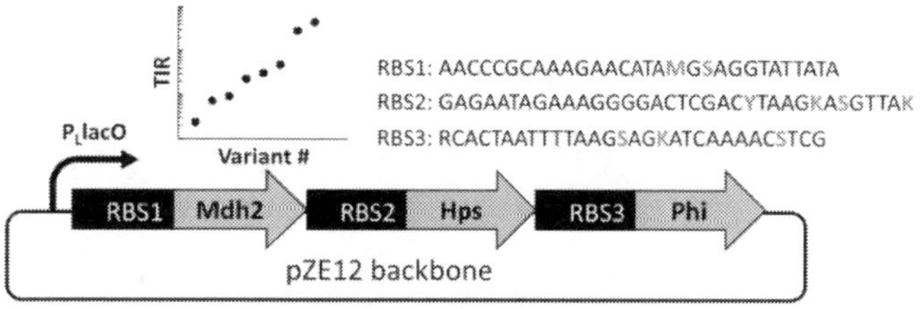

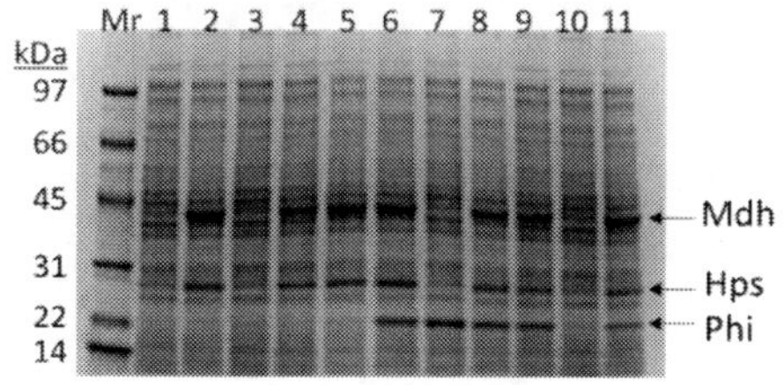

32

Construction of synthetic methanol auxotrophy -Overexpress Mdh, Hps, Phi in Δ*rpe* strain-

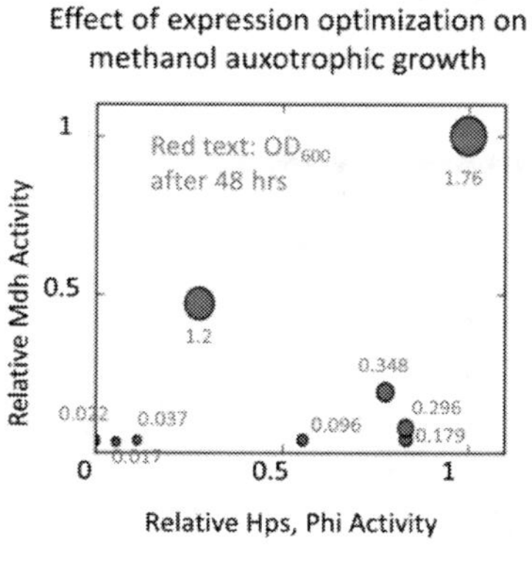

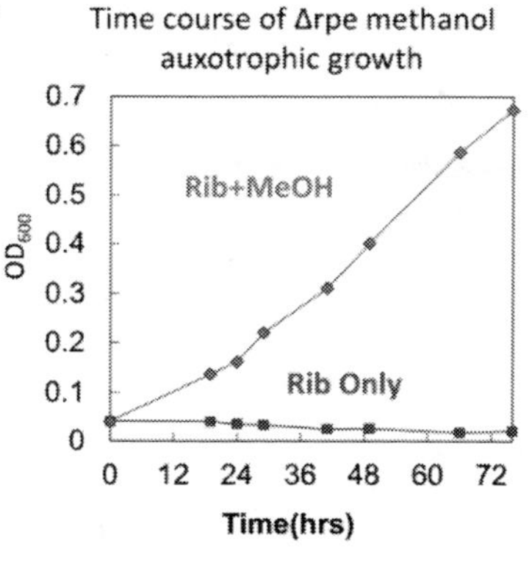

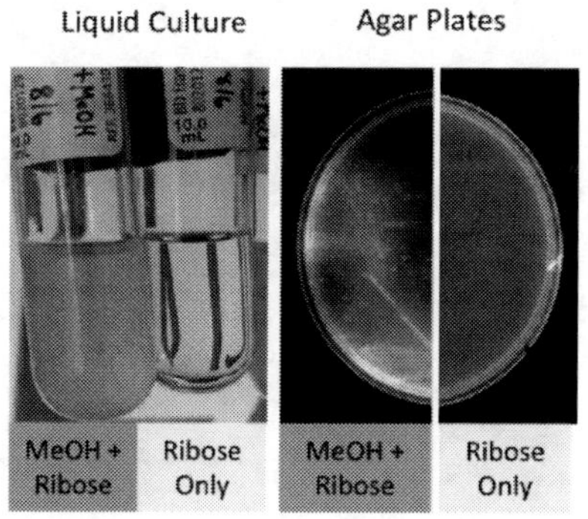

33

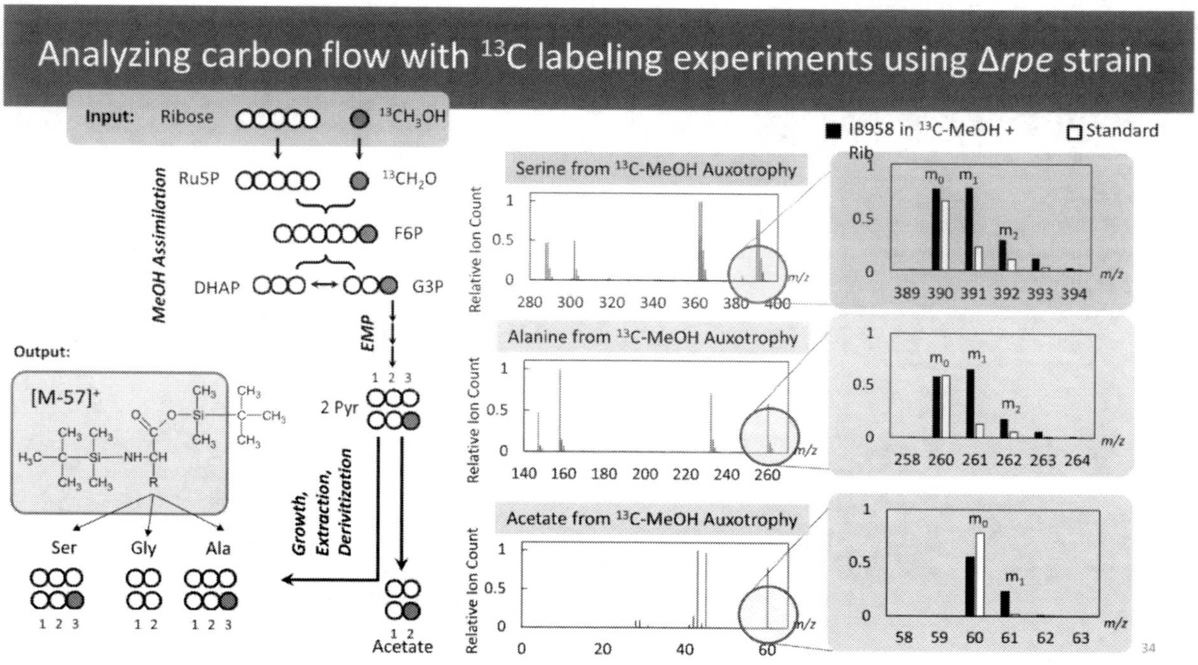

Analyzing carbon flow with 13C labeling experiments using Δrpe strain
Input: Ribose 13CH3OH
MeOH Assimilation
Ru5P 13CH2O
F6P
DHAP G3P
EMP
1 2 3
2 Pyr
Growth, Extraction, Derivitization
Output:
[M-57]+
Ser
Gly
Ala
1 2 3
1 2
1 2 3
1 2
Acetate
IB958 in 13C-MeOH + Rib
Standard
Serine from 13C-MeOH Auxotrophy
Relative Ion Count
280 300 320 340 360 380 400
m/z
389 390 391 392 393 394
Alanine from 13C-MeOH Auxotrophy
140 160 180 200 220 240 260
258 260 261 262 263 264
Acetate from 13C-MeOH Auxotrophy
0 20 40 60
58 59 60 61 62 63
34

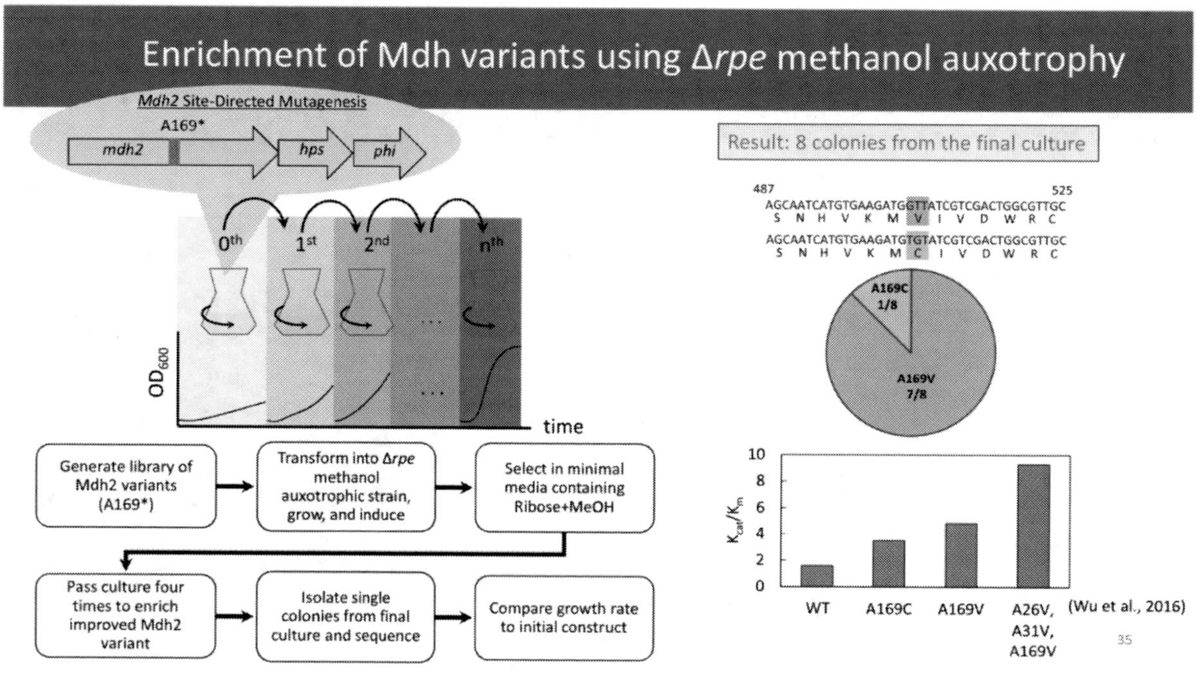

Enrichment of Mdh variants using Δrpe methanol auxotrophy
Mdh2 Site-Directed Mutagenesis
A169*
mdh2
hps
phi
0th
1st
2nd
nth
OD600
time
Generate library of Mdh2 variants (A169*)
Transform into Δrpe methanol auxotrophic strain, grow, and induce
Select in minimal media containing Ribose+MeOH
Pass culture four times to enrich improved Mdh2 variant
Isolate single colonies from final culture and sequence
Compare growth rate to initial construct
Result: 8 colonies from the final culture
487
525
AGCAATCATGTGAAGATGGTTATCGTCGACTGGCGTTGC
S N H V K M V I V D W R C
AGCAATCATGTGAAGATGTGTATCGTCGACTGGCGTTGC
S N H V K M C I V D W R C
A169C 1/8
A169V 7/8
Kcat/Km
WT
A169C
A169V
A26V, A31V, A169V
(Wu et al., 2016)
35

Production of ethanol and *n*-butanol using Δ*rpe* methanol auxotrophy

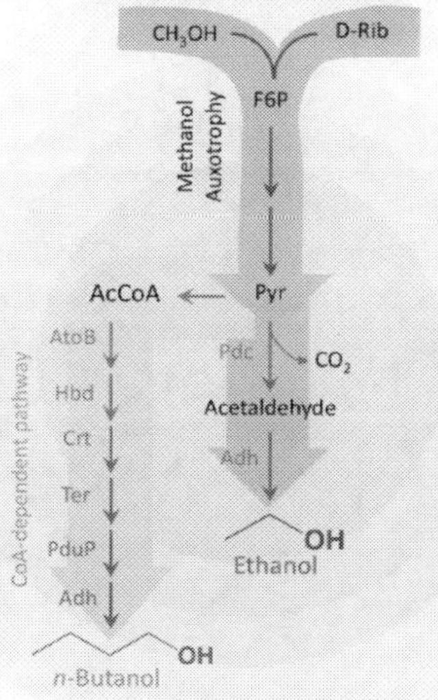

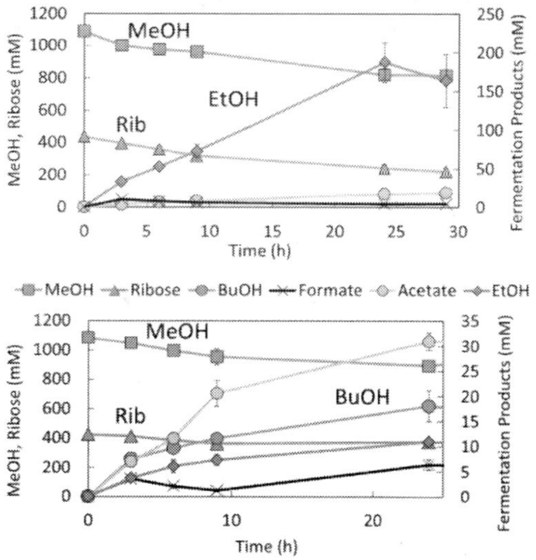

EtOH production

8.6 g/L
(187 mM)

Carbon yield
30.5%

BuOH production

1.3 g/L
(18 mM)

Carbon yield
5.45%

36

Construction of synthetic methanol auxotrophy
-Overexpress Mdh, Hps, Phi in Δ*rpiAB* strain-

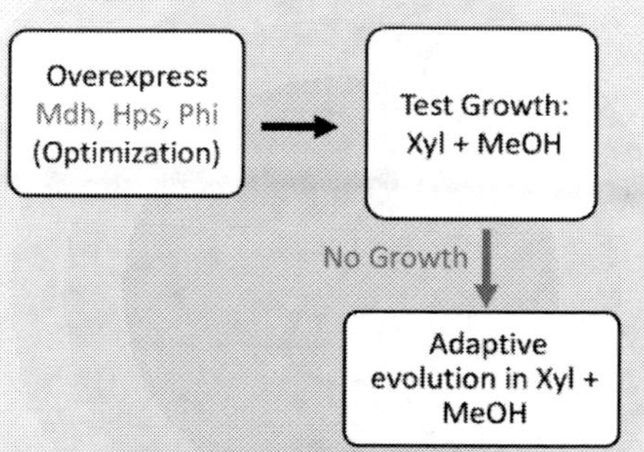

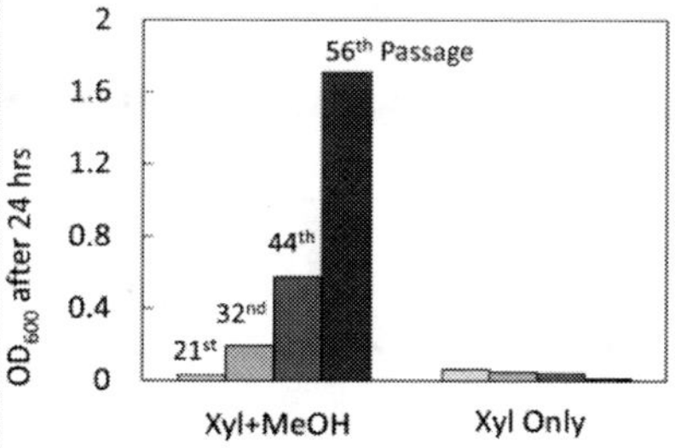

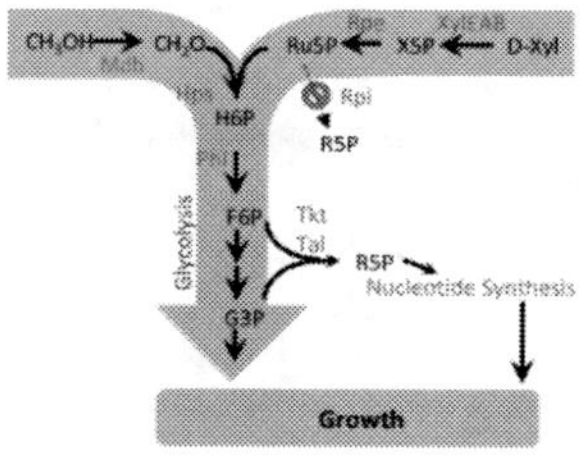

37

Confirming methanol assimilation of Δ*rpiAB* strain with ^{13}C labeling

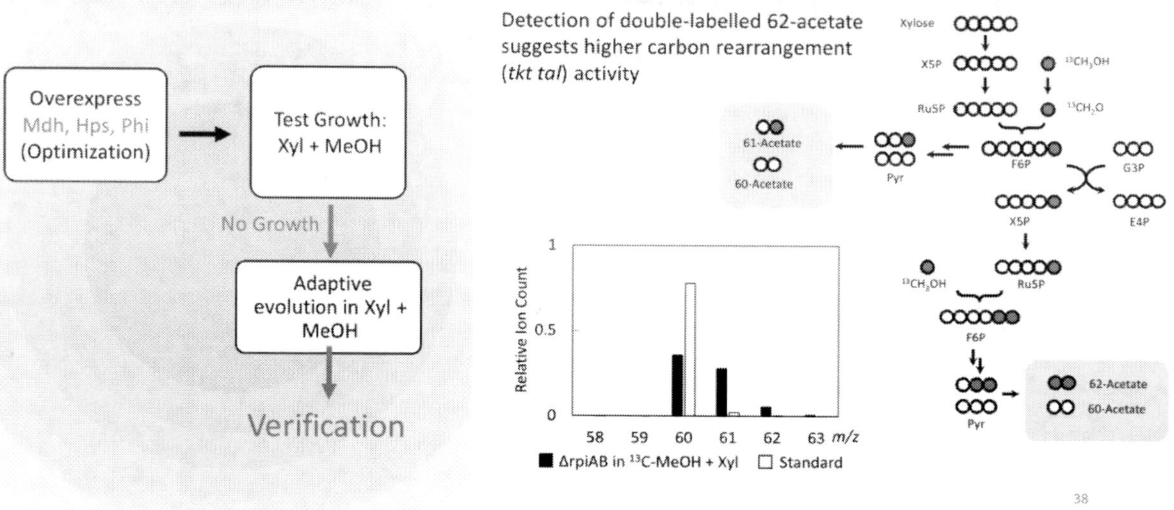

Further evolution required for EtOH production from Δ*rpiAB* strain

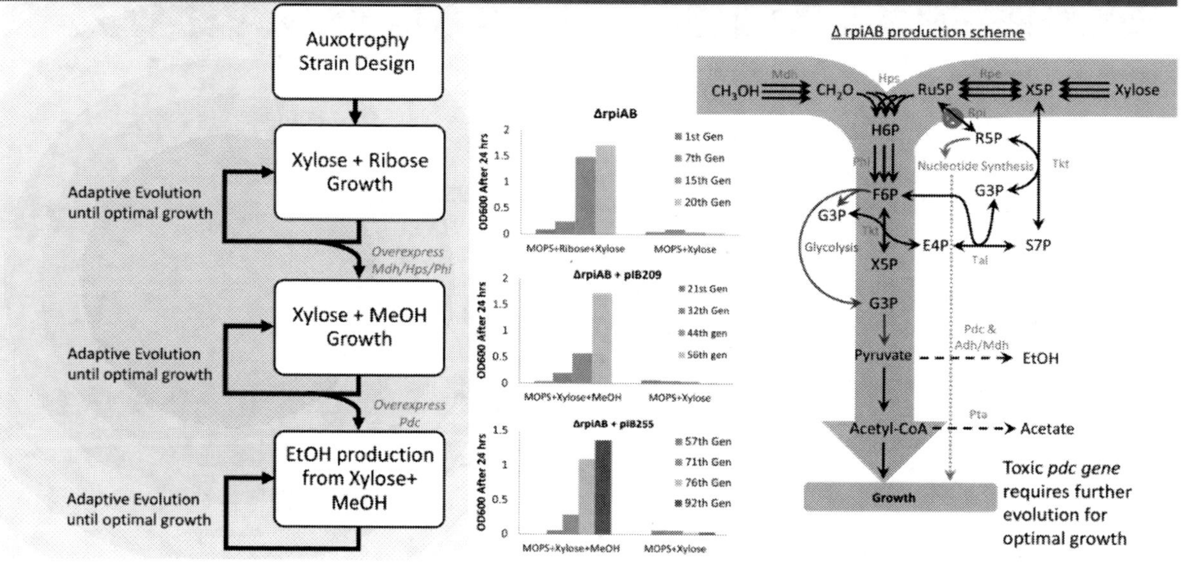

EtOH production from MeOH+Xylose with ΔrpiAB strain

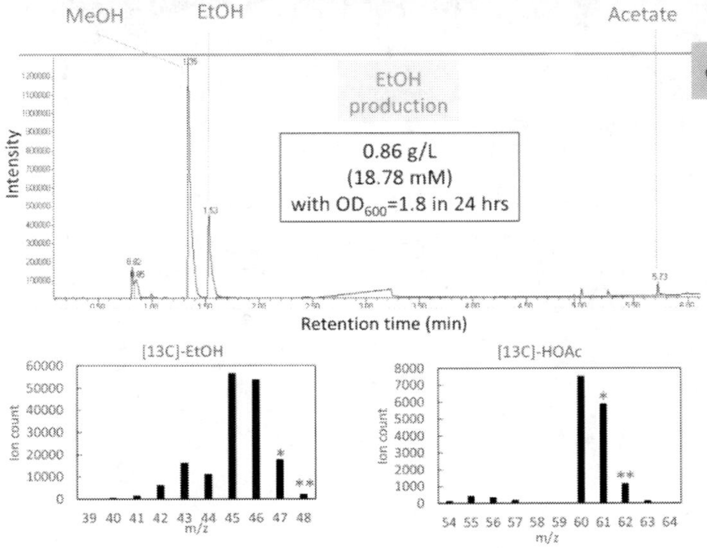

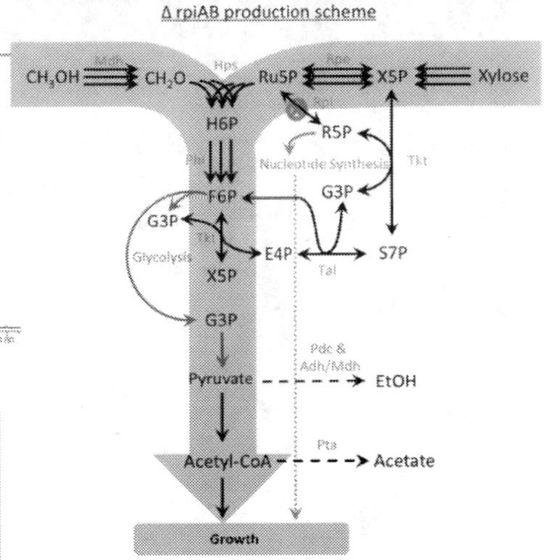

Acknowledgement

- DOE, ARPA-E REMOTE project (Award: DE-AR0000430)
- REMOTE team:
- Dr. Tung-Yun Wu, Charlie Chang-Ting Chen, Dr. Igor Bogorad, Freddy Yu-Hsiao Chen. Po-Wei Chen

41

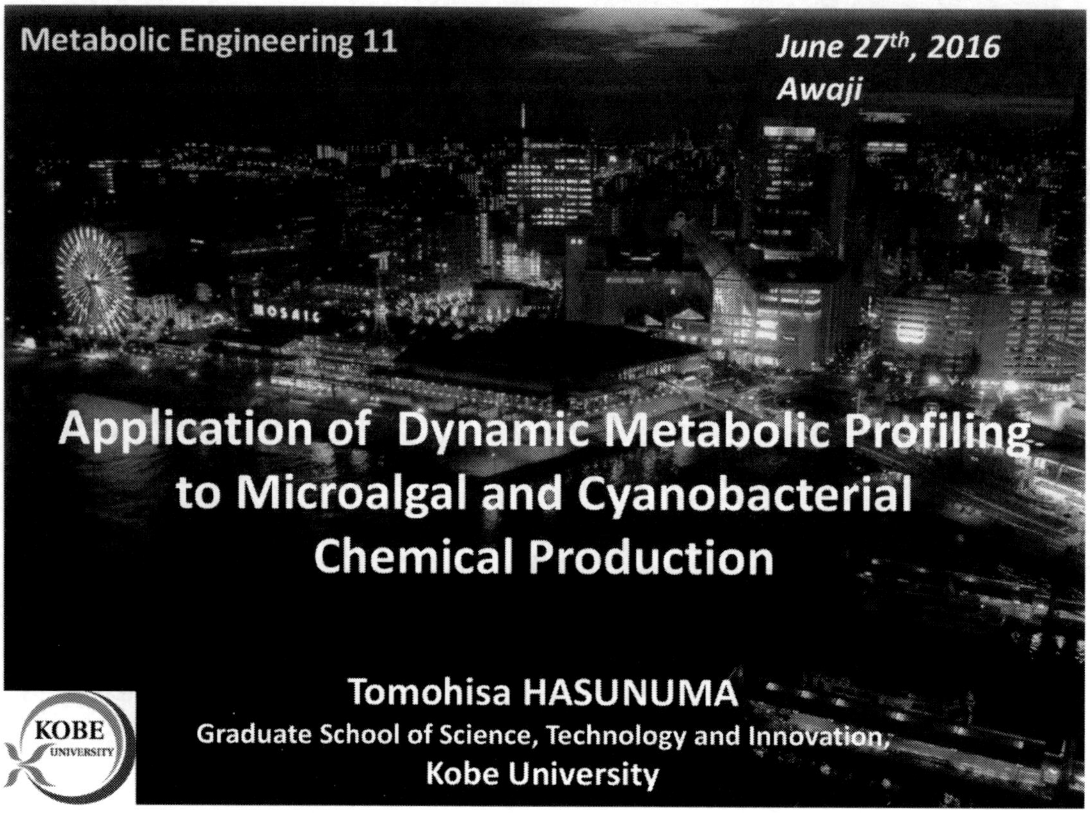

Energy supply from marine biomass

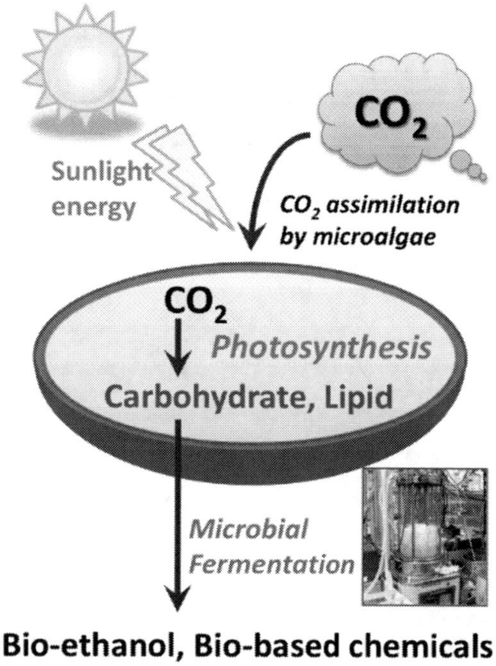

Algae as an ideal feedstock

- **To use as renewable energy source**
  - Products; directly converted from CO_2
- **To avoid issues of land biomass utilization**
  - Full-year harvesting
  - No competition with food supply
  - No limitation of farm land and fresh water
- **To utilize unused aquatic resource**
  - Marine biomass; 30-50% of total amount
  - Marine area; 12 times land area (Japan)

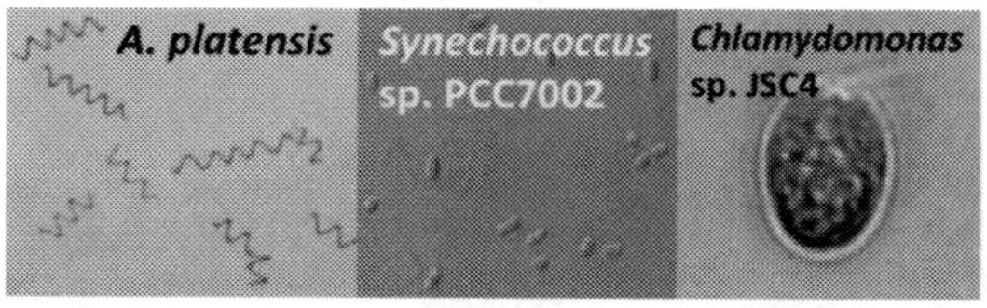

Microalgae and cyanobacteria represent a major portion of aquatic biomass, which has been expected to be utilized as renewable energy source.

2

Outline

1. Development of oil production process with the marine green alga, *Chlamydomonas* sp. JSC4

2. Improvement of photosynthetic carbon assimilation by *Flv3* overexpression in the cyanobacteria *Synechocystis* sp. PCC6803

3. Sugar-free succinate production in the cyanobacteria, *Synechocystis* sp. PCC6803 following identification of the limiting steps in glycogen catabolism

3

Outline

1. Development of oil production process with the marine green alga, *Chlamydomonas* sp. JSC4

2. Improvement of photosynthetic carbon assimilation by *Flv3* overexpression in the cyanobacteria *Synechocystis* sp. PCC6803

3. Sugar-free succinate production in the cyanobacteria, *Synechocystis* sp. PCC6803 following identification of the limiting steps in glycogen catabolism

4

Development of oil production process with marine green algae

Chlamydomonas sp. JSC4

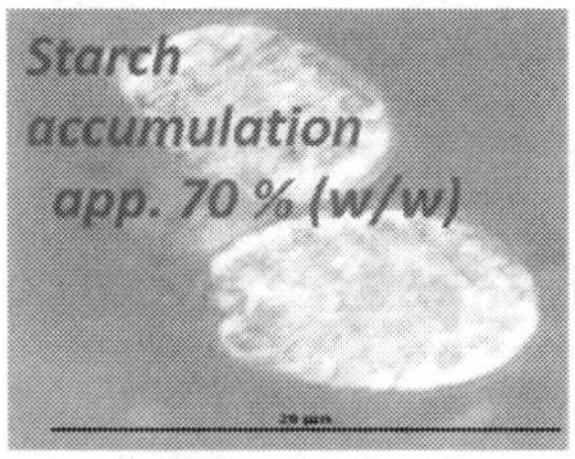

Metabolic perturbation

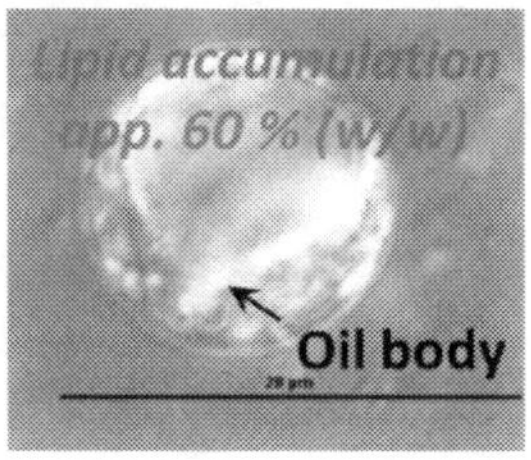

NIBB

EARTHRISE

Development of cultivation process for optimizing oil production

2 L Photobioreactor (PBR)

30 L Tubular PBR

Scale up

Feed back

Outdoor cultivation (50 m²)

Laboratory-scale cultivation system of *Chlamydomonas* sp. JSC4

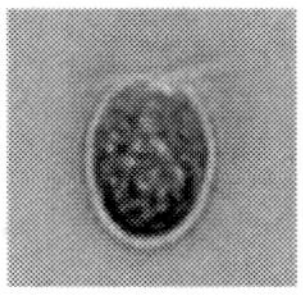

***Chlamydomonas* sp. JSC4**

isolated from brackish-water region of Taiwan

Experimental condition

Light intensity: 200 μE m^{-2} s^{-1}

Temperature: 28 °C

Medium: MN3N with sea salt

Photobioreactor (PBR) design

6

Effect of sea salt addition on lipid production

- 0.5%, 2.0%, 3.5%, and 5.0% sea salt were investigated. With increase in sea salt concentration, lipid content increased. The highest lipid content was 60% under 5% sea salt condition. However, high sea salt inhibited the cell growth.
- After 3-day nitrogen depletion, the highest lipid productivity of 158 mg/L/d was obtained in 2% sea salt condition, along with the lipid content of 42%.

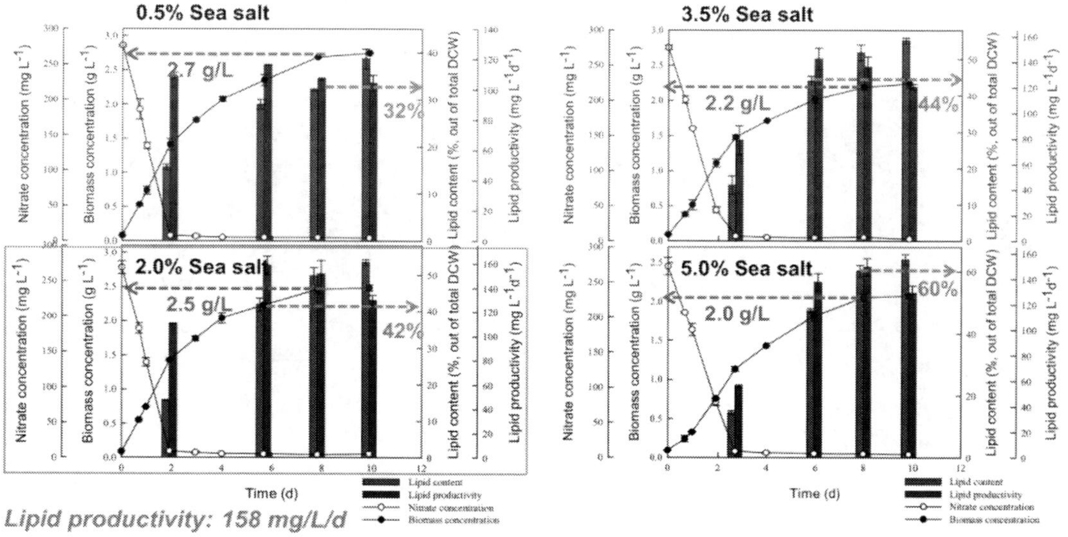

Lipid productivity: 158 mg/L/d

Development of Salinity Gradient Operation System

- ➢ 2% sea salt was selected for cell growth in the beginning.
- ➢ When nitrogen exhausted, the stepwise addition of sea salt (0.5%, 1.0%, and 1.5% per day) was studied for lipid accumulation.

- The stepwise addition of sea salt enhanced the lipid content dramatically, along with the higher cell growth rate.
- Small amounts of sea salt addition (0.5% per day) was enough for lipid accumulation. It is a very good news for the practical outdoor cultivation.

Maximum lipid productivity: 223 mg/L/d
Lipid content: 59%

Ho *et al.* *Biotechnology and Biofuels* 7:97 (2014)

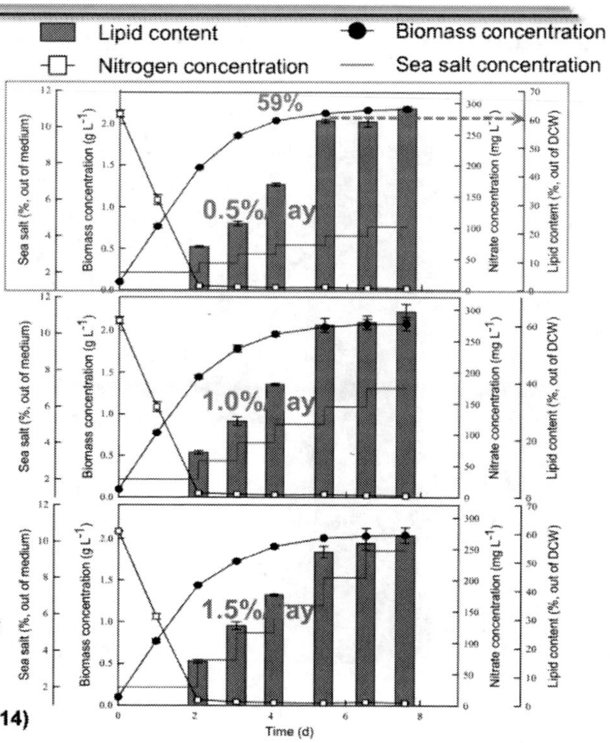

Development of oil production process with marine green alga

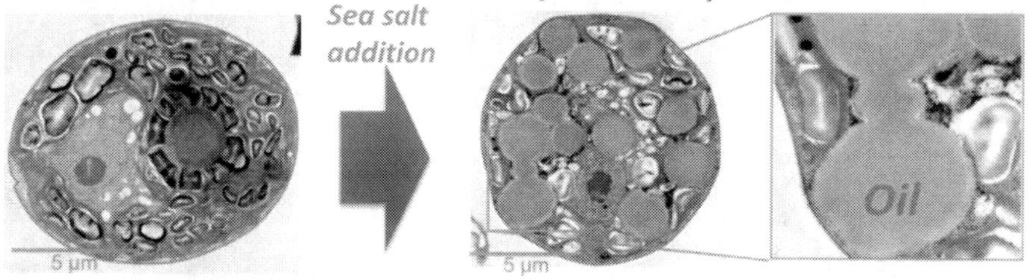

Species	Lipid productivity (mg/L/d)	Lipid content (%)	Reference
Chlorella sorokiniana	140.8	57.7	Chen et al., 2010
Chlorella vulgaris	79.1	17.3	Talebi et al., 2013
Nannochloropsis sp.	110.1	68.5	Bondioli et al., 2012
Isochrysis zhangjiangensis	136.2	46.0	Feng et al., 2011
Nannochloropsis gaditana	51	73.1	San Pedro et al., 2013
Chlamydomonas sp. JSC4	223.2	59.4	Ho et al., 2014
Chlamydomonas sp. JSC4	312.0	43.1	Ho et al., 2015

Ho *et al., Biotechnol. Biofuels* ***2015***, **8, 48**

Outdoor cultivation of *Chlamydomonas* sp. JSC4

Outdoor cultivation set up based on laboratory experiments

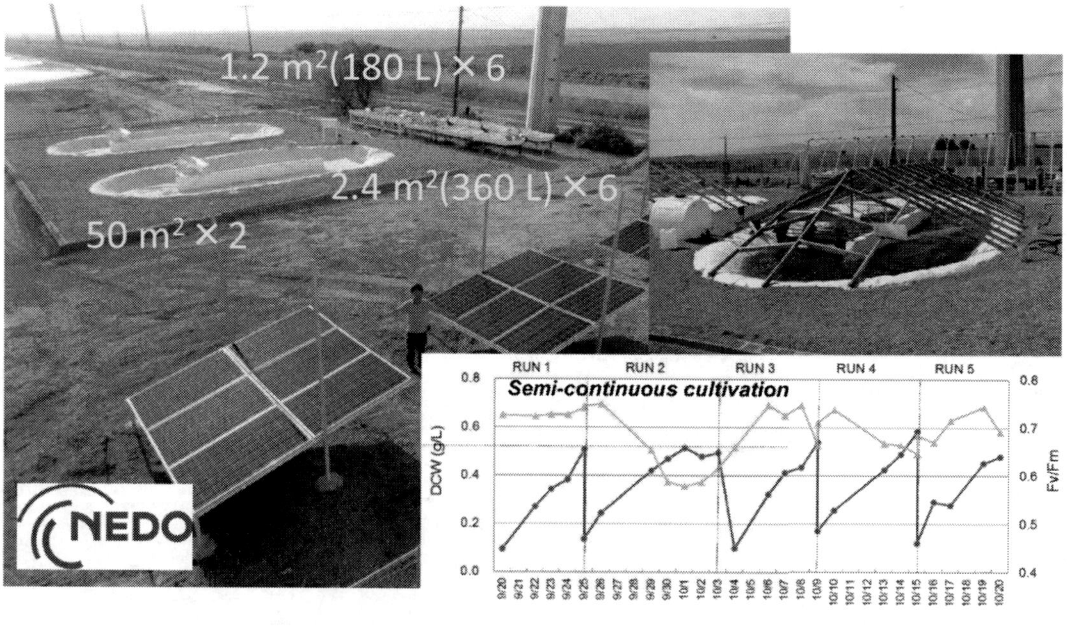

Improvement of oil production ability based on multi-omics approach

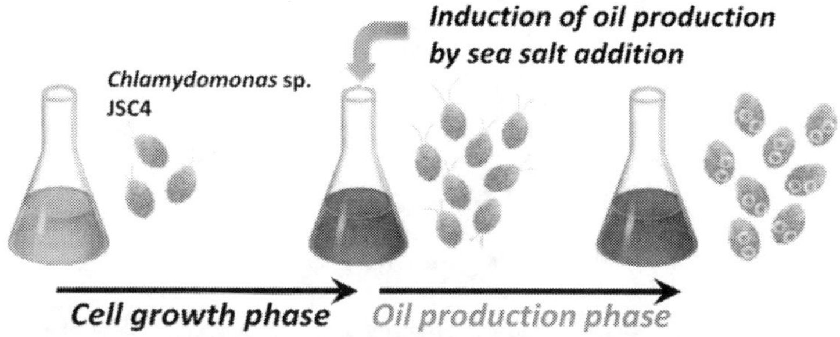

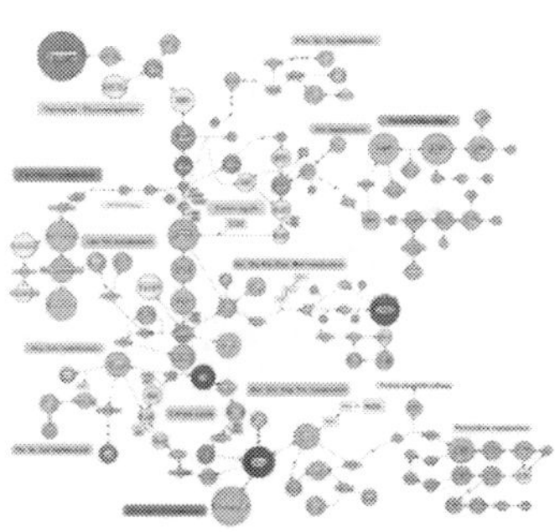

Comparison analysis of cellular metabolism

Global gene expression analysis

Metabolite profiling

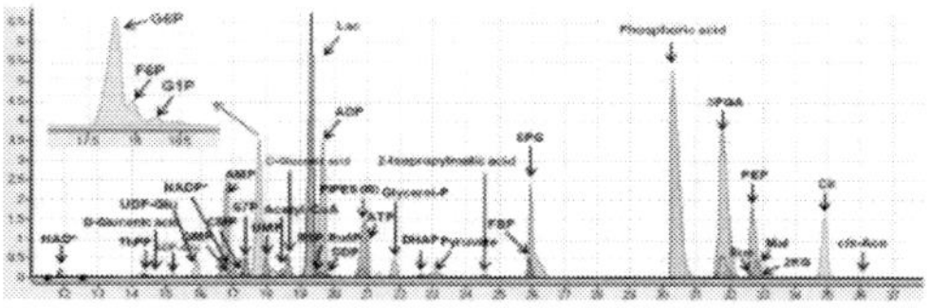

Identification of key reactions in lipid biosynthesis

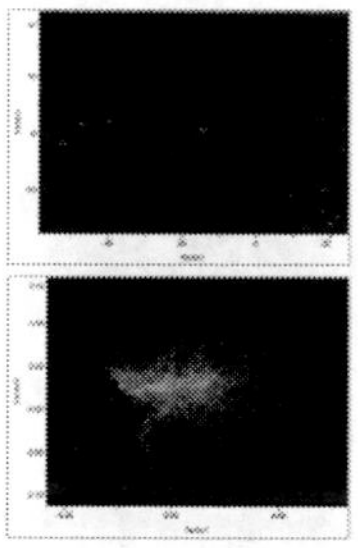

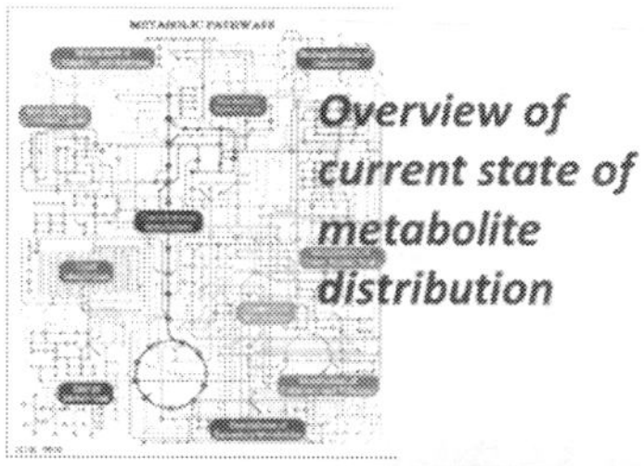

Metabolomic approach is not enough to account for dynamic metabolic flux.

To observe dynamic metabolic flow, analysis of turnover of metabolic intermediates is needed.

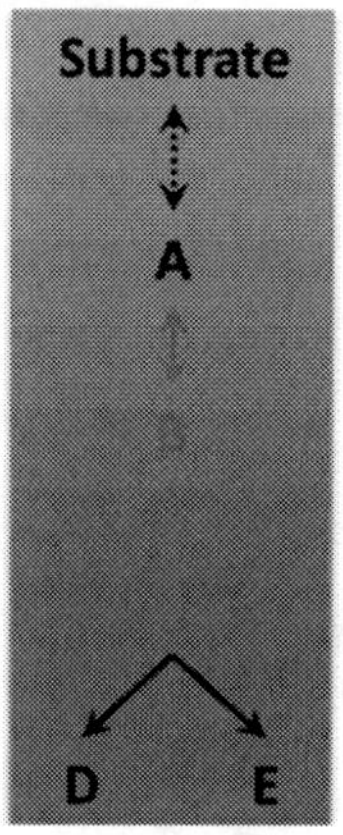

Development of Dynamic Metabolic Profiling

In vivo ^{13}C labeling

$$CO_2 + H_2O \xrightarrow{\textit{Photosynthesis}} (CH_2O)_n + O_2$$

Replacement of carbon source

In vivo ^{13}C-labeled

$$^{13}CO_2 + H_2O \xrightarrow{\textit{Photosynthesis}} (^{13}CH_2O)_n + O_2$$

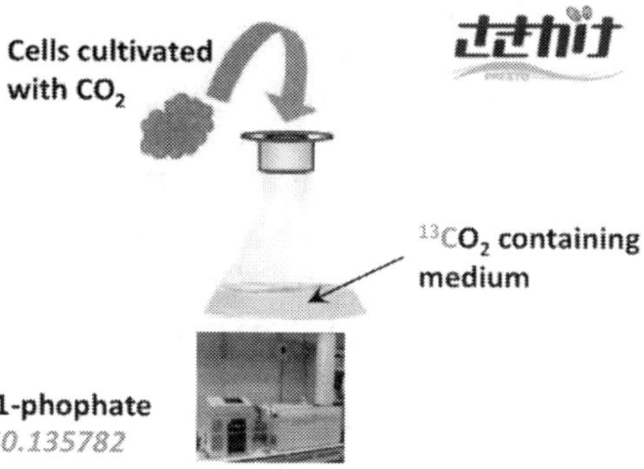

Metabolomics

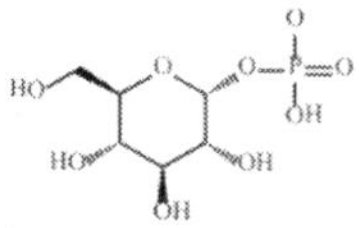

Glucose-1-phophate
Mw: 260.135782

Mass distribution analysis

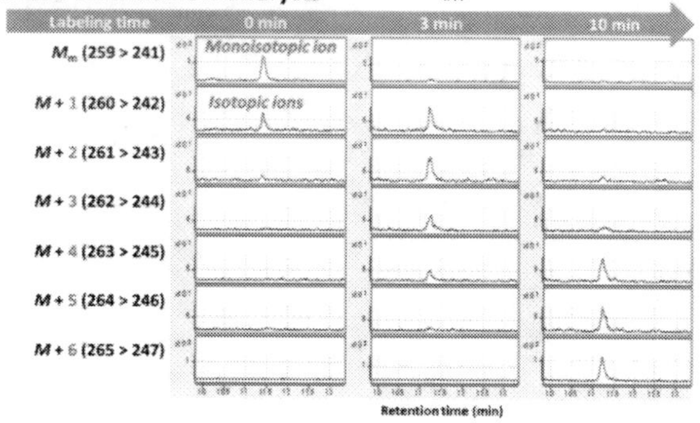

Kinetic analysis of ^{13}C fraction

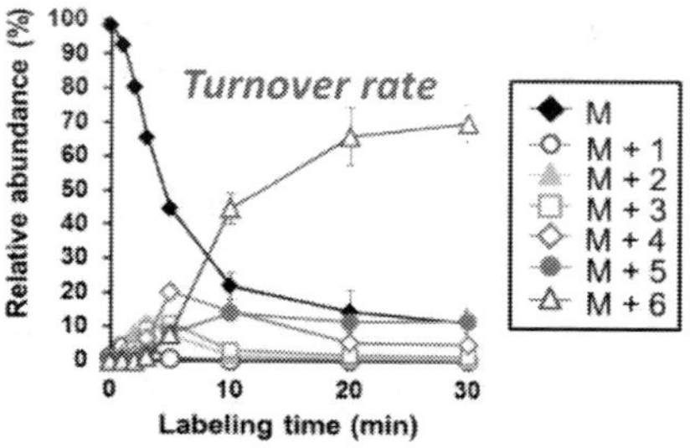

Hasunuma *et al.*, *J. Exp. Bot.*, 61, 1041-1051 (2010)
Hasunuma *et al.*, *J. Exp. Bot.*, 64, 2943-2954 (2013)

Effect of sea salt addition on *Chlamydomonas* sp. JSC4

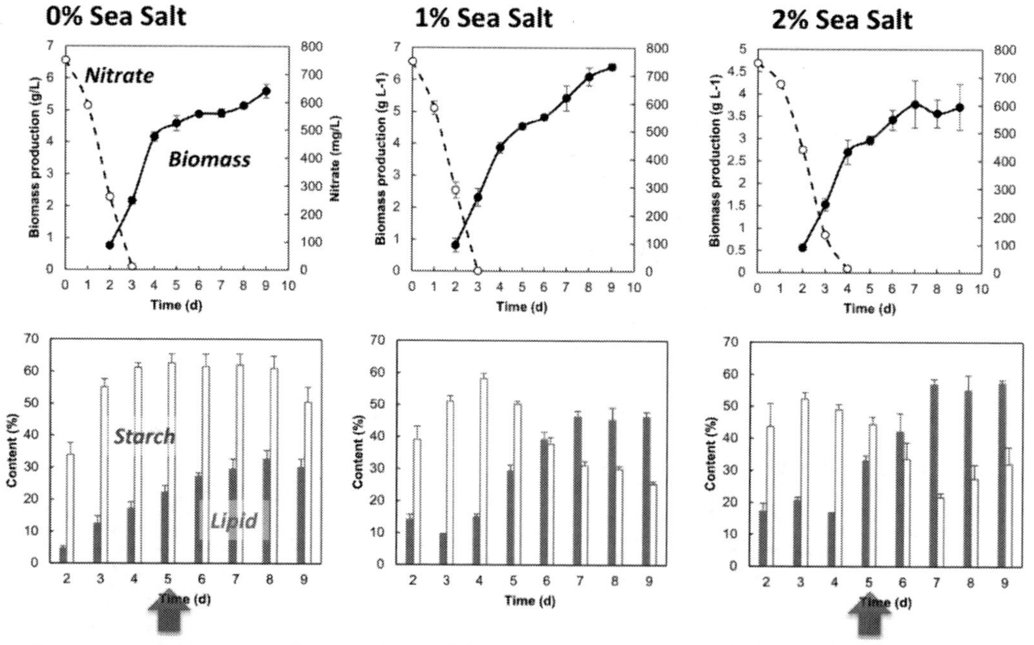

Dynamic metabolic profiling of *Chlamydomonas* was performed after 5-day cultivation. 14

Outline

1. **Development of oil production process with the marine green alga, *Chlamydomonas* sp. JSC4**

2. Improvement of photosynthetic carbon assimilation by *Flv3* overexpression in the cyanobacteria *Synechocystis* sp. PCC6803

3. **Sugar-free succinate production in the cyanobacteria, *Synechocystis* sp. PCC6803 following identification of the limiting steps in glycogen catabolism**

15

Enhancement in ATP level by *Flv3* overexpression

Photosynthetic electron transport

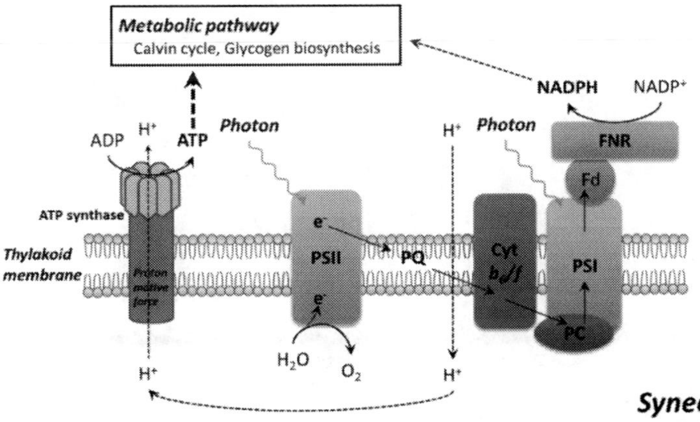

***Synechocystis* sp. PCC6803**

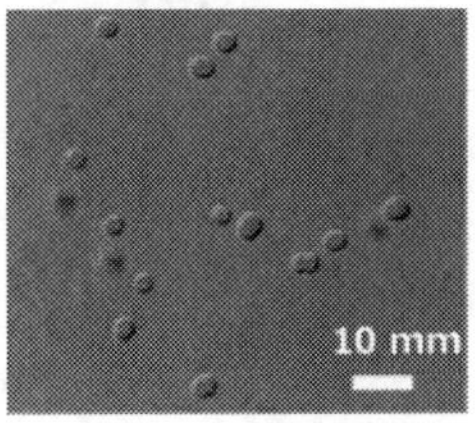

Enhancement in ATP level by *Flv3* overexpression

Photosynthetic electron transport

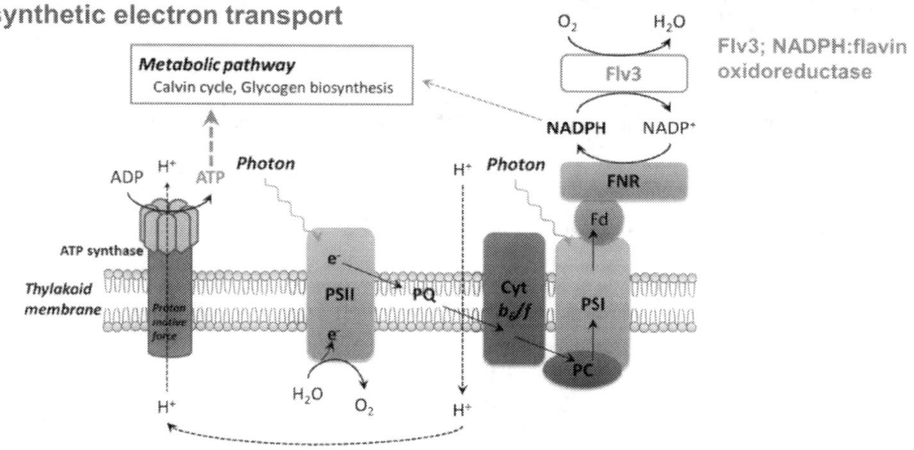

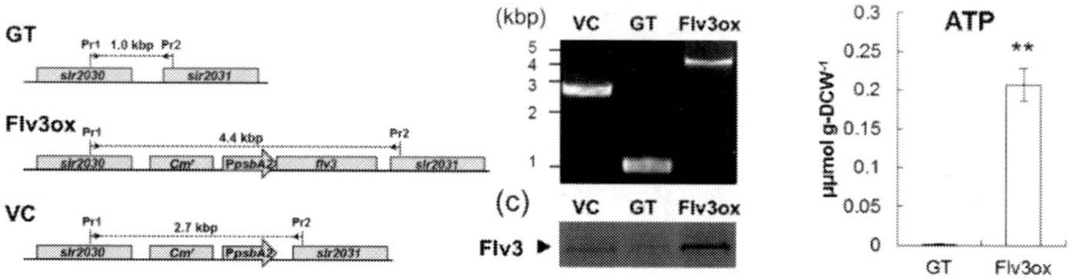

Improvement of photosynthesis by *Flv3*

Light intensity;
200 μmol photons m^{-2} s^{-1}
Temperature;
30 °C
CO_2 concentration;
10,000 ppm

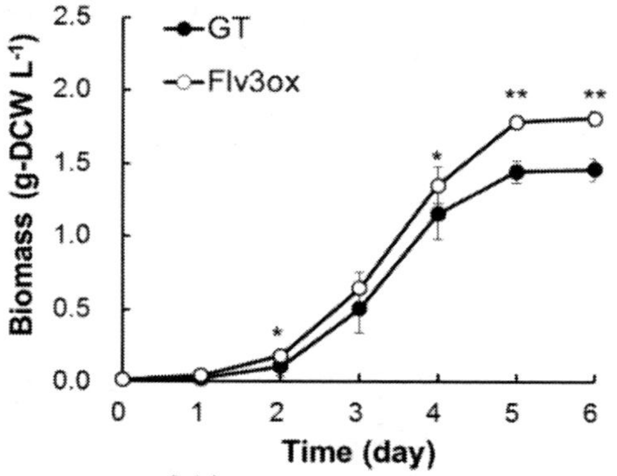

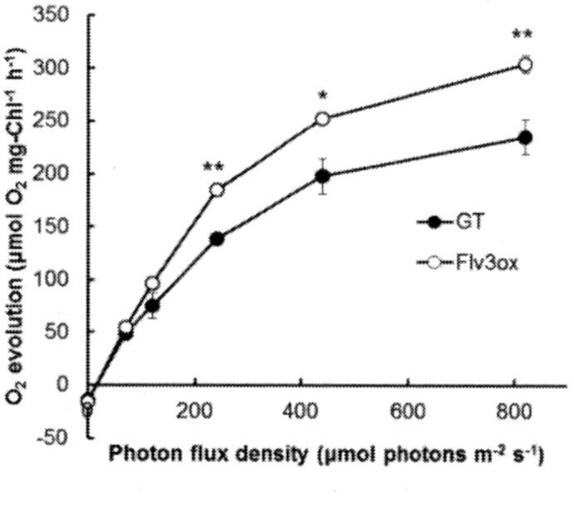

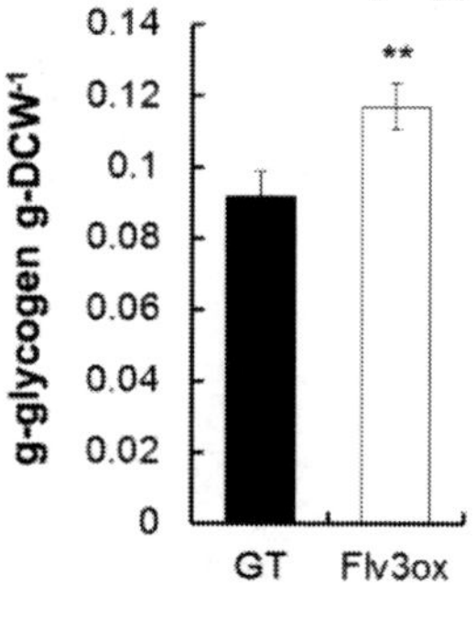

Improvement of carbon assimilation by *Flv3* overexpression in *Synechocystis*

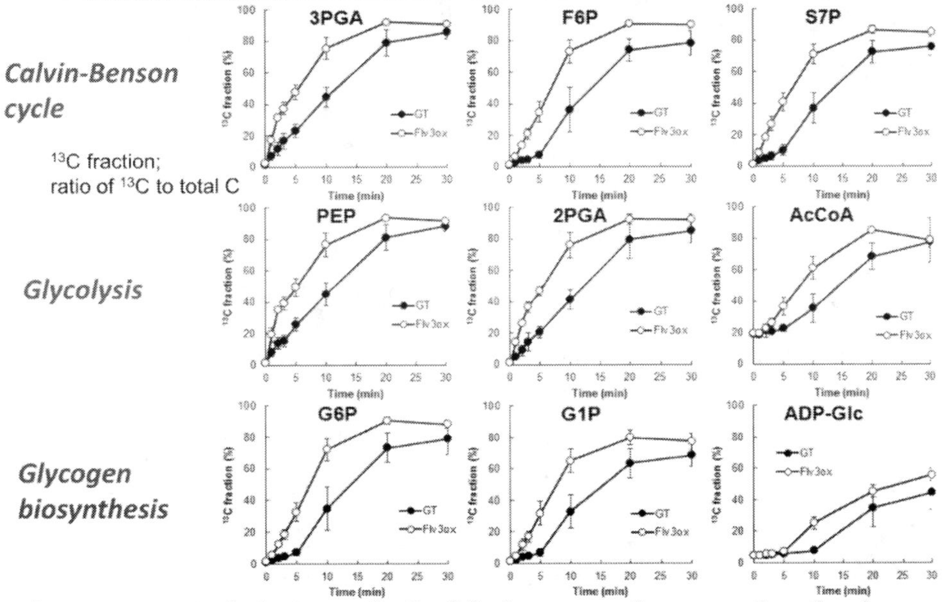

Improvement of photosynthetic CO_2 incorporation was directly observed in recombinant cyanobacterial strains. **Hasunuma *et al.*, *Biotechnol. Biofuels*, 7, 493 (2014)**

Outline

1. **Development of oil production process with the marine green alga, *Chlamydomonas* sp. JSC4**
2. **Improvement of photosynthetic carbon assimilation by *Flv3* overexpression in the cyanobacteria *Synechocystis* sp. PCC6803**
3. **Sugar-free succinate production in the cyanobacteria, *Synechocystis* sp. PCC6803 following identification of the limiting steps in glycogen catabolism**

20

Carbon dioxide Capture and Utilization by cyanobacteria

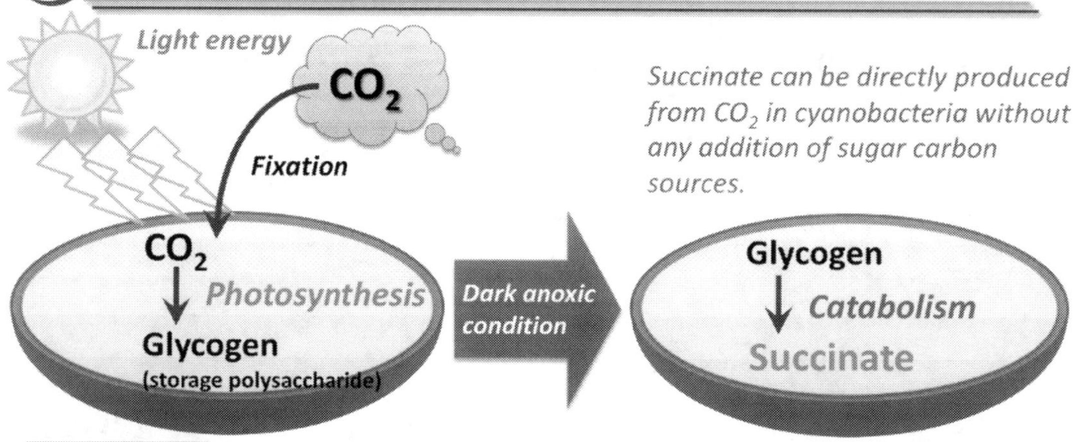

What is metabolic perturbation suitable for succinate production?

Metabolic mechanism is investigated through metabolomics approach.

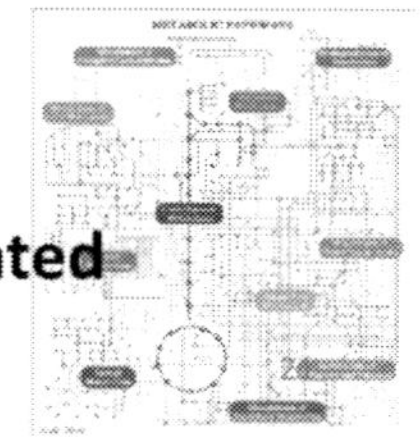

Glycogen degradation under dark anoxic condition

Phototrophic cultivation (3 days)
Medium; BG11, 5 mM $NH_4Cl/NaNO_3$, 50 mM Hepes-KOH (pH7.8)
CO_2; 1% (v/v), Light; 120 μE, Initial OD_{750}; 0.4

Dark anoxic cultivation
Medium; 50 mM Hepes-KOH (pH7.8)
Initial OD_{750}; 20

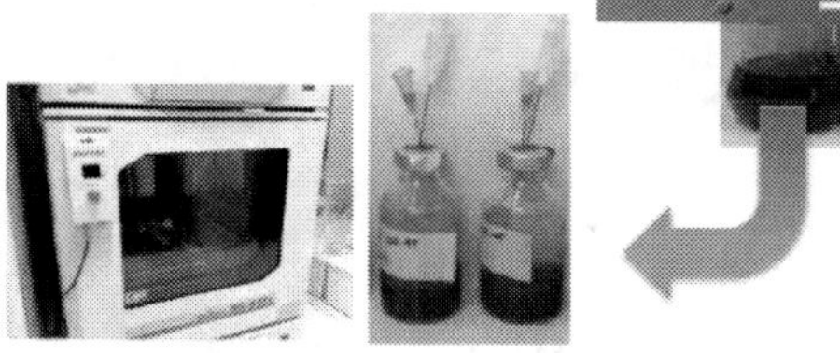

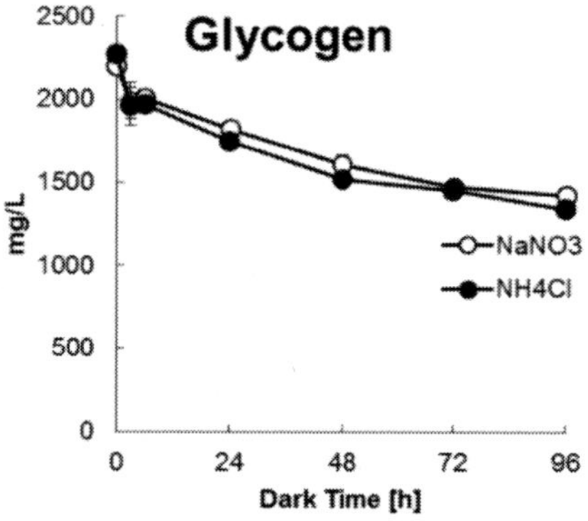

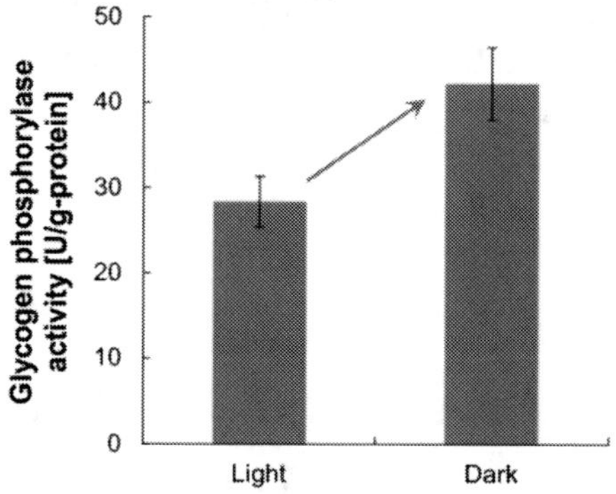

Organic acid produced into medium under dark anoxic condition

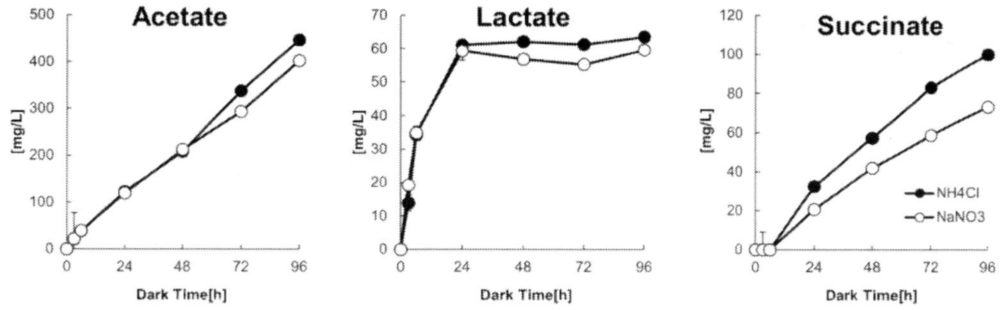

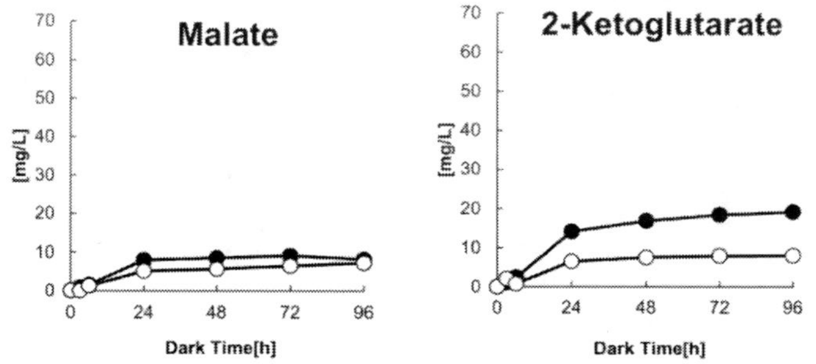

Succinate production was initiated from 6 h.

More than 100 mg/L succinate was produced under dark anoxic condition.

Time course of metabolome during autofermentation

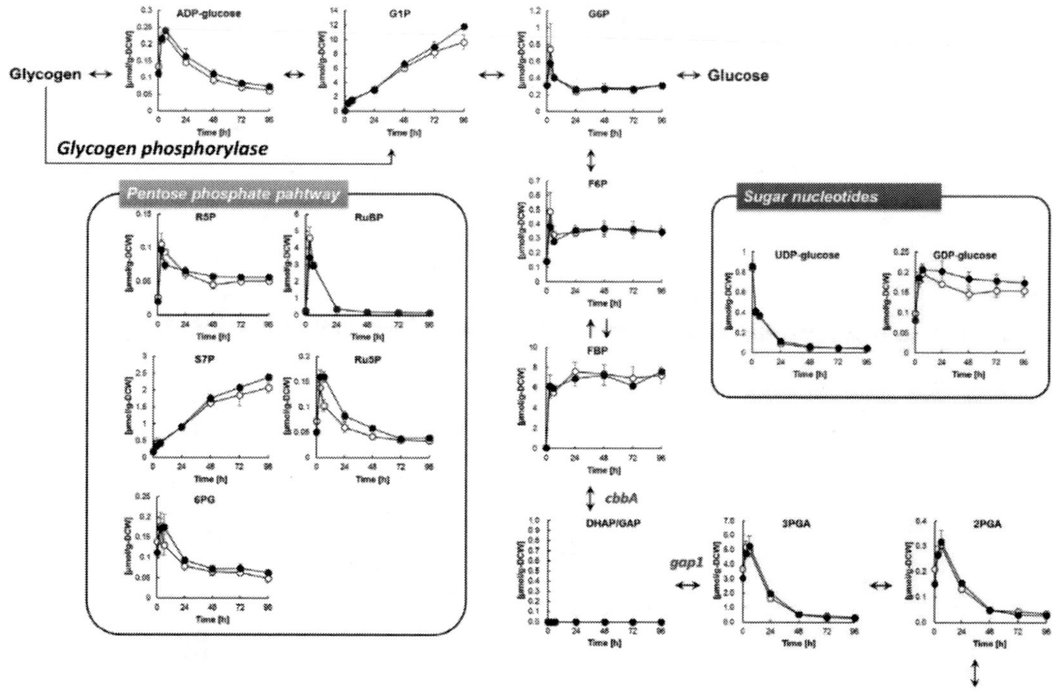

Time course of metabolome during autofermentation

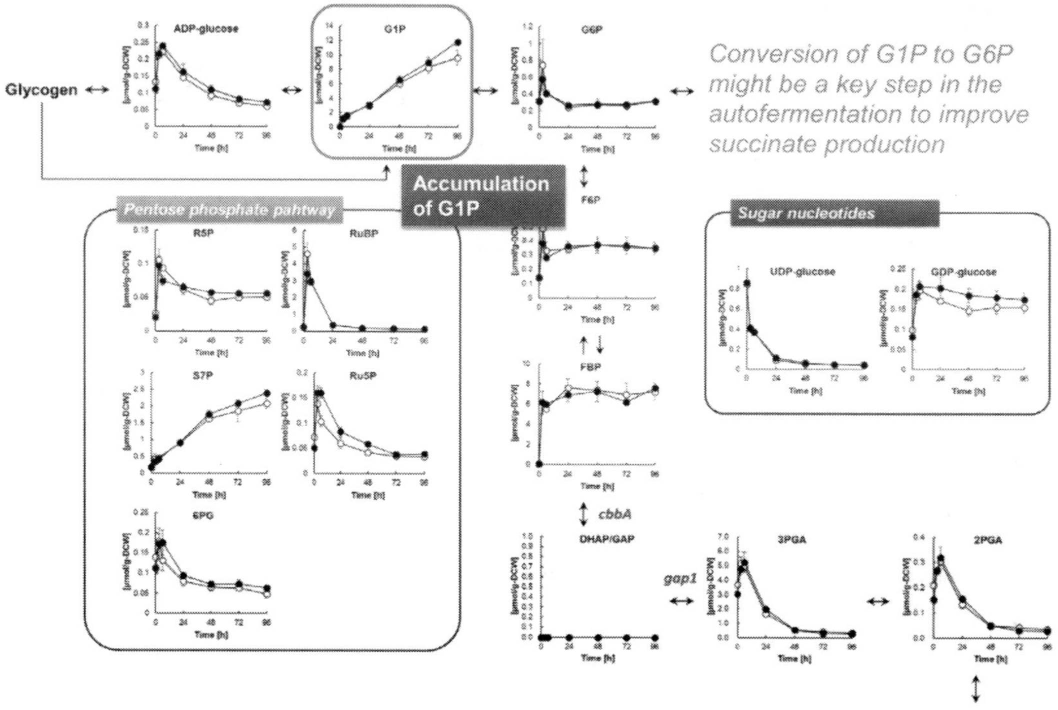

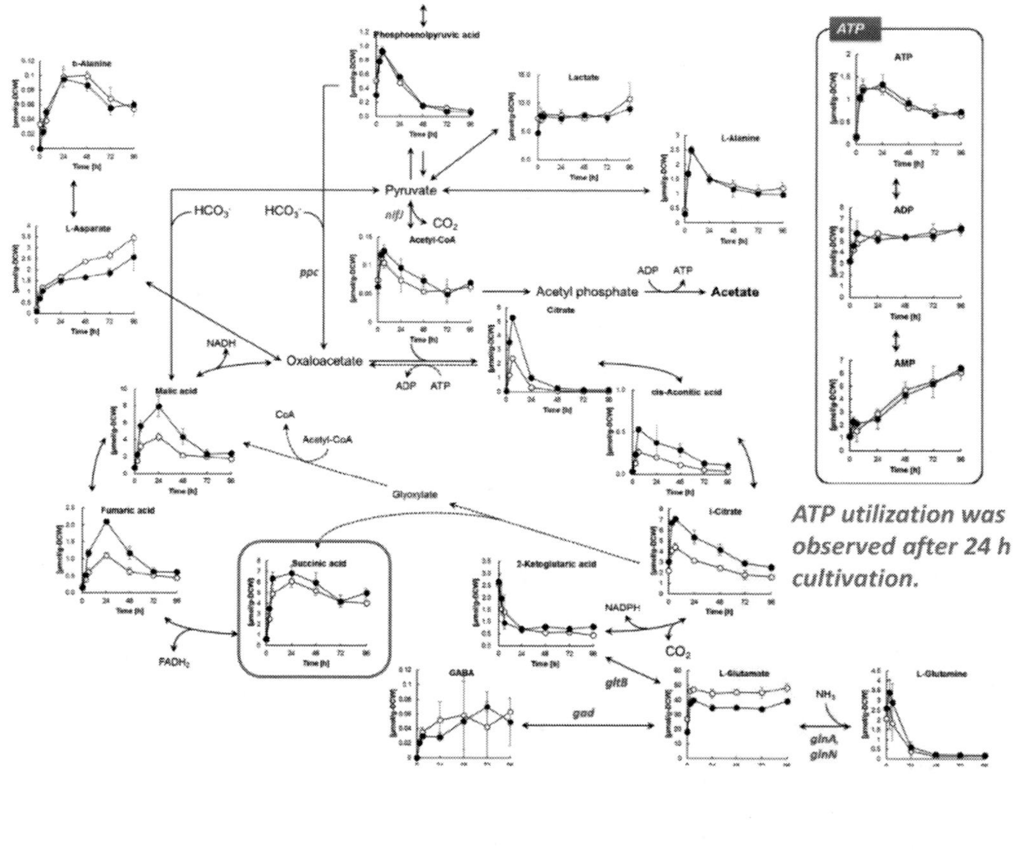

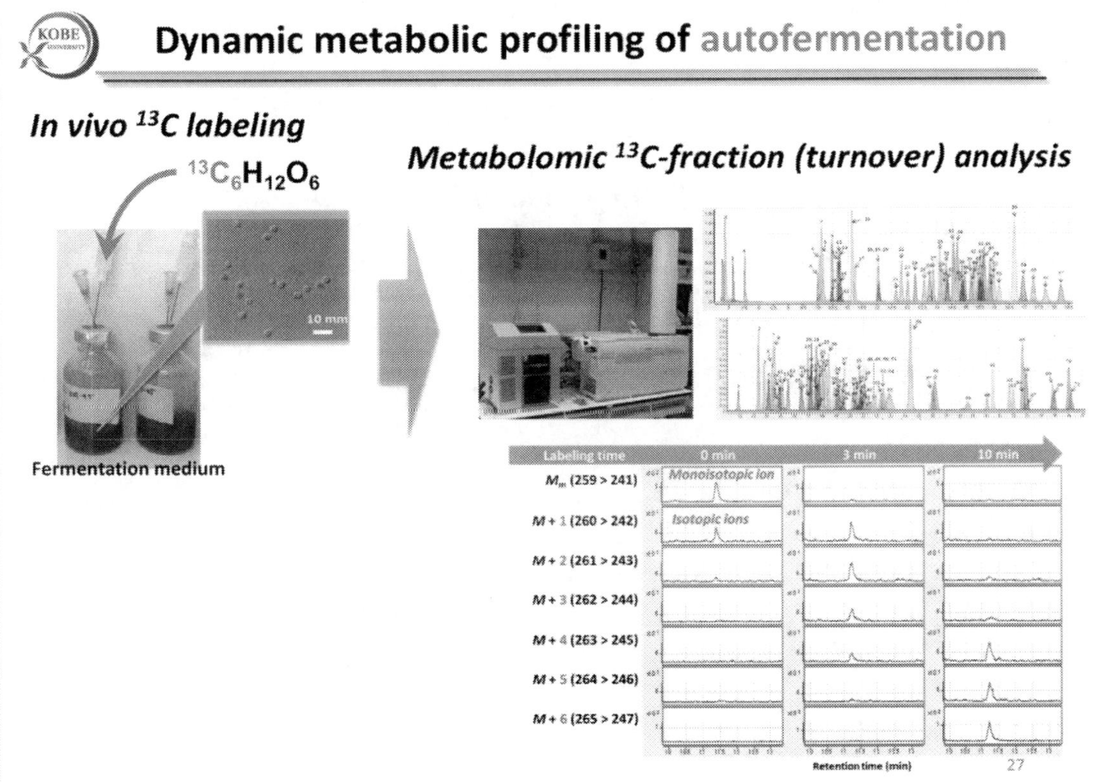
KOBE UNIVERSITY
Dynamic metabolic profiling of autofermentation
In vivo ^{13}C labeling
$^{13}C_6H_{12}O_6$
10 mm
Fermentation medium
Metabolomic ^{13}C-fraction (turnover) analysis
Labeling time
0 min
3 min
10 min
M_m (259 > 241)
Monoisotopic ion
M + 1 (260 > 242)
Isotopic ions
M + 2 (261 > 243)
M + 3 (262 > 244)
M + 4 (263 > 245)
M + 5 (264 > 246)
M + 6 (265 > 247)
Retention time (min)
27

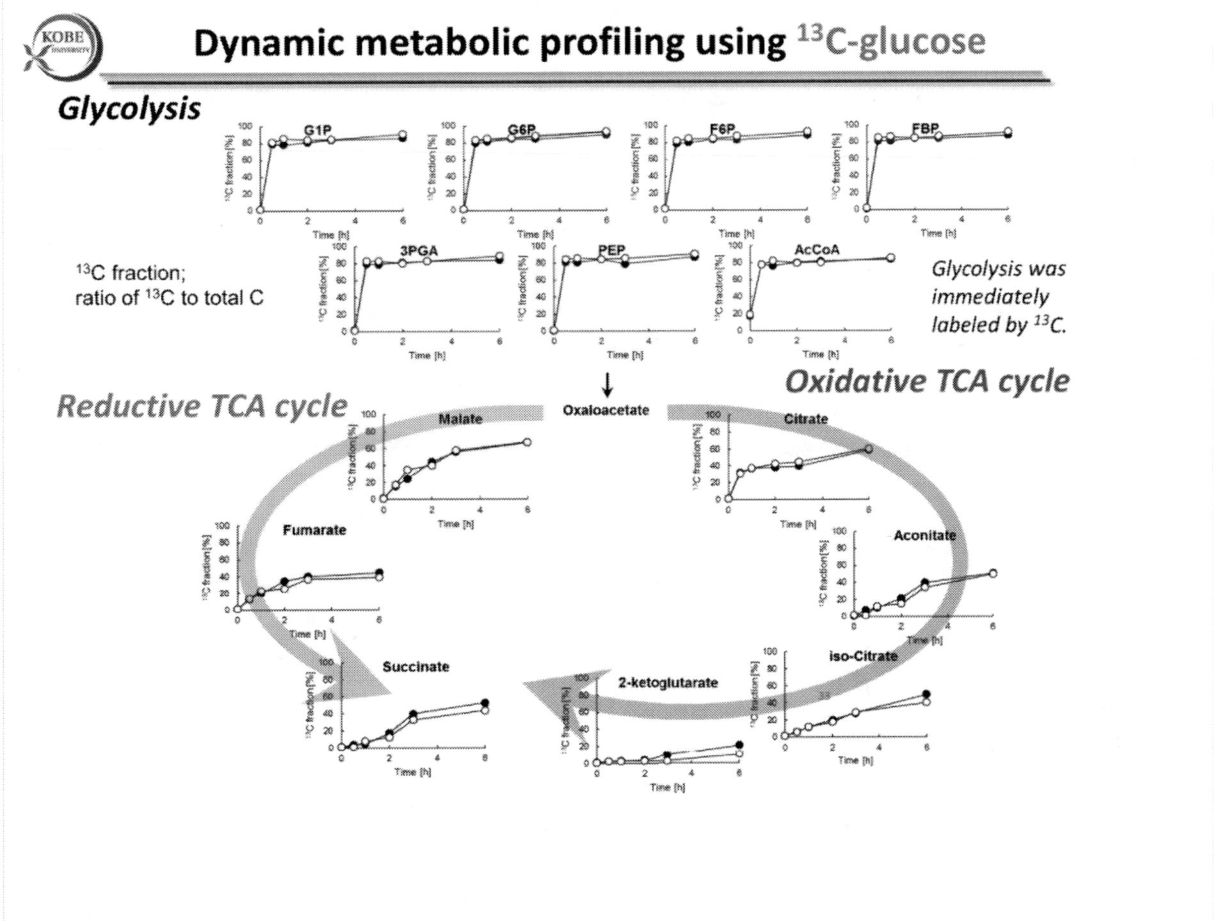
KOBE UNIVERSITY
Dynamic metabolic profiling using ^{13}C-glucose
Glycolysis
G1P
G6P
F6P
FBP
3PGA
PEP
AcCoA
^{13}C fraction [%]
Time [h]
^{13}C fraction;
ratio of ^{13}C to total C
Glycolysis was immediately labeled by ^{13}C.
Reductive TCA cycle
Oxidative TCA cycle
Oxaloacetate
Malate
Citrate
Fumarate
Aconitate
Succinate
2-ketoglutarate
iso-Citrate

Dynamic metabolic profiling using ^{13}C-glucose

Glycolysis

- **There is a bottle neck between glycolysis and TCA cycle.**
- **Succinate is biosynthesized via the reductive TCA cycle**
- **Conversion of fumarate to succinate would be a limiting step.**

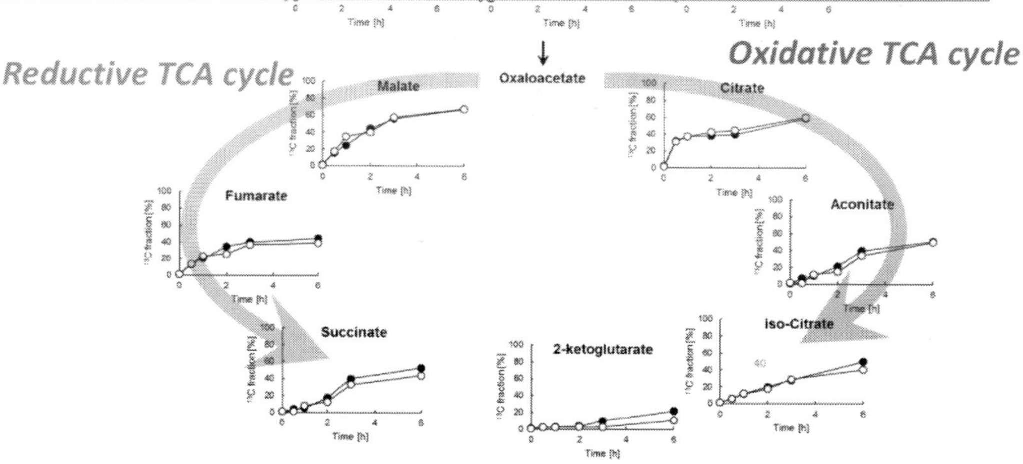

Dynamic metabolic profiling using ^{13}C-glucose

Glycolysis

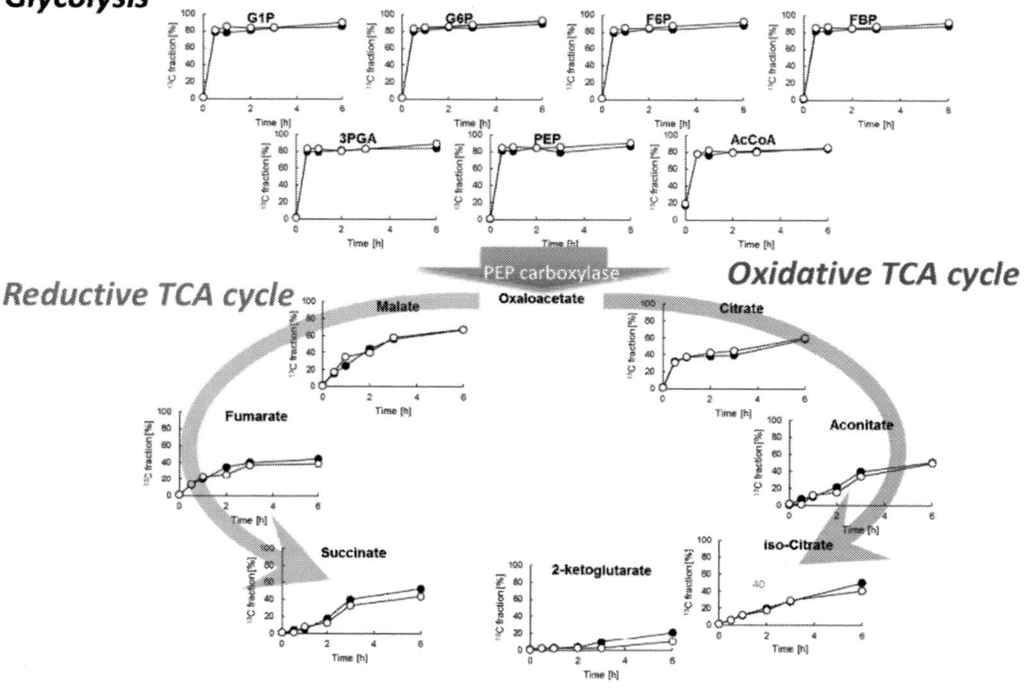

Improvement of succinate production by *ppc* overexpression

PEP carboxylase (Ppc)

Phospho*enol*pyruvate ⟷ Oxaloacetate (HCO_3^-)

Construction of *ppc* overexpressing strain

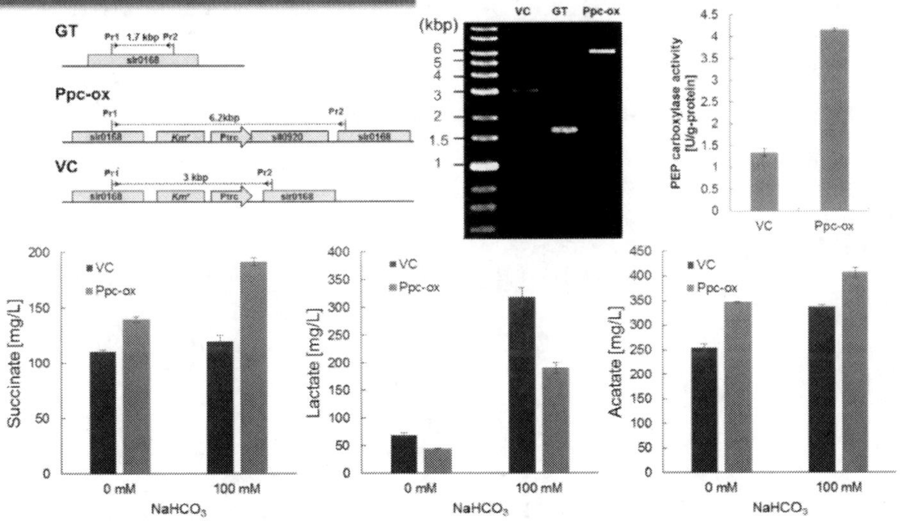

Overexpression of *ppc* improves succinate production.

Hasunuma *et al.*, *Metab. Eng. Commun.* 2016, In press

Carbon Capture and Utilization (CCU) for Organic Acid Production

Improvement of organic acid production by addition of $NaHCO_3$

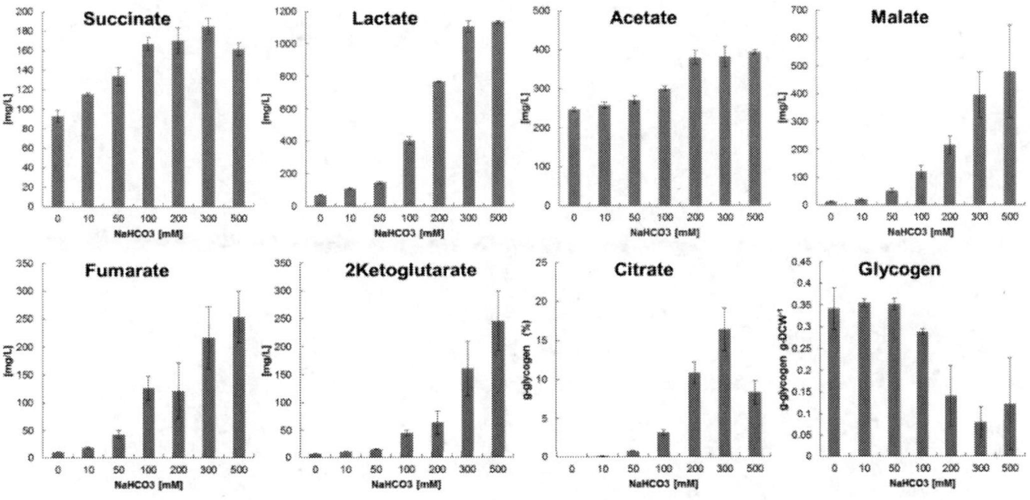

Equilibrium state $CO_2 \xleftrightarrow{pK_{a1} = 6.35} HCO_3^- \xleftrightarrow{pK_{a2} = 10.33} CO_3^{2-}$

32

Thanks to ...

Kobe University

Prof. Akihiko Kondo

Dr. Shimpei Aikawa
Dr. Akihito Nakanishi
Dr. Shih-Hsin Ho
Ms. Mami Matsuda

Assoc. Prof. Chikahiro Miyake

Kyushu UNIVERSITY

Assoc. Prof. Yoshihiro Izumi

National Cheng Kung University

Prof. Jo-Shu Chang

National Institute of Basic Biology

Prof. Jun Minagawa

DIC Corporation

Dr. Haruo Nishie
Dr. Hiroyuki Taroda

Meiji UNIVERSITY

Dr. Takashi Osanai

RIKEN

Dr. Tomokazu Shirai

FUNDING

33

A Robust and Flexible Production Platform Utilizing Dynamic Metabolic Control

Lynch, M., Duke University
Ye, Z.,
Reed, B.,
Burg, J.,
Cooper, C.,
Rodriguez, D.,
Menacho-Melgar, R.,
Hennigan, J.,
Moreb, E.,
Originally presented on: 6/27/2016 9:30:00 - 9:50:00

We have developed a robust and generalizable bioprocess platform, which decouples microbial growth and metabolic production. We are able to simultaneously eliminate combinations of key enzymes in central metabolism, minimizing the metabolic network and design space. Metabolic flux is effectively redirected from growth to production using synthetic metabolic "valves" which rely on a combination of controlled protein degradation and gene silencing, in a phosphate limited stationary phase.
This methodology has been successfully demonstrated in the two important platform microbes, *Escherichia coli* and *Saccharomyces cerevisiae.* To date we have demonstrated the utility of this approach with over 50 different production strains, resulting in the commercially relevant production of a wide variety of products ranging from biofuels, and commodity and specialty chemicals to natural products, and even lipidated peptides useful as biologics or in the production of nanostructures. Speed of strain optimization is increased due to the robustness of the minimal metabolic networks to key process conditions such as oxygen and glucose concentration. This robustness has enabled predictable scale up from screening scale (96 well plates) to controlled benchtop bioreactors.

Exploring Conventional and Nonconventional Yeasts for Producing Bioplastic and Pharmaceutical Precursors Via Shikimate Pathway

Shao, Z., Iowa State University
Suastegui, M., Iowa State University
Cao, M., Iowa State University
Gao, M., Iowa State University
Originally presented on: 6/27/2016 9:50:00 - 10:10:00

Shikimate pathway serves an essential metabolic role in all living organisms. Not only are the three aromatic amino acids synthesized through this pathway, but also many bioactive alkaloid and flavonoid types of secondary metabolites are derived from here. In addition to the ability to express membrane-bound cytochrome P450 enzymes, yeast is generally recognized as safe (GRAS) for producing compounds used as nutraceutical and pharmaceutical ingredients. On the other hand, the synthesis of muconic acid, the precursor of nylon 6,6 and polyethylene terephthalate (PET), was also established recently in yeast as a branch-out pathway from shikimate biosynthesis to replace the environmentally harmful petrochemical synthetic processes. However, the intrinsically complicated genetic and metabolic regulations involved in yeast central carbon metabolism prohibit the production of high-value compounds derived from this treasure at levels of commercial interest.

Here we report our recent progress on establishing yeast platforms to produce shikimate and shikimate-derived bioplastic and pharmaceutical precursors through three aspects: (1) developing a seamless integration with electrochemical hydrogenation at ambient temperature and pressure, which allows the creation of a novel nylon 6,6 replacement polymer with tunable properties[1]; (2) deciphering the bottlenecks of shikimate biosynthesis in both conventional and nonconventional yeast species and achieving the highest production among all the compounds derived from this pathway using yeast platforms[2,3]; (3) establishing platform technologies to enable rapid functional modifications of a series of high-potential nonconventional yeast species[4]. Our work represents the new explorations in expanding the current collection of microbial factories and bridging the gap between biological and chemical catalysis to nurture a sustainable biorenewable chemical industry.

References:

1. M. Suastegui, J. E. Matthiesen, J. M. Carraher, N. Hernandez, N. R. Quiroz, A. Okerlund, E. W. Cochran, Z. Shao, and J.-P. Tessonnier, "**Combining Metabolic Engineering and Electrocatalysis: Application to the Production of Polyamides from Sugar**", *Angewandte Chemie International Edition* DOI: 10.1002/anie.201509653 (featured as the front cover, 2016).

2. M. Suastegui, W. Guo, X. Feng, and Z. Shao, "**Investigating Strain Dependency in the Production of Aromatic Compounds in *Saccharomyces cerevisiae***" (submitted, 2016).

3. M. Gao, M. Cao, M. Suastegui, Y. Wu, J. Shanks, and Z. Shao, "**Innovating a Nonconventional Yeast Platform for Producing Shikimate as the Building Block of High-Value Aromatics**" (submitted, 2016).

4. M. Cao, M. Gao, C. Lopez, Y. Wu, A. Seetharam, A. Severin, and Z. Shao. "**Rapid Isolation of the Hubs of Chromosomes to Facilitate Nonconventional Yeast Engineering**" (submitted, 2016).

Cyanide-free Nitrile Synthesis and "Metabolic Tuning" of Molecular Parts

Asano, Y., Toyama Prefectural University
Originally presented on: 6/27/2016 10:40:00 - 11:05:00

We are exploiting new enzymes from microorganisms, plants and animals and utilizing them for the synthesis of various useful chemicals, and L-amino acid determination, etc. In this presentation, I will describe how we are learning from Nature to get the molecular parts and combine them to synthesize chemicals and improve the biocatalysts for applications.

1. Cyanide-free nitrile synthesis: *In microorganisms*, we have been characterizing the â??aldoxime-nitrile pathwayâ?□, comparing with those found in plants and animals. We have shown that aldoxime dehydratase from *Bacillus* sp. strain OxB-1catalyzes not only the dehydration reaction of Î±-substituted Î²-arylaldoximesÂ to from the corresponding chiral nitriles (1), but also â??Kemp eliminationâ?□ reaction as the first natural enzyme. *In plants*, we identified two cytochrome P450s CYP79D16 (L-Phe to phenylacetaldoxime) and CYP71AN24 (phenylacetaldoxime to mandelonitrile) in *Prunus mume* (2). We also identified a new P450 from a plant *Fallopia sachalinensis*, catalyzing the formation of phenylacetonitrile (PAN) from phenylacetaldoxime (the same reaction as aldoxime dehydratase) (3). *Millipedes* use the â??aldoxime-nitrile pathwayâ?□ to synthesize poisonous HCN precursor cyanohydrin in a quite elegant way (4). **Nitriles as a new target of metabolic synthesis**: An artificial hybrid biosynthetic pathway for phenylacetonitrile (PAN) production was constructed in *E. coli*. In the pathway, L-Phe was converted to PAN by a combination of CYP79A21 from *Arabidopsis thaliana* and *Bacillus*aldoxime dehydratase (5).

2. â??Metabolic tuningâ?□ of molecular parts: **(i) Growth-dependent molecular selection of stable L-tryptophan dehydrogenase:** A mutant enzyme TrpDH L59F/D168G/A234D/I296N with thermal stability was simply obtained by cultivating *E. coli* transformants harboring various mutant genes in an L-tryptophan auxotroph of *E. coli* (6). **(ii) Soluble expression by directed evolution using a fusion reporter method**:Â For the heterologous production of L-lysine Îµ-oxidase (LodA), a plasmid carrying LodA gene fused in-frame with a phleomycine resistant gene was randomly mutated and the plasmids were transformed into *E. coli*Â BL21 (DE3) harboring *lodB*, which acts in the posttranslational modification of LodA. A soluble LodA variant isolated by this method contained six silent mutations and one missense mutation (7). These enzymes are used for selective determination of L-amino acids.

References

1. R. Metzner et al, *ChemCatChem*, **6**, 3105 (2014).
2. T. Yamaguchi et al, *Plant Molec. Biol.*, **86**, 215 (2014).
3. T. Yamaguchi et al, *Plant Molec. Biol.*, published online (2016).
4. M. Dadashipour et al, *Proc. Natl. Acad. Sci. USA*, **112**, 10605 (2015).
5. Y. Miki and Y. Asano, *Appl. Environ. Microbiol.*, **80**, 6828 (2014)
6. D. Matsui et al, *J. Biotechnol.*, **196-197**, 27-32 (2015).

7. D. Matsui and Y. Asano, *Biosci. Biotechnol. Biochem.*, **79**, 11473 (2015).

Synthetic Methylotrophy: Engineering Escherichia coli for Methanol Utilization and Metabolite Production

Papoutsakis, E. T., University of Delaware
Whitaker, W. B.,
Bennett, R. K., University of Delaware
Palmer, M. A.,
Gonzalez, J., University of Delaware
Antoniewicz, M. R., University of Delaware
Jones, J. A., Rensselaer Polytechnic Institute
Koffas, M. A. G., University at Buffalo, the State University of New York
Originally presented on: 6/27/2016 11:30:00 - 11:55:00

Methylotrophy is the ability of microorganisms to utilize one carbon compounds such as methane and methanol as growth and energy sources. The recent discovery of large natural gas reserves has prompted considerable interest in utilizing methane or methanol as a substrate or co-substrate with sugars in the industrial fermentation of fuels and chemicals. Increased biomass and product yields are expected from methane and methanol since both are more reduced than sugars. Native methylotrophs are poor industrial microorganisms since many are strict aerobes, produce few metabolites and lack genetic tools. Therefore, the development of synthetic methylotrophy is of considerable interest. Herein, we describe the development of *Escherichia coli* as a platform microorganism for methanol metabolism. By incorporating native methylotrophic genes, we demonstrate the ability of *E. coli* to metabolize methanol into biomass and metabolites. We further discuss how to overcome the challenges involved in synthetic methylotrophy, including unfavorable methanol oxidation, carbon conservation and regulatory limitations. We also discuss the importance of the pentose phosphate pathway during methanol metabolism in *E. coli* and its regulatory tuning for optimal methanol assimilation. Finally, we provide data demonstrating the use of methanol as a substrate for cell growth, energy generation and metabolite production in *E. coli*.

SUPPORTED by the US DOE ARPA-E agency through contract no. DE-AR0000432.

Synthetic Regulation of Metabolic Pathways at the Single Cell Level

Zhang, F., Washington University in St. Louis

Originally presented on: 6/27/2016 11:55:00 - 12:15:00

Cell-to-cell metabolic variations among isogenetic cell populations are ubiquitous and can substantially affect the ensemble bioproduction of cell cultures. We showed that engineered cells grown from a single colony could have ten-fold variations between individual cells in their metabolic activities.[1] Methods to control non-genetic variation are lacking, such that low-performing variants arise in all cell cultures and their proportions cannot be regulated, resulting in suboptimal ensemble production. We have developed a synthetic regulatory tool, termed *in vivo* population quality control (PopQC), which employs an intracellular, product-responsive biosensor to regulate the expression of a survival gene – permitting continuous enrichment of high-performing, non-genetic variants under a given selection pressure and leading to enhanced ensemble production. We validated PopQC in two biosynthetic pathways and tested two alternative design principles, yielding enhanced bioproduction of both free fatty acid (FFA) and tyrosine. We further confirmed that PopQC improves ensemble performance by selecting for non-genetic high-performers, an avenue unexplored by existing methods. Given the ubiquity of non-genetic phenotype variation, PopQC should be applicable to a variety of metabolic pathways, providing a unique approach to realizing efficient bioproduction for metabolic engineering.

[1] Xiao Y., Bowen C.H., Liu D., Zhang F., Exploiting non-genetic, cell-to-cell variation for enhanced biosynthesis. ***Nat Chem Biol*** Accepted

Design and Application of Transcription Factor Based Biosensors

Siedler, S., Technical University of Denmark
Khatri, N.,
Vogt, M., Forschungszentrum Juelich
Marienhagen, J., Forschungszentrum Juelich
Sommer, M.,
Jönsson, H., KTH Royal Institute of Technology
Originally presented on: 6/27/2016 12:15:00 - 12:35:00

Transcription-factor-based biosensors are an emerging technology to accelerate cell factory production and optimization. We describe several successful applications of those biosensors, e.g. for the optimization of an NADPH-dependent enzyme (Siedler et al. 2014a) and detection of flavonoid production in *Escherichia coli* (Siedler et al. 2014b). Furthermore, a lysine biosensor was combined with the recombineering technology in *Corynebacterium glutamicum* for the efficient identification of beneficial mutations at the genome level by detection of the lysine concentration *in vivo* (Binder et al. 2013). Direct transplantation of heterologous transcriptional regulators into new hosts for the generation of functional biosensors is often challenging and expression of the regulator needs to be fine-tuned. We analyzed the effect of different translational strengths, introduced by variation of the RBS, to the dynamic range of a biosensor. The best variant showed an up to 40-fold increased fluorescent signal upon induction. Another challenge of biosensor application is that intracellular concentration of small molecules is not always correlating with the extracellular concentration. We describe a novel screening method for identification of a producer strain in co-cultivation experiments of a producer strain and a biosensor containing strain in microfluidic droplets.

References

Binder, S; Siedler, S; Marienhagen, J; Bott, M; Eggeling, L (2013) Recombineering in Corynebacterium glutamicum combined with optical nanosensors: a general strategy for fast producer strain generation. In: *Nucleic Acids Res.*41(12): 6360–6369.

Siedler, S; Schendzielorz, G; Binder, S; Eggeling, L; Bringer, S; Bott, M (2014a) SoxR as a Single-Cell Biosensor for NADPH-Consuming Enzymes in Escherichia coli. In: *ACS Synth Biol* (3), S. 41–47.

Siedler, S; Stahlhut, S. G.; Malla, S; Maury, J; Neves, A. R. (2014b) Novel biosensors based on flavonoid-responsive transcriptional regulators introduced into Escherichia coli. In: *Metab. Eng.* 21, S. 2–8.

General Purposing Synthetic Biology: a Biofoundry in a Box

Lipscomb, T., Muse Biotechnology, Inc.
Gill, R. T., University of Colorado at Boulder
Garst, A., University of Colorado Boulder
Originally presented on: 6/27/2016 14:05:00 - 14:30:00

What could you build with unlimited DNA writing? Productivity of DNA synthesis continues to increase at exponential rates. The oligonucleotide synthesis market alone is forecast to reach nearly $2B USD by 2020 with an annual growth rate in excess of 10%. This increase in oligo synthesis capacity and the associated reduction in cost is transforming synthetic biology; cheap oligo synthesis will remake the impossible to the new routine. Muse's foundational technology is a proprietary synthetic biology platform that scales directly with multiplex oligo synthesis, enabling designer DNA construction at 100X lower cost than the current state of the art. To date, Muse has demonstrated the ability to generate and track over 50,000 designer mutants simultaneously. Applications of this technology include sequence to activity mapping for site saturation mutagenesis, protein and pathway engineering, and strain reconstruction of complex evolved strains to name a few. Gene editing capabilities have been demonstrated in various industrially relevant organisms, including applications across biotechnology. Further, this technology is being automated in the form of a bench-top device that allows a single user to perform a broad range of molecular biology endeavors at rates 1000x beyond the current state of the art.

Our Scenarios to Scrap and Build Bacterial Genomes by a Bacillus subtilis Genome Vector

Itaya, M., Keio University
Originally presented on: 6/27/2016 14:55:00 - 15:20:00

All cells/microbes possess their own genome DNA. Besides their robustness in structure, recent technologies on giant DNA synthesis can reproduce existing genomes and related gene networks. These technologies seem to break the first step to produce novel chassis run by synthesized genomes. We have long developed a unique molecular cloning system that allows whole genome synthesis using Bacillus subtilis as a cloning host. In addition to the first successful case for a whole genome cloning of Synechosystis PCC6803 [1], many elaborate techniques have been under study to produce fully designed genomes and activate them in appropriate chassis. I would like not only to recap our original systems but also present scenarios with on-going works aiming at production of cells with designed genomes [2]. They might lead to new pipelines through which functional genomes are tested for novel synthetic biology fields.

References:
[1] Itaya, et. al., PNAS 102, 15971-15976 (2005).
[2] Itaya, M. In Synthetic Biology tools and applications (ed. Zhao, H), Academic Press, Elsevier Inc. pp. 225-242 (2013).

Sugar and Spice in Mammalian Cell Cultures

Michael J. Betenbaugh

Metabolic Engineering
June 2016

Introduction

Mammalian Cell Factories
Ultimate Goal

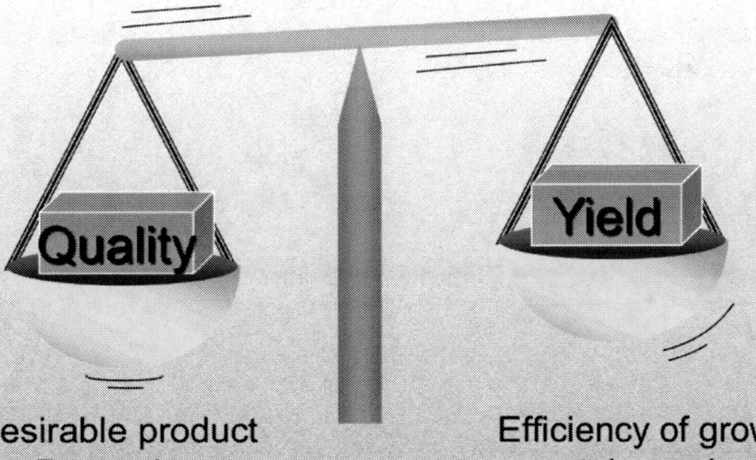

Desirable product
Properties
"Sugar"

Efficiency of growth &
protein production
"Spice"

Mammalian Cell Culture Insults

Cell Viability: Comparison of Depleted (Spent) and Nutrient Rich (Fresh) Media

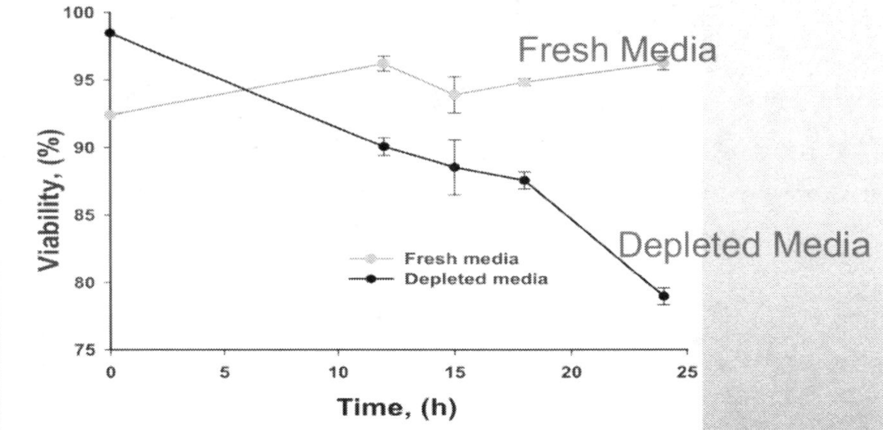

Rapid decline (hours) in cell viability and productivity for cells in spent media

Alex Druz and Yossi Shiloach, NIH

Mammalian Cell Death by Apoptosis

Healthy Chinese Hamster Ovary (CHO) cell

Apoptotic CHO cell

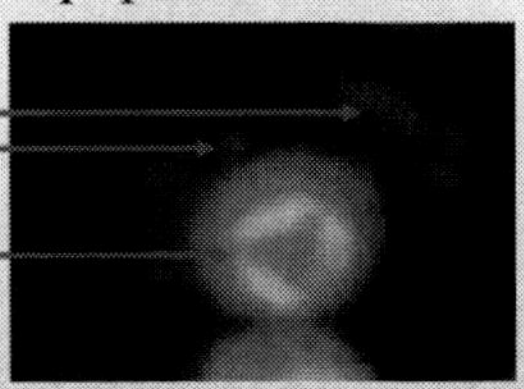

Acridine Orange Staining

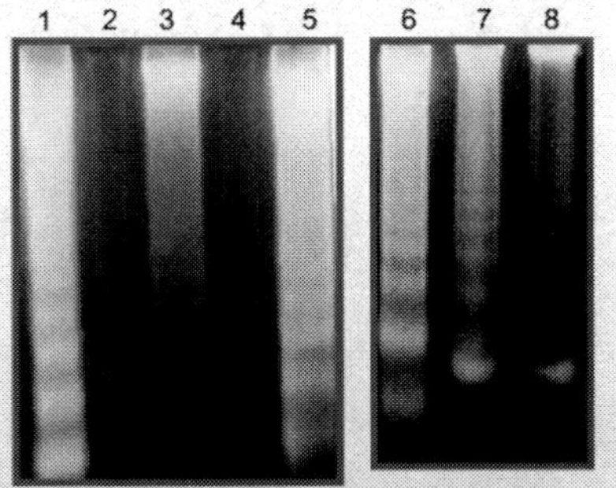

CHO cells (3 and 7) and Baby Hamster Kidney (BHK) (5 and 8) undergo DNA fragmentation

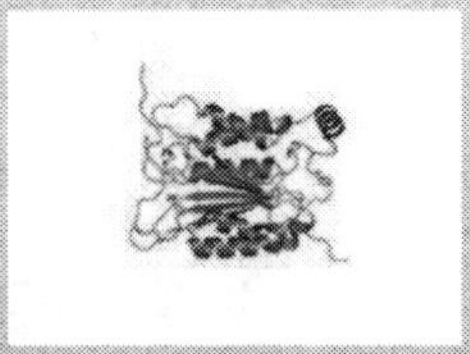

Apoptosis stimulated by activation of intracellular proteins:

cysteine-**aspartic** prote**ases** =**CASPASES**

Genome-wide RNA interference screening

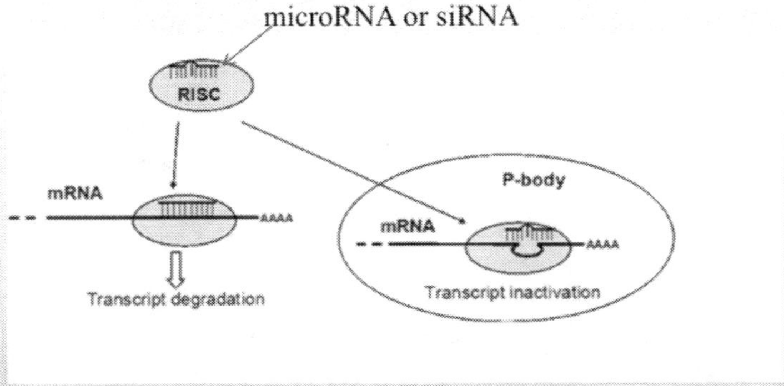

- MicroRNA (miRNA) - silence multiple genes
- Works together RISC complex
- Small interfering RNA (siRNA) - silence one gene

Do miRNA play a role in Apoptosis Activaton?

Large Scale Screening of miRNA for apoptosis?

1,038 mouse microRNA mimics transfected into mammalian cell lines: mmu-mir-….

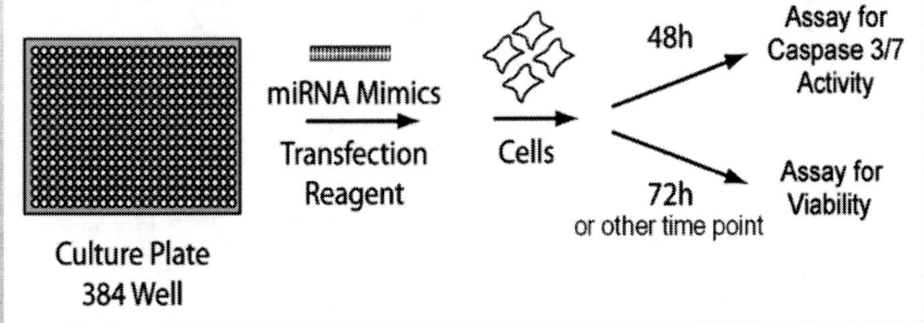

Apoptotic miRNA mimics screen 7

Caspase-3/7 **Viability**

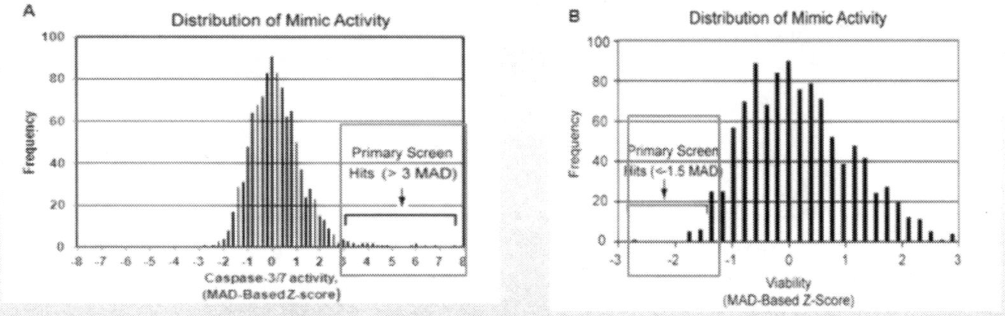

Mouse MicroRNAs associated with Caspase Activity

- mmu-miR-15a-3p
- mmu-miR-706
- mmu-miR-1948-5p
- mmu-miR-1948-3p

Viability reduction in mammalian HeLa and AsPc-1 cells overexpressing mmu-miR-15a-3p

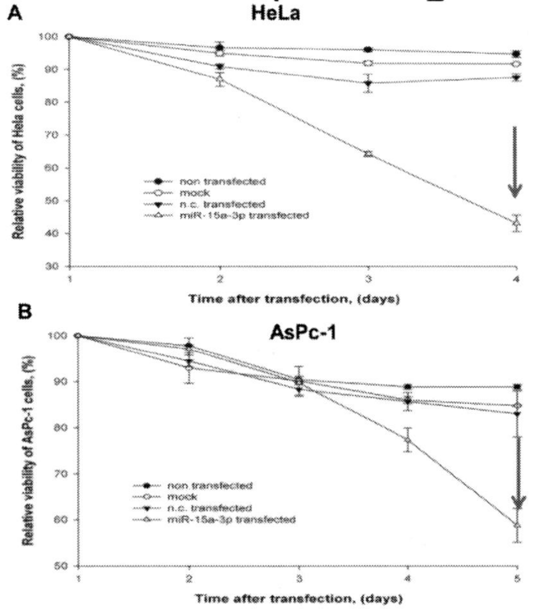

- Viability of HeLa miR-15a-3p was 43% on day 4 compared to 87.5% for n.c. transfected HeLa on day 4
- Viability of AsPc-1 transfected with miR-15a-3p to 59% on day compared to 83% for n.c. transfected AsPc1 on day 5
- Loss of viability in both cells lines followed Caspase-3/7 activation

8

Identification of mmu-miR-15a-3p targets in mice

gene symbol	binding sites	fold down-regulation in miR-15a-3p samples	P-value	Anti-apoptotic role of targeted gene
bcl-x_L	CAG:AUGGCCU (2-8:10-12) GU:UGGCCU (2-7:9-10) U:UGGCCU (2-7:9) 4 of UGGCCU (2-7)	3.34± 0.23	0.041	Bcl2l1(bcl-x_L) inhibits formation of permeability transition pore and subsequent release of cytochrome C and other pro-apoptotic molecules from mitochondria by binding to bax and bak
naip5	AUGGCCU (2-8)	2.40± 0.15	0.013	Also known as naip1, birc1a. Baculoviral IAP domain-containing protein. IAP domains are known to inhibit apoptosis by direct inhibition of Caspases
fgfr2	2 of UGGCCU (2-7)	3.89± 0.44	0.007	Fibroblast growth factor receptor 2 which binds fibroblast growth factor-2 ligand and initiates signaling cascade leading to over-expression of anti-apoptotic genes bcl-X_L and XIAP
mybl2	AUGGCCU (2-8) CAGU:UGGCCU (2-7:9-12) UGGCCU (2-7)	2.65± 0.17	0.032	Mybl2 belongs to the family of transcription factors which are involved in regulation of cell survival by induction of bcl2 transcription

Pro- and Anti-Apoptosis Pathway

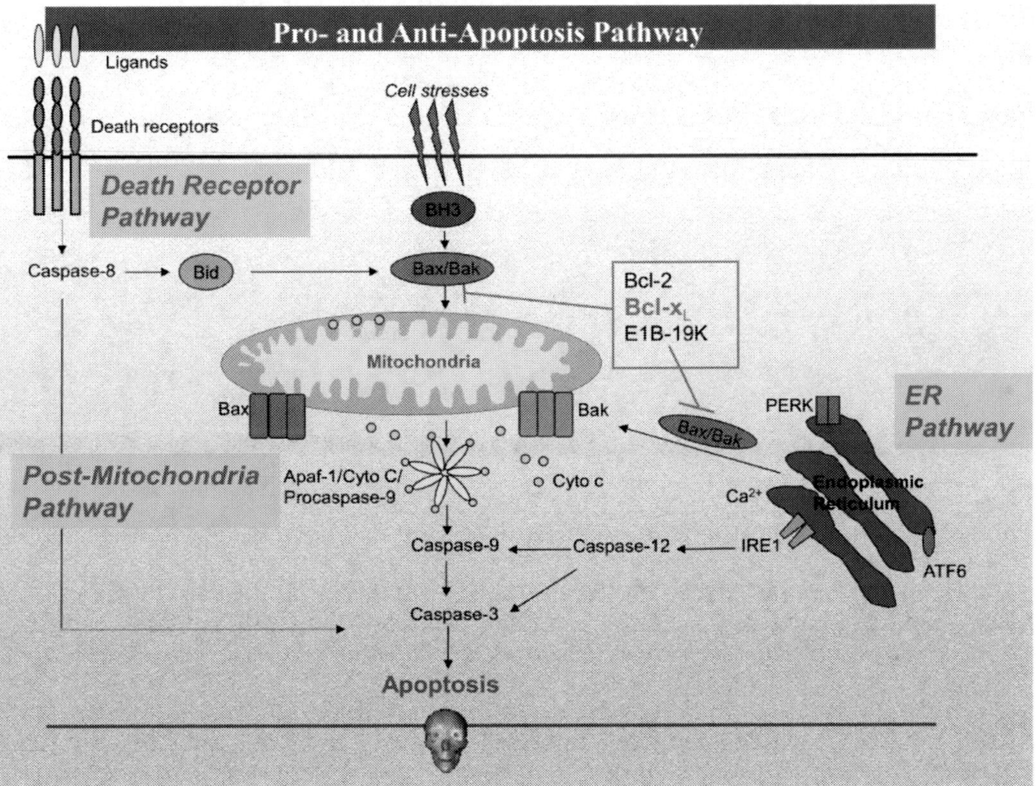

Impact of microRNA mmu-miR-15a-3p on Bcl-xL expression levels in mammalian cells?

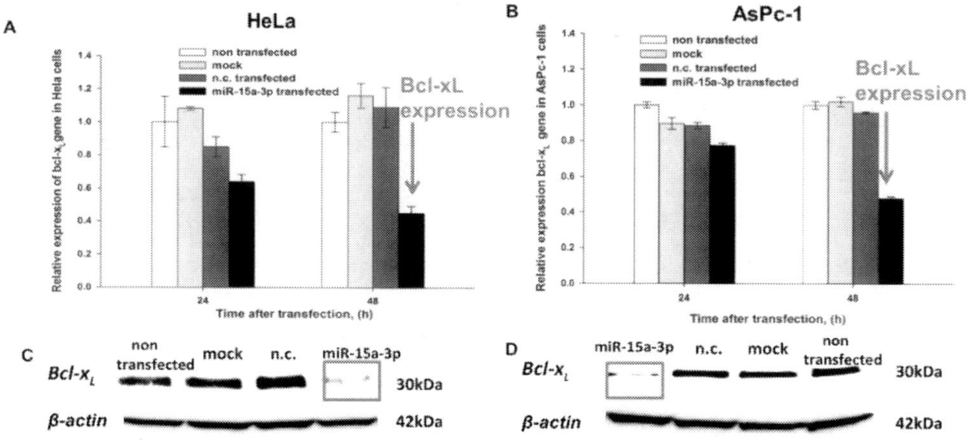

- Expression levels of *bcl-x$_L$* gene in miR-15a-3p transfected HeLa (A) and AsPc-1 (B) cells lowered to 45 and 48% respectively compared with non transfected cells at 48h after transfection
- Protein levels of Bcl-x$_L$ were reduced in both miR-15a-3p transfectedHeLa (C) and AsPc-1 (D) at 72h
- Other putative anti-apoptotic targets of miR-15a-3p in mouse (fgfr2, mybl2, naip5) not affected in both HeLa and AsPc-1

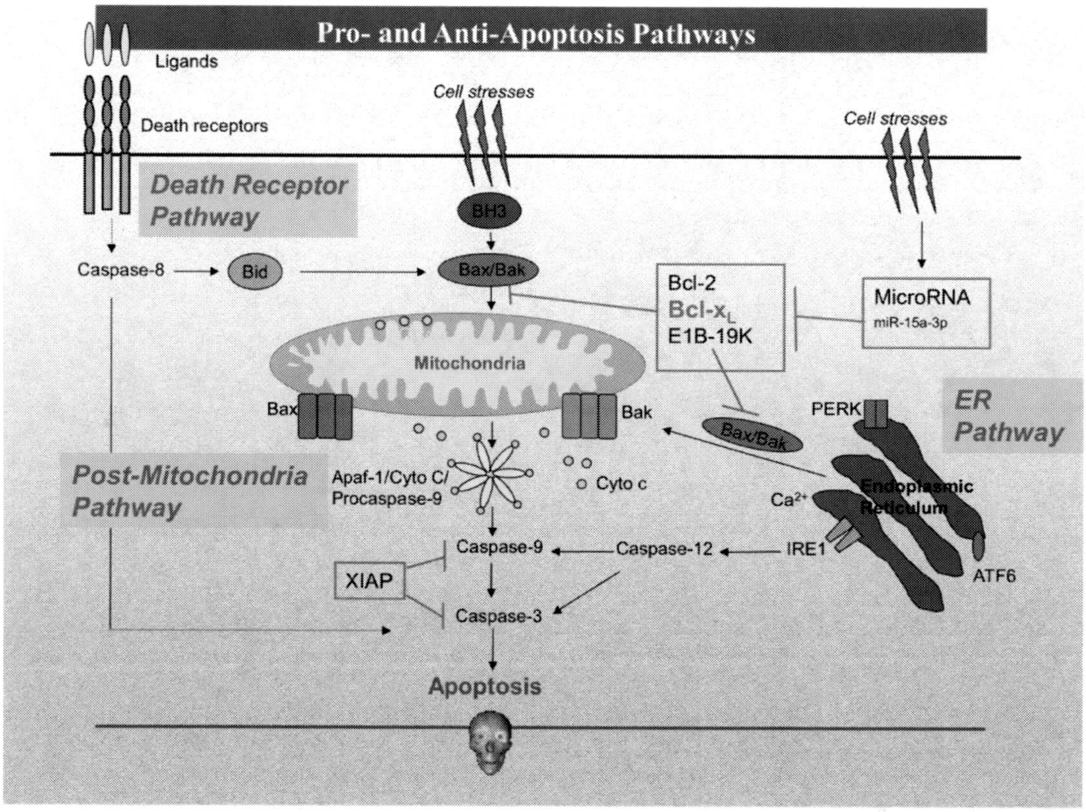

How does miR-15a-3p target bcl-x_L ?

eGFP gene Bcl-x 3'UTR

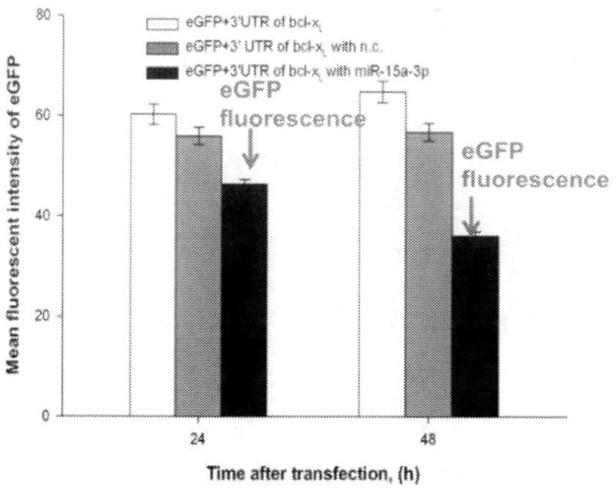

- Binding of miR-15a-3p to 3'UTR of *bcl-x_L* gene added to the terminus of eGFP was tested in HeLa cells
- GFP expression in miR-15a-3p transfected cells was reduced at 24h after transfection and further reduced at 48h

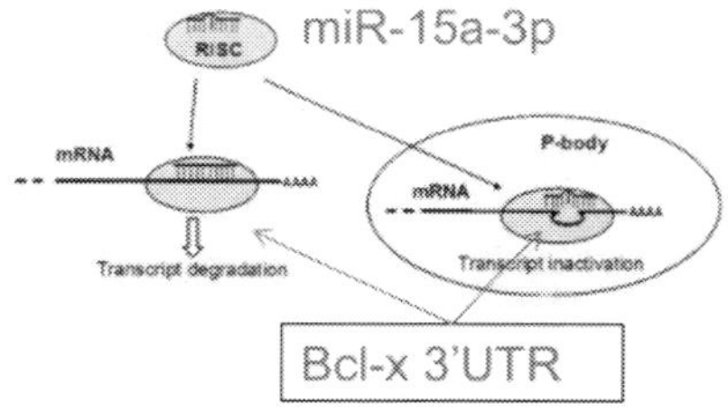

13

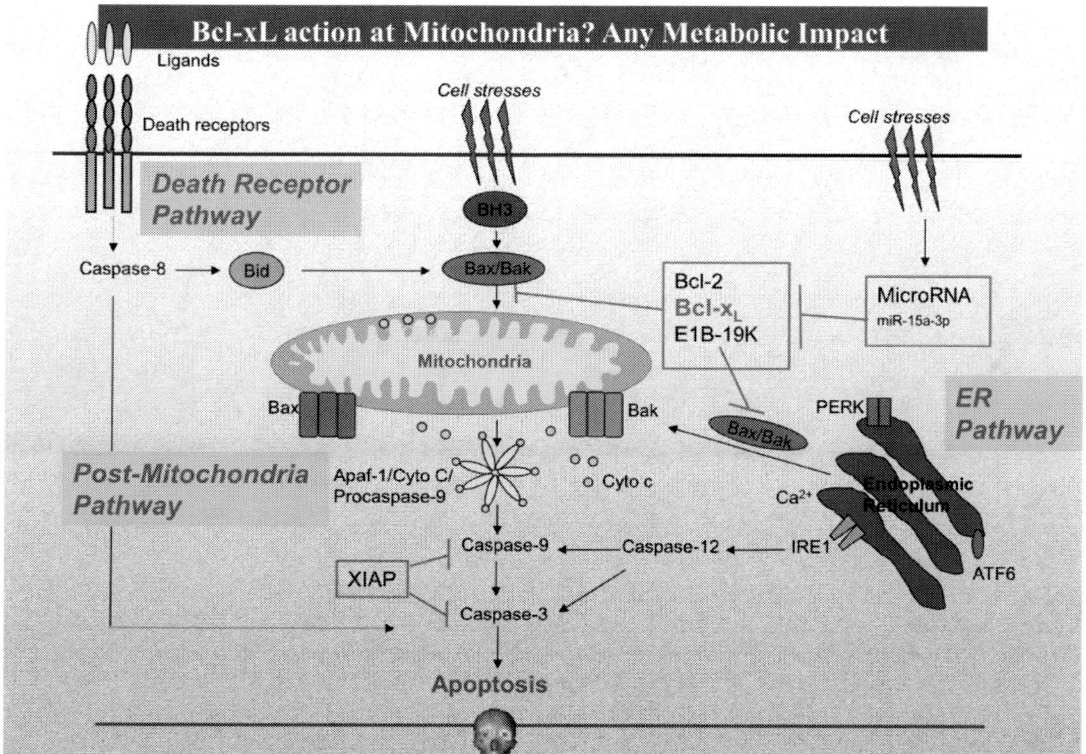

Where is Bcl-x_L in relation to the Mitochondria and ER Mitochondrial Associated Membrane (MAM)

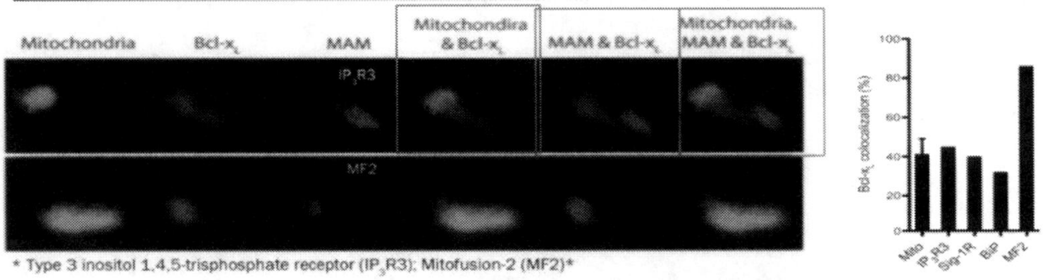

* Type 3 inositol 1,4,5-trisphosphate receptor (IP_3R3); Mitofusion-2 (MF2)*

Some Bcl-x_L in red clusters between ER and next to mitochondria and associates with the ER MAM

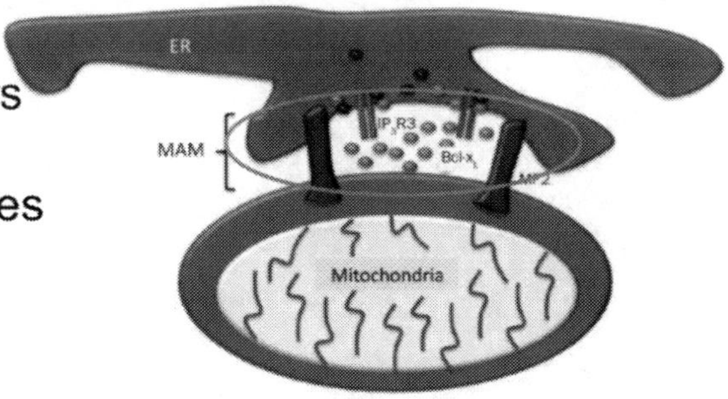

Abasha Lewis and Tsung Ping Su, NIH

Is Bcl-x_L involved with Calcium Ca^{2+} Transport??

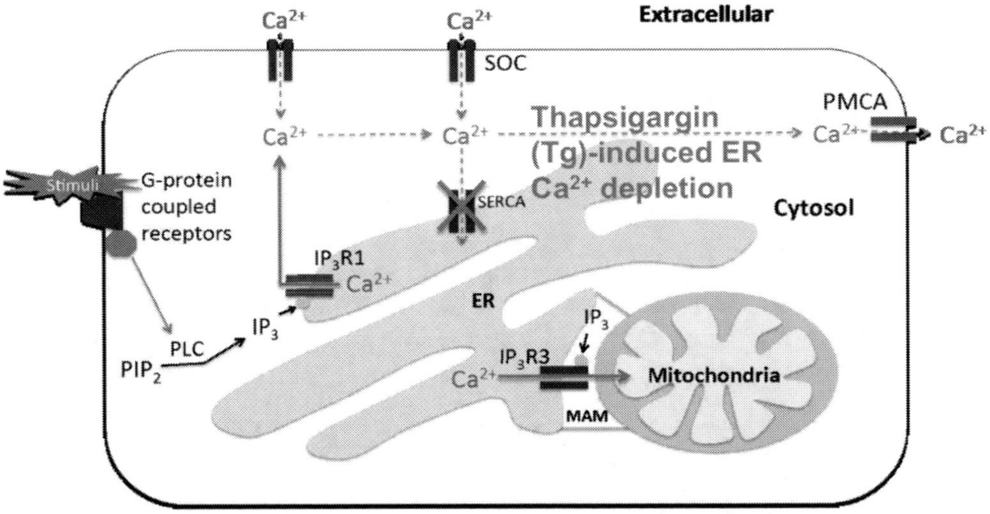

Calcium stored in ER has 2 possible destinations: Cytosol and Mitochondria

Abasha Lewis and Tsung-Ping Su, NIH

INTRACELLULAR Ca^{2+} FLUX: IMPACT OF Bcl-x_L

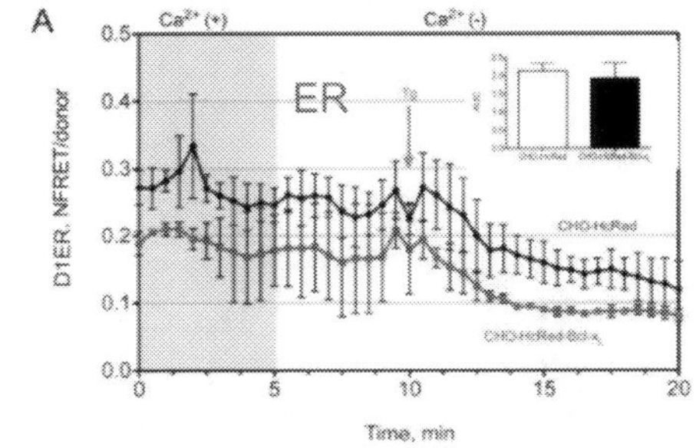

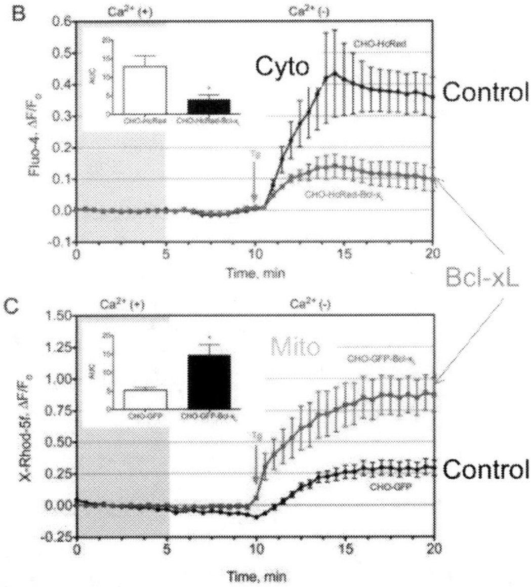

- Bcl-x_L expression:
  - Reduces cytosolic Ca^{2+} efflux
  - Enhances mitochondrial Ca^{2+} efflux

Abasha Lewis and Tsung Ping Su, NIH

POSSIBLE ROLE OF Bcl-x_L IN Ca^{2+} TRANSPORT

Control Mammalian (CHO-K1)

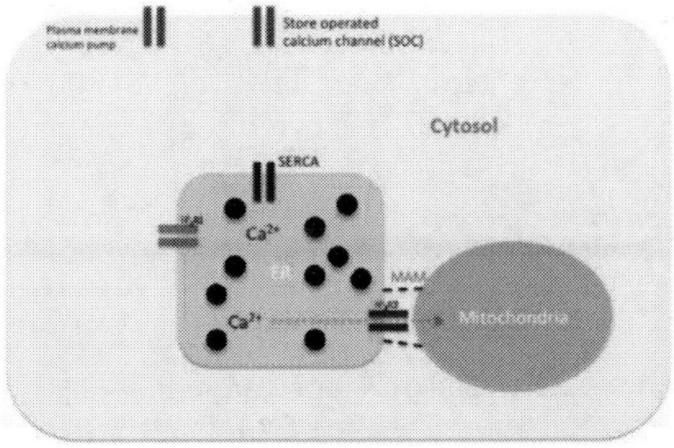

Overexpressing Bcl-x_L

CHO-Bcl-x_L

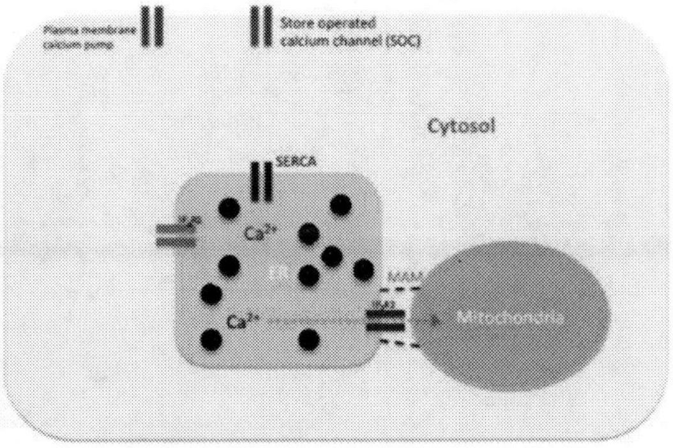

-Bcl-x_L Enhances Ca^{2+} transport into the mitochondria, possibly through the MAM

-Does Ca^{2+} transport change cellular metabolism?

What about downstream metabolic effects of Increased Ca^{2+} in Mitochondria?

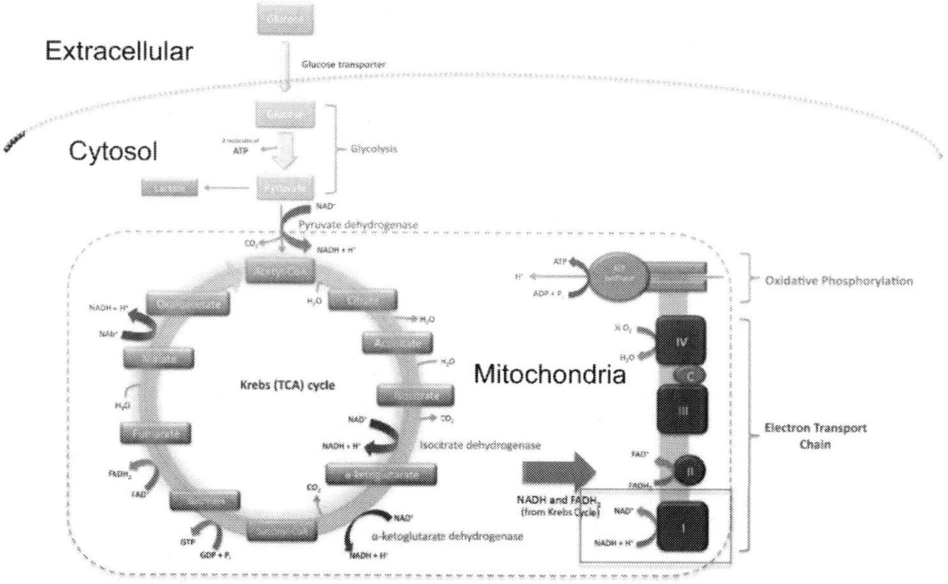

Part 3 of 5: Bcl-x_L and Ca^{2+} Signaling

NADH oxidase activity of Complex I of the electron transport chain

- Bcl-x_L and intracellular Ca^{2+} enhances Complex I activity

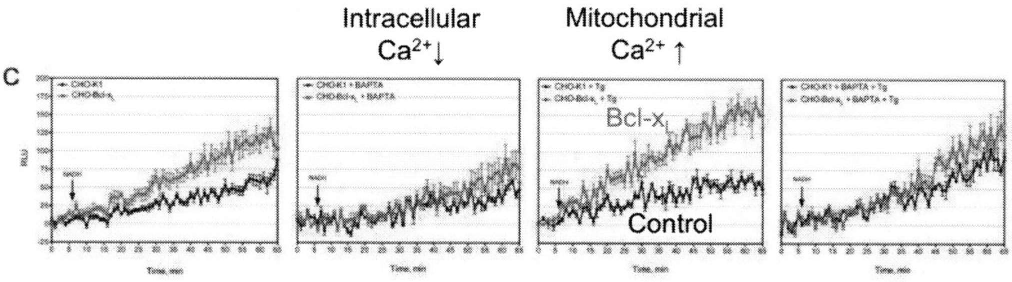

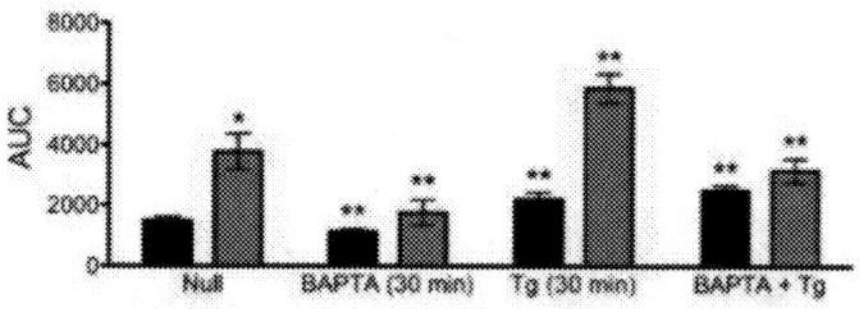

Tg=Thapsigargin to Enhance Ca releast

BAPTA= Ca++ Chelator

Abasha Lewis and Tsung Ping Su

BCL-X_L REDUCES LACTATE ACCUMULATION AND INCREASES NADH/NAD RATIOS

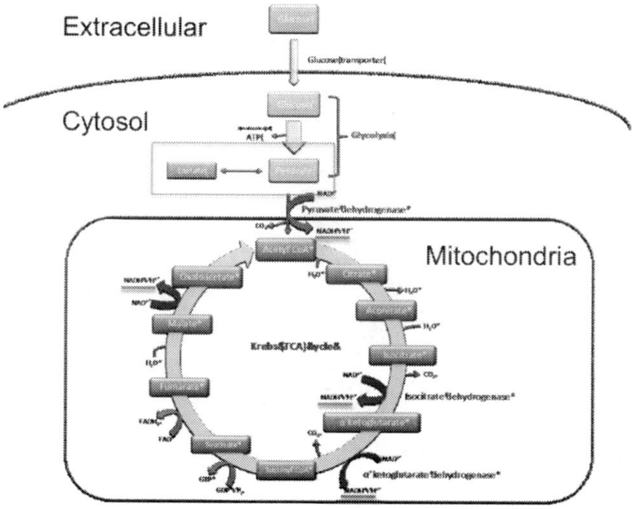

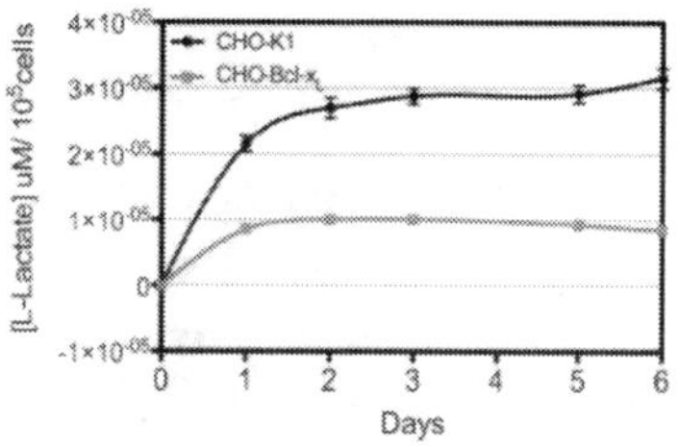

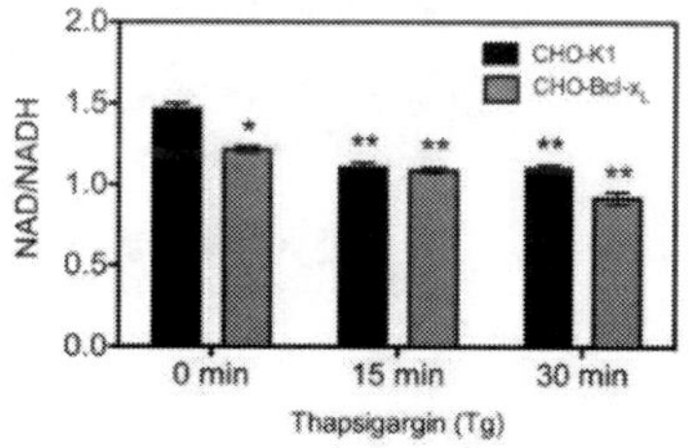

Abasha Lewis and Tsung Ping Su

^{13}C Metabolic Flux Analysis: Increased flux into TCA for mammalian cells overexpressing anti-apoptosis (Bcl-2Δ) and antibody (M1) genes

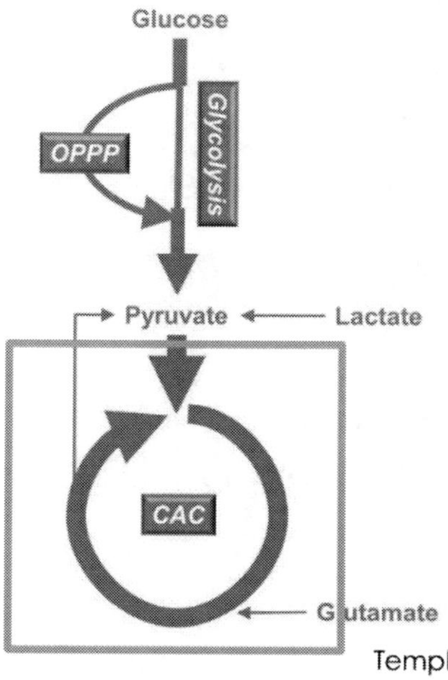

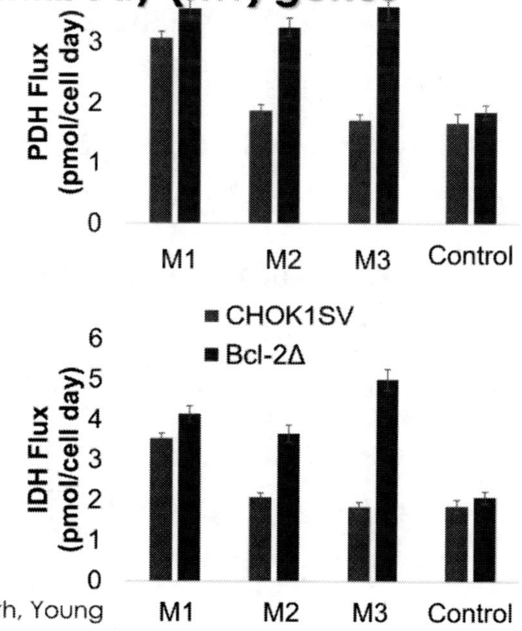

Templeton, Smith, Young

Impact of anti-apoptosis protein expression on specific productivity

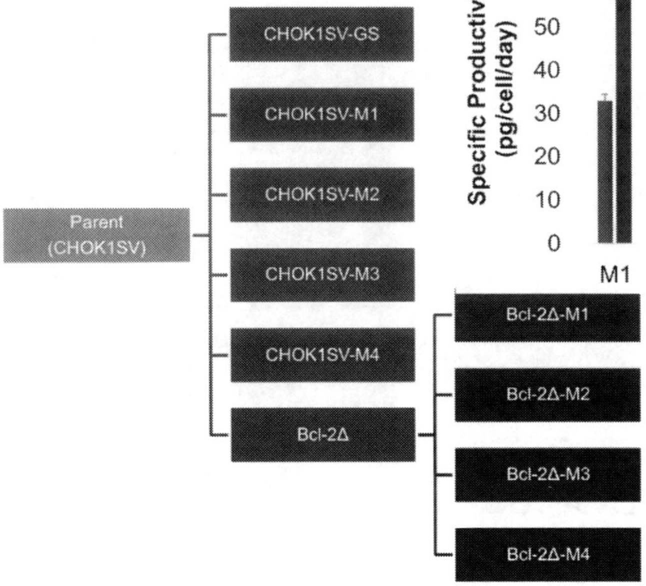

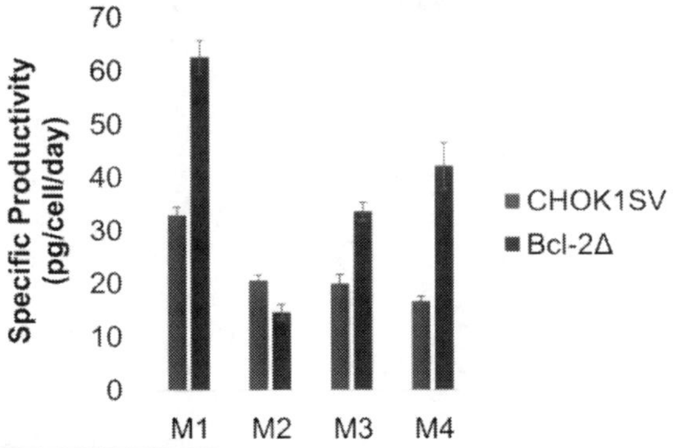

Increased productivity for 3 out of 4 cell lines with anti-apoptosis protein expression

Summary

- **High-throughput screening delivers miRNAs with enhanced pro-apoptotic properties**
- **MicroRNAs reduce apoptosis resistance by degrading Bcl-xL in mammalian cells**
- **Overexpression of Bcl-xL alters Ca flux in mammalian cells**
- **Altering Bcl-xL and Bcl-2delta levels alters the intracellular metabolic state as well as cellular productivity in mammalian cells**

Acknowledgements

Collaborators

- Yossi Shiloach, NIH
- Tsung-Ping Su, NIH
- Kevin Smith, Steve Lang, (Janssen R&D),
- Neil Templeton, Jamey Young (Vanderbilt)
- Haimanti Dorai

Funding

- NSF GOALI award CBET-1067766

Caspase-3/7 activation by mmu-miR-15a-3p in two Mammalian Cell Lines: HeLa and AsPc-1

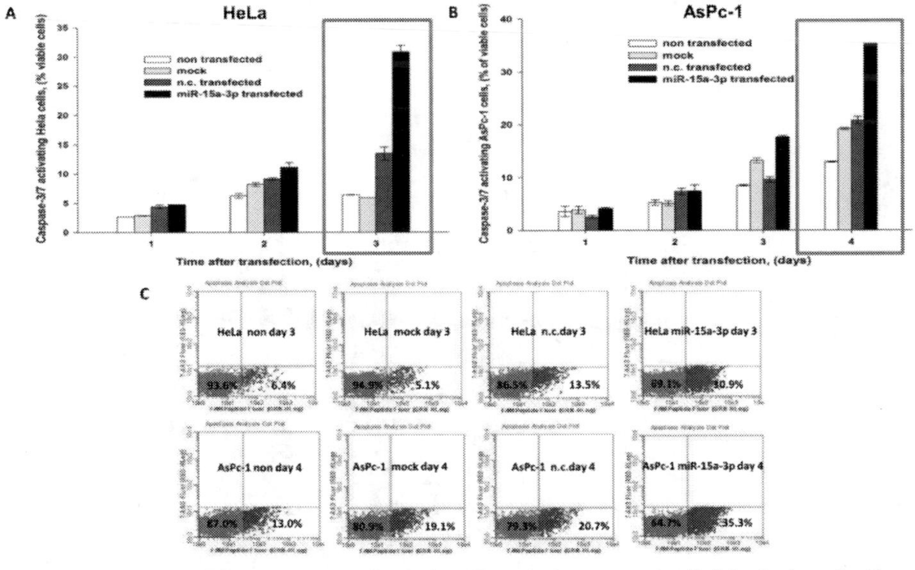

- Tested for both cell lines: non transfected cells , mock, n.c. and miR-15a-3p transfection
- Caspase-3/7 activated in miR-15a-3p transfected HeLa on day 3 : 30.9% of transfected were apoptotic
- Caspase-3/7 activated in miR-15a-3p transfected AsPc-1 on day 4 : 35.3% of transfected were

^{13}C : Increased Lactate Consumption and Antibody Production (qP) in Anti-Apoptosis Cells

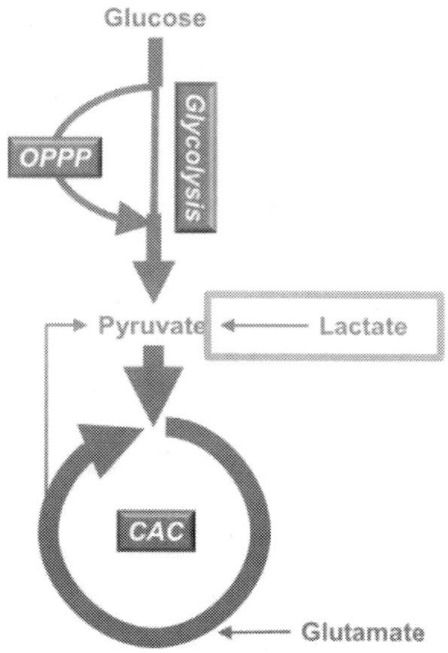

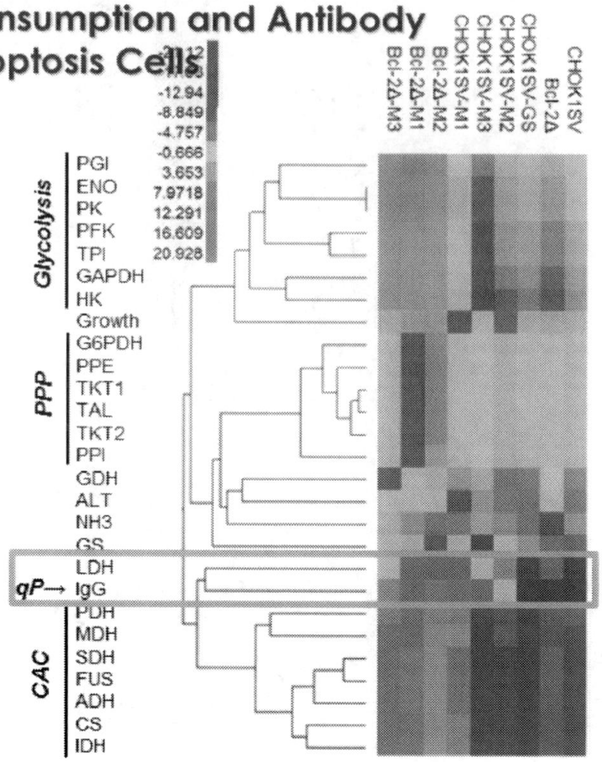

IgG-producing lines channel pyruvate carbon into CAC

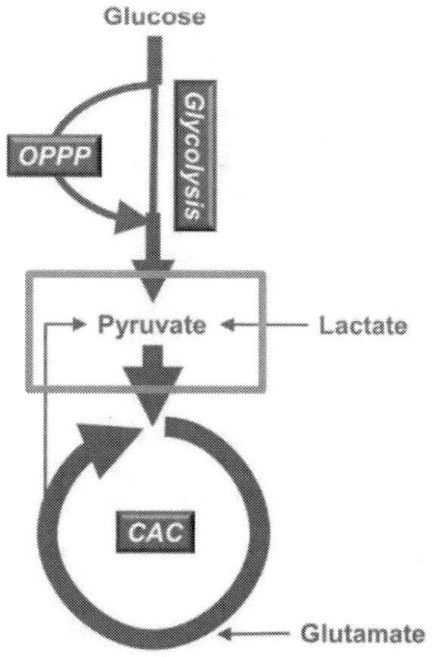

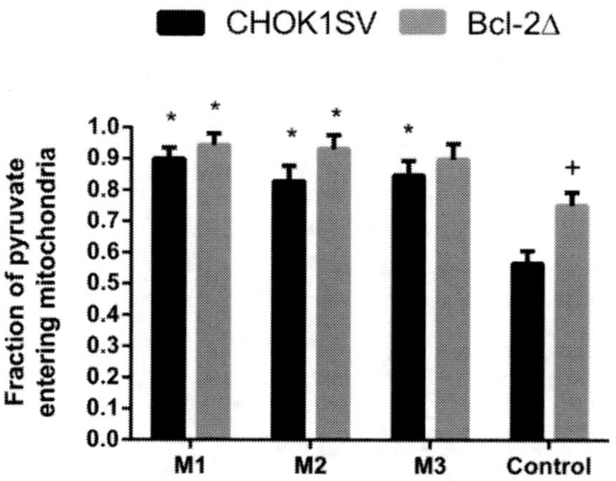

^{13}C MFA results

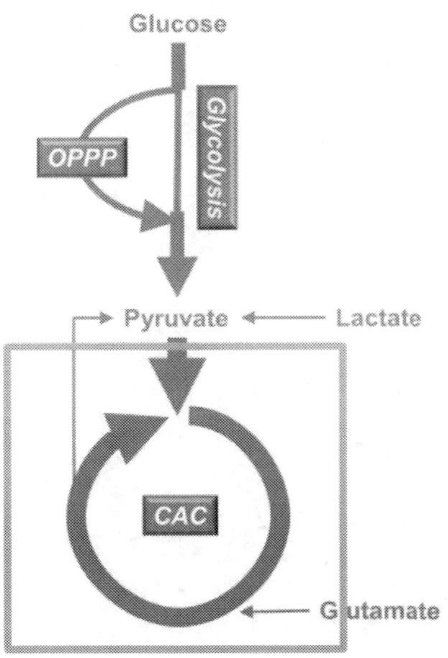

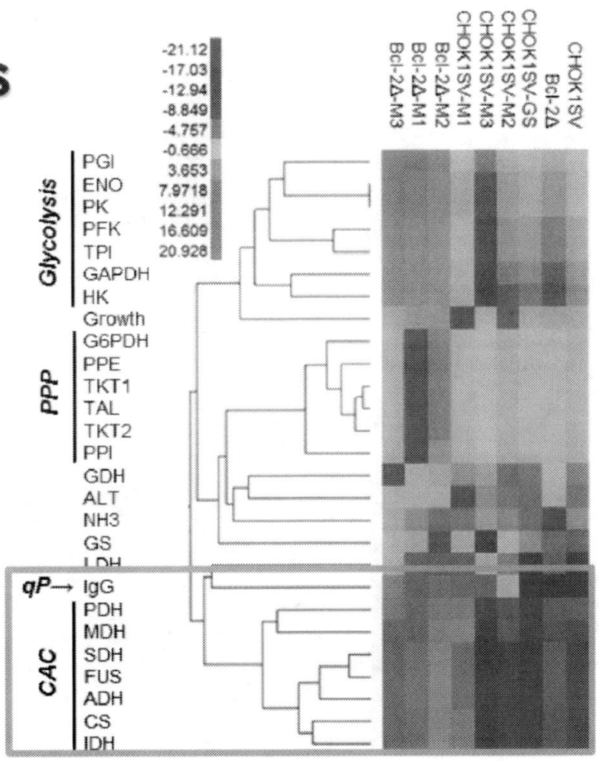

Glucose Deprivation of Chinese Hamster Ovary Cells

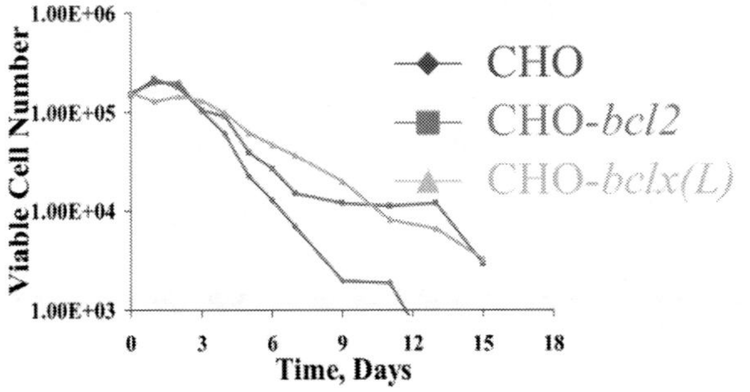

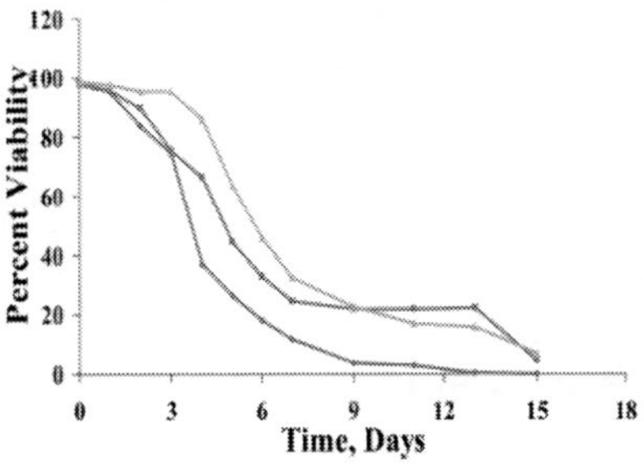

- Expression of anti-apoptosis proteins Bcl-2 and Bcl-X improves survival
- Viable cell number and percentage viability increase with anti-apoptosis
- CHO cells growing on medium with serum subsequently deprived of glucose

Mastrangelo et al., Biotech. Bioeng. 2000

More pyruvate directed to mitochondria for cells overexpressing anti-apoptosis genes (more oxidative metabolism)

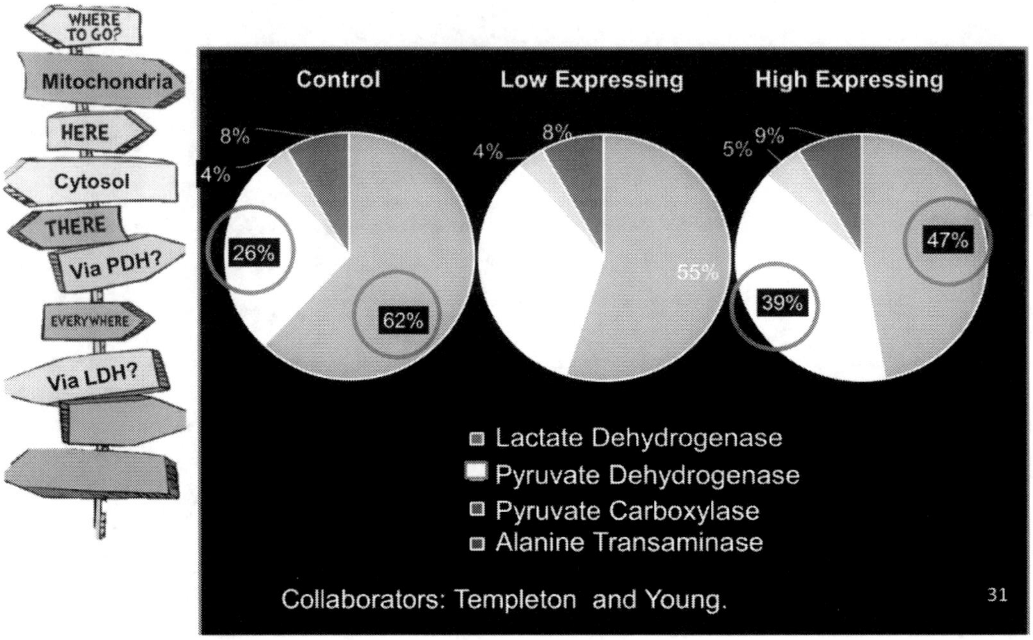

Rate controlling reaction of Isocitrate Dehydrogenase (IDH)

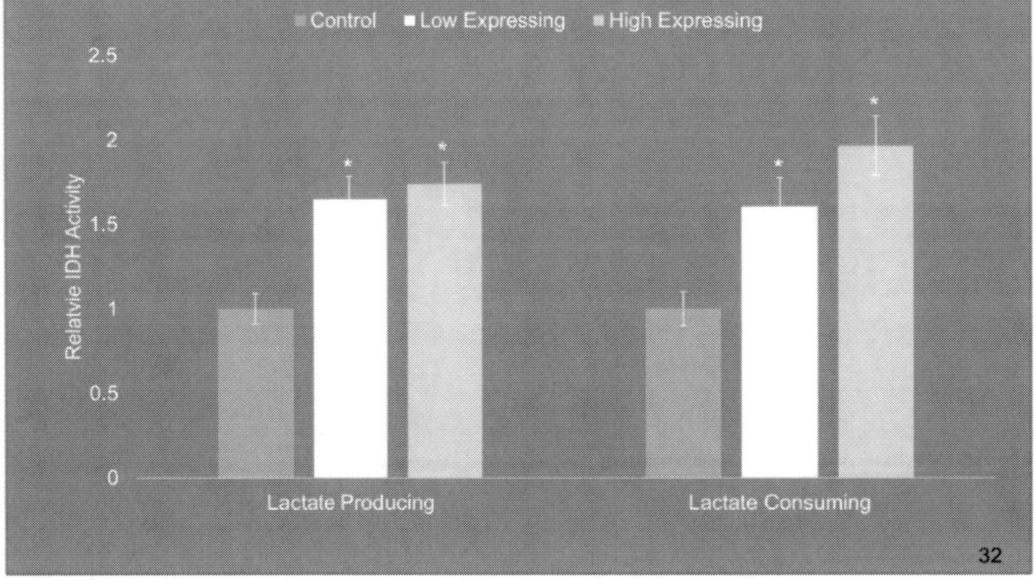

Confocal Imaging of Bcl-x_L at Mitochondria

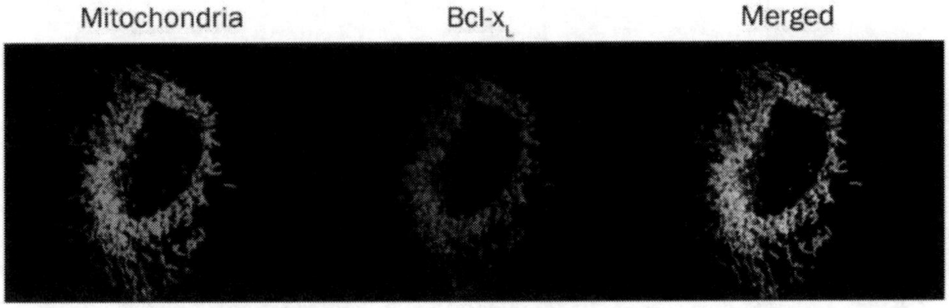

- CHO-Bcl-x_L cells:
  - Transiently transfected with GFP-mito vector
  - Anti-Bcl-x_L antibody, followed by Alexa Fluor-conjugated secondary antibody

Where is Bcl-x_L in relation to the ER Mitochondrial Interface (MAM)?

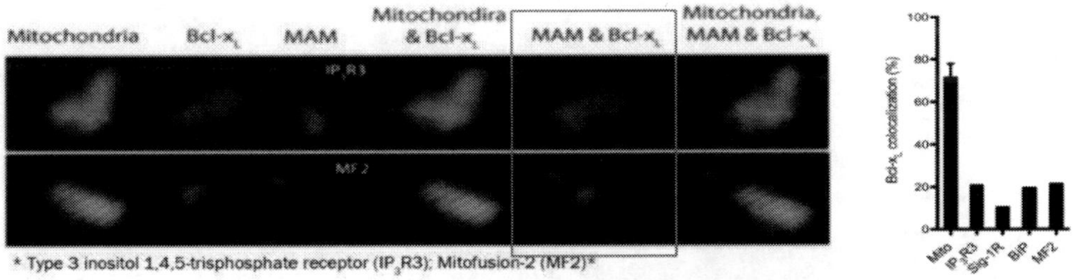

* Type 3 inositol 1,4,5-trisphosphate receptor (IP_3R3); Mitofusion-2 (MF2)*

Type 1:
Bcl-x_L clusters on mitochondria adjacent to the MAM

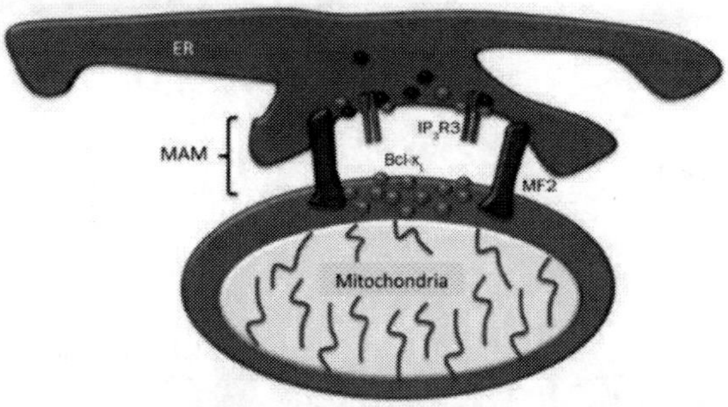

Conclusions for miRNA Screening

- miR-15a-3p was identified to have a pro-apoptotic role in mammalian cells (human and mouse cells were tested)
- This miRNA has a pro-apoptotic role in multiple carcinoma cell lines
- The role of miR-15a-3p was verified in HeLa and AsPc-1 cells: miR-15a-3p binds to bcl-x_L directly and reduces its expression which results in activation of Caspase-3/7 and loss of cell viability
- The pro-apoptotic role of miR-15a-3p in human cancers may be dependant on levels of bcl-x_L expression
- Large-scale screening methodology was validated and can be utilized for identify additional microRNAs targets with potential anti- and pro-apoptotic capabilities

Apoptotic miRNA mimics screen

Caspase-3/7 activation by miR 15a-3p transfection

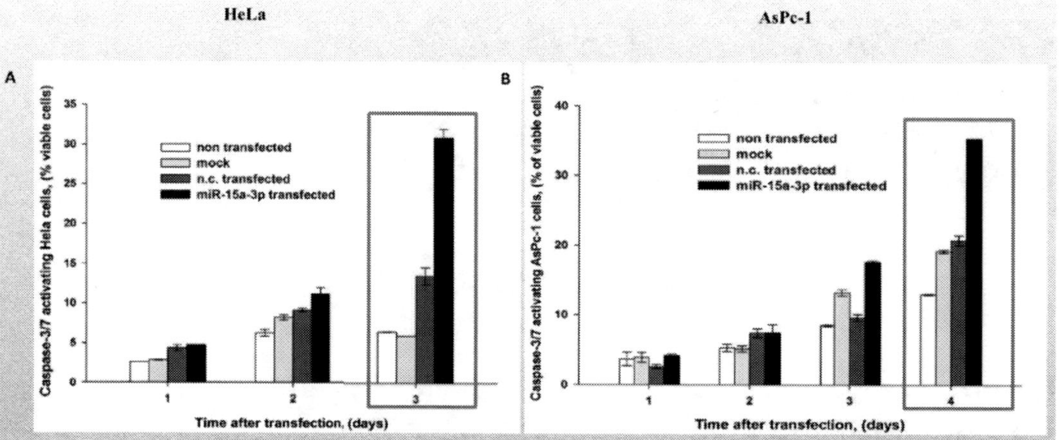

Apoptotic miRNA mimics screen

Viability reduction by miR 15a-3p

HeLa AsPc-1

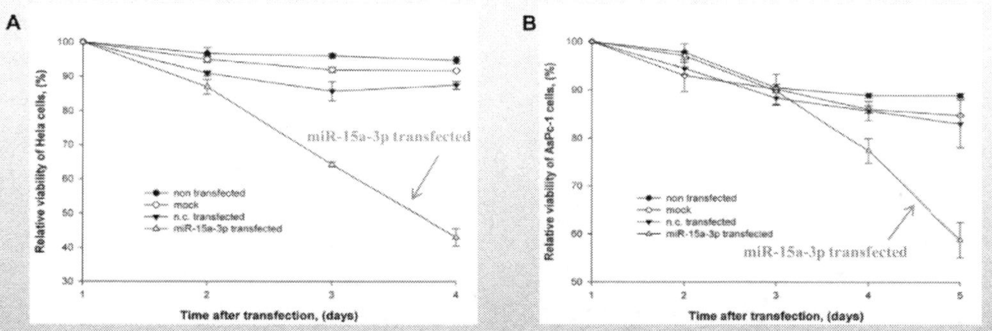

Apoptotic miRNA mimics screen

miR-15a-3p targets bcl-x_L

HeLa AsPc-1

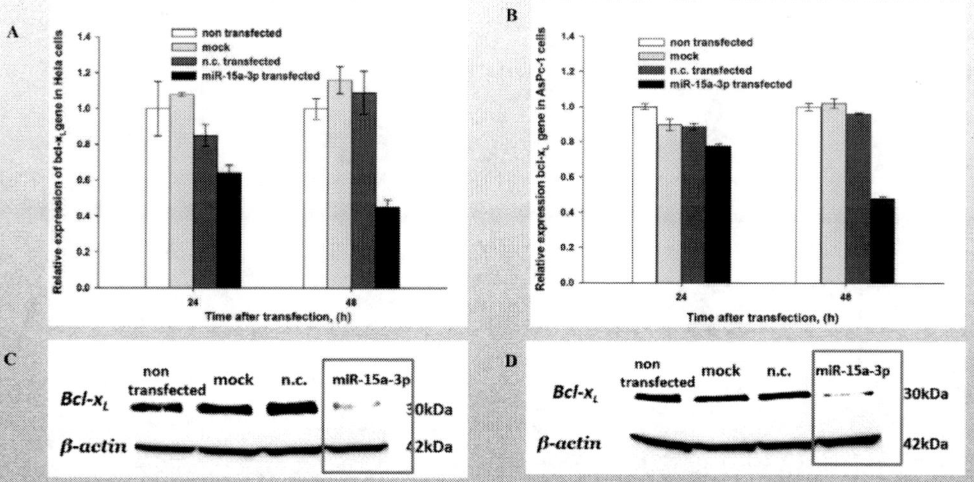

miR-15a-3p binds to bcl-x_L in the 3'UTR and reduces its expression which results in activation of Caspase-3/7 and loss of cell viability

MicroRNA analysis of depleted medium

Mouse miR ID (mmu-)	Variant	Mature miR sequence	Fresh media relative fluorescent signal	Depleted media relative fluorescent signal	Up-regulation in depleted media (times)	p-value
miR-297a	-	AUGUAUGUGUGCAUGUGCAUGU	14±4	1412±220	101±13	0.039
miR-466a	3p	UAUACAUACACGCACACAUAAGA	8±1	205±30	25±1	0.050
miR-466b	3-3p	AAUACAUACACGCACACAUAAGA	5±1	209±39	42±1	0.035
miR-466d	3p	UGUGUGUGCGUACAUGUACAUG	10±1	648±122	65±6	0.039
	5p	UAUACAUACACGCACACAUAG	8±2	552±92	69±6	0.050
miR-466f	3p	CAUACACACACACAUACACAC	19±1	5056±105	266±8	0.009
	5p	UACGUGUGUGUGCAUGUGCAUG	102±13	22200±5982	218±31	0.036
miR-466g	-	AUACAGACACAUGCACACACA	51±9	14935±4286	293±32	0.038
[illegible]	-	[illegible]	29±4	[illegible]	[illegible]	[illegible]
miR-467a	minor	AUAUACAUACACACACCUACAC	44±1	9567±2001	217±40	0.025
miR-467b	minor	AUAUACAUACACACACCAACAC	72±8	14451±2817	201±17	0.027
miR-467c	minor	AUAUACAUACACACACCUAUAU	14±2	1056±38	75±8	0.022
miR-669a	-	AGUUGUGUGUGCAUGUUCAUGU	7±0	57±3	8±0	0.016
miR-669b	-	AGUUUUGUGUGCAUGUGCAUGU	8±0	46±1	6±0	0.003
[illegible]	-	[illegible]	[illegible]	[illegible]	[illegible]	[illegible]
miR-669d	-	ACUUGUGUGUGCAUGUAUAUGU	7±0	1511±463	216±66	0.038
miR-669e	-	UGUCUUGUGUGUGCAUGUUCAU	12±0	4090±971	341±81	0.027
miR-669f	-	CAUAUACAUACACACACACGUAU	44±1	13444±4267	306±90	0.037
miR-669h	3p	UAUGCAUAUACACACAUGCACA	8±0	326±6	41±1	0.007
miR-669i	-	UGCAUAUACACACAUGCAUAC	18±1	39±3	2±0	0.048

- 18 members of mouse miR 297-669 cluster were up-regulated in CHO cells in nutrient-depleted conditions

Pro- and Anti-Apoptosis Genes

Upregulation of genes and microRNA in glucose-deprived media

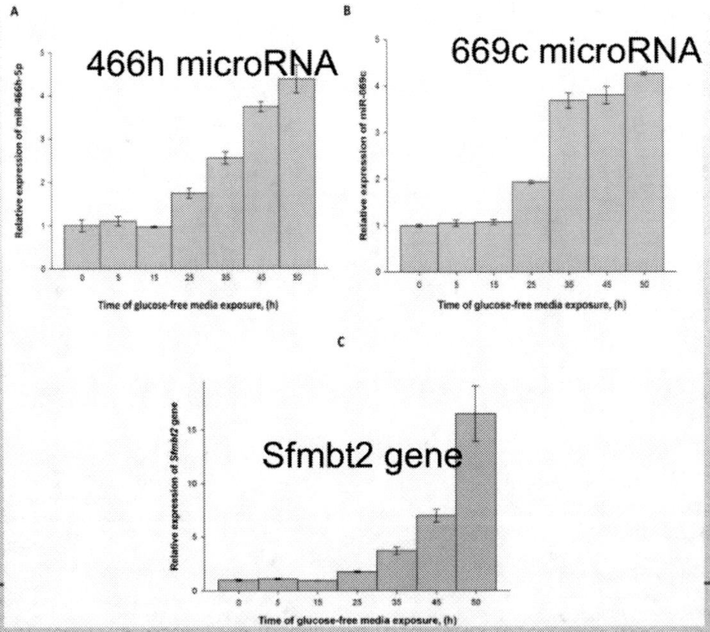

Druz et al. Nucleic Acids Res. 2012

Pro- and Anti-Apoptosis Genes

miR 297-669 cluster in mice is intronic and co-expressed with Sfmbt2 gene

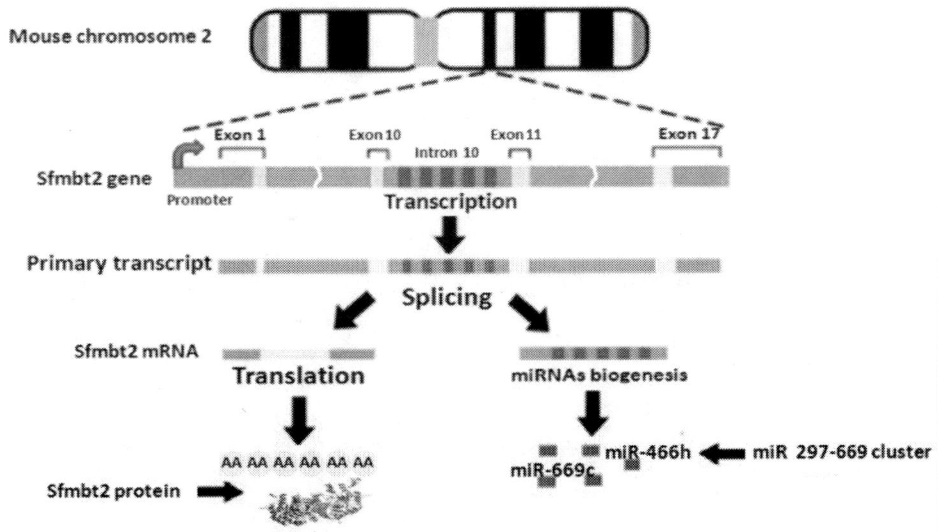

Pro- and Anti-Apoptosis Genes

Examine Potential targets of mmu-mir-466h

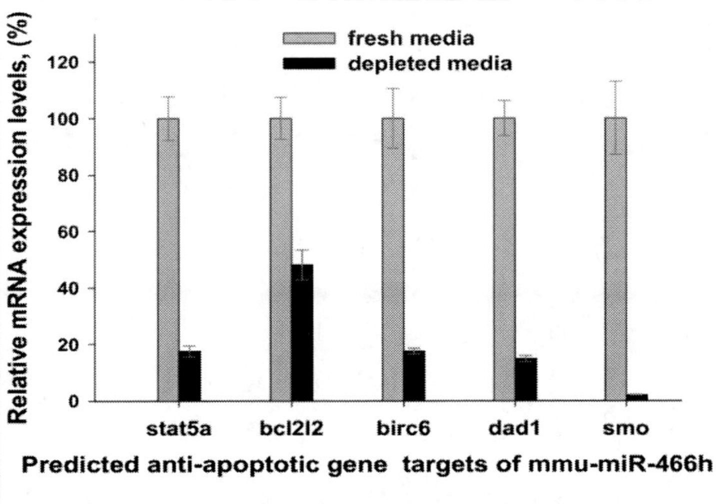

Pathway for MicroRNA Activation

Enrichment
Of
Acetylated
Histones

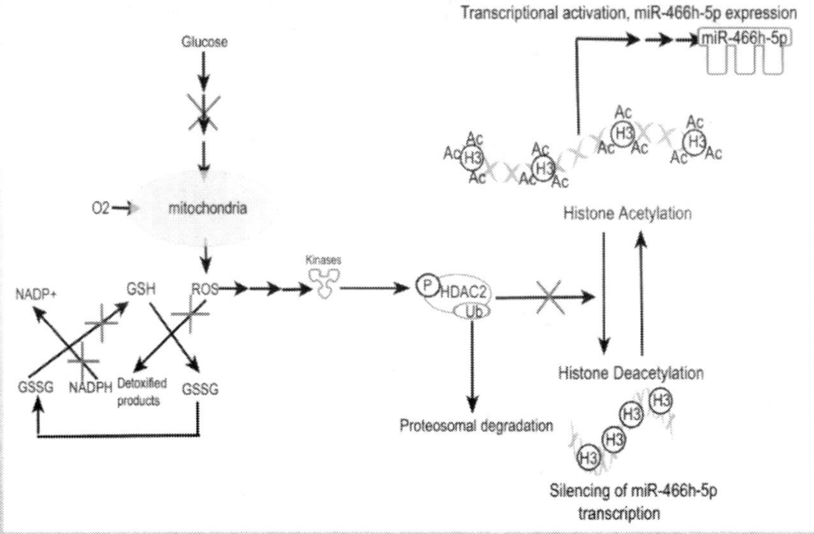

Glucose deprivation leads to MicroRNA activation through ROS generation and Acetylated Histone Generation

Pro- and Anti-Apoptosis Genes

Levels of miR-466h-5p gene targets in anti-miR-466h and n.c. CHO

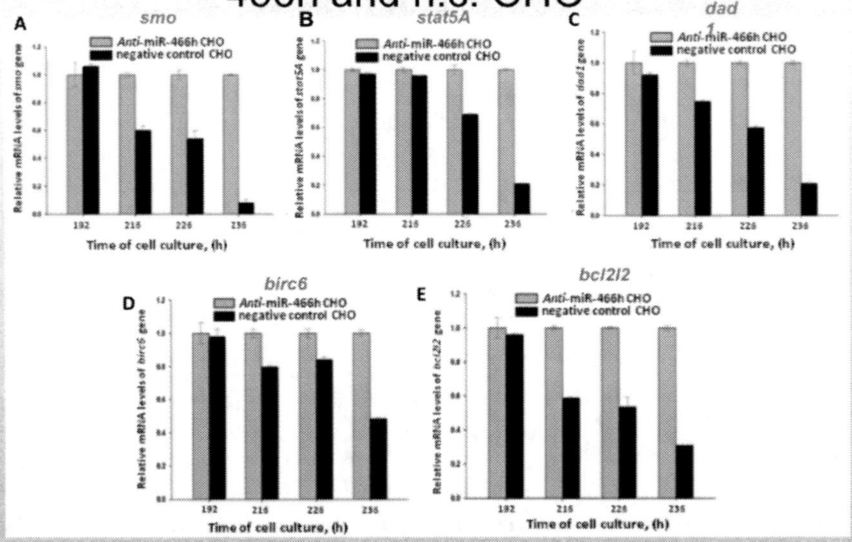

- reduced expression of all genes in the n.c. CHO cells compared to anti-miR-466h CHO cells at 216, 226, and 236h: 2.1÷12.5-fold reduction at 236h

Reduction of Pro-apoptosis MicroRNA

Viability and Protein Production of *anti*-miR-466h, n.c. and parental CHO

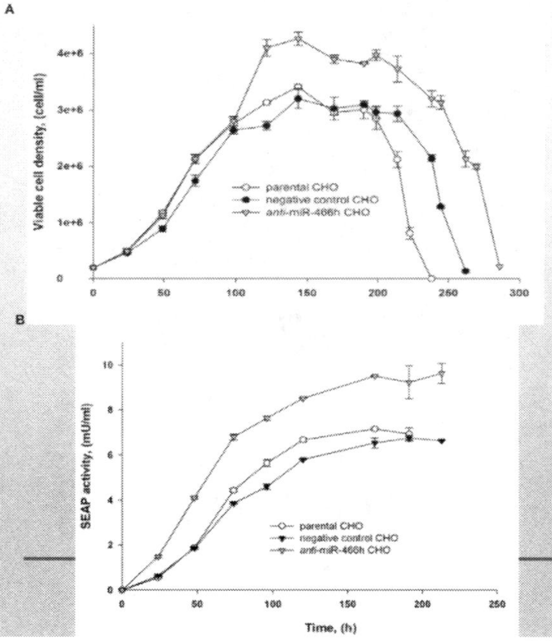

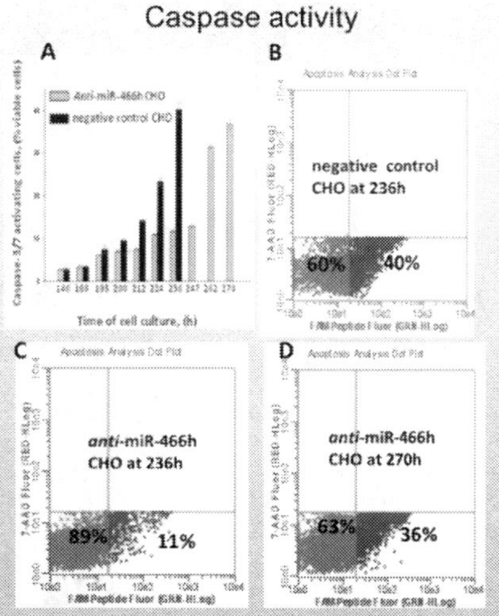

Druz et al., Met. Eng., 2013

MicroRNA screen for recombinant protein expression

High-throughput RNAi facilities

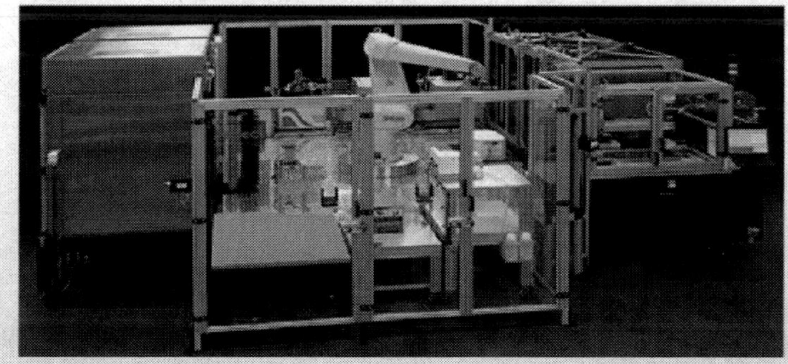

Dr. Scott Martin

370 Plates for siRNA screen

- MiRNA library (~900 miRNAs)

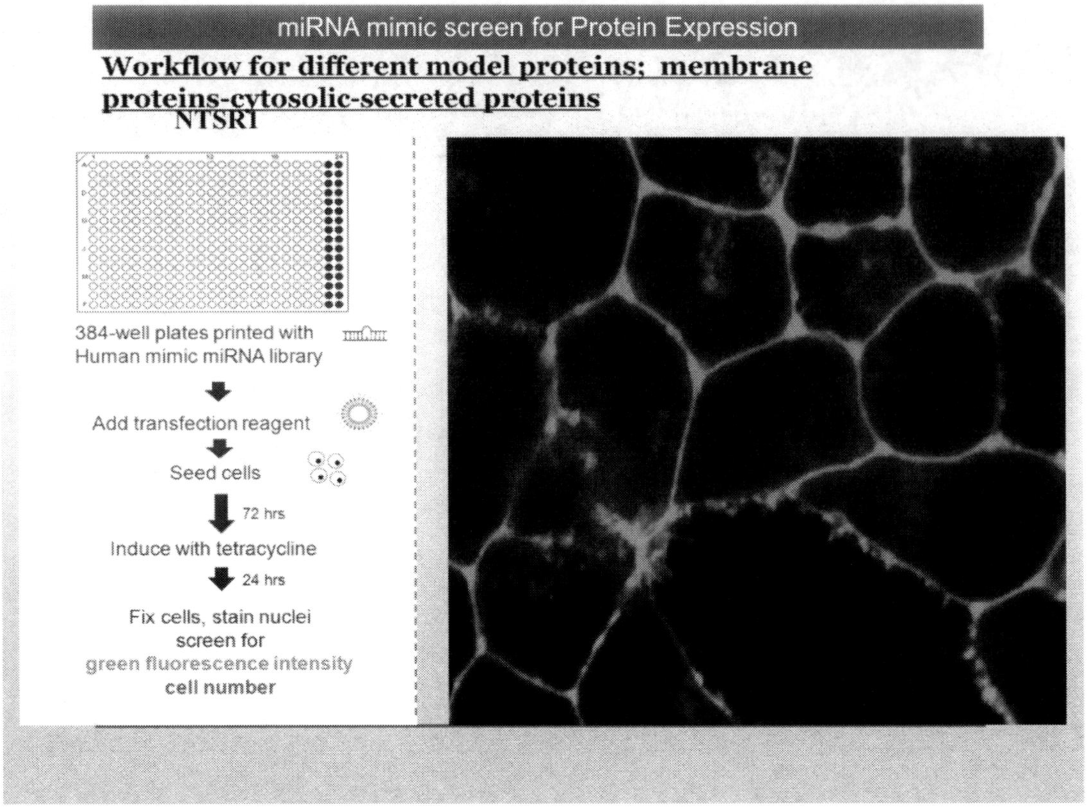

miRNA mimic screen for Protein Expression
Workflow for different model proteins; membrane proteins-cytosolic-secreted proteins
NTSR1
384-well plates printed with Human mimic miRNA library
Add transfection reagent
Seed cells
72 hrs
Induce with tetracycline
24 hrs
Fix cells, stain nuclei
screen for
green fluorescence intensity
cell number

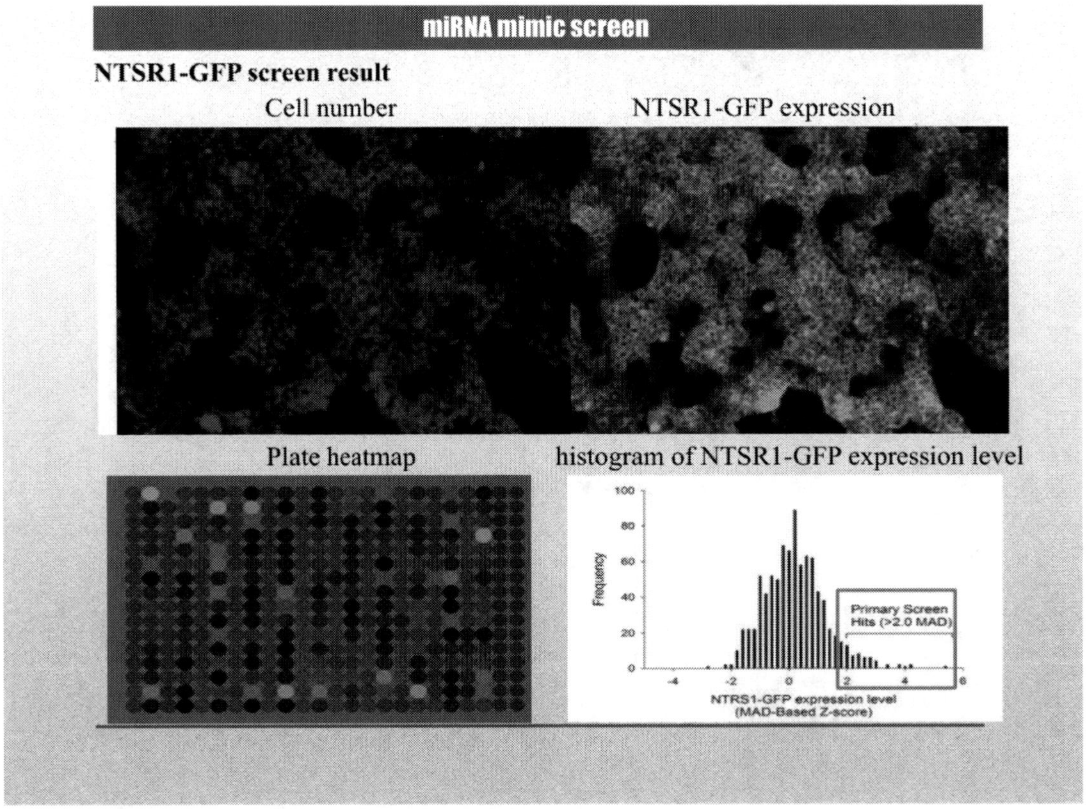

miRNA mimic screen
NTSR1-GFP screen result
Cell number
NTSR1-GFP expression
Plate heatmap
histogram of NTSR1-GFP expression level
100
80
60
40
20
0
Frequency
Primary Screen Hits (>2.0 MAD)
-4
-2
0
2
4
6
NTRS1-GFP expression level (MAD-Based Z-score)

miRNA mimic screen for recombinant protein expression

Functional NTSR1 expression
(Ligand-binding assay)

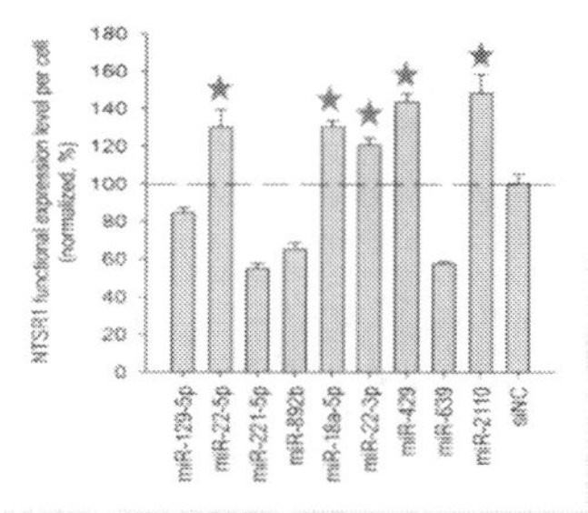

Luciferase (cytosolic protein)

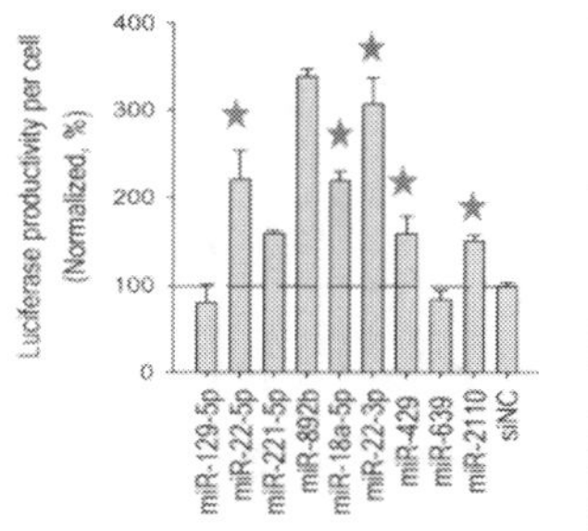

miR-22-5p
miR-18a-5p
miR-22-3p
miR-429
miR-2110

SERT (membrane protein)

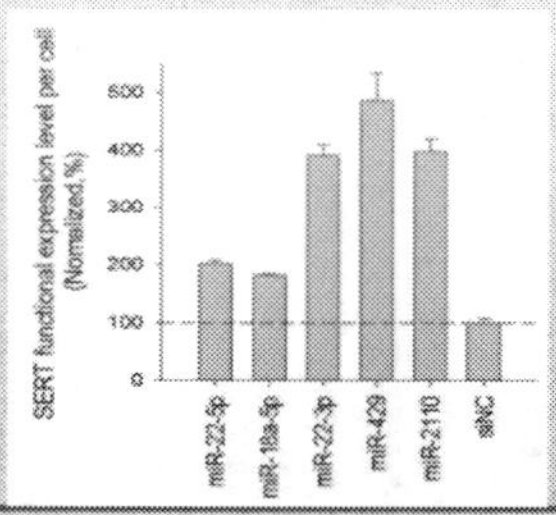

GPC3-hFc (secreted protein)

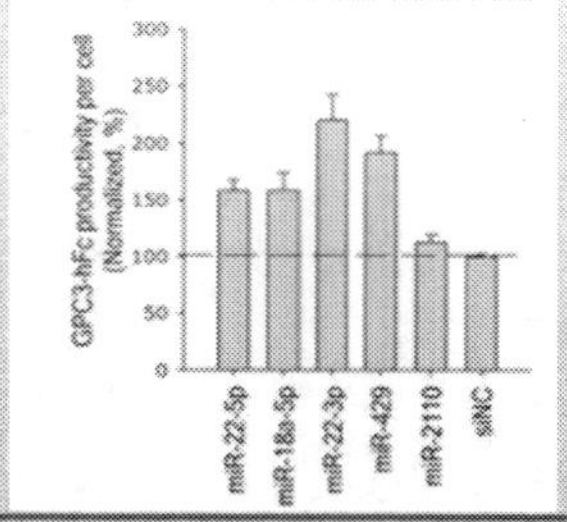

Whole-geome siRNA screen

50

- **<u>Primary siRNA screen:</u>** Whole-genome siRNA library (3 siRNAs ×21,000 genes → 65,000 siRNAs)

Specific productivity
vs.
viable cell number

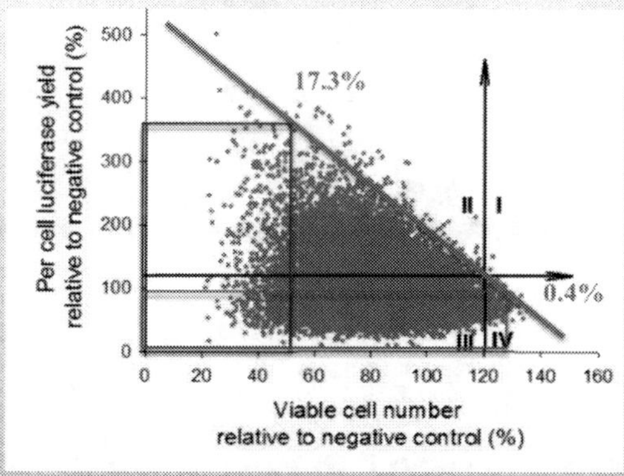

Histogram for luciferase expression

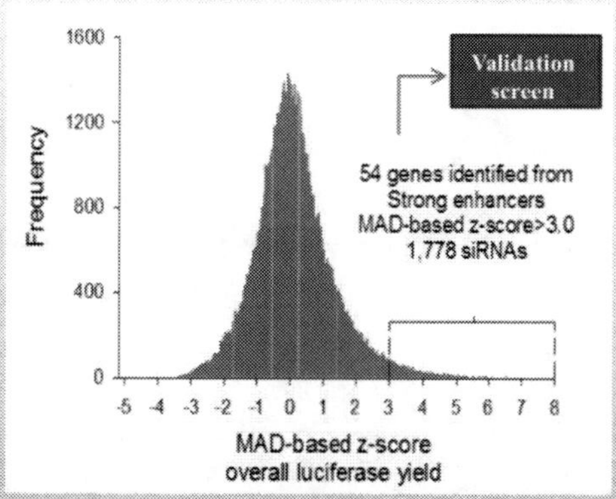

Overal expression:
300%*40%=120%

Test of Top 10 hits with different proteins

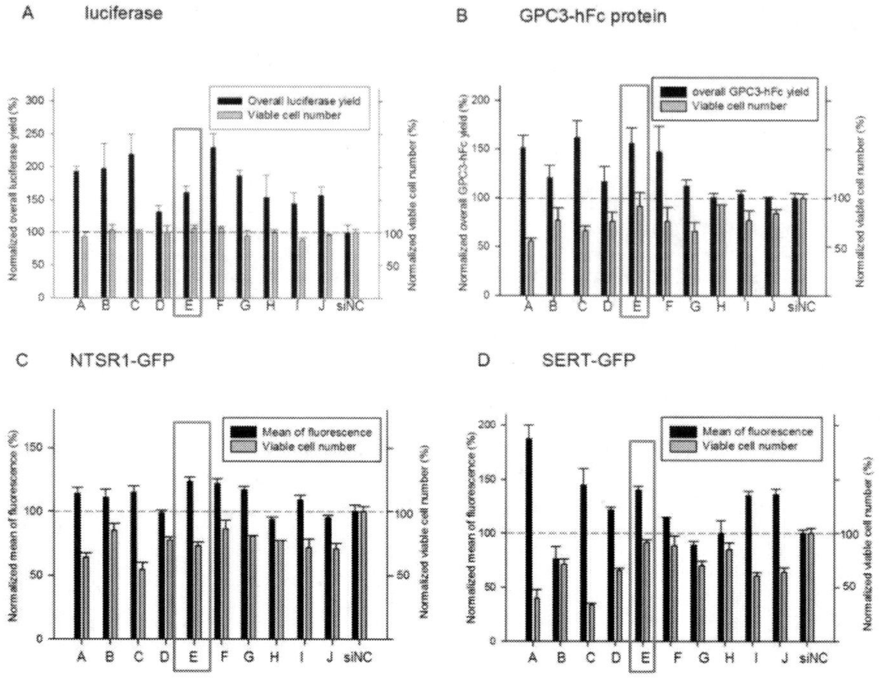

Transfecting 2 or more siRNAs for luciferase production

	A	B	C	D	E	F	G	H	I	J
A	186±8	174±22	346±31	153±16	209±26	253±21	189±35	185±36	149±12	152±16
B		159±26	342±45	144±24	194±28	294±29	202±31	130±14	124±16	144±14
C			299±30	260±33	304±37	435±13	327±23	296±39	210±41	260±23
D				121±14	149±25	236±23	146±15	111±9	136±16	94±10
E					151±18	311±14	172±4	152±8	159±3	106±9
F						251±35	285±41	222±27	197±20	204±27
G							179±13	151±11	132±19	123±19
H								120±10	134±15	105±17
I									127±16	154±14
J										170±11

Best combination: C+F → 435 ± 13
3 siRNAs: A+C+F → **463±22**

Acknowledgements and Collaborators

53

This work was supported by the Intramural Research Program of the National Institutes of Health and partially supported by NIH grant 5R01GM095685-03.

- National Institutes of Health

  NIDDK Biotechnology Core Laboratory
  Dr. Joseph Shiloach Sarah Inwood

  NCATS NIH Chemical Genomics Center
  Dr. Scott E. Martin Yu-Chi Chen

  NINDS Membrane Protein Structure and Function Unit
  Dr. Reinhard Grisshammer Jim F. White

- MRC Laboratory of Molecular Biology, Cambridge, UK

  Dr. Christopher G. Tate Dr. Juni Andréll

Pro- and Anti-Apoptosis Pathway

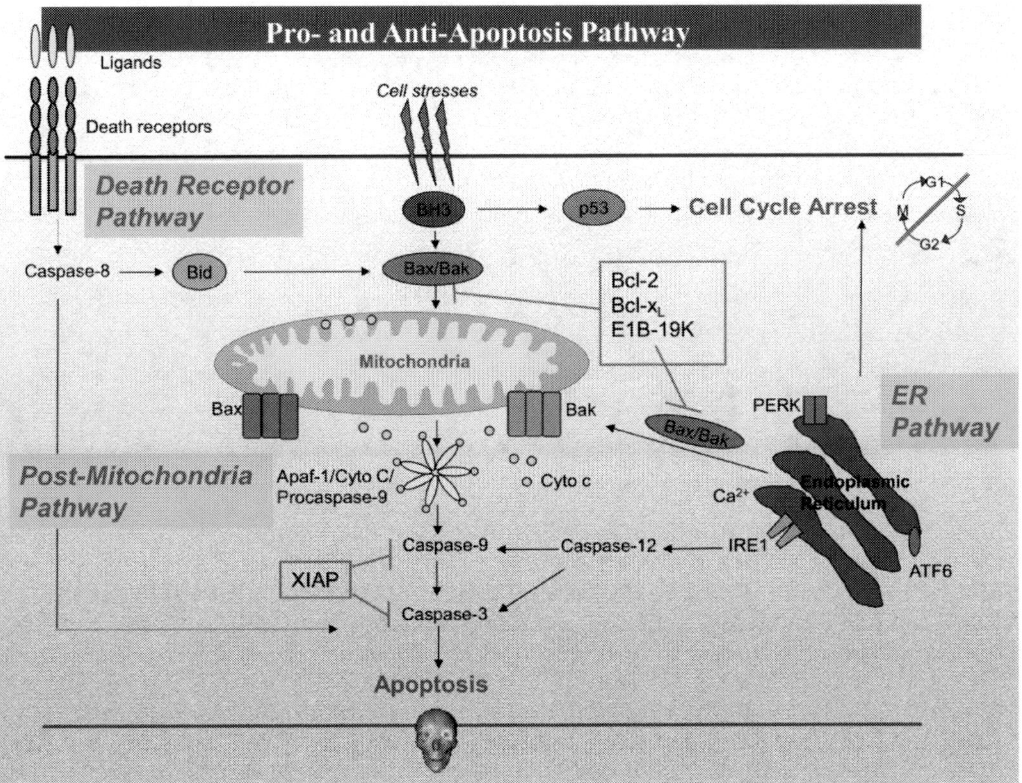

siRNA screen workflow

Primary siRNA screen
3 siRNAs × 21,585 genes

↓

Data analysis

↓

Identify top genes

↓

Validation siRNA screen
Another 3 siRNAs × 54 genes

↓

Confirm top genes

↓

Test top 10 hits with different proteins

luciferase

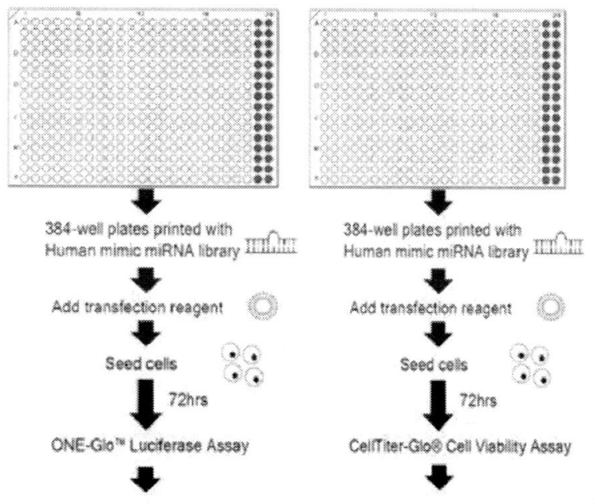

Calculate per cell luciferase expression

Top miRNA hits from NTSR1-GFP screen

Human miR ID (hsa-)	Variant	Mature miRNA sequence	MAD-based z-score	Signal relative to negative control (%)
miR-221	5p	ACCUGGCAUACAAUGUAGAUUU	5.3	248
miR-429	-	UAAUACUGUCUGGUAAAACCGU	4.2	212
miR-22	5p	AGUUCUUCAGUGGCAAGCUUUA	4.0	215
miR-892b	-	CACUGGCUCCUUUCUGGGUAGA	3.7	201
miR-210	3p	CUGUGCGUGUGACAGCGGCUGA	3.2	183
let-7f-2	3p	CUAUACAGUCUACUGUCUUUCC	3.0	178
miR-130b	5p	ACUCUUUCCCUGUUGCACUAC	2.9	178
miR-188	5p	CAUCCCUUGCAUGGUGGAGGG	2.9	177
miR-301a	3p	CAGUGCAAUAGUAUUGUCAAAGC	2.9	176
miR-129	5p	CUUUUUGCGGUCUGGGCUUGC	2.7	172
let-7c	5p	UGAGGUAGUAGGUUGUAUGGUU	2.6	168
miR-1909	5p	UGAGUGCCGGUGCCUGCCCUG	2.6	169
miR-138-1	3p	GCUACUUCACAACACCAGGGCC	2.5	167
miR-193b	3p	AACUGGCCCUCAAAGUCCCGCU	2.5	166
miR-639	-	AUCGCUGCGGUUGCGAGCGCUGU	2.4	165
miR-10b	3p	ACAGAUUCGAUUCUAGGGGAAU	2.4	162

Whole-geome siRNA screen 57

Summary

- Top 10 genes involved in improving luciferase expression was identified from human whole-genome
- Pathways where top genes enriched in were identified
- *E* gene is a promising engineering target for improving recombinant protein production and cell growth. Mechanism study has been achieved.
- On-going efforts: Evaluation of gene E in CHO cell line, construction of stable knock-out cell line.

Whole-geome siRNA screen 58

Confirmed top 10 gene hits

Gene	Description	Overall luciferase yield (%)	MAD-based z-score	Function
A	A Subunit 1	172	4.55	
B	A Subunit 2	165	4.17	
C	CA	163	4.10	transcription
D	D Subunit 2	156	3.70	signal transduction
E	E	153	3.57	
F	F Subunit A	153	3.56	Apoptosis
G	A Factor xx	147	3.27	protein translation
H	H Subunit A	138	2.80	Protein folding
I	I Subunit 2	126	2.23	etc.
J	J Subunit 2	124	2.13	

KEGG pathway enrichment analysis

59

Whole-genome siRNA screen to improve recombinant protein expression

---

miRNA mimic screen 60

Summary

- Developed a method for NTSR1-GFP screening

  cost-effective, image-based, high-throughput
  applicable to other membrane/cytosolic protein

- Identified 5 novel miRNAs improving NTSR1 functional expression

- The 5 miRNAs can improve cytosolic, secreted and membrane protein expression up to 4 fold in HEK293 cells

- **Xiao S**, Chen YC, Betenbaugh MJ, Martin SE, Shiloach J: MiRNA screen for improved functional expression of neurotensin receptor. Biotechnology and Bioengineering 2015

Introduction- projects overview

61

What are the phenotypes being pursued?

Dr. Su Xiao

Dr. Alex Druz

recombinant protein productivity ↑

Viable cell number ↑

Apoptosis onset

loss of viability

Introduction- projects overview

62

What are the phenotypes being pursued?

Dr. Su Xiao

Cell line:	HEK293
Screen library:	human miRNA and siRNA
Model proteins:	luciferase neurotensin receptor type I

recombinant protein productivity ↑

Viable cell number ↑

Introduction- NTSR1

63

- Neurotensin receptor type I (NTSR1): **Neurotensin (agonist)**
  Parkinson's disease
  pathogenesis of schizophrenia
  modulation of dopamine neurotransmission
  hypothermia
  promoting the growth of cancer cells

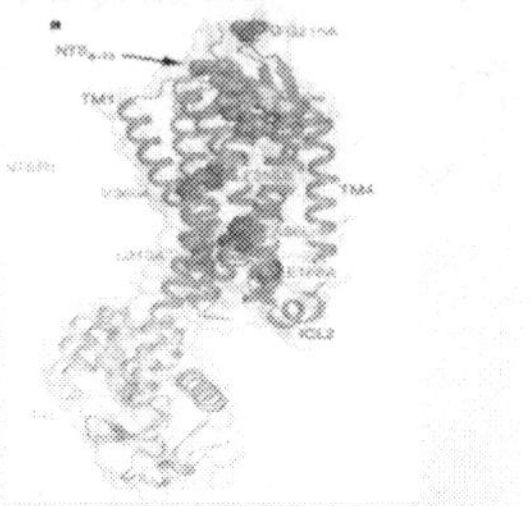

- **G-protein-coupled receptor (GPCR)**
  most medicines sold worldwide target GPCR
  hard-to-express integral membrane protein

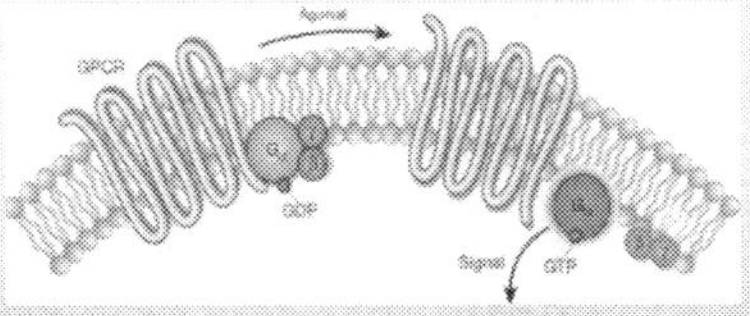

Not naturally abundant in native host (mammalian)	
Poor functional expression in common expression systems	
bacteria and yeast:	aggregates, non-functional
insect cells:	considerate mis-folding rate
mammalian cells:	poor cell growth
	loss of expression upon prolonged cell culture

White JF, Noinaj N, Shibata Y, Love J, Kloss B, et al. (2012) Structure of the agonist-bound neurotensin receptor. Nature 490: 508–513. doi: 10.1038/nature11558.

64

miRNA screen to improve recombinant protein expression

- **Xiao S et al** Biotechnology and Bioengineering 2015

miRNA mimic screen
65

NTSR1 screen

Construct NTSR1-GFP cell line
Assay development

↓

Primary miRNA screen
(875 miRNA mimics)

↓

Data analysis

↓

Identify top miRNAs

↓

Follow-up assay: Ligand binding assay

luciferase screen

Assay development

↓

Primary miRNA screen

↓

Luciferase screen data

↓

Identify common hits

↓

Screen common hits with different proteins

MicroRNA screen for recombinant protein expression
66

High-throughput RNAi facilities

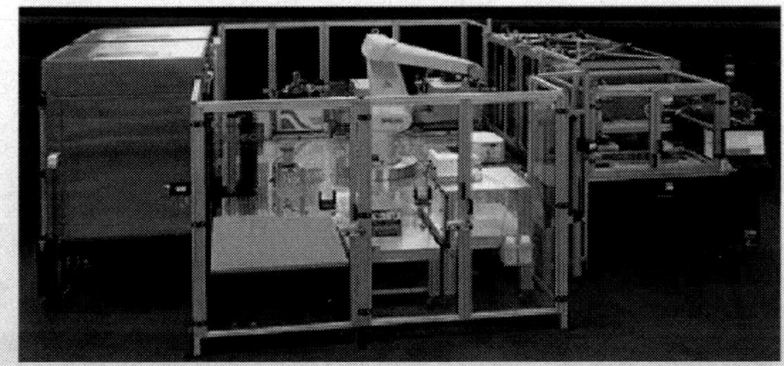

Dr. Scott Martin

370 Plates for siRNA screen

- MiRNA library
  (~900 miRNAs)

- Whole-genome siRNA library
  (3 siRNAs ×21,000 genes → 65,000 siRNAs)

What are the phenotypes being pursued?

Dr. Su Xiao

Dr. Alex Druz

recombinant protein productivity ↑

Viable cell number ↑

Apoptosis onset

loss of viability

What are the phenotypes being pursued?

Dr. Su Xiao

Cell line:	HEK293
Screen library:	human miRNA and siRNA
Model proteins:	luciferase neurotensin receptor type I

recombinant protein productivity ↑

Viable cell number ↑

Introduction- NTSR1

69

- Neurotensin receptor type I (NTSR1): **Neurotensin (agonist)**
  - Parkinson's disease
  - pathogenesis of schizophrenia
  - modulation of dopamine neurotransmission
  - hypothermia
  - promoting the growth of cancer cells

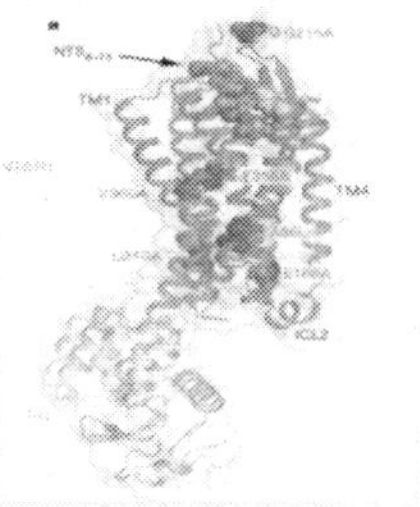

- **G-protein-coupled receptor (GPCR)**
  - most medicines sold worldwide target GPCR
  - hard-to-express integral membrane protein

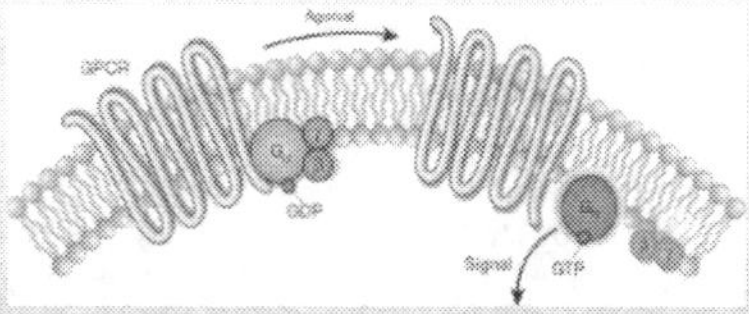

Not naturally abundant in native host (mammalian)
Poor functional expression in common expression systems

bacteria and yeast:	aggregates, non-functional
insect cells:	considerate mis-folding rate
mammalian cells:	poor cell growth
	loss of expression upon prolonged cell culture

White JF, Noinaj N, Shibata Y, Love J, Kloss B, et al. (2012) Structure of the agonist-bound neurotensin receptor. Nature 490: 508–513. doi: 10.1038/nature11558.

70

miRNA screen to improve recombinant protein expression

- **Xiao S et al** Biotechnology and Bioengineering 2015

miRNA mimic screen

71

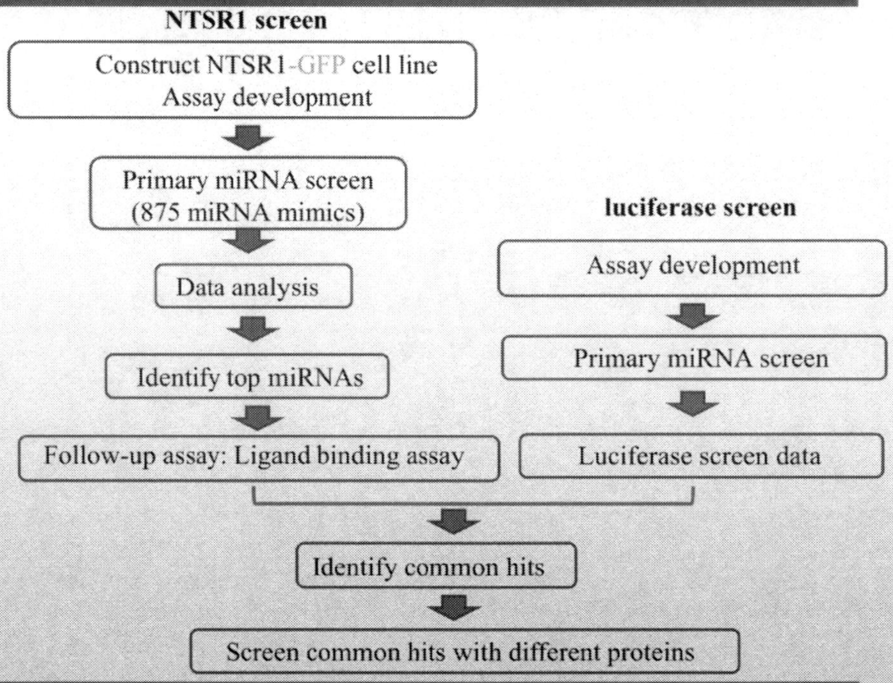

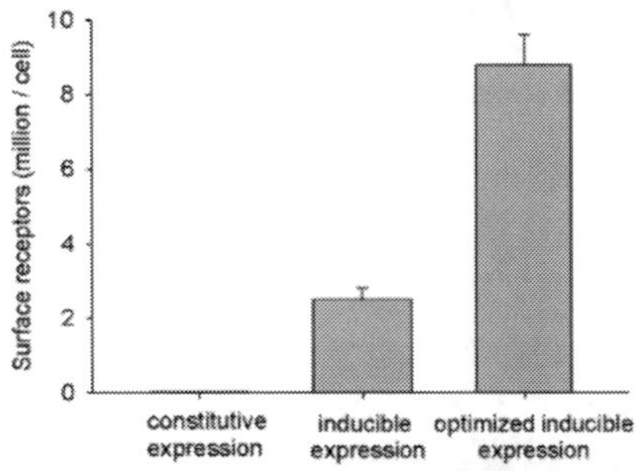

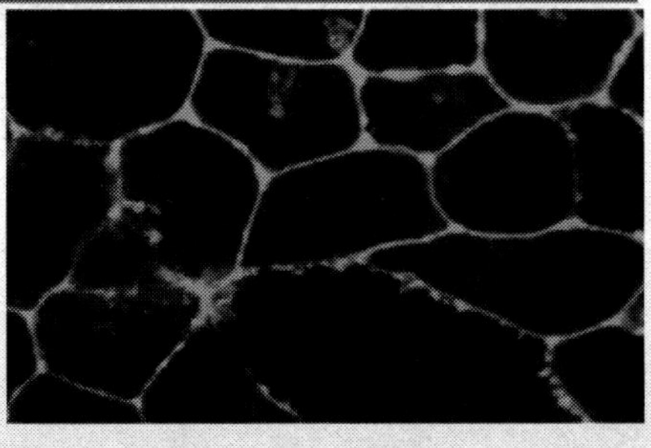

Expression of Target Membrane Protein in HEK Cells

Fig. 2 Confocal microscopy of tetracycline-induced T-REx-293-NTSR1-GFP cells with NTSR1-GFP fusion protein located on plasma membrane.NTSR1 is the neurotensin receptor membrane protein.

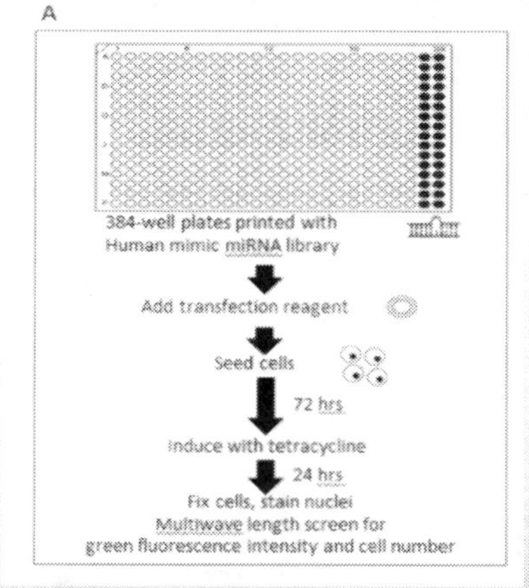

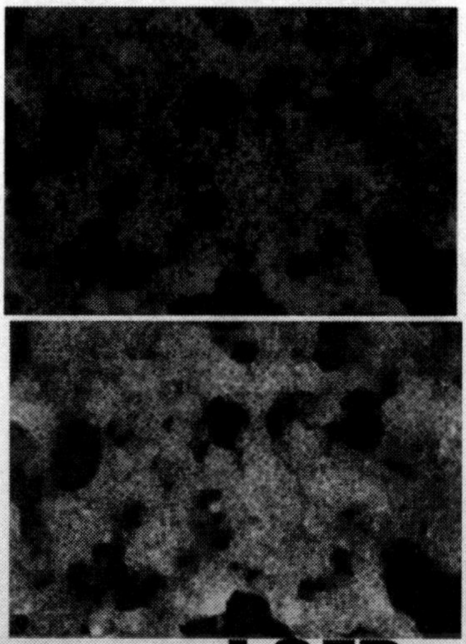

Nuclei staining and GFP expression captured.

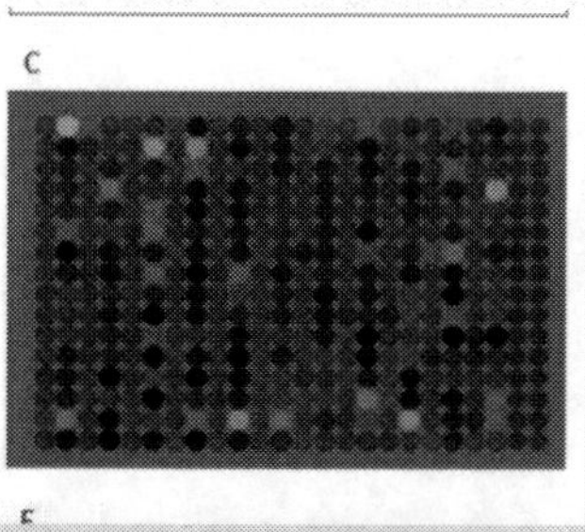

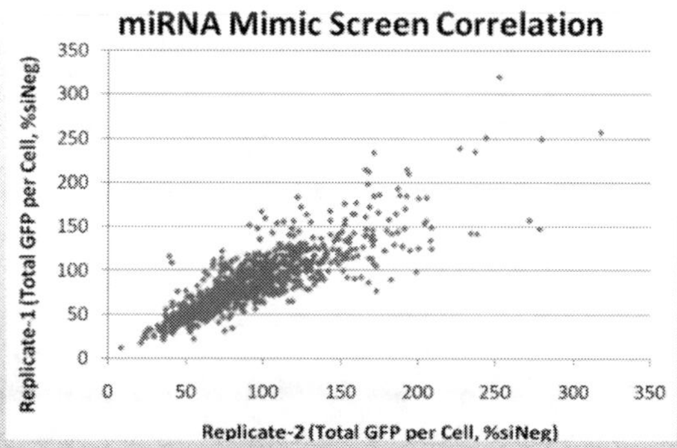

miRNA mimic screen and reproducibility

Apoptotic miRNA mimics screen

Large Scale Screening of miRNA for apoptosis?

1,038 mouse miRNA Mimics

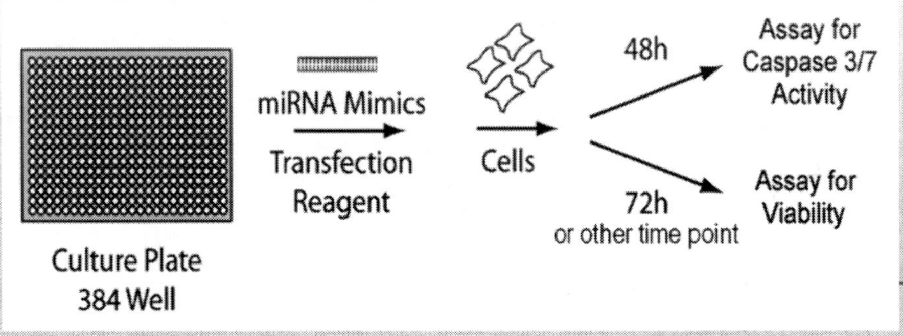

Apoptotic miRNA mimics screen **76**

Caspase-3/7

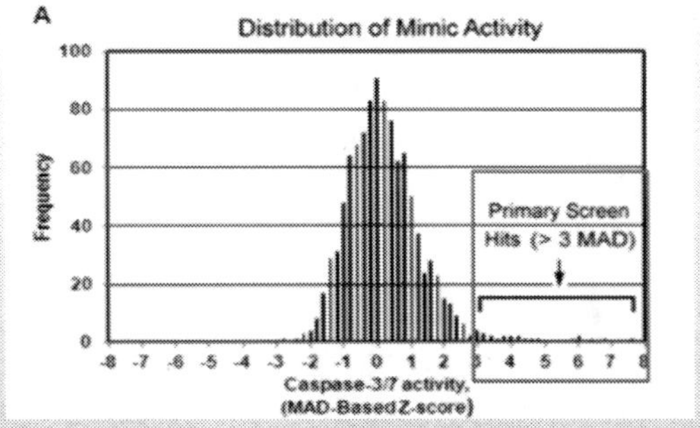

Viability

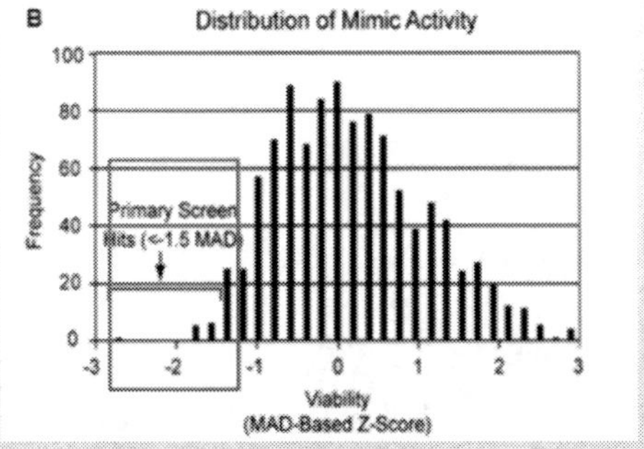

MicroRNA associated with Caspase Activity

- mmu-miR-15a-3p
- mmu-miR-706
- mmu-miR-1948-5p
- mmu-miR-1948-3p

Apoptotic miRNA mimics screen

Caspase-3/7 activation by miR 15a-3p transfection

HeLa AsPc-1

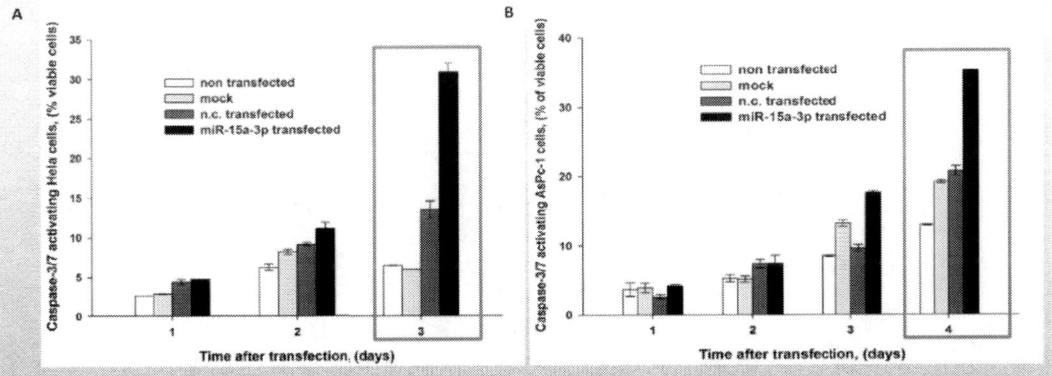

Apoptotic miRNA mimics screen

Viability reduction by miR 15a-3p

HeLa AsPc-1

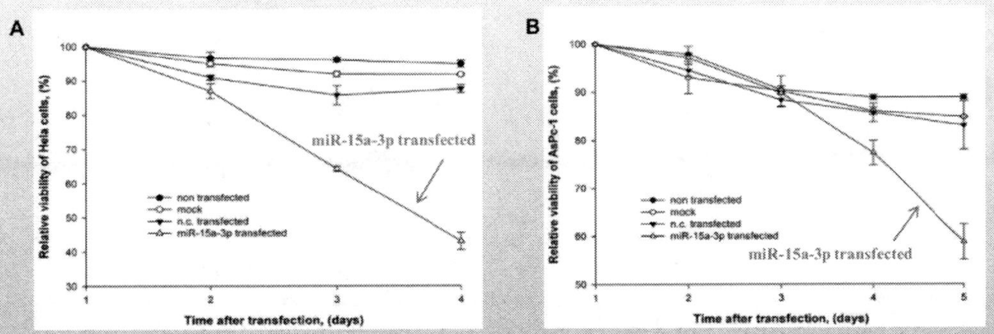

Apoptotic miRNA mimics screen

Identification of miR-15a-3p targets in mice

gene symbol	binding sites	fold down-regulation in miR-15a-3p samples	P-value	Anti-apoptotic role of targeted gene
bcl-x_L	CAG:AUGGCCU (2-8:10-12) GU:UGGCCU (2-7:9-10) U:UGGCCU (2-7:9) 4 of UGGCCU (2-7)	3.34± 0.23	0.041	Bcl2l1(bcl-x_L) inhibits formation of permeability transition pore and subsequent release of cytochrome C and other pro-apoptotic molecules from mitochondria by binding to bax and bak
naip5	AUGGCCU (2-8)	2.40± 0.15	0.013	Also known as naip1, birc1a. Baculoviral IAP domain-containing protein. IAP domains are known to inhibit apoptosis by direct inhibition of Caspases
fgfr2	2 of UGGCCU (2-7)	3.89± 0.44	0.007	Fibroblast growth factor receptor 2 which binds fibroblast growth factor-2 ligand and initiates signaling cascade leading to over-expression of anti-apoptotic genes bcl-X_L and XIAP
mybl2	AUGGCCU (2-8) CAGU:UGGCCU (2-7:9-12) UGGCCU (2-7)	2.65± 0.17	0.032	Mybl2 belongs to the family of transcription factors which are involved in regulation of cell survival by induction of bcl2 transcription

Apoptotic miRNA mimics screen

miR-15a-3p targets bcl-x_L

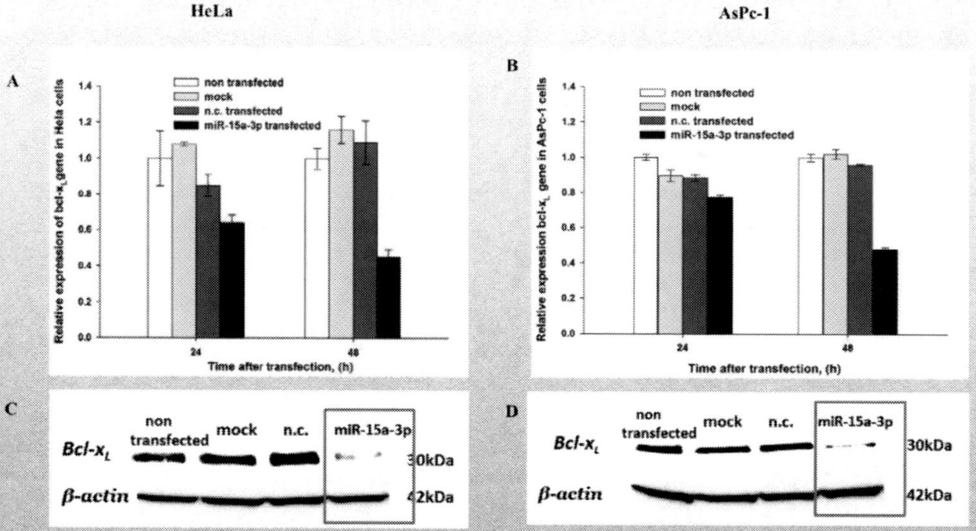

miR-15a-3p binds to bcl-x_L in the 3'UTR and reduces its expression which results in activation of Caspase-3/7 and loss of cell viability

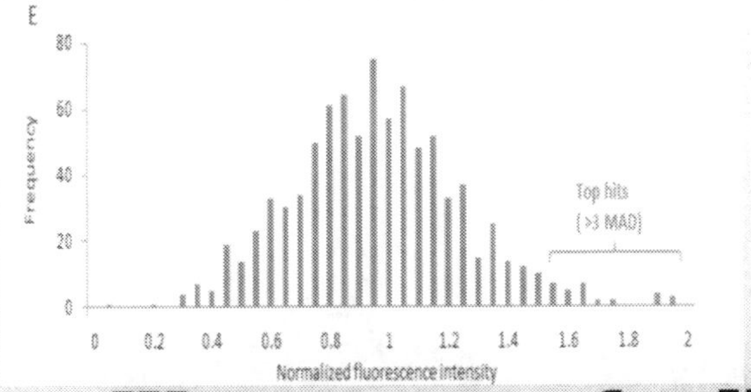

Histogram of miRNA hits for GFP Expression and Flow Cytometry

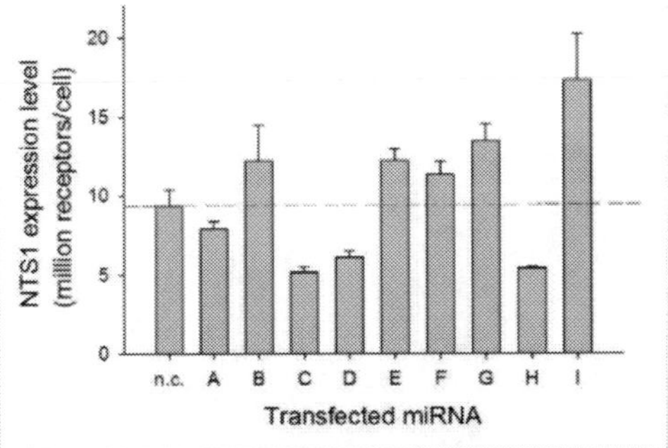

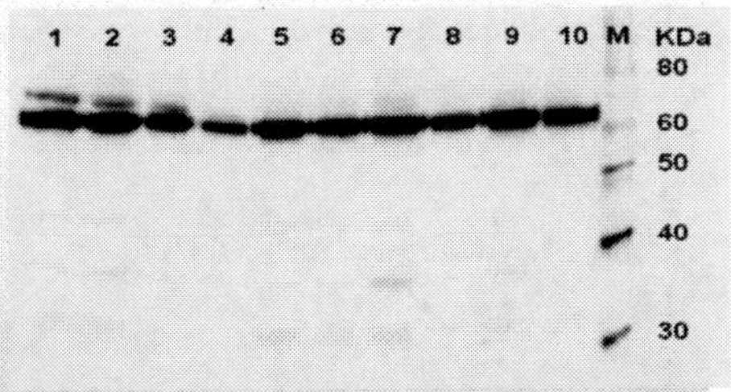

Relative Membrane Protein Expression for different miRNA hits

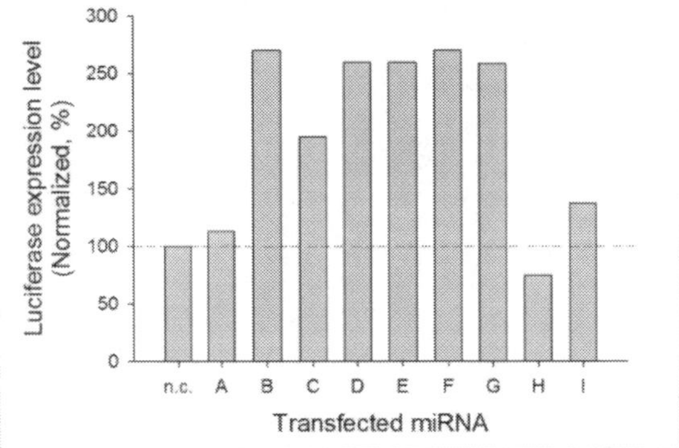

Effect of transfected miRNAs on Expression of Luciferase Model Protein

- MicroRNA's in clusters can be activated upon environmental insults leading to apoptosis
- These genes can target intracellular anti-apoptotic targets of importance to survival
- Control of microRNA's can be used to alter viability in mammalian
- A number of additional microRNA identified to have a pro-apoptotic role in human and mouse cells
- The pro-apoptotic role of miR-15a-3p in human cancers may involve control of bcl-x_L expression
- These microRNAs may have potential anti-cancer applications in therapy
- Large-scale screening methodology can be utilized to identify microRNAs targets with anti- and pro-apoptotic capabilities as well as effects on recombinant protein production

Conclusions: miRNA Screening

Summary

- miR-15a-3p was identified to have a pro-apoptotic role in mammalian cells (human and mouse cells were tested)
- miR-15a-3p binds to bcl-x_L directly and reduces its expression which results in activation of Caspase-3/7 and loss of cell viability

Druz, A., et al., *Large-scale screening identifies a novel microRNA, miR-15a-3p, which induces apoptosis in human cancer cell lines.* RNA Biol, 2013. **10**(2): p. 287-300.

Large-scale screening in mouse cells

BCMBA/OV mouse cells

Transfected with mouse miRNAs mimics library (1028 mimics)

Assayed for Caspase-3/7 activation 48h after transfection

Identified 19 candidates (p<0.001, q<0.05)

Assayed for loss of viability 72h after transfection

Identified 20 candidates (p<0.1)

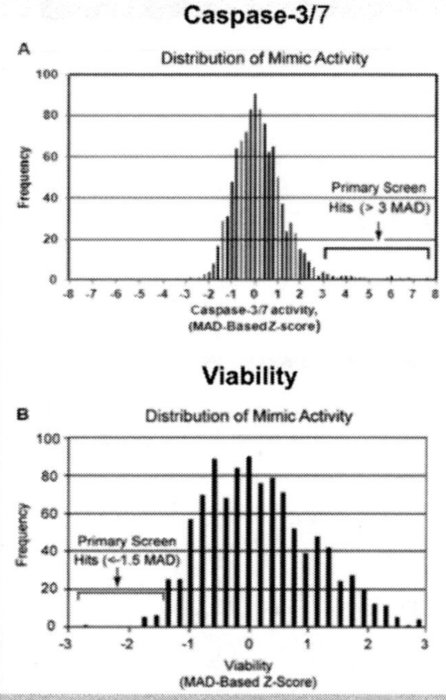

86

TARGET MICRORNA:

miR-15a-3p conserved in both humans and mice

Four mouse miRNA identified:

PUTATIVE TARGETS:

anti-apoptotic genes (*bcl2l1, naip5, fgfr2, mybl2, bcl-xL*).

- mmu-miR-15a-3p
- mmu-miR-706
- mmu-miR-1948-5p
- mmu-miR-1948-3p

Significant Caspase 3/7 and Viability Loss in Mouse Cells

What about mir-15a-3p in humans?

miRNA screen for viability in human lines

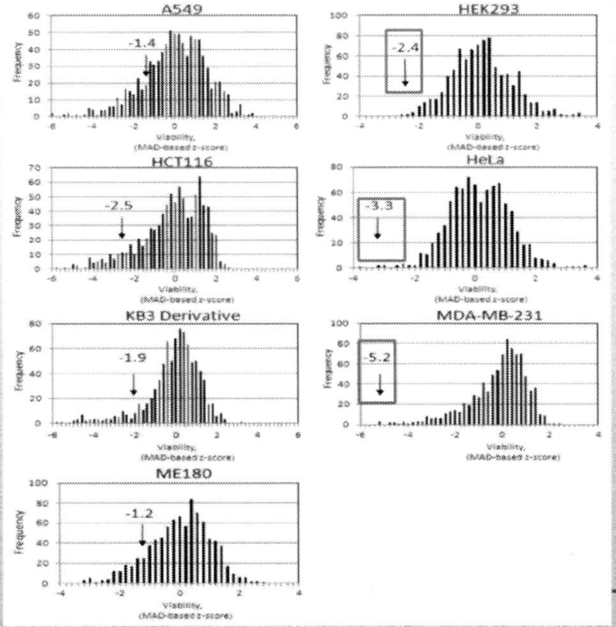

- miRNA mimics activities for viability in human cells based on screening with complete library (1212 miRNAs)
- hsa-miR-15a-3p activity is shown on diagrams with arrows
- hsa-miR-15a-3p activity was trending toward viability reduction in all of the analyzed cell lines (scores ranging from -5.2 to -1.2)
- hsa-miR-15a-3p was significant for viability loss in a numbe of lines including HEK 293, MDA-MB-231, and HeLa cells

88

Caspase-3/7 activation in HeLa and AsPc-1 cells by miR-15a-3p

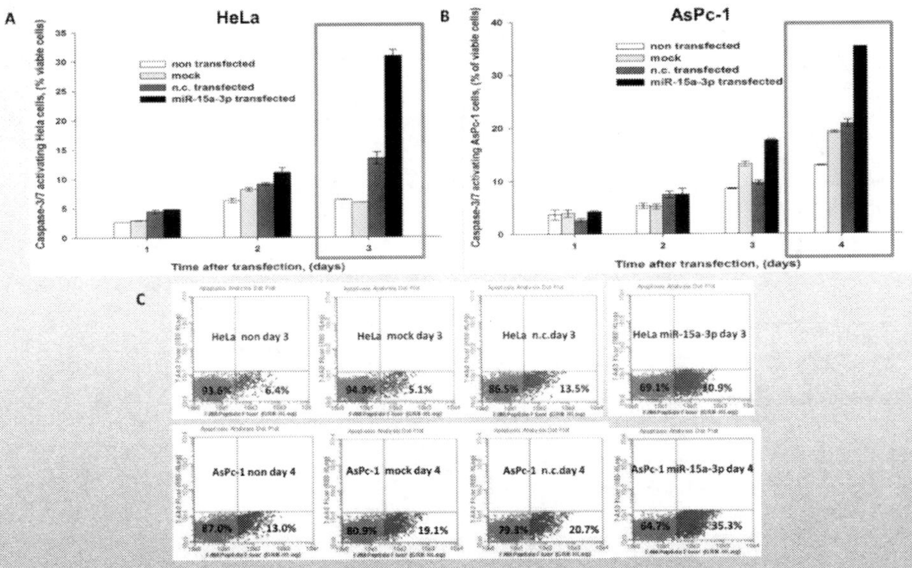

- Tested for both cell lines: non transfected cells , mock, n.c. and miR-15a-3p transfection
- ~~Caspase-3/7 activated in miR-15a-3p transfected HeLa on day 3 : 30.9% of transfected were apoptotic~~
- Caspase-3/7 activated in miR-15a-3p transfected AsPc-1 on day 4 : 35.3% of transfected were apoptotic

Viability reduction in HeLa and AsPc-1 cells

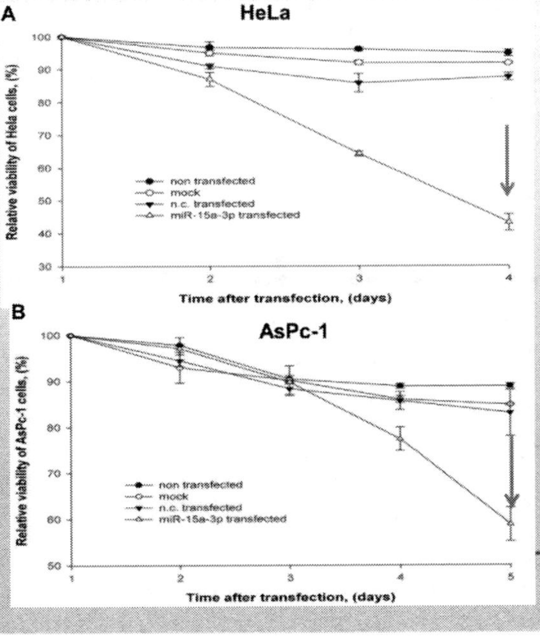

- Viability of HeLa miR-15a-3p was 43% on day 4 compared to 87.5% for n.c. transfected HeLa on day 4
- Viability of AsPc-1 transfected with miR-15a-3p to 59% on day compared to 83% for n.c. transfected AsPc1 on day 5
- Loss of viability in both cells lines followed Caspase-3/7 activation

90

What are miR-15a-3p targets in HeLa and AsPc-1 cells?

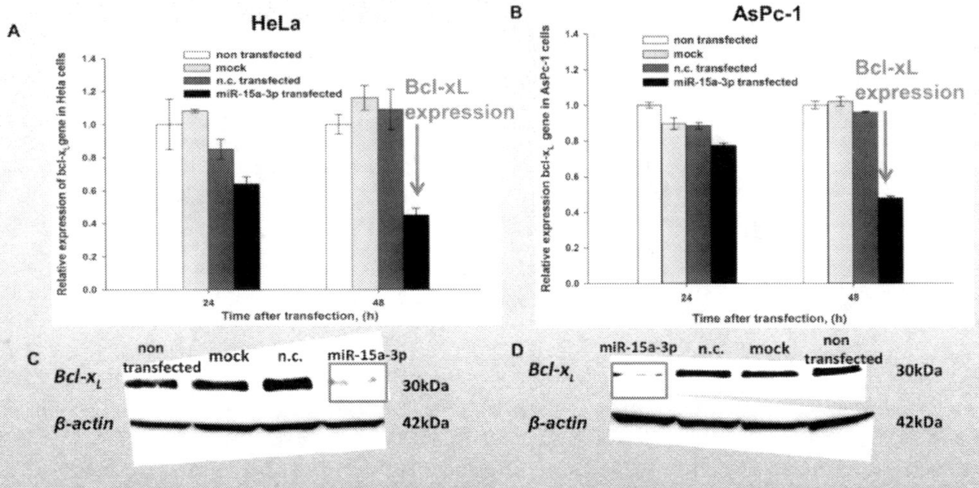

- Expression levels of *bcl-x$_L$* gene in miR-15a-3p transfected HeLa (A) and AsPc-1 (B) cells lowered to 45 and 48% respectively compared with non transfected cells at 48h after transfection
- Protein levels of Bcl-x$_L$ were reduced in both miR-15a-3p transfectedHeLa (C) and AsPc-1 (D) at 72h
- ~~Other putative anti-apoptotic targets of miR-15a-3p in mouse (fgfr2, mybl2, naip5) not affected in both HeLa and AsPc-1~~

target bcl-xL

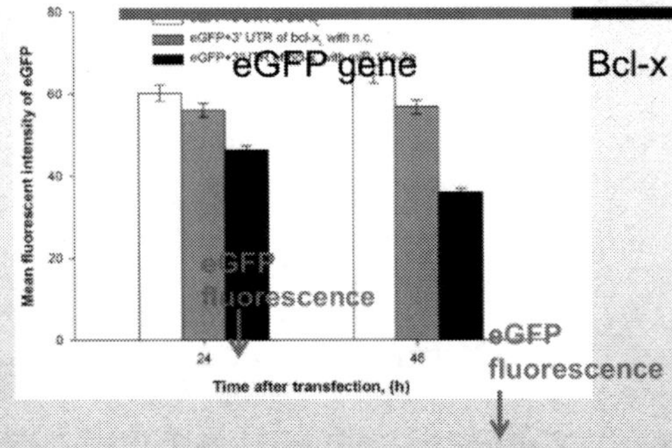

- Binding of miR-15a-3p to 3'UTR of *bcl-x$_L$* gene added to the terminus of eGFP was tested in HeLa cells
- HeLa co-transfected with eGFP+3'UTR of bcl-x$_L$ and miR-15a-3p (or n.c.)
- GFP expression in miR-15a-3p transfected cells was reduced at 24h after transfection and further reduced at 48h

92

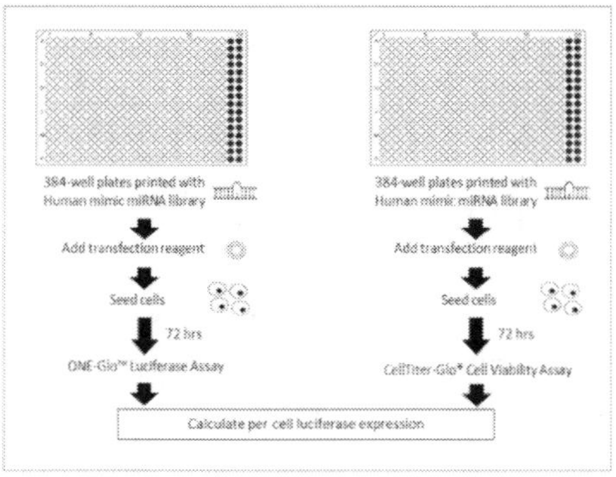

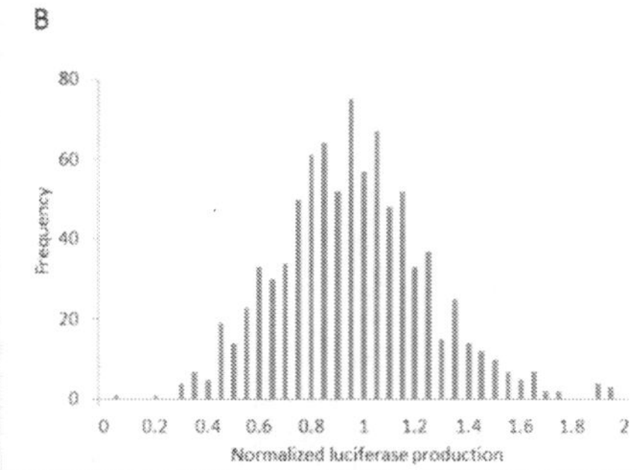

MiRNA screening for luciferase production

miR-15a-3p targets bcl-x_L by direct binding

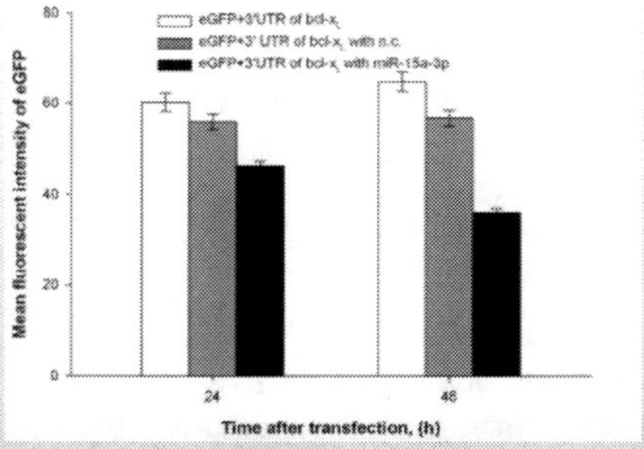

- Binding of miR-15a-3p to 3'UTR of *bcl-x_L* gene added to the terminus of eGFP was tested in HeLa cells
- HeLa were co-transfected with eGFP+3'UTR of bcl-x_L and miR-15a-3p (or n.c.)
- GFP expression in miR-15a-3p transfected cells was reduced at 24h after transfection and further reduced at 48h

Apoptotic miRNA mimics screen 95

Verification of pro-apoptotic activity of miR-15a-3p in human cell lines

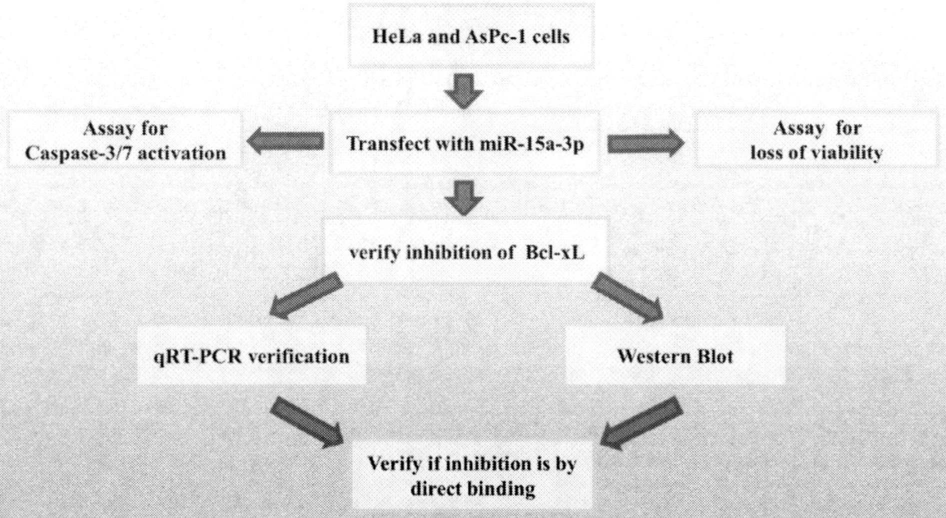

Apoptotic miRNA mimics screen 96

Identification of miR-15a-3p and its role in apoptosis activation in cancer cells

Dr. Alex Druz

Apoptosis onset

loss of viability

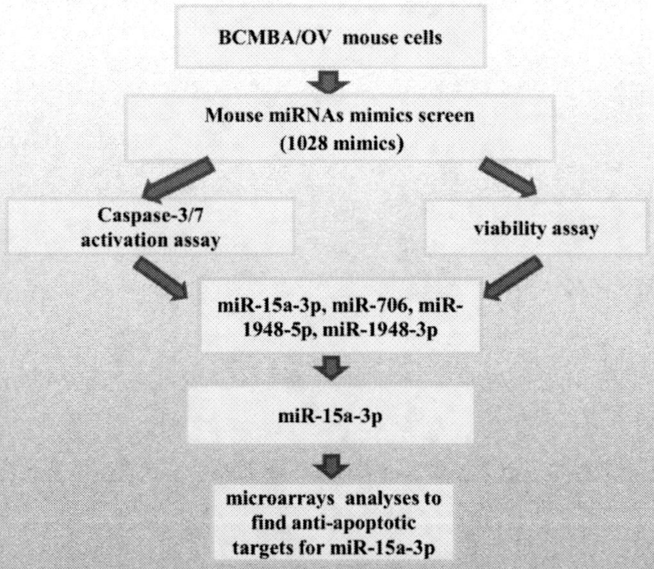

Druz, A., et al. RNA Biol, 2013. **10**(2): p. 287-300.

Our strategy **Genome-wide RNA interference screening**

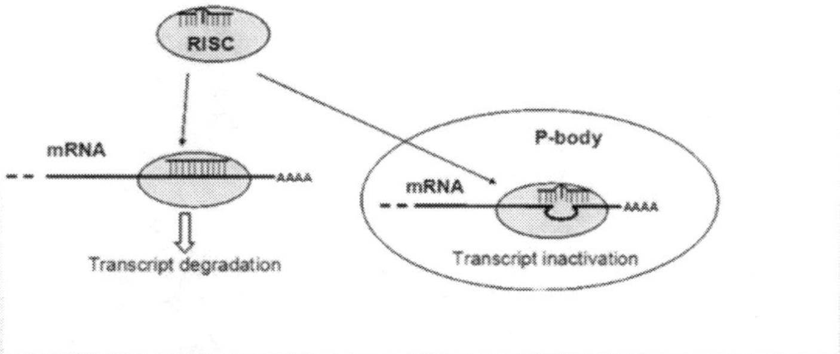

- miRNA - silence multiple genes
- siRNA - silence one gene

bottom-up method
bottleneck/target identification
provide valuable dataset for understanding the host organism

Recombinant SEAP protein production in *anti*-miR-466h, negative control and parental CHO

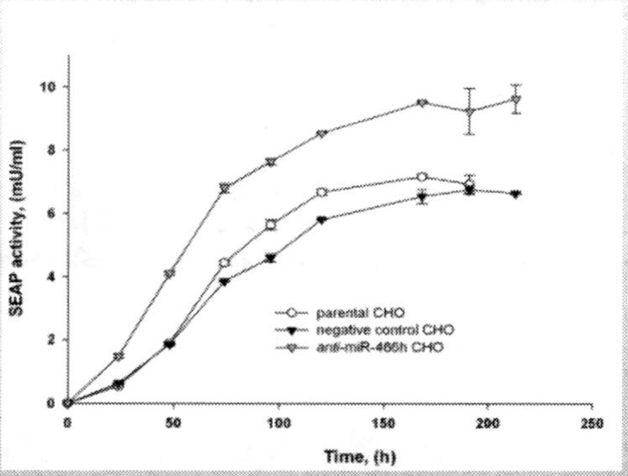

- Maximum SEAP activity in *anti*-miR-466h CHO was increased by 35 and 43% compared to negative control and parental CHO

Druz et al., Met. Eng. ,2013

Examine chemical anti-mir466h

- A chemically modified single stranded oligonucleotide specific for mmu-miR-466h was added to the cells
- The levels of miR-466h in fresh media and in the depleted media, with or without anti-miR-466h, were measured
- **Compared with fresh media, miR-466h levels were 10 times higher in depleted media in the absence of the inhibitor**
- **Mir-466h levels were reduced by a factor of 4.5 in depleted media containing anti-miR-466h.**

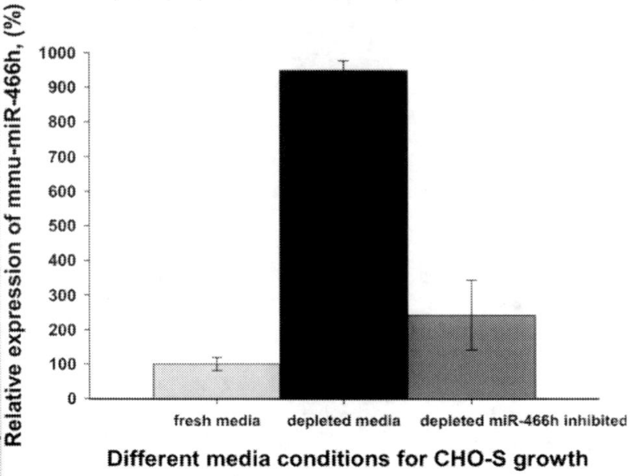

Increases in Mmu-mir-466h expression levels in depleted media are inhibited by oligonucleotide

Does chemical inhibition of mmu-mir-466h affect cell viability

- Cell viability in the depleted media started to decline after about 18 hours and fell to 81% by 23 hours.
- **When the cells were chemically treated with anti -miR-466h oligonucleotide, the cell viability was higher at both the 20 and 23 hours time points.**

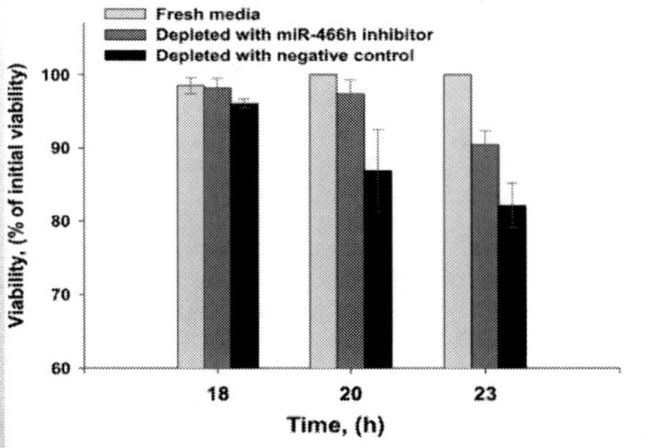

Addition of anti-mmu-mir-466h chemical to cells enhances viability following nutrient depletion

Effect of glucose deprivation on Acetylated Histones

Enrichment Of Acetylated Histones

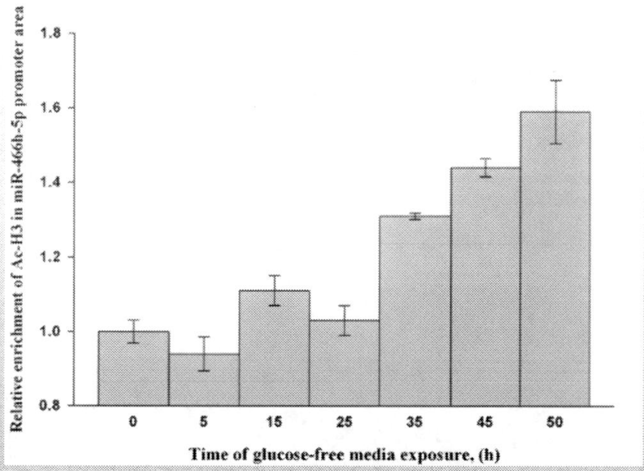

Time in glucose free medium

Acetylated Histones Increase following glucose deprivation

Location of microRNAs within cluster

Mouse miR ID (mmu-)	Variant	Mature miR sequence	Fresh media relative fluorescent signal	Depleted media relative fluorescent signal	Up-regulation in depleted media (times)	p-value
miR-297a	-	AUGUAUGUGUGCAUGUGCAUGU	14±4	1412±220	101±13	0.039
miR-466a	3p	UAUACAUACACGCACACAUAAGA	8±1	205±39	25±1	0.030
miR-466b	3-3p	AAUACAUACACGCACACAUAAGA	5±1	209±39	42±1	0.035
miR-466d	3p	UGUGUGUGCGUACAUGUACAUG	10±1	648±122	65±6	0.039
	5p	UAUACAUACACGCACACAUAG	8±2	552±92	69±6	0.030
miR-466f	3p	CAUACACACACACAUACACAC	19±1	5056±305	266±8	0.009
	5p	UACGUGUGUGUGCAUGUGCAUG	102±13	22208±5982	218±31	0.036
miR-466g	-	AUACAGACACAUGCACACACA	51±9	14935±4286	293±32	0.038
miR-466h		[illegible]	[illegible]	[illegible]	[illegible]	[illegible]
miR-467a	minor	AUAUACAUACACACACCUACAC	44±1	9567±2081	217±40	0.025
miR-467b	minor	AUAUACAUACACACACCAACAC	72±8	14451±2817	201±17	0.027
miR-467c	minor	AUAUACAUACACACACCUAUAU	14±2	1056±38	75±8	0.022
miR-669a	-	AGUUGUGUGUGCAUGUUCAUGU	7±0	57±3	8±0	0.016
miR-669b	-	AGUUUGUGUGUGCAUGUGCAUGU	8±0	46±1	6±0	0.003
miR-669c	-	[illegible]	[illegible]	[illegible]	[illegible]	[illegible]
miR-669d	-	ACUUGUGUGUGCAUGUAUAUGU	7±0	1511±463	216±66	0.038
miR-669e	-	UGUCUUGUGUGUGCAUGUUCAU	12±0	4080±971	341±81	0.027
miR-669f	-	CAUAUACAUACACACACACGUAU	44±1	13444±4267	306±90	0.037
miR-669h	3p	UAUGCAUAUACACACAUGCACA	8±0	326±6	41±1	0.007
miR-669i	-	UGCAUAUACACACAUGCAUAC	18±1	39±1	2±0	0.048

- 18 detected members of mouse miR 297-669 cluster were up-regulated in CHO cells in nutrient-depleted conditions

The End– Thank you

"Strategies based on speculation about where bottlenecks lie, can be successful on a case-by-case basis, but also do not reveal the relevant mechanisms of a high-yielding cell."

Bill, R.M., et al., *Overcoming barriers to membrane protein structure determination.* Nat Biotechnol, 2011. **29**(4): p. 335-40.

Microarray Analysis of MicroRNA: Fresh and Depleted Media

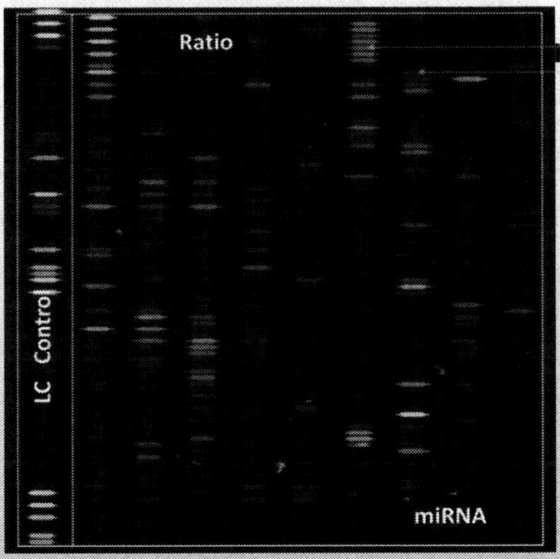

(A) Microarray map of overlaid images of miRs in fresh (labeled red) and depleted (labeled green) conditions.

MicroRNAs including mouse (mmu) microRNA (mir)-466h and mmu-mir-669c upregulated in depleted media

Location of microRNAs within cluster

Mouse miR ID (mmu-)	Variant	Mature miR sequence	Fresh media relative fluorescent signal	Depleted media relative fluorescent signal	Up-regulation in depleted media (times)	p-value
miR-297a	-	AUGUAUGUGUGCAUGUGCAUGU	14±4	1412±220	101±13	0.039
miR-466a	3p	UAUACAUACACGCACACAUAAGA	8±1	203±30	25±1	0.030
miR-466b	3-3p	AAUACAUACACGCACACAUAAGA	5±1	209±39	42±1	0.035
miR-466d	3p	UGUGUGUGCGUACAUGUACAUG	10±1	648±122	65±6	0.039
	5p	UAUACAUACACGCACACAUAG	8±2	552±92	69±6	0.030
miR-466f	3p	CAUACACACACACAUACACAC	19±1	5056±305	266±8	0.009
	5p	UACGUGUGUGUGCAUGUGCAUG	102±13	22200±5982	218±31	0.036
miR-466g	-	AUACAGACACAUGCACACACA	51±9	14935±4286	293±32	0.038
miR-466h	-	[illegible]	[illegible]	[illegible]	[illegible]	[illegible]
miR-467a	minor	AUAUACAUACACACACCUACAC	44±1	9567±2001	217±40	0.025
miR-467b	minor	AUAUACAUACACACACCAACAC	72±8	14451±2817	201±37	0.027
miR-467c	minor	AUAUACAUACACACACCUAUAU	14±2	1056±38	75±8	0.022
miR-669a	-	AGUUGUGUGUGCAUGUUCAUGU	7±0	57±3	8±0	0.016
miR-669b	-	AGUUUGUGUGCAUGUGCAUGU	8±0	46±1	6±0	0.005
miR-669c	-	[illegible]	[illegible]	[illegible]	[illegible]	[illegible]
miR-669d	-	ACUUGUGUGUGCAUGUAUAUGU	7±0	1511±463	216±66	0.038
miR-669e	-	UGUCUUGUGUGUGCAUGUUCAU	12±0	4090±971	341±81	0.027
miR-669f	-	CAUAUACAUACACACACACGUAU	44±1	13444±4267	306±90	0.037
miR-669h	3p	UAUGCAUAUACACACAUGCACA	8±0	326±6	41±7	0.007
miR-669i	-	UGCAUAUACACACAUGCAUAC	18±1	39±1	2±0	0.048

- 18 detected members of mouse miR 297-669 cluster were up-regulated in CHO cells in nutrient-depleted conditions

Engineering and Application of Prokaryotic Transcriptional Regulators As Biosensors in Yeast

Keasling, J. D., Lawrence Berkeley National Laboratory
Snoek, T., Technical University of Denmark
Skjoedt, M., Novo Nordisk
Jensen, M. K., Technical University of Denmark
Originally presented on: 6/27/2016 15:45:00 - 16:05:00

Bio-based production of chemicals and fuels is an attractive avenue in order to reduce dependence on fossil fuels. However, the current efficiency of genome engineering methods and parts prospecting allow for unprecedented genotype diversity that vastly outstrips our ability to screen for best cell performance. Here we describe the systematic characterization of parameters important for the successful engineering of biosensors for high-throughput screening of yeast cell factories. Our multi-parametric engineering of a ligand-binding prokaryotic transcription factor for aromatic compound catabolism allowed us to develop a biosensor for dicarboxylic acid accumulation in yeast cell factories. Using the optimal design of the dicarboxylic acid biosensor, we tested a number of other candidate biosensors derived from ligand-binding prokaryotic transcriptional regulators, and here report successful examples of their direct transplantation into yeast cell factories. Most importantly, the biosensors engineered and characterized in our laboratory are applicable for *in vivo* screening and selection of optimal cell factory performance in a high-throughput manner within relevant operational ranges. Furthermore, omics data analysis suggests that they have no off-target imprint on the native transcriptome. Acknowledging the vast repertoire of ligand-binding transcriptional regulators in prokaryotes, these findings should be of interest to the metabolic engineering community in order to develop future high-throughput screening and selection tools of cell factory libraries targeting bio-based production of a wide range of both fine and bulk chemicals.

New Methods for Computational Strain Design

Klamt, S.,

Originally presented on: 6/28/2016 8:00:00 - 8:25:00

Computational strain design is a subfield of metabolic engineering which uses mathematical models and dedicated algorithms to identify metabolic intervention strategies optimizing the production performance of microbial strains. In particular, stoichiometric and constraint-based modeling approaches have been used to rationally design microbial cell factories. In my talk I will present recent developments in this field achieved by us and our collaborators:

1. Minimal cut set (MCS) based algorithms for the enumeration of suitable intervention strategies in genome-scale metabolic networks [1].
2. Generalization of MCS to *regulatory* MCS allowing combinations of gene/reaction knockouts with flux up- and downregulations [2].
3. Metabolic network properties enabling growth-coupled product synthesis [3]. We found that a growth-coupled production of almost all metabolites in *E. coli* is possible.
4. By a rigorous use of the above methods in a realistic application we designed an *E. coli* strain for high-yield production of itaconic acid [4]. In a fed-batch cultivation, this strain produced 32 g/l itaconic acid with an overall yield of 0.68 mol/(mol glucose) and a peak productivity of 0.45 g/l/h. These values are by far the highest that have ever been achieved for heterologous itaconic acid production.

References:

[1] von Kamp A, Klamt S (2014) Enumeration of smallest intervention strategies in genome-scale metabolic networks. PLoS Computational Biology 10(1):e1003378.
[2] Mahadevan R, von Kamp A, Klamt S (2015) Genome-scale strain designs based on regulatory minimal cut sets. Bioinformatics 31:2844-2851.
[3] Klamt S, Mahadevan R (2015) On the feasibility of growth-coupled product synthesis in microbial strains. Metabolic Engineering 30:166-178.
[4] Harder B-J, Bettenbrock K, Klamt S (2016) Model-based metabolic engineering enables high yield itaconic acid production by *Escherichia coli*. Under revision.

Development of Databases and Computational Methods for Exploring and Managing Metabolic Complexity

Hatzimanikatis, V., Swiss Federal Institute of Technology of Lausanne (EPFL)
Originally presented on: 6/28/2016 8:25:00 - 8:50:00

Metabolic engineering has benefited from the application of mathematical methods and the availability of biological databases. Two of the central problems in metabolic engineering have been the expansion of our current knowledge of biochemical reactions and compounds and the identification of targets for manipulating metabolism with the minimum possible impact on cells physiology. We will present and discuss our work in addressing these two problems.

In the first case we will present and discuss the ATLAS of biochemistry, a database of all the biochemically plausible reactions between compounds reported to occur in living organisms. ATLAS uses KEGG as the reference database with 16,000 metabolites and 10,000 reactions and it contains more than 130,000 new reactions. The mining of ATLAS will undoubtedly offer new opportunities for the design of synthetic pathways and metabolic engineering.

In the second case, we will introduce an optimisation framework that (i) uses a kinetic representation of genome-scale metabolic networks, and (ii) identifies enzyme manipulations that can optimize a metabolic function while maintaining metabolite levels and reaction rates within physiological ranges.

Finding Enzymes for Novel Synthetic Pathways Constructions to Generate High Value Added Chemicals

Ghosh, D.,
Yunus, I. S.,
Jones, P. R., Imperial College London
Originally presented on: 6/28/2016 9:15:00 - 9:35:00

The low-hanging fruit for metabolic engineering targets have in many cases been picked and challenges in commercialization of low-value fuel production systems have resulted in a gradual shift in attention towards higher value products with expanding potential diversity. Synthetic biology allows us to create new metabolic pathways for industrial biocatalysis by transfer of existing "parts" available in nature, or to produce pathways towards molecules that naturally do not exist at all. A limiting factor for the creativity solution space is the availability of genes encoding suitable enzymes. Towards this objective, we have developed a protocol for structure-based prediction of the promiscuity of enzyme substrate specificity with the aim to narrow potential targets for secondary experimental trials in the construction of novel pathways. The protocol includes a phylogenetic search at the primary level, followed by structural model refinement (Ramachandran, ERRAT, GDT-HA, RMSD, Molprobity) and molecular docking using three complementary prediction algorithms and tools (Phyre2, Swiss Model, Autodock). The combined score provides an estimate of the predicted affinity (Kcal/mol) of the tested substrates. *In vitro* assays with purified recombinant novel enzymes are used to validate the molecular predictions. We will present case studies where we (i) have carried out *in silico* screening for substrate specificity of novel enzymes involved in synthetic pathways by molecular docking; (ii) evaluated these novel enzymes *in vitro* activities to validate the computational predictions.

Toward Next-Generation Antibiotics: Model-Guided Dissection of Bacterial Nitric Oxide Stress Networks

Brynildsen, M. P., Princeton University

Originally presented on: 6/28/2016 9:35:00 - 9:55:00

Nitric oxide (NO) is an important antimicrobial produced by the immune system to neutralize pathogens. Many pathogens, such as *Mycobacterium tuberculosis*, *Neisseria meningitides*, *Pseudomonas aeruginosa*, and pathogenic *Escherichia coli*, harbor NO detoxification and repair systems that have been linked to their virulence, and thus represent attractive targets toward which novel anti-infective therapies could be identified. Unfortunately, the few compounds that have been found to inhibit bacterial NO detoxification systems suffer from poor transport or toxicity, which warrants searches for other compounds or strategies to impair those processes. To complicate matters, the broad reactivity of NO and its reaction products give rise to an expansive and interconnected biochemical reaction network whose experimental measurables are difficult to interpret. To facilitate quantitative analyses of bacterial NO stress networks, we constructed a kinetic model of the *E. coli* NO biochemical network, including relevant processes such as enzymatic NO detoxification, iron-sulfur cluster damage and repair, DNA deamination and repair, transcriptional regulation, and NO autoxidation. Here, I will discuss how we have used this computational tool to mechanistically investigate how a variety of chemical, genetic, and environmental perturbations alter the *E. coli* NO stress response, and how this tool could facilitate the development of next-generation antibiotics that inhibit bacterial NO defenses.

Engineered Polyketide Synthases for Production of Commodity and Specialty Chemicals

Keasling, J. D., Lawrence Berkeley National Laboratory
Originally presented on: 6/28/2016 10:25:00 - 10:50:00

Engineered modular polyketide synthases (PKSs) have the potential to be an extraordinarily effective retrosynthesis platform. Native PKSs assemble and tailor simple, readily available cellular acyl-CoAs into large, complex, chiral molecules. By successfully rearranging existing polyketide modules and domains, one could exquisitely control chemical structure from DNA sequence alone. As an example of the diverse biosynthetic potential of PKSs, we have concluded that approximately 20 of the roughly 150 commodity chemicals tracked by the petrochemical market information provider ICIS could be produced by mixing and matching naturally occurring PKS domains. To form these chemicals, engineered PKSs would load acyl- CoAs, perform a programmed number of extension reactions, and then release products using previously published mechanisms. However, this potential has only just begun to be realized as the compounds that have been made using engineered PKSs represent a small fraction of the potentially accessible chemical space. We envision a future in which a single design algorithm, using a molecule of interest as input, successfully combines natural PKS sequences to produce the desired molecule. This kind of algorithm could also be used to make analogs of existing natural products. In my talk, I will highlight work from our laboratory where we have engineered polyketide synthases to produce a variety of commodity and specialty chemicals.

Reprogramming Halomonas for Enhanced Production of Chemicals

Chen, G. G. Q.,

Originally presented on: 6/28/2016 10:50:00 - 11:15:00

Due to the disadvantages of bioprocesses including energy consuming sterilization, high fresh water consumption, discontinuous fermentation to avoid microbial contamination, highly expensive stainless steel fermentation facilities and competing substrates for human consumption, industrial biotechnology is less competitive compared with chemical processes. Recently, halophiles have shown promises to overcome these shortcomings. Due to their unique halophilic properties, some halophiles are able to grow in high pH and high NaCl containing medium, practically, seawater, under higher temperature, allowing fermentation processes to run contamination free under unsterile conditions and continuous way in seawater. We have developed genetic manipulation methods for halophiles. In this lecture, halophiles were reprogrammed to produce bioplastics polyhydroxyalkanoates (PHA), ectoines, enzymes, and bio-surfactants with improved efficiency including morphology engineering to change the shapes of *Halomonas* spp to allow convenient downstream processing. This lecture will focus on bioplastics polyhydroxyalkanoates (PHA) as an example to illustrate how *Halomonas* can be reprogrammed for various applications.

Construction of a Hydrogenase minus Mutant of Clostridium acetobutylicum : a Platform Strain for the Continuous Production of Fuels and Chemicals

Soucaille, P.,
Nguyen, N. P. T.,
Originally presented on: 6/28/2016 11:15:00 - 11:40:00

Currently, there is a resurgence of interest in *Clostridium acetobutylicum*, the biocatalyst of the historical Weizmann process, to produce bulk chemicals like 1, 3 propanediol and fuels like n-butanol. To develop commercial processes via a metabolic engineering approach, it is necessary to better characterize both the primary metabolism of *C. acetobutylicum* and its regulation. Here we developed an improved and biochemically validated genome-scale model, (iCac967) and apply a quantitative system-scale analysis to acidogenic, solventogenic and alcohologenic steady-state *C. acetobutylicum* cells to report for the first time quantitative transcriptomic, proteomic and fluxomic data. This approach allows for a better understanding of the regulation of primary metabolism and for the functional characterization of numerous enzymes involved in primary metabolism. Based on these data a metabolic engineering strategy was developed to engineer *C. acetobutylicum* for the continuous production of several bulk chemicals at high yield. A new method was first developed to simultaneously inactivate several genes. Using this method we first engineered a â??hydrogenase minus platform strainâ? producing lactate as the major product without any production of either butyrate, acetate, acetone, ethanol or butanol. When used in continuous culture a titer of 28 g/l with a yield on glucose of 0.92 g/g and a productivity of 1.4 g/l.h were obtained.

This platform strain was further engineered to continuously convert raw glycerin to 1, 3 propanediol at a yield of 0.61 g/g, a titer of 60g/l and a productivity of 3 g/l.h with acetate as the only by-product. Finally the platform strain was engineered to produce ethanol as the major product. Used in continuous culture a titer of 39 g/l with a yield on glucose of 0.48 g/g and a productivity of 2 g/l.h were obtained. Stability of the three processes was demonstrated for several months and the cultures were only stopped due to biofilmsâ?? formation on probes and problems of pH and level control.

Work in our laboratory is currently concentrating on engineering the â??hydrogenase minus platform strainâ? strain for the continuous production of n-butanol at high yield.

KOBE UNIVERSITY

FUJICCO Good Food, Good Balance

Production of rare inositols: conversion of agricultural wastes into value added products

Kobe University
Department of Science, Technology and Innovation
Ken-ichi Yoshida, Ph.D.

APPLMIC
Applied Microbiology

ME11(26-30 June 2016)

Converting Waste Agricultural Biomass into a Resource

UNEP

Globally, 140 billion metric tons of biomass is generated every year from agriculture. This volume of biomass can be converted to an enormous amount of energy and raw materials.

Equivalent to approximately 50 billion tons of oil, agricultural biomass waste converted to energy can substantially displace fossil fuel, reduce emissions of greenhouse gases and provide renewable energy to some 1.6 billion people.

As raw materials, biomass wastes have attractive potentials for large-scale industries and community-level enterprises. Biomass takes the form of residual stalks, straw, leaves, roots, husk, nut or seed shells, waste wood and animal husbandry waste. Widely available, renewable, and virtually free, waste biomass is an important resource.

ME11(26-30 June 2016)

Rice as the MI-resource plant…? "Phytic acid " in rice bran

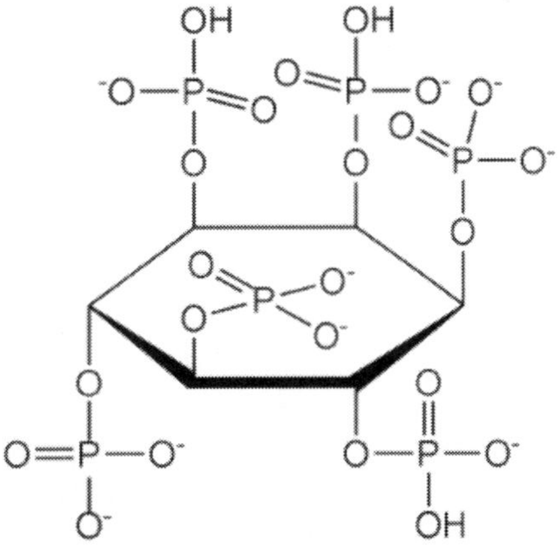

Phytic acid (known as *myo*-inositol hexakisphosphate (IP_6), or phytate when in salt form) is the principle storage form of phosphorus in many plant tissues, especially bran and seeds.

Phytase

Releasing phosphorus from phytate

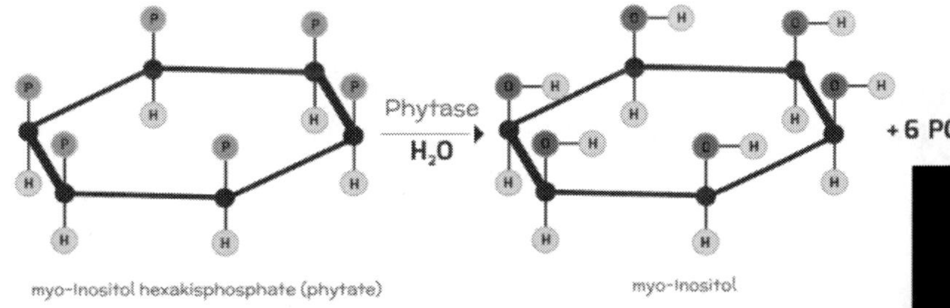

Four distinct classes of phytase have been characterized so far:

- Beta-propeller phytases (including "natto" starter enzymes)
- Histidine acid phosphatases (HAPS)
- purple acid phosphatases
- protein tyrosine phosphatase-like phytases (PTP-like phytases)

To optimize phytase secretion in *Bacillus subtilis*

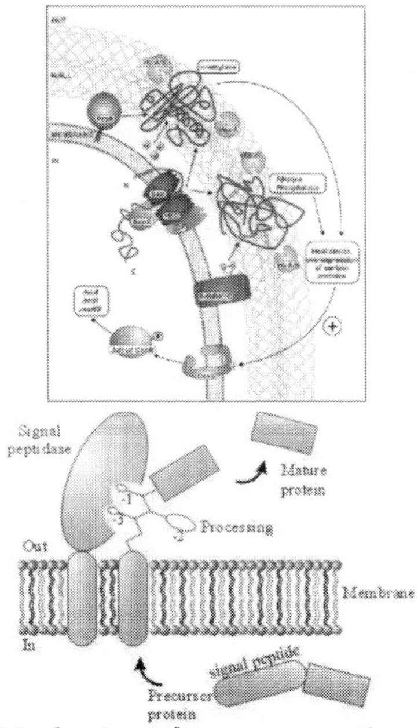

Mechanism of enzyme secretion

Addition of an appropriate signal peptide...

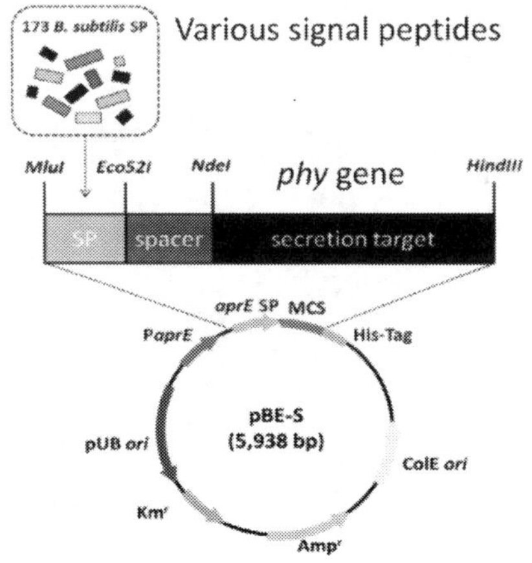

... enables the best efficiency in secretion of target enzyme.

Phytase secretion enhanced by the customized signal peptide

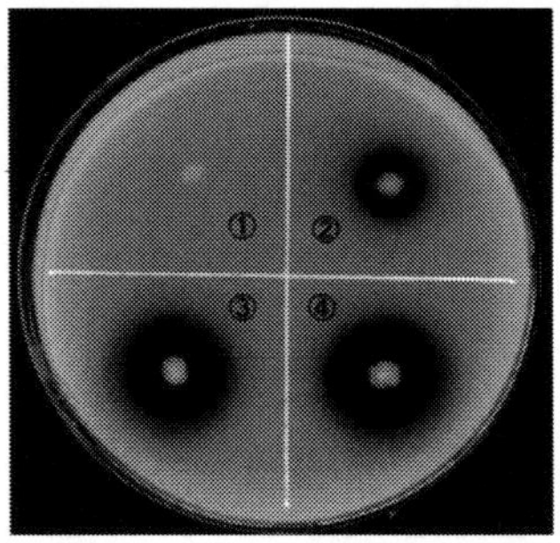

Out of 173 signal peptides identified in *B. subtilis* genome, the best one was selected to double the secreted activity.

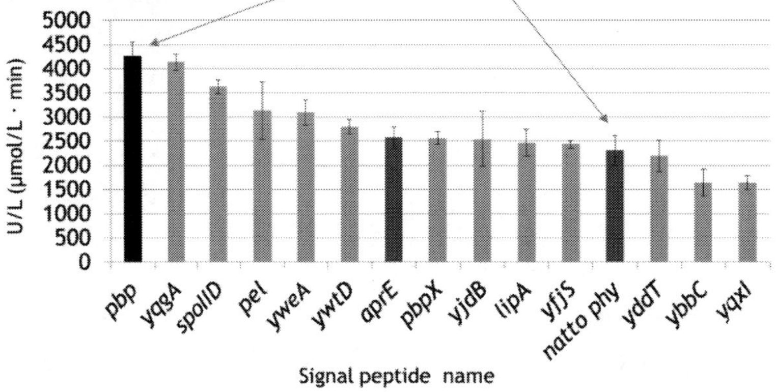

Inositol isomers

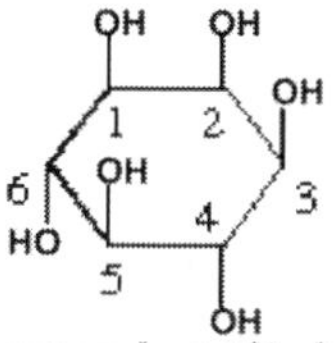

myo-inositol (MI)

- 1,2,3,4,5,6-Cyclohexanehexol
- Nine stereoisomers (MI is the most abundant in nature.)
- Natural ingredients found in plants and animals
- Various biological functions

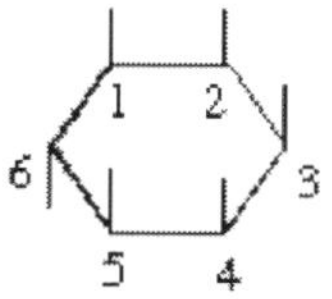

***epi*-inositol**

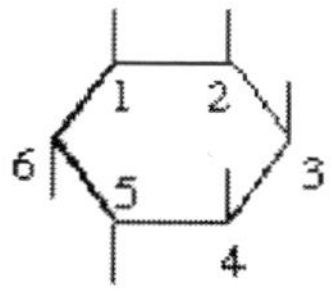

***allo*-inositol**

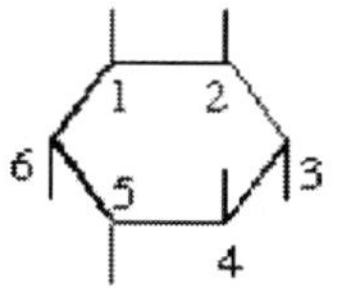

L-*chiro*-inositol

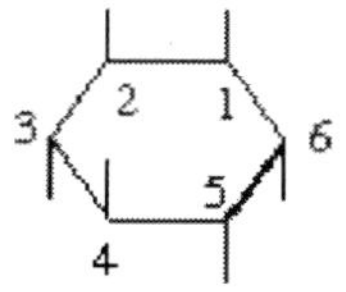

D-*chiro*-inositol

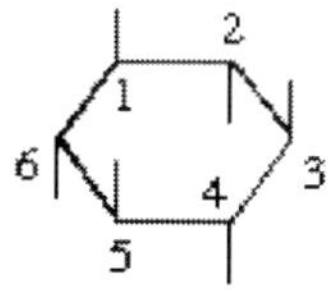

scyllo-inositol

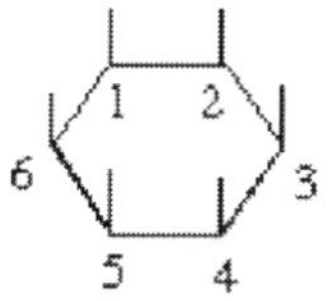

***cis*-inositol**

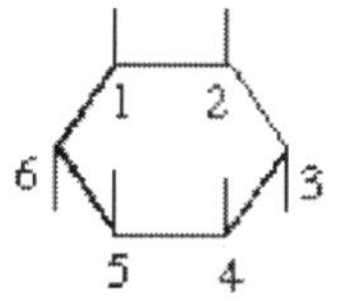

***muco*-inositol**

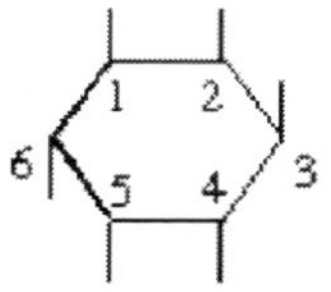

***neo*-inositol**

scyllo-Inositol (SI)

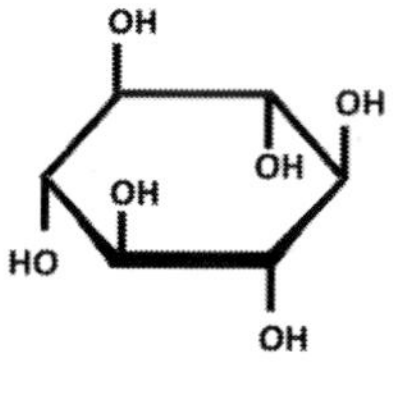

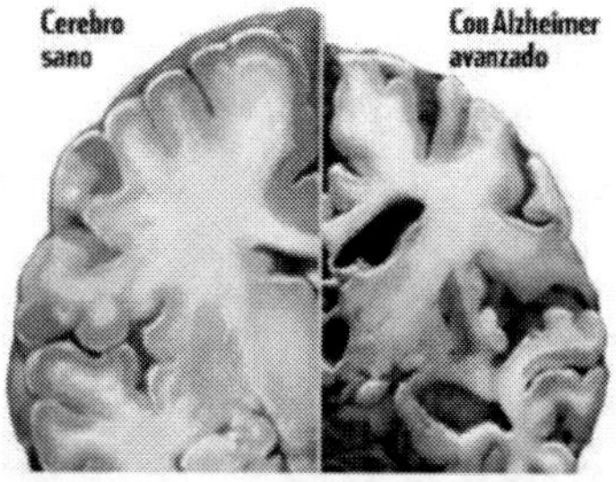

Some types of inositol prevented the accumulation of amyloid â deposits, a hallmark of Alzheimer's disease. *scyllo*-Inositol treatment also improved cognitive abilities in the mice and allowed them to live a normal lifetime. (McLaurin *et al.*, 2006. Nature Medicine 12, 801-808.)

Modified metabolism for conversion from MI to SI

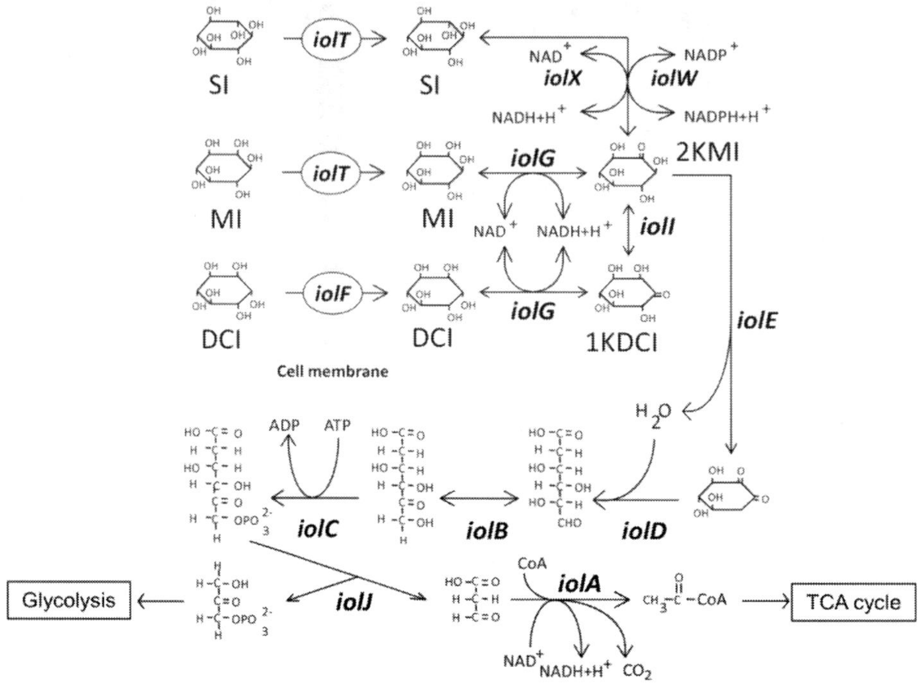

Modified metabolism for conversion from MI to SI

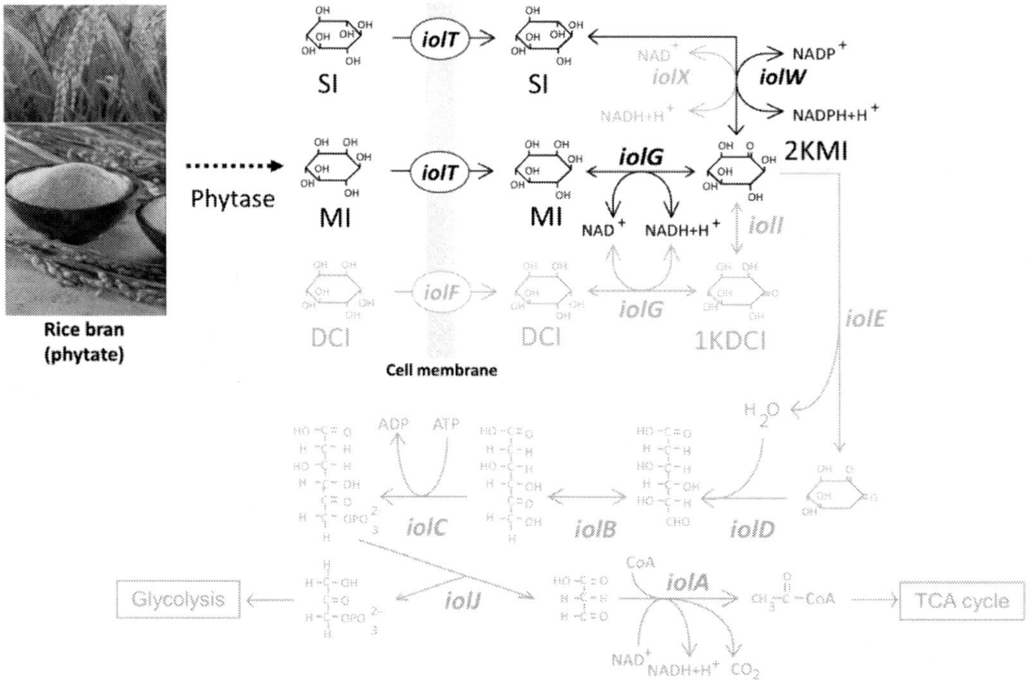

Improvement in SI production

• Parental strain (168)

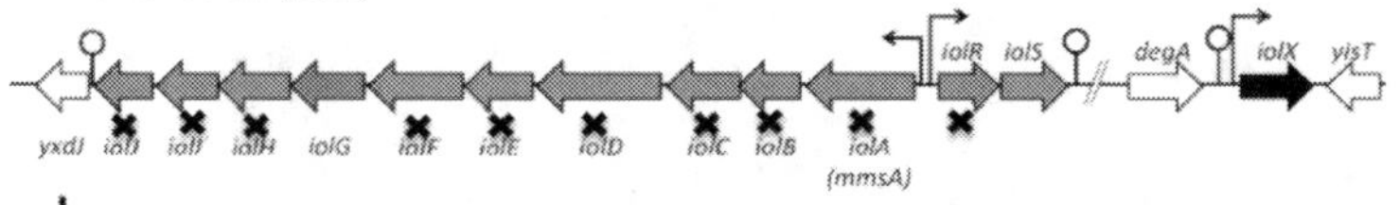

↓ Deletion ⟶ Addition (Overexpression) SI production

• MYI04

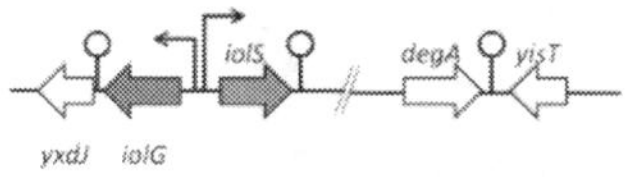

•KU106 10 g/L

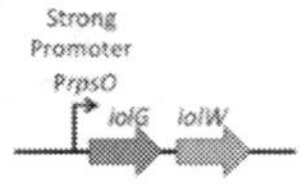

+

NADPH •KU108 19 g/L

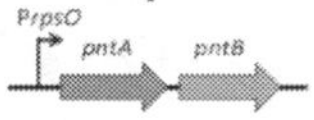

Substrate MI •KU301 21 g/L

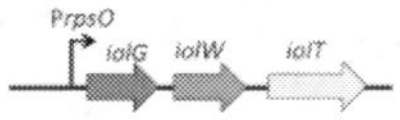

+

Both NADPH and substrate MI •KU302 <u>25 g/L</u>

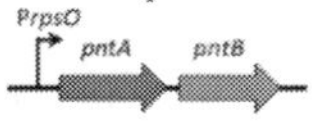

Poster Session 3 (this afternoon, p. 130, #80): Kosei Tanaka

D-Pinitol (3-*O*-methyl-D-*chiro*-inositol)

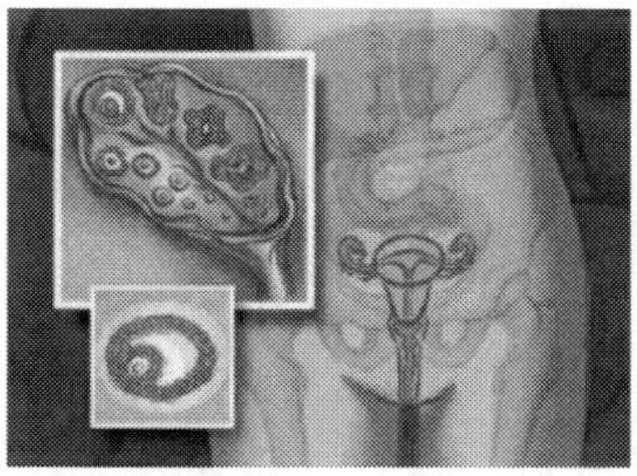

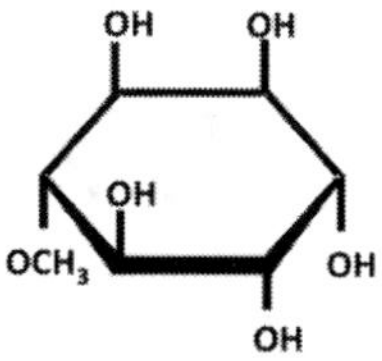

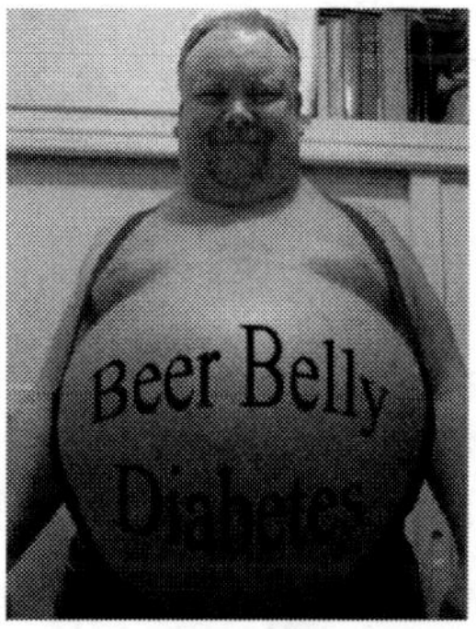

D-Pinitol lowers serum glucose in STZ rats, diabetes-model animals (Kawa *et al.*, 2003).

A drug candidate for the treatment of type 2 diabetes and polycystic ovary syndrome.

2-deoxy-D-[^{3}H]-glucose (2DG) uptake into L6 myotubes

Collaboration with Prof. Ashida

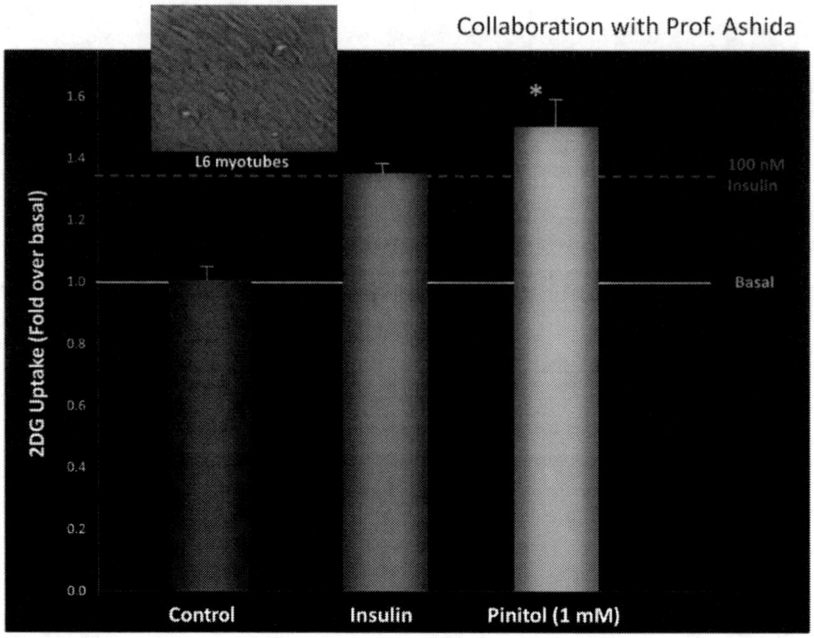

* Significant difference from control cells, P<0.05. ‡ Significant difference from insulin-treated cells, P<0.05.

D-Pinitol does not stimulate 2DG uptake into 3T3-L1 adipocyte as insulin does.

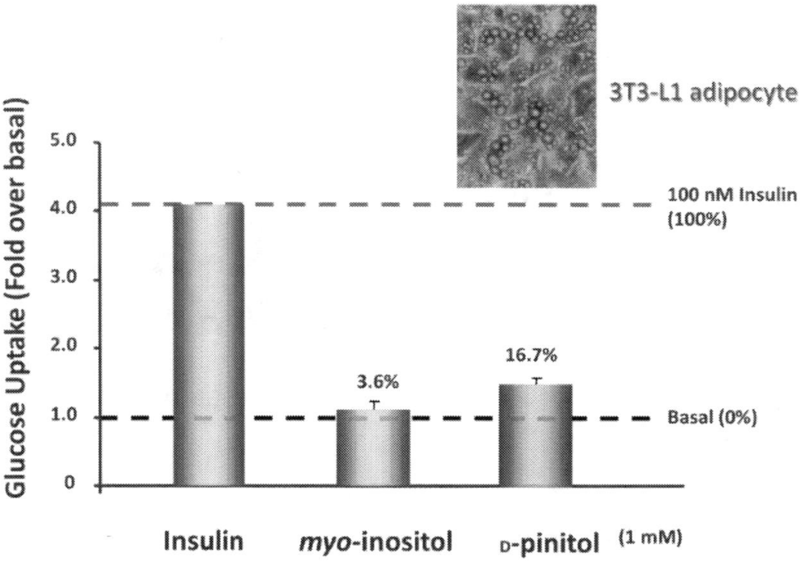

Soybean plant is rich in D-pinitol.

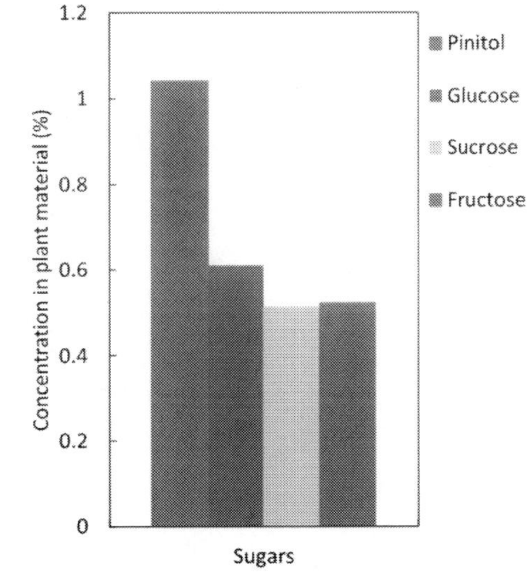

What is GK?

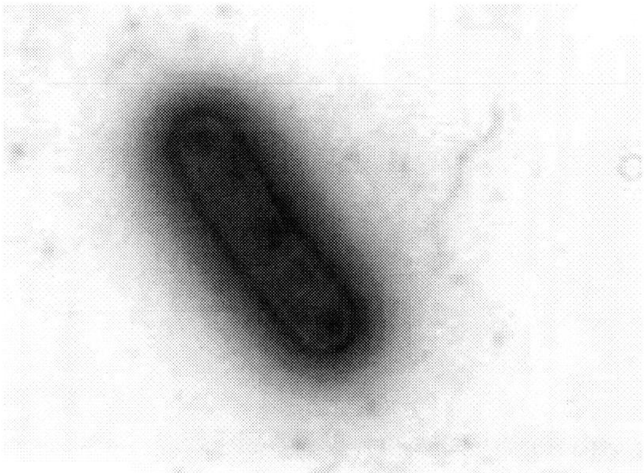

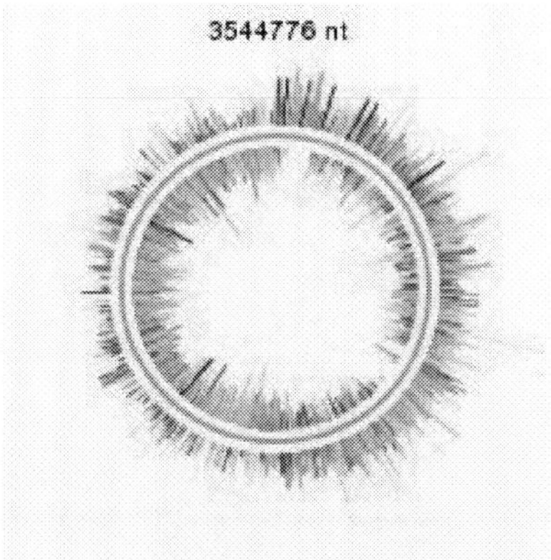

Geobacillus kaustophilus HTA426

- isolated from the deep-sea sediment of the Mariana Trench.
- a thermophilic *Bacillus*-related species whose upper temperature limit for growth is 74°C (optimally 60°C).
- its genome is composed of a 3.54 Mb chromosome and a 47.9 kb plasmid, and is an aerobic endospore-forming Gram-positive *Bacillus*-related bacterium.

Predicted inositol metabolism in GK

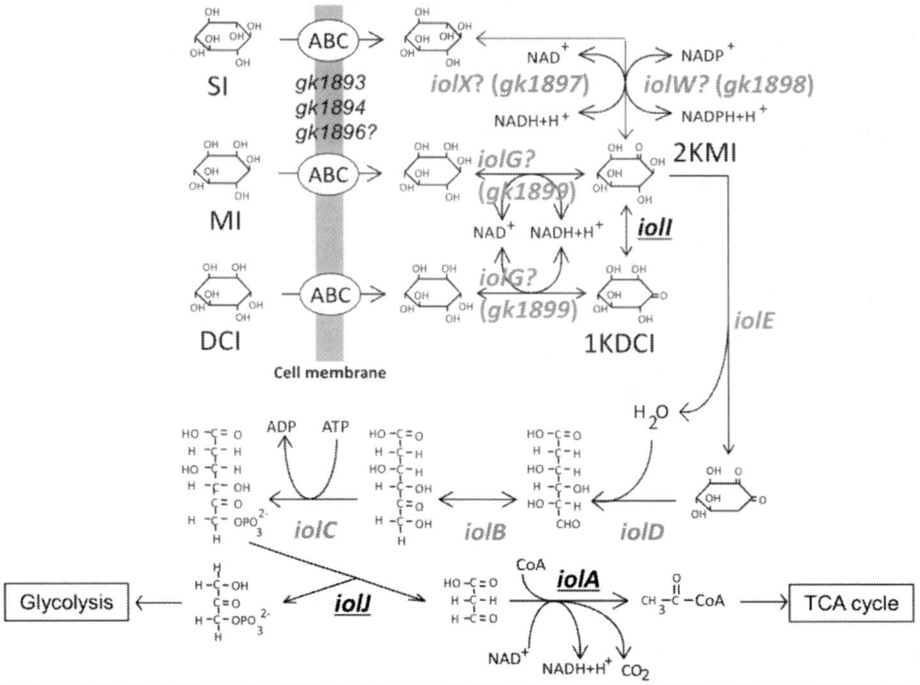

Pilot-scale PI production

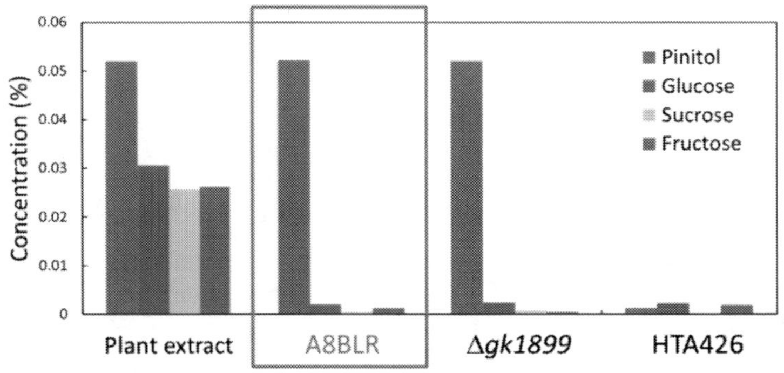

Collaboration with Fujicco

350 L fermenter

Centrifuge

Filtration apparatus

Safety and Functionality

Prof. Osawa

Prof. Mizuno

Safety testing

Grad Sch Agricultural Science
In vitro and animal studies

Collaboration

Prof. Azuma

Prof. Yoshida

Functionality testing

Grad Sch Medicine
Clinical studies

The poster to recruit borderline diabetes volunteers

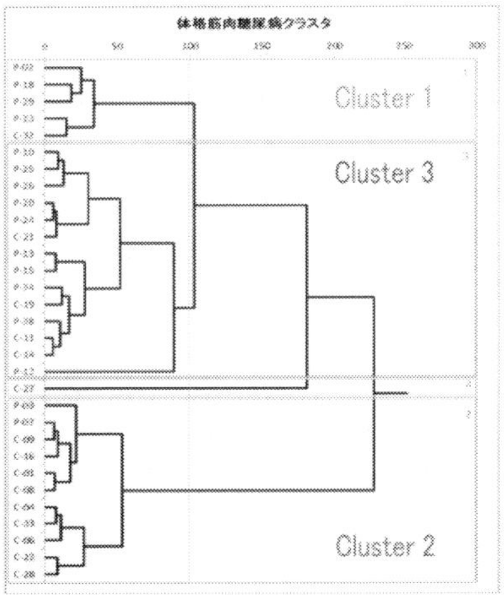

Cluster.	BMI	Total Fat	Muscle	Visceral fat	CK(CPK)	Insulin (IRI)	HbA1c(NGSP)	Glucose level
1	0.178	1.940	-1.004	0.400	48.600	-2.920	0.040	5.800
2	0.442	1.464	-0.682	0.455	-32.818	-1.245	0.118	0.909
3	0.021	0.500	-0.168	0.107	11.214	2.029	0.014	8.071
4	-1.200	-3.600	-0.500	-2.000	152.000	-1.100	0.000	1.000

ME11(26-30 June 2016)

Acknowledgements

Good Food, Good Balance
FUJICCO
Fujicco
Mr. Katsushige Yamada
Mr. Toshiaki Kamitani

Grad Sch Medicine
Prof. Takeshi Azuma
Prof. Masaru Yoshida

Grad Sch Agricultural Science
Prof. Hitoshi Ashida
Prof. Ro Osawa
Prof. Masashi Mizuno

Grad Sch Sci Tech Innov
Prof Akihiko Kondo
Prof Tomohisa Hasunuma

KOBE UNIVERSITY

iBioK

APPLMIC
Applied Microbiology

ME11(26-30 June 2016)

Methanol-Based Production of Value-Added Products like 1,5-Diaminopentane
Wendisch, V. F., Bielefeld University
Originally presented on: 6/28/2016 12:00:00 - 12:20:00

Methanol, a one-carbon compound, is regarded as an attractive substrate for biotechnological production of value-added bulk products, such as amino acids and polyamines. Methanol can be utilized by a variety of bacteria and other organisms as carbon and energy source. Two strategies were followed to utilize methanol for cadaverine production, namely using non-methylotrophic *Corynebacterium glutamicum* and the methylotrophic *Bacillus methanolicus.*

A strain of non-methylotrophic *C. glutamicum* was engineered to produce the polyamide building block 1,5-diaminopentane (cadaverine) as non-native product. Moreover, this strain was engineered further for co-utilization of methanol. Expression of the gene encoding NAD^+-dependent methanol dehydrogenase (Mdh) from the natural methylotroph *B. methanolicus* increased methanol oxidation. Deletion of the endogenous aldehyde dehydrogenase genes *ald* and *fadH* prevented methanol oxidation to carbon dioxide and formaldehyde detoxification via the linear formaldehyde dissimilation pathway. Heterologous expression of genes for the key enzymes hexulose-6-phosphate synthase and 6-phospho-3-hexuloisomerase of the ribulose monophosphate (RuMP) pathway in this strain restored growth in the presence of methanol or formaldehyde, which suggested efficient formaldehyde detoxification involving RuMP key enzymes. While growth with methanol as sole carbon source was not observed, the fate of ^{13}C-methanol added as co-substrate to sugars was followed and the isotopologue distribution indicated incorporation into central metabolites and in vivo activity of the RuMP pathway. In addition, ^{13}C-label from methanol was traced to the secreted product cadaverine. Thus, this synthetic biology approach led to a *C. glutamicum* strain that converted the non-natural carbon substrate methanol at least partially to the non-native product cadaverine.

B. methanolicus, a thermophilic methylotrophic bacterium, was engineered into a microbial cell factory for the production of the platform chemical 1,5-diaminopentane (cadaverine) from methanol. This was achieved by the heterologous expression of the *Escherichia coli* genes *cadA* and *ldcC* encoding two different lysine decarboxylase enzymes, and by increasing the overall L-lysine production levels in this host. Both CadA and LdcC were functional in *B.□methanolicus* cultivated at 50°C and expression of *cadA* resulted in cadaverine production levels up to 500□mg/l during shake flask conditions. A volume-corrected concentration of 11.3□g/l of cadaverine was obtained by high-cell density fed-batch methanol fermentation. Our results demonstrated that efficient conversion of L-lysine into cadaverine presumably has severe effects on feedback regulation of the L-lysine biosynthetic pathway in *B.□methanolicus*. By also investigating the cadaverine tolerance level, *B.□methanolicus* proved to be an exciting alternative host and comparable to the well-known bacterial hosts E.□*coli* and *C. glutamicum*. This study represents the first demonstration of microbial production of cadaverine from methanol.

Rewiring Metabolism to Maximize Lipid Overproduction in Yarrowia lipolytica

Qiao, K., Massachusetts Institute of Technology
Wasylenko, T., Massachusetts Institute of Technology
Xu, P., MIT
Stephanopoulos, G., Massachusetts Institute of Technology
Originally presented on: 6/28/2016 13:50:00 - 14:15:00

Production of lipids by microbial fermentation of carbohydrate feedstocks outcompetes oil crops in terms of productivity, however, presently achievable carbohydrate-to-lipid process yields have not yet reached the point that can support cost-effective production of lipids and biodiesel. To improve process yields, one needs to maximize lipid content and capture as many of the electrons generated from the catabolism of substrates as possible. Guided by a quantitative model that predicts process yield, Y from the non-lipid biomass yield, lipid content, and yield of lipid synthesis pathway, we show that Y is significantly improved by introducing synthetic pathways that recycle glycolytic NADHs into cytosolic NADPH or acetyl-CoA. We experimentally built and evaluated thirteen engineered strains, with the best strain achieving a lipid titer of 99 g/L with a productivity of 1.2 g/L/h and an overall process yield 0.269 g-FAME/g-glucose. These figures of merit advance the commercialization opportunities of carbohydrate-based lipids production. In addition, these Yarrowia strains show promise in lipid production from other feedstocks, especially when integrated with fermentation of gases fixing CO_2 with hydrogen or carbon hydroxide to produce volatile fatty acids. Results on an integrated system converting syngas to lipids will be presented as a model for a biological Gas-to-Liquids (GTL) process.

Engineering Yeast Central Metabolism for Improved Yields of Fuels and Chemicals

Pronk, J. T., Department of Biotechnology, Delft University of Technology, Julianalaan 67, 2628 BC Delft, The Netherlands
Daran-Lapujade, P.,
Daran, J. M. G., Department of Biotechnology, Delft University of Technology, Julianalaan 67, 2628 BC Delft, The Netherlands
van Maris, T., Delft University of Technology
Originally presented on: 6/28/2016 14:15:00 - 14:40:00

Breath-taking advances in synthetic biology, including the recent addition of CRISPR-Cas9 to the metabolic engineering toolbox, greatly accelerate introduction and optimization of pathways towards native and non-native products in the yeast *Saccharomyces cerevisiae.*

Especially for transport fuels and commodity chemicals, economic constraints require that product yields on the carbohydrate substrate approach the theoretical limits set by conservation laws and thermodynamics. For many products, achieving such near-theoretical yields represents a formidable challenge that not only requires 'tuning' of product pathways but also a reconfiguration of central metabolism, which plays pivotal roles in precursor supply, energy conservation and redox-cofactor regeneration.

With the aim to improve ethanol yields on sugar substrates, we have explored new engineering strategies to engineer redox metabolism in anaerobic *S. cerevisiae* cultures. Implementation of acetic-acid reduction pathways based on a bacterial acetylating acetaldehyde dehydrogenase enabled improved ethanol yields on lignocelllosic hydrolysates, in which acetic acid is a key inhibitor of yeast performance. Functional expression of Calvin-cycle enzymes enables increased ethanol yields by the reduction of carbon dioxide, which is abundantly present in all yeast-based anaerobic fermentation processes.

Cytosolic acetyl-coenzyme A is required for the formation of many heterologous products made by engineered *S. cerevisiae* strains, including lipids, isoprenoids and flavonoids. In *S. cerevisiae*, formation of this precursor is notoriously expensive in terms of ATP requirements due to the involvement of acetyl-CoA synthetase. We have demonstrated functional expression and assembly of a bacterial pyruvate-dehydrogenase complex in the yeast cytosol, which offers a new and attractive pathway towards cytosolic acetyl-CoA. The mitochondrial carnitine shuttle in another pathway that, theoretically, could lead to ATP-efficient provision of cytosolic acetyl-CoA. However, previous research indicated that, *in vivo*, the yeast carnitine shuttle operates unidirectionally and does not allow for export of acetyl-CoA from the mitochondrial matrix to the cytosol. By a combination of metabolic engineering, laboratory evolution and reverse engineering of laboratory-evolved strains, we identified mutations that enable the 'reversal' of this shuttle mechanism in glucose-grown yeast cultures.

Institute of Bio- and Geosciences
IBG-1: Biotechnology

JÜLICH
FORSCHUNGSZENTRUM

Mitglied der Helmholtz-Gemeinschaft

Metabolic engineering of bacterial cell factories and use of transcriptional regulators as single-cell biosensors for strain and enzyme development

Michael Bott

ME XI, Awaji Island, June 29, 2016

Aim of our work: establishment of knowledge-based bioeconomy

JÜLICH
FORSCHUNGSZENTRUM

Present

Petrochemical products

Petroleum

Fuels & energy

Future

Bio-based products
- Commodity chemicals
- Fine chemicals
- Food, feed and pharmaceutical ingredients
- Proteins

Biomass

Food & feed

Fuels & energy

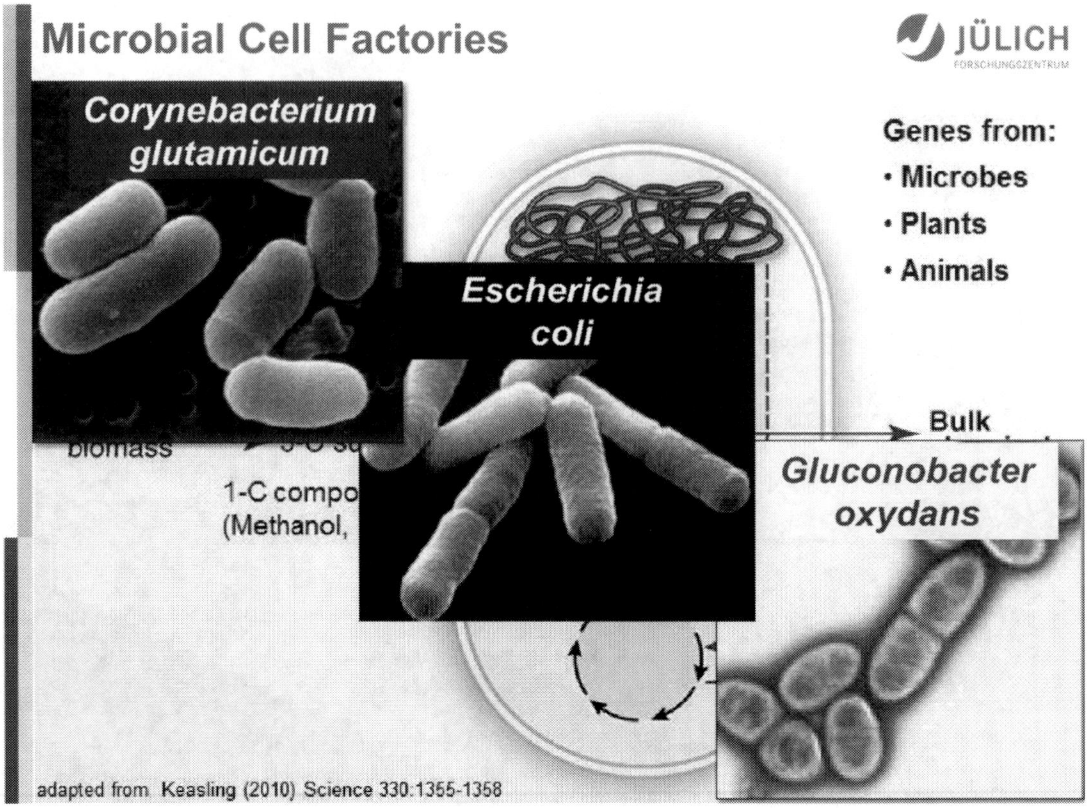

Microbial Cell Factories
JÜLICH
FORSCHUNGSZENTRUM
Corynebacterium glutamicum
Escherichia coli
Gluconobacter oxydans
Genes from:
• Microbes
• Plants
• Animals
Bulk
1-C compo
(Methanol,
adapted from Keasling (2010) Science 330:1355-1358

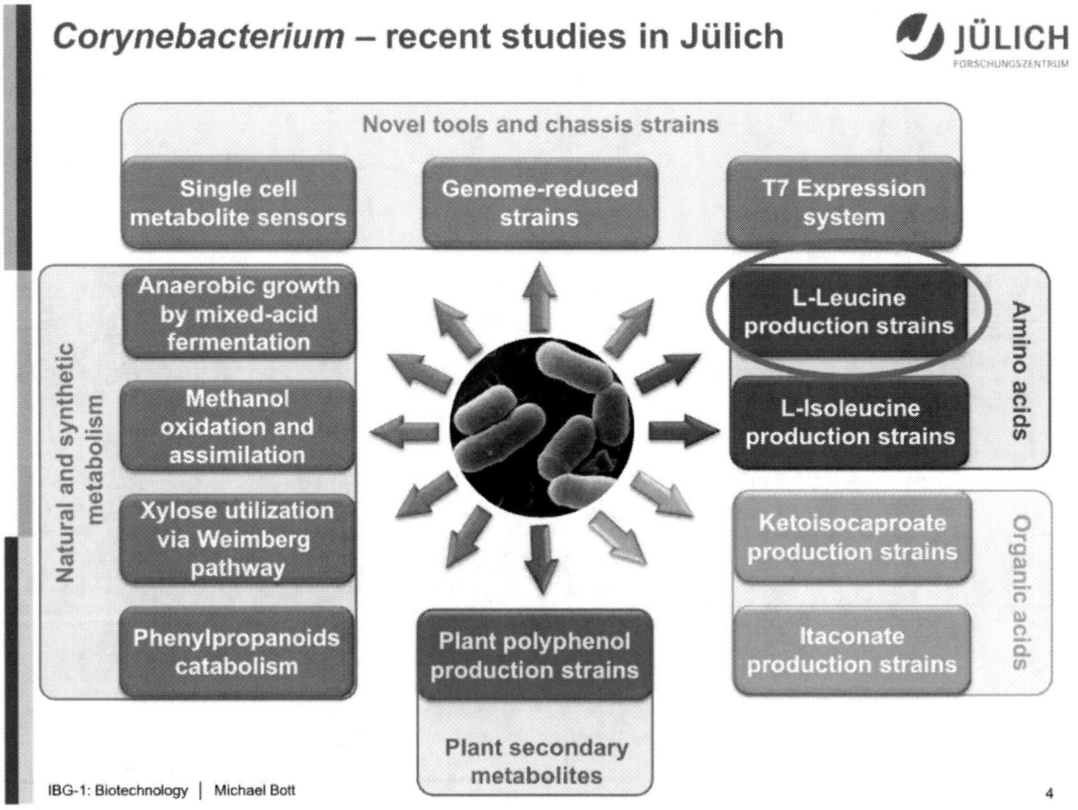

Corynebacterium – recent studies in Jülich
JÜLICH
FORSCHUNGSZENTRUM
Novel tools and chassis strains
Single cell metabolite sensors
Genome-reduced strains
T7 Expression system
Natural and synthetic metabolism
Anaerobic growth by mixed-acid fermentation
Methanol oxidation and assimilation
Xylose utilization via Weimberg pathway
Phenylpropanoids catabolism
Amino acids
L-Leucine production strains
L-Isoleucine production strains
Organic acids
Ketoisocaproate production strains
Itaconate production strains
Plant polyphenol production strains
Plant secondary metabolites
IBG-1: Biotechnology | Michael Bott
4

Corynebacterium as host for amino acid production

JÜLICH FORSCHUNGSZENTRUM

Glucose + Ammonium →

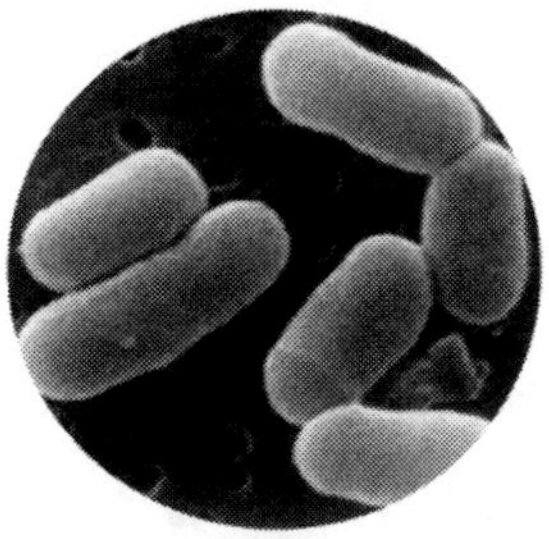

↓

L-Leucine

Michael Vogt

Jan van Ooyen

Simon Klaffl

L-Leucine

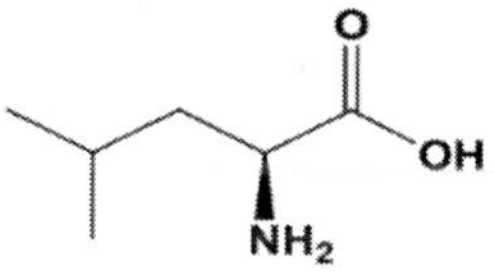

JÜLICH FORSCHUNGSZENTRUM

- essential amino acid
- various applications in the food and pharma industry
- price about 20 US-$ /kg
- production volume about 2000 – 3000 tons/year, steadily increasing

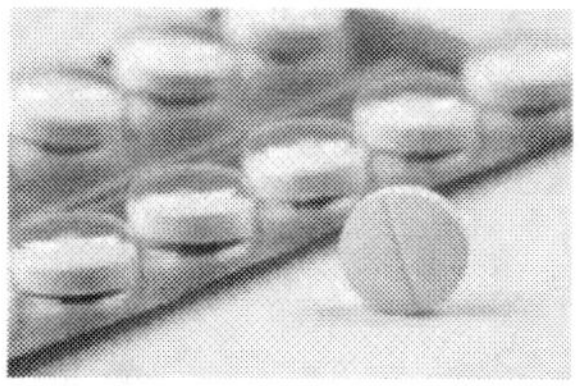

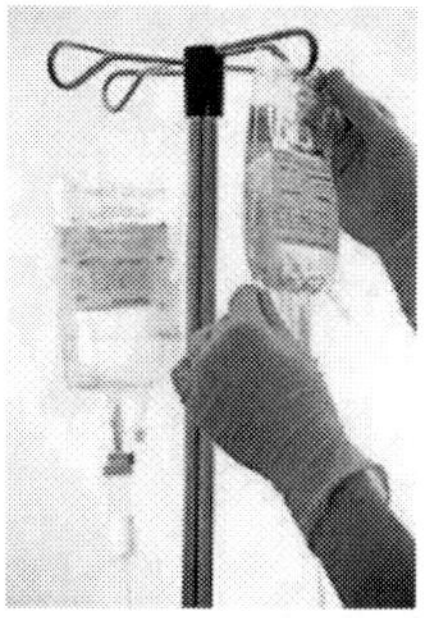

L-Leucine biosynthesis in *C. glutamicum*

JÜLICH
FORSCHUNGSZENTRUM

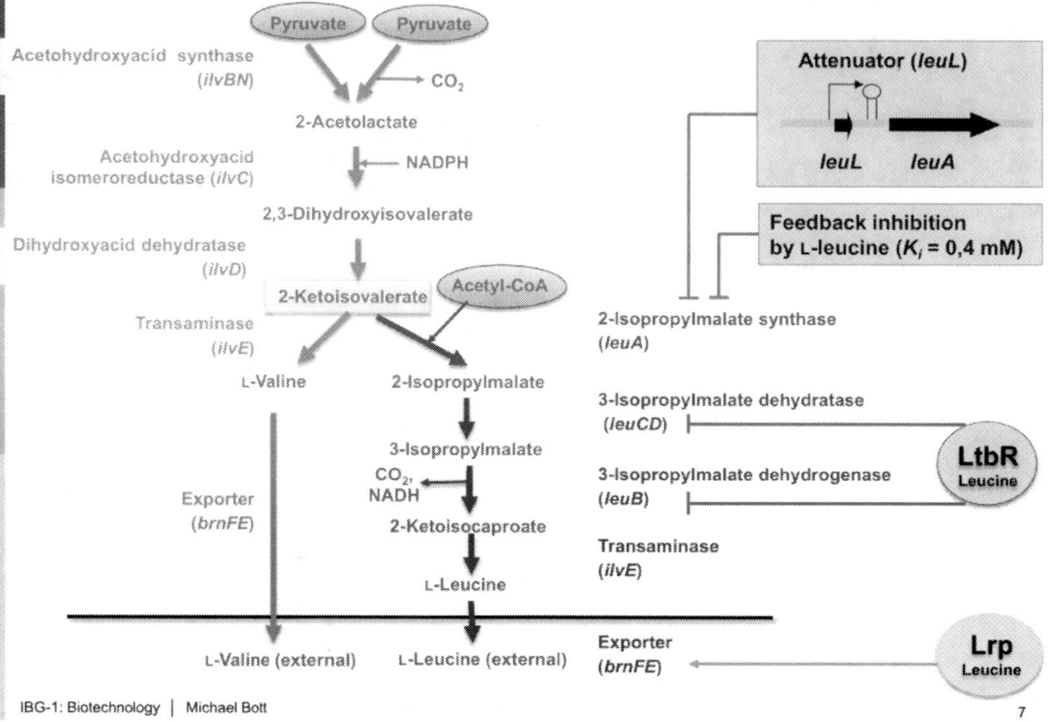

Feedback-resistent isopropylmalate synthase

JÜLICH
FORSCHUNGSZENTRUM

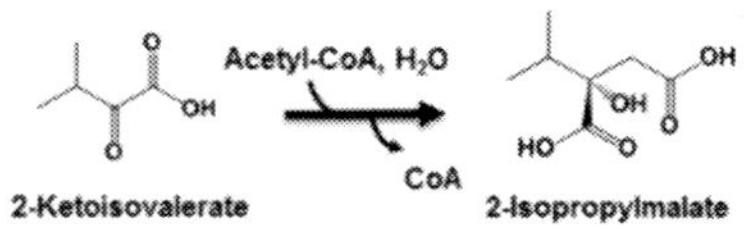

Sequence analysis of *leuA* gene from a weak leucine producer B018 obtained by random mutagenesis and screening

→ two amino acid exchanges identified (R529H and G532D), located in leucine-binding region

Koon *et al.* (2004) PNAS 101:8295-300
LeuA structure from *M. tuberculosis*

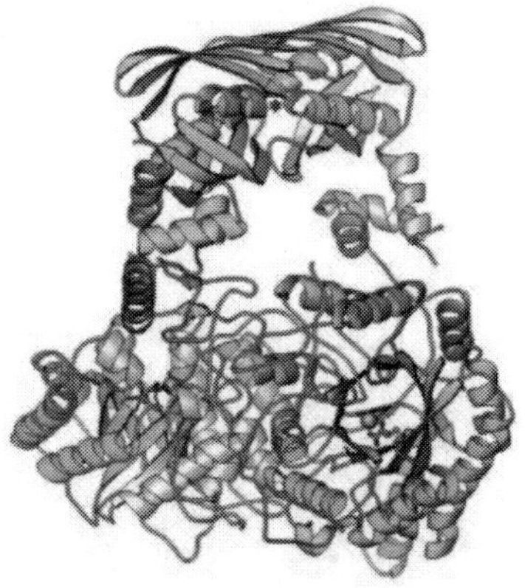

	LeuA of wild type	LeuA of strain B018
K_m 2-Ketoisovalerate (μM)	6.4 ± 0.1	10.0 ± 0.2
K_m Acetyl-CoA (μM)	9.6 ± 0.4	9.7 ± 0.4
K_i L-Leucin (mM)	0.4 ± 0.1	>40

Strain	Specific IPMS activity ($\mu mol\ min^{-1}\ mg_{protein}^{-1}$)	L-Leucine titer[d] (mM)
C. glutamicum wild type	0.06 ± 0.01	0
C. glutamicum pAN6-*leuA*_wt	0.59 ± 0.02	0.8 ± 0.1
B018	0.03 ± 0.01[c]	7.8 ± 2.0
C. glutamicum pAN6-*leuA*_B018	0.63 ± 0.04	40.7 ± 0.4

Strain development: *C. glutamicum* MV-Leu2

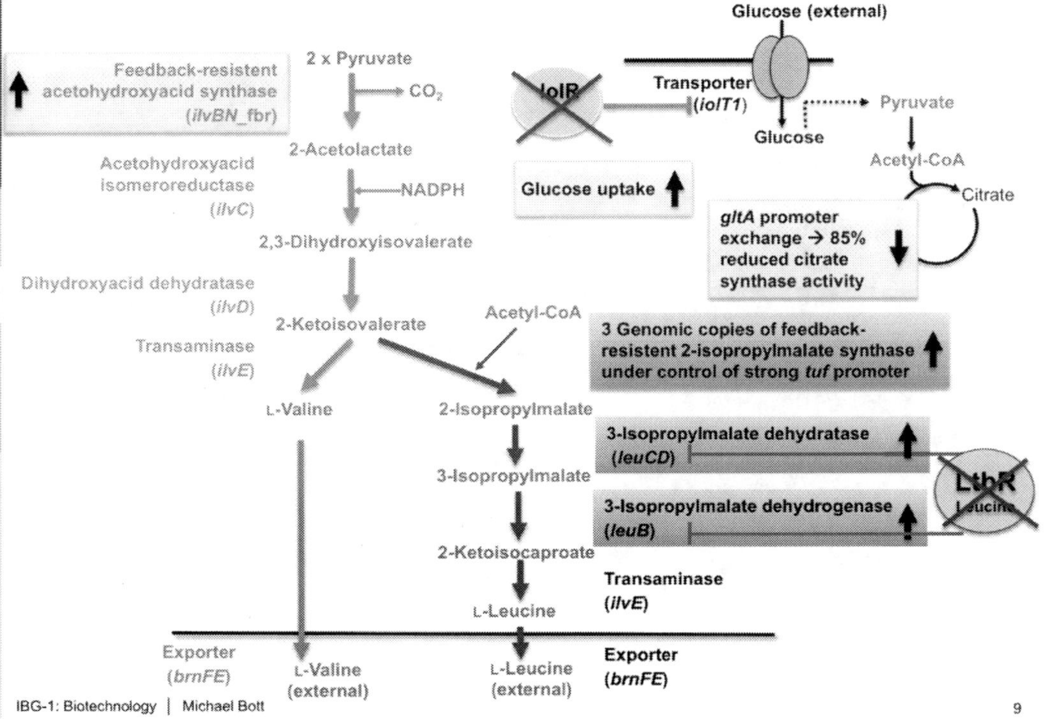

IBG-1: Biotechnology | Michael Bott 9

Fed-batch cultivation of strain MV-Leu2

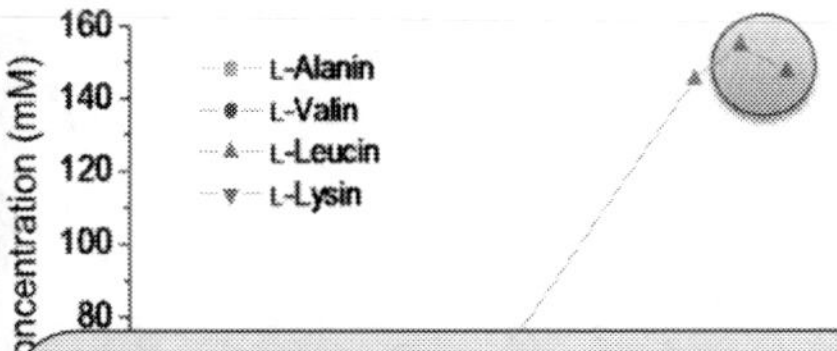

C. glutamicum MV-Leu2 represents highly efficient plasmid-free non-GMO L-leucine production strain suitable for industrial application

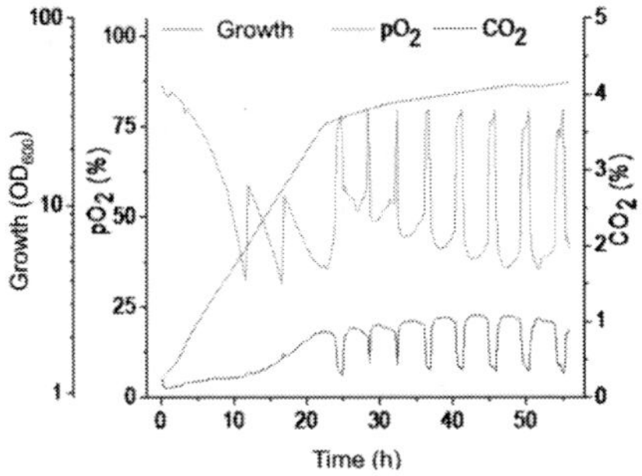

L-Leucine titer: 56 h: 152 mM = 19.9 g/l
72 h: 181 mM = 23.7 g/l
(Solubility: 24 g/l)

Productivity: 4.4 mmol l^{-1} h^{-1} (0.6 g l^{-1} h^{-1})
(cultivation time <72h)

Yield: 30% mol L-Leucine/mol Glucose
(theor. maximum = 60%)

IBG-1: Biotechnology | Michael Bott 10

Corynebacterium – recent studies in Jülich

JÜLICH FORSCHUNGSZENTRUM

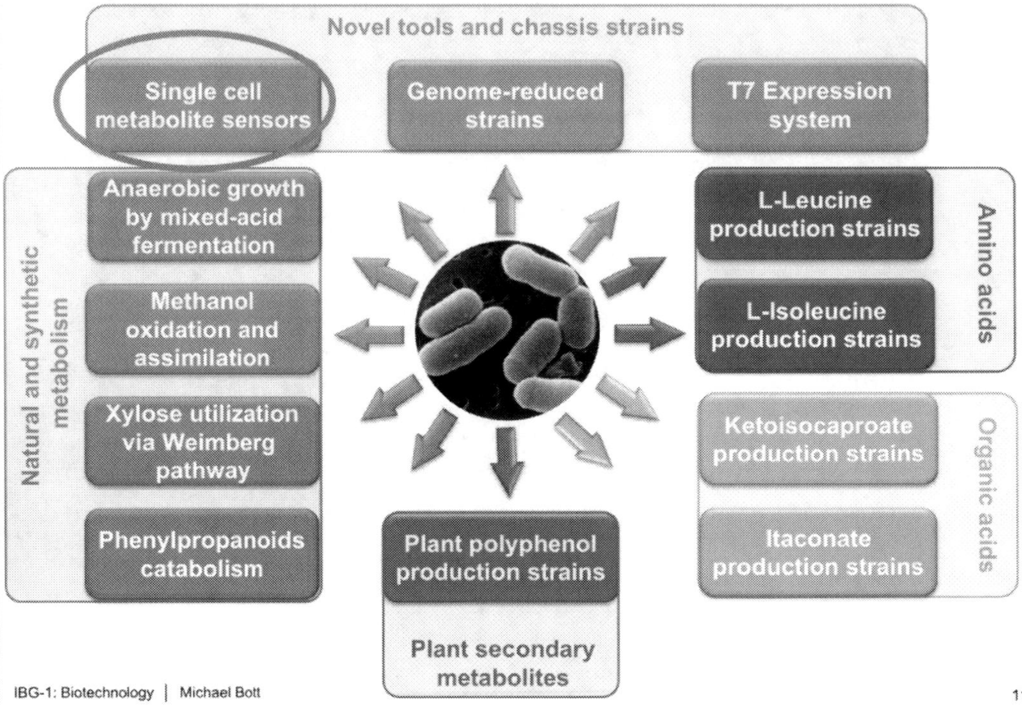

Metabolite sensors based on transcriptional regulators

JÜLICH FORSCHUNGSZENTRUM

Target molecule	Sensor	Transcription factor	Organism	Reference
L-Lysine	pSenLys	LysG	*C. glutamicum*	Binder et al. 2012
L-Arginine	pSenLys	LysG	*C. glutamicum*	Schendzielorz et al. 2013
L-Histidine	pSenLys	LysG	*C. glutamicum*	Schendzielorz et al. 2013
L-Methionine, L-Valine, L-Leucine, L-Isoleucine	pJC1-lrp-brnF'-eyfp	Lrp	*C. glutamicum*	Mustafi et al. 2012 Mustafi et al. 2014
O-Acetyl-serine	pSenOAS	CysR	*C. glutamicum*	Binder et al. 2012, Hoffmann et al. 2013
L-Arginine	pSenArg	ArgP	*E. coli*	Binder et al. 2012
NADPH	**pSenSox**	**SoxR**	***E. coli***	

For reviews see: Eggeling et al. 2015 Curr Opin Biotechnol 35: 30-36, Mahr & Frunzke 2016 AMB 100: 79-90

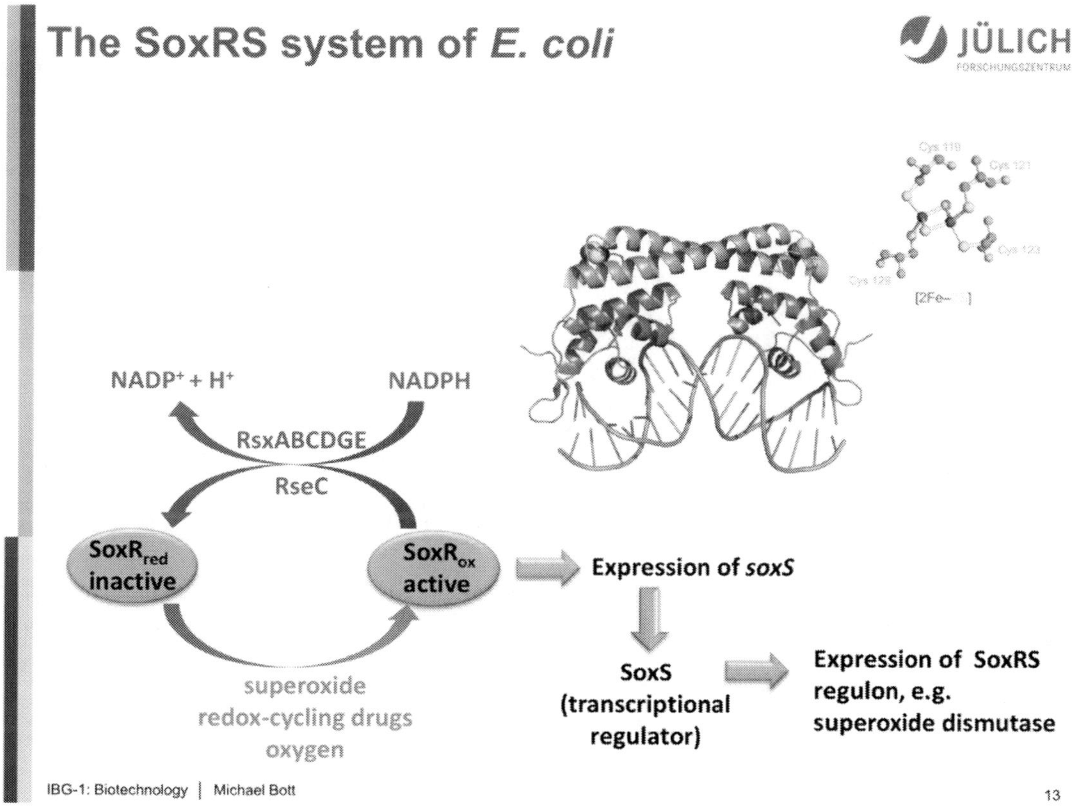
The SoxRS system of *E. coli*
JÜLICH
FORSCHUNGSZENTRUM
NADP⁺ + H⁺
NADPH
RsxABCDGE
RseC
SoxR red inactive
SoxR ox active
superoxide
redox-cycling drugs
oxygen
Expression of *soxS*
SoxS (transcriptional regulator)
Expression of SoxRS regulon, e.g. superoxide dismutase
IBG-1: Biotechnology | Michael Bott
13

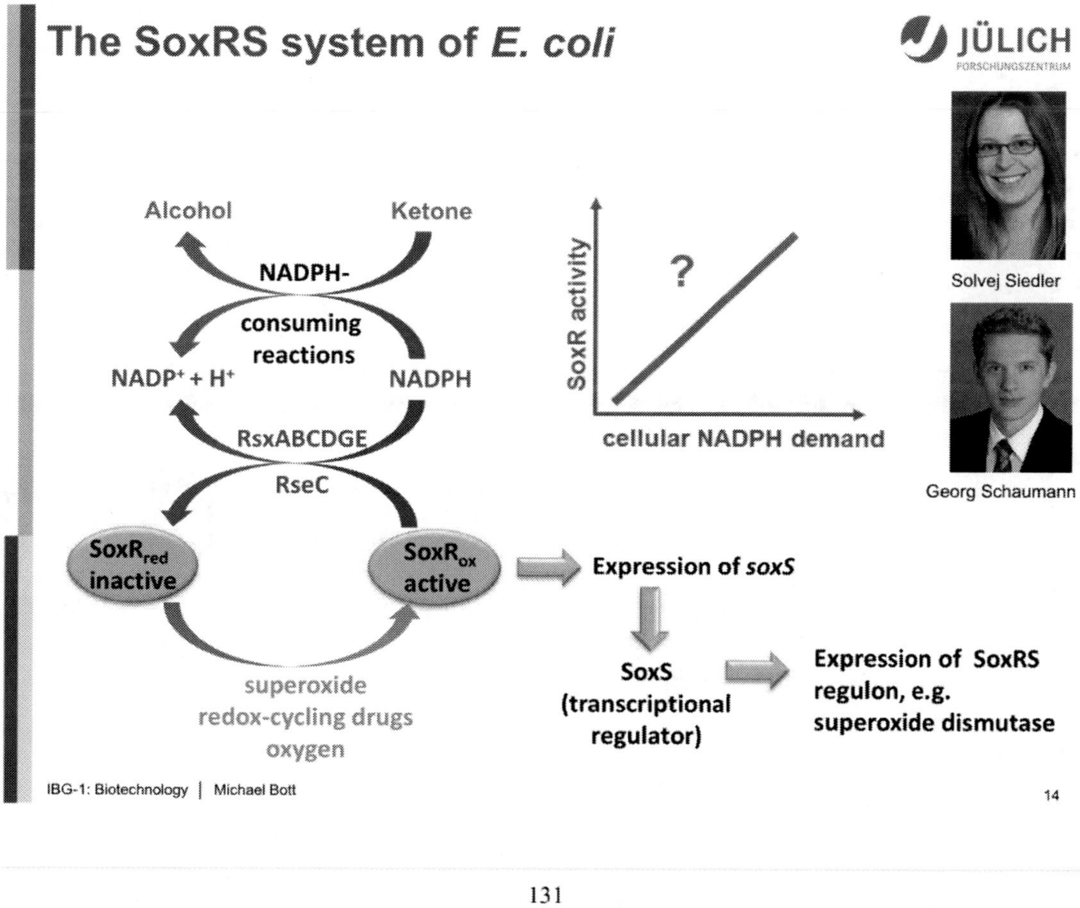
The SoxRS system of *E. coli*
JÜLICH
FORSCHUNGSZENTRUM
Alcohol
Ketone
NADPH-
consuming
reactions
NADP⁺ + H⁺
NADPH
RsxABCDGE
RseC
SoxR red inactive
SoxR ox active
superoxide
redox-cycling drugs
oxygen
SoxR activity
?
cellular NADPH demand
Solvej Siedler
Georg Schaumann
Expression of *soxS*
SoxS (transcriptional regulator)
Expression of SoxRS regulon, e.g. superoxide dismutase
IBG-1: Biotechnology | Michael Bott
14

The SoxR-based NADPH sensor

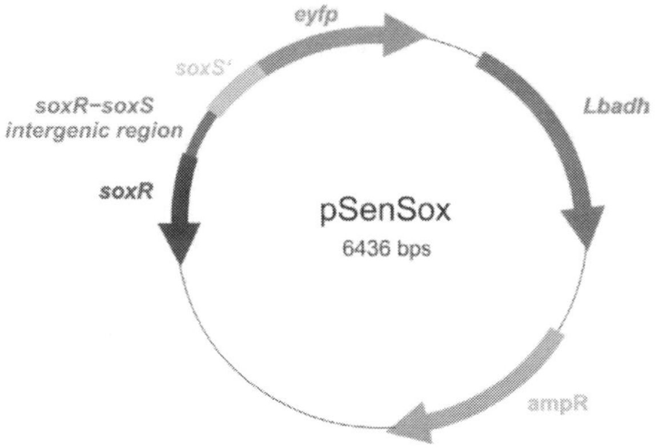

*Lb*Adh: NADPH-dependent alcohol dehydrogenase of *Lactobacillus brevis*

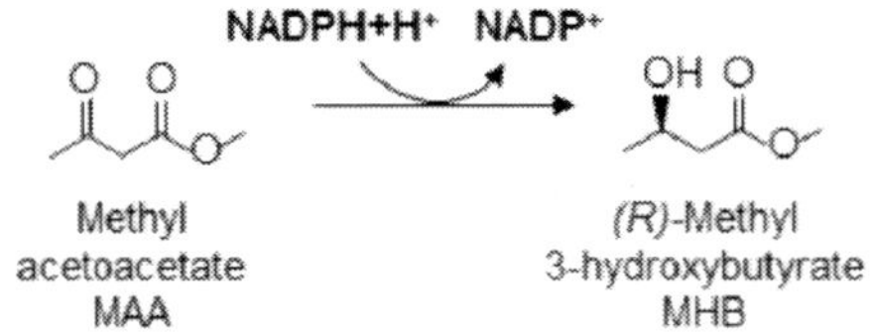

Variation of substrate concentration (MAA) at constant *Lb*ADH specific activity

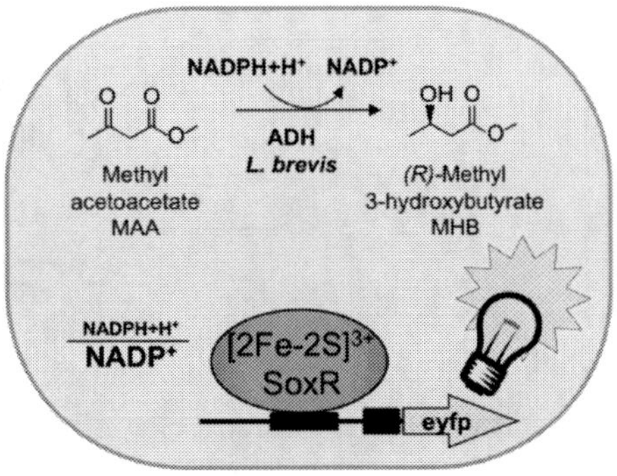

Escherichia coli

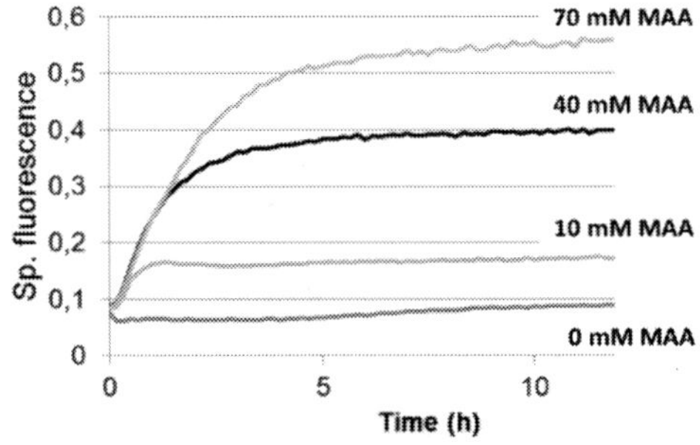

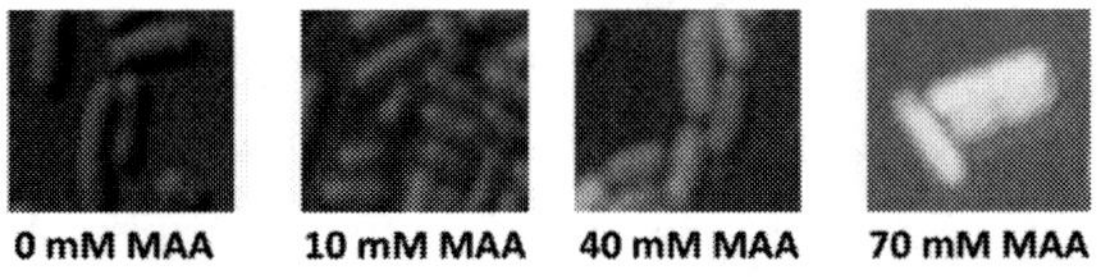

Increased NADPH demand results in increased fluorescence

Variation of specific *Lb*ADH activity at constant substrate concentration

JÜLICH FORSCHUNGSZENTRUM

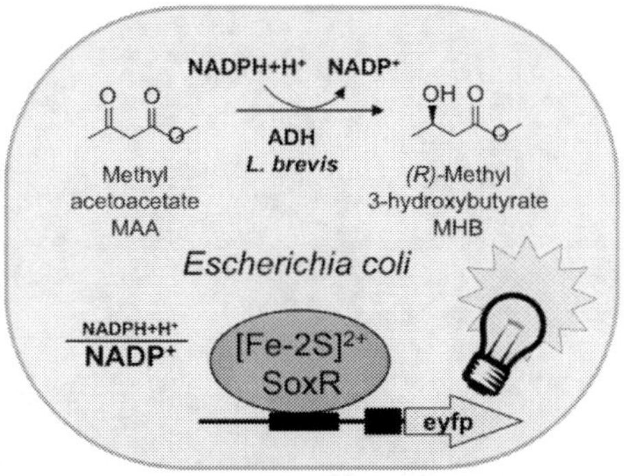

Escherichia coli	LbADH activity (U mg^{-1})	Sp. culture fluorescence (AU)
BL21 pSenNeg	0.03 ± 0.01	**0.06**
BL21 pSenSox pET28a	0.50 ± 0.1	**0.09**
BL21 pSen-L194S	0.70 ± 0.3	**0.11**
BL21 pSen-L194A	2.70 ± 0.6	**0.17**
BL21 pSenSox	6.28 ± 0.6	**0.38**
BL21 pSenSox; IPTG	**15.2 ± 2.0**	**0.45**

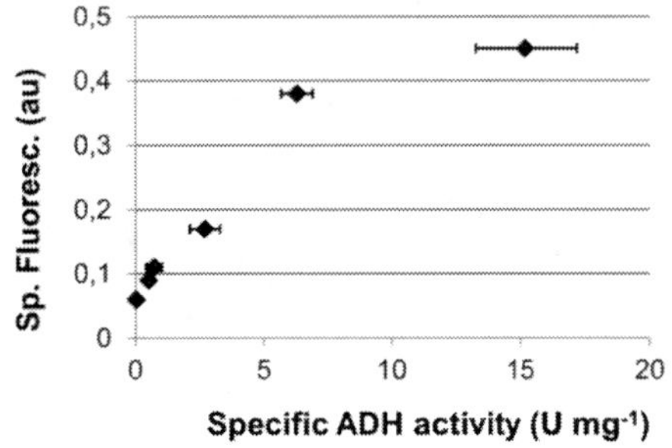

→ **single cell fluorescence correlates with specific ADH activity**

→ **use of FACS screening for isolation of ADH variants with increased activity?**

FACS-based screening of *Lb*ADH library for variants with improved activity for a selected substrate

JÜLICH FORSCHUNGSZENTRUM

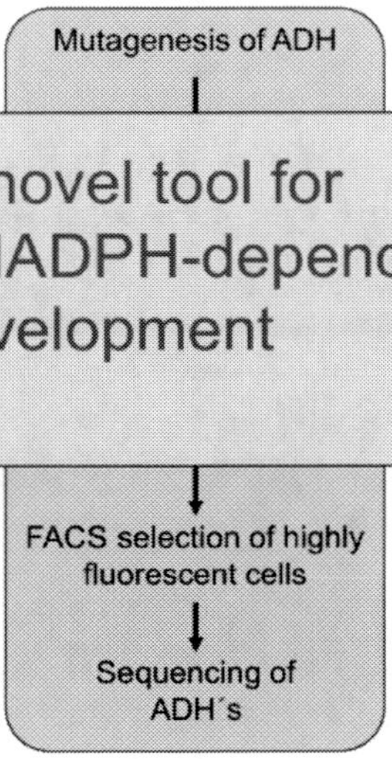

pSenSox is a novel tool for

- analyzing NADPH-dependent reactions
- enzyme development
- screening

ADH-A93M	2.62 ± 0.03	135%

pSenSox allows FACS screening for isolation of ADH variants with increased activity

Siedler et al 2014
ACS Synth. Biol. 3: 41-47

Recent studies in Jülich

JÜLICH FORSCHUNGSZENTRUM

Novel tools and chassis strains	Natural and synthetic metabolism	New production strains
Single cell metabolite sensors Genome Biol (2012) 13 R40 Metab Eng (2012) 14: 449 NAR (2013) 41: 6360 ACS Synth Biol (2014) 3: 21 ACS Synth Biol (2014) 3: 41	**Anaerobic growth by mixed-acid fermentat.** AEM (2015) 81: 7496	**L-Leucine** Metab Eng (2014) 22: 40
Genome-reduced strains AEM (2013) 79: 6006 Biotechnol J (2015) 10: 290	**Xylose utilization via Weimberg pathway** J Biotechnol (2014) 192:156	**L-Isoleucine** AMB (2015) 99: 791
T7 Expression system Microb Biotechnol 8: 253	**Methanol oxidation and assimilation** AEM (2013) 79: 6974 AEM (2015) 81: 2215	**Ketoisocaproate** Microb Biotech (2015) 8:351
	Phenylpropanoid catabolism AMB (2016) 100: 1871	**Itaconate** Metab Eng (2015) 30: 156
		Plant polyphenols Metab Eng (2016) in press

Further information

JÜLICH FORSCHUNGSZENTRUM

Presentation Regina Mahr (Wednesday, 3.00 pm):
A high-throughput approach for the identification of suitable parts for biosensor design

Poster Nicolai Kallscheuer (page 11):
Construction of a *Corynebacterium* platform strain for the production of high-value plant secondary metabolites

Poster Julia Frunzke (page 110):
Biosensor-driven adapative laboratory evolution of amino acid production in *Corynebacterium glutamicum*

Poster Jan Marienhagen (page 137):
Metabolic engineering of *Corynebacterium glutamicum* for the utilization of biomass-derived carbon & energy sources

Acknowledgements

Lothar Eggeling
Jan Marienhagen
Julia Frunzke
Stephan Binder
Georg Schaumann
Nurije Mustafi

Wolfgang Wiechert
Stephan Noack
Dietrich Kohlheyer
Alex. Grünberger

Volker Wendisch, Jörn Kalinowski — Universität Bielefeld

Reinhard Krämer, Gerd Seibold — Universität zu Köln

Ralf Takors — Universität Stuttgart

Kay Marin — Evonik Industries

Lutz Thomas — Amino

Helmholtz Association | BioSC Bioeconomy Science Center | Seventh Framework Programme | Bundesministerium für Bildung und Forschung

21

An Orthogonal and Modular Pathway for the Efficient Synthesis of Functionalized Small Molecules

Gonzalez, R., Rice University
Originally presented on: 6/28/2016 15:05:00 - 15:30:00

Anabolic metabolism can produce an array of small molecules, but yields and productivities areÂ often limited by carbon and energy inefficiencies and slow kinetics. Catabolic and fermentativeÂ pathways, on the other hand, are carbon and energy efficient but support only a limited productÂ range. To address these limitations, we engineered an orthogonal pathway for the synthesis ofÂ functionalized small molecules based on non-decarboxylative Claisen condensation reactions (andÂ subsequent Î²-reductions) that uses functionalized primers and functionalized extender units andÂ operates in an iterative manner. This carbonâ??carbon elongation mechanism was selected because ofÂ its ability to support iterative condensation reactions at high energy (ATP) efficiency, as previouslyÂ demonstrated in our laboratory (*Nature 476, 355-359, 2011*). The orthogonality of the newlyÂ developed platform enables predictable, tunable, and programmable operation of a pathway thatÂ retains the high product diversity, modularity, and combinatorial capabilities of anabolism. UsingÂ different Ï?- and Ï?-1-functionalized primers and Î±-functionalized extender units in combination withÂ various termination pathways, we engineered the synthesis of 18 products from 10 classes inÂ *Escherichia coli*, including Ï?-phenylalkanoic, Î±,Ï?-dicarboxylic, Ï?-hydroxy, Ï?-1-oxo, Ï?-1-methyl, 2-methyl, 2-methyl-2-enolic and 2,3-dihydroxy acids, Î²-hydroxy-Ï?-lactones, and Ï?-1-methyl alcoholsÂ (*Nature Biotechnology, 2016, 34 (5): 556-561.*

Bio-Based Isobutene

Chayot, R., Global Bioenergies

Originally presented on: 6/28/2016 15:30:00 - 15:50:00

As of today, most industrial bioproduction processes are based on naturally existing metabolic pathways, limiting the scope of industrial biology, and preventing the access to many of the chemistry's largest markets. The purpose of Global Bioenergies is to develop innovative metabolic pathways for the production of light olefins from renewable resources, by direct fermentation.Light olefins (ethylene, propylene, linear butylene, isobutylene and butadiene) are the core of the petrochemical industry. However, these volatile compounds are not naturally produced by microorganisms and no bioprocess to convert renewable resources to these molecules has been industrialized so far.

Global Bioenergies has developed artificial metabolic pathways including all the necessary enzymatic reactions from feedstock to isobutene. In contrast with most former approaches, the metabolic route leading to isobutene includes non-naturally occurring reactions as key steps, for example the decarboxylation of hydroxyisovaleric acid into isobutene. Global Bioenergies' current production strain can directly convert glucose, sucrose and second generation feedstocks into Isobutene without any chemical treatment. In 2015, the bio-Isobutene process has reached 70% of target commercial yield.

Importantly, light olefins are gaseous and therefore spontaneously volatilize from the fermentation medium, making the process less energy intensive and cheaper. Commercialization efforts took a step forward with the announcement of the successful production of isobutene at industrial pilot scale in Pomacle (France), where operations started since November 2014. Samples of bio-based isobutene or of isooctane derived exclusively from bio-isobutene where delivered to Arkema and Audi and to CFBP an industry organisation representing several gas supply distribution companies. Furthermore, a Demo Plant is currently being built in Leuna (Germany).

Finally, while isobutene production moves towards industrial scale, Global Bioenergies is also developing new artificial pathways enabling direct bio-production of Butadiene and Propylene.

Engineering Competitive Advantage in Industrial Microbial Bioprocesses

Shaw, A. J.,
Lam, F., Massachusetts Institute of Technology
Hamilton, M. A.,
Consiglio, A., Novogy, Inc.
MacEwen, K., Novogy, Inc.
Brevnova, E., Novogy, Inc.
Greenhagen, E., Novogy, Inc.
LaTouf, W. G.,
South, C. R., Novogy, Inc
van Dijken, J. P., Delft University of Technology
Rajgarhia, V.,
Stephanopoulos, G., Massachusetts Institute of Technology
Originally presented on: 6/28/2016 15:50:00 - 16:10:00

Microbial contamination is a liability for industrial bioprocesses affecting yield, productivity, and operability. Mitigation often increases overall cost or results in undesired process steps, such as antibiotic use. For scale-up of new bioprocesses, contamination can be the single largest barrier to successful operation.

Traditional control methods, such as sterilization, operation at low pH, or application of antimicrobial compounds kill or inhibit undesired microbes. The ROBUST principle focuses instead on creating an environment where only the desired microorganism has access to essential growth nutrients, leaving contaminants unable to reproduce. To create such an environment, we have engineered metabolic pathways that enable degradation of nitrogen and phosphorous containing chemicals not commonly encountered in nature. When cultured in media with the desired chemical as sole nitrogen or phosphorous source, engineered organisms rapidly outcompete contaminants, resulting in a broadly applicable alterative to antibiotic use for creation of selective fermentation environments.

Metabolomics as a Hypothesis Generator

Sauer, U., ETH Zurich

Originally presented on: 6/29/2016 8:00:00 - 8:25:00

The omics revolution has vastly expanded our capability to monitor and quantify molecular events in cells and tissues. Unfortunately, this technological advance has not (yet) been matched with conceptual advances in our ability to interpret and understand such data or to systematically generate hypotheses from them. In particular the rapidly developing metabolomics technologies generate data that are even harder to interpret mechanistically than transcript and protein abundances, simply because the direct connection to the genome is lost and many effects overlap. This is particularly problematic for metabolomics data in a cellular context, where multivariate statistics alone is typically not sufficient to obtain biologically meaningful hypotheses in fundamental science and metabolic engineering alike.

To temporally delineate overlapping influence on the metabolome, we developed technologies for broad coverage dynamic metabolomics at high-throughput. Based on previously developed high-throughput flow injection TOF mass spectrometry that enables to measure 400-800 metabolite in one sample per minute (1, 2), we developed automated, near real-time metabolome profiling directly from living cells at a resolution of 10 sec (3). We thus effectively removed limitations on data generation in our lab, shifting the challenge even more to the data interpretation side. Here I will present two unpublished examples to illustrate recent advances in generating hypotheses on molecular functions from metabolomics data. In the first I will address the question how specific enzymes really are by *in vitro* metabolomics on about 1000 purified *E. coli* enzymes. We find a surprisingly large number of promiscuous enzymes and assess their evolutionary potential by FBA. In the second example, I will exploit the potential of dynamic metabolomics data to systematically unravel bottlenecks in synthetic production pathways.

1. Fuhrer et al. 2011 **Anal Chem** 83: 7074;
2. Sevin & Sauer 2014, **Nature Chem Biol** 10: 266
3. Link H, Fuhrer T, Gerosa L, Zamboni N & U. Sauer. **Nature Methods**, 12: 1091-1079.

Conference Proceedings

Browse My Content

me11
An IMES Conference

Metabolic Engineering 11
Home | Author & Chair Index
General Submissions
Advances in 13c Metabolic Flux Analysis: Co-Culture MFA and New Pathway Discovery
Antoniewicz, M. R., University of Delaware
Originally presented on: 6/29/2016 8:25:00 - 8:50:00

Measuring intracellular fluxes by ^{13}C metabolic flux analysis (^{13}C-MFA) has become a key activity in metabolic engineering, biotechnology and medicine. In this talk, I will present several important new advances in ^{13}C-MFA that extend the scope of this technology to more complex biological systems. First, I will describe a new ^{13}C-flux analysis technology that we have developed for measuring fluxes in microbial communities using GC-MS analysis. The advantages of our approach are: 1) no seperation of cells or proteins is required; 2) no genetic manipulations of cells is required; 3) no advanced analytical instruments are required; 4) we can simultaneously measure metabolic fluxes in multiple species, population dynamics, and inter-species metabolite exchange in a single step. The second ^{13}C-MFA advance is the use of parallel labeling experiments for metabolic pathway discovery. In this talk, I will describe several surprising new metabolic pathways that we have discovered in central carbon metabolism in microbes, including *E. coli*, as well as mammalian cells. These metabolic pathways have been "hidden" so far because they cannot be estimated using traditional tracer experiments. Only through the use of carefully selected tracers in parallel labeling experiments are we able, for the first time, to visualize these pathways. Implications of these new pathways will be discussed.

The Secret of Good Taste: Novel Approaches for Metabolic Flux Analysis of Microorganisms and Plants in Agro-biotechnology

Wittmann, C., Saarland University

Originally presented on: 6/29/2016 8:50:00 - 9:15:00

One of the most important tasks for biotechnology is ensuring food security and raw material availability over the next fifty years. The increasing world population and the shortage of arable land are major drivers to develop next-generation plants and processes in agro-biotechnology in order to cope with the growing demand for food, feed and raw materials. Without doubt, the enormous complexity of the underlying biological and technological systems displays a huge challenge, but, at the same time holds the key for the implementation of knowledge-based engineering and optimization strategies (1).

Here, we have taken approaches from microbial systems biology and systems biotechnology into the area of agro-biotechnology. In particular, the lecture will demonstrate the application of metabolic flux analysis to support safe and efficient production of human nutrition and illustrate its potential via two examples. First, it will describe the analysis of mixed microbial fermentations of cocoa, the core piece of chocolate manufacturing. A comprehensive approach of experimental design, parallel isotope studies and metabolic network modelling allowed studying the cooperation of different bacterial species to bring out what we all love: chocolate flavor (2, 3). Second, novel concepts to quantify metabolic fluxes plants will be discussed. The power of *in silico* network modelling will be demonstrated fort the model plant Arabidopsis (4), while novel technological approaches for *in vivo* metabolic flux quantification will be shown for whole rice plants, staple food for more than 3 billion people worldwide (5).

Literature

1. Dersch L, Beckers V, Wittmann C (2016) Green pathways: Metabolic network analysis of plant systems. Metab. Eng. 34:1-24.
2. Adler P, Frey LJ, Berger A, Bolten C J, Hansen C E, Wittmann C (2014) The key to acetate: Metabolic fluxes of acetic acid bacteria under cocoa pulp fermentation simulating conditions.
3. Adler P, Bolten CJ, Dohnt K, Hansen CE, Wittmann C (2013) Core fluxome and meta-fluxome of lactic acid bacteria under cocoa pulp fermentation simulating conditions. Appl. Environ. Microb. 79: 5670-5681.
4. Beckers V, Dersch LM, Lotz K, Melzer G, Bläsing OE, Fuchs R, Ehrhardt T, Wittmann C (2016) In silico metabolic network analysis of *Arabidopsis* leaves. Under revision.
5. Dersch L, Beckers V, Rasch D, Melzer G, Bolten CJ, Kiep K, Becker H, Blaesing H, Fuchs R, Ehrhardt T, Wittmann C (2016) Novel approach for high-throughput metabolic screening of whole plants by stable isotopes. Plant Physiol. 171:25-41.

Integrated Genomics for Enzyme and Pathway Optimization and Stability

Cardinale, S., Technical University of Denmark
Ye, L., Technical University of Denmark
Genee, H., Technical University of Denmark
Herrgård, M. J., Technical University of Denmark
Sommer, M.,
Originally presented on: 6/29/2016 9:15:00 - 9:35:00

The engineering of microbial strains with synthetic genes and pathways is often hindered by the fact that cellular metabolism and regulation is complex and small perturbations of this context can impact the function of the cell or the synthetic genes. For example, modifications of microbial central metabolism for increasing cellular levels of NAD(P)H cofactors, such as a deletion of the gene encoding phosphoglucoisomerase (*pgi*), can cause deleterious growth defects and impact their application in metabolic engineering.

Our study presents an integrated genomics approach to improve rational and intuitive cell factory engineering via whole-genome screening and metabolic modeling, and pathway regulation and integration via synthetic molecular regulators. We apply this approach to identify cellular functions that directly influence the availability of NADPH and the yield and stability of a 7-gene biosynthetic pathway for the production of the vitamin Thiamine. NADPH is a crucial cofactor in reductive enzymatic reactions to synthesize valuable bioactive compounds

We show that our methodology, based on the screening of a very large number of barcoded transposon insertion mutants ($\sim 10^6$) coupled with product biosensors and synthetic genes expressed from plasmids, allows a direct identification of key genes whose deletion improves performance during NADPH consumption and increases pathway stability and yield during vitamin production. We identified a number of deletions of oxidoreductases which we confirm lead to an increased NADPH/$NADP^+$ ratio ranging from 12 to 62% compared to a wild type strain. We further constructed and applied a RNA riboswitch-based molecular biosensor for specific vitamins to select mutant genotypes with better growth and higher stability of vitamin production.

We are continuing to validate the role of identified cellular factors, enzymes and metabolic fluxes to build a system-wide map of cellular redox capabilities and pathway stability to assist targeted and robust metabolic engineering.

me11
An IMES Conference
Awaji, June-29-2016

Integrated omics-based approach toward understanding and engineering of valuable secondary metabolism in medicinal plants

Junfeng Chen

Department of Pharmacy, Changzheng Hospital
Department of Pharmaceutical Botany, School of Pharmacy
Modern Research Centre for TCM

第二军医大学
The Second Military Medical University

1

What we investigate

Natural products from Traditional Chinese Medicinal (TCM) herbs

Discovery of active compounds
Cardiovascular activation, Hypoglycemic, Anti inflammatory

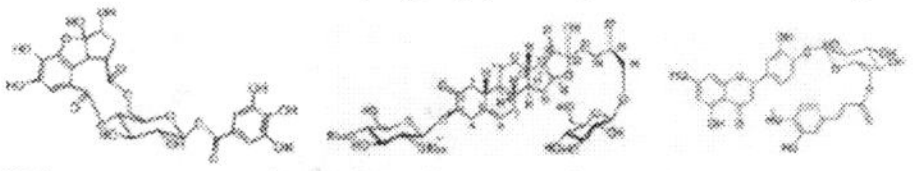

Pharmacological mechanism
Pharmacology, Pharmacokinetics

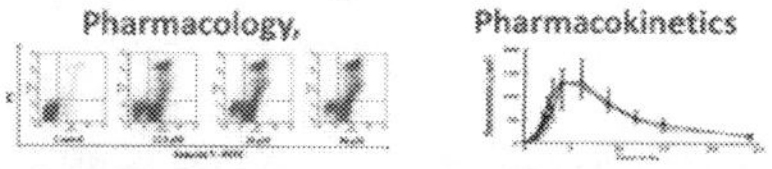

Identification and quality control of herb
Chemical finger print

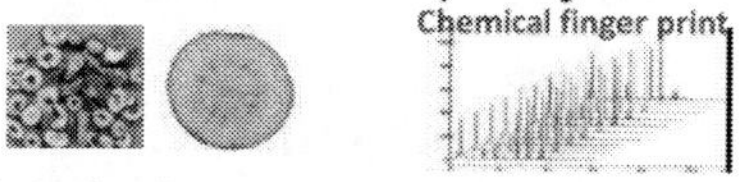

Metabolic engineering of TCM plants
Plant cell cultures, Synthetic manchism, Metabolic engineering

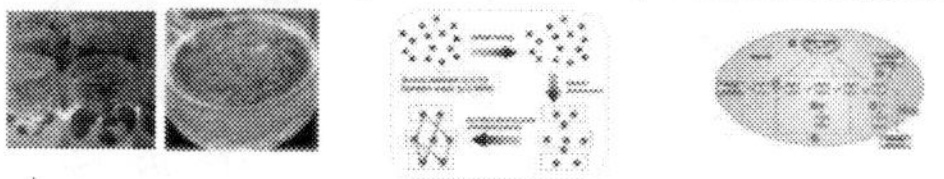

For a better use of Traditional Chinese Medicinal

2

Synthetic mechanism of NP is the basic for metabolic engineering

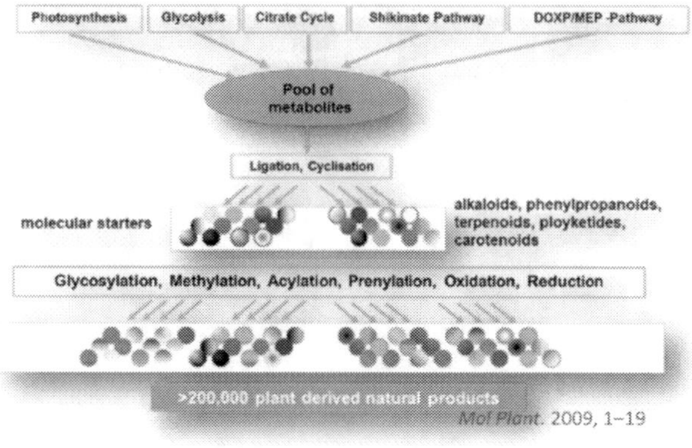

Mol Plant. 2009, 1–19

Biosynthesis pathways
Regulatory mechanism

⇩ **Basically knowledge**

Cell cultures
Secondary metabolic engineering
Synthetic biology

⇩

Improved yields

Synthetic mechanism for most natural products in specific species is unknown

3

Challenges of understanding NPs biosynthesis in plant

Metabolic flux

The most biologically important description of metabolism is the set of metabolic fluxes that it generates.
——*Plant Cell*, 2013, 25: 694

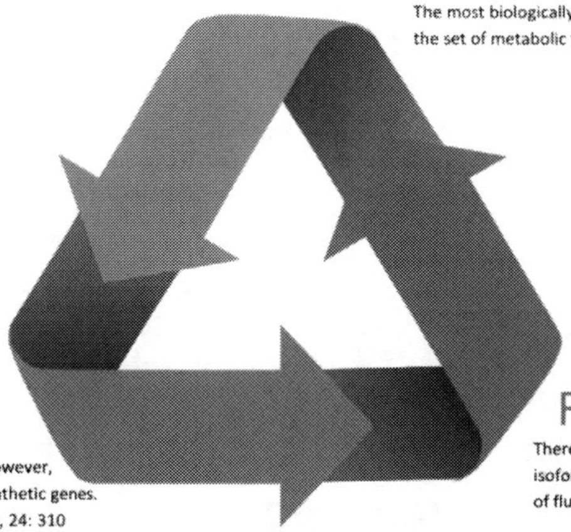

Regulators

The first obstacle to this approach, however, is knowledge of the underlying biosynthetic genes.
——*Curr Opin Biotech*, 2013, 24: 310

Rate-limiting steps

There are several branch-points, many enzyme isoforms involved, thus a quantitative understanding of flux control is required.
——*Mol Syst Biol*, 2009, 5: 1

4

Platform I: Metabolic flux analysis

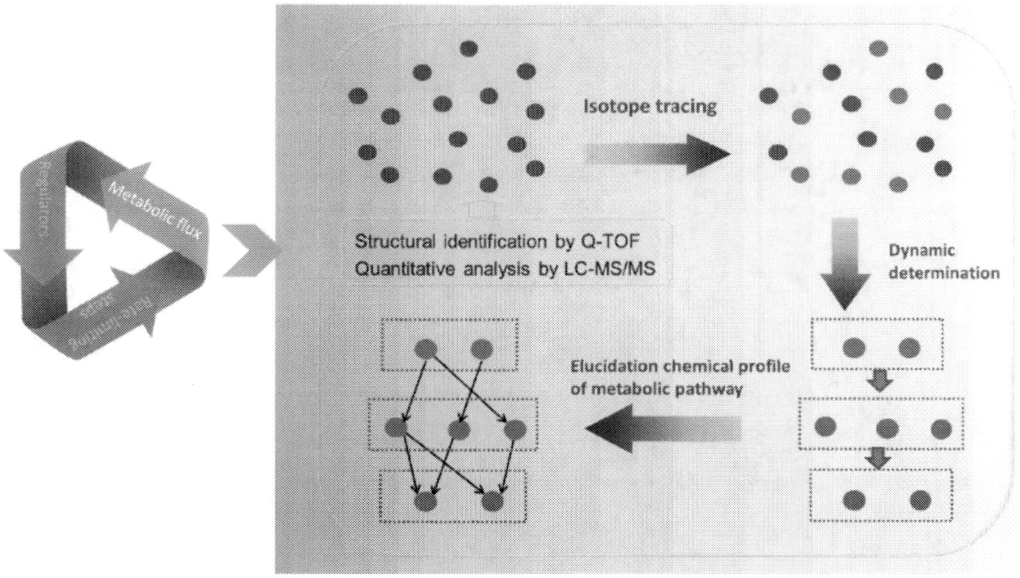

5

Platform II: rate-limiting steps and regulators

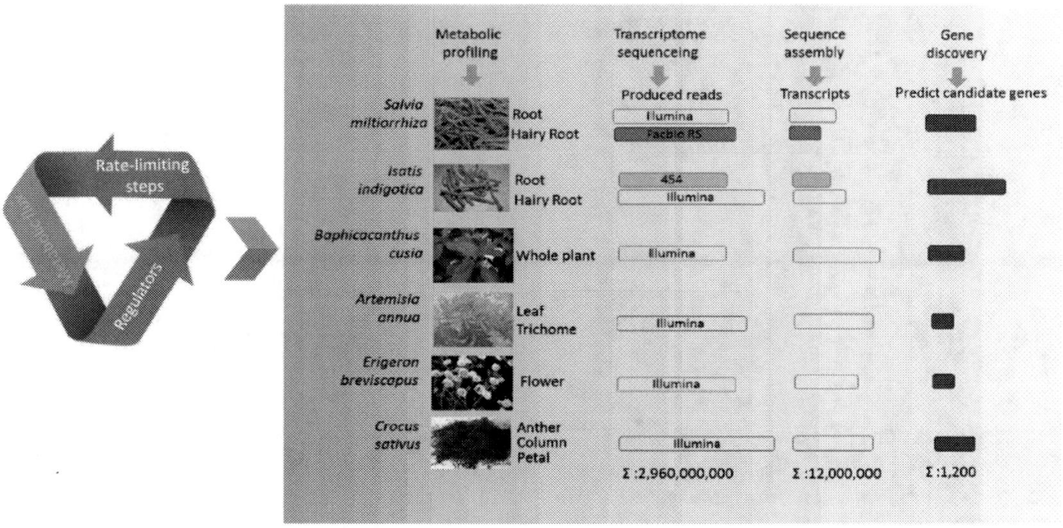

6

Platform III: increasing productivity via diverse transgenic strategies for secondary metabolic engineering

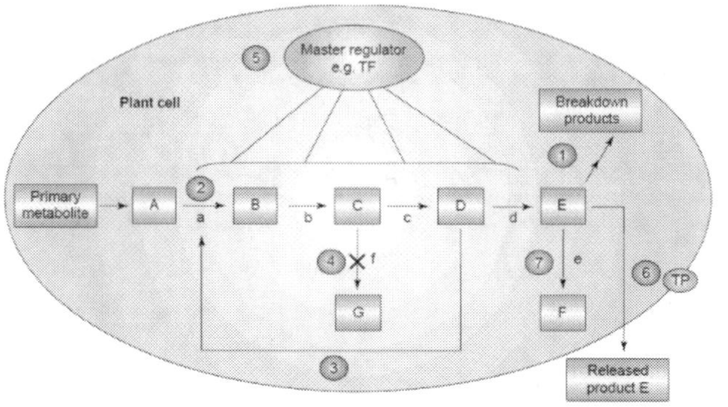

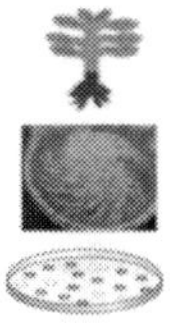

Plantlet

Hairy root culture

Cell suspension culture

Regulation of growth cycle.

Regulation at organ / tissue / subcellular compartmentation level

Abiotic / biotic induced regulation

7

Case study I: ^{13}C Tracer Reveals Phenolic Acids Biosynthesis in Hairy Root Cultures of *Salvia miltiorrhiza*

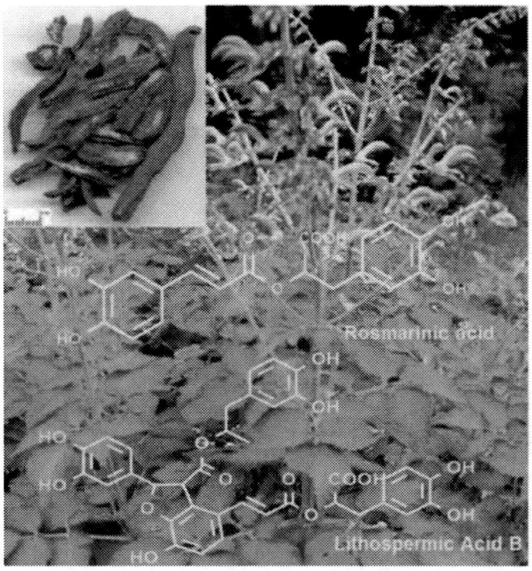

- Promoting blood circulation
- Removing blood stasis
- Activate collaterals

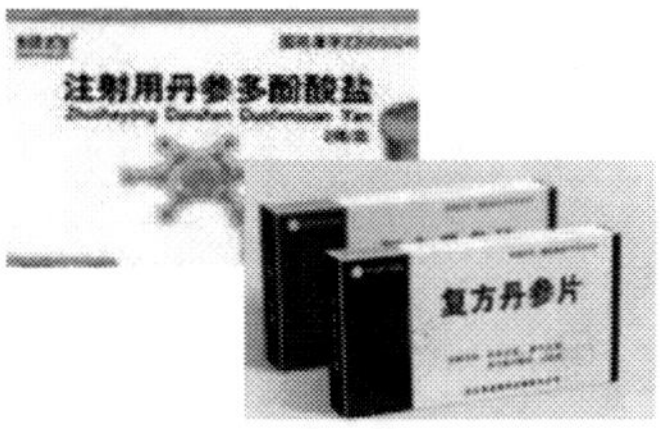

8

Phenolic acids show specific accumulation pattern in *Salvia miltiorrhiza*

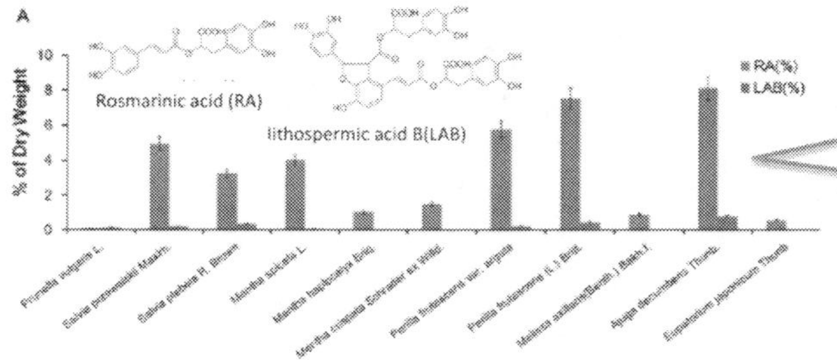

Superior accumulation of RA in other species

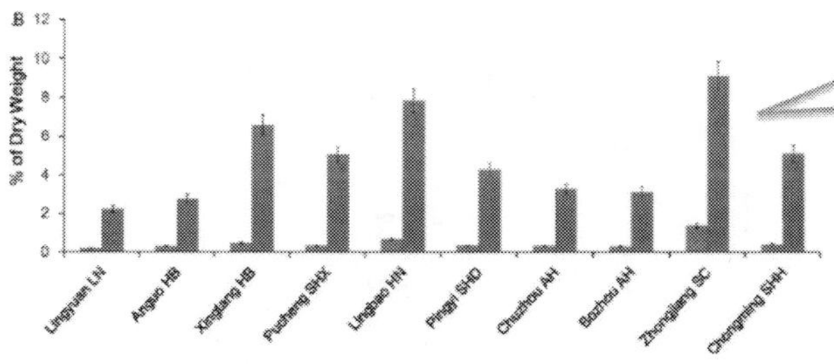

Superior accumulation of LAB in *S. miltiorrhiza*

Is there a different biosynthesis pathway?

9

Identification of ^{13}C labeled phenolic acids and precursors using UPLC-QTOF

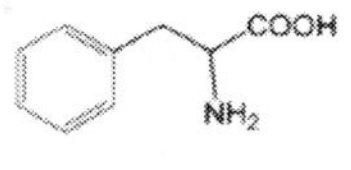

^{13}C labled L-Phenylalanine

Salvia miltiorrhiza hairy roots

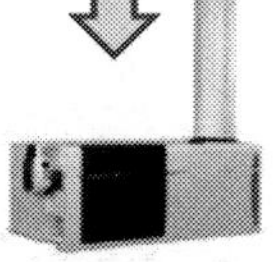

Identification and Profiling

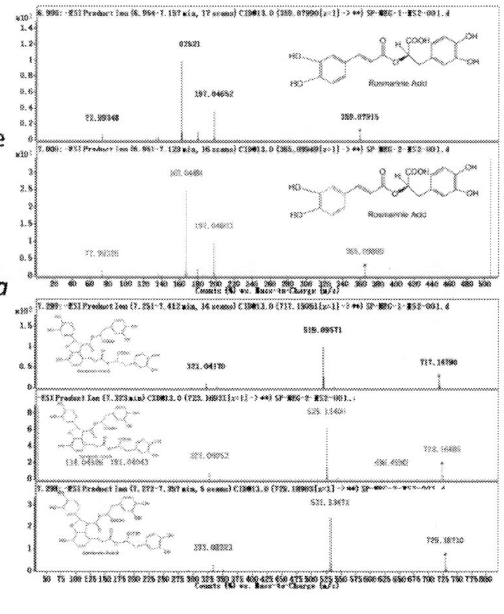

10 ^{13}C labled compounds in RA pathway were identified :

- L-phenylalanine
- 4-coumaric acid
- Rosmarinic acid
- lithospermic acid B
- Caffeic acid
- Caffeoyl-4'-hydroxyphenyllactic acid
- 4-Coumaroyl-4'-hydroxyphenyllactic acid
- Salvianolic acid E
- Feruloyl-3',4'-dihydroxy-phenyllactic acid
- Methyl lithospermic acid B

10

Metabolic Flux Analysis Reveals RA and LAB pathway in *Salvia miltiorrhiza*

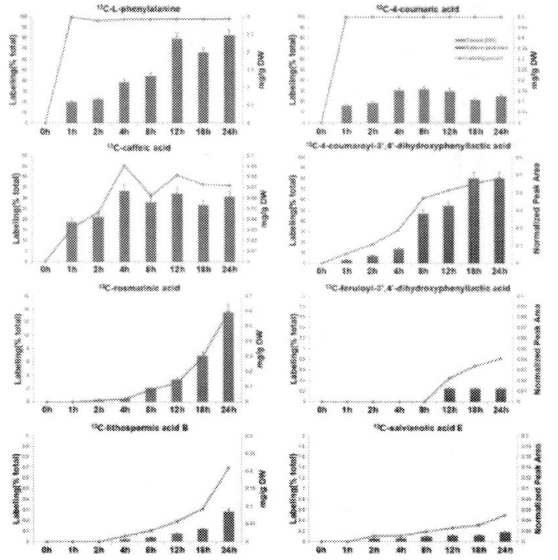

The order in which ^{13}C-compounds appeared

+

Accumulation patterns of ^{13}C-compounds

Metabolic flux of identified phenolic acids

Synthesis pathway

A Specific RA Biosynthesis Pathway in *Salvia miltiorrhiza*

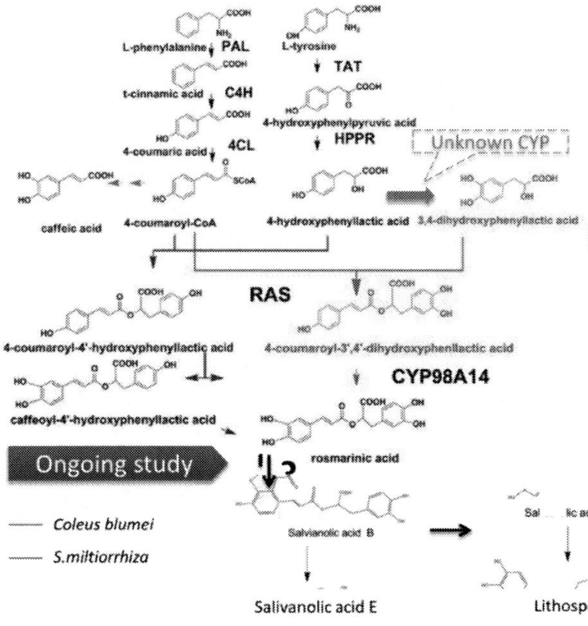

More efficient pathway for LAB synthesis

Di *et al., ACS Chem. Biol.*, 2013

Secondary Metabolic Engineering of *Salvia miltiorrhiza*

Target genetic regulation of RA/LAB pathway

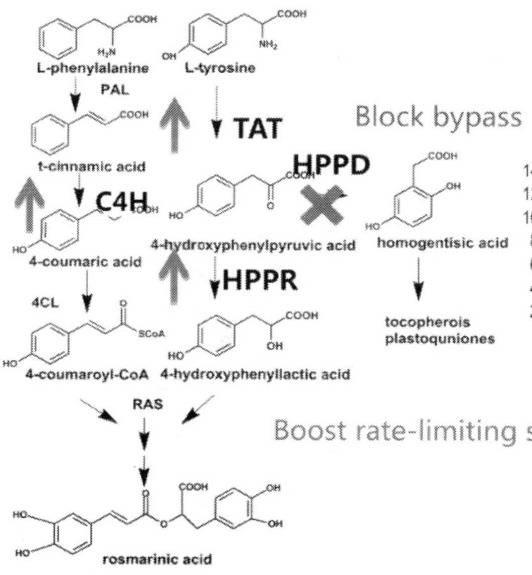

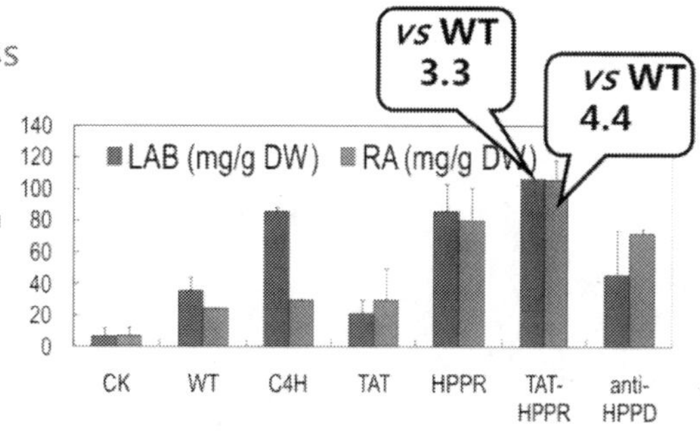

Xiao *et al.*, *PLoS ONE*, 2012

13

Case study II: Dynamic metabolic and transcriptomic profiling reveals biosynthesis of lignans in *Isatis indigotica* Fort.

- Famous TCM ——"Ban-Lan-Gen"
- Anti-inflammatory, antiviral and detoxification
- Active substances: Lignans

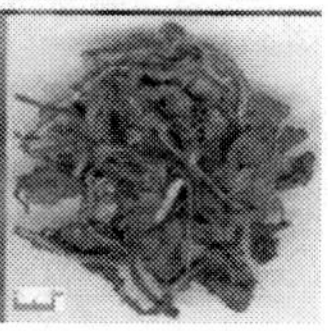

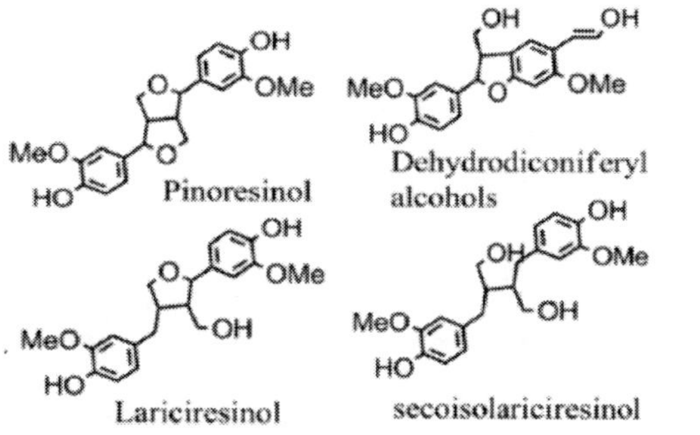

- Lariciresinol is the material basis for antiviral activity
- The content in *I. indigotica* is only 68 μg/g DW

Li *et al.*, *Acta Pharm Sin* (2000)

14

Methyl jasmonate improved lignan synthesis in *I. indigotica*

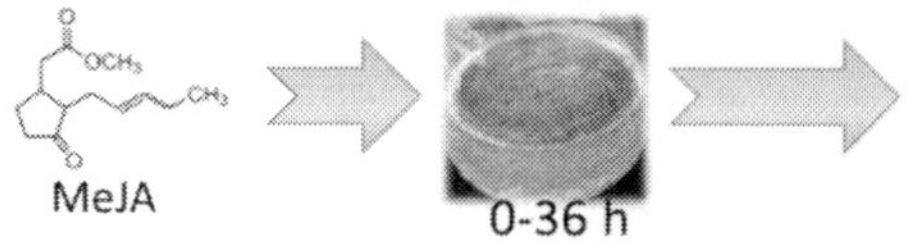

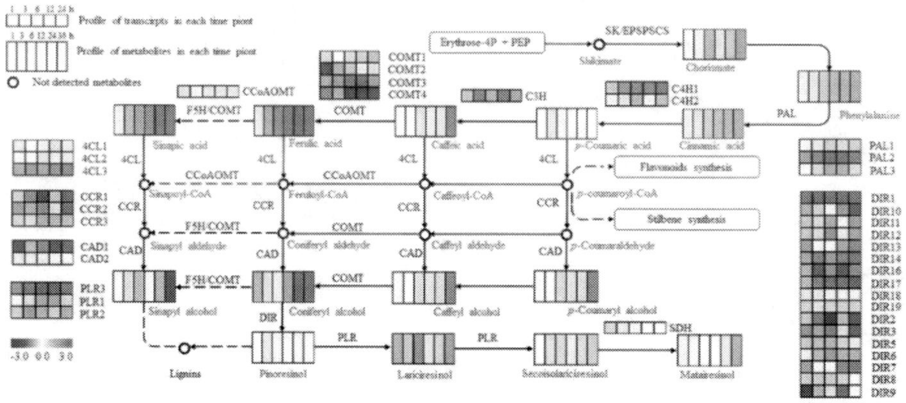

MeJA caused perturbation of entire lignan synthesis pathway

Chen *et al., Plant Biotechnol J,* 2016

15

Transcriptional and Metabolic shifts of MeJA treated *I. indigotica*

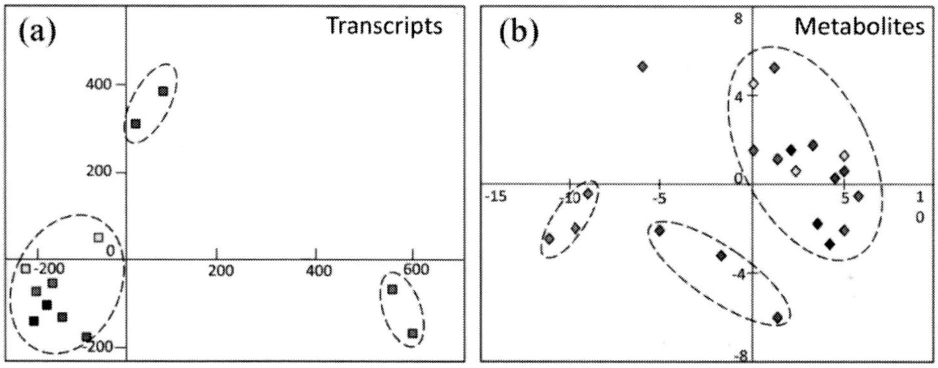

PCA analysis showed biological process from transcriptional variation to metabolites in response to MeJA:

- Major transcriptional variations were observed at early stages
- Metabolic differences were detected at late stages

16

Trends tendency narrowed abundance of different expressed genes (DEGs)

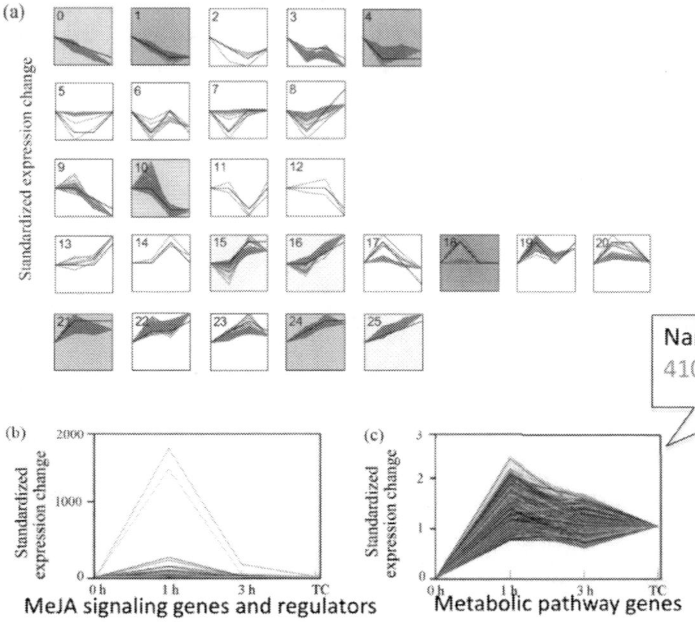

MeJA signaling genes and regulators

Metabolic pathway genes

Narrowed gene pool
410 candidate genes

17

Co-expression network represent hubs of lignan synthesis

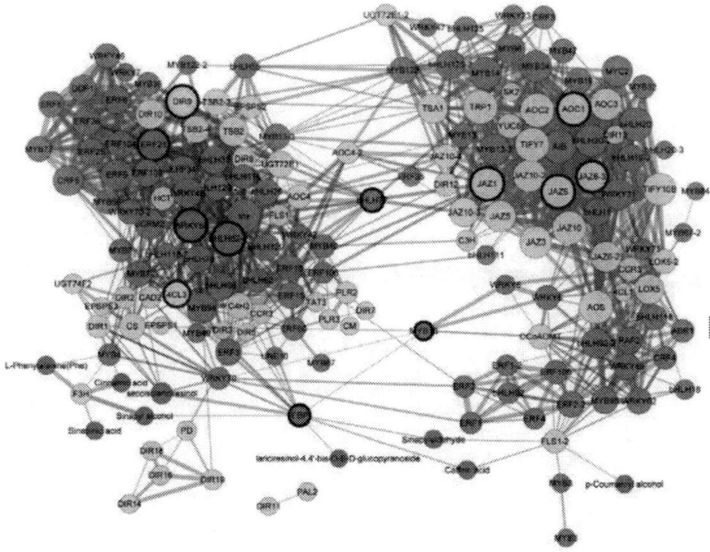

175 nodes were connected with 1956 edges

11 hub genes: 8 TF genes and 3 synthetic genes

⇩

Modeling the Pathway :

Synthetic pathway

Regulators

18

4-coumarate:CoA ligase (4CL3) plays distinct role in lignan biosynthesis

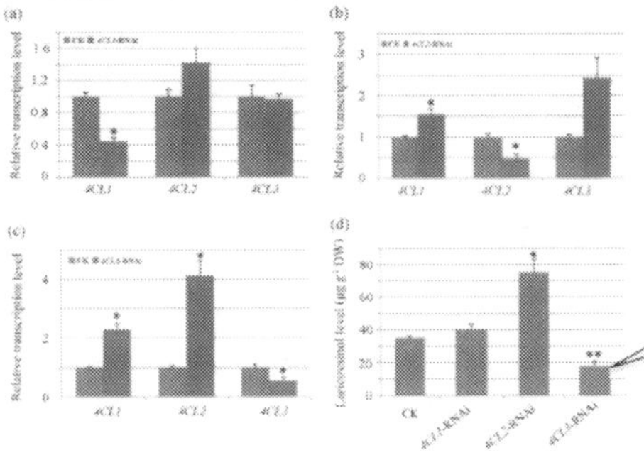

Only *4CL3-RNAi* leads to dramatic decrease of lignan synthesis, suggesting it's distinct role for lignan synthesis

Substrates

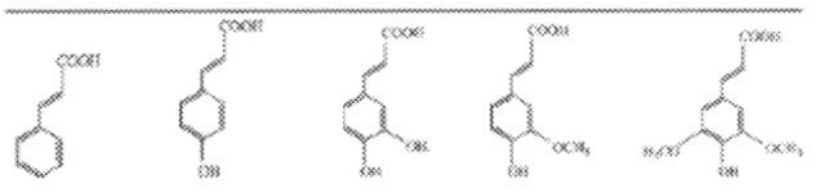

Protein name	Kinetic parameter	Cinnamic acid	p-Coumaric acid	Caffeic acid	Ferulic acid	Sinapic acid
4CL1	K_m (μM)	NDA	NDA	NDA	NDA	NDA
	V_{max} (nmol mg^{-1} protein S^{-1})	–	–	–	–	–
	K_{cat} (S^{-1})	–	–	–	–	–
	K_{cat}/K_m	–	–	–	–	–
4CL2	K_m (μM)	NDA	24 ± 10	8 ± 0.7	6 ± 0.9	12 ± 1.5
	V_{max} (nmol mg^{-1} protein S^{-1})	–	2.9 ± 0.09	3 ± 0.11	1 ± 0.02	1.0 ± 0.08
	K_{cat} (S^{-1})	–	0.22	0.19	0.06	0.08
	K_{cat}/K_m	–	8300	26 400	9600	8100
4CL3	K_m (μM)	137 ± 2.5	NDA	3 ± 0.4	16 ± 0.8	NDA
	V_{max} (nmol mg^{-1} protein S^{-1})	9 ± 0.6	–	44.20 ± 0.34	40.60 ± 1.60	–
	K_{cat} (S^{-1})	0.61	–	3.12	2.76	–
	K_{cat}/K_m	4400	–	1 276 000	167 500	–

Each 4CL have different catalytic property:
4CL3 showed the highest affinity towards caffeic acid

19

Primary synthesis route of lignan among the complex phenylpropanoids network

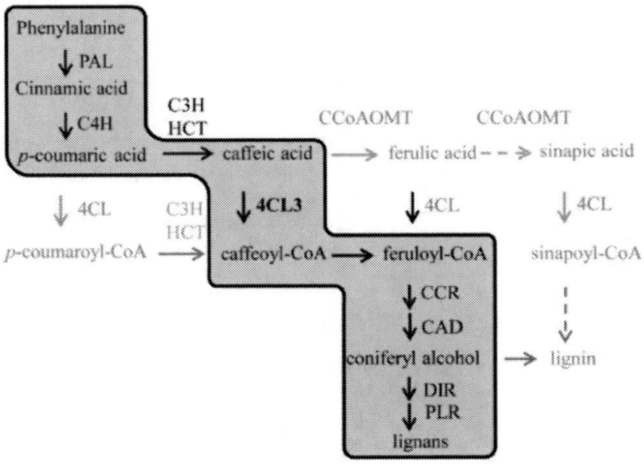

4CL3 is an ideal target for metabolic engineering of lignan

20

Engineering lignan pathway via *ovx PLR1*

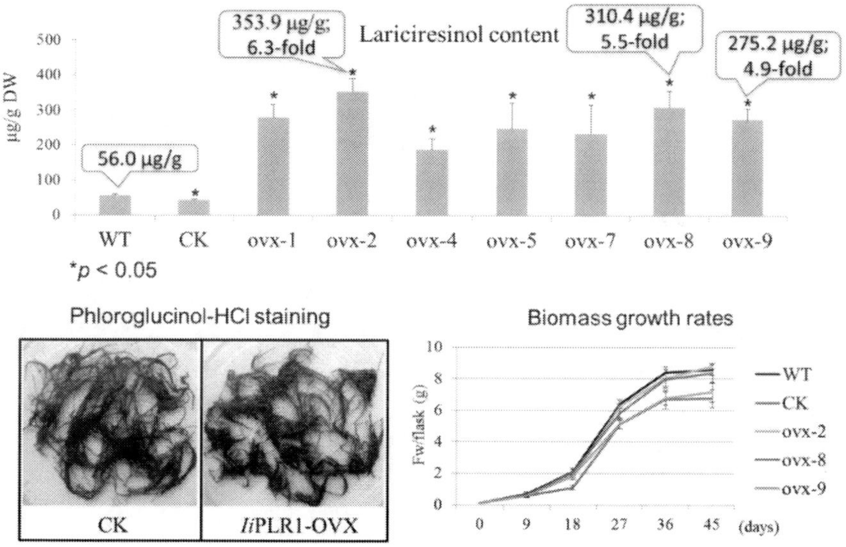

Xiao *et al.*, *J Exp Bot* (2015)

21

Phenotype analysis of AP2-49-*ovx* hairy root

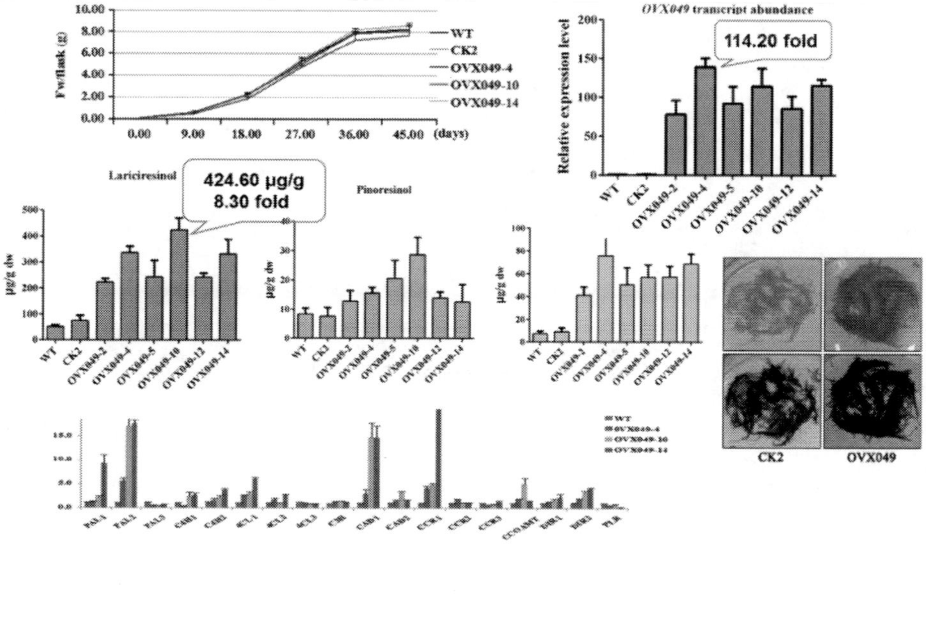

22

Case study III: Metabolic engineering of artemisinin by regulating development of trichome in *Artemisia annua*

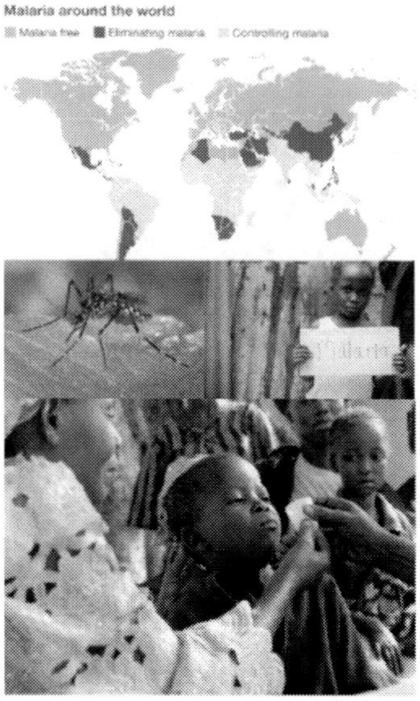

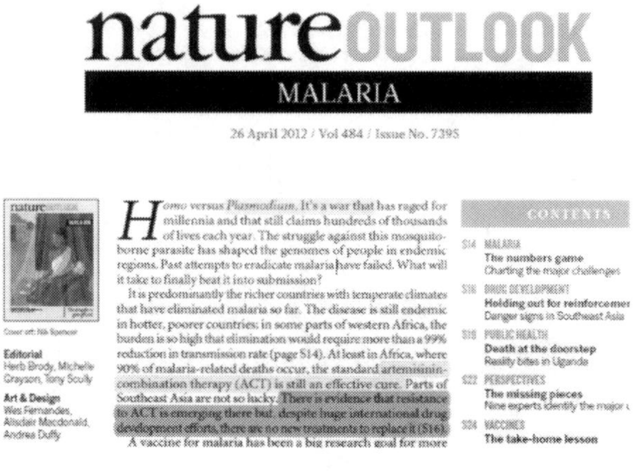

nature OUTLOOK

MALARIA

26 April 2012 / Vol 484 / Issue No. 7395

Cover art: Nik Spencer

Editorial
Herb Brody, Michelle Grayson, Tony Scully

Art & Design
Wes Fernandes, Alisdair Macdonald, Andrea Duffy

*H*omo versus *Plasmodium*. It's a war that has raged for millennia and that still claims hundreds of thousands of lives each year. The struggle against this mosquito-borne parasite has shaped the genomes of people in endemic regions. Past attempts to eradicate malaria have failed. What will it take to finally beat it into submission?

It is predominantly the richer countries with temperate climates that have eliminated malaria so far. The disease is still endemic in hotter, poorer countries: in some parts of western Africa, the burden is so high that elimination would require more than a 99% reduction in transmission rate (page S14). At least in Africa, where 90% of malaria-related deaths occur, the standard artemisinin-combination therapy (ACT) is still an effective cure. Parts of Southeast Asia are not so lucky. There is evidence that resistance to ACT is emerging there but, despite huge international drug development efforts, there are no new treatments to replace it (S16).

A vaccine for malaria has been a big research goal for more

CONTENTS

23

Plant extraction is still the main source of artemisinin

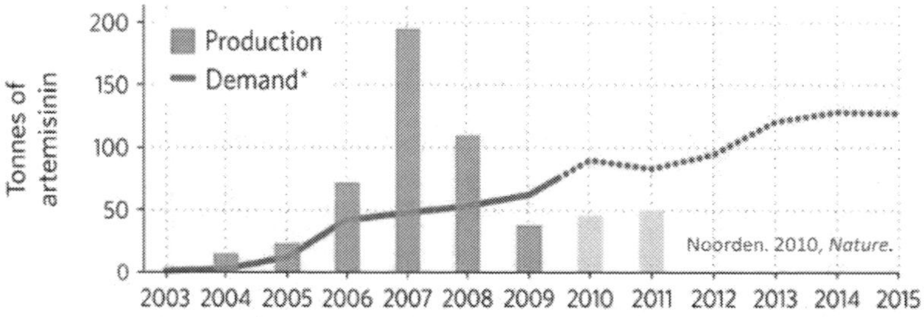

*One tonne gives about two million doses of artemisinin-based combination therapy.

Source	Plant extraction	Semi synthesis
Supply quantity 200 tons	120-140 tons 2013 140 tons 2014	35 tons 2013 60 tons 2014

Peplow. 2013, *Nature*.

24

Glandular secreting trichomes (GSTs) is the organ for synthesis and storage of artemisinin

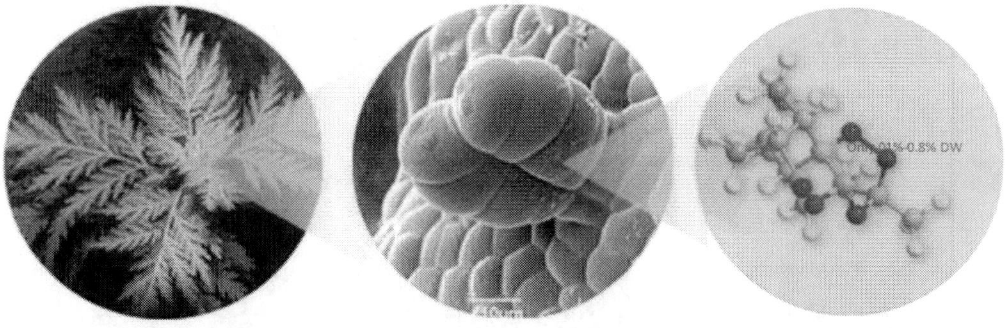

Artemisia annua Glandular trichomes Artemisinin

Trichome is a promising target for metabolic engineering to improve artemisinin yield

The schematic to engineering *Aa*GSTs

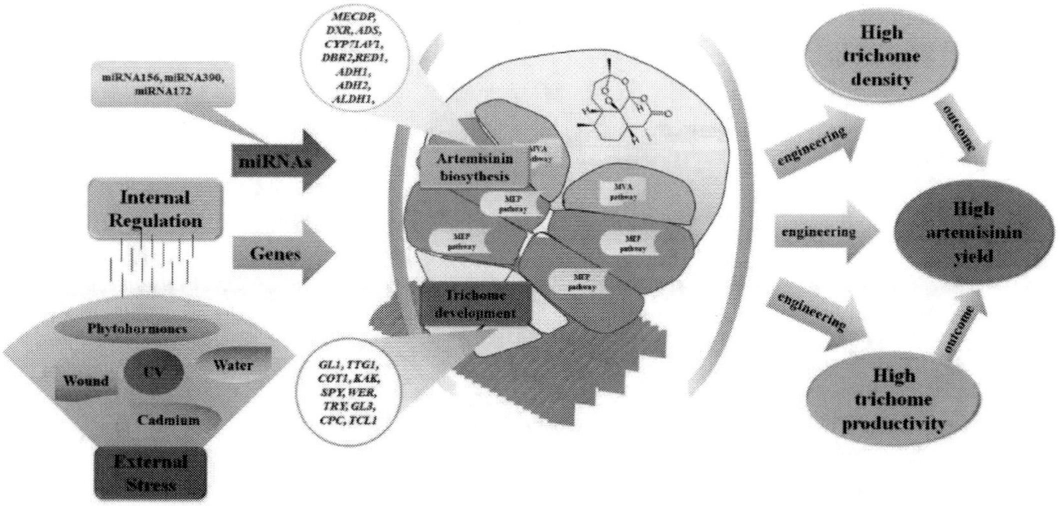

Xiao *et al., Sci Bull* (2015)

Target on a candidate Trichome and Artemisinin Regulator gene (*AaTAR1*)

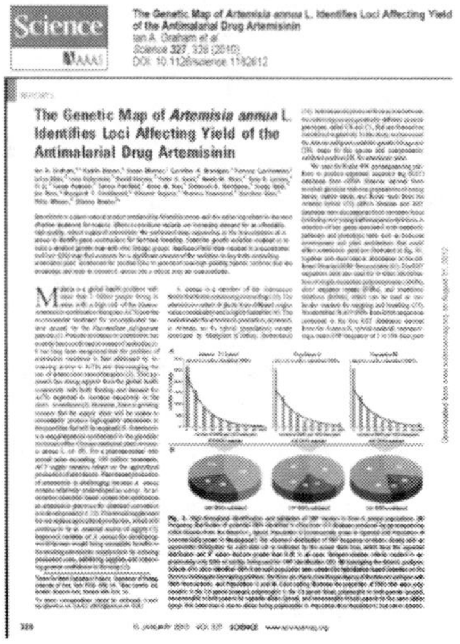

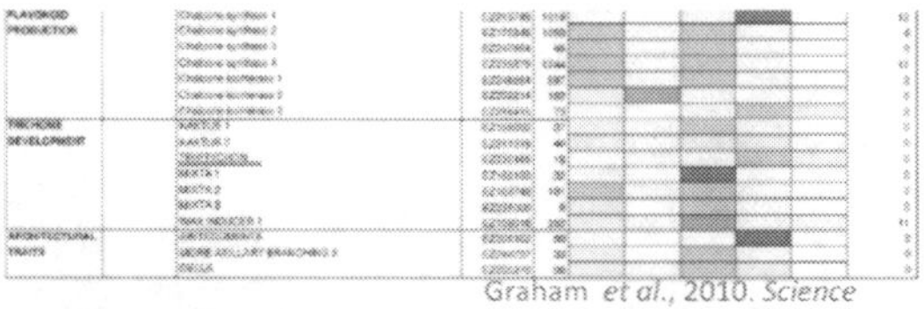

Graham *et al*., 2010. *Science*

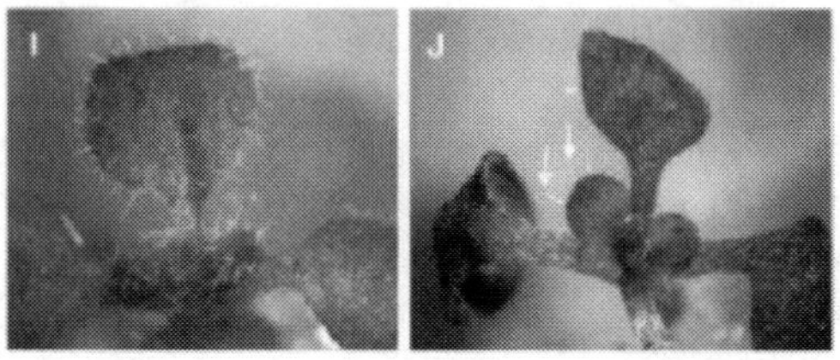

Asaph *et al*., 2004, ***Plant Cell***

TAR1 is a AP2/ERF transcription factor

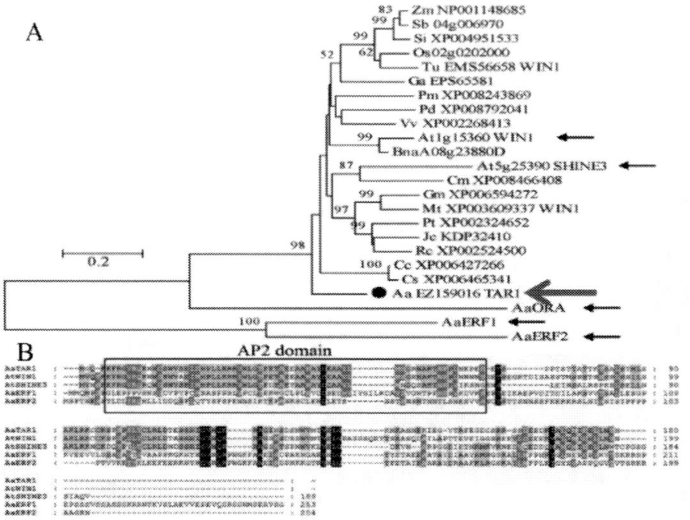

Tan *et al*., *Mol Plant* (2015)

TAR1 is a AP2/ERF transcription factor

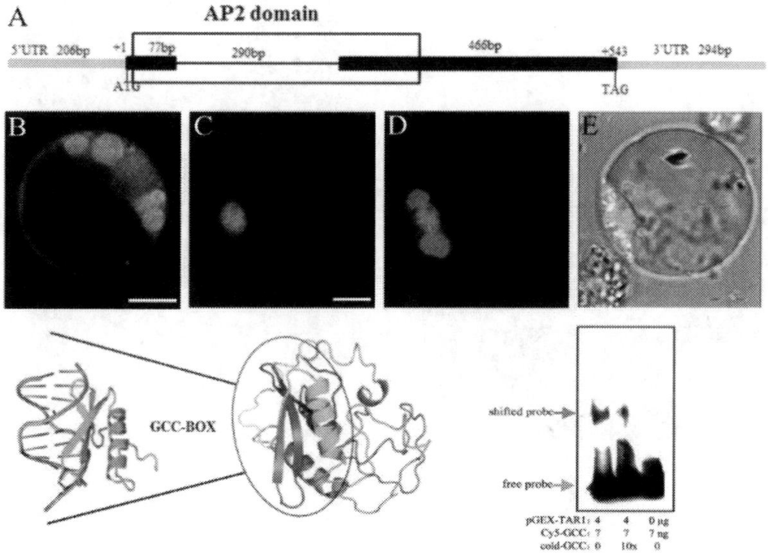

Tan *et al., Mol Plant* (2015)

29

TAR1 is mainly expressed in trichomes of young tissues

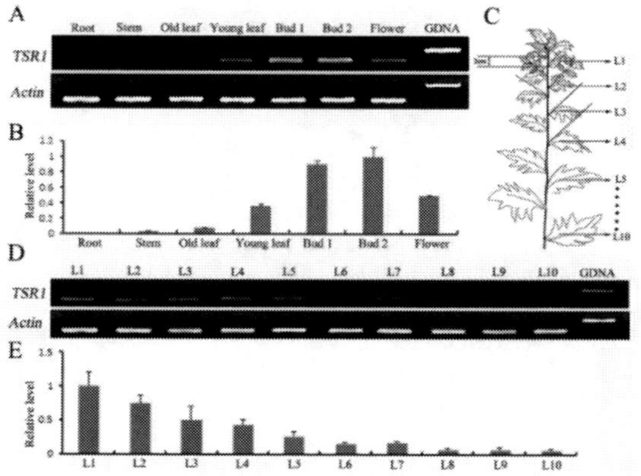

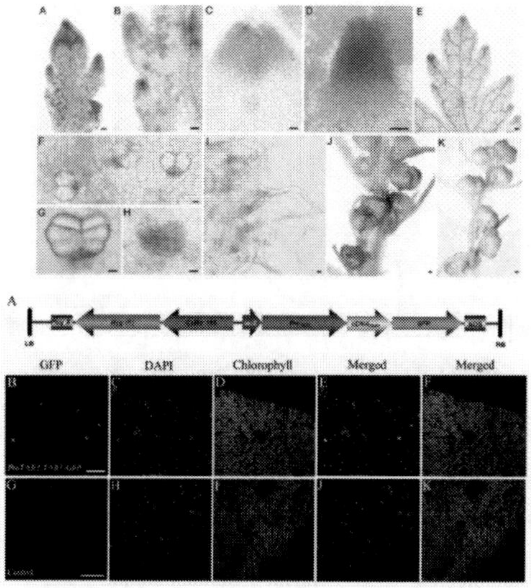

Tan *et al., Mol Plant* (2015)

30

Phenotypic analysis of *TAR1-RNAi* transgenic plants

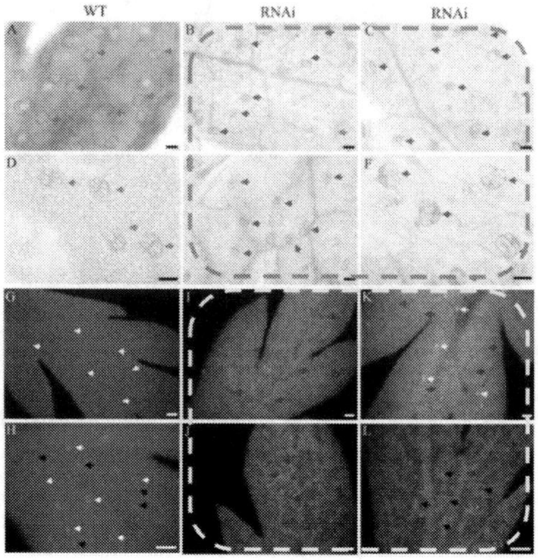

Morphological changes of glandular secreting trichomes (GSTs)

No effect on trichome density

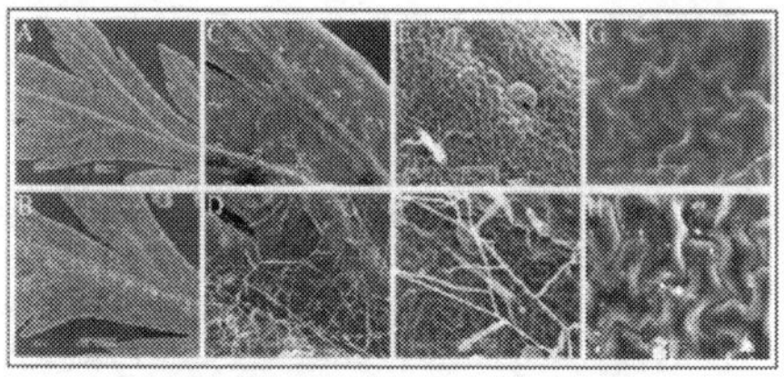

Morphological changes of T-shaped non-glandular trichomes (TNGs)

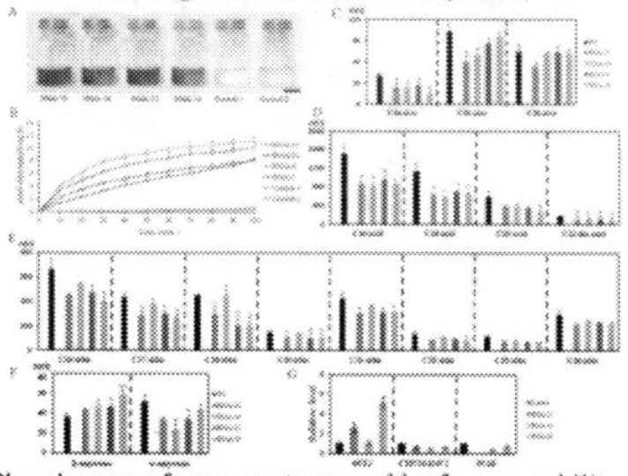

The change of wax content and leaf permeability

Influence development of trichomes by regulating the synthesis of wax

Tan *et al., Mol Plant* (2015)

31

Chemical analysis of *TAR1*-RNAi lines

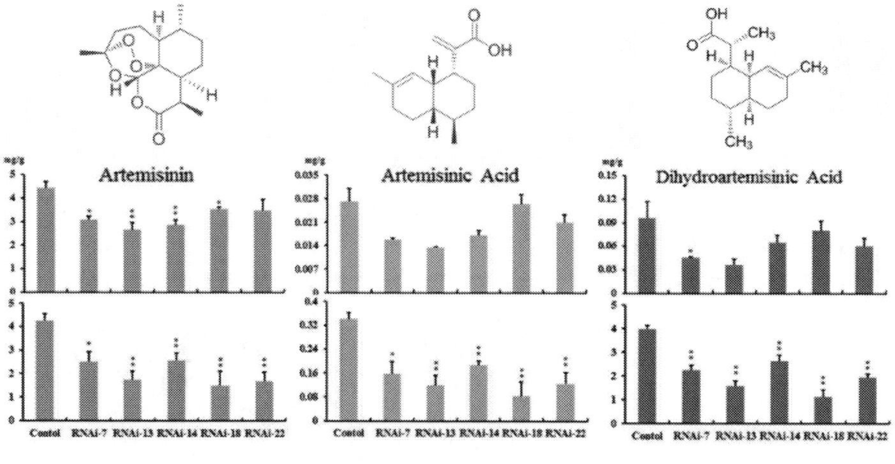

Tan *et al., Mol Plant* (2015)

32

Metabolic engineering via *ovx-TAR1*

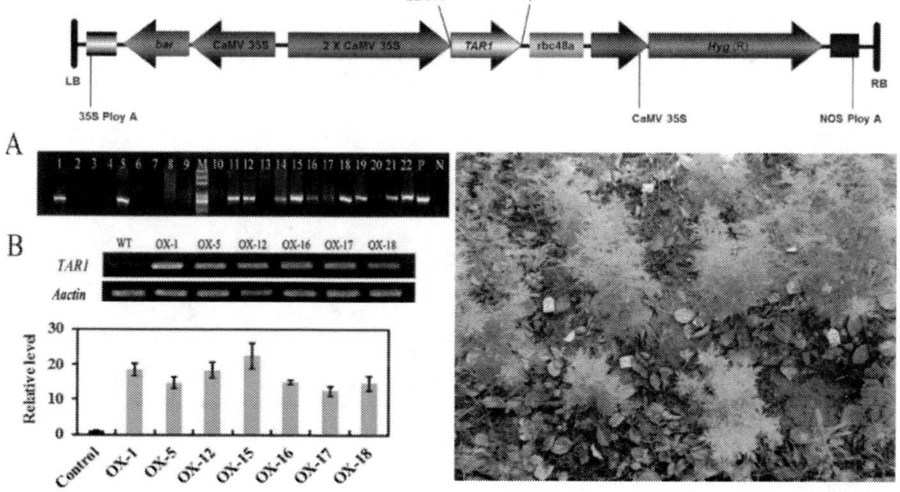

Tan *et al., Mol Plant* (2015)

33

Chemical analysis of *ovx-TAR1* materials

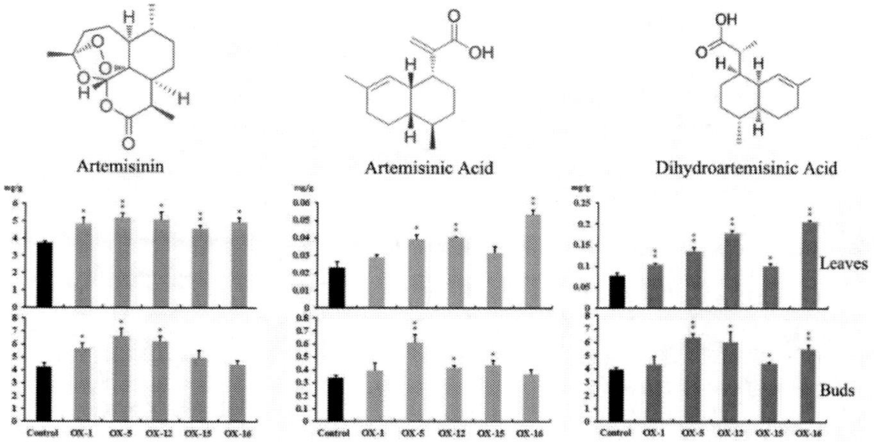

Tan *et al., Mol Plant* (2015)

34

Mechanism of the regulation by *AaTAR1*

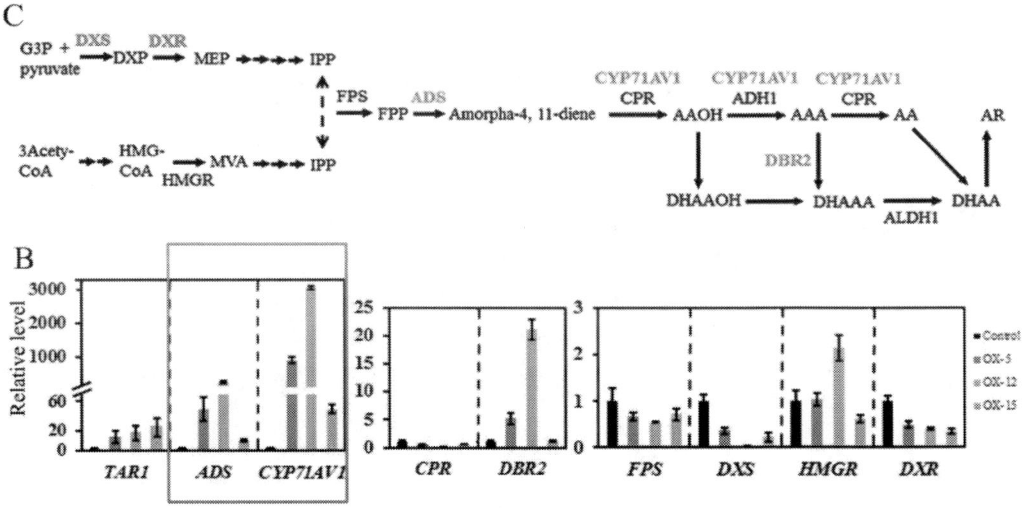

Tan *et al., Mol Plant* (2015)

35

Mechanism of the regulation by *AaTAR1*

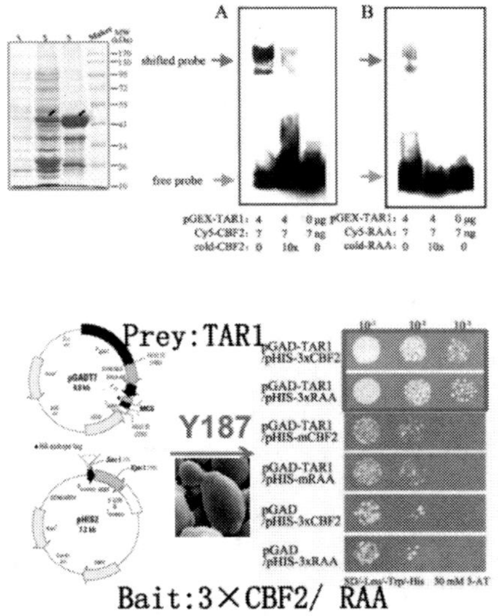

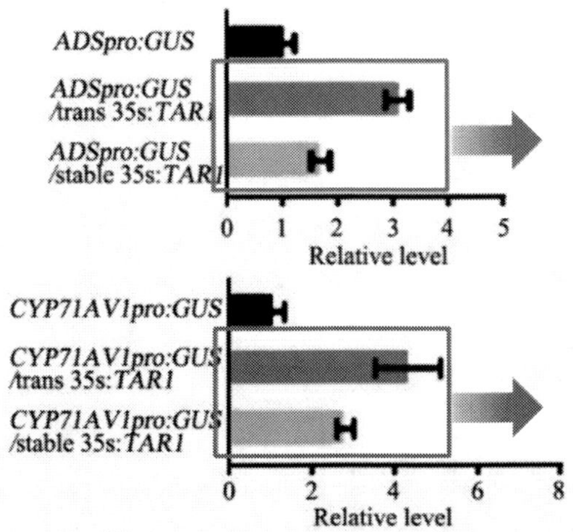

Tan *et al., Mol Plant* (2015)

36

Summary

Knowledge of NPs in plant

- Valuable target chemicals
- Potential elements

- Calculating models
- Potential target genes

Metabolic Engineering in Microbe

Desire for cooperation !

37

Acknowledgements

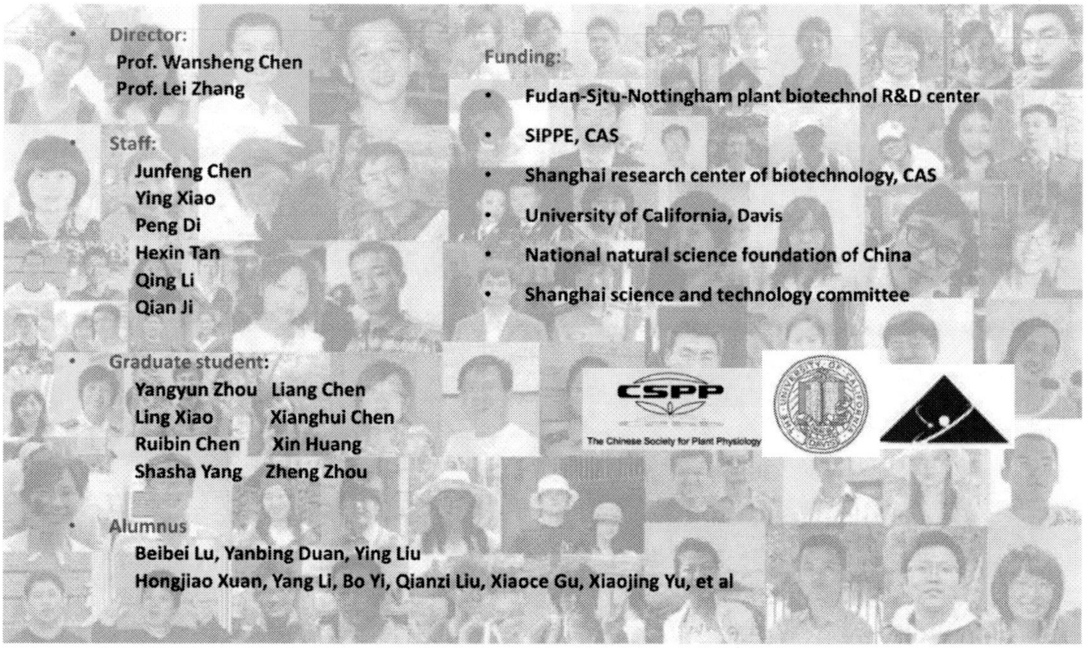

- Director:
  Prof. Wansheng Chen
  Prof. Lei Zhang
- Staff:
  Junfeng Chen
  Ying Xiao
  Peng Di
  Hexin Tan
  Qing Li
  Qian Ji
- Graduate student:
  Yangyun Zhou Liang Chen
  Ling Xiao Xianghui Chen
  Ruibin Chen Xin Huang
  Shasha Yang Zheng Zhou
- Alumnus
  Beibei Lu, Yanbing Duan, Ying Liu
  Hongjiao Xuan, Yang Li, Bo Yi, Qianzi Liu, Xiaoce Gu, Xiaojing Yu, et al

Funding:

- Fudan-Sjtu-Nottingham plant biotechnol R&D center
- SIPPE, CAS
- Shanghai research center of biotechnology, CAS
- University of California, Davis
- National natural science foundation of China
- Shanghai science and technology committee

CSPP The Chinese Society for Plant Physiology

38

End and thanks indeed

39

Cell cultivation…history

Only 100 year history for animal cell culture!

Animal cell culture world

tadpole

1907…*in vitro* cultivation **Harrison**

1916…Trypsinization *Rous*

1958…Construction of ***CHO (Chinese hamster ovary)*** *cell line*

1959…*Eagle*'s MEM, *Dulbecco*'s MEM

1970s… Vaccine production by large-scale cell culture

1980s…Development of serum-free medium

Production of biophamaceuticals using recombinant cells

1990s… Regenerative medicine, cell therapy, tissue engineering, stem cell, bio-artificial organs

Human iPS

2006… start of iPS

21 century is Animal cell century

Microbial world

BC Beer, Wine

1560 Starting Japanese "Sake" brewing

1600s Microbial world *Leeuwenhoek*

1800s The concept for "Fermentation" *Pasteur*

1895 Finding for Sake yeast

1914 Acetone-butanol fermentation

1943 Penicillin fermentation

1963 Biochemical Engineering textbook

1972 Recombinant DNA technologies

Development of recombinant DNA technologies.

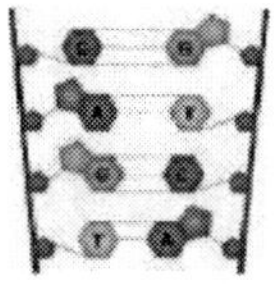

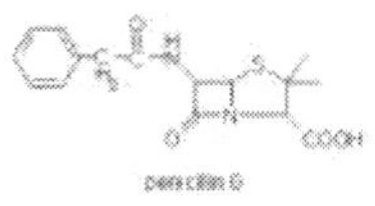

2

Application of mammalian cell culture is classified into two categories

(1)Mammalian cell is a mediator for production.

Global workhorse for pharma-industry

Market is growing up

Product is produced material by mammalian cell….Biopharmacesuticals

i.e. *Biologics* (antibody, vaccine, cytokine, etc.)
Using cell line (CHO, BHK, HEK293…..), mainly recombinant cell line

(2)Mammalian cell is the product itself.

People hope and expected big market but it is still small.

Product is mammalian cell

i.e. cell therapy, tissue engineering, regenerative medicine, diagnosis usage, cell for gene-therapy, evaluation for chemicals, drugs, food etc.
Using primary cell ; stem cell, ES cell, iPS cell etc. mainly not-recombinant cell line

大政 日本生物工学会誌(照井賞受賞論文) (2005)

World top 10 selling drugs

No.	製品名(Product)	一般名 (General Name)	主な薬効/クラス	Company in Word and Japan	FY2012	Change	FY2011
1	ヒュミラ(Humira)	アダリムマブ (Adalimumab)	Rheumatoid arthritis	AvvVie(Abbot)/ Eisai	9,603	17%	8,242
2	レミケード (Remicade)	インフリキシマブ (Infliximab)	Arthritis / Crohn's disease	J & J / Merck / Mitsubishi Tanabe	9,071	1%	9,016
3	エンブレル(Embrel)	エタネルセプト (Etanercept)	Rheumatoid arthritis	Amgen / Pfizer / Takeda	8,476	7%	7,902
4	アドエア／セレタイド (Advair Diskus, Seretide)	サルメテロール／フルチカゾン (Salmeterol/Fluticasone)	Anti-asthma drugs	GSK / Almirall	8,216	4%	7,891
5	クレストール (Crestor)	ロスバスタチン (Rosuvastatin)	Hyperlipidemia	Shionogi / AstraZeneca	7,430	-6%	7,919
6	リツキサン (Rituxan)	リツキシマブ (Rituximab)	Non-Hodgkin's lymphoma	Roche / Biogen Idec	7,227	-2%	7,386
7	**ランタス (Lantus)**	**インスリングラルギン (Insulin glargine)**	**Diabetes**	**Sanofi**	**6,555**	**19%**	**5,071**
8	ハーセプチン (Herceptin)	トラスツズマブ (Trastsuzumab)	Breast cancer	Roche / Chugai	6,444	11%	5,589
9	アバスチン (Avastin)	ベバシズマブ (Bevacizumab)	Colon cancer	Roche / Chugai	6,307	6%	5,631
10	ジャヌビア (Januvia)	シタグリプチン／配合剤(Sitagliptin)	Type 2 diabetes / DPP4	Merck / Ono / Almirall	6,208	22%	5,095

7 products(Yellow) are biopharmaceuticals **and 6 (Blue) was produced by mammalian cells**

Data from CSD (Cegedim Strategic Data)

More than 50% therapeutic antibodies are produced by **CHO cells**

Therapeutic antibodies approved in US, EU, and JP

平成27年8月22日 国立医薬品食品衛生研究所 生物薬品部

Antibody name	Product name	Structure	Target	主な適応疾患	承認年 US	承認年 EU	承認年 Japan	Host
mouse								
muromonab-CD3	Orthoclone OKT3	IgG2a	CD3	腎移植後の急性拒絶反応	1986	NA	1991	mouse hybridoma
ibritumomab tiuxetan	Zevalin	IgG1κ (MX-DTPA:⁹⁰Y標識)	CD20	B細胞性非ホジキンリンパ腫	2002	2004	2008	CHO
	Zevalin	IgG1κ (MX-DTPA:¹¹¹In標識)		イブリツモマブ チウキセタンの集積部位の確認			2008	CHO
iodine 131 Tositumomab	Bexxar	IgG2aλ (¹³¹I標識)	CD20	非ホジキンリンパ腫	2003	NA	NA	mammalian cell
catumaxomab	Removab	mIgG2aκ (EpCAM), rIgG2bλ (CD3	EpCAM, CD3	癌性腹水	NA	2009	NA	rat/mouse bridoma
blinatumomab	Blincyto	scFv-scFv	CD19, CD3	急性リンパ性白血病	2014	NA	NA	CHO
Chimera								
abciximab	ReoPro	IgG1(Fab)	GPIIb/IIIa	心筋虚血	1994	NA	NA	mammalian cell
rituximab	Rituxan/MabThera	IgG1κ	CD20	B細胞性非ホジキンリンパ腫	1997	1998	2001	CHO
basiliximab	Simulect	IgG1κ	CD25	腎移植後の急性拒絶反応	1998	1998	2002	Sp 2/0
infliximab	Remicade	IgG1κ	TNFα	関節リウマチ	1998	1999	2002	Sp 2/0
cetuximab	Erbitux	IgG1κ	EGFR	頭頸部癌, 結腸・直腸癌	2004	2004	2008	Sp 2/0
brentuximab vedotin	Adcetris	IgG1(MMAE修飾)	CD30	ホジキンリンパ腫	2011	2012	2014	CHO
siltuximab	Sylvant	IgG1κ	IL-6	キャッスルマン病	2014	NA	NA	CHO
dinutuximab	Unituxin	IgG1κ	GD2	神経芽細胞腫(小児)	2015	NA	NA	SP2/0
humanized								
daclizumab	Zenapax	IgG1κ	CD25	腎移植後の急性拒絶反応	1997	~~1999~~	NA	
palivizumab	Synagis	IgG1κ	RSV F protein	RSウイルス感染	1998	1999	2002	NS0
trastuzumab	Herceptin	IgG1κ	HER2	転移性乳癌	1998	2000	2001	CHO
gemtuzumab ozogamicin	Mylotarg	IgG4κ (カリケアマイシン修飾)	CD33	急性骨髄性白血病	~~2000~~	refused	2005	NS0
alemtuzumab	Campath	IgG1κ	CD52	B細胞性慢性リンパ性白血病	2001	2001	2014	CHO
omalizumab	Xolair	IgG1κ	IgE	喘息	2003	2005	2009	CHO
efalizumab	Raptiva	IgG1κ	CD11	尋常性乾癬	~~2003~~	~~2004~~	NA	CHO
bevacizumab	Avastin	IgG1κ	VEGF	結腸・直腸癌	2004	2005	2007	CHO
natalizumab	Tysabri	IgG4κ	α4integrin	多発性硬化症	2004	2006	2014	mouse myeloma cell
tocilizumab	Actemra	IgG1κ	IL-6R	キャッスルマン病、関節リウマチ	2010	2009	2005	CHO
ranibizumab	Lucentis	IgG1κ Fab	VEGF-A	加齢黄斑変性	2006	2007	2009	E.Coli
eculizumab	Soliris	IgG2/4κ	C5	発作性夜間血色素尿症	2007	2007	2010	NS0
certolizumab pegol	Cimzia	Fab'+PEG	TNFα	関節リウマチ、重症クローン病	2008	2009	2012	E.Coli
mogamulizumab	Poteligeo	IgG1κ	CCR4	CCR4陽性成人T細胞白血病リンパ腫	NA	NA	2012	CHO
pertuzumab	Perjeta	IgG1κ	HER2	HER2陽性手術不能または再発乳癌	2012	2013	2013	CHO
trastuzumab emtansine	Kadcyla	IgG1κ (メイタンシン修飾)	HER2	HER2陽性転移・再発乳癌	2013	2013	2013	CHO
obinutuzumab	Gazyva	IgG1	CD20	慢性リンパ性白血病	2013	2014	NA	CHO
vedolizumab	Entyvio	IgG1	α4β7 integrin	クローン病	2014	2014	NA	CHO
pembrolizumab	Keytruda	IgG4κ	PD-1	黒色腫	2014	2015	NA	
human								
adalimumab	Humira	IgG1κ	TNFα	関節リウマチ	2002	2003	2008	CHO
panitumumab	Vectibix	IgG2κ	EGFR	結腸・直腸癌	2006	2007	2010	CHO
golimumab	Simponi	IgG1κ	TNFα	関節リウマチ	2009	2009	2011	Sp 2/0-Ag14
ustekinumab	Stelara	IgG1κ	IL12, IL23-p40	乾癬	2009	2009	2011	Sp 2/0
canakinumab	Ilaris	IgG1κ	IL-1β	クリオピリン関連周期性症候群	2009	2009	2011	Sp 2/0-Ag14
ofatumumab	Arzerra	IgG1κ	CD20	慢性リンパ性白血病	2009	2010	2013	NS0
denosumab	Prolia/Xgeva ランマーク	IgG2	RANKL	骨病変, 骨粗鬆症	2010	2010	2012	CHO
ipilimumab	Yervoy	IgG1κ	CTLA4	黒色腫	2011	2011	2015	CHO
belimumab	Benlysta	IgG1λ	BlyS	SLE	2011	2011	NA	NS0
raxibacumab	Raxibacumab	IgG1λ	B. anthracis to	吸入炭疽, 肺炭疽	2012	NA	NA	murine cell
ramucirumab	Cyramza	IgG1	VEGFR2	胃癌	2014	2014	2015	NS0
nivolumab	Opdivo	IgG4	PD-1	悪性黒色腫	2015	2015	2014	CHO
secukinumab	Cosentyx	G1/κ	IL17-A	尋常性乾癬, 関節症性乾癬	2015	2015	2014	CHO
evolocumab	Repatha	IgG2	PCSK9	高コレステロール血症	NA	2015	NA	CHO
alirocumab	Praluent	IgG1	PCSK9	高コレステロール血症	2015	NA	NA	CHO

NA: Not approved (未承認)

5

Champion data for mammalian cell culture in fed-batch

Now
20,000L reactor
(Genentech)
10g/L antibody
(Wyeth, Takeda)
for industrial mammalian cell culture

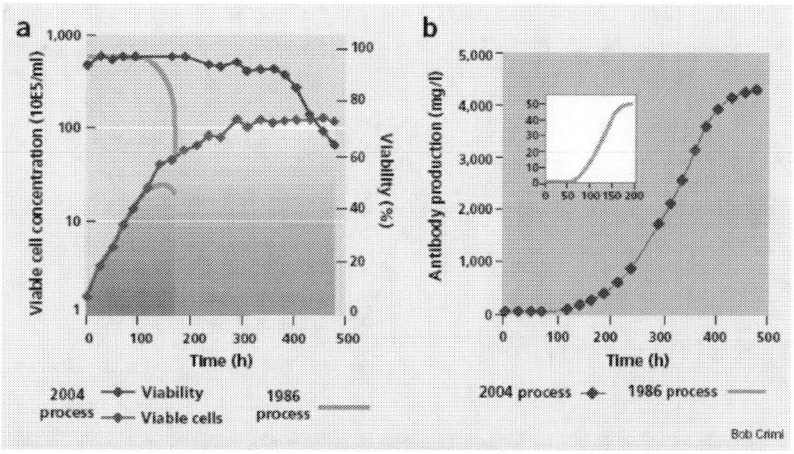

Figure 2 Comparison of cell culture processes from 1986 and 2004. (**a**,**b**) The viable cell mass and viability (**a**) and the productivity (**b**) of both cell culture processes are shown. The examples are derived from processes in suspension cultures of CHO cells. The data from 2004 was generated from a process for the expression of a human recombinant antibody at the 10-liter scale. The hypothetical process from 1986 is based on unpublished results. For clarity the product accumulation data for the hypothetical 1986 process is also represented in the insert in **b**. Data courtesy of Lonza Biologics, Slough, UK.

Nature Biotech. (2004)

Transgenic animals i.e. clone sheep milk ~ 10g /L
Transgenic plant 10 mg/1kg-wet-plant
Recombinant yeast 5-15 g/L

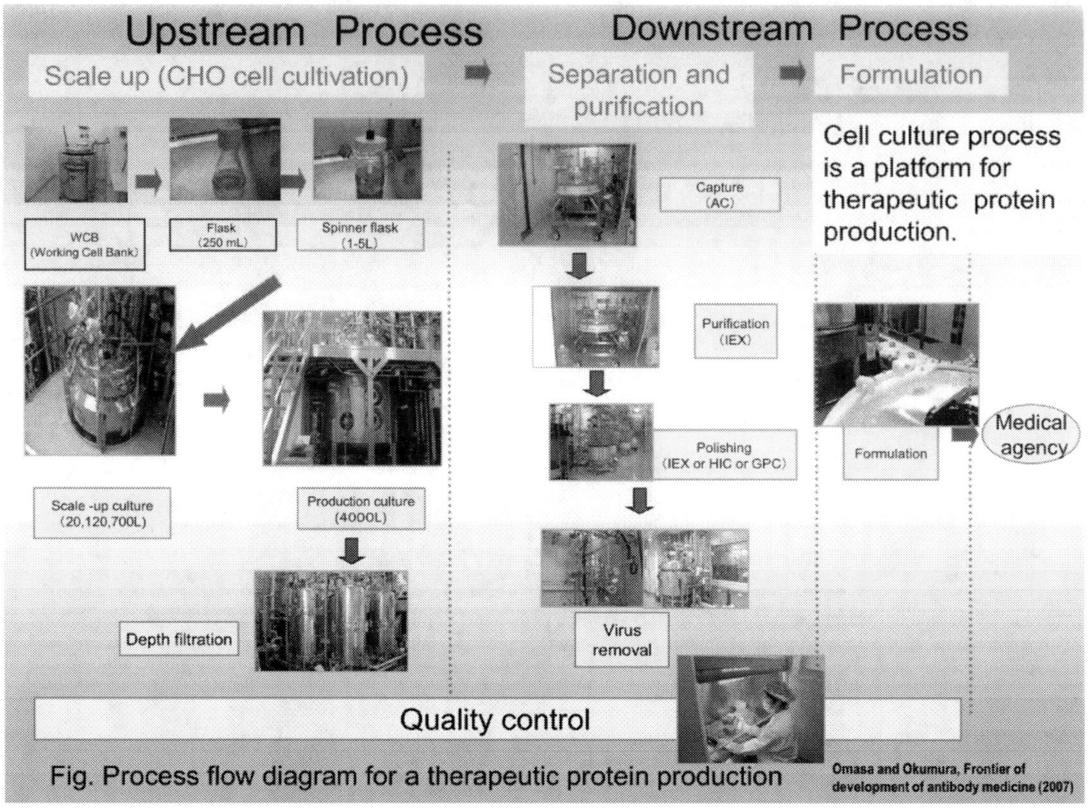
Upstream Process
Downstream Process
Scale up (CHO cell cultivation)
Separation and purification
Formulation
Cell culture process is a platform for therapeutic protein production.
WCB (Working Cell Bank)
Flask (250 mL)
Spinner flask (1-5L)
Capture (AC)
Purification (IEX)
Polishing (IEX or HIC or GPC)
Scale -up culture (20,120,700L)
Production culture (4000L)
Depth filtration
Virus removal
Formulation
Medical agency
Quality control
Fig. Process flow diagram for a therapeutic protein production
Omasa and Okumura, Frontier of development of antibody medicine (2007)

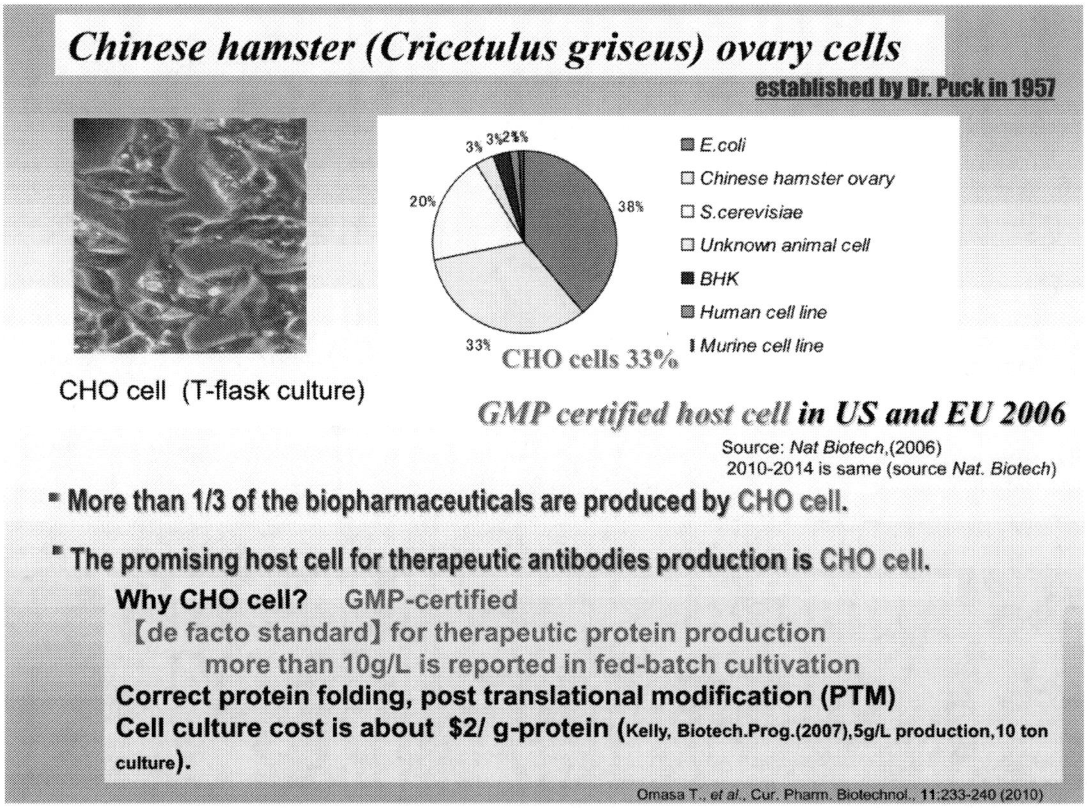
Chinese hamster (Cricetulus griseus) ovary cells
established by Dr. Puck in 1957
CHO cell (T-flask culture)
38%
33%
20%
3%
E.coli
Chinese hamster ovary
S.cerevisiae
Unknown animal cell
BHK
Human cell line
Murine cell line
CHO cells 33%
GMP certified host cell in US and EU 2006
Source: Nat Biotech,(2006)
2010-2014 is same (source Nat. Biotech)
More than 1/3 of the biopharmaceuticals are produced by CHO cell.
The promising host cell for therapeutic antibodies production is CHO cell.
Why CHO cell? GMP-certified
【de facto standard】 for therapeutic protein production
more than 10g/L is reported in fed-batch cultivation
Correct protein folding, post translational modification (PTM)
Cell culture cost is about $2/ g-protein (Kelly, Biotech.Prog.(2007),5g/L production,10 ton culture).
Omasa T., et al., Cur. Pharm. Biotechnol., 11:233-240 (2010)

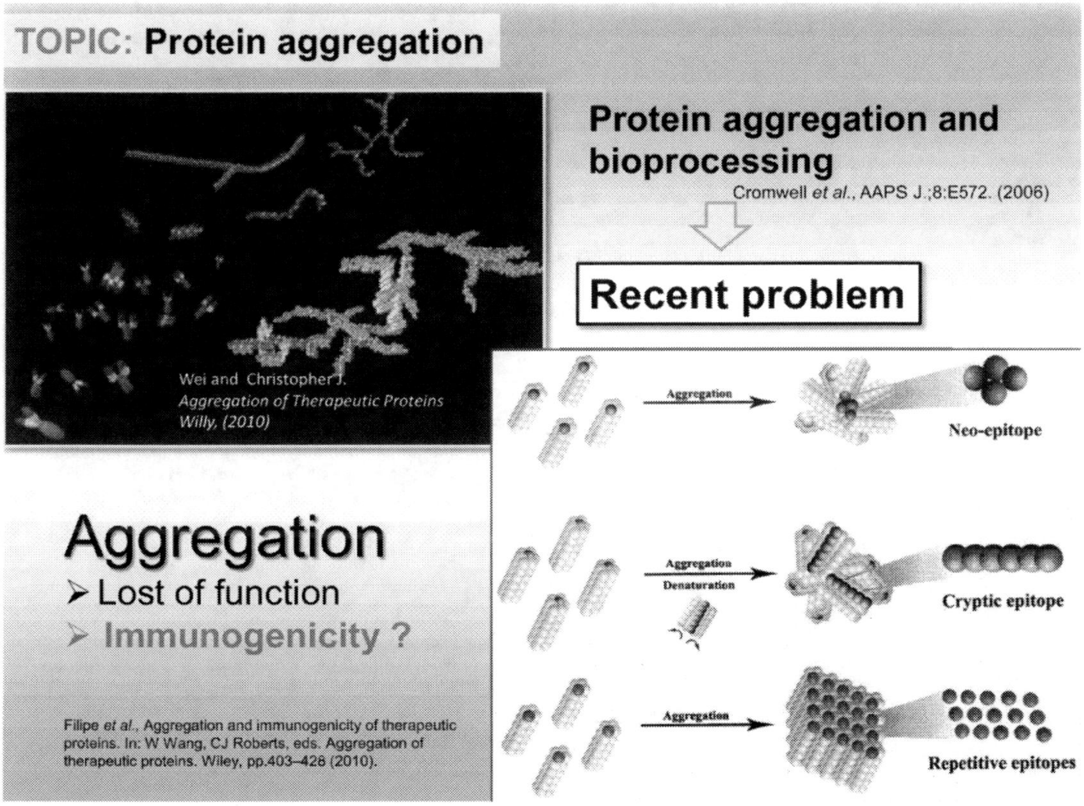
TOPIC: Protein aggregation
Protein aggregation and bioprocessing
Cromwell et al., AAPS J.;8:E572. (2006)
Recent problem
Wei and Christopher J.
Aggregation of Therapeutic Proteins
Willy, (2010)
Aggregation
➢ Lost of function
➢ Immunogenicity ?
Filipe et al., Aggregation and immunogenicity of therapeutic proteins. In: W Wang, CJ Roberts, eds. Aggregation of therapeutic proteins. Wiley, pp.403–428 (2010).
Aggregation
Neo-epitope
Aggregation
Denaturation
Cryptic epitope
Aggregation
Repetitive epitopes

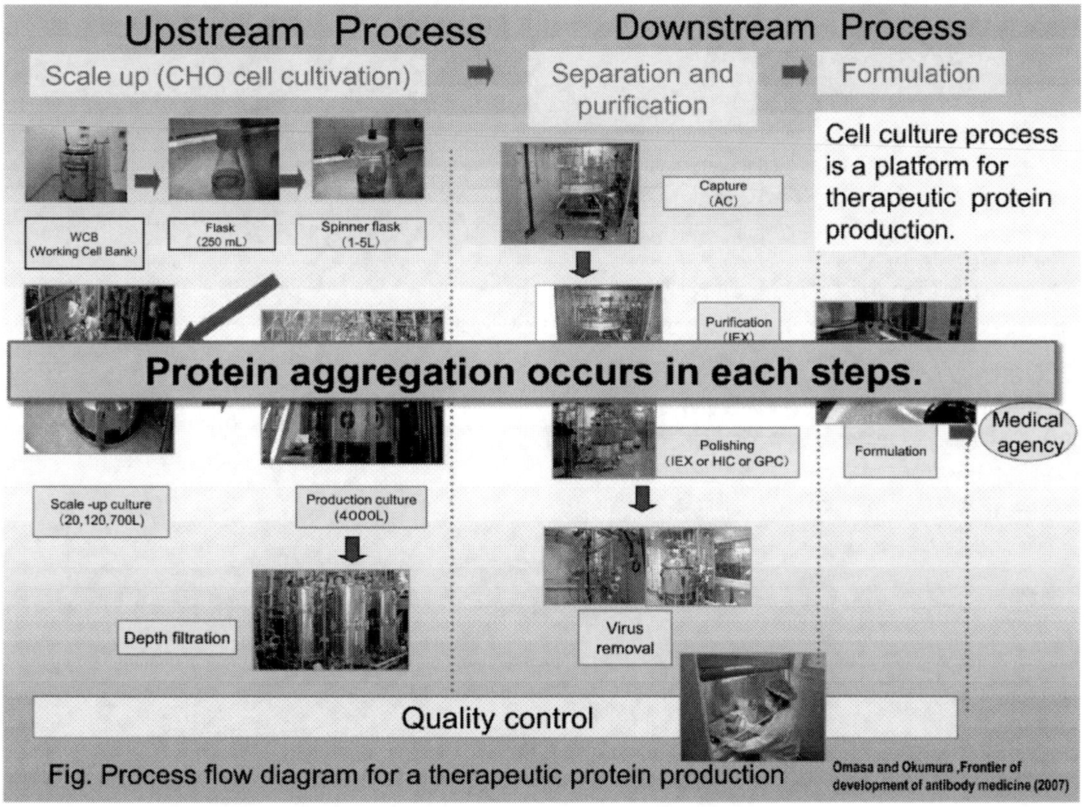
Upstream Process
Scale up (CHO cell cultivation)
Downstream Process
Separation and purification
Formulation
Cell culture process is a platform for therapeutic protein production.
WCB (Working Cell Bank)
Flask (250 mL)
Spinner flask (1-5L)
Capture (AC)
Purification (IEX)
Protein aggregation occurs in each steps.
Polishing (IEX or HIC or GPC)
Formulation
Medical agency
Scale -up culture (20,120,700L)
Production culture (4000L)
Depth filtration
Virus removal
Quality control
Fig. Process flow diagram for a therapeutic protein production
Omasa and Okumura ,Frontier of development of antibody medicine (2007)

Application of chemical chaperon

Anti-aggregate chemical chaperon: Sugar, polyole, amino acids etc...

Trehalose

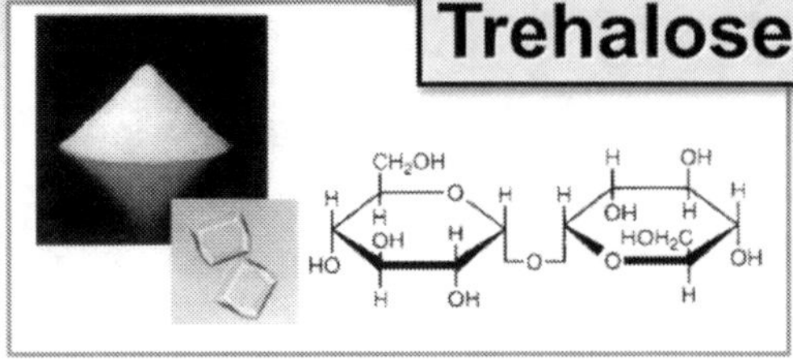

- Effective anti-aggregates
- Used for formulation

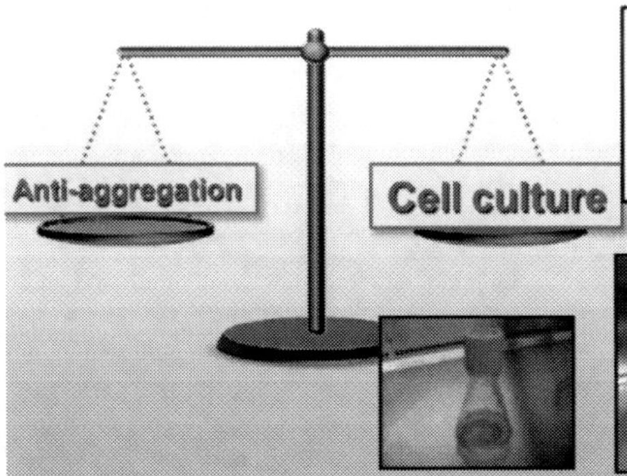

Trehalose containing cell culture is effective for anti-aggregation.

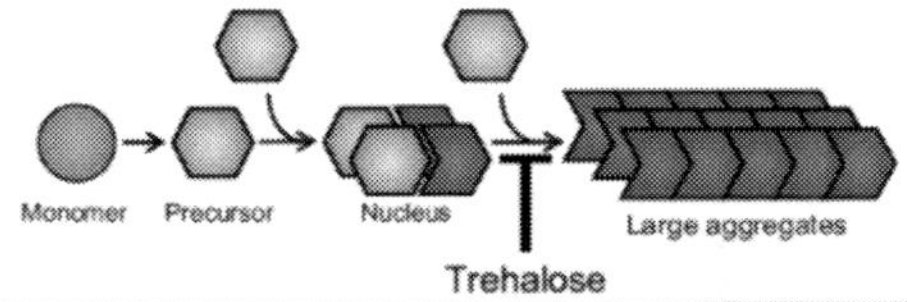

Onitsuka *et al.*, J.Biosci.Bioeng. 117:632-638 (2014)

Materials and Methods

Cell culture

- Cell line: CHO-IgG1, batch-culture, Erlenmeyer Flask (working 100mL),
- Medium: TOp2 (JX), supplemented with 8mM L-Gln and 15 mg/mL Puromycin, Trehalose addition (100mM)
- Shaker: Climo-shaker ISF1-X (Kuhner), 80 rpm, 5% CO_2, 80% humidity
- Cell counting: Vi-CELL XR (Beckman Coulter)
- Extracellular metabolites: Bioprofile 400 (Nova Biomedicals)

- IgG1 detection: AlphaLISA Human IgG Immunoassay kit
- CHO protein estimation: Cell lysate from $4.0x10^6$ cells with M-PER (ThermoFisher), Calibrator: BSA
- Enspire Alpha plate reader (PerkinElmer)
- IgG1 purification: HiTrap rProtein A
- SEC analysis: SuperdexS200

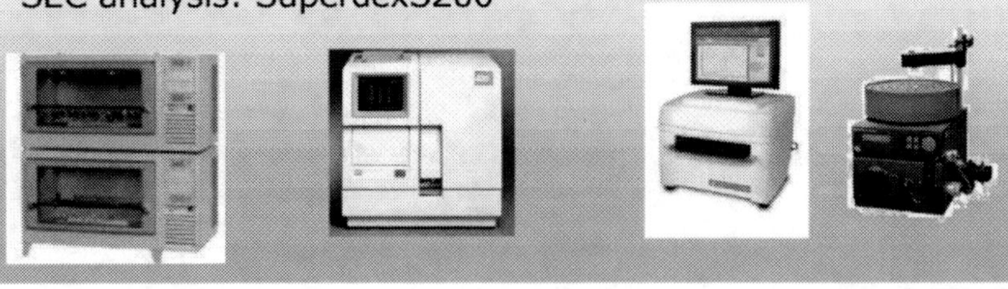

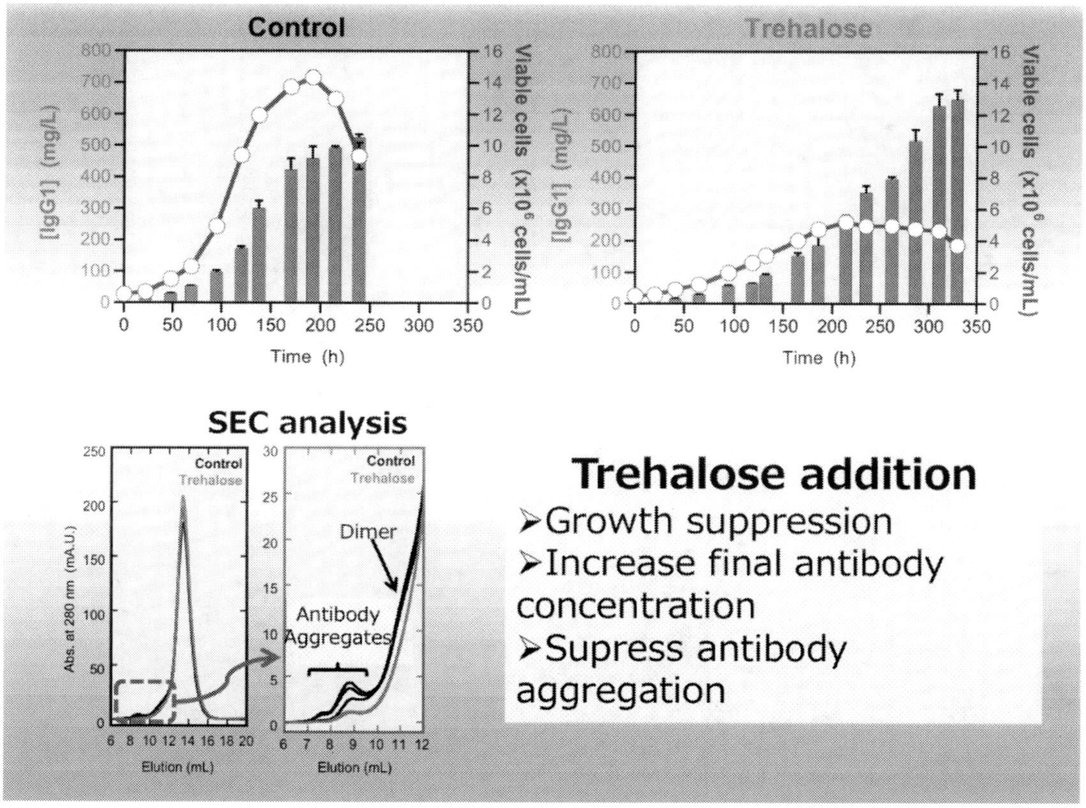
Control
Trehalose
[IgG1] (mg/L)
Viable cells (x10^6 cells/mL)
Time (h)
SEC analysis
Abs. at 280 nm (mA.U.)
Elution (mL)
Dimer
Antibody Aggregates
Trehalose addition
➢Growth suppression
➢Increase final antibody concentration
➢Supress antibody aggregation

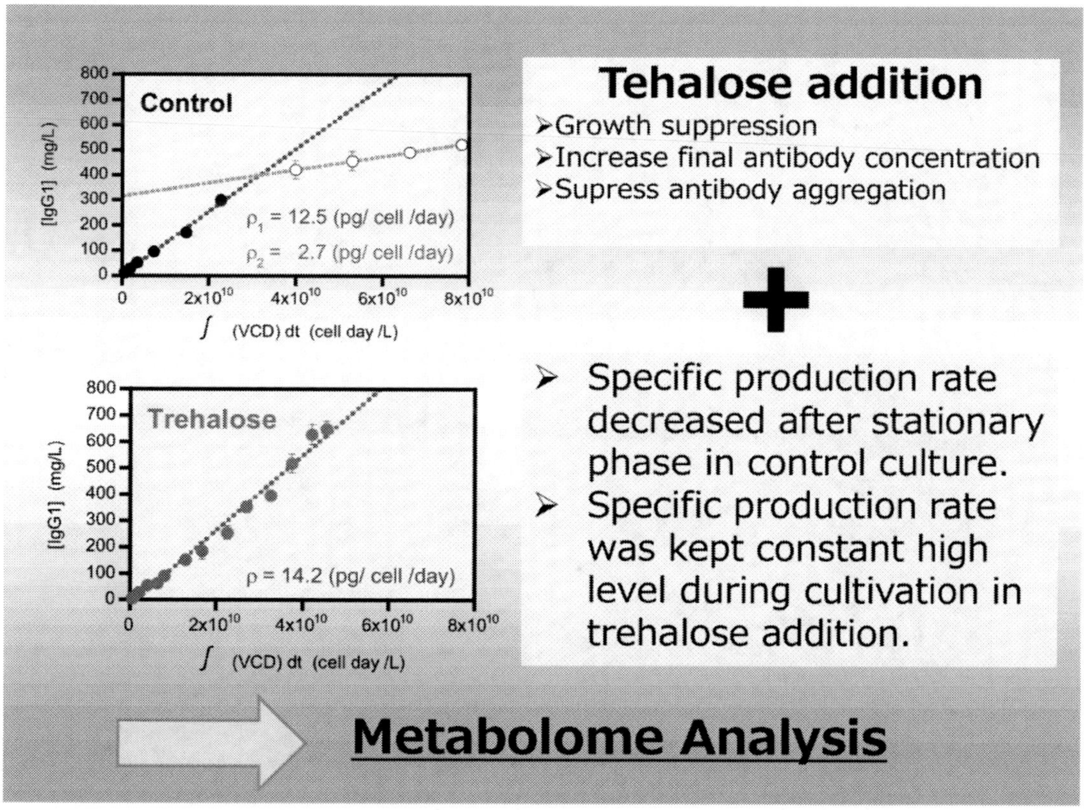
Control
[IgG1] (mg/L)
ρ_1 = 12.5 (pg/ cell /day)
ρ_2 = 2.7 (pg/ cell /day)
$\int$ (VCD) dt (cell day /L)
Trehalose
ρ = 14.2 (pg/ cell /day)
Tehalose addition
➢Growth suppression
➢Increase final antibody concentration
➢Supress antibody aggregation
➢ Specific production rate decreased after stationary phase in control culture.
➢ Specific production rate was kept constant high level during cultivation in trehalose addition.
Metabolome Analysis

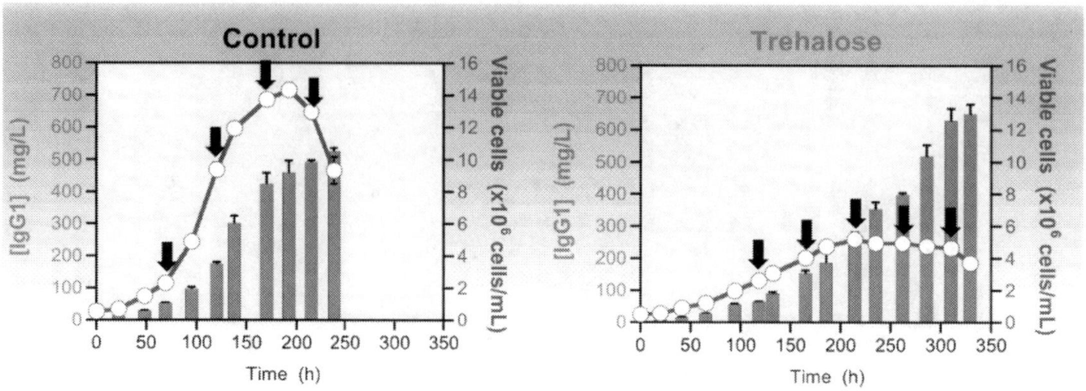

⬇ : Extract intracellular contents

Metabolome analysis supported by

HMT Human Metabolome Technologies

- Methanol quenching, $4.0x10^6$ cells
- HMT internal standard
- Filtration with cellulose membrane
- Cation measurements
  CE-MS : Agilent CE-TOFMS system (Agilent Technologies)
  Capillary : Fused silica capillary, i.d. 50 μm × 80 cm
- Anion measurements
  CE : Agilent CE system
  MS : Agilent 6460 TripleQuad LC/MS system
  Capillary : Fused silica capillary, i.d. 50 μm × 80 cm

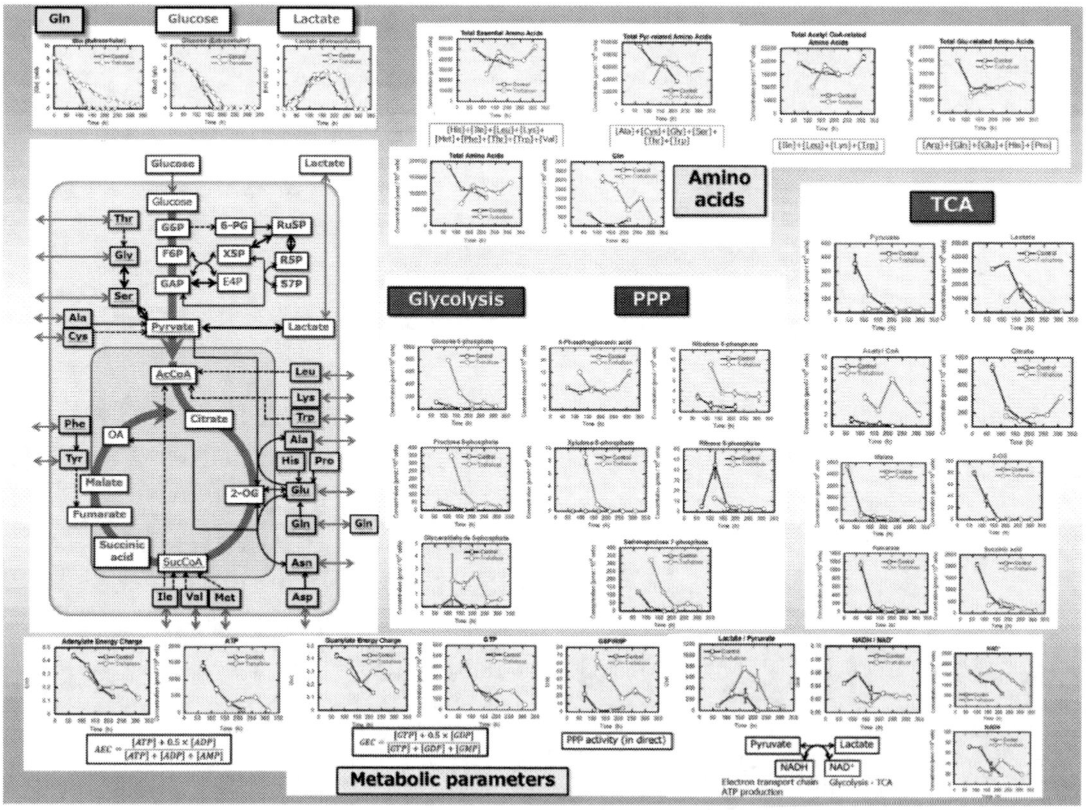

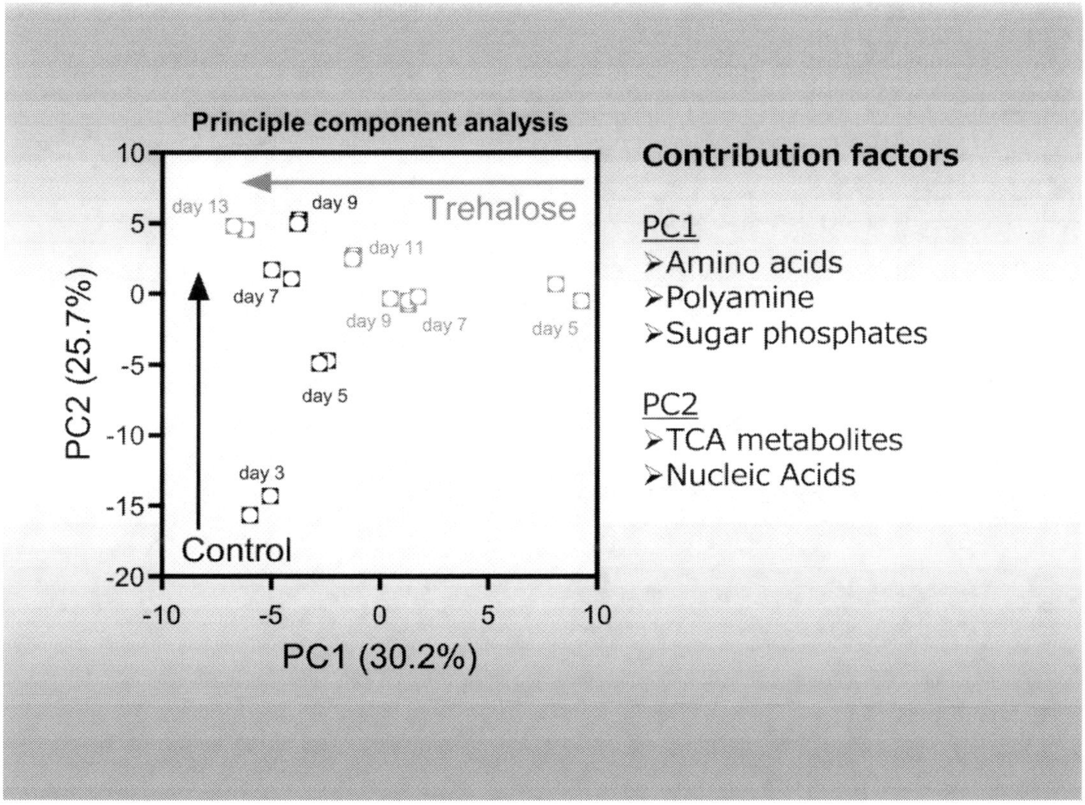
Principle component analysis
Trehalose
Control
day 13
day 9
day 11
day 7
day 9
day 7
day 5
day 5
day 3
PC2 (25.7%)
PC1 (30.2%)
Contribution factors
PC1
➢Amino acids
➢Polyamine
➢Sugar phosphates
PC2
➢TCA metabolites
➢Nucleic Acids

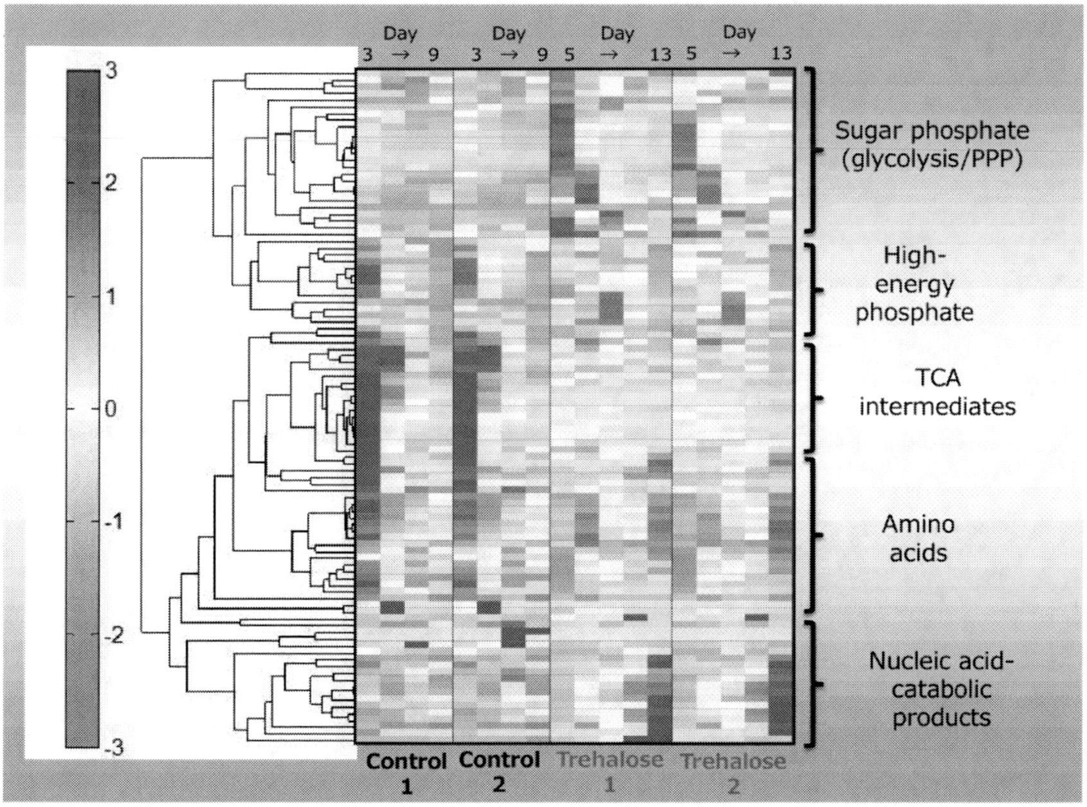
Day
3 → 9
Day
3 → 9
Day
5 → 13
Day
5 → 13
Sugar phosphate (glycolysis/PPP)
High-energy phosphate
TCA intermediates
Amino acids
Nucleic acid-catabolic products
Control 1
Control 2
Trehalose 1
Trehalose 2

Summary

Control culture

Amino acid metabolism is enhanced in early stage of cultivation. From late-growth phase, these AAs are consumed and energy metabolism is inactivated.

Trehalose addition

In early stage, metabolites in glycolysis are accumulated and energy metabolism is not so active. However, from late-growth phase, AAs are consumed and energy metabolism is activated.

Estimated mechanism

Trehalose addition suppressed the cell growth and energy metabolism. Consequently, AAs may be used for antibody production.

References
1. M. Onitsuka et al., "Trehalose suppresses antibody aggregation during the culture of Chinese hamster ovary cells" J. Biosci. Bioeng., 117: 632-638 (2014).
2. M. B. Badsha et al., "Metabolic analysis of antibody producing CHO cell culture under different stresses conditions" J. Biosci. Bioeng.,122:117-124 (2016).

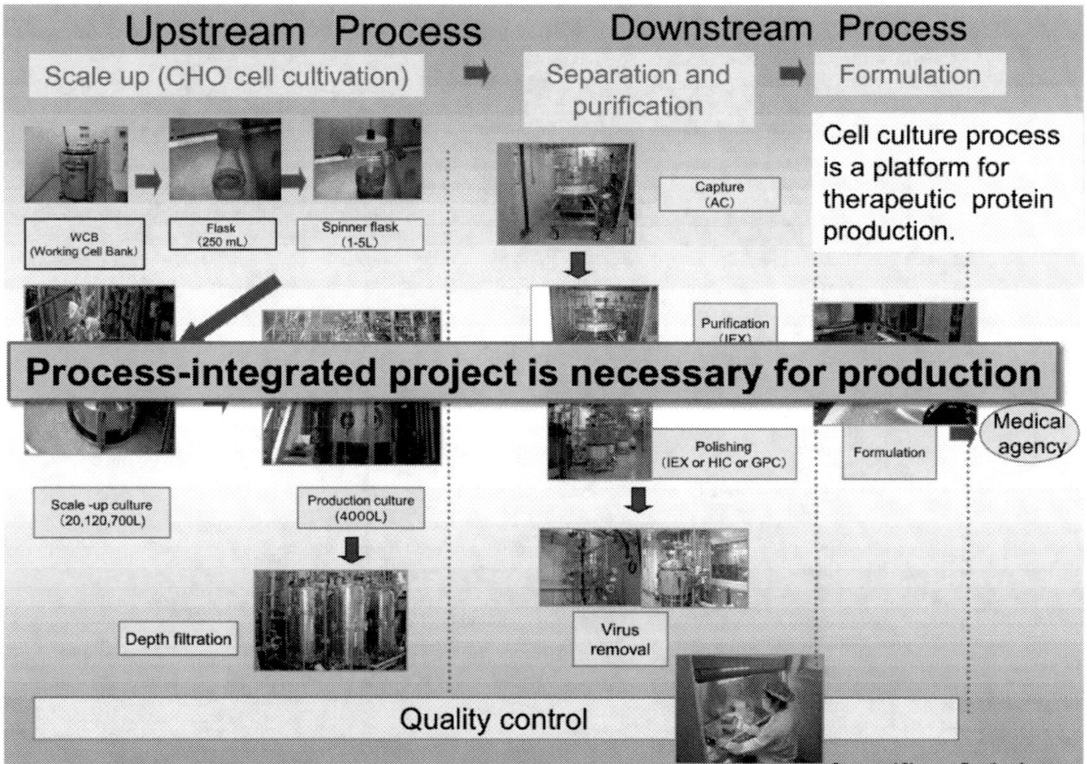

Fig. Process flow diagram for a therapeutic protein production

Omasa and Okumura ,Frontier of development of antibody medicine (2007)

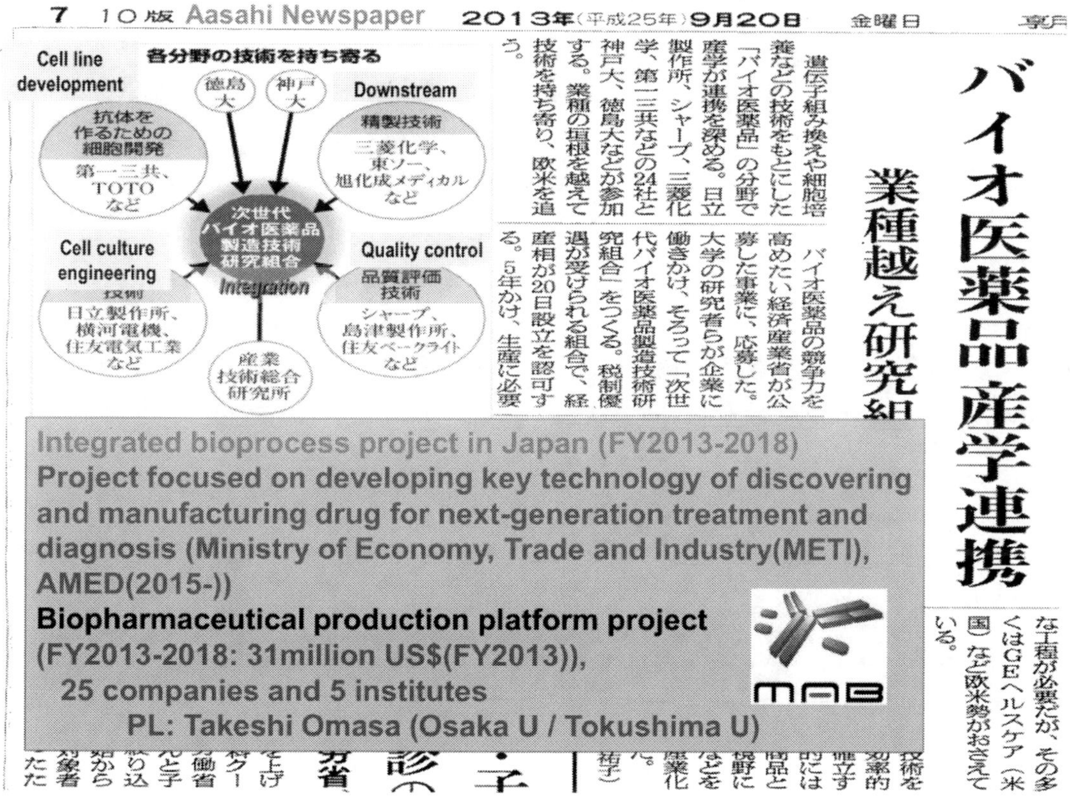

7 10版 Aasahi Newspaper 2013年（平成25年）9月20日 金曜日

バイオ医薬品 産学連携

業種越え研究組

遺伝子組み換えや細胞培養などの技術をもとにした「バイオ医薬品」の分野で産学が連携を深める。日立製作所、シャープ、三菱化学、第一三共などの24社と神戸大、徳島大などが参加する。業種の垣根を越えて技術を持ち寄り、欧米を追う。

バイオ医薬品の競争力を高めたい経済産業省が公募した事業に、応募した。大学の研究者らが企業に働きかけ、そろって「次世代バイオ医薬品製造技術研究組合」をつくる。税制優遇が受けられる組合で、経産相が20日設立を認可する。5年かけ、生産に必要

な工程が必要だが、その多くはGEヘルスケア（米国）など欧米勢がおさえている。

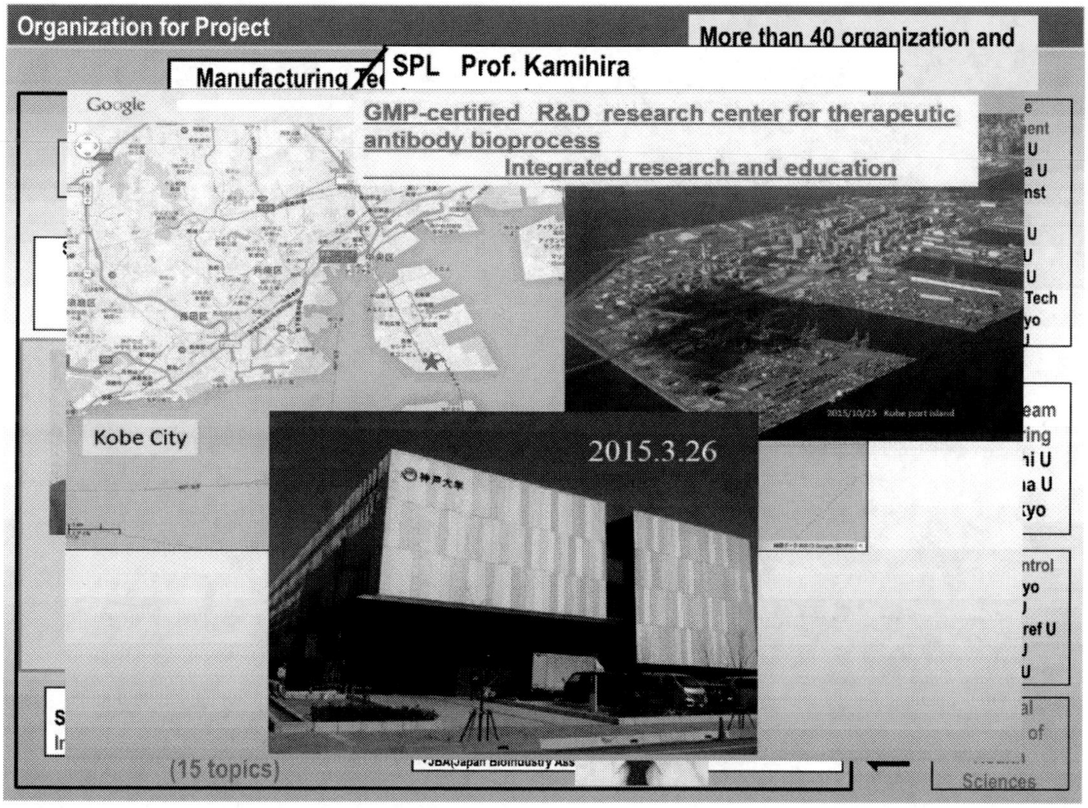

The 29th Annual and International Meeting of
the Japanese Association for Animal Cell Technology
JAACT
JAACT2016 Kobe
9 – 12 November 2016
Kobe International Conference Center
Kobe, Japan
Chair: Prof. Yamaji
(Kobe U)
http://www.aeplan.co.jp/jaact2016/
Vice-Chair:
Prof. Omasa (Osaka U)
MAB
About ~400 attendees are expected.
Site visit tour for GMP-facility in Kobe U.

Thank you for your attention.
Acknowledgements
for my students
Ms. Miki Tatsuzawa
Mr. Masahiro Noda
Mr. Daiki Oka
Mr. Ryonosuke Harata
and
MAB
Metabolome analysis supported by
HMT
Human Metabolome Technologies

Novel Stable Isotope Method to Determine Liver Metabolic Fluxes In Vivo

Hasenour, C. M., Vanderbilt University
Young, J. D., Vanderbilt University
Originally presented on: 6/29/2016 11:10:00 - 11:30:00

Mouse models designed to examine hepatic metabolism are critical to diabetes and obesity research. Thus, a microscale method to quantitatively assess hepatic glucose and intermediary metabolism in conscious, unrestrained mice was developed. [$^{13}C_3$]propionate, [$^{2}H_2$]water, and [6,6-$^{2}H_2$]glucose isotopes were delivered intravenously in short- (9 h) and long-term-fasted (19 h) C57BL/6J mice. GC-MS and mass isotopomer distribution (MID) analysis were performed on three 40-microliter arterial plasma glucose samples obtained during the euglycemic isotopic steady state. Model-based regression of hepatic glucose and citric acid cycle (CAC)-related fluxes was performed using a comprehensive isotopomer model to track carbon and hydrogen atom transitions through the network and thereby simulate the MIDs of measured fragment ions. Glucose-6-phosphate production from glycogen diminished, and endogenous glucose production was exclusively gluconeogenic with prolonged fasting. Gluconeogenic flux from phosphoenolpyruvate (PEP) remained stable, whereas that from glycerol modestly increased from short- to long-term fasting. CAC flux was reduced with long-term fasting. Interestingly, anaplerosis and cataplerosis increased with fast duration. We demonstrate for the first time that the infusion of [$^{13}C_3$]propionate does not significantly perturb gluconeogic fluxes. We also found evidence of significant $^{13}CO_2$ recycling in liver, which may bias flux calculations using prior methods that do not take $^{13}CO_2$ recycling into account. This new approach utilizes state-of-the-art *in vivo* methodology and comprehensive isotopomer modeling to quantify hepatic glucose and intermediary fluxes during physiological stress in mice. The small plasma requirements permit serial sampling without stress and the affirmation of steady-state glucose kinetics. Furthermore, the approach can accommodate a broad range of modeling assumptions, isotope tracers, and measurement inputs without the need to introduce ad hoc mathematical approximations. I will discuss the development and validation of this approach, as well as our initial forays into applying this approch to assess the liver response to genetic knockouts, exercise, and diet manipulation in mice.

Split-GFP and Droplet Microfluidics for High-Throughput Screening of Mammalian Cell Factories

Volk, A. L., KTH - Royal Institute of Technology
Lundqvist, M., KTH- Royal Institute of Technology
Hammar, P.,
Gram Hansen, H.,
Uhlen, M., KTH- Royal Institute of Technology
Faustrup Kildegaard, H.,
Jönsson, H., KTH Royal Institute of Technology
Rockberg, J., KTH- Royal Institute of Technology
Originally presented on: 6/29/2016 11:30:00 - 11:50:00

The industrial production of monoclonal antibodies and other protein therapeutics is currently mainly achieved by heterologous overexpression of recombinant genes in Chinese hamster ovary (CHO) cells. These cells are popular for a multitude of reasons, in particular for their ability to generate folded and glycosylated proteins at high yield at large scale. To meet the growing demands of high-yielding and cost-effective production strains numerous rational, undirected and combinatorial approaches have been employed to enhance the level of protein secretion. The lack of access to scalable and fast ways of monitoring product secretion is however limiting the ability for a complete and efficient screening of such large libraries of cells.

Here we present a high throughput screening system allowing for single cell monitoring of protein secretion and fluorescent sorting using a combination of droplet microfluidics and split-GFP complementation. Briefly a short tag, comprising the 11-strand of GFP, was genetically fused to either the monoclonal antibody Herceptin or the glycoprotein EPO and expressed in CHO as fusion proteins. The expressed fusion protein will remain non-fluorescent until complemented with the remaining part of the GFP protein (GFP 1-10). We demonstrate the ability to encapsulate and cultivate single CHO cells in mono-disperse droplets containing cultivation medium supplemented with GFP 1-10 and that we are able to monitor protein secretion of single cells by fluorescence and to sort CHO cell factory libraries based on their protein secretion level. Further we show the ability to de-tag the 11th strand from the high producing clones once identified using Crispr/Cas9. We believe this generic technology will allow for a better ability to perform thorough and fast screenings of CHO cell factories for the identification of high-secreting production clones.

Using Metabolomics to Identify Gut Microbiota-Derived Metabolites As Host Inflammation Modulators

Lee, K., Tufts University
Krishnan, S., Tufts University
Choi, M.,
Ding, Y.,
Saedi, N.,
Jayaraman, A., Texas A&M University
Yarmush, M. L., Harvard Medical School
Originally presented on: 6/29/2016 11:50:00 - 12:10:00

Non-alcoholic fatty liver disease (NAFLD) has emerged as the most prevalent chronic liver disease in Western countries, paralleling the rise in obesity. It is characterized by steatosis (fat deposition in the liver), which can remain a benign condition. However, in ~20% of patients, NAFLD progresses from steatosis to steatohepatitis, which can lead to cirrhosis and liver cancer. Despite accumulating evidence that the gut microbiota plays a mechanistic role, the molecular mediators and cellular pathways involved remain unclear. Recent findings, including our own work [1], demonstrated that microbiota species produce bioactive metabolites that engage host cellular pathways. The present study investigates the hypothesis that gut microbiota dysbiosis perturbs the balance of immunomodulatory microbiota metabolites, exacerbating liver inflammation in steatosis.

We compared the metabolite profiles in the cecum of germ-free (GF) and conventionally raised (CONV-R) mice, and identified metabolites whose levels in the mouse intestine depend on the presence of the microbiota. Focusing on metabolites depleted in GF mice, we found derivatives of aromatic amino acids (AAAs), several of which we had previously [1] shown to activate the aryl hydrocarbon receptor (AHR), a transcriptional regulator of host-microbiota interactions. To investigate a potential role for these metabolites in NAFLD, we compared their profiles in serum and liver samples from 14-week old male C57BL/6J mice raised for 8 weeks either on high-fat diet (HFD) and low-fat diet (LFD) mice. Untargeted metabolomics detected microbiota-dependent AAA derivatives in serum and liver samples from both HFD and LFD mice. We selected a pair of representative metabolites from this set, tryptamine (TA) and indole-3-acetate (I3A), for further analysis using *in vitro* culture models. Cultured HepG2 and AML12 cells were treated with 10-500 mM TA or I3A for 24 h, and then exposed to the pro-inflammatory cytokine TNFα for another 24 h in the presence of the metabolites. We focused on bile and fatty acids as functional readouts, as dysregulation of lipid metabolism is a hallmark of NAFLD. Compared to vehicle control, TNFα treatment significantly increased both intra- and extracellular levels of several bile acids. Pretreatment with either TA or I3A significantly attenuated the increases in bile acid levels. Similarly, TNFα treatment significantly increased both intra- and extracellular levels of several free fatty acids, notably palmitate. Pretreatment with I3A, but not TA, significantly attenuated these increases. To further delineate the signaling and regulatory pathways affected by the metabolites, we performed an untargeted proteomics analysis. Cytokine treatment significantly decreased the level of albumin, while increasing the expression of several acute phase proteins. Pretreatment with I3A, but not TA, attenuated this trend.

In summary, we show that the levels of AAA-derived microbiota metabolites are significantly depleted in a diet model of liver steatosis, and that these metabolites can act directly on hepatocytes, possibly through the AHR, to modulate inflammatory pathways. Ongoing work utilizes metabolomics and proteomics to investigate potential crosstalk between AHR and other pathways regulating liver lipid metabolism. Insights into the mechanisms linking dysbiosis and NAFLD could provide novel strategies to treat fatty liver diseases through pre-/pro-/postbiotics.

Enzyme Screening and Engineering for Improvement of Novel Biosynthetic Pathways

Prather, K. L. J., Massachusetts Institute of Technology
Originally presented on: 6/29/2016 13:50:00 - 14:15:00

Biological synthesis is increasingly being pursued as an alternative to traditional organic synthesis for the production of chemical compounds. As known biochemical pathways are transferred to non-native hosts and new pathways are designed, access to genes encoding desired enzymatic activities with appropriately high reaction rates is often a limiting factor. In this presentation, I will discuss our efforts to identify enzymes with appropriate activity and selectivity in the context of two novel biosynthetic pathways. In the first example, we have exploited the promiscuous activity of four different enzymes in order to build a versatile platform pathway for the synthesis of various 3-hydroxyalkanoic acids. These products can be subsequently dehydrated to yield the unsaturated acids, or hydrogenated to produce the saturated product, but only if enzymes with the appropriate specificity and promiscuity can be found. Examining the product profiles in vivo has led to greater understanding of the constituent enzymes in the absence of access to the substrates needed for in vitro analysis. I will discuss our findings to date with this pathway, including an analysis of enzyme variants obtained through bioprospecting and rational design, as well as challenges that we face towards linking in vitro characterization to in vivo performance. Using a second model pathway for the production of glucaric acid, I will also describe opportunities to use experimental data to guide the selection of additional enzyme variants that may have higher target activities through network analysis of protein sequences.

Optimization and Evolution of Pathways and Metabolism in Yeast
Alper, H., The University of Texas at Austin
Originally presented on: 6/29/2016 14:15:00 - 14:40:00

Advances in metabolic engineering are enabling the production of nearly any organic molecule of interest—from biofuels to biopolymers to pharmaceuticals. However, these feats require the ability to "hijack" native cellular machinery and metabolism and navigate the complexity inherent in cellular regulation. One particular challenge in the field involves the design of new metabolic pathways that properly interface within the context of cellular metabolism. In this regard, introduced pathways must be properly assembled and evolved to work within a given cellular context. In addition, local and global metabolic genes must be altered to interface with these pathways and improve production levels. Both of these tasks traditionally rely on extensive experimentation. This talk will describe two recent advances in speeding the design process for pathway optimization and evolution. First, we develop an *in vivo* mutagenesis approach in yeast by synthetically optimizing the retrotransposon Ty1. This approach is used in the directed evolution of global transcriptional regulators, single enzymes, and multi-gene pathways. In each case described here, we obtain rapid and significant improvements in performance over days instead of weeks, a clear advantage over traditional *in vitro* approaches, especially in fungal systems. Additionally, we demonstrate how this approach can be coupled with microdroplet technologies to enable rapid screening and selection of pathway variants. Second, we develop an approach to multiplex gene expression within pathways and broader metabolism to identify key bottlenecks limiting flux. By utilizing a CRISPR-Cas9 system, we demonstrate a rapid method for assessing the importance of these metabolic targets in a manner that is rapidly scalable and combinatorial. We demonstrate how this approach can be used to both identify and tune rate limiting steps in several pathway case studies of interest. Together, these two approaches enable rapid pathway design and optimization in yeast.

Synthetic Quorum Sensing As a Tunable Cell Density Sensor-Regulator for Dynamic Metabolic Engineering

Hanai, T., Kyushu University
Soma, Y., Kyushu University

Originally presented on: 6/29/2016 14:40:00 - 15:00:00

Metabolic engineering have enabled improvements in titer and yield for a variety of chemicals produced naturally in microorganisms, as well as those produced via the synthetic metabolic pathways consist of heterogeneous enzymes. Almost all precursors for the synthetic metabolic pathways are served from the endogenous intermediates in glycolysis, such as phosphoenolpyruvate, pyruvate and acetyl-CoA. However, endogenous central metabolism also requires these key intermediates for cell growth. Accordingly, there is a trade-off relationship between the cell growth and the target chemicals production, which decrease the total productivity of target chemical. To achieve the drastic improvement in productivity, it is necessary to improve both of final cell density and the specific productivity for efficient bioprocess.

For this end, we previously constructed a synthetic genetic circuit in *E. coli* for control the local metabolic fluxes around acetyl-CoA, which was termed as synthetic metabolic toggle switch (MTS). The MTS was designed to redirect the metabolic flux from TCA cycle toward target chemical production in the response to the addition of the exogenous inducer IPTG. This circuit succeeded to improve titer and yield of isopropanol production with ensuring the final cell density. However, it required strict optimization of timing for the addition of IPTG in order to achieve the adequate cell density and the improvement in productivity. To solve such problem, it is desired that the engineered *E. coli* would sense their population and self-trigger the function of the genetic circuit in the response to appropriate cell density.

Here, we reconstructed *Vibrio fischeri*'s quorum sensing (QS) system, *lux* system, in *E. coli* so that it may work as a tunable cell density sensor-regulator. In native *lux* system, the gene expression under the QS promoter (P_{lux}) would be activated by transcription factor LuxR in the response to the concentration of auto-inducer AHL reaching to threshold level. AHL is produced by LuxI under control of P_{lux} promoter. Even though the *lux* system is the most well studied model QS, there is no report for its application to the microbial chemical production because the cell density upon the self-induction dependent on the QS (threshold cell density) is too much low to induce the genes' expression for the chemical production. To elevate the threshold cell density, we constructed a synthetic *lux* system in *E. coli* by using a synthetic *lux* promoter (P_{lux}lacO) which has dual responsiveness to the natural QS activation and the repression by LacI. In this system, the self-induction of target gene expression is driven by QS-signal, and its threshold cell density can be tuned depending on the concentration of IPTG which was added at the beginning of the fermentation. We demonstrated the self-redirection of metabolic flux by the MTS integrated with the synthetic *lux* system at a desired cell density, which resulted in a significant increase in isopropanol production. This study demonstrated that the dynamic gene expression profiles which is designed by using the genetic circuit allow regulating the microbial phenotype plastically and can help to manage the trade-offs between the growth and the production.

Member of the Helmholtz Association

JÜLICH
FORSCHUNGSZENTRUM

A high-throughput approach for the identification of suitable parts for biosensor design

Regina Mahr

Institute for Bio- and Geosciences, IBG-1
Forschungszentrum Jülich, Germany

June 29th, 2016

Multiple applications of biosensors

JÜLICH
FORSCHUNGSZENTRUM

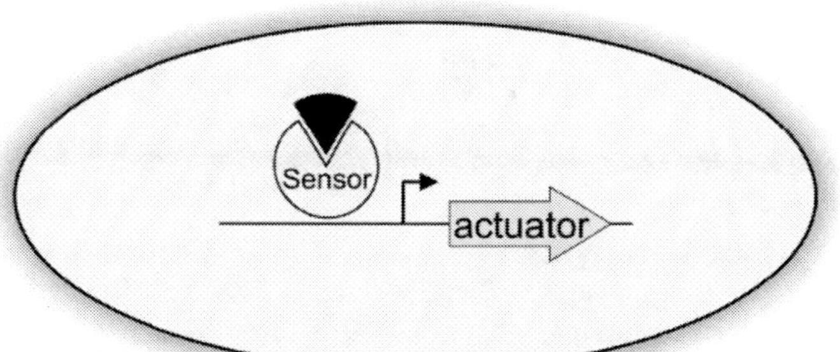

June 29th, 2016 | Adapted to Mahr and Frunzke (2016) *Appl Microbiol Biotechnol* | Slide 2

Multiple applications of biosensors

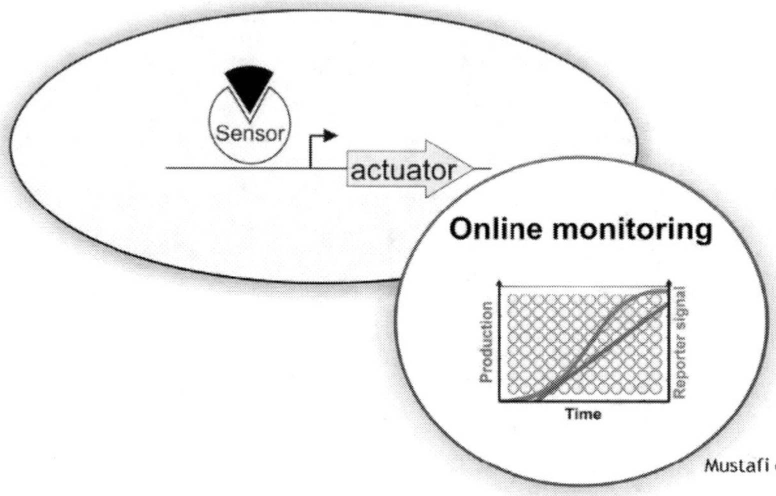

Mustafi et al. (2014) *PLoS One*

June 29th, 2016 Adapted to Mahr and Frunzke (2016) *Appl Microbiol Biotechnol* Slide 2

Multiple applications of biosensors

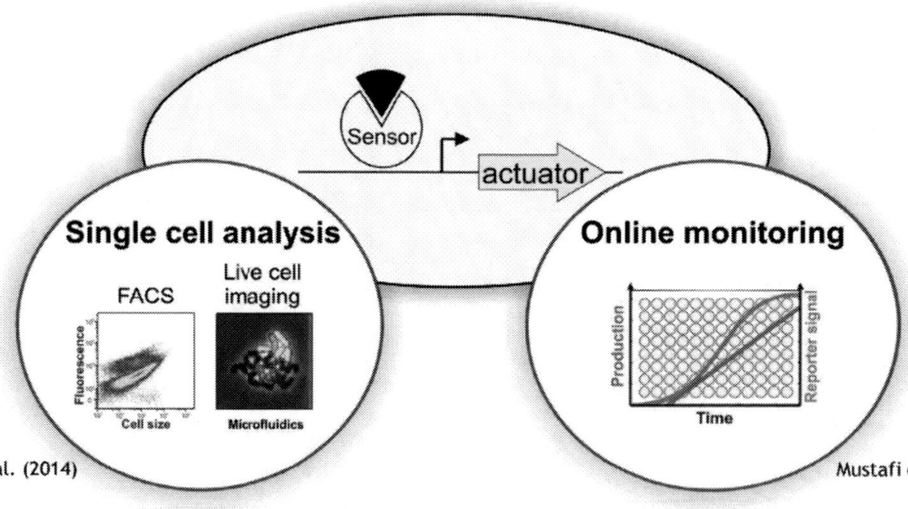

Mustafi et al. (2014) *PLoS One*

Mustafi et al. (2014) *PLoS One*

June 29th, 2016 Adapted to Mahr and Frunzke (2016) *Appl Microbiol Biotechnol* Slide 2

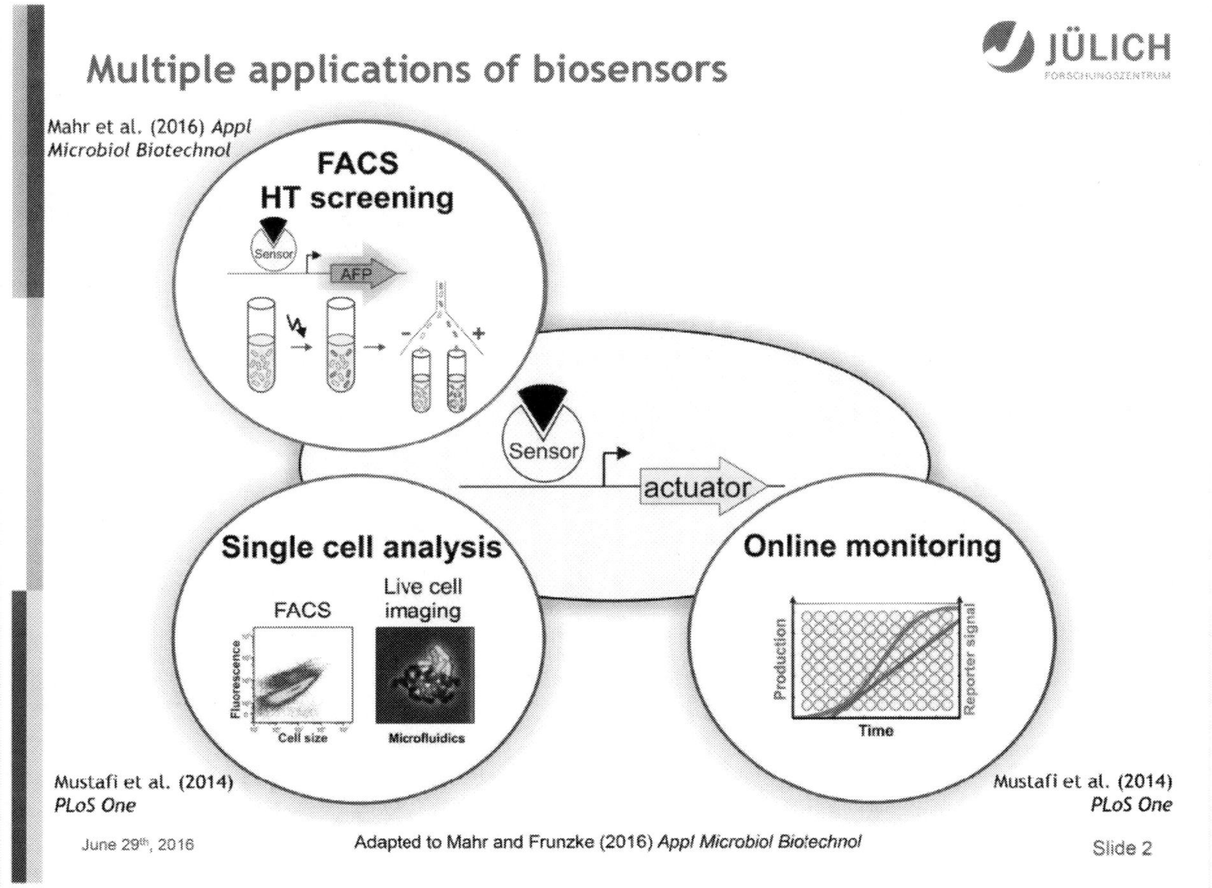
JÜLICH
FORSCHUNGSZENTRUM
Multiple applications of biosensors
Mahr et al. (2016) *Appl Microbiol Biotechnol*
FACS
HT screening
Sensor
AFP
-
+
Sensor
actuator
Single cell analysis
FACS
Live cell imaging
Fluorescence
Cell size
Microfluidics
Online monitoring
Production
Reporter signal
Time
Mustafi et al. (2014) *PLoS One*
Mustafi et al. (2014) *PLoS One*
June 29th, 2016
Adapted to Mahr and Frunzke (2016) *Appl Microbiol Biotechnol*
Slide 2

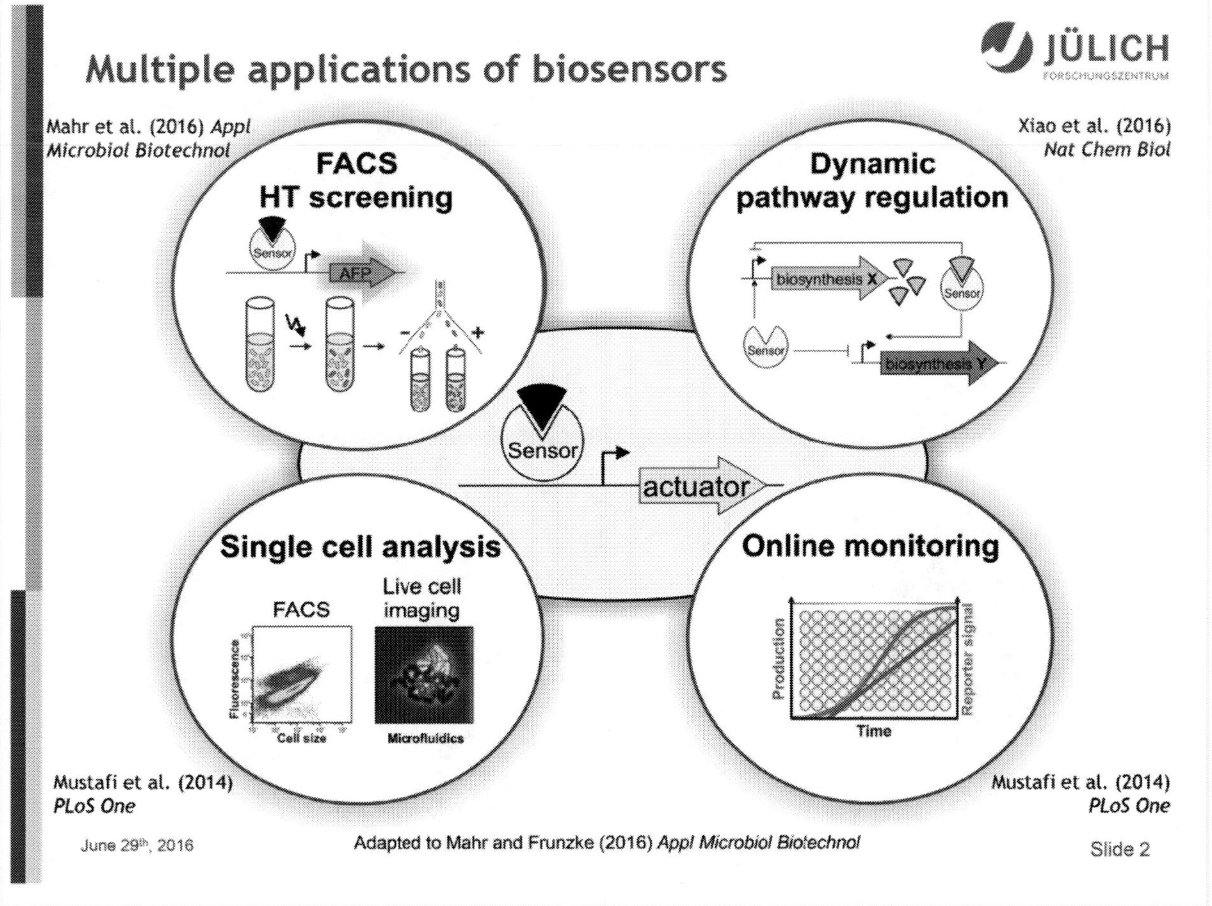
JÜLICH
FORSCHUNGSZENTRUM
Multiple applications of biosensors
Mahr et al. (2016) *Appl Microbiol Biotechnol*
Xiao et al. (2016) *Nat Chem Biol*
FACS
HT screening
Sensor
AFP
-
+
Dynamic pathway regulation
biosynthesis X
Sensor
Sensor
biosynthesis Y
Sensor
actuator
Single cell analysis
FACS
Live cell imaging
Fluorescence
Cell size
Microfluidics
Online monitoring
Production
Reporter signal
Time
Mustafi et al. (2014) *PLoS One*
Mustafi et al. (2014) *PLoS One*
June 29th, 2016
Adapted to Mahr and Frunzke (2016) *Appl Microbiol Biotechnol*
Slide 2

Overview

- HT approach for the **identification of suitable parts** for biosensor design
- Influence of biosensor design on **performance characteristics**
- **Biosensor-driven adaptive laboratory evolution** of metabolite production

June 29th, 2016

Slide 3

Identification of new biosensor constructs

Uri Alon library: about 2000 different promoter-*gfpmut2* fusions in *Escherichia coli* K-12 MG1655 Zaslaver et al. (2006) *Nat Methods*

Raphael von Boeselager

June 29th, 2016

Adapted to Mahr et al. (2016) *Appl Microbiol Bictechnol*

Slide 4

Screening for phenylalanine-inducible promoters

JÜLICH FORSCHUNGSZENTRUM

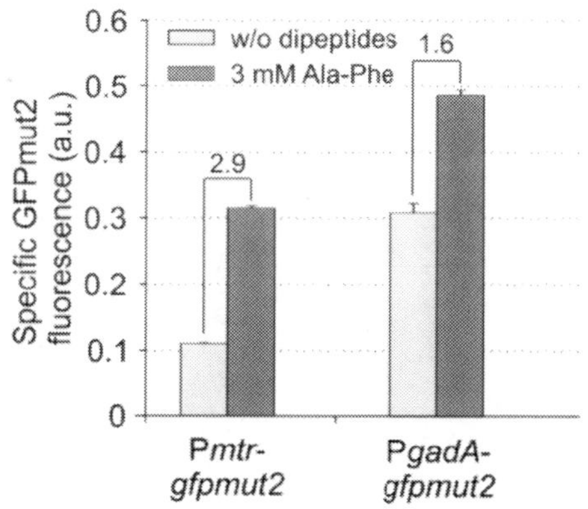

23 isolated and sequenced clones

⇨ 12 clones with the fusion of *mtr* promoter and *gfpmut2*

June 29th, 2016 | Adapted to Mahr et al. (2016) *Appl Microbiol Bictechnol* | Slide 5

Influence of the sensor's architecture

JÜLICH FORSCHUNGSZENTRUM

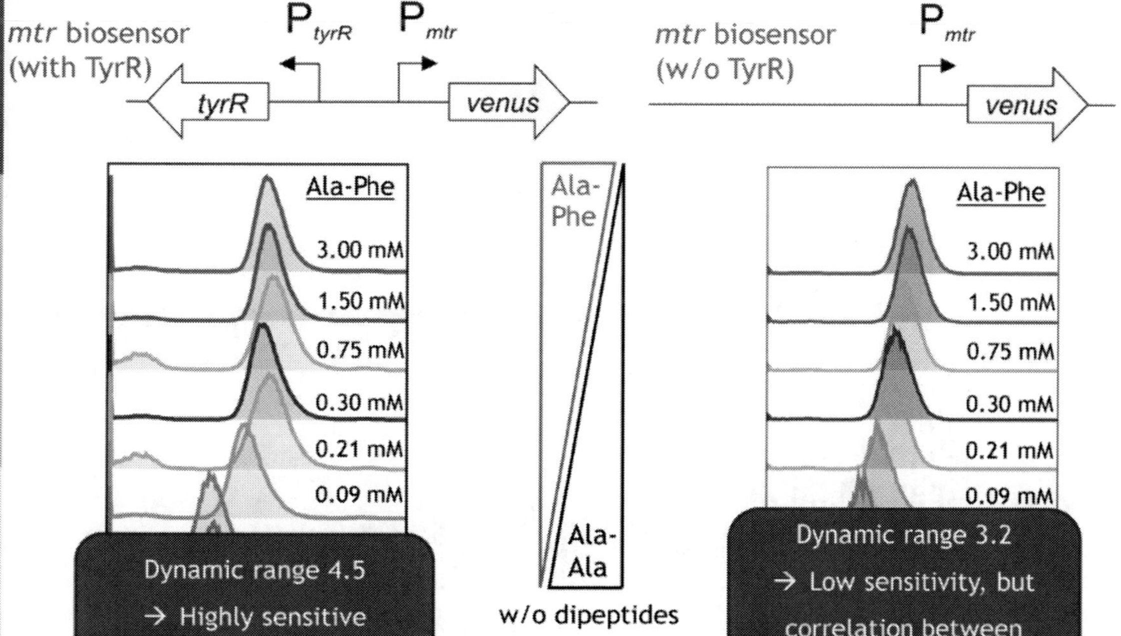

Dynamic range 4.5
→ Highly sensitive in the low µM range

Dynamic range 3.2
→ Low sensitivity, but correlation between effector and output

June 29th, 2016 | Adapted to Mahr et al. (2016) *Appl Microbiol Bictechnol* | Slide 6

Parameters influencing the sensor performance

JÜLICH FORSCHUNGSZENTRUM

- Copy number of gene encoding the regulator
- Number of genomic targets

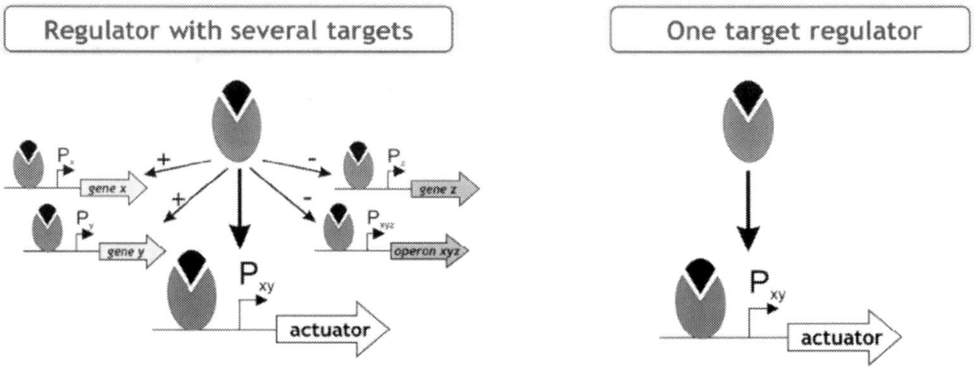

- Interaction with competing regulatory networks

June 29th, 2016

Slide 7

FACS high-throughput screening

JÜLICH FORSCHUNGSZENTRUM

E. coli K-12 MG1655 pJC1-*mtr*Sensor (w/o TyrR)

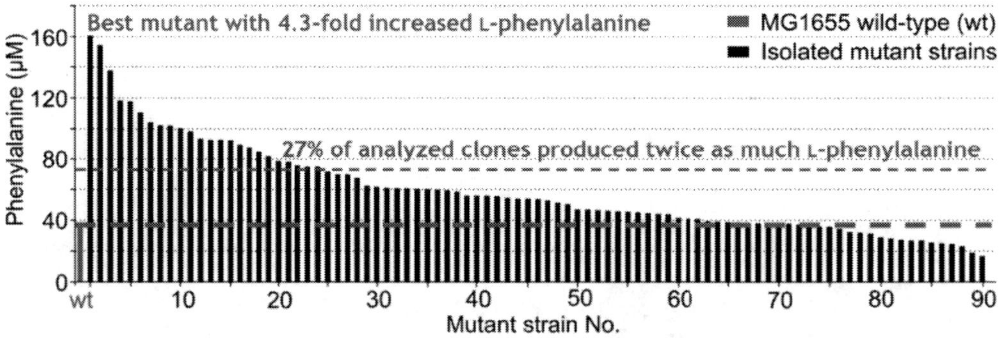

June 29th, 2016

Adapted to Mahr et al. (2016) *Appl Microbiol Biotechnol*

Slide 8

Identification of mutations

- **538 *SNPs* identified in five different strains**
- **310 *SNPs* lead to an amino acid exchange or a stop codon**

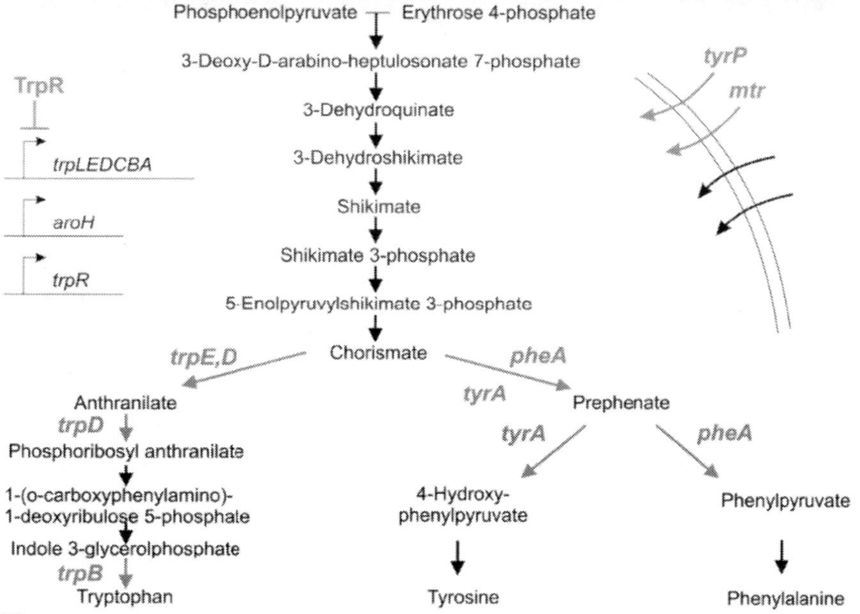

June 29th, 2016

Slide 9

Challenge

Identification of (non-intuitive) mutations, which are highly valuable for strain engineering.

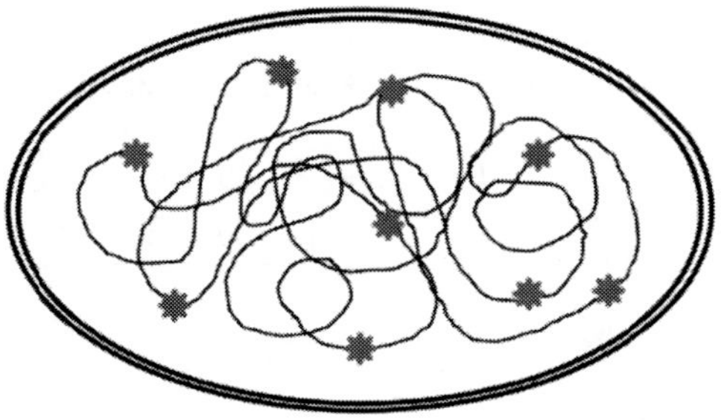

June 29th, 2016

Slide 10

Adaptive laboratory evolution

JÜLICH FORSCHUNGSZENTRUM

Evolution	
Mutation	Selection
Natural mutation frequency (10^{-10}-10^{-9} mutations per base pair per replication cycle) Barrick and Lenski (2014), *Nature Reviews*	**Selection pressure by environmental impact** → Influences **fitness** of organisms

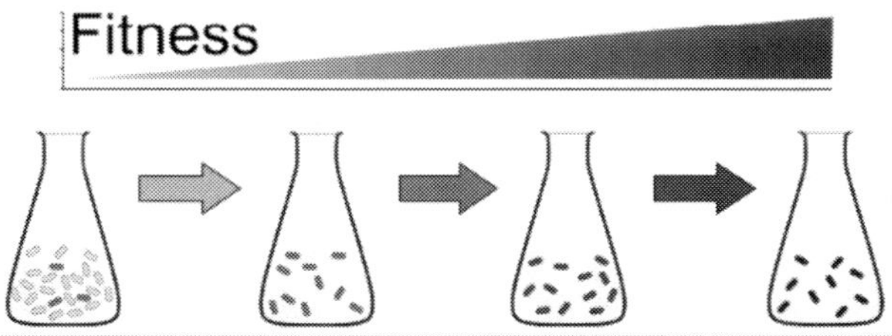

June 29th, 2016

Slide 11

Biosensor-driven adaptive laboratory evolution

JÜLICH FORSCHUNGSZENTRUM

Evolution	
Mutation	Artificial selection pressure
Natural mutation frequency (10^{-10}-10^{-9} mutations per base pair per replication cycle) Barrick and Lenski (2014), *Nature Reviews*	e.g. Lrp-biosensor (for the detection of L-methionine, L-valine, L-isoleucine and L-leucine) Mustafi et al. 2012, *Metab Eng*

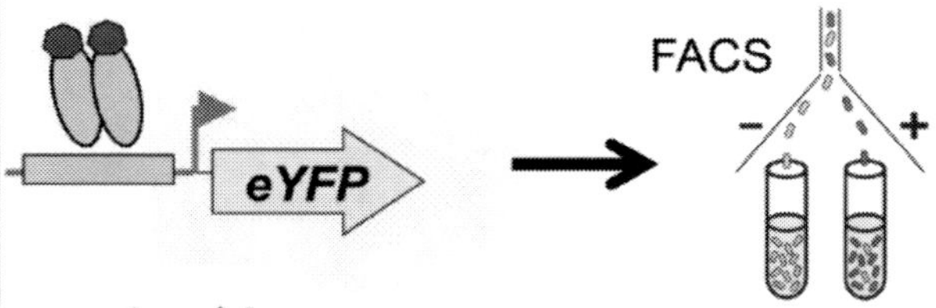

Mahr et al. (2015) Biosensor-driven adaptive laboratory evolution of L-valine production in *Corynebacterium glutamicum*. *Metab Eng* 32: 184-94

June 29th, 2016

Slide 12

Corynebacterium glutamicum ΔaceE - a basal L-valine producer strain

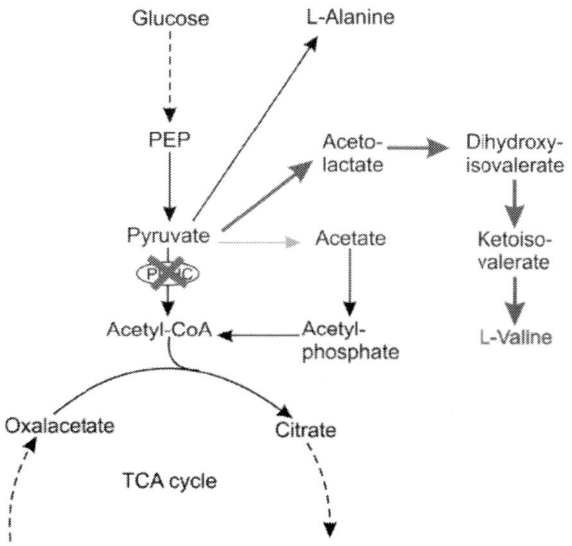

June 29th, 2016

Blombach et al. (2007) *Appl Environ Microbiol*
Schreiner et al. (2005) *J Bacteriol*

Slide 13

Evolution - Isolation and recultivation

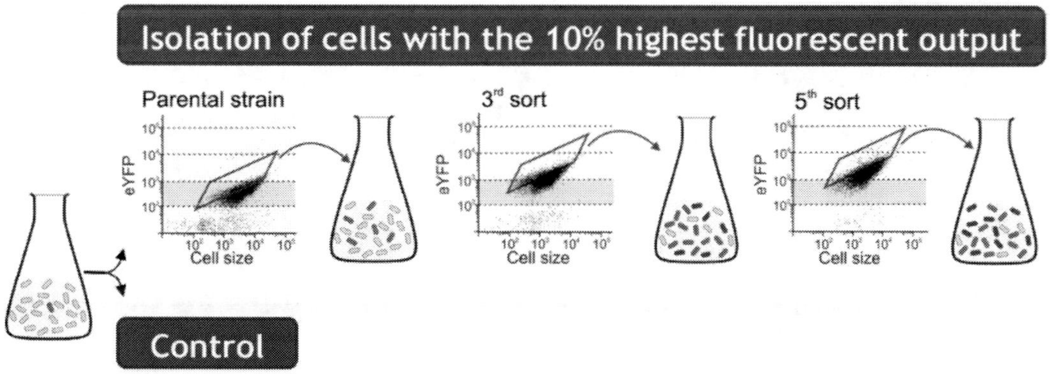

The biosensor signal imposes a selective pressure on product formation.

June 29th, 2016

Mahr et al. (2015) *Metab Eng*

Slide 14

Improvement of growth

JÜLICH
FORSCHUNGSZENTRUM

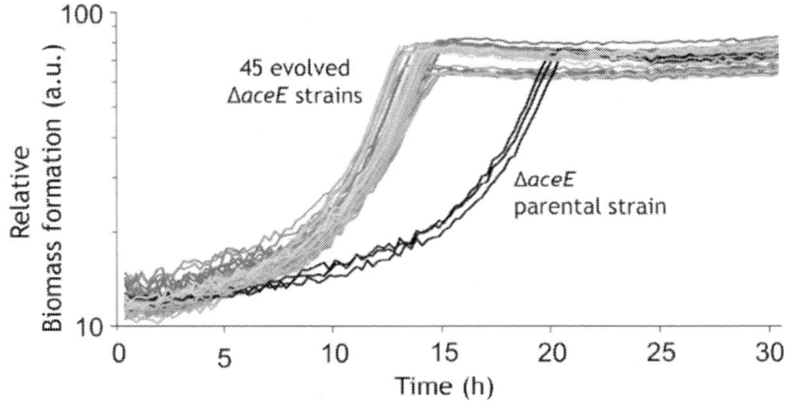

Shortened lag-phase

Increased growth rate (μ_{max})

evolved *ΔaceE* strains: μ_{max} = **0.41 ±0.019 h^{-1}**

Non-evolved *ΔaceE* strain: μ_{max} = **0.36 ±0.007 h^{-1}**

Mahr et al. (2015) *Metab Eng*

Improvement of L-valine production

JÜLICH
FORSCHUNGSZENTRUM

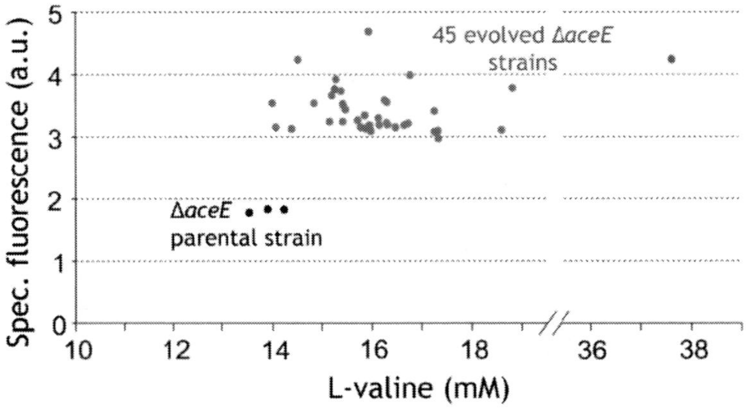

Ø 20% increased production

Best clone produces **38 mM L-valine**
→ **2.7-fold** more than the parental strain

Mahr et al. (2015) *Metab Eng*

Genome sequencing of evolved strains

- 7 different SNPs
- 4 SNPs in annotated genes

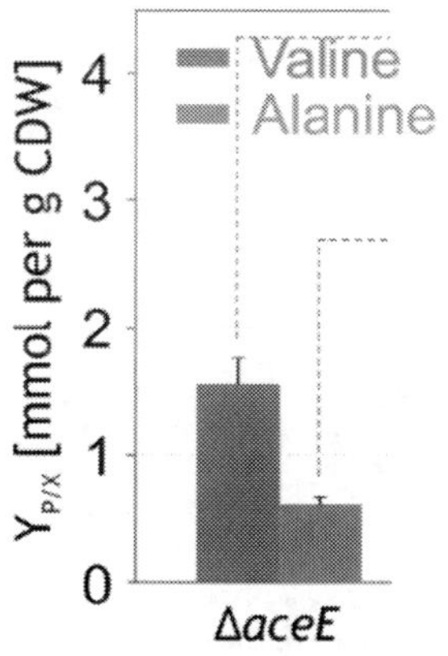

Mahr et al. (2015) Biosensor-driven adaptive laboratory evolution of L-valine production in *Corynebacterium glutamicum*. *Metab Eng* 32: 184-94

Summary

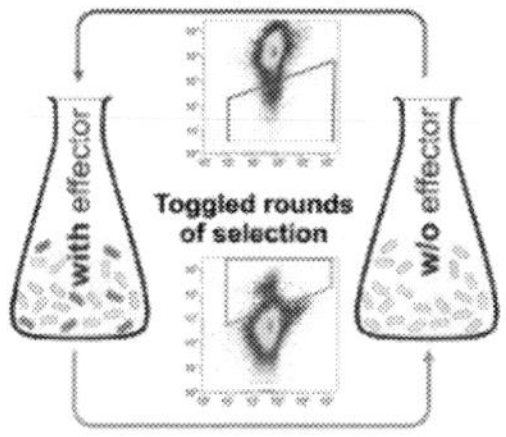

Identification of suitable parts for biosensors based on FACS.

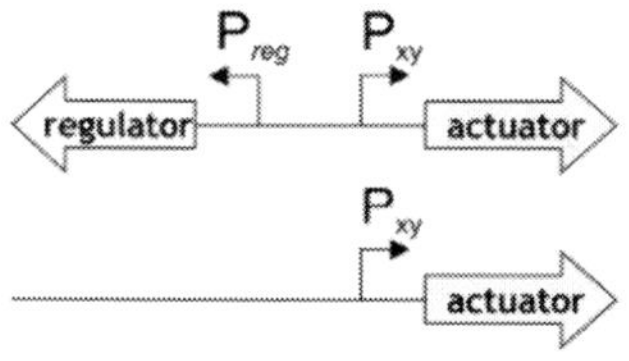

Biosensor design with strong impact on performance characteristics.

Biosensor-driven adaptive laboratory evolution as powerful approach to identify non-intuitive targets for metabolic engineering.

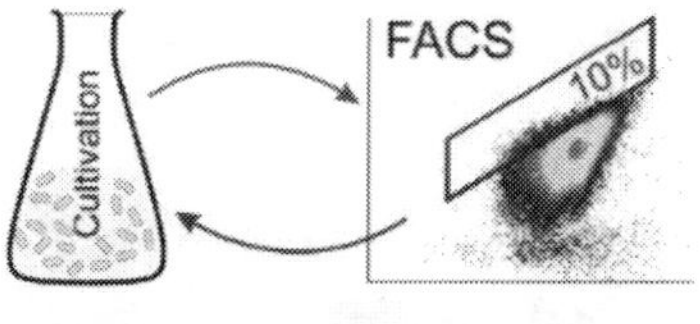

Evolutionary strategies and the combinatory assembly of circuits can complement **Synthetic Biology** where classical, rational approaches are limited.

JÜLICH
FORSCHUNGSZENTRUM

Julia Frunzke

Michael Bott

Bastian Blombach

Jörn Kalinowski

Raphael Frhr. v. Boeselager
Johanna Wiechert
Cornelia Gaetgens
Tino Polen

Thank you for your attention!

Federal Ministry of Education and Research

HELMHOLTZ ASSOCIATION

JÜLICH
FORSCHUNGSZENTRUM

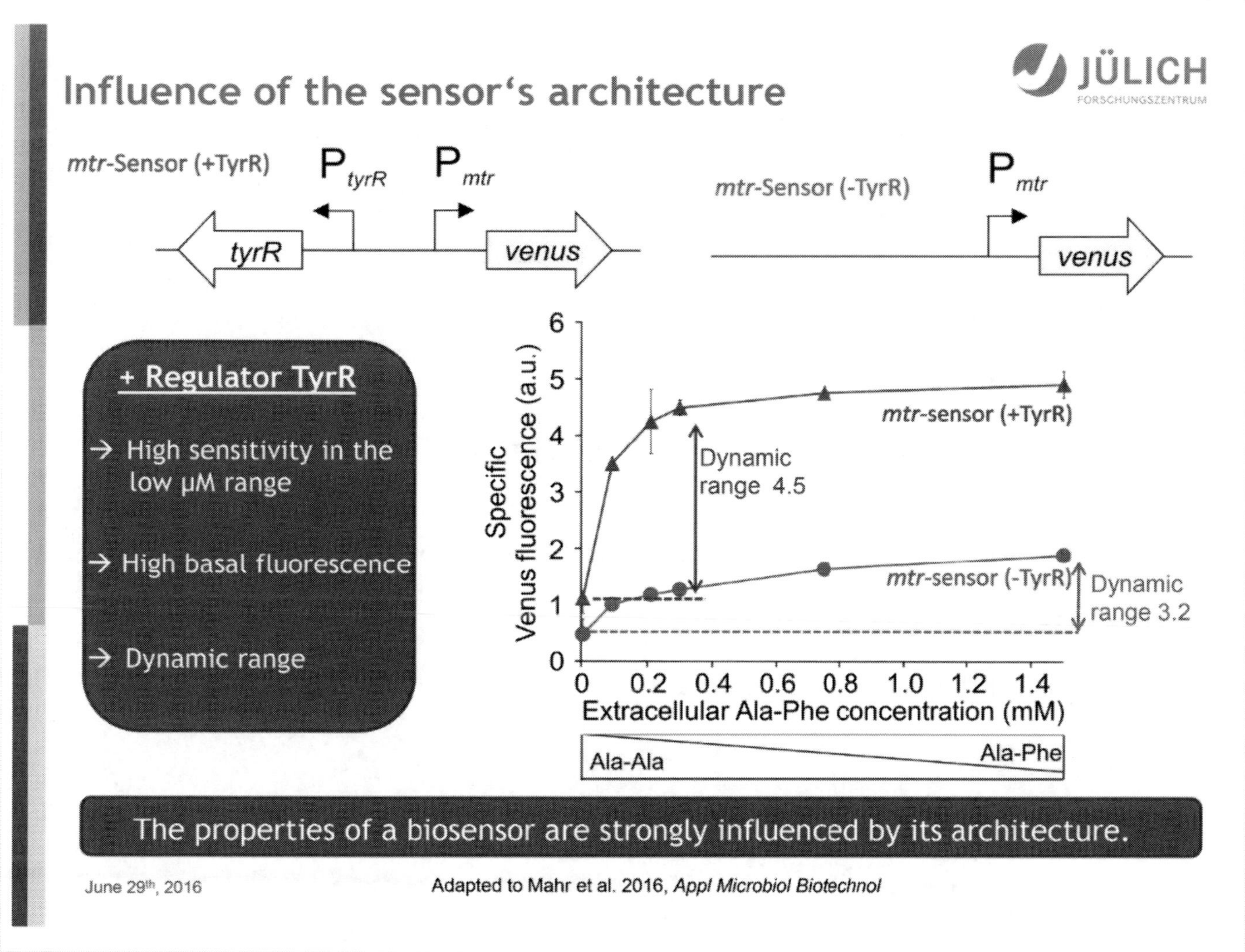
Influence of the sensor's architecture
JÜLICH
FORSCHUNGSZENTRUM
mtr-Sensor (+TyrR)
P tyrR
P mtr
tyrR
venus
mtr-Sensor (-TyrR)
P mtr
venus
+ Regulator TyrR
→ High sensitivity in the low µM range
→ High basal fluorescence
→ Dynamic range
Specific Venus fluorescence (a.u.)
6
5
4
3
2
1
0
mtr-sensor (+TyrR)
Dynamic range 4.5
mtr-sensor (-TyrR)
Dynamic range 3.2
0 0.2 0.4 0.6 0.8 1.0 1.2 1.4
Extracellular Ala-Phe concentration (mM)
Ala-Ala
Ala-Phe
The properties of a biosensor are strongly influenced by its architecture.
June 29th, 2016
Adapted to Mahr et al. 2016, Appl Microbiol Biotechnol

Developing a Commercial-Ready Production Microbe and Process - The 1,4-Butanediol Story

Nagarajan, H., University of California - San Diego
Burgard, A., Genomatica, Inc.
Originally presented on: 6/29/2016 15:50:00 - 16:10:00

Genomatica has established an integrated computational and experimental technology platform[1] to design, build, and optimize high-producing microorganisms and fermentation-based processes for the production of a broad variety of chemicals from renewable feedstocks[2, 3]. The platform enables the development of cost-effective, sustainable processes that will compete with and ultimately displace existing processes that rely on fossil fuel raw materials. Our first product, 1,4-butanediol (BDO)[4], is a 1.8 billion kg/year industrial chemical used in numerous applications including polymers such as spandex (PTMEG), poly(butyleneterephthalate) (PBT), and polyurethanes. This presentation will highlight the process-first approach that led to a final production microbe that produces BDO at a high titer, rate, and yield with minimal soluble byproducts to simplify and improve the economics of downstream recovery. Key recent advances include increasing ATP and NADPH availability, identifying the rate-limiting pathway enzyme, improving its activity and substrate specificity, rebalancing sugar uptake with increased pathway capacity, eliminating sources of genome instability, and developing a robust scalable fermentation process that accentuates the strengths of the final production microbe. To our knowledge, this is the first example of a successfully developed commercial-scale process (validated at scale within 6 years of project initiation) for a bio-based â??exact sameâ?□ commodity chemical that will compete directly with existing petro-based processes[5].

[1] Barton, et al., 2015, J Ind Microbiol Biotechnol, 42(3), 349-60
[2] Burk, 2010, Int Sugar J, 112(1333)
[3] Van Dien, 2013, Curr Opin Biotechnol, 24(6), 1061-1068
[4] Yim, et al., 2011, Nat Chem Biol, 7(7), 445-52
[5] Burgard, et al., 2016, Curr Opin Biotechnol, 42, 118-125

Contribution to Create KAITEKI Society Based on Bio-related Technologies

Mizunashi, W.,

Originally presented on: 6/29/2016 16:10:00 - 16:30:00

Chemical Industry has contributed human society and human life through the providing various bulk or functional chemicals to the industry of agriculture, medical, mobility, information communication, etc. However, we have fallen out of balance with the global environment, and are confronted by complex problems spanning from climate change to natural resource/energy depletion and an uneven distribution of food and water.

By contributing to resolving environmental and social issues, Mitsubishi Chemical Holdings Corporation (MCHC) will build a sustainable society together with stakeholders toward the realization of KAITEKI. KAITEKI means "a sustainable condition which is comfortable for people, society and the Earth, transcending time and generations." It is an original concept of the MCHC Group that proposes a way forward in the sustainable development of society and the planet, in addition to serving as a guide for solving environmental and social issues.

Today, I would discuss with you through the introduce our some works for contribution to create KAITEKI society based on Bio-related technology, such as bio-based chemicals from sustainable resources, environmental friendly production process.

Small Molecule Glycosylation – a Powerhouse in Synthetic Biology
Hansen, J.,
Originally presented on: 6/29/2016 16:30:00 - 16:50:00

Small molecule glycosyltransferases or UGTs are the "awesome office managers" of Nature, making sure everything gets made in the right way, gets to the right location and gets stored for easy retrieval when needed. UGT stands for "UDP-glucose dependent Glycosyl Transferase" and close to 10,000 such enzymes were annotated by the start of 2016 (www.cazy.org). Plants are particularly rich sources, each having from 100-300 UGTs, but most if not all living organisms have them. UGTs are employed for de-toxification before either excretion (animals) or storage (plants), for stabilization of unstable molecules (e.g. cyanogenic glycosides and anthocyanins) or for attaining precise functionality (e.g. rebaudioside sweeteners and saponins).

The de-toxification aspect is particularly interesting as this allows for biological production of molecules otherwise harmful to the producing organism but the often accompanying increase in watery solubility can be of equal importance. Obtaining the right functionality of molecules by complex glycosylation is of obvious importance but requires deep knowledge on UGT specificity. Evolva has for many years been developing a state-of-the-art UGT platform and used this for current or near-to-market products such as Vanillin and Stevia sweeteners, but also for a large number of pipeline products.

The use of small molecule glycosylation in current and future Evolva ingredients will be discussed as will a number of the detailed studies that led to the success of the UGT platform.

Hansen, E.H., Møller, B.L., Koch, G., Bünner, C.M., Kristensen, C., Jensen, O.R., Motawia, M.S., Olsen, C.E., Okkels, F.T., & Hansen, J. 2009. De novo biosynthesis of vanillin in fission yeast (*Schizosaccharomyces pombe*) and baker's yeast (*Saccharomyces cerevisiae*). Appl. Env. Microbiol. 75: 2765-2774.
Hansen, E.H., Osmani S.A., Kristensen, C., Møller, B.L. & Hansen, J. 2009. Substrate specificities of family 1 UGTs gained by domain swapping. Phytochemistry 70: 473-482.
Olsson, K., Simon Carlsen, S., Semmler, A., Simon, E., Mikkelsen, M., & Møller, B.L. Microbial production of next-generation *Stevia* sweeteners. Submitted for publication.

Development of Next Generation Yeast Strains for Ethanol Production from Lignocellulosic Feedstocks

de Waal, P., DSM Biotechnology Center, Alexander Fleminglaan 1, 2613AX, Delft, The Netherlands
Klaassen, P., DSM
Driessen, A. J. M.,
Pronk, J. T., Department of Biotechnology, Delft University of Technology, Julianalaan 67, 2628 BC Delft, The Netherlands
van Maris, T., Delft University of Technology
de Bruijn, H., DSM
van Rijsewijk, B. R. B. H., DSM Biotechnology Center
Originally presented on: 6/29/2016 16:50:00 - 17:10:00

DSM is a frontrunner in the development of bioconversion technologies for the production of lignocellulosic ethanol, being the first company to commercialize C5-fermenting *Saccharomyces cerevisiae* and proving this technology in pilot and commercial facilities worldwide. Since then, next generation advanced yeast strains have been developed, commercialized and produced at industrial scale, offering significantly accelerated C5-fermentation rates, improved robustness and an expanded substrate range to the market.

In taking fermentation of lignocellulosic biomass streams to the next level, DSM has established a unique IP and technology position. Breakthroughs in C5 uptake by sugar transporter engineering, by-product reduction and conversion of previously untouched carbon sources, such as acetic acid and biorefinery waste streams, into additional ethanol are currently finding their way to commercial implementation.

We will link these proprietary technologies to industrial needs through the development of commercial yeast strains.

Metabolic Engineering of Gas Fermenting Organisms for Commercial Scale Production of Fuels and Commodity Chemicals from Waste and Low Cost Resources

Köpke, M., LanzaTech, Inc.
Behrendorff, J. B. Y. H., LanzaTech, Inc.
Simpson, S. D., LanzaTech
Originally presented on: 6/29/2016 17:10:00 - 17:25:00

In December 2015, 195 countries adopted the Paris Agreement at the United Nations Climate Change Conference. The agreement "aims to strengthen the global response to the threat of climate change" and seeks to hold the increase in global average temperature to "well below 2°C above pre-industrial levels". Though non-binding, this underscores the urgency for action that will radically limit global greenhouse gas emissions. Specifically, recent data indicates that this will require leaving a third of oil reserves, half of gas reserves, and over 80% of coal reserves unused until 2050.

LanzaTech has developed novel technology that recycles carbon from local, highly abundant, waste and low cost resources. The process converts gas containing carbon monoxide or CO_2 (e.g. from industrial sources like steel mills or processing plants) or syngas generated from any biomass resource (e.g. municipal solid waste, organic industrial waste or agricultural waste) into a range of sustainable fuels and commodities. Such feedstock flexibility and product diversification are important parts of ensuring favorable economics for this breakthrough technology.

At the heart of the process is a gas fermenting, acetogenic microbe *Clostridium autoethanogenum* that is able to autotrophically grow on a range of C1 substrates. This exceptional feedstock flexibility relies on unique redox chemistry and energy conservation mechanisms that do not exist in standard hosts such as *E. coli* or yeast. LanzaTech has evolved a high performing chassis strain through extensive natural selection and developed a complete genetic toolbox for the organism, including genome editing tools and a comprehensive genetic parts library.

In order to maximize the value that can be added to the array of gas resources the process can use as an input, LanzaTech has developed several lines of metabolically engineered strains. This includes strains with optimized flux and tolerance for native products ethanol or 2,3-butanediol, as well as strains for synthesis of an array of non-native products via both natural and synthetic routes. Examples include production of acetone, isopropanol or butanol via combinatorial testing and optimization of genes from the commercial ABE fermentation process. Yields and production rates achieved on gas have exceeded those of native producers utilizing sugars or published examples of *E. coli* or yeast engineered for these products. In addition, the company is working on novel synthetic routes to products including 1,3-butanediol, butadiene and isobutylene. Effective engineering of gas fermenting organisms requires a deep understanding of energy and redox metabolism which dictates the product spectrum and is vastly different between gas and sugar fermenting organisms. To guide the metabolic engineering efforts, we have developed a systems biology platform and a genome-scale model that has been validated against hundreds of fermentation runs.

Scalable reactor designs and optimized process chemistry ensure efficient, continuous, single-pass gas conversion. The process has been successfully scaled-up to fully integrated 100,000-gallon/year demonstration plants with over 40,000 hours on stream. Such diverse gas streams as by-product gases from steel making and syngas produced from gasified biomass or municipal solid waste. Currently, LanzaTech is constructing its first commercial plants in Belgium, China and Taiwan.

Production of the Human Milk Oligosaccharide 2'-O-Fucosyllactose Using *E. coli* As a Bacterial Cell Factory

GLYCOM

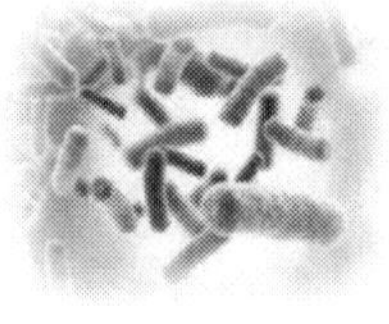

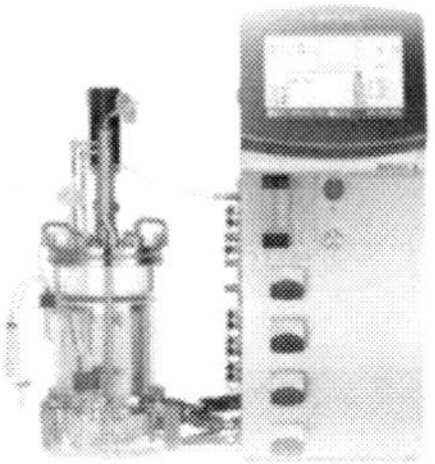

Metabolic Engineering 11
An IMES conference
Wednesday 29th of May 2016

Katrine Bych
Head of Strain Development, Ph.D.
Glycom A/S

Glycom A/S and Glycom Manufacturing: From Chemistry to Biotechnology

GLYCOM

- Privately owned company based in Copenhagen, Denmark
  - Operations in Denmark
  - Founded in 2005, Manufacturing facility acquired late 2016
  - Nestlé SA a major investor
- About + 90 employees
  - Of which about 75% are scientists and technologists
- Focused on synthesis and commercialisation of human milk sugars:
  - Especially oligosaccharides (HMOs)
  - Single HMO products and HMO-blends
- Successfully developed different HMO technology platforms
  - Production of HMOs in high titers and low cost

Human milk

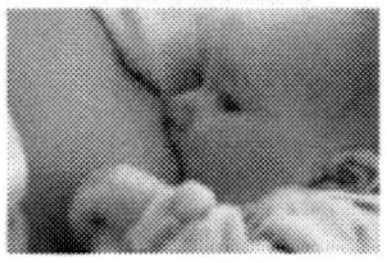

HMOs ⇨

Infant nutrition formulas

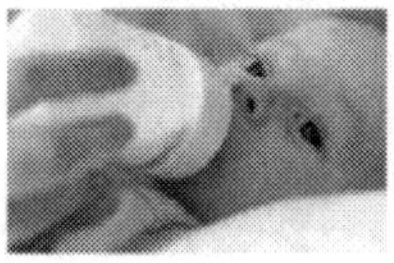

Facts about human milk oligosaccharides (HMOs)

GLYCOM

- HMOs are the 3rd largest solid component in human milk
- HMOs are highly abundant in and unique to human milk

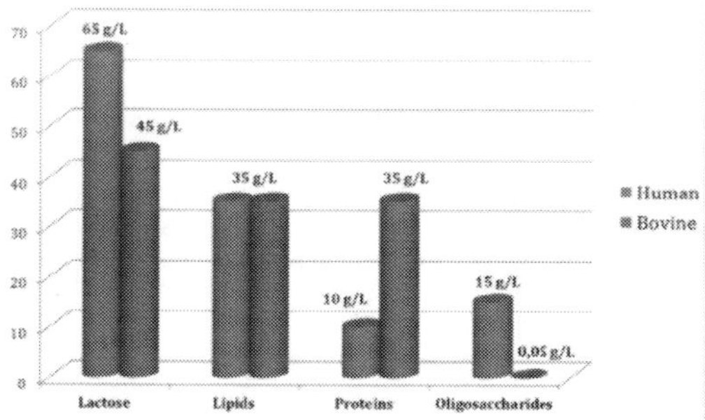

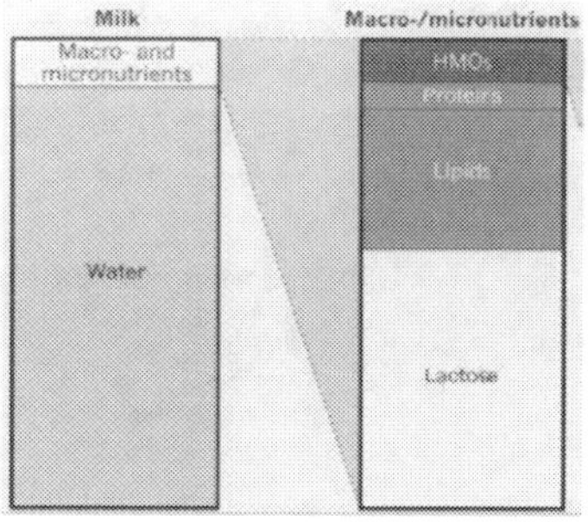

Component	Concentration (g/L)
Oligosaccharides	**~ 10 – 23**
-- colostrum	20 – 23
-- mature milk	10 – 15
Proteins	**~ 8**
Lipids	**~ 38**
Lactose	**~ 70**

Facts about human milk oligosaccharides (HMOs)

GLYCOM

- >200 structurally different (unique) HMOs – Structure determines biological function of the HMO
- 12 most abundant HMOs ~75% of total

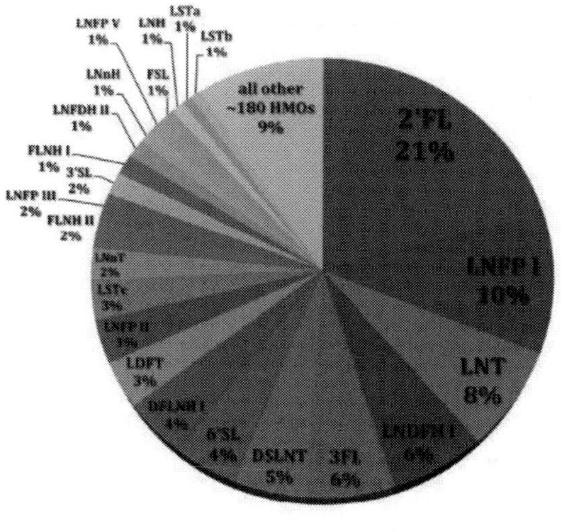

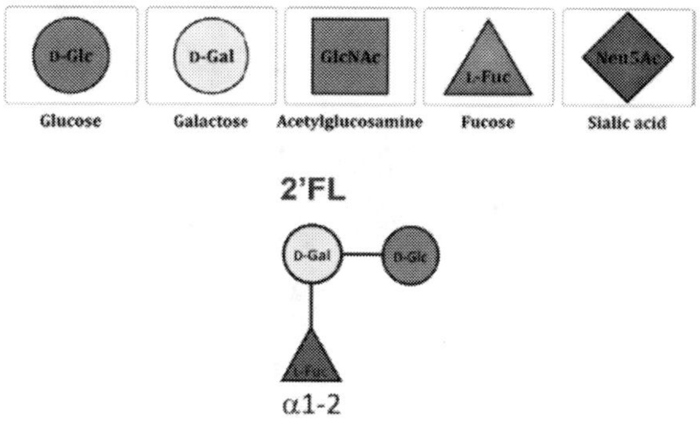

Functions of HMOs – Health benefits of breastfeeding

GLYCOM

- Biology reasonably undeveloped due to **lack of availability** and **high costs**
  - Primarily comparisons between breast fed and non-breast fed infants
  - Preclinical studies

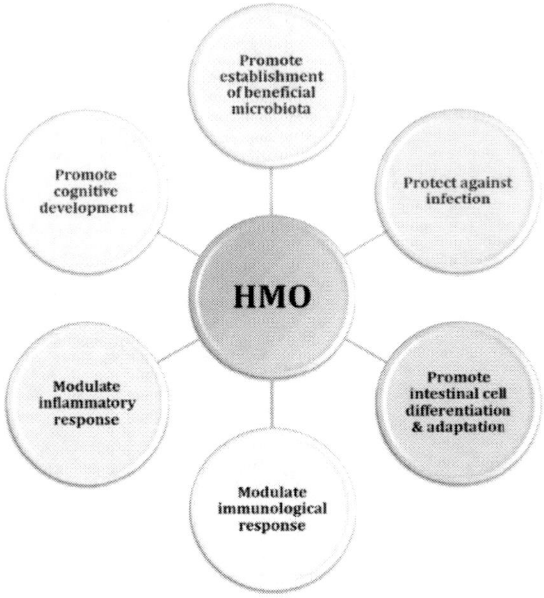

Our technology – Biochemical synthesis platform

GLYCOM

Development of bacterial cell factories for HMO production

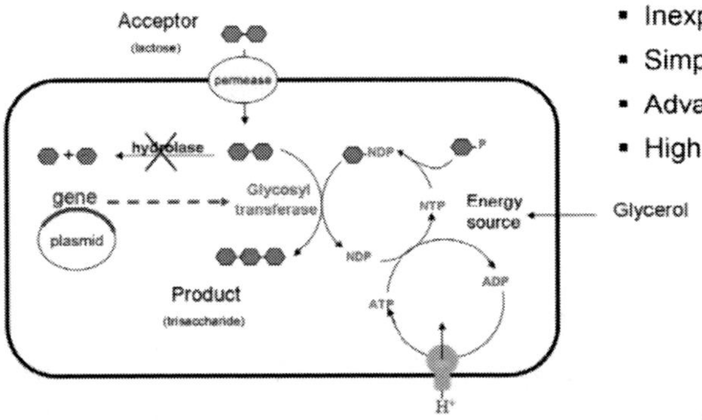

- Inexpensive energy sources and feed saccharides
- Simple fermentation process
- Advanced downstream processing strategies
- High yields and high purity at low production costs

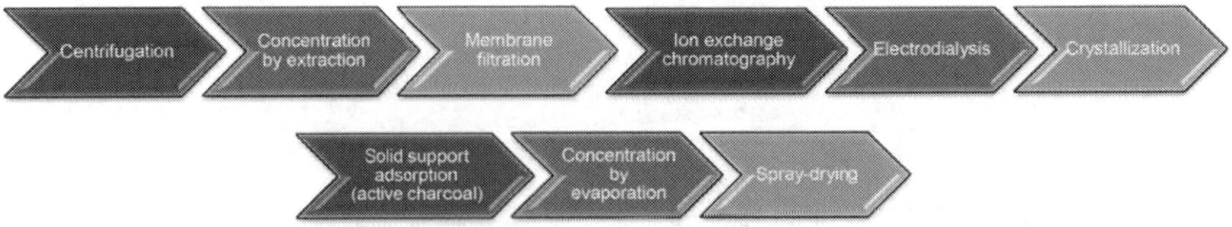

Designing bacterial cell factories: Metabolic engineering and systems biology

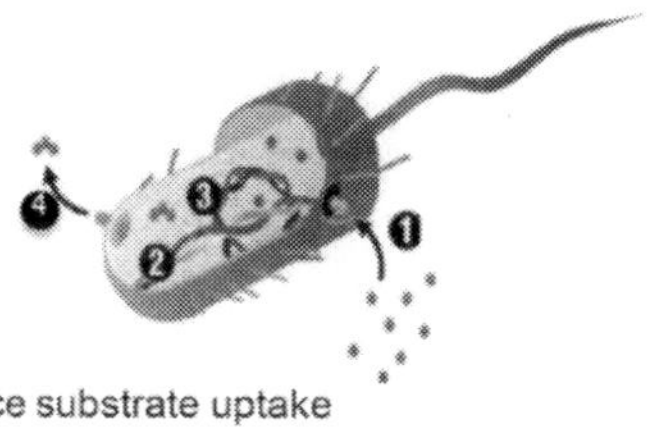

Systems biology toolbox
-- Genomics
-- Transcriptomics
-- Metabolome analyses

1. Enhance substrate uptake
2. Flux reduction or enhancement
3. Heterologous pathway expression and optimization – Enzyme engineering
4. Enhance product export

Gene targeting, Recombineering

Heterologous gene expression optimization

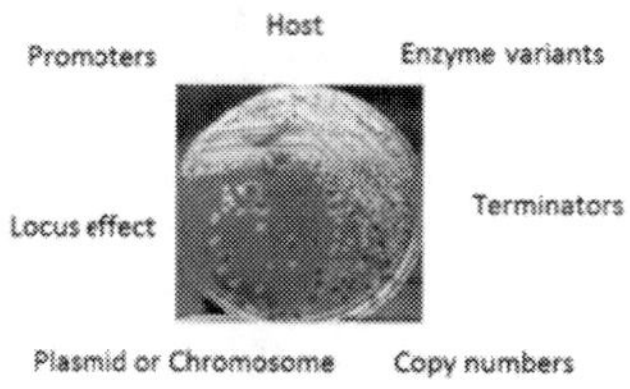

Antibiotic-free plasmid systems

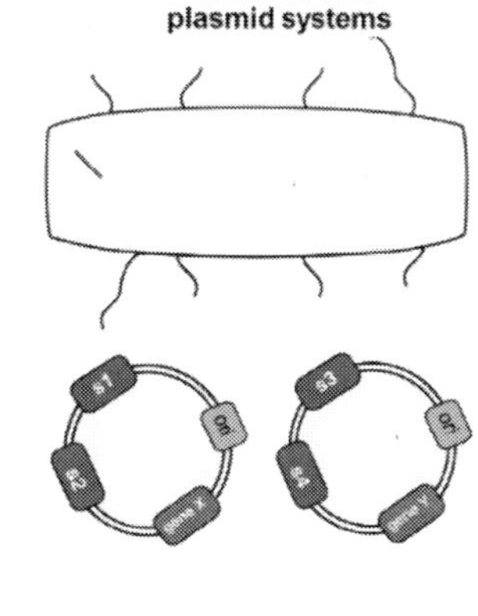

Large-scale production of fucosyllactose (2'FL) in *E. coli*

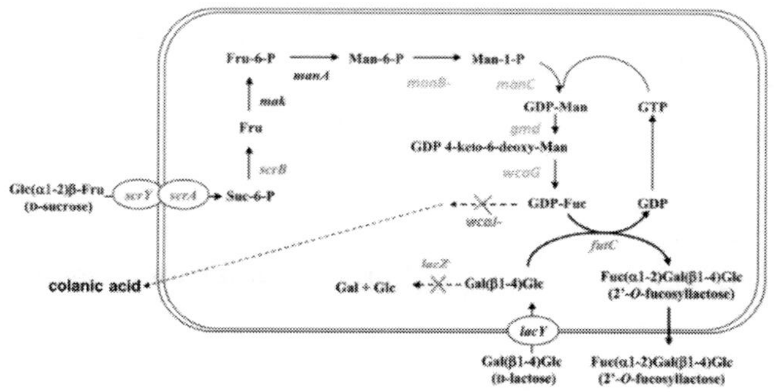

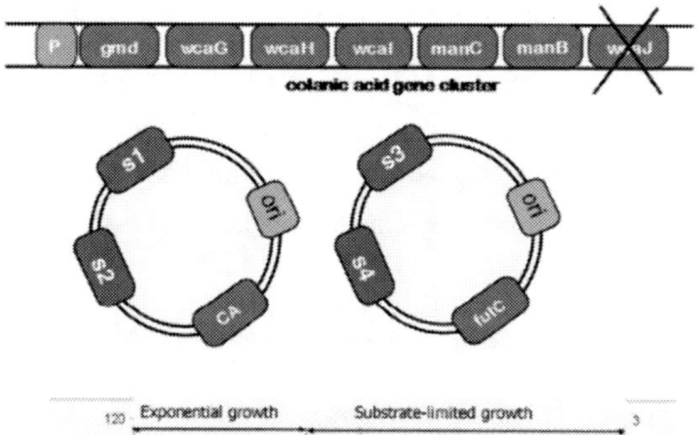

Strain engineering – 2-plasmid system
- ❑ Number of host modifications - suitability for HMO production
- ❑ Overexpression of endogenous genes involved in GDP-fucose synthesis
- ❑ Expression of a-1,2-fucosyltransferase gene (FutC) from *Helicobacter pylori*

Simple fermentation process
- ❑ In high cell density cultures, with lactose feed
- ❑ High production yields

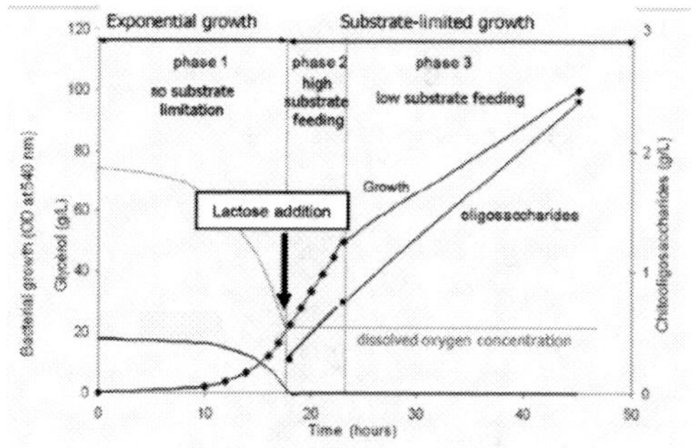

Wang et al, 1999, Microbiology 145(11):3245-3253

Our pipeline

- ❑ Our lead compounds are
  - 2'-Fucosyllactose
  - Lacto-*N*-neotetraose
- ❑ Both have been produced on multi-100 kg scale
  - Under GMP conditions
- ❑ Safety regulatory is well advanced for both
  - Tox studies completed and published
  - Novel food and GRAS processes well advanced
- ❑ Both have been used in recently completed clinical studies – **Excellent tolerance** and **potential health benefits**

-- Nestlé Nutrition **growth and safety** study in infants

-- Glycom **tolerance dose ranging** study in healthy adults

2'FL

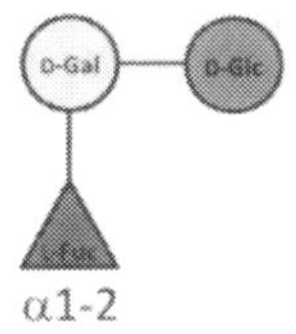

Lacto-*N*-tetraose

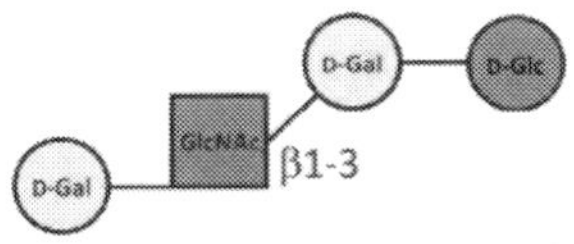

From science to application: Glycom's paradigm

❑ Leader in the microbial production of HMOs on large scale and at commercially viable cost

1. Conversion of *E. coli* into a beneficial cell factory for the **large-scale production** of HMOs (biotechnology)
2. Advanced methodologies for the **isolation** and **purification** of complex compounds (chemistry)

❑ Provides **evidence** confirming benefits of HMOs to human health – Innovative solutions for **infant** and **adult nutrition**

❑ Encourages widespread investigations on biological significance of HMOs and precursors – Promotion and support of **research** in glycobiology and HMO biology

❑ Lead compounds are well on the way to the market and other compounds are entering the pilot plant stage of development

Plenty of pure and safe HMOs for babies

Thank you for your attention !

Metabolic Engineering of E. coli for the Production of Non-Natural Polymers

Lee, S. Y., Korea Advanced Institute of Science and Technology (KAIST)

Originally presented on: 6/30/2016 8:00:00 - 8:25:00

Polyhydroxyalkanoates are natural polyesters accumulated by numerous microorganisms. The key enzymes involved are an enzyme that generates hydroxyalkanoyl-CoAs and a synthase that polymerizes hydroxyalkanoyl-CoAs. Non-natural polymers such as poly(lactic acid) and other lactate-containing polymers are of great interest, but they cannot synthesized by natural microorganisms. In this lecture, I will describe our recent works on metabolic engineering of E. coli for the production of various lactate-containing polymers by one-step fermentation. Strategies for establishing new pathways and merging fluxes coming from different sources will be described.

Yeast Cell Factories for Chemicals

Irina Borodina, Senior Scientist, Group Leader

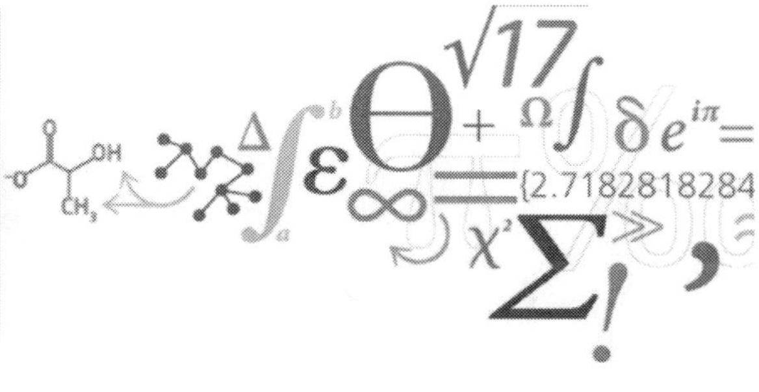

DTU Biosustain
The Novo Nordisk Foundation Center for Biosustainability

Biotech sector growth

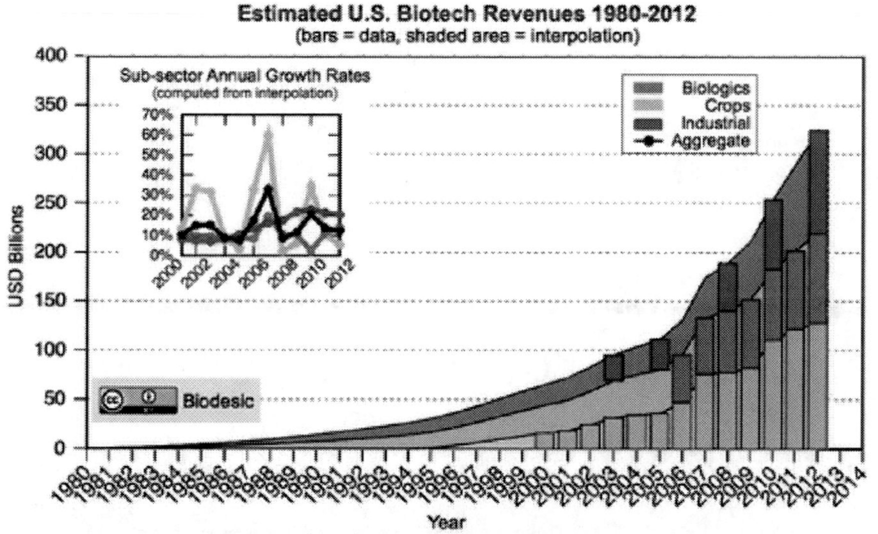

As published in R. Carlson, Nature Biotechnology, March 2016. Source data, figure, and Octave code available at www.biodesic.com

Carlson R (2016). "Estimating the biotech sector's contribution to the US economy". *Nat Biotechnol*, 34:247–255.

Creating a cell factory

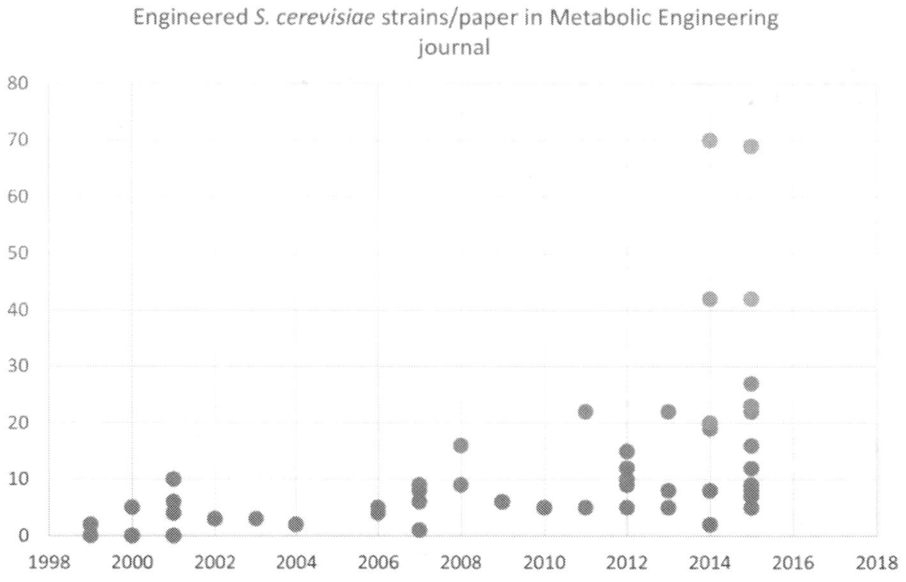

* Not complete

EasyClone – for iterative stable integration of multiple genes in *S. cerevisiae*

Editor's Choice: Ian Dawes

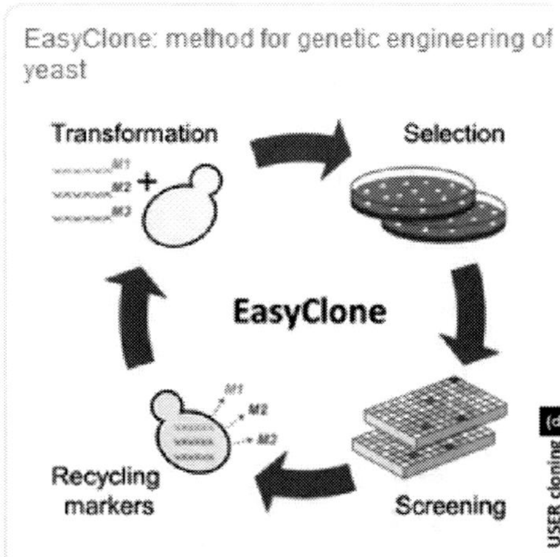

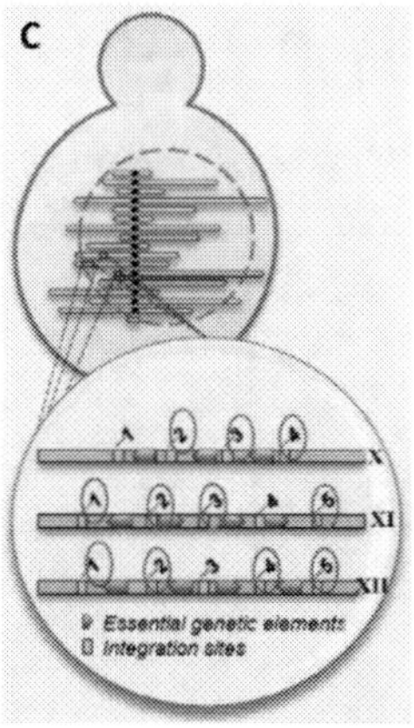

Niels Bjerg Jensen

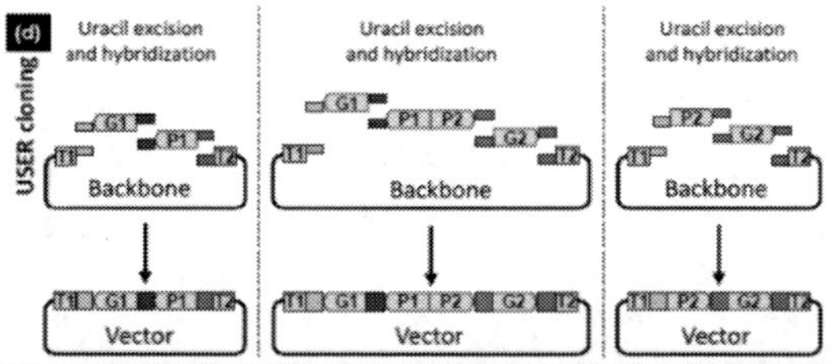

Jensen et al. (2014) "EasyClone: method for iterative chromosomal integration of multiple genes in *Saccharomyces cerevisiae*". *FEMS Yeast Res* 14:238-248.

EasyClone ensures high expression level of multiple genes and low population heterogeneity

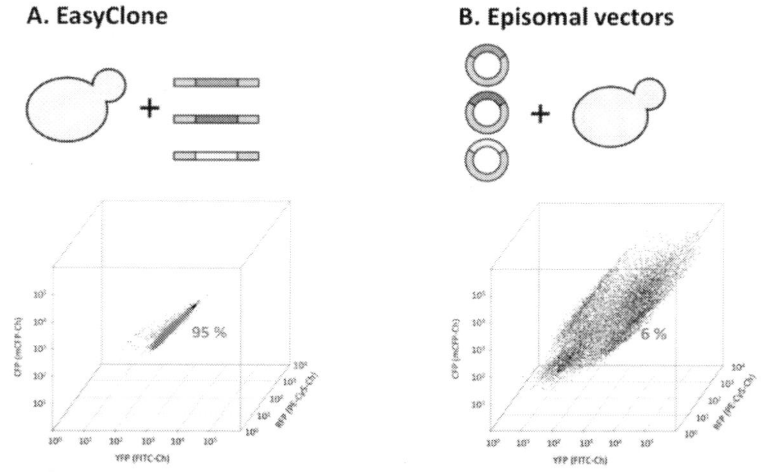

Jensen et al. (2014) "EasyClone: method for iterative chromosomal integration of multiple genes in *Saccharomyces cerevisiae*". *FEMS Yeast Res* 14:238-248.

EasyCloneMulti – for multi-genomic insertions

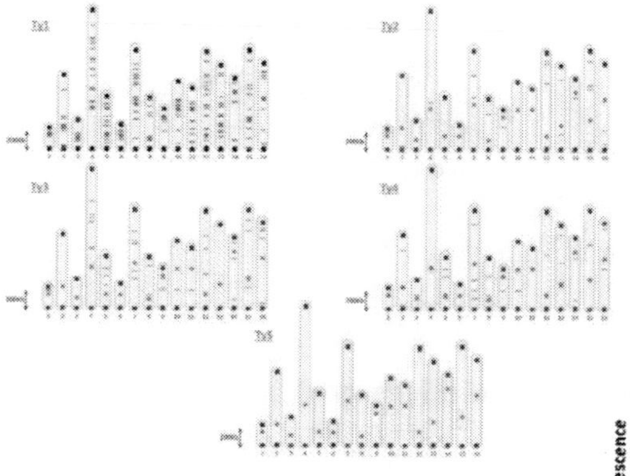

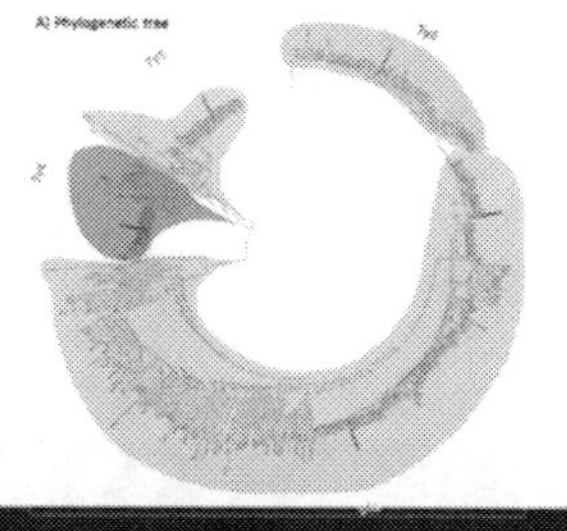

Jérôme Maury

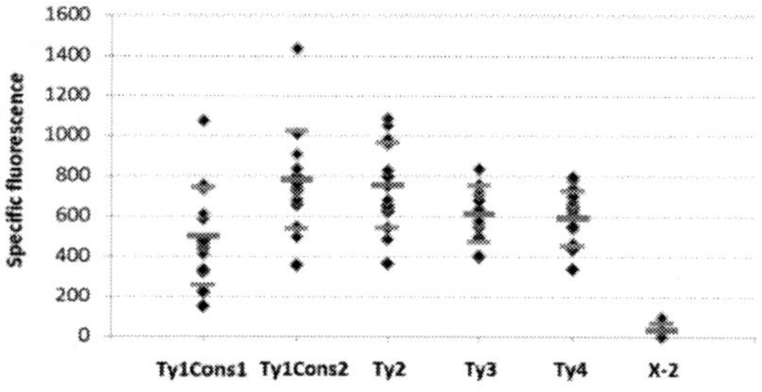

Maury et al (2016) "EasyCloneMulti: A Solution For Simultaneous Multi-Genomic Insertions of Expression Cassettes in *Saccharomyces cerevisiae*". *PLOS One*, 11(3):e0150394.

EasyClone 2.0 – for engineering industrial strains

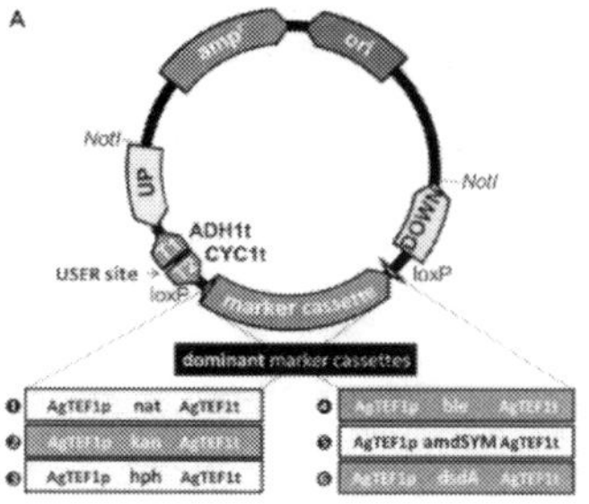

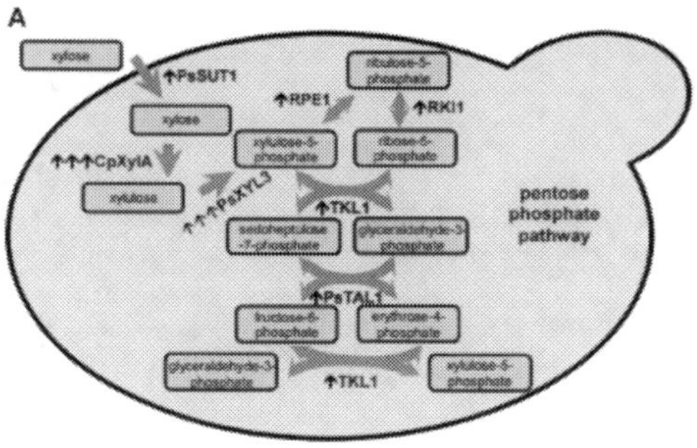

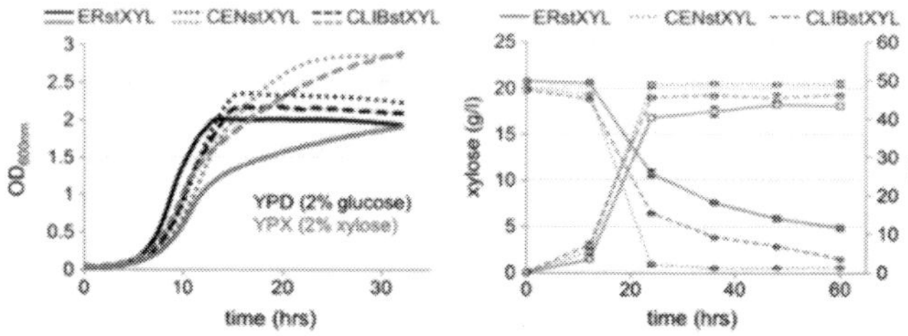

Stovicek et al. "EasyClone 2.0: Expanded toolkit of integrative vectors for stable gene expression in industrial *S. cerevisiae* strains". *J Ind Microbiol Biotechnol*, 1:13.

CRISPR-Cas9-mediated knock-ins/knock-outs in laboratory and industrial *S. cerevisiae*

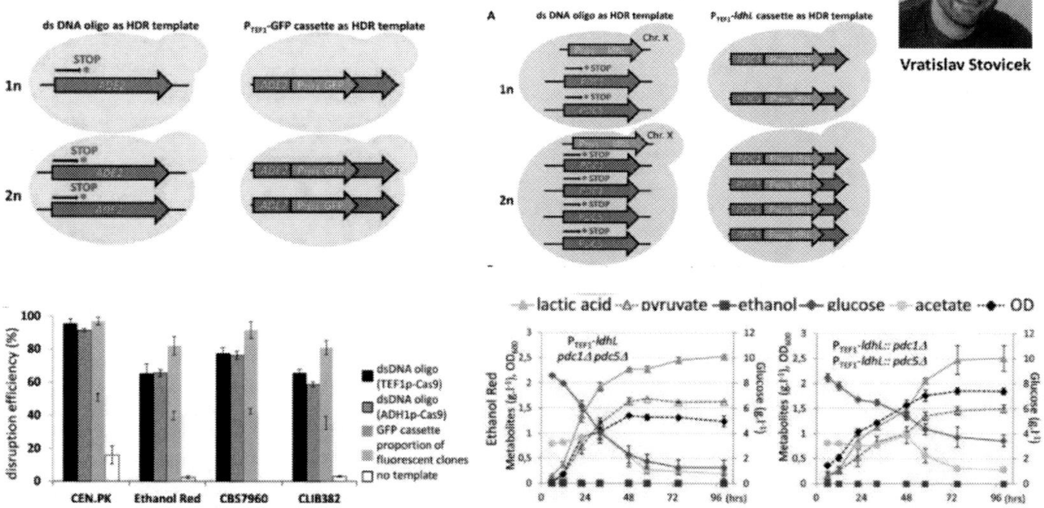

Stovicek et al. (2015) "CRISPR-Cas system enables fast and simple genome editing of industrial *Saccharomyces cerevisiae* strains". *Metab Eng Commun* 2:13-22.

EasyClone-MarkerFree via CRISPR-Cas9

DTU

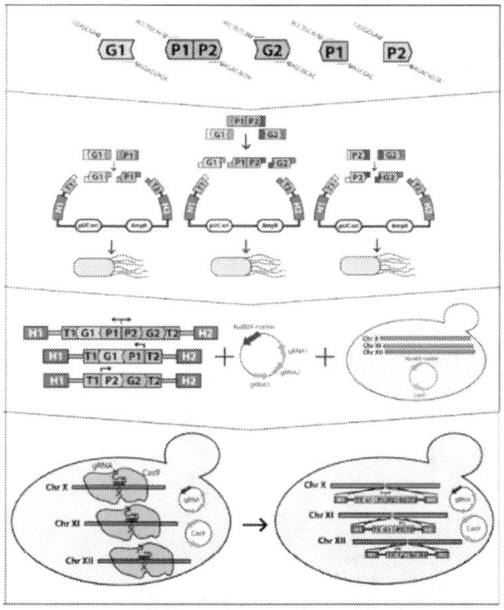

Mathew MJ Fabre

Tadas Jakočiūnas

95-100% integration efficiency

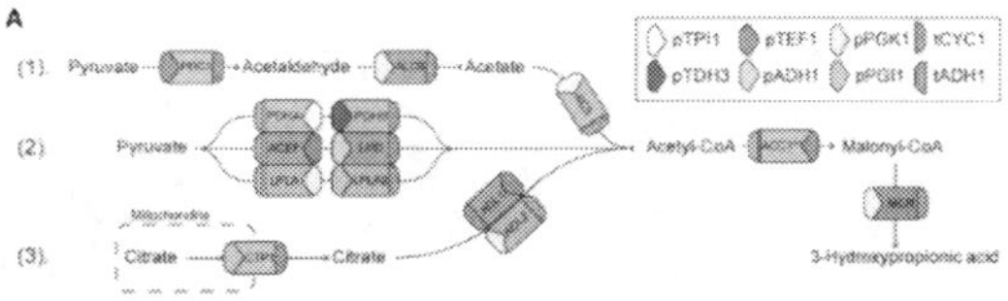

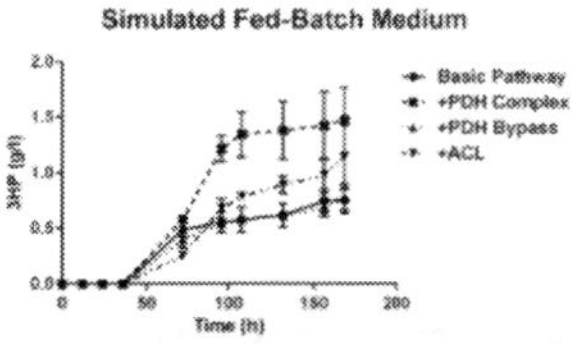

Fabre & Jakočiūnas (2016) "EasyClone-MarkerFree: A vector toolkit for marker-less integration of genes into *Saccharomyces cerevisiae* via CRISPR-Cas9", *Biotechnol J*, in press.

Multiplex metabolic pathway engineering using CRISPR/Cas9

DTU

Tadas Jakočiūnas

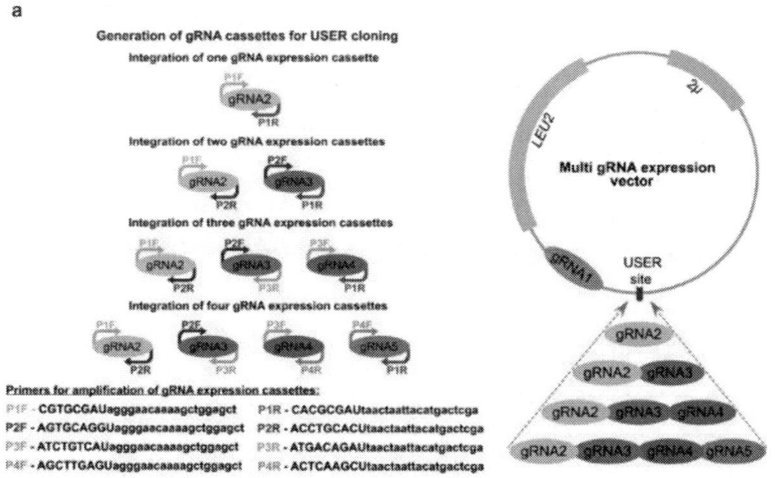

Jakočiūnas T et al. (2015) "Multiplex metabolic pathway engineering using CRISPR/Cas9 in *Saccharomyces cerevisiae*". *Metab Eng*, 28:213-222

CfB robotic platform

Screening robot

Cloning robot

2 x ALE robots

Cultivation robot

Applications: DIY yeast

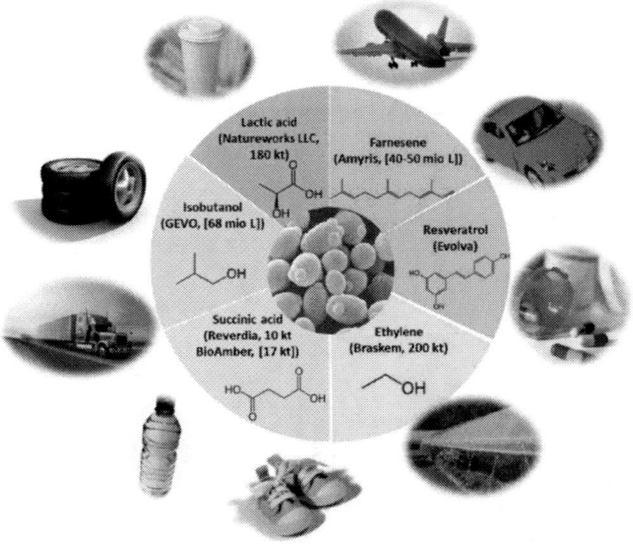

" FRANK IS INTO D.I.Y. ... 'DESTROY IT YOURSELF!' "

Borodina I & Nielsen J (2014) "Advances in metabolic engineering of yeast *Saccharomyces cerevisiae* for production of chemicals". *Biotechnol J* 9(5):609-20.

Engineering yeast for sustainable acrylates

Acrylic acid market: $11.5 billion
Annual growth rate: 4.7%

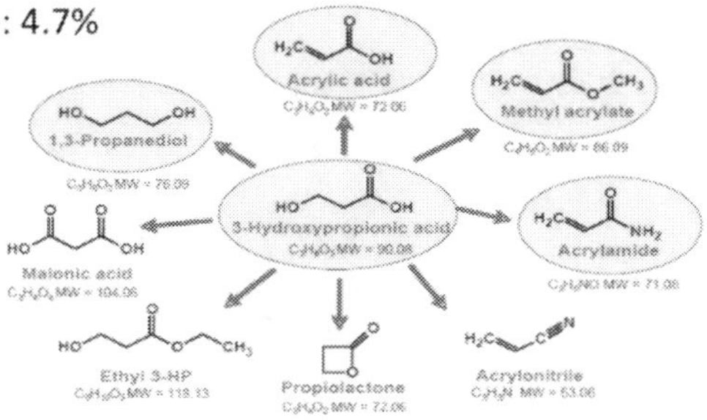

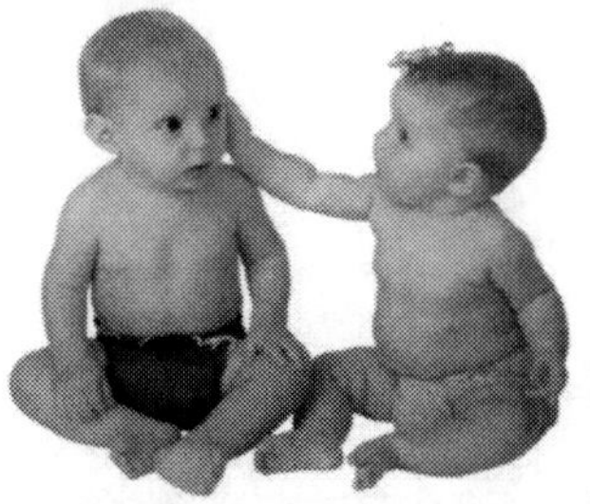

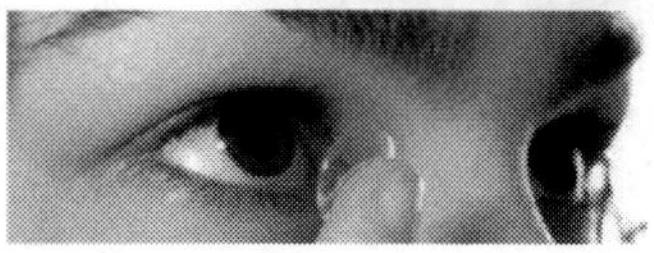

Metabolic engineering team

novonordiskfonden

ENGINEERED YEAST YIELDS SUSTAINABLE PLASTIC SUBSTITUTE

FRIDAY, 20 FEBRUARY 2015

Pathways towards 3HP

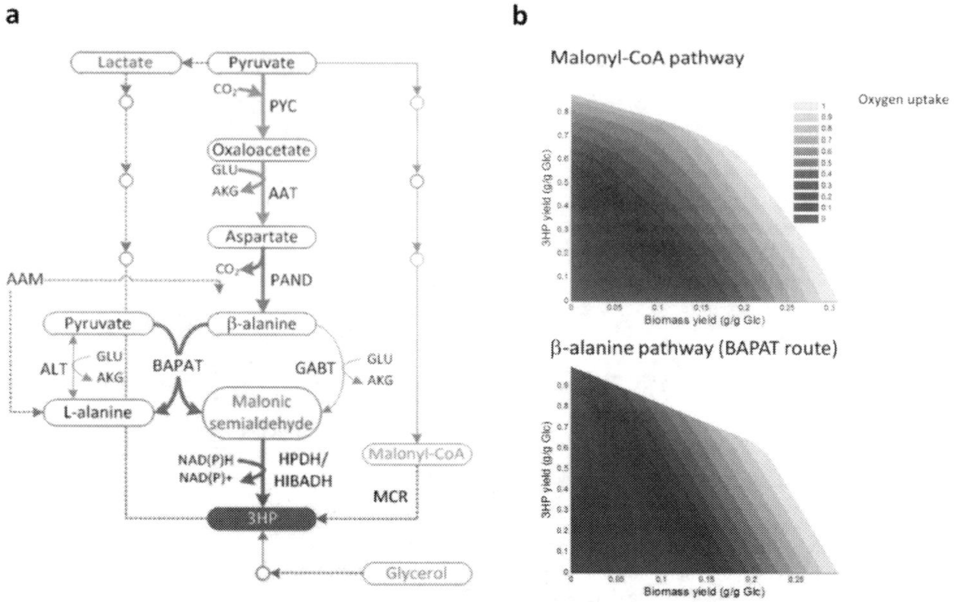

Borodina et al. (2015). "Establishing a synthetic pathway for high-level production of 3-hydroxypropionic acid in *Saccharomyces cerevisiae* via β-alanine". *Metab Eng*, 27:57-64.

Malonyl-CoA pathway to 3HP: optimization of gene expression and precursor supply

Overexpressed genes	
$ACC1^{Ser659, Ser1157}$	*CaMCR*
episomal plasmid	
integrative plasmid	
multiple integrative plasmid	

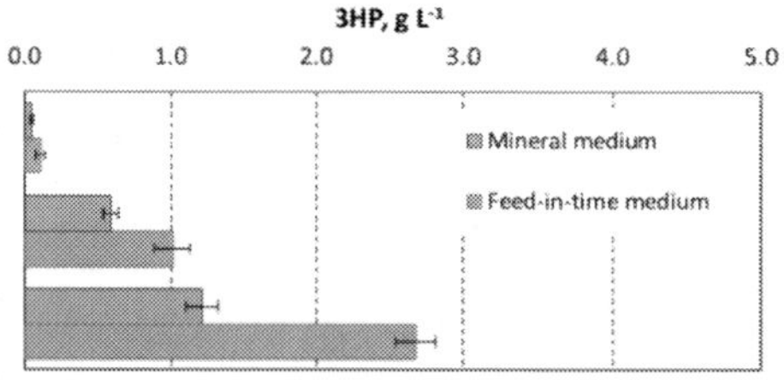

Overexpressed genes			
ACC1**↑↑	CaMCR↑↑	ACS_{SE}-ALD6	PDC1
+	+		
+	+	+	
+	+	+	+

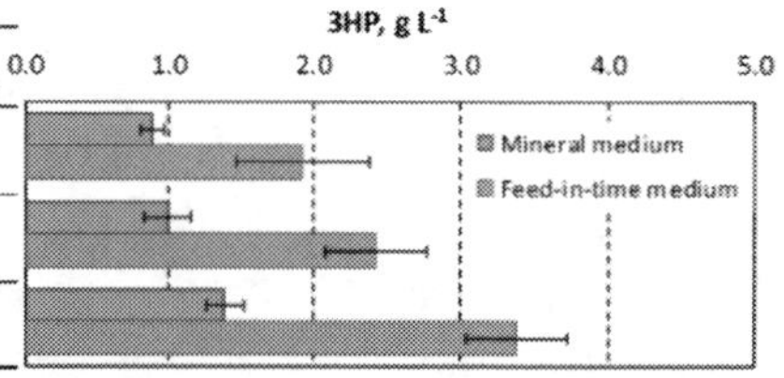

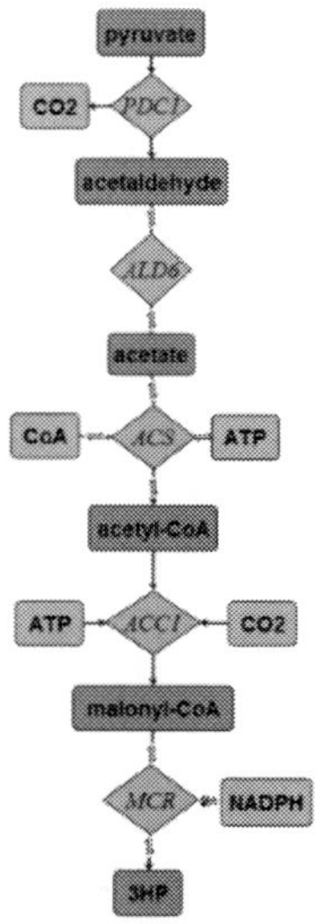

Kildegaard, Jensen et al (2016) "Engineering of *Saccharomyces cerevisiae* for production of 3-hydroxypropionic acid via malonyl-CoA reductase-dependent pathway". *Microb Ce*
Fact 15:53. Jensen et al. "Microbial production of 3-hydroxypropionic acid". WO2014198831 (A1).

Optimization of redox co-factor supply

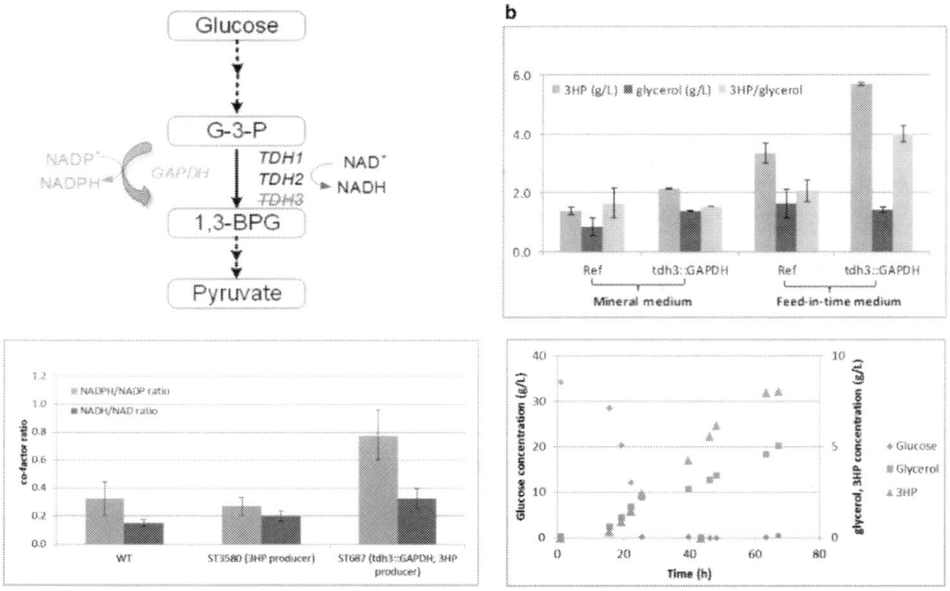

Kildegaard, Jensen et al (2016) "Engineering and systems-level analysis of *Saccharomyces cerevisiae* for production of 3-hydroxypropionic acid via malonyl-CoA reductase-dependent pathway". *Microb Cell Fact* 15:53. Jensen et al. "Microbial production of 3-hydroxypropionic acid". WO2014198831 (A1).

Synthetic pathway via β-alanine

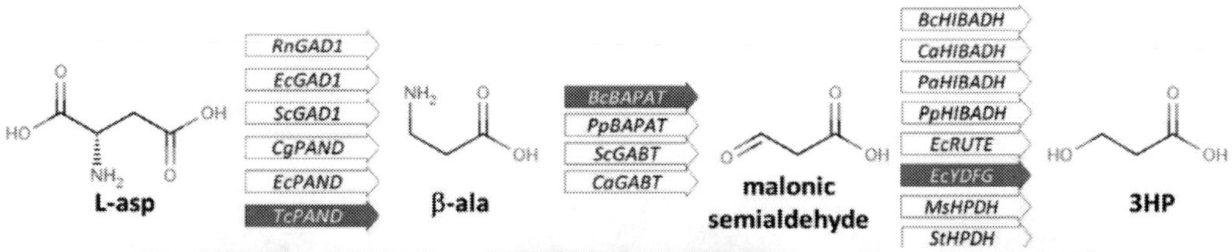

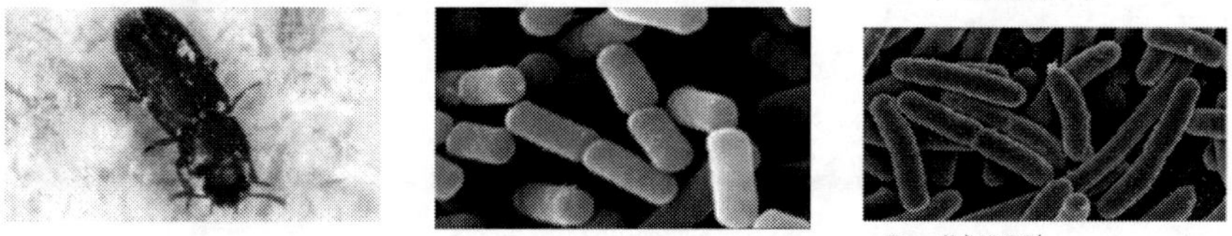

Tribolium castaneum (PAND) | *Bacillus cereus* (BAPAT) | *E. coli* (YDFG)

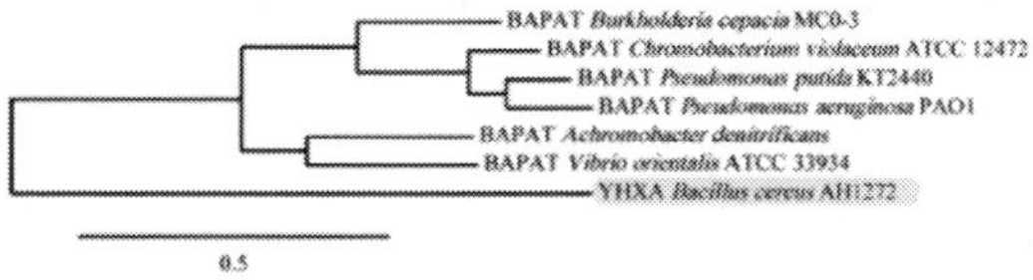

Borodina et al. (2015). "Establishing a synthetic pathway for high-level production of 3-hydroxypropionic acid in *Saccharomyces cerevisiae* via β-alanine". *Metab Eng*, 27:57-64. Borodina et al. "Genetically engineered yeast". WO2014057036 (A1).

β-alanine pathway to 3HP: optimization of gene expression and precursor supply

Strain name	Overexpressed genes				
	BcBAPAT	EcYDFG	TcPAND	ScAAT2	ScPYC1/PYC2
SCE-R2-188	+	+	+		
SCE-R2-180	+	+	+	+	
SCE-R2-204	+	+	+		+
SCE-R2-196	+	+	+	+	+
SCE-R2-190	+++	+++	+		
SCE-R2-182	+++	+++	+	+	
SCE-R2-206	+++	+++	+		+
SCE-R2-198	+++	+++	+	+	+
SCE-R2-192	+	+	+++		
SCE-R2-184	+	+	+++	+	
SCE-R2-208	+	+	+++		+
SCE-R2-200	+	+	+++	+	+

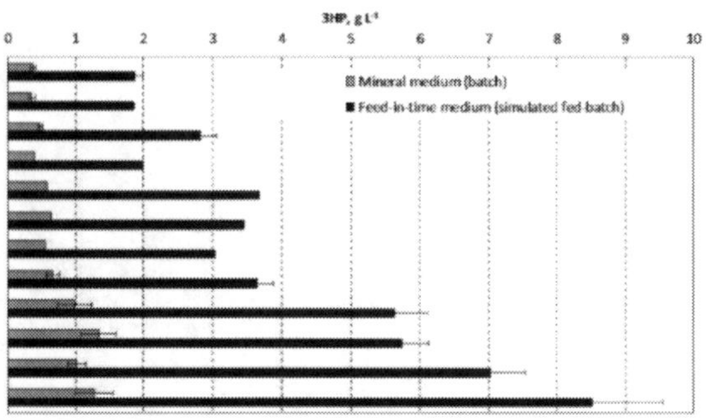

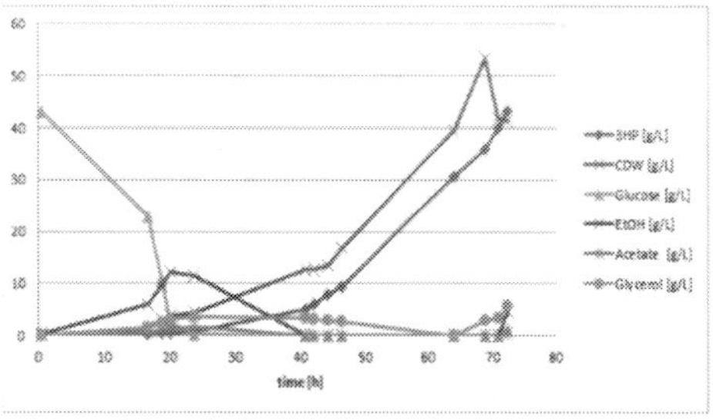

Titer	43.4 g/L
Rate (FB)	0.82 g/L/h
Yield	14% C-mol/C-mol

Borodina et al. (2015). "Establishing a synthetic pathway for high-level production of 3-hydroxypropionic acid in *Saccharomyces cerevisiae* via β-alanine". *Metab Eng*, 27:57-64. Borodina et al. "Genetically engineered yeast". WO2014057036 (A1). Schneider (*unpublished*).

3HP production from xylose

Kanchana R. Kildegaard

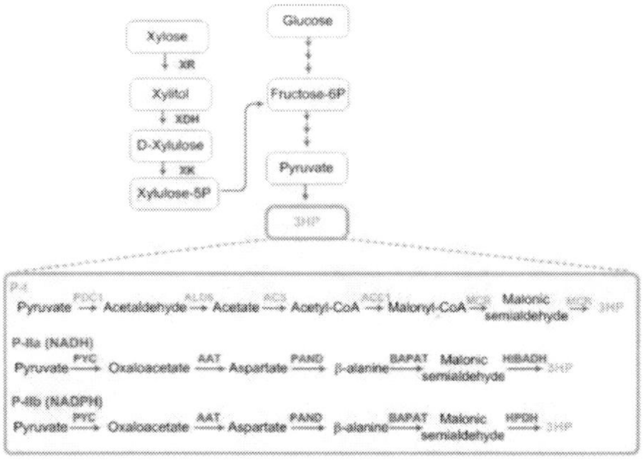

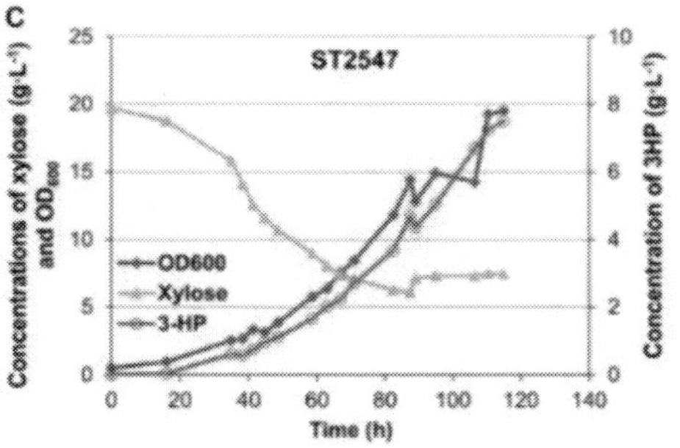

Kildegaard KR, Wang Z, Chen Y, Nielsen J, Borodina I (2015) "Production of 3-hydroxypropionic acid from glucose and xylose by metabolically engineered *Saccharomyces cerevisiae*". *Metab Eng Commun*, 2:132-136.

Adaptive Laboratory Evolution for 3HP tolerance

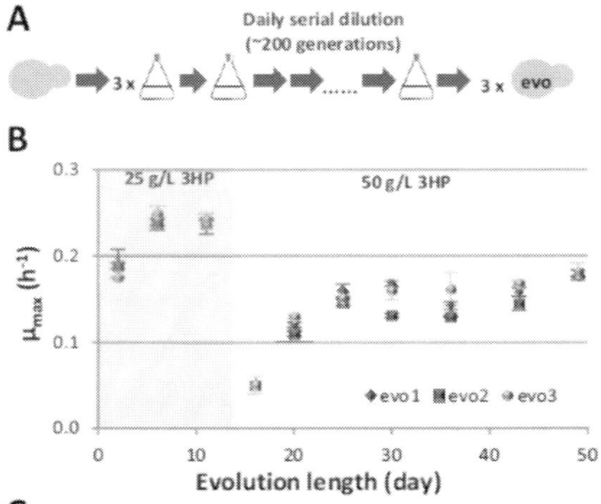

Kanchana R. Kildegaard

C

Gene	Strain	d2	d6	d11	d16	d20	d25	d30	d36	d43	d49	Mutation
SFA1	evo1	G	G	G	Gc	gC	gC	C	C	C	C	Cys276Ser
	evo2	G	G	G	G	GA	gA	A	A	A	A	Met283Ile
	evo3	G	G c	Gc	gC	C	C	C	C	C	C	Cys276Ser

Kildegaard et al. (2014) "Evolution reveals a glutathione-dependent mechanism of 3-hydroxypropionic acid tolerance". *Metab Eng*, 26C:57-66.
Kildergaard et al. "3HP tolerance". WO2014198841(A1).

3HP tolerance is dependent on *SFA1* and glutathione

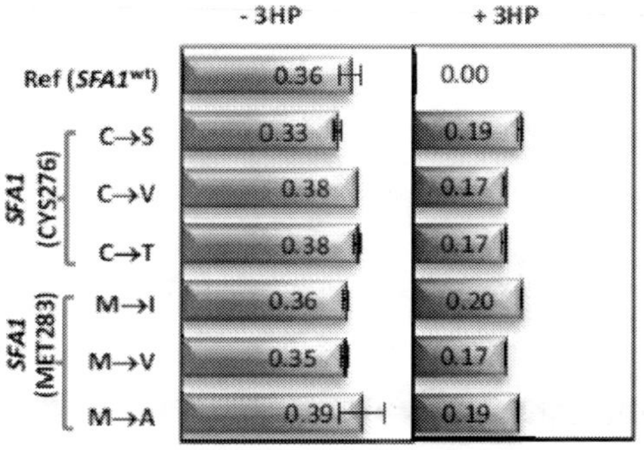

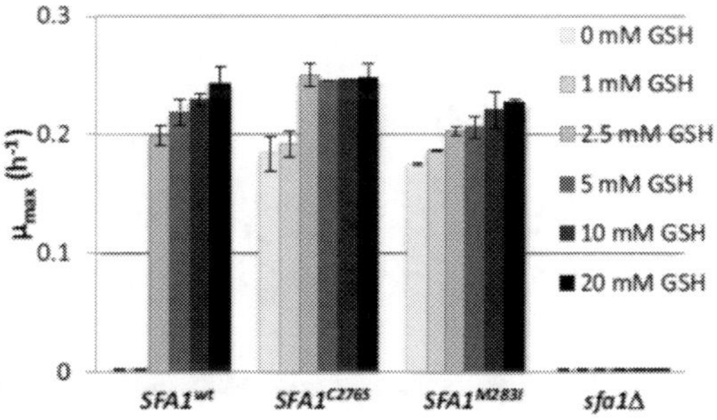

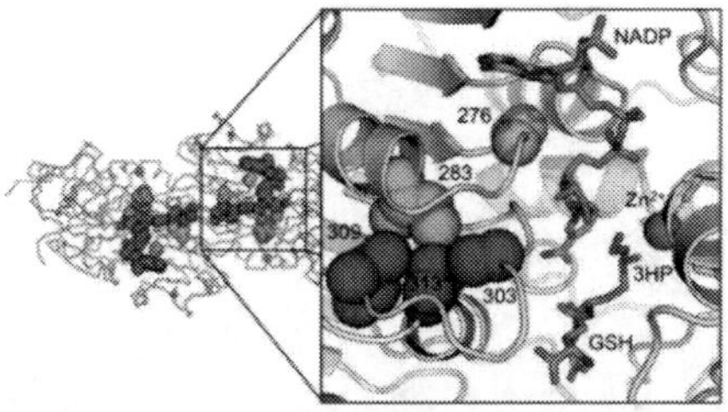

SFA1p
formaldehyde dehydrogenase

Kildegaard et al. (2014) "Evolution reveals a glutathione-dependent mechanism of 3-hydroxypropionic acid tolerance". *Metab Eng*, 26C:57-66.
Kildergaard et al. "3HP tolerance". WO2014198841(A1).

Effect of glutathione on 3HP tolerance in yeast and *E. coli*

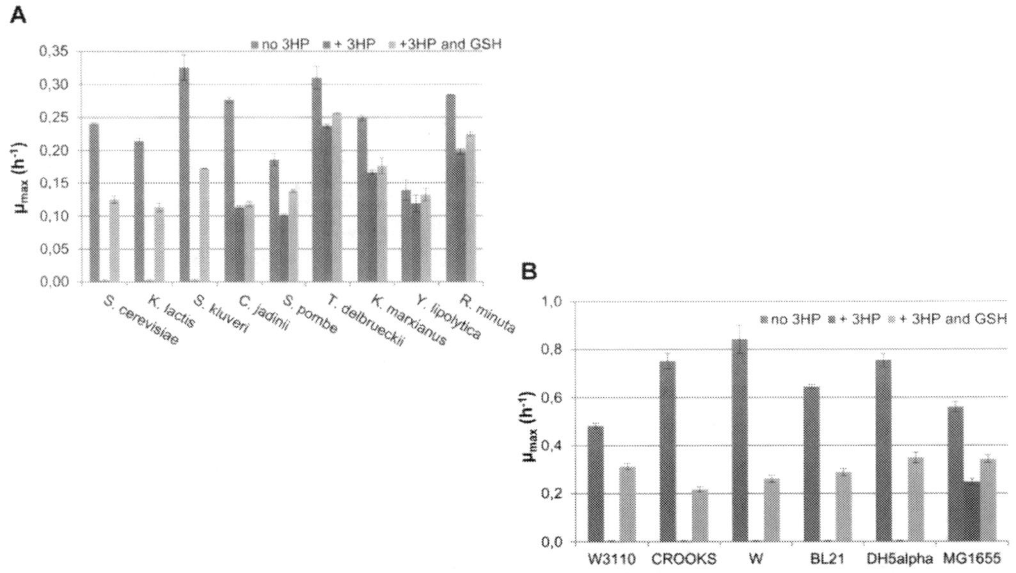

Kildegaard et al. (2014) "Evolution reveals a glutathione-dependent mechanism of 3-hydroxypropionic acid tolerance". *Metab Eng*, 26C:57-66.
Kildergaard et al. "3HP tolerance". WO2014198841(A1).

Proposed mechanism for 3HP toxicity and tolerance

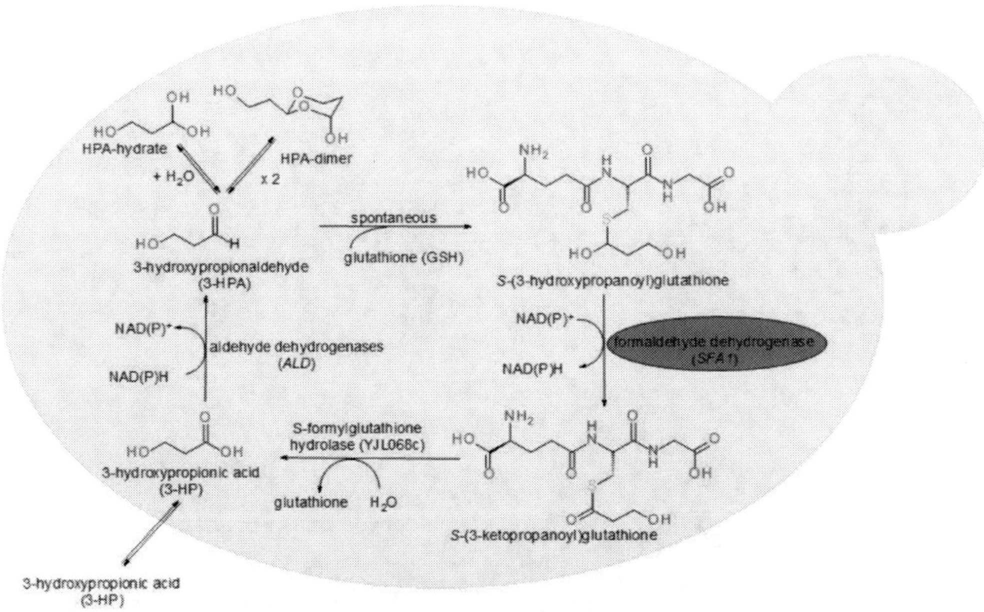

Kildegaard et al. (2014) "Evolution reveals a glutathione-dependent mechanism of 3-hydroxypropionic acid tolerance". *Metab Eng*, 26C:57-66.
Kildergaard et al. "3HP tolerance". WO2014198841(A1).

Yeast as cell factory for aromatic products

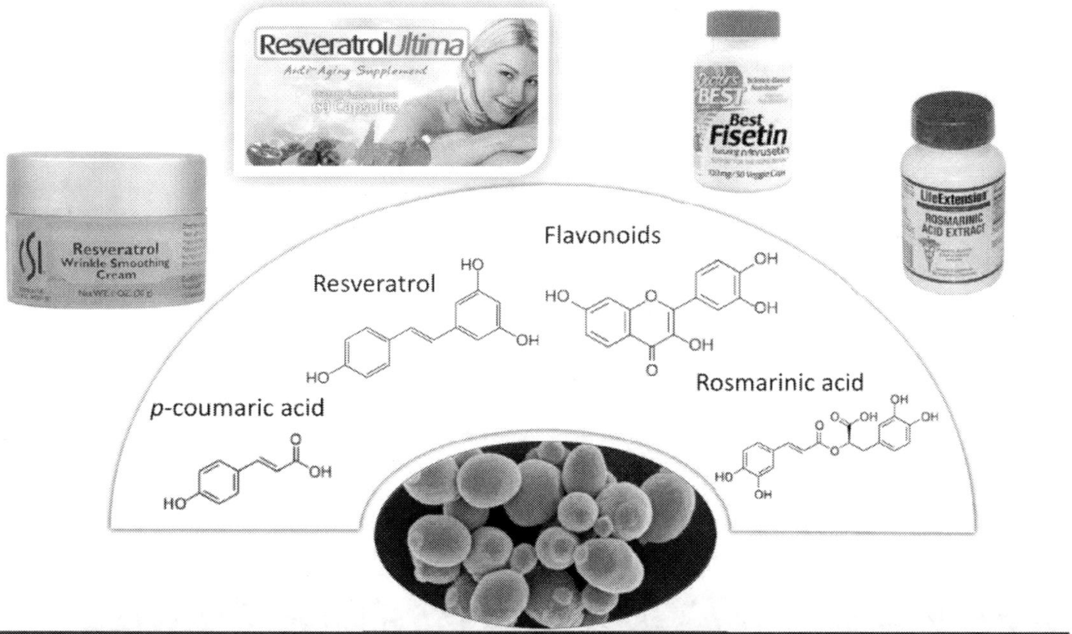

p-coumaric acid production in yeast

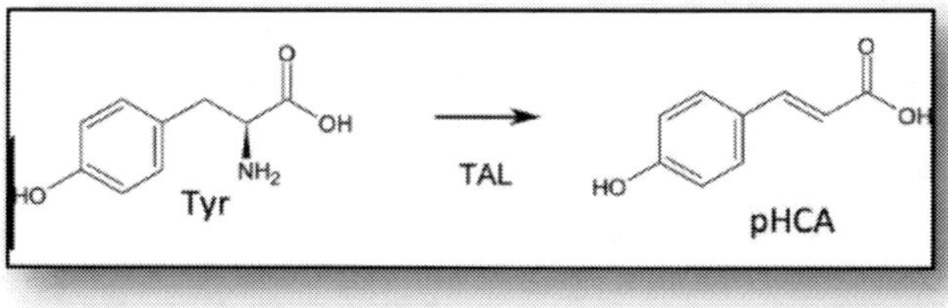

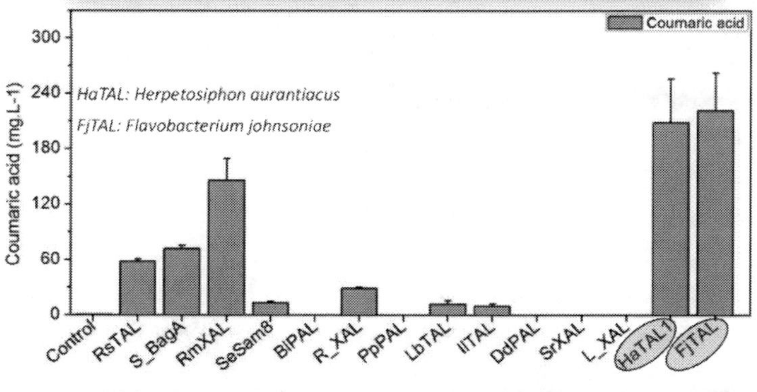

Jendresen CB, Stahlhut SG, Li M, Gaspar P, Siedler S, Förster J, Maury J, Borodina I, Nielsen AT (2015) "Novel highly active and specific tyrosine ammonia-lyases from diverse origins enable enhanced production of aromatic compounds in bacteria and yeast". *Appl Environ Microbiol* 81(13):4458-76.

p-coumaric acid production in yeast

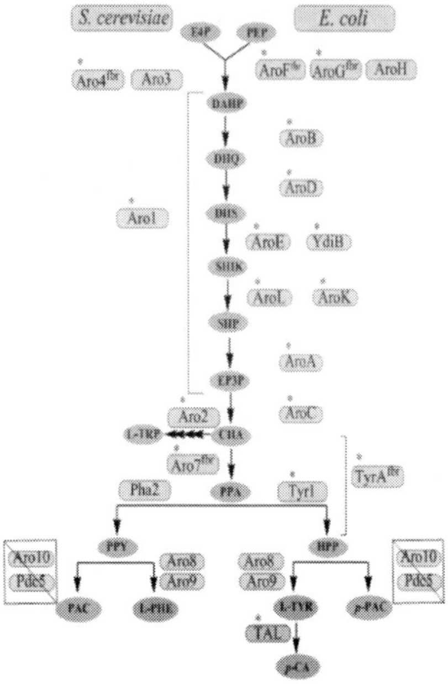

Angelica Rodriguez

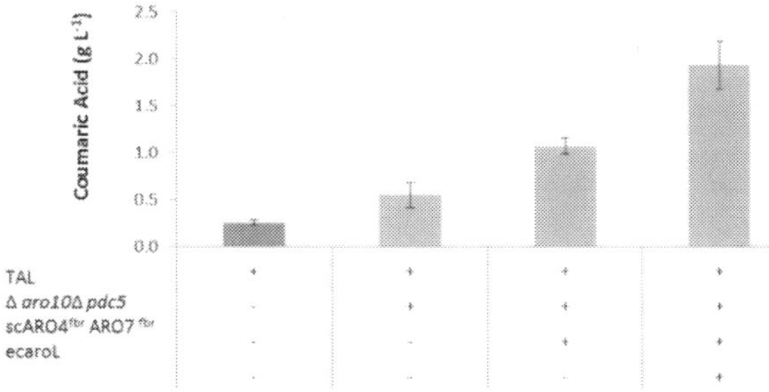

Rodriguez A, Kildegaard KR, Li M, Borodina I, Nielsen J (2015). "Establishment of a yeast platform strain for production of *p*-coumaric acid through metabolic engineering of aromatic amino acid biosynthesis". *Metab Eng* 31:181-188.

Resveratrol biosynthesis via L-Tyr

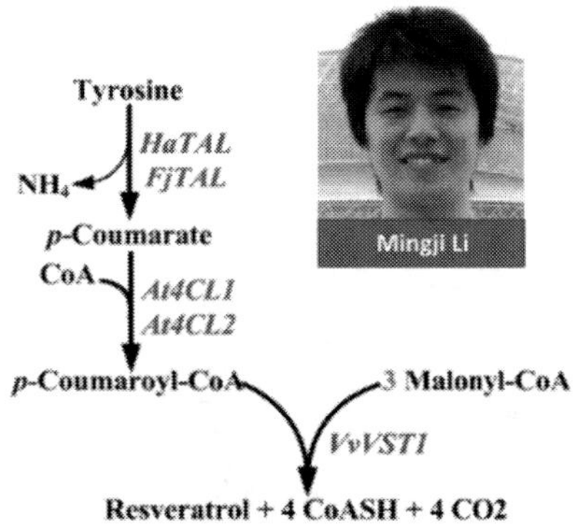

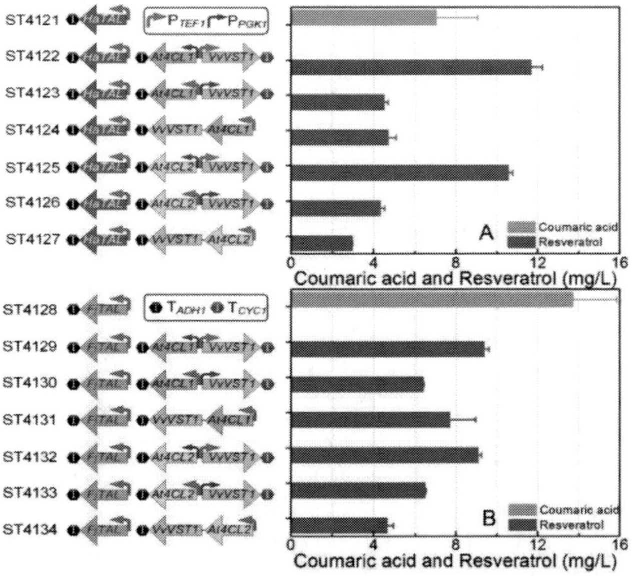

HaTAL: Herpetosiphon aurantiacus
FjTAL: Flavobacterium johnsoniae
At4CL1: Arabidopsis thaliana
At4CL2: Arabidopsis thaliana
VvVST1: Vitis vinifera

Li et al (2015) "*De novo* production of resveratrol from glucose or ethanol by engineered *Saccharomyces cerevisiae*". *Metab Eng*, 32:1-11.

Improvement of precursor supply

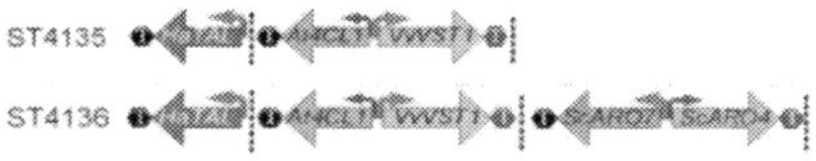

2.8 mg L^{-1}

4.9 mg L^{-1}

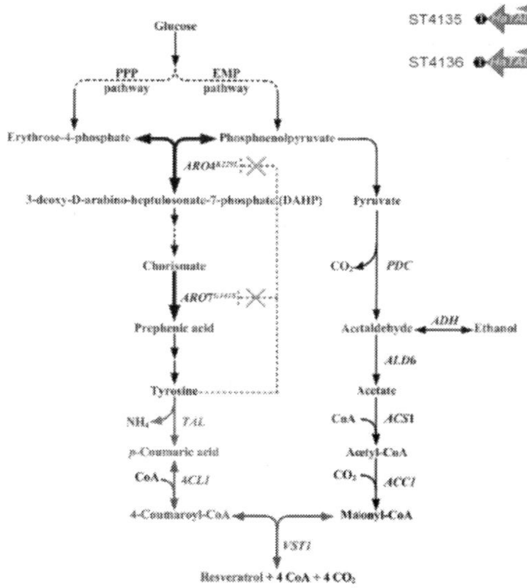

Li et al (2015) "*De novo* production of resveratrol from glucose or ethanol by engineered *Saccharomyces cerevisiae*". *Metab Eng*, 32:1-11.

Improvement of precursor supply

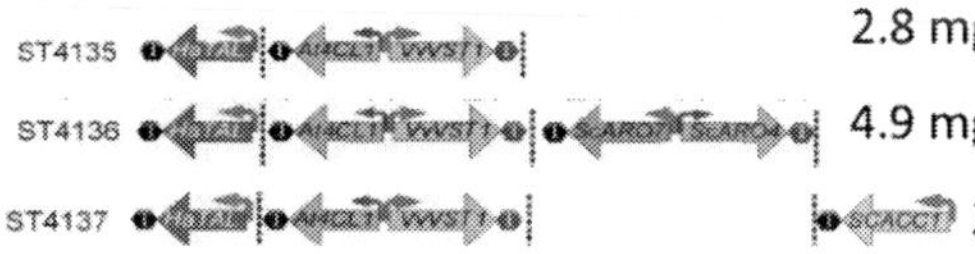

2.8 mg L^{-1}

4.9 mg L^{-1}

3.6 mg L^{-1}

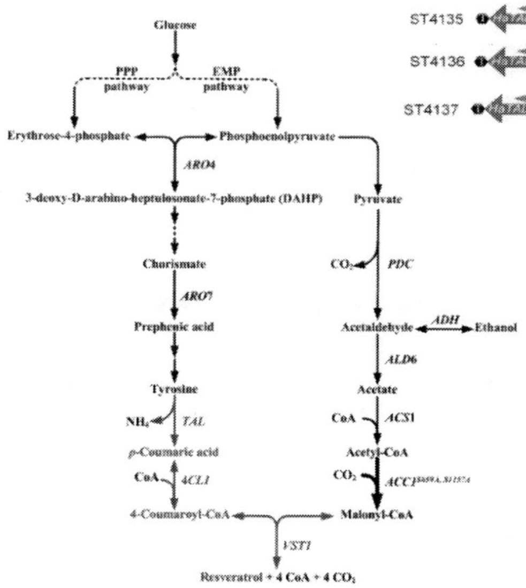

Li et al (2015) "*De novo* production of resveratrol from glucose or ethanol by engineered *Saccharomyces cerevisiae*". *Metab Eng*, 32:1-11.

Improvement of precursor supply

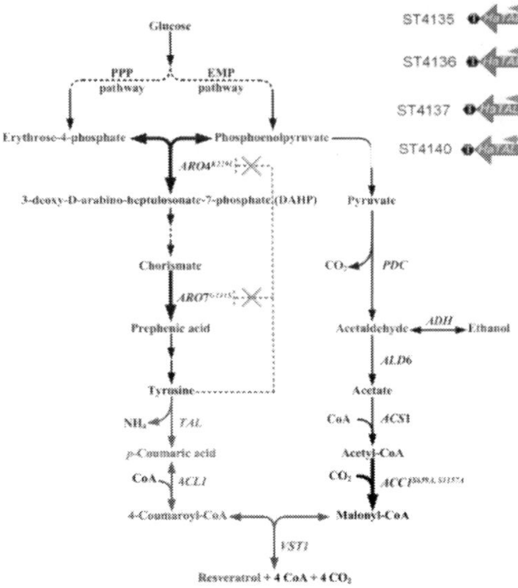

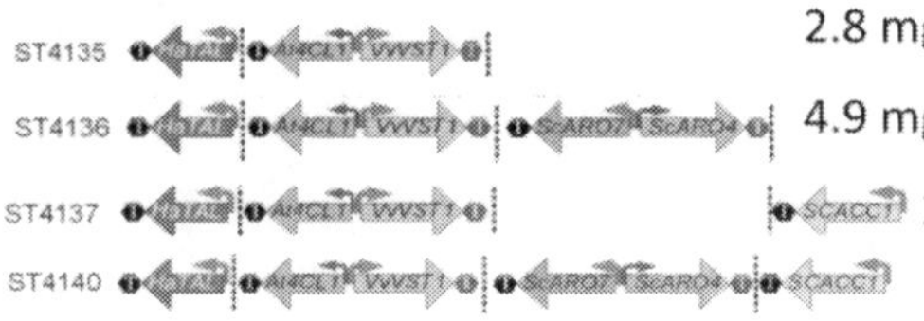

2.8 mg L^{-1}
4.9 mg L^{-1}
3.6 mg L^{-1}
6.4 mg L^{-1}

Li et al (2015) "*De novo* production of resveratrol from glucose or ethanol by engineered *Saccharomyces cerevisiae*". *Metab Eng*, 32:1-11.

Multiple integration of resveratrol pathway

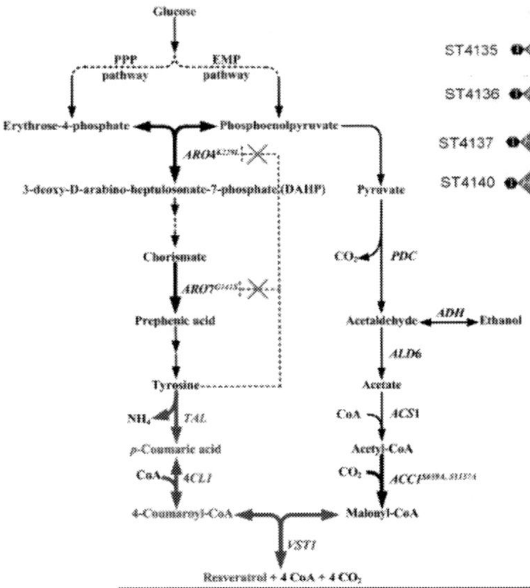

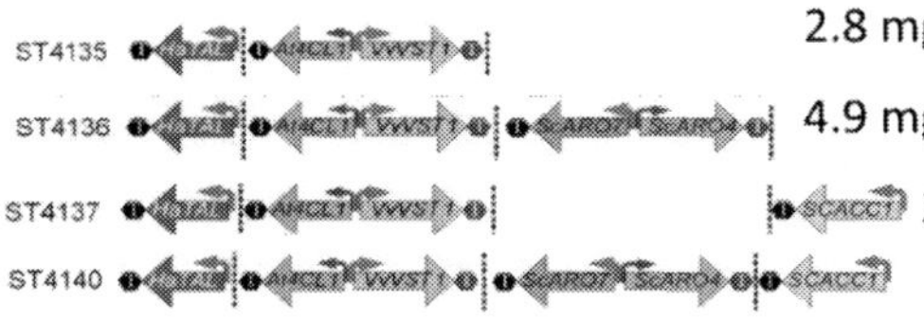

2.8 mg L^{-1}
4.9 mg L^{-1}
3.6 mg L^{-1}
6.4 mg L^{-1}

	HaTAL	*At4CL1*	*VvVST1*
Copy number	8.43±0.85	8.64±1.60	11.16±1.23

236 mg L^{-1}

Li et al (2015) "*De novo* production of resveratrol from glucose or ethanol by engineered *Saccharomyces cerevisiae*". *Metab Eng*, 32:1-11.

Resveratrol production in fed-batch fermentation

Glucose-feed | Ethanol-feed

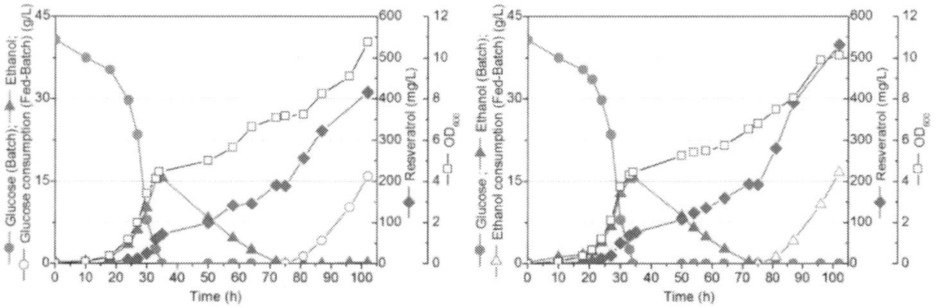

0.5 g/L resveratrol

Li et al (2015) "*De novo* production of resveratrol from glucose or ethanol by engineered *Saccharomyces cerevisiae*". *Metab Eng*, 32:1-11.

Summary

- **Synthetic biology tools** enable rapid iterative strain construction.
  - Vectors & manuals available from AddGene
  - Strains available upon request

- **Adaptive laboratory evolution** followed by systems biology analysis is an excellent tool for development of tolerance and deciphering the mechanism behind.

- Via rational metabolic engineering, novel yeast cell factories have been created for **production of 3HP and aromatic secondary metabolites**.

Acknowledgements

DTU

Group:

Kanchana R Kildegaard
Carina Holkenbrink
Vratislav Stovicek
Behrooz Darbani
Marie Inger Dam
Gheorghe MB Zamfir
Mathew MJ Fabre

Alumni:

Niels B Jensen
Jérôme Maury
Svetlana Sherstyk
Johannes Beder
Angelica Rodriguez
Mingji Li

novo nordisk fonden

bioREFINE-2G

SEVENTH FRAMEWORK PROGRAMME

Novo Seeds

Collaborators:

Jens Nielsen
Nikolaus Sonnenschein
Jochen Förster
Yun Chen
Tomas Strucko
Florian David
Birgitta Ebert
Vratislav Stovicek

Konstantin Schneider
Scott James Harrison
Emre Özdemir
Il-Kwon Kim
Agnieszka Juncker
Christian B. Jendresen
Christer Löfstedt
Xiao Chen

Inger Rosenstand
Jens William Kindtler
Alex T. Nielsen
Thomas Holberg Blicher
Björn Hallström
Steen G. Stahlhut
Baojian Ding

Markus J. Herrgård
Susanne M. Germann
Verena Siewers
Uffe Mortensen
Pedro Lamos
Lars Blank
Hong-Lei Wang

Metabolic Engineering Approaches Harnessing the CRISPR/dCas9 CRISPRi System and Polyculture Strategies for Rapid Assessment and Improvement of Natural Product Production in Recombinant Hosts

Koffas, M. A. G., Rensselaer Polytechnic Institute
Originally presented on: 6/30/2016 8:45:00 - 9:05:00

The field of metabolic engineering is entering a new era in which synthetic transcription factors such as RNAguided-dCas9 are providing the means for dynamic regulation and for rapid assessment of multiplexed genetic interventions to alter the cellular metabolic landscape.

In this talk we will first present a modular assembly method for traditional restriction-ligation cloning of type IIA CRISPR array libraries for multiplex, combinatorial dCas9-mediated transcriptional repression and its application in metabolic engineering of *E. coli* for enhancing flavonoid titers through simultaneous repression of an endogenous transcription factor and several central carbon enzymes, including partial downregulation of a synthetic lethal pair. We will also demonstrate how this method can be expanded by adapting Golden Gate Shuffling for assembly of natural type IIA CRISPR arrays, enabling randomized one-pot assembly of array libraries simultaneously repressing all combinations of a user-defined set of target genes. This approach was tested for improving production of glycosylated natural products—including the pharmaceutically and nutraceutically valuable polysaccharides heparin and chondroitin and the colorful class of plant natural pigments known as anthocyanins—through repression of novel knockdown targets and partial downregulation of an essential glycolytic enzyme. Direct comparison of CRISPRi strains with analogous deletion strains (obtained through λ-red recombineering) shows that CRISPRi can lead to better production improvements compared to gene deletions. Finally, we will demonstrate the construction of orthogonal variants of the classic T7lac promoter using site-directed mutagenesis, generating a panel of inducible hybrid promoters regulated by both LacI and dCas9 and covering a wide expression range. Remarkably, dCas9 orthogonality in our system is mediated by only 23 nucleotide mismatches in a narrow window of the RNA:DNA hybrid, neighboring the protospacer adjacent motif (PAM). We demonstrate that, contrary to many reports, one PAMproximal mismatch is insufficient to abolish dCas9-mediated repression. A subset of these refactored promoters were incorporated into the highly branched violacein biosynthetic pathway, where they act as orthogonal, dCas9-dependent valves capable of throttling and selectively redirecting carbon flux in *E. coli*.

Finally, the presentation will also cover our work on the development and optimization of polycultures (three or more strains in co-culture) for the extension of recombinant pathways, such as the flavonoid branch pathway, *in vivo*. This technology has enabled, for the first time, the *de novo* production of flavan-3-ols and anthocyanins in *E. coli*. Utilizing a computationally guided optimization approach, we were able to demonstrate up to a 970-fold improvement over previously published monoculture titers.

In Vitro Reconstitution of Mevalonate Pathway Guide to Build the Terpenoids Overproduction Cell Factories

Liu, T., Wuhan Institute of Biotechnology
Originally presented on: 6/30/2016 9:05:00 - 9:25:00

Terpenoids are very important natural products, like fuel chemical farnesene, nutritional compounds lycopene, β-carotene, astaxanthin, and drugs like artemisinin and taxol are belong to this big family. However, due to the lack of understanding of the pathway and enzymes, to build a high efficient synthetic pathway is still a huge challege. In order to sovle this problem to accelerate the progress of pathway engineering, we have develop a strategy called "Targeted Synthetic Metabolism". In this strategy, the biosynthetic pathway will be fully reconstituted or partially reconstituted in vitro firstly, and then the contributions of cofactors, substrates and each component and best ratio of components will be analyzed systematically. Second, we will choose the right promoters to control each gene in vivo and also prepare the chasis host with the guidance from in vitro study to generate our first generation mutant. Third, we used our Mass-spec platform to quantitatively analyze the intermediates accumulation. And then, we will adjust the targeted point by replacing of different genes or different promoters to improve the titer step by step under our predictions. Here, we have successfully in vitro reconstituted the mevalonate pathway, and these informations guided us to build the cell factories for farnesene, lycopene and astaxanthin in E. coli, and also we can copied the success for lycopene, β-carotene, astaxanthin in Saccharomyces cerevisiae and other important plant natural products in microbial system. The robust IPP and DMAPP chassis can be used for key enzymes evaluation since most of intermediates are not stable. Furthermore, this strategy also can be applied into fatty acid derived compound overproduction and polyketide compounds overproduction and we have successfully increased the performances of some current industrial strains by using this strategy.

jbei
Joint BioEnergy Institute

BERKELEY LAB

U.S. DEPARTMENT OF ENERGY

Pathway Optimization in *E. coli* through Systematic Design-Build-Test-Learn (DBTL) Cycle and Crispr-Cas9 Mediated Chromosomal Integration

Jorge Alonso-Gutierrez, PhD

Metabolic Engineering XI
Awaji Yumebatai, Japan
June 30, 2016

Joint BioEnergy Institute (JBEI)
Lawrence Berkeley National Laboratory (LBNL)

Metabolic Engineering Workflow: DBTL cycle

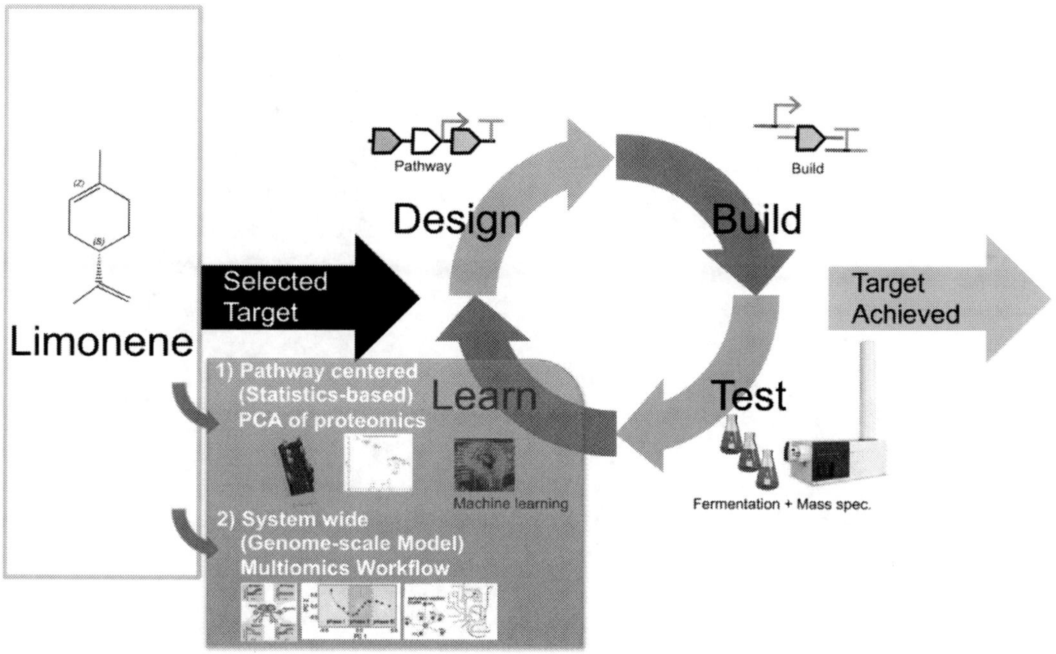

Complexity of the MVA pathway

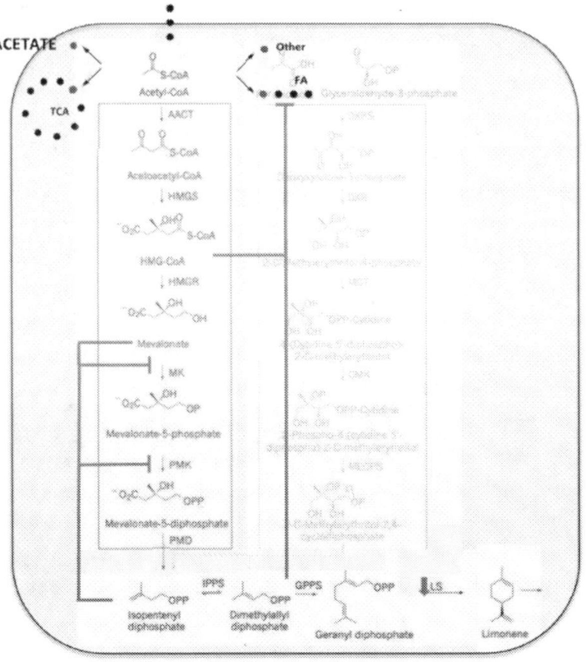

1) Pathway-focused / Statistics-based approach (PCAP)

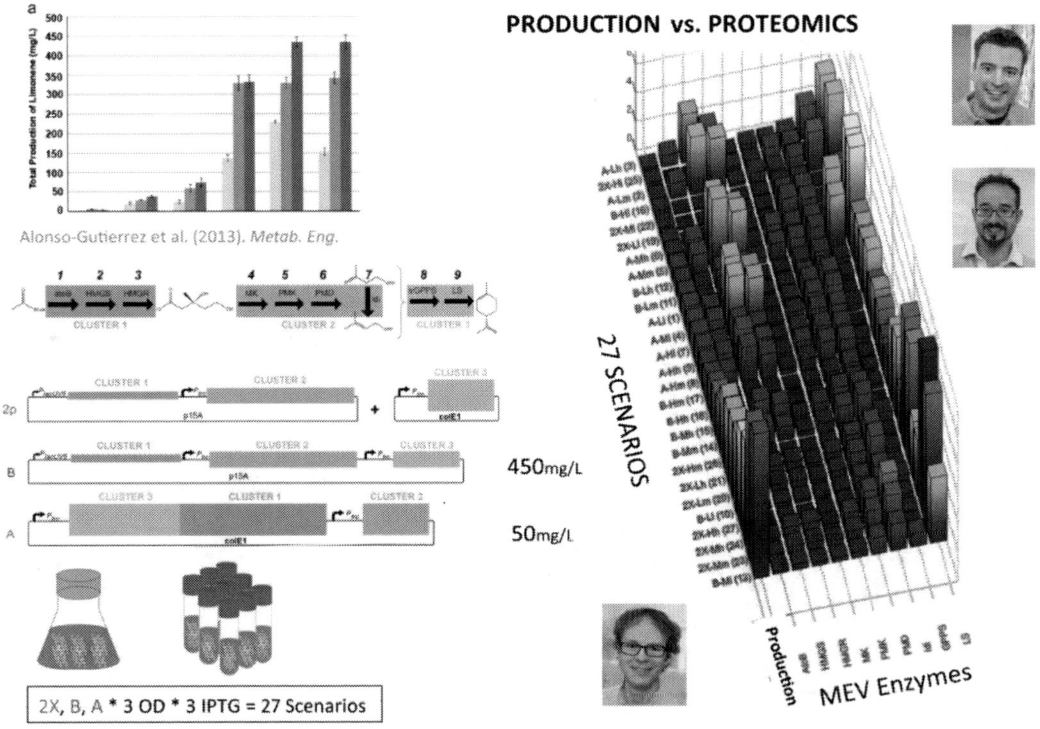

1) Pathway-focused / Statistics-based approach (PCAP)

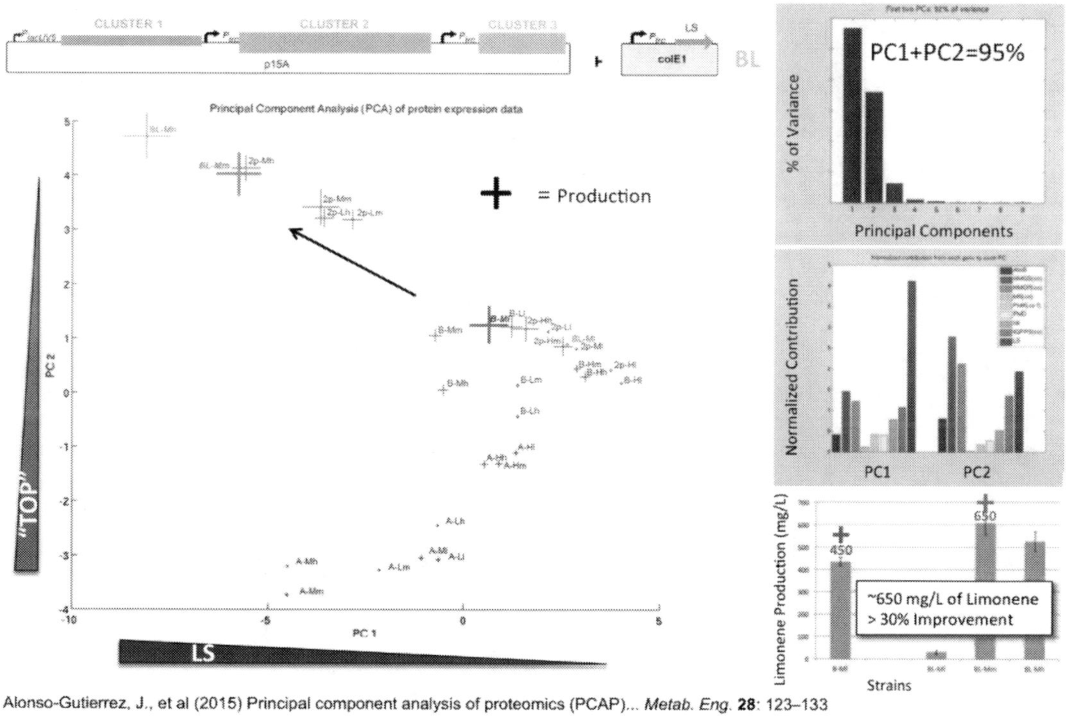

Alonso-Gutierrez, J., et al (2015) Principal component analysis of proteomics (PCAP)... *Metab. Eng.* **28**: 123–133

1) Pathway-focused / Statistics-based approach (PCAP)

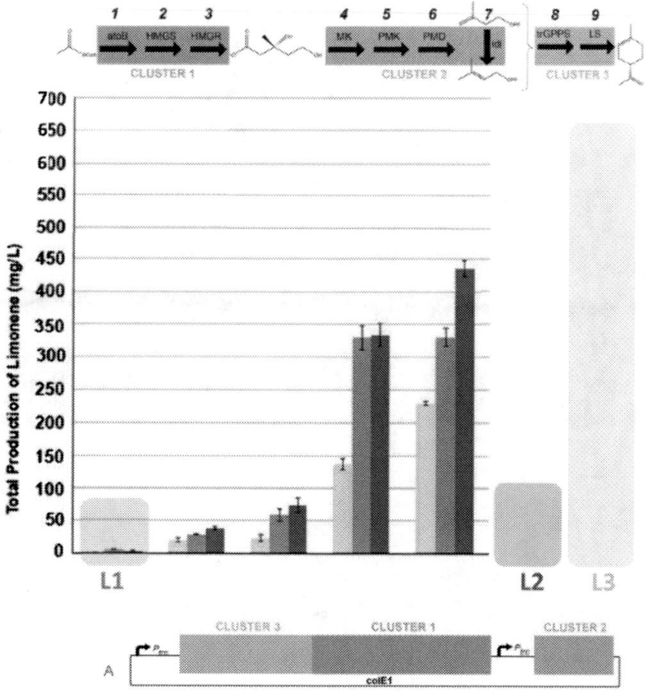

2) Host-focused / Genome scale model-based approach

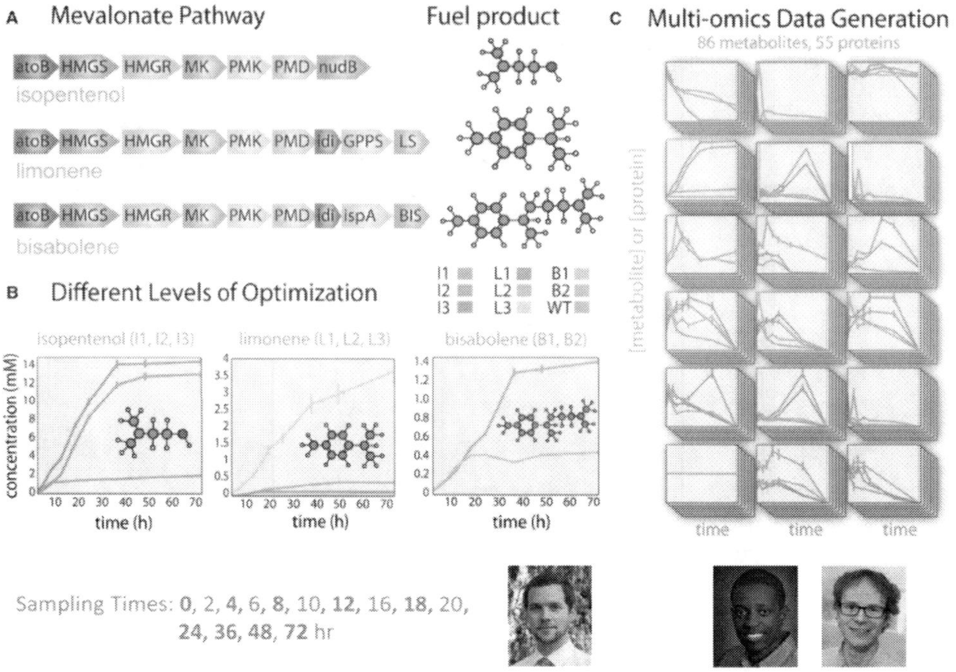

Brunk, E., et al. (2016) Characterizing Strain Variation in Engineered E. coli Using a Multi-Omics-Based Workflow. *Cell Syst.* 1–12

2) Host-focused / Genome scale model-based approach

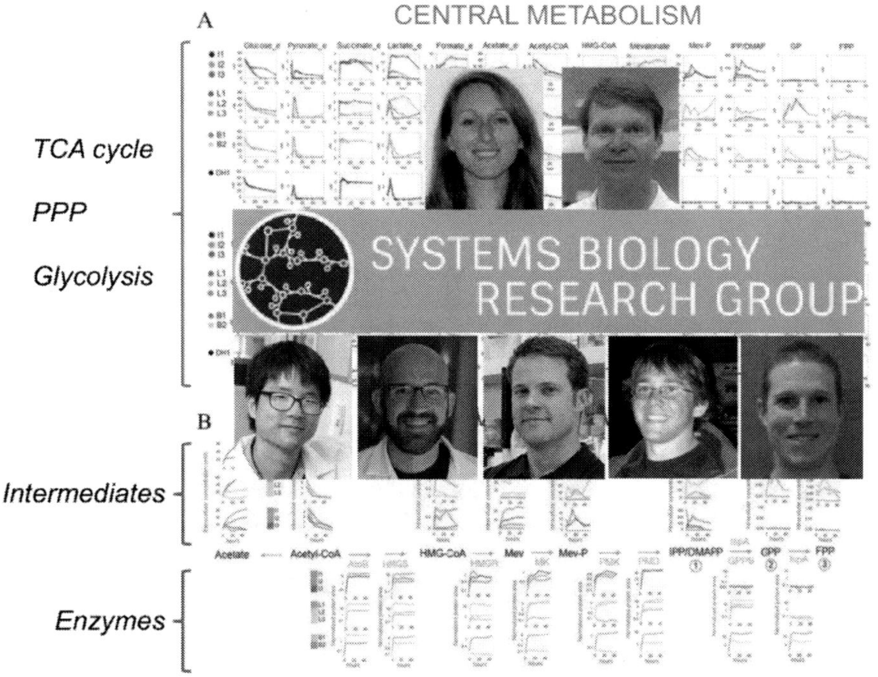

Brunk, E., et al. (2016) Characterizing Strain Variation in Engineered E. coli Using a Multi-Omics-Based Workflow. *Cell Syst.* 1–12

2) Host-focused / Genome scale model-based approach

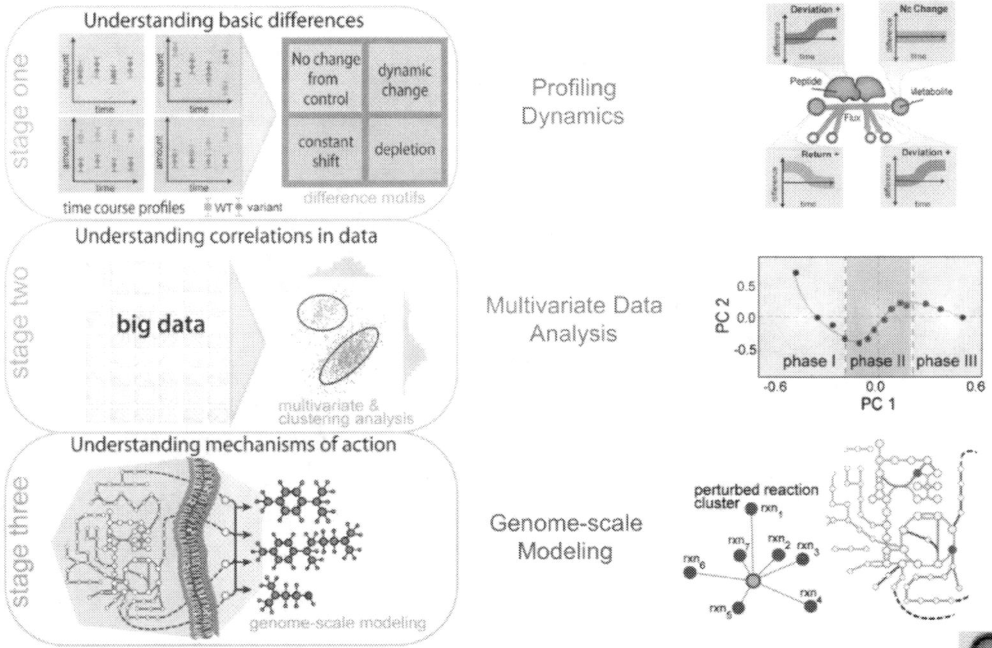

Brunk, E., et al. (2016) Characterizing Strain Variation in Engineered E. coli Using a Multi-Omics-Based Workflow. *Cell Syst.* 1–12

2) Host-focused / Genome scale model-based approach

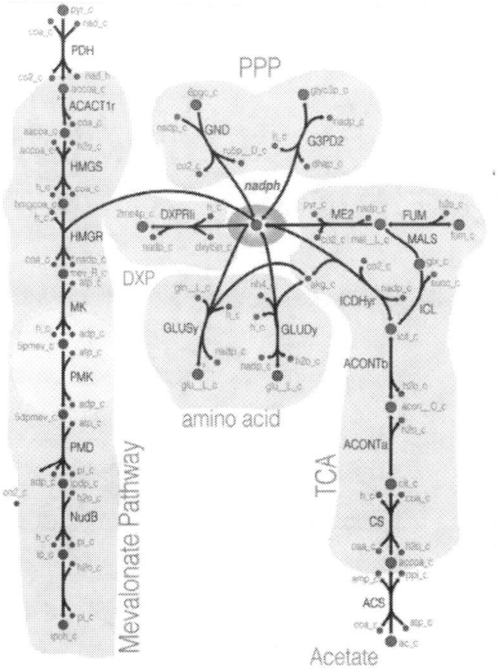

Fitness characteristic	*Gene knock-outs*	*Gene names*
maximized flux through NADPH-producing reactions	b0197, b0198, b0199, b4238	metQ, metL, metN, nrdD
maximized flux through pentose-phosphate pathway	b1378, b0197, b0198, b0199, b4238	**ydbk**, metQ, metL, metN, nrdD
minimized flux through NADPH-consuming reactions	b4209, b1748, b1747, b2501, b4468	ytfE, astC, astA, ppk, glcE

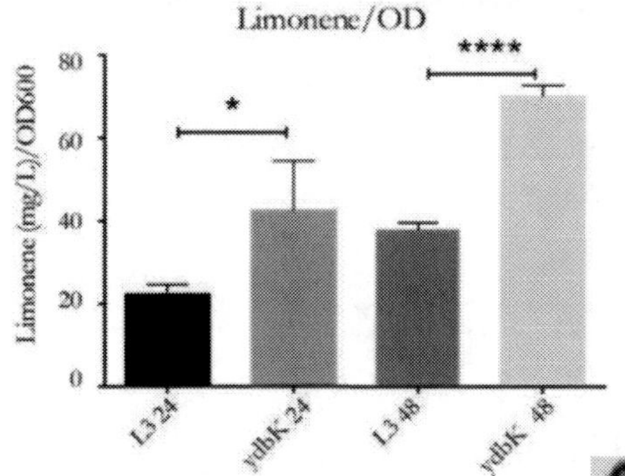

Brunk, E., et al. (2016) Characterizing Strain Variation in Engineered E. coli Using a Multi-Omics-Based Workflow. *Cell Syst.* 1–12

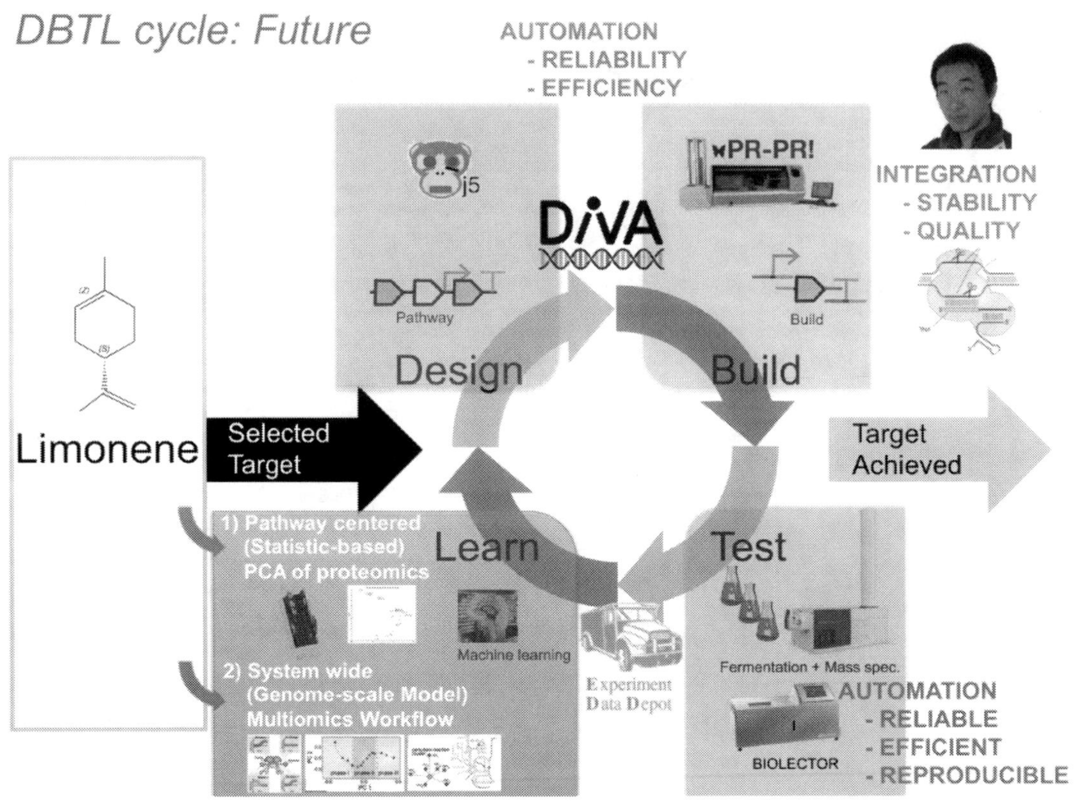
DBTL cycle: Future
AUTOMATION
- RELIABILITY
- EFFICIENCY
j5
DiVA
Pathway
Build
Design
Build
INTEGRATION
- STABILITY
- QUALITY
Limonene
Selected Target
Target Achieved
1) Pathway centered
(Statistic-based)
PCA of proteomics
Learn
Test
Machine learning
2) System wide
(Genome-scale Model)
Multiomics Workflow
Experiment Data Depot
Fermentation + Mass spec.
AUTOMATION
- RELIABLE
- EFFICIENT
- REPRODUCIBLE
BIOLECTOR

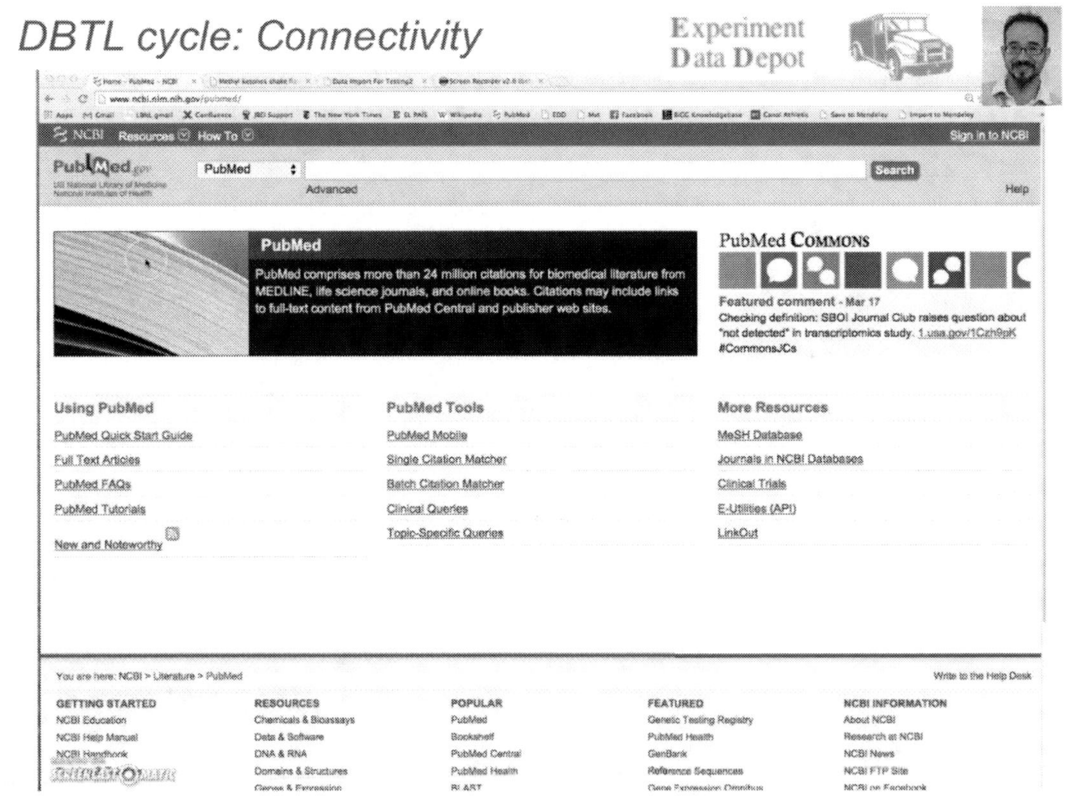
DBTL cycle: Connectivity
Experiment Data Depot
NCBI
Resources
How To
Sign in to NCBI
PubMed
Search
Advanced
Help
PubMed
PubMed comprises more than 24 million citations for biomedical literature from MEDLINE, life science journals, and online books. Citations may include links to full-text content from PubMed Central and publisher web sites.
PubMed Commons
Featured comment - Mar 17
Checking definition: SBOI Journal Club raises question about "not detected" in transcriptomics study. 1.usa.gov/1Czh9pK #CommonsJCs
Using PubMed
PubMed Quick Start Guide
Full Text Articles
PubMed FAQs
PubMed Tutorials
New and Noteworthy
PubMed Tools
PubMed Mobile
Single Citation Matcher
Batch Citation Matcher
Clinical Queries
Topic-Specific Queries
More Resources
MeSH Database
Journals in NCBI Databases
Clinical Trials
E-Utilities (API)
LinkOut
You are here: NCBI > Literature > PubMed
Write to the Help Desk
GETTING STARTED
NCBI Education
NCBI Help Manual
RESOURCES
Chemicals & Bioassays
Data & Software
DNA & RNA
Domains & Structures
POPULAR
PubMed
Bookshelf
PubMed Central
PubMed Health
FEATURED
Genetic Testing Registry
PubMed Health
GenBank
Reference Sequences
NCBI INFORMATION
About NCBI
Research at NCBI
NCBI News
NCBI FTP Site

Summary (take home messages)

A statistical-based method (**PCA of Proteomics**) has been proved useful to direct pathway optimization and **substantially increase titers** (from 450 to 650mg/L of Limonene and from **250 to 1200mg/L** of bisabolene)

A systems-level, model-driven **multi-omics-based workflow** successfully i) **reconciles** metabolomics data, proteomics data, and predictions from genome-scale models ii) **explains** host response to pathway engineering iii) suggest SKO that lead to **increase productivity** of both limonene and isopentanol

Automation of Build-Test increases **reproducibility**

Chromosomal Integration of heterologous pathways is key to provide **Quality data** to feed the DBTL cycle

Share all results derived from every strain in a **standardized** form (EDD)

Acknowledgements

Jay Keasling
Taek Soon Lee

Mevalonate Pathway
Kevin George

Computational biology
Hector Garcia-Martin
Nathan Hillson

Metabolomics
Edward Baidoo

Proteomics
Chris Petzold

CRISPR/Cas9 MVA optimization
Daisuke Koma
Sarah Richardson
Qijun (Daniel) Hu
Yuchen Yang
Luis Iniesta

Palsson's Lab (UCSD)
Elizabeth Brunk
Bernhard Palsson
Douglas McCloskey
Jonathan Monk
Laurence Yang
Edward J. O'Brien
Adam Feist

All of you for listening !!!
ありがとうございます。

jbei Joint BioEnergy Institute

BERKELEY LAB

U.S. DEPARTMENT OF ENERGY | Office of Science

This work conducted by the Joint BioEnergy Institute was supported by the Office of Science, Office of Biological and Environmental Research, of the U.S. Department of Energy under Contract No. DE-AC02-05CH11231

Gene Circuit Boosts High-Temperature Fermentation with Low Energy Consumption

Jia, H., Beijing Institute of Technology
Xu, K., Beijing Institute of Technology
Li, C., Beijing Institute of Technology
Originally presented on: 6/30/2016 9:40:00 - 9:55:00

The growth and production of microorganisms in bioconversion are often hampered by heat stress and product feed-back repression. In this study, the intelligent microbial heat regulating engines (IMHeRE) which are novel gene circuits were developed and customised to improve the thermo-robustness of *Escherichia coli* and *Saccharomyces cerevisiae* via the integration of a thermotolerant system and a quorum regulating system. At the cell level, the thermotolerant system composed of different heat shock proteins and RNA thermometers hierarchically expands the optimum temperature by sensing heat changes in both strains. At the community level, the quorum regulating system dynamically programs the altruistic sacrifice of individuals to reduce metabolic heat release by sensing the temperature and cell density in *Escherichia coli*. Using this hierarchical, dynamical and multilevel regulation, the IMHeRE are able to significantly improve cell growth and production. In real applications, the production of lysine was increased five-fold at 40°C using the IMHeRE. Additionally, both tolerant to thermo stress and high concentration of ethanol in *Saccharomyces cerevisiae* has recently become much useful as the industry moves toward the use of simultaneous sacchrification and fermentation (SSF). So double-tolerance gene circuits were rationally designed and successfully obtained through one-pot parts random assembly via Golden Gate Shuffling. The cooling water and energy consumption are all significantly reduced of 31% and 24% as well as ethanol production and productivity enhanced by 5% and 12% by engineered *Saccharomyces cerevisiae*, respectively, in 35-37°C fermentation. Our work provides new potential for the development of bioconversion by conserving energy and increasing productivity.

Keywords: gene circuit, hierarchical thermotolerance, ethanol tolerance, bioconversion, microorganism, synthetic biology

Production of 2,3-Butanediol By Engineered Saccharomyces cerevisiae

Seo, J. H., Seoul National University

Originally presented on: 6/30/2016 10:10:00 - 10:35:00

2,3-Butanediol (2,3-BD) is a platform chemical with wide applications including a precursor chemical to 1,2-butadiene, a monomer for synthetic rubber. *Saccharomyces cerevisiae* is an attractive host strain for the production of 2,3-BD. While *S. cerevisiae* is known to produce trace amounts of 2,3-BD naturally, it is necessary to engineer *S. cerevisiae* to produce 2,3-BD with high yield and productivity. To increase 2,3-BD production through elimination of ethanol production, a pyruvate decarboxylase (Pdc)-deficient mutant was constructed. Also the *alsS* and *alsD* encoding acetolactate synthase and acetolactate decarboxylase from *Bacillus subtilis* and endogenous *BDH* coding for 2,3-BD dehydrogenase were overexpressed. The engineered Pdc-deficient *S. cerevisiae* with the 2,3-BD biosynthetic pathway has two main problems for efficient production of 2,3-BD: redox imbalance induced by accumulation of cytosolic NADH, shortage of C_2-compounds for cell growth. To overcome these problems, NADH oxidase from *Lactococcuslactis*(*noxE*) was introduced and the expression levels of *PDC* were optimized by combination of heterologous *PDC* genes and native promoters. The reduced intracellular $NADH/NAD^+$ ratio by the expression of *noxE*increased 2,3-BD yield and decreased glycerol yield as a byproduct. In addition, the optimized expression of *PDC* relieved C_2-auxotropy and increased 2,3-BD productivity. Through the fed-batch fermentation with the BD5_Ctnox strain with co-expression of *noxE*and optimized *PDC*, 154.3 g/L of 2,3-BD was produced with 1.98 g/L/h of productivity. The overall yield of 2,3-BD was 0.404 g/g which corresponds to 80.7% of theoretical yield. The metabolic engineering strategy used in this study will be applicable to production of 2,3-BD as well as other chemicals by using Pdc-deficient *S. cerevisiae* as a host strain.

Metabolic Engineering 11
Kobe, Japan

From Cyanochemicals to Cyanofactories

Yin Li

Tianjin Institute of Industrial Biotechnology, CAS
Institute of Microbiology, CAS

MPSysBio
Lab of Molecular Physiology and Systems Biotechnology

Industrial biotechnology

Raw materials → Bio-based processes → Bio-based Products

2

Raw materials

Easiness for utilization | **Availability**

- Sugars/Starches
- Non-grain Starches
- Low-Value Chemicals
- Cellulosic Biomass
- Syngas/Offgas
- Methanol/Methane
- Carbon dioxide

3

Photosynthesis

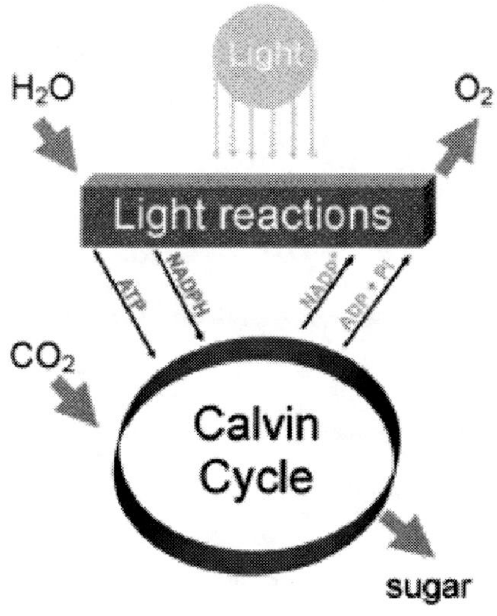

Oxygenic Photosynthesis

4

Cyanobacteria

- Photoautotrophic prokaryotes, evolutionary precursor of chloroplasts
- 20-30% organic carbon derived from cyanobacteria
- Photosynthesis efficiency 10-fold that higher plants
- Fastest doubling time 2 hours (*Synechococcus elongatus* UTEX 2973)
- Genetic manipulation easier than higher plants/eukaryotic algae

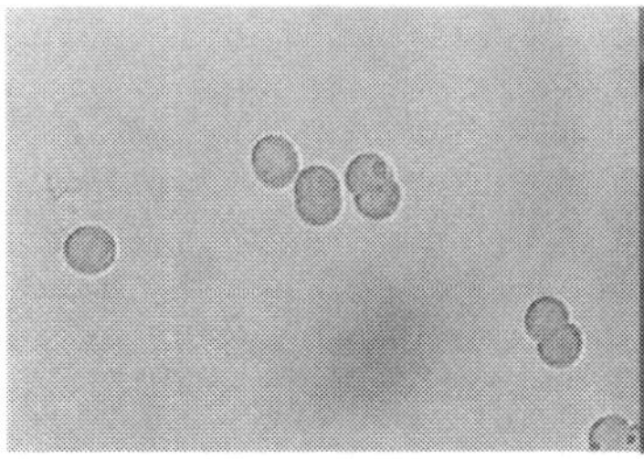

Unicellular Form

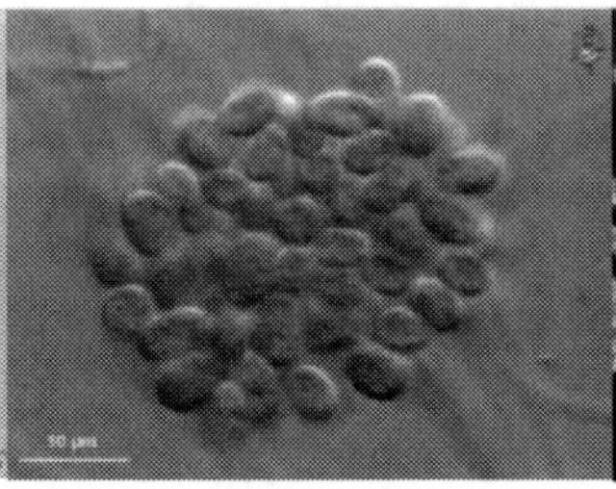

Aggregates

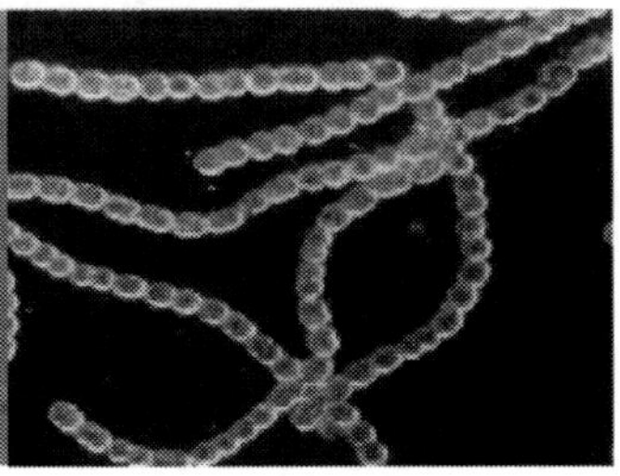

Filamentous form

5

CO_2 to fuels / chemicals

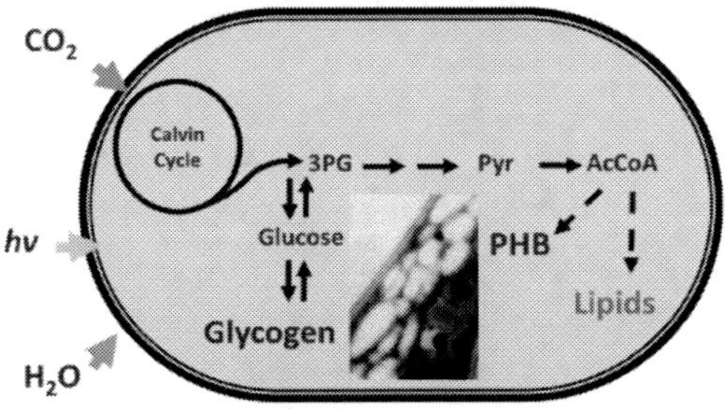

- **Biomass concentration limitation**

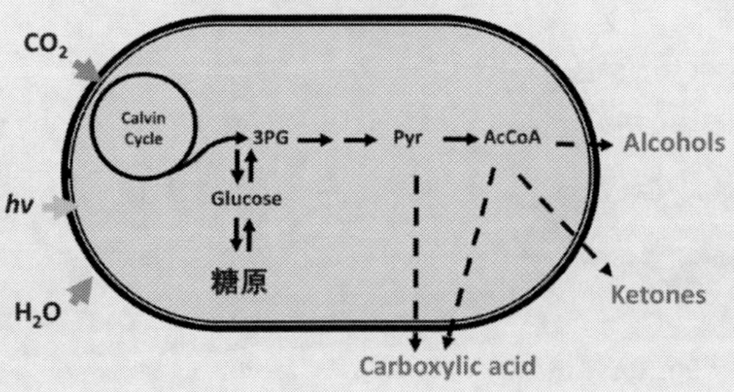

- **Extracellular small molecules**
- **Photobiocatalysts**

Cyanochemicals

2009-2014: > 20 chemicals produced by cyanobacteria

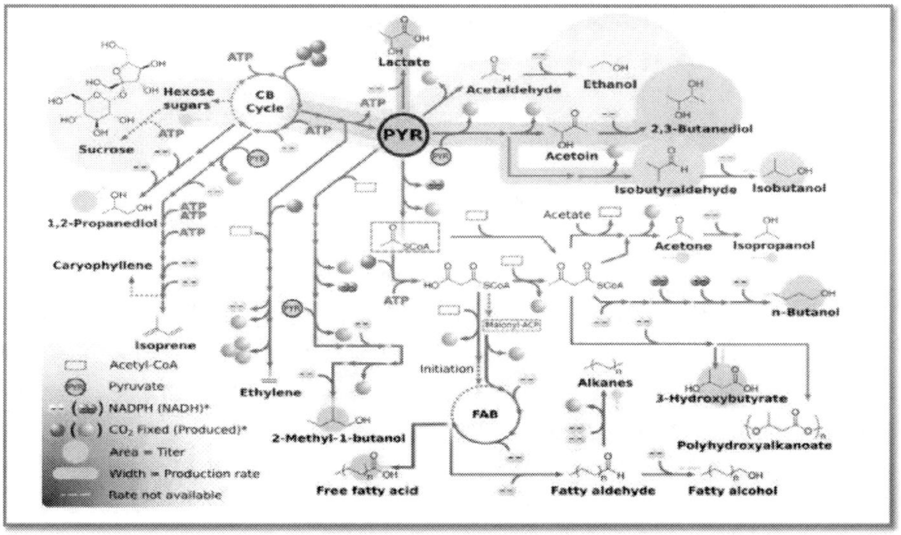

Oliver & Atsumi, *Photosynth. Res.* 2014, 120:249-261

7

Cyanochemicals > 1 g/L

Host	Chemical	Titer (g/L)	Strategies toward increased production
7942	Isobutyraldehyde [8]	1.10	Overexpression of *rbcLS* genes; Supplement of 50 mM $NaHCO_3$.
6803	D-lactate [14]	1.06	Introducing a more efficient enzyme DldhL with a super-strong promoter; Blocking poly-3-hydroxybutyrate and acetate pathways from acetyl-CoA node.
6803	D-lactate [15]	1.14	Introducing a more efficient enzyme GlyDH*; Expression of a soluble transhydrogenase.
7942	2,3-butanediol [9]	2.38	Selection of key enzyme sADH with higher efficiency; Using NADPH-dependent sADH; Supplement of 50 mM $NaHCO_3$.
7942	sucrose [15]	3.50	Expressing a symporter of protons and sucrose; Minimizing glucose or sucrose consuming reactions.
6803	ethanol [4]	5.50	Selection of key enzymes with higher activity; Blocking poly-3-hydroxybutyrate synthetic pathway.

8

Each cell is a tiny factory

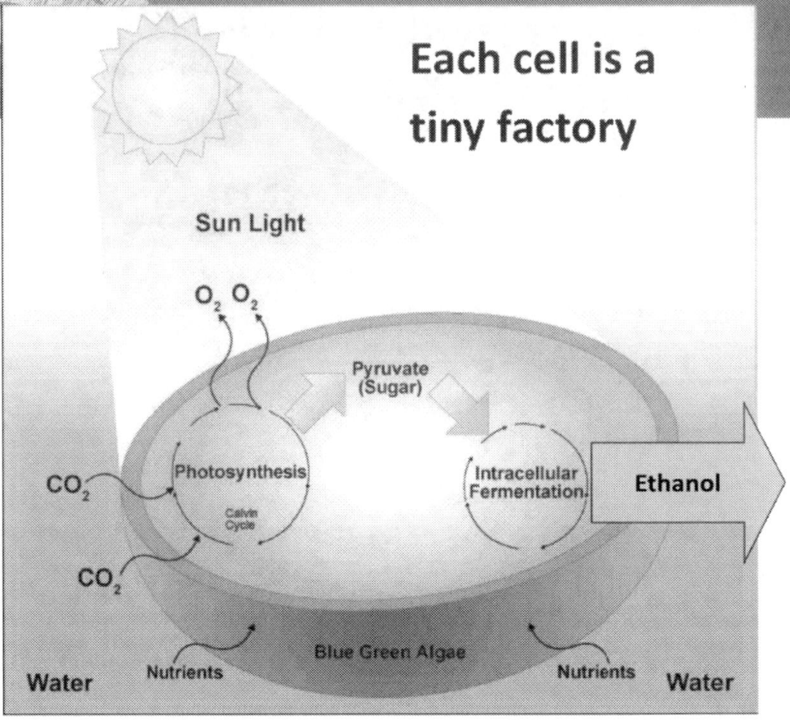

www.algenol.com

9

Algenol pilot facility

Florida, US

10

Strategies for cyanochemicals

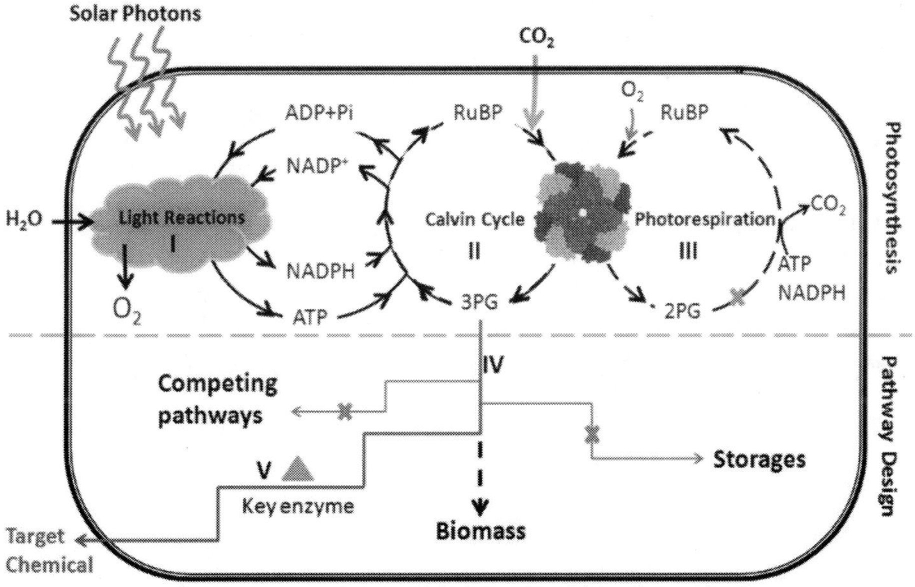

11

Common challenges

- **Efficiency for increasing CO_2-fixation**
- **Efficiency for genetic operation**
- **Efficiency for redirecting carbon flux**

12

Engineering Rubisco

Rubisco

- -- Pivotal role in crop production
- -- Hexadecamer: L_8S_8
- -- Bifunctional
- -- Extremely low CO_2 activity (k_{cat}: 0.3-13.4 s^{-1})
- -- Poor CO_2/O_2 selectivity ($S_{C/O}$: 10-240)

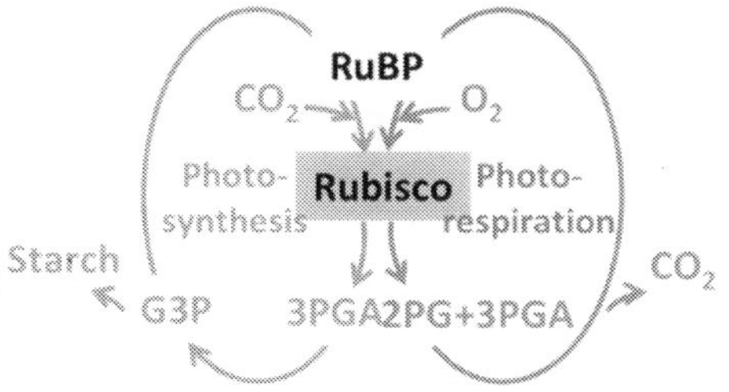

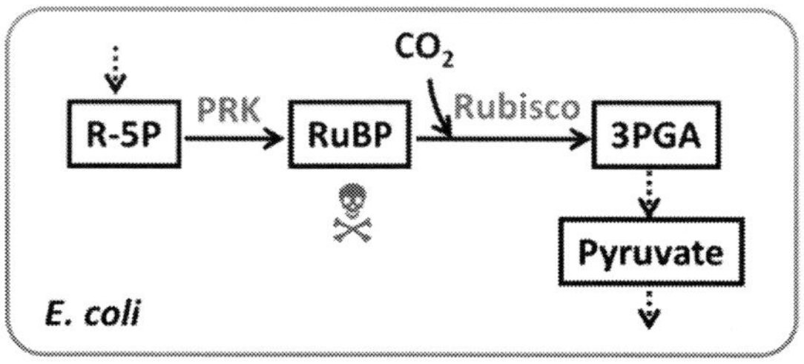

Previous molecular engineering improved expression rather than activity

Heterotrophic CO_2-fixation

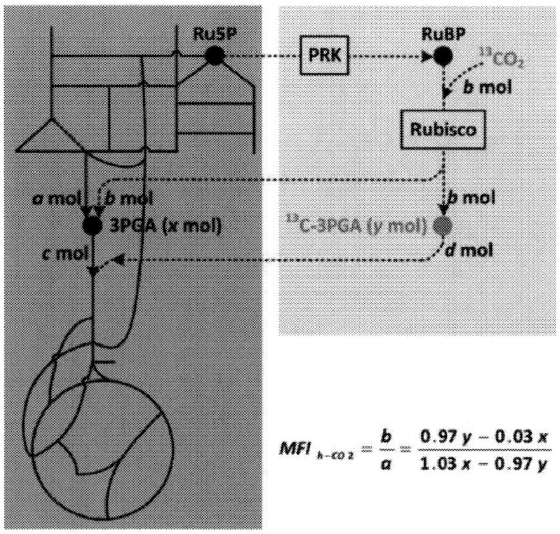

$$MFI_{h\text{-}CO2} = \frac{b}{a} = \frac{0.97y - 0.03x}{1.03x - 0.97y}$$

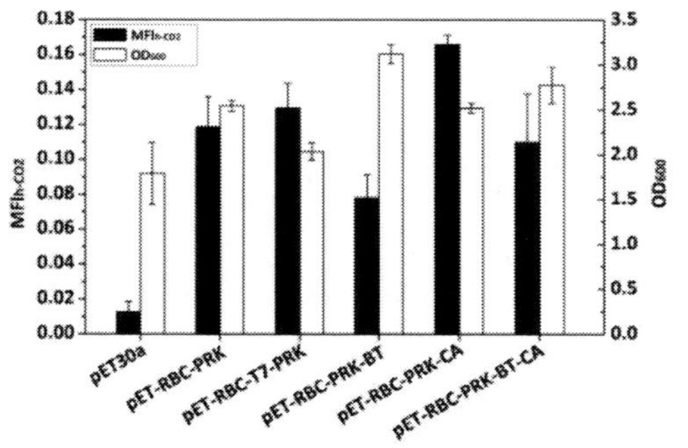

RBC: Rubisco

PRK: Phosphoribulokinase

CA: Carbonic anhydride

BT: Bicarbonate transporter

Improved Rubisco

Saturation functional expression

-- Overexpression chaperon RbcX

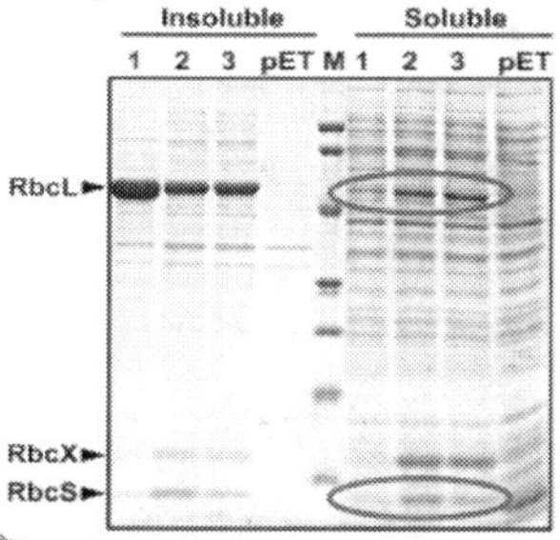

13 ↗ 65 mg/L

0 ↗ 53 mg/L

Activity-directed selection system

-- Cell growth ~ Rubisco activity

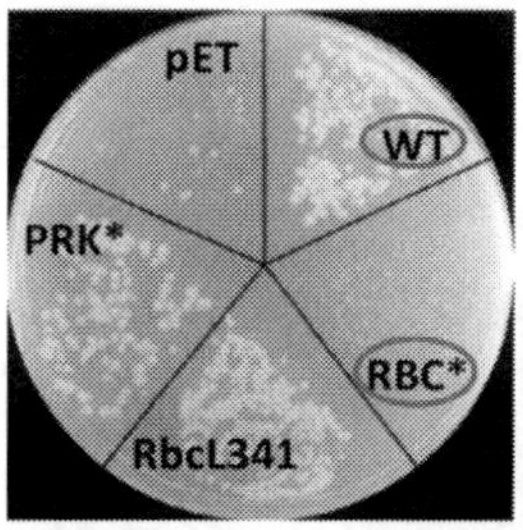

RBC* (^{RbcL}K197M)
PRK* (K20M/S21A)
RbcL341 (F341I)

	Specific activity (U/mg)	k_{cat} (s^{-1})	K^{C}_{M} (mM)	k_{cat}/K_{M}	K^{RuBp}_{M} (mM)
WT	0.91±0.04	6.66±0.58	0.131±0.005	173.33±21.84	0.052±0.01
M6-5	1.68±0.14	14.64±1.08	0.198±0.012	264.17±32.34	0.049±0.01
	↑85%			↑52%	

15

Cyano-gene expression

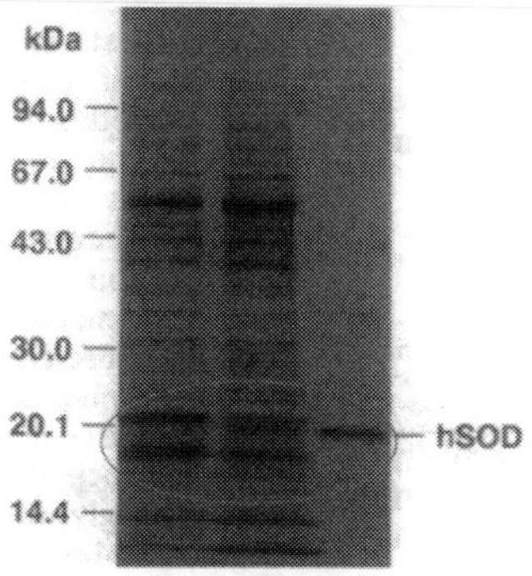

Takeshima, et al., *PNAS*. 1994, 91: 9685-9689

- Native PrbcL promoter and optimized SD sequence
- High level expression SOD in cyanobacteria (3%)

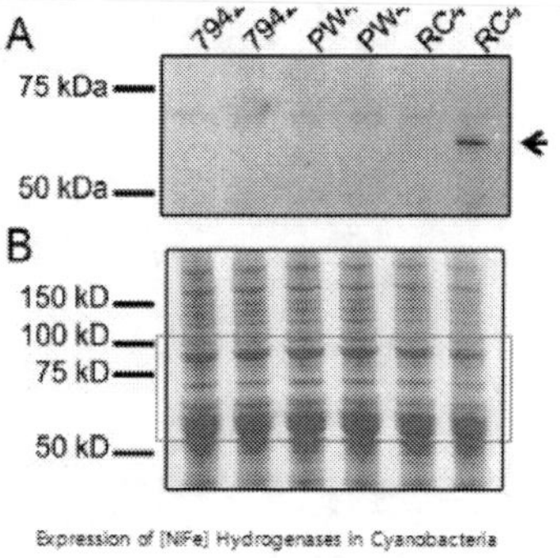

Expression of [NiFe] Hydrogenases in Cyanobacteria

Weyman, et al., *PLoS One*. 2011, 5:1-6

- Identification of gene expression by western blot
- Target protein bands are invisible in the gel

Cyano-gene expression

- ➢ **Lack of strong promoters**

- ■ *E. coli* strong promoter, Ptrc, Ptac...
- ■ Native short promoter (200-300 bp), PrbcL, PpsbA1...

- ➢ **Strategy for improving expression**

Strain	Modification on Ter	NSI	NSII	Ter activity (U/mg)	Butanol production (mg/L)
EL1	Original	Ptrc	P_LlacO_1 T.d-ter	0.0021	3.04±0.41
EL7	Codon optimized for *S. elongatus* 7942	Ptrc	P_LlacO_1 $T.d\text{-}ter_{So}$	0.019	3.67±0.33
EL13	Gene rearrangement	Ptrc T.d-ter	P_LlacO_1	0.057	11.58±0.58
EL14	Gene rearrangement + His-tag	Ptrc His-T.d-ter	P_LlacO_1	0.41	13.10±0.40

Berla et al., *Front Micorbiol.* 2013, 4:246
Lan & Liao, 2011. *Metab Eng.*, 13, 353-363

- ■ Adding His tag
- ■ Using multiple promoters
- ■ Gene multiplication

17

Strong promoter Pcpc560

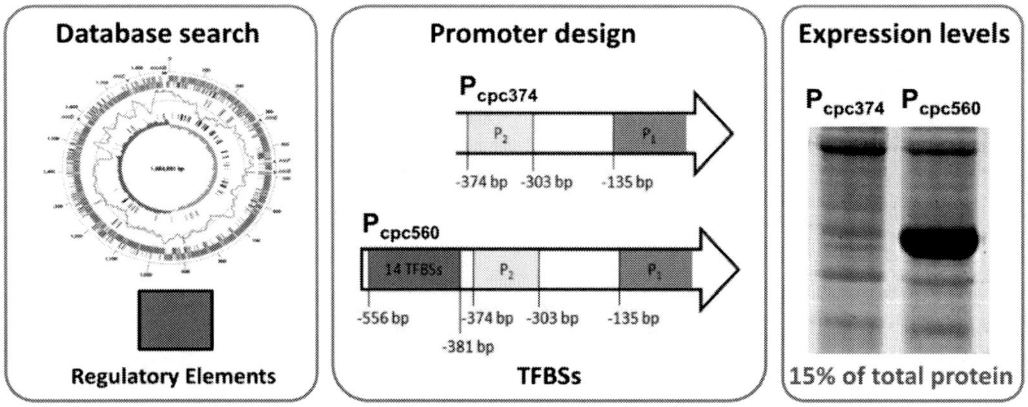

Zhou et al., *Sci. Rep.* 2014, 4:4500

18

Verification of Pcpc560

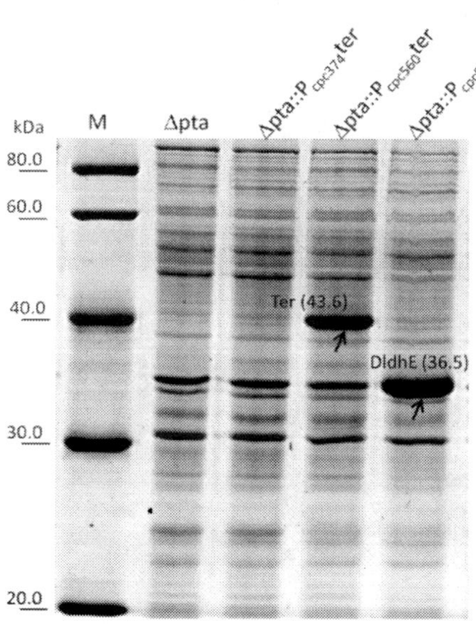

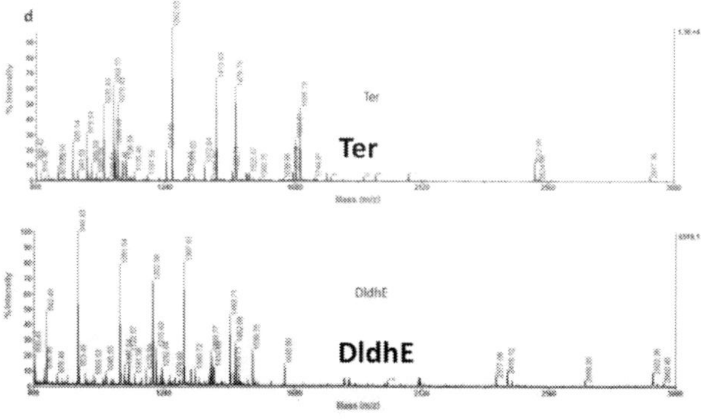

19

Cyano-D-lactate

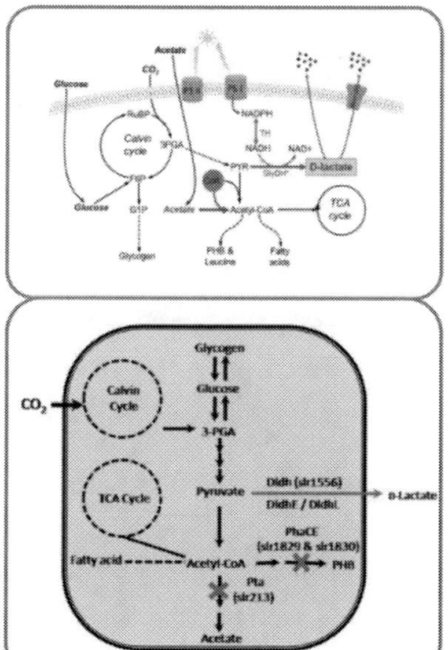

>1 g/L

Strains	Genome modification	Metabolic production		
		PHB (% (w/w) of DW)	Acetate (g/L)	Lactate (g/L)
WT	phaE up, phaE phaC, phaC down; pta up, pta, pta down	5.59 ± 0.57	1.67 ± 0.30	0.04 ± 0.02
ΔphaCE	phaE up, Cm^r, phaC down	0.00 ± 0.05	0.88 ± 0.13	0.06 ± 0.01
Δpta	pta up, Km^r, pta down	9.85 ± 1.34	0.56 ± 0.05	0.08 ± 0.03
ΔphaCE Δpta	phaE up, Cm^r, phaC down; pta up, Km^r, pta down	0.00 ± 0.01	0.41 ± 0.08	0.36 ± 0.04
ΔphaCE Δpta::DldhE	phaE up, Cm^r, phaC down; pta up, Km^r, DldhE, pta down	0.00 ± 0.01	0.20 ± 0.06	0.68 ± 0.1
ΔphaCE Δpta::DldhL	phaE up, Cm^r, phaC down; pta up, Km^r, DldhL, pta down	0.00 ± 0.01	0.10 ± 0.1	1.06 ± 0.07

Step-wise optimization

20

Cyano-acetone

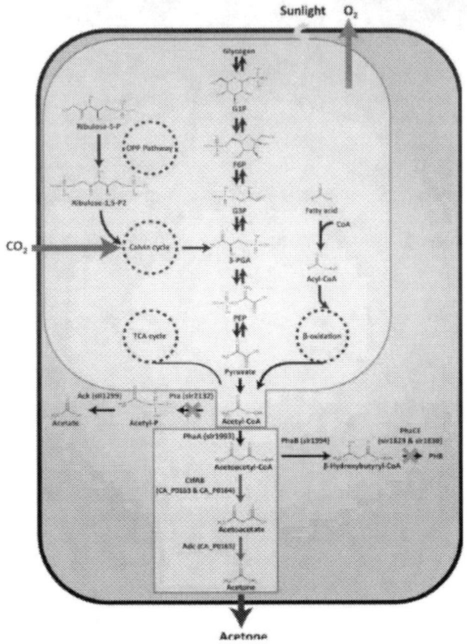

Strain	Insertion	Acetone production (mg l^{-1})
SM1	*phaE* up – Cm^r – *phaC* down	0
SM2	*phaE* up – Cm^r – *phaC* down; *pta* up – Km^r – *pta* down	0
SM3	*phaE* up – PphaE – *adc* – *ctfAB* – PrbcL – Cm^r – *phaC* down	3.0 ± 0.4
SM4	*phaE* up – Pcpc – *adc* – *ctfAB* – PrbcL – Cm^r – *phaC* down	5.0 ± 1.0
SM5	*pta* up – Km^r – *pta* down; *phaE* up – Pcpc – *adc* – *ctfAB* – PrbcL – Cm^r – *phaC* down	36.0 ± 1.5

Zhou et al., *Matob. Eng.* 2012, 14:394-400

21

Low productivity

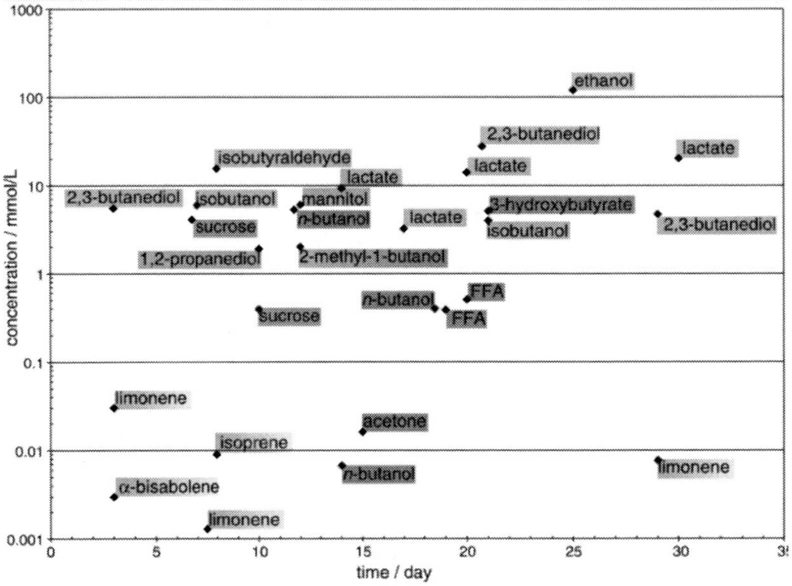

Savakis & Hellingwerf, *Curr Opin Biotechnol*. 2015, 33:8–14.

22

Energy conversion efficiency

- Visible region 400-700 nm
- Blue light heat loss
- Photorespiration
- Delivery of CO_2

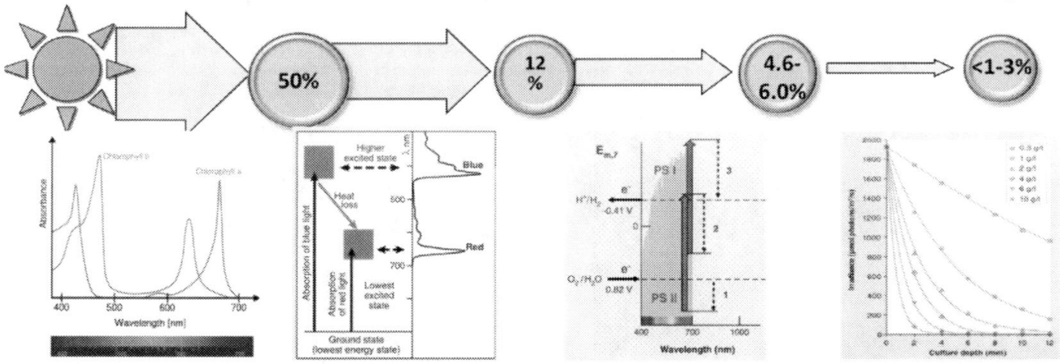

Data from *Science*, 2011, 332: 805

23

Future challenges

- Low efficiency for CO_2 fixation
- Low efficiency for energy conversion/growth
- Ability to use concentrated CO_2
- Effective carbon repartitioning

24

Acknowledgements

Cyano-chemicals

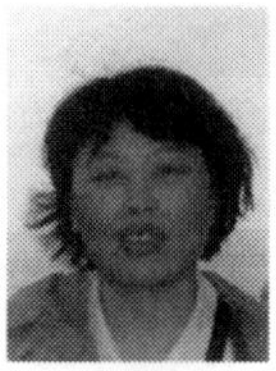

Jie Zhou

Haifeng Zhang

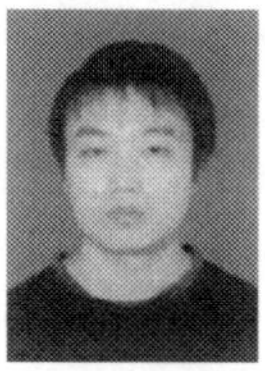

Fuliang Zhang

Hengkai Meng

Rubisco

Zhen Cai

Guoxia Liu

Fuyu Gong

Junli Zhang

- **Ministry of Science and Technology of China, 973 program**
- **Chinese Academy of Sciences, Knowledge Innovation Program**
- **National Science Foundation of China**

25

Zhou *et al. Microb Cell Fact* (2016) 15:2
DOI 10.1186/s12934-015-0405-3

Microbial Cell Factories

REVIEW Open Access

From cyanochemicals to cyanofactories: a review and perspective

Jie Zhou, Taicheng Zhu, Zhen Cai and Yin Li*

Abstract

Engineering cyanobacteria for production of chemicals from solar energy, CO_2 and water is a potential approach to address global energy and environment issues such as greenhouse effect. To date, more than 20 chemicals have been synthesized by engineered cyanobacteria using CO_2 as raw materials, and these studies have been well reviewed. However, unlike heterotrophic microorganisms, the low CO_2 fixation rate makes it a long way to go from cyanochemicals to cyanofactories. Here we review recent progresses on improvement of carbon fixation and redistribution of intercellular carbon flux, and discuss the challenges for developing cyanofactories in the future.

Keywords: Cyanobacteria, Cyanochemicals, Cyanofactories, Rubisco, Metabolic Engineering, Rerouting carbon flux, Strong promoter, Cofactor balance

26

Thank you for your attention!
MPSysBio
Lab of Molecular Physiology and Systems Biotechnology

Engineering Bacterial Organelles for the Sequestration of Synthetic Metabolic Pathways

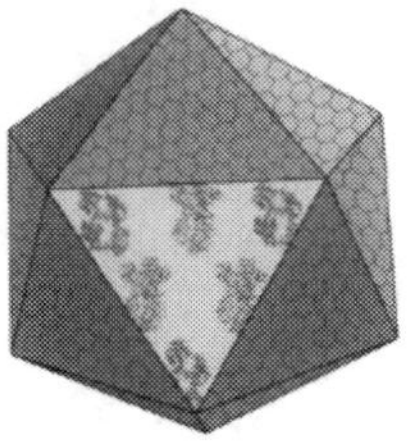

Marilyn Slininger
Christopher Jakobson, Prof. Danielle Tullman-Ercek
UC Berkeley
Dept. of Chemical and Biomolecular Engineering
Metabolic Engineering 11 – June 30, 2016

Engineering complex protein systems to control the movement of molecules across membranes

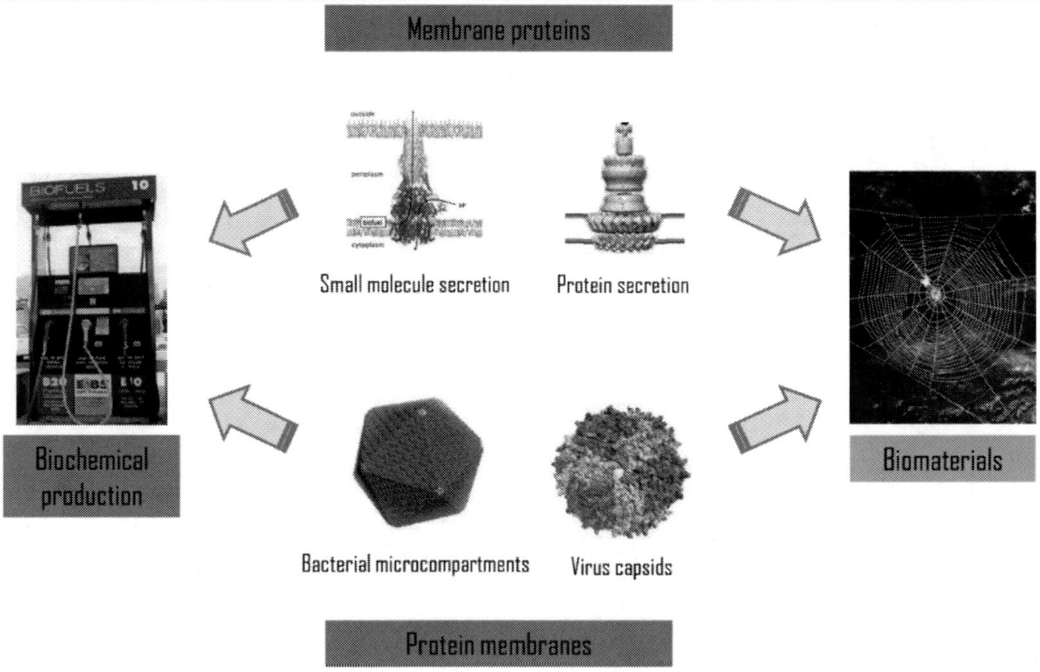

Engineering complex protein systems to control the movement of molecules across membranes

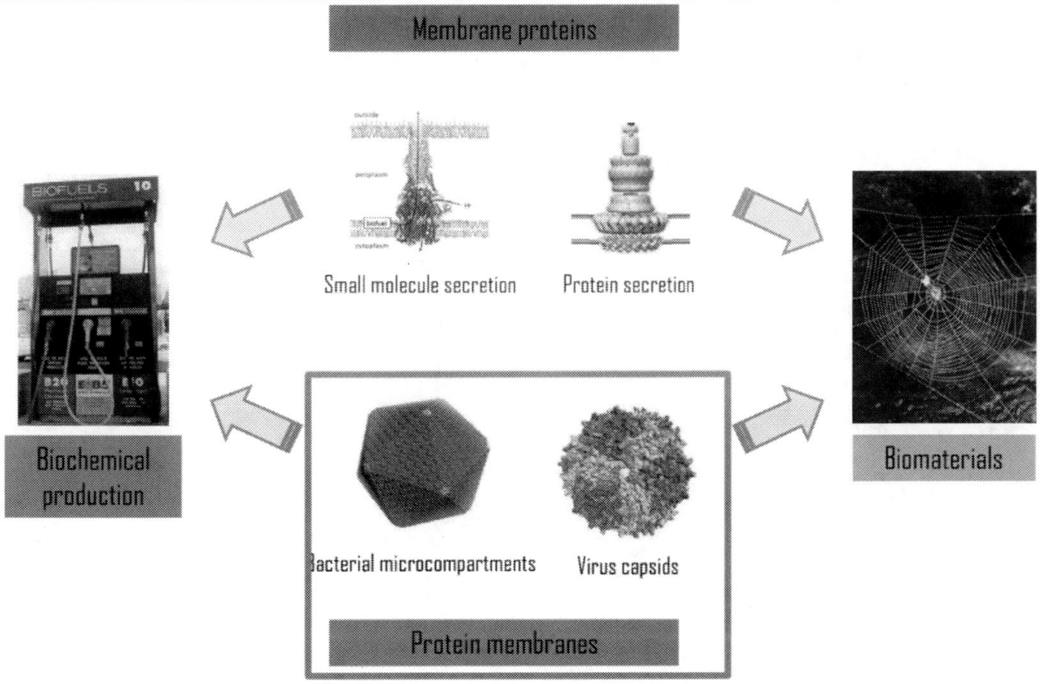

Designing bacterial organelles

Goal: Create designer compartments using protein-based membranes in bacteria

Application: Sequester synthetic metabolic pathways (enzymes and intermediates) for select chemical production

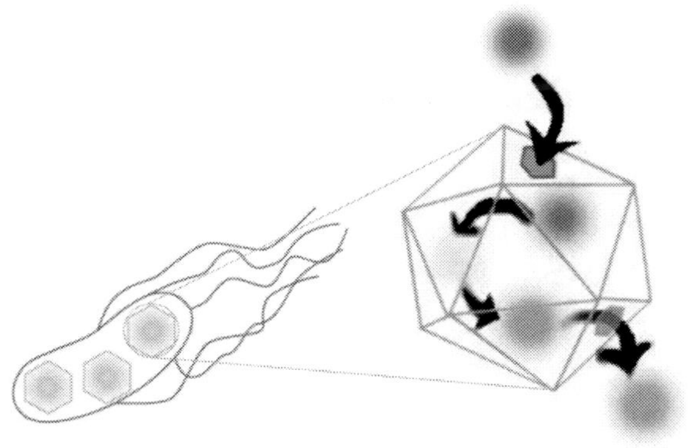

Toxicity and bottlenecks occur in metabolic engineering due to:

Cofactor imbalances or low local reactant concentrations

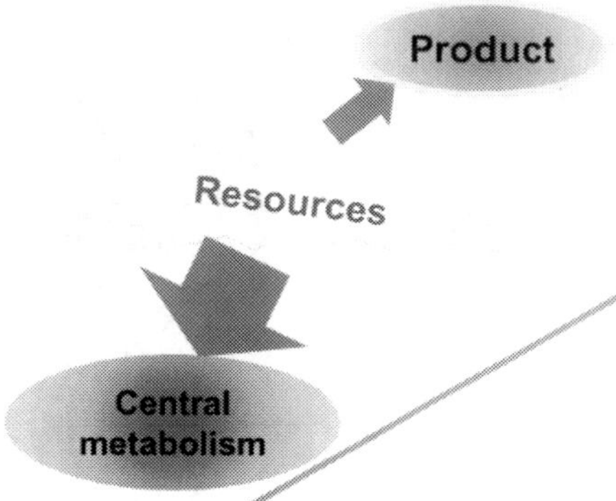

Toxicity and bottlenecks occur in metabolic engineering due to:

Cofactor imbalances or low local reactant concentrations

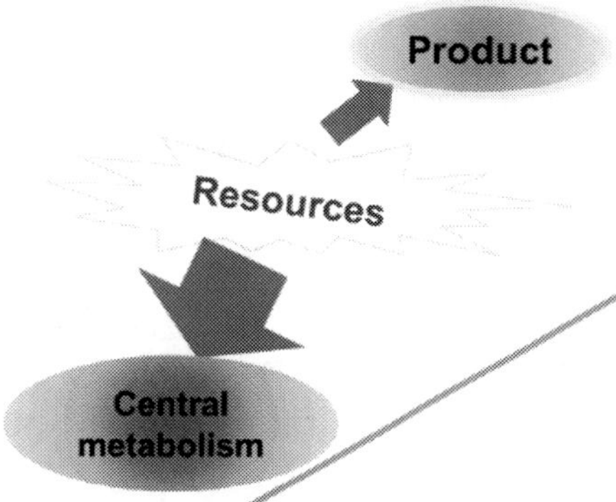

Undesired reactions and byproducts

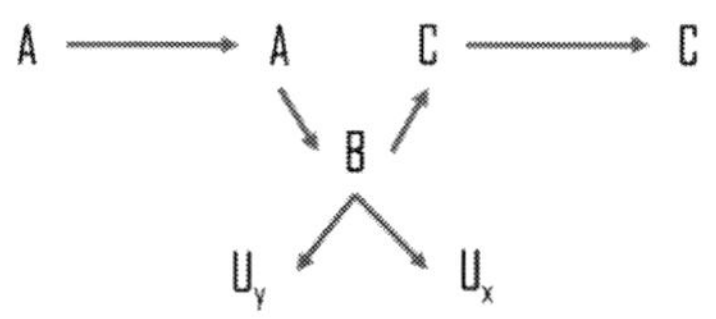

Toxicity and bottlenecks occur in metabolic engineering due to:

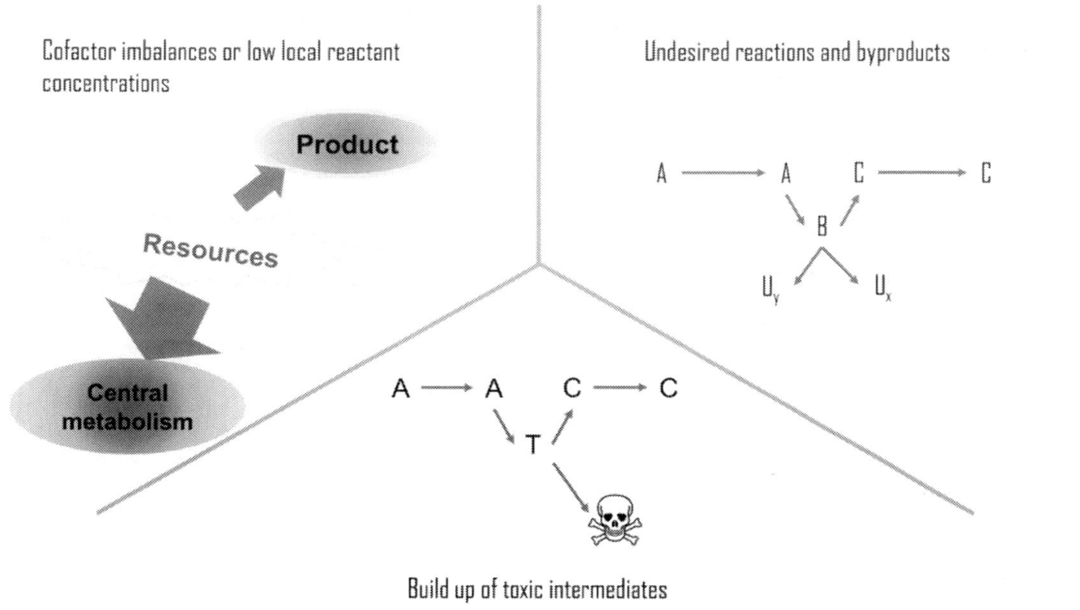

Intracellular organization addresses toxicity and bottlenecks by:

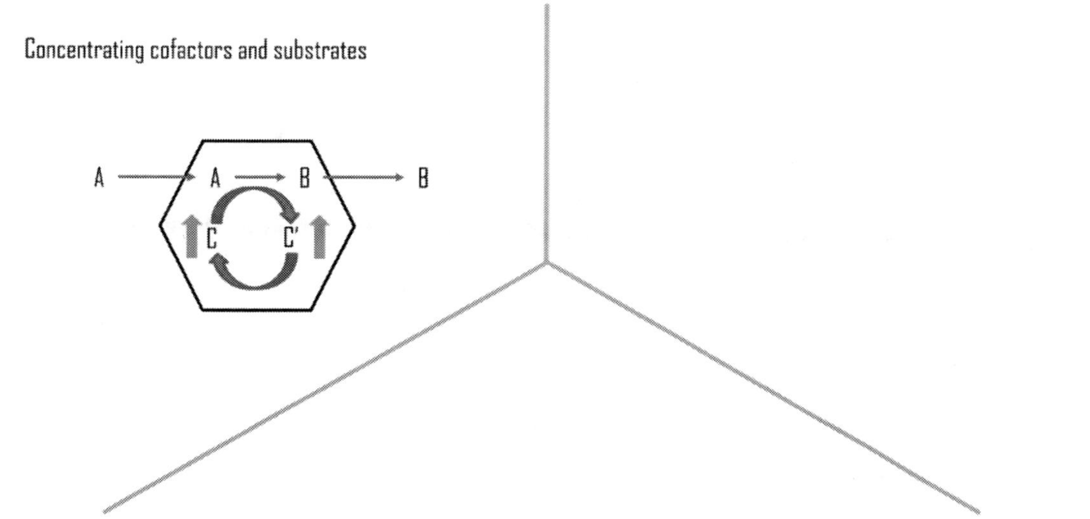

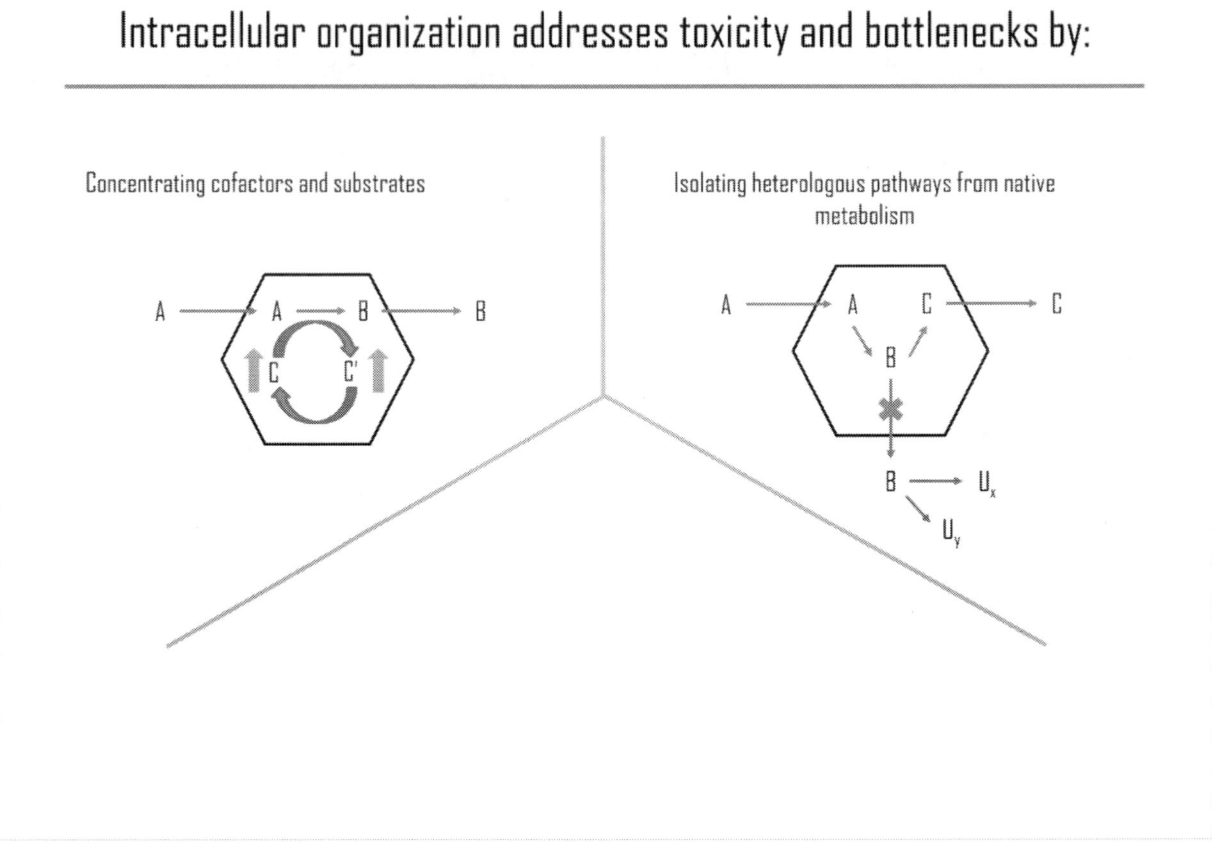
Intracellular organization addresses toxicity and bottlenecks by:
Concentrating cofactors and substrates
A
A
B
B
C
C'
Isolating heterologous pathways from native metabolism
A
A
C
C
B
B
U_x
U_y

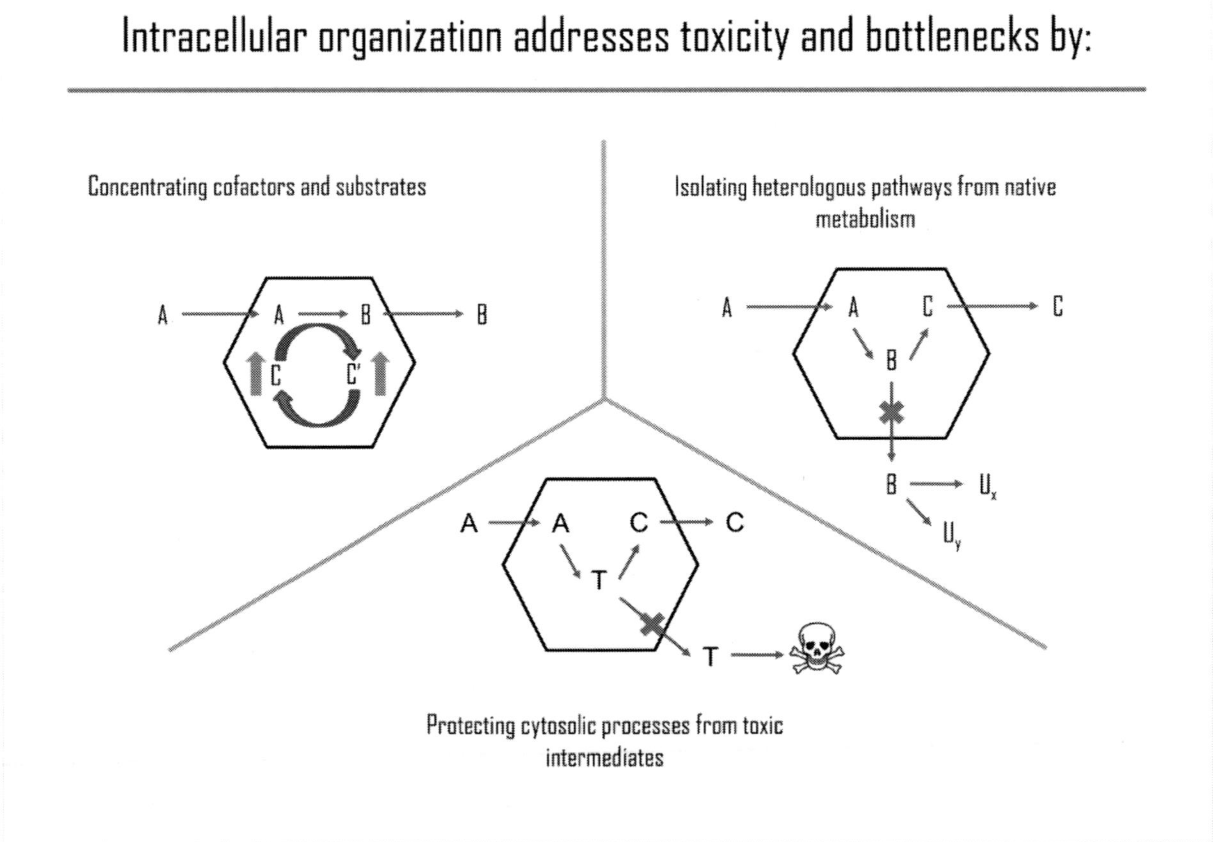
Intracellular organization addresses toxicity and bottlenecks by:
Concentrating cofactors and substrates
A
A
B
B
C
C'
Isolating heterologous pathways from native metabolism
A
A
C
C
B
B
U_x
U_y
A
A
C
C
T
T
Protecting cytosolic processes from toxic intermediates

The propanediol utilization (Pdu) bacterial microcompartment (MCP) is an excellent starting point

- Found in enteric bacteria
- Encapsulates enzymes involved in 1,2-propanediol metabolism

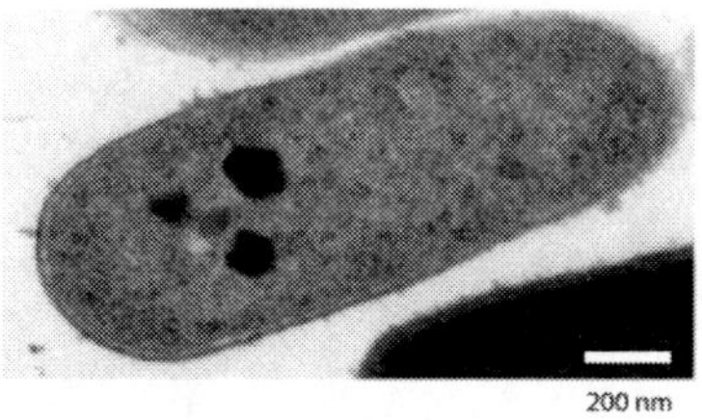

TEM of cell sections

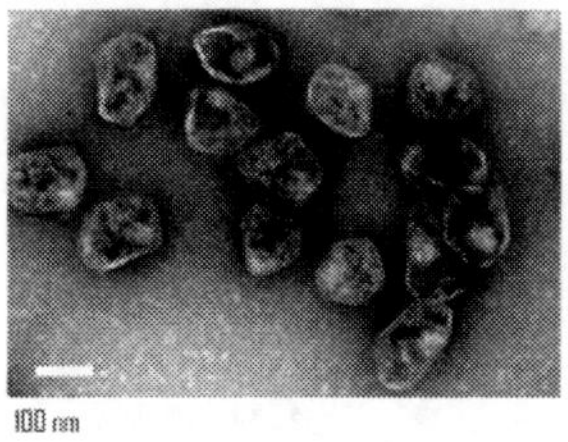

Purified Pdu MCPs

The Pdu operon encodes a protein shell and enzymatic cargo

Enzyme Cargo
PduCDE
PduGH
PduL
PduO
PduP
PduQ
PduS
PduW

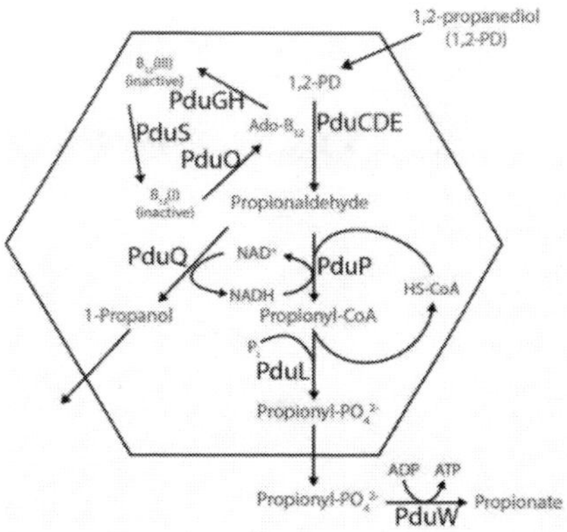

structural proteins
PduA
PduB
PduJ
PduK
PduM
PduN
PduT
PduU

Crowley et al., J. Biol. Chem., 2010;
Fan et al., PNAS, 2012.

Pdu MCP constructed from shell proteins with pores

- Protein shell composed of thousands of subunits

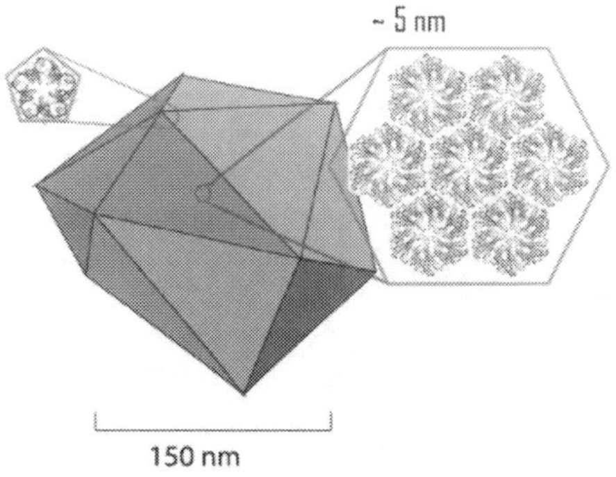

1,2-propanediol utilization (Pdu) MCP

Fan et al (2012) *PNAS*

Pdu MCP constructed from shell proteins with pores

- Protein shell composed of thousands of subunits

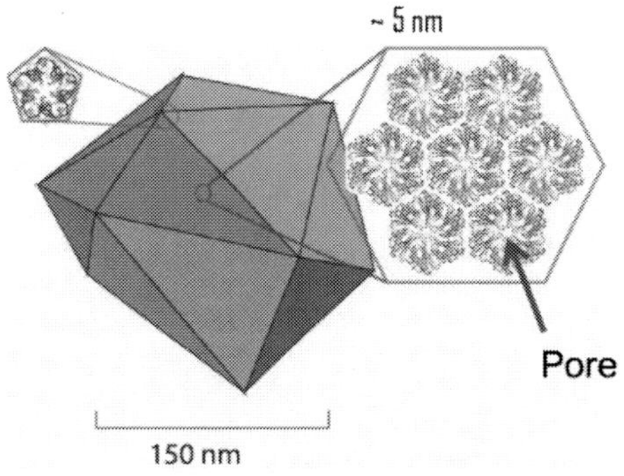

1,2-propanediol utilization (Pdu) MCP

Fan et al (2012) *PNAS*

Pdu MCP encapsulates cargo proteins via peptide interactions

- Protein shell composed of thousands of subunits
- Proteins are encapsulated by peptide tag interactions with the shell.

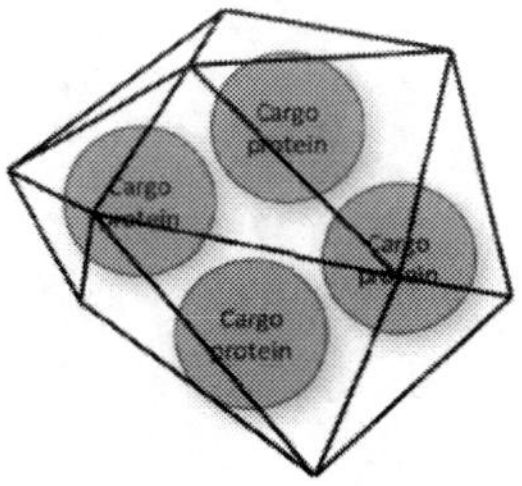

1,2-propanediol utilization (Pdu) MCP from S. enterica

Fan et al (2012) *PNAS*

Pdu MCP encapsulates cargo proteins via peptide interactions

- Protein shell composed of thousands of subunits
- Proteins are encapsulated by peptide tag interactions with the shell.

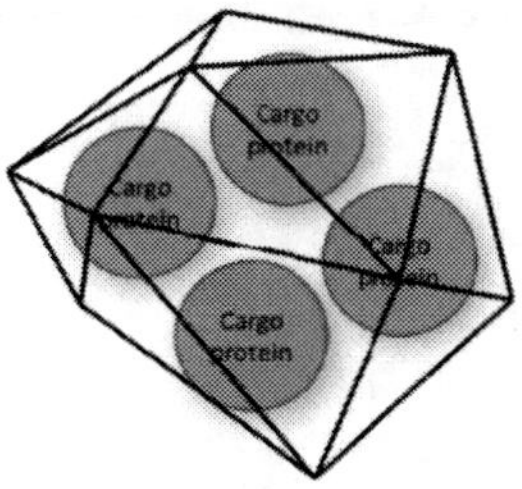

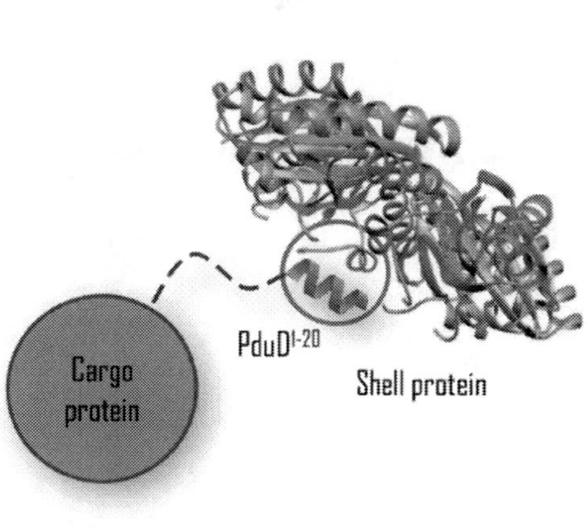

1,2-propanediol utilization (Pdu) MCP from S. enterica

Fan et al (2012) *PNAS*

Addition of a degradation-tagged GFP enables an assay compatible with microscopy and flow cytometry

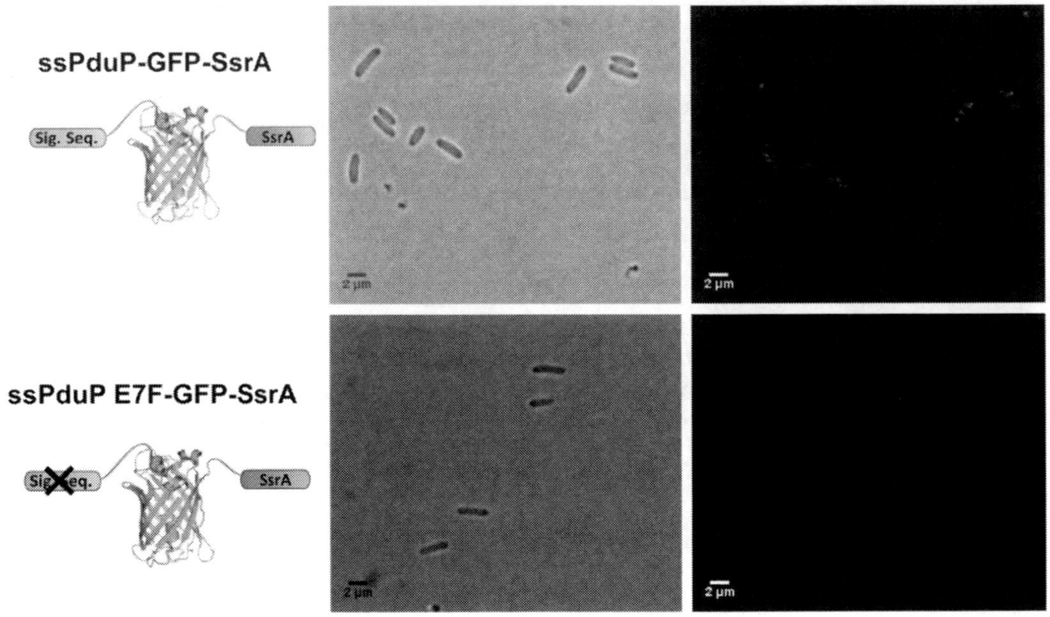

Kim and Tullman-Ercek, *Biotech J.* 2014.

Engineering custom subcellular organelles

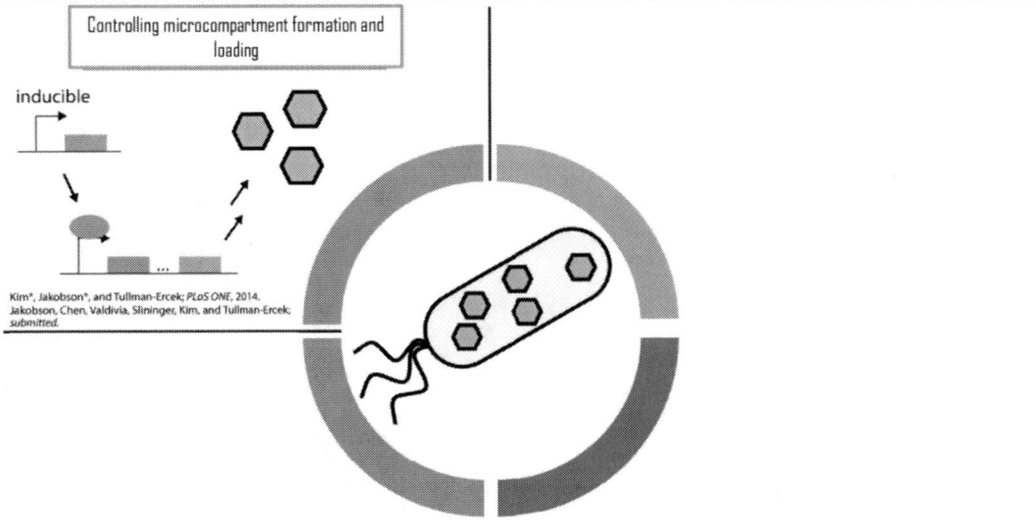

Native microcompartment formation is controlled via a complex regulatory network

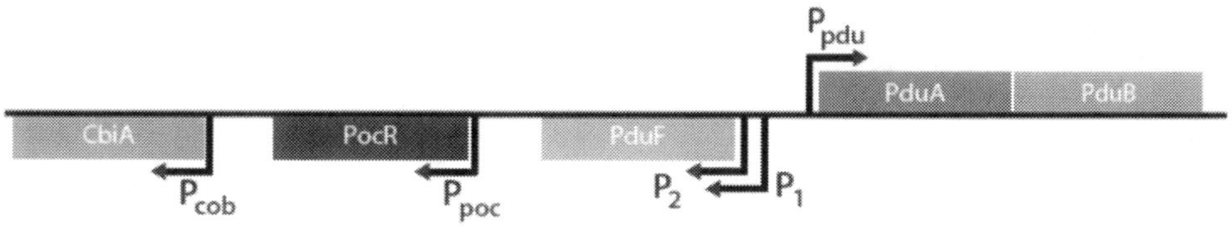

- 1,2-propanediol activates the Pdu operon
- Glucose represses the Pdu operon
- PocR is a known transcriptional regulator

We hypothesized that we could use PocR production in place of 1,2-propanediol addition and bypass the need for special media to create native microcompartments.

We can inducibly control native microcompartment formation in standard rich media...

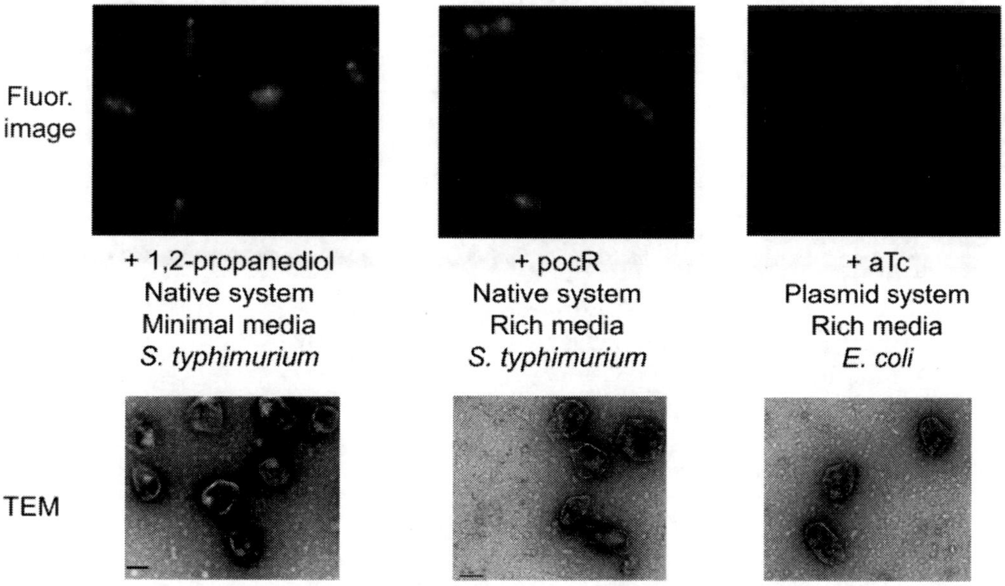

...and in *E. coli* using a minimal operon

Kim, Jakobson, and Tullman-Ercek, *PLoS ONE, 2014.*

Altering MCP induction increases loading

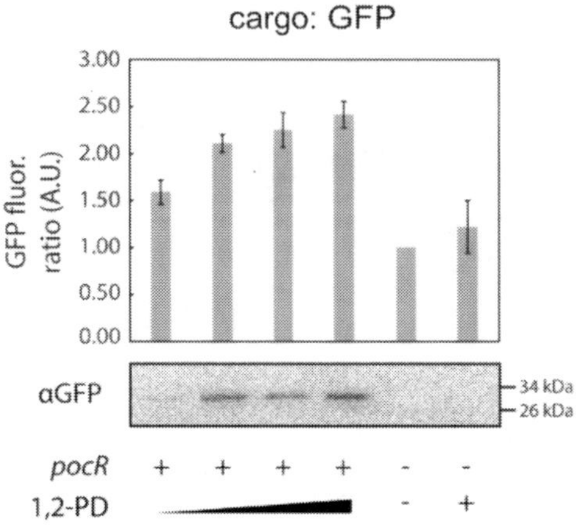

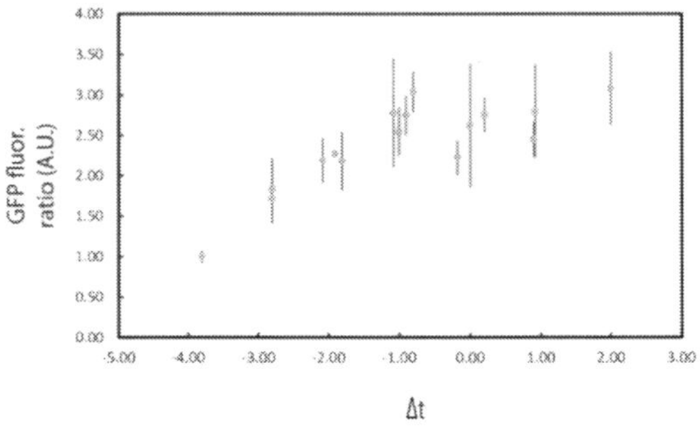

Engineering custom subcellular organelles

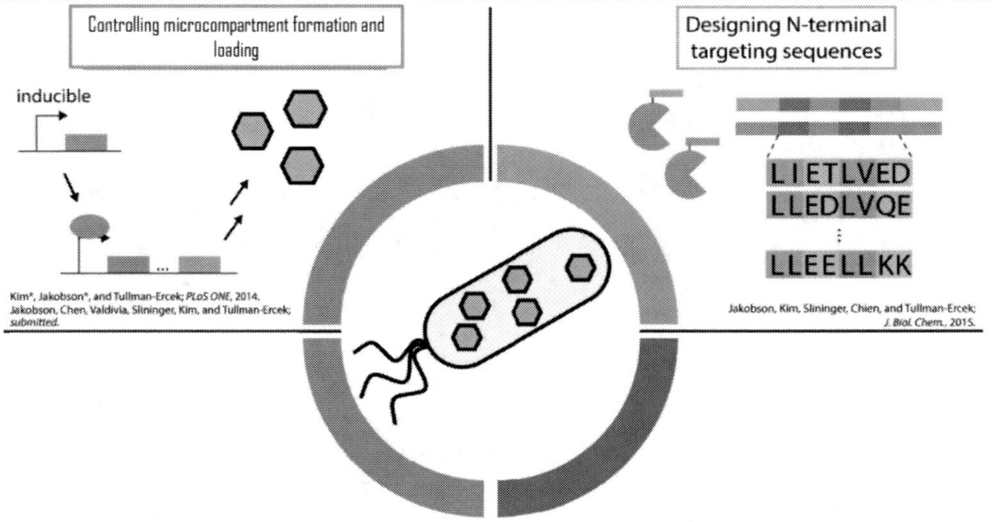

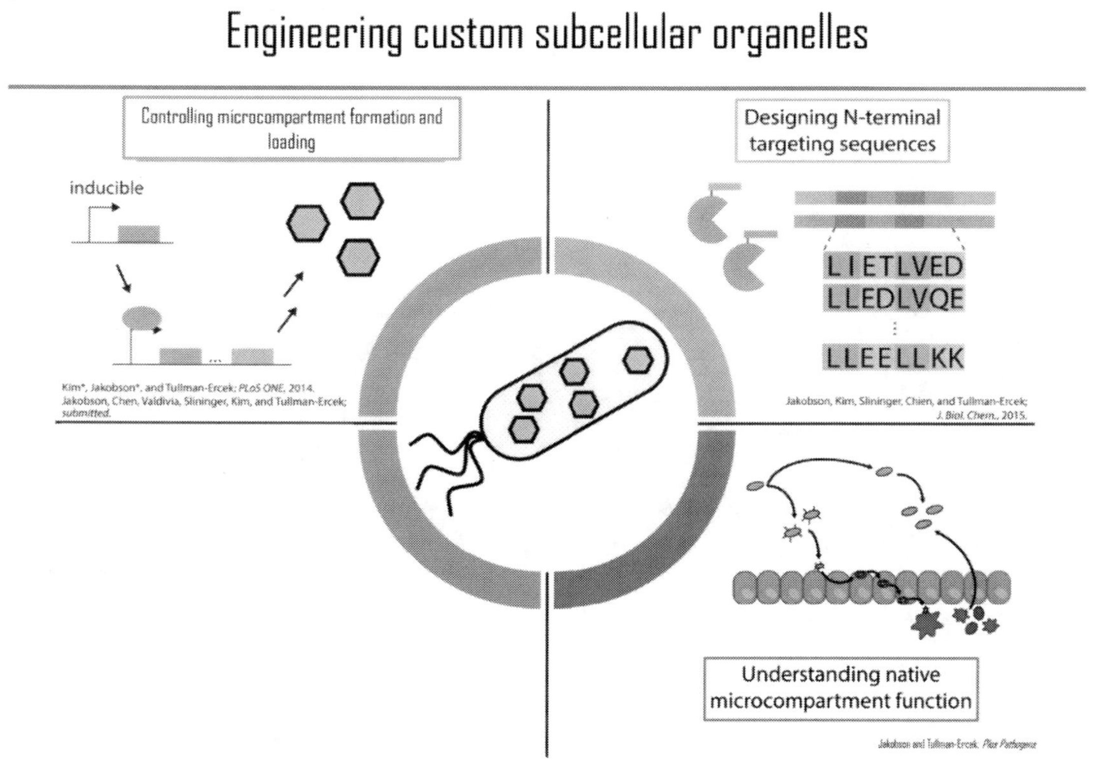

Engineering custom subcellular organelles
Controlling microcompartment formation and loading
inducible
Kim*, Jakobson*, and Tullman-Ercek; PLoS ONE, 2014.
Jakobson, Chen, Valdivia, Slininger, Kim, and Tullman-Ercek; submitted.
Designing N-terminal targeting sequences
LIETLVED
LLEDLVQE
LLEELLKK
Jakobson, Kim, Slininger, Chien, and Tullman-Ercek; J. Biol. Chem., 2015.
Understanding native microcompartment function
Jakobson and Tullman-Ercek. Plos Pathogens

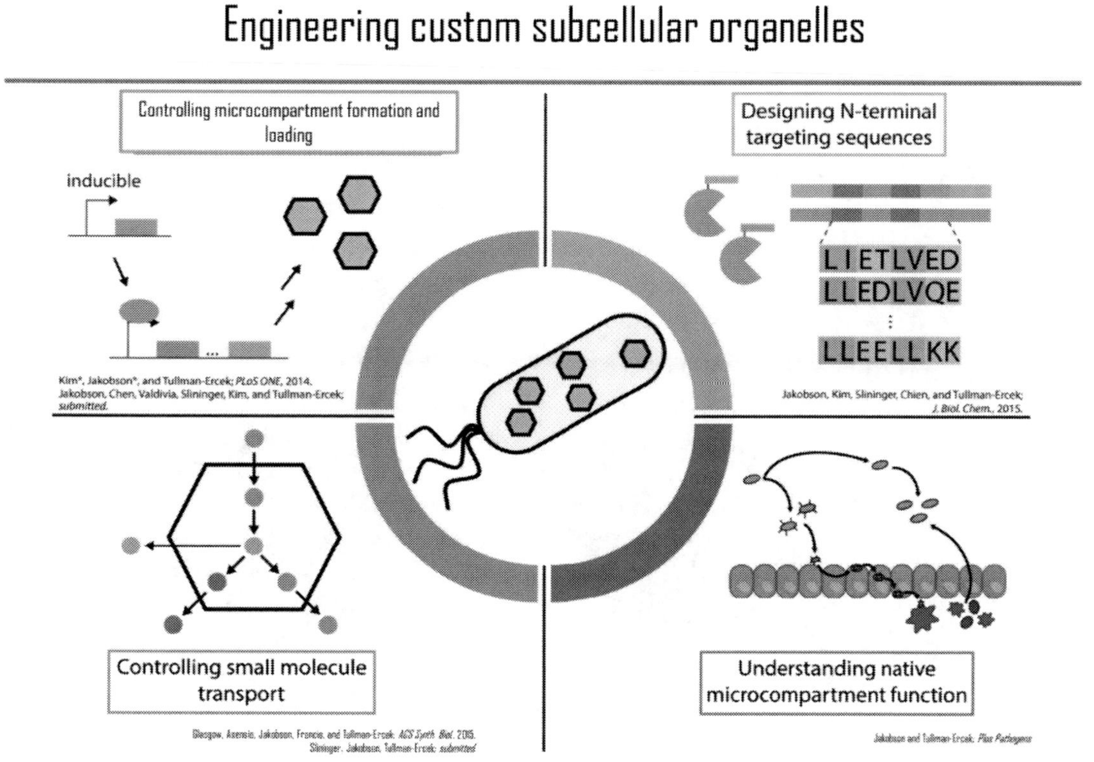

Engineering custom subcellular organelles
Controlling microcompartment formation and loading
inducible
Kim*, Jakobson*, and Tullman-Ercek; PLoS ONE, 2014.
Jakobson, Chen, Valdivia, Slininger, Kim, and Tullman-Ercek; submitted.
Designing N-terminal targeting sequences
LIETLVED
LLEDLVQE
LLEELLKK
Jakobson, Kim, Slininger, Chien, and Tullman-Ercek; J. Biol. Chem., 2015.
Controlling small molecule transport
Glasgow, Asensio, Jakobson, Francis, and Tullman-Ercek. ACS Synth. Biol. 2015.
Slininger, Jakobson, Tullman-Ercek; submitted
Understanding native microcompartment function
Jakobson and Tullman-Ercek. Plos Pathogens

Engineering custom subcellular organelles

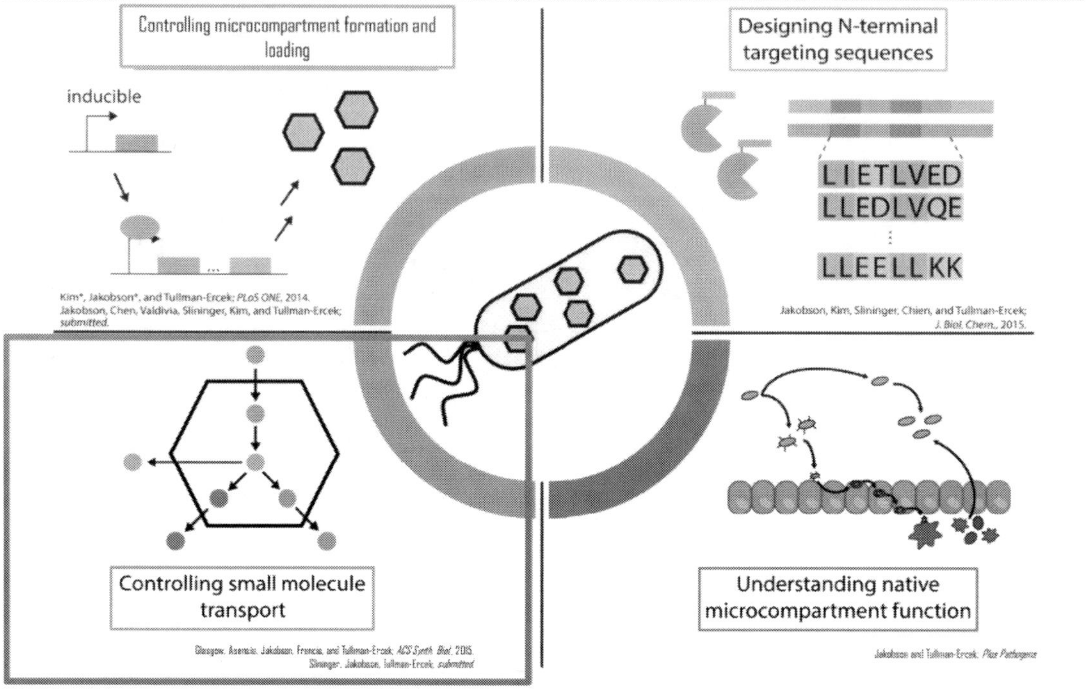

We investigate the effect of pore structure on function of encapsulated enzymes

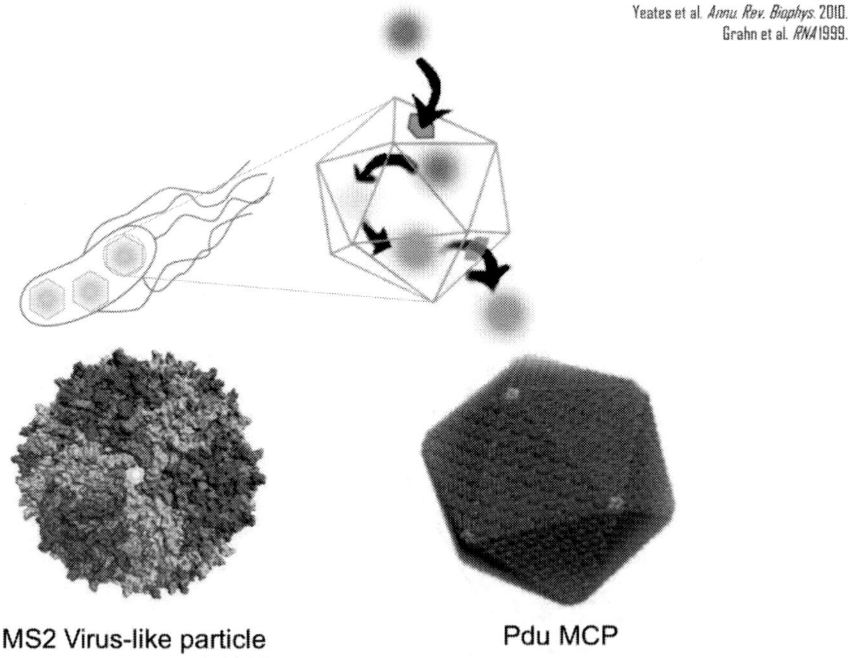

Virus-like particles: analogous protein membrane systems

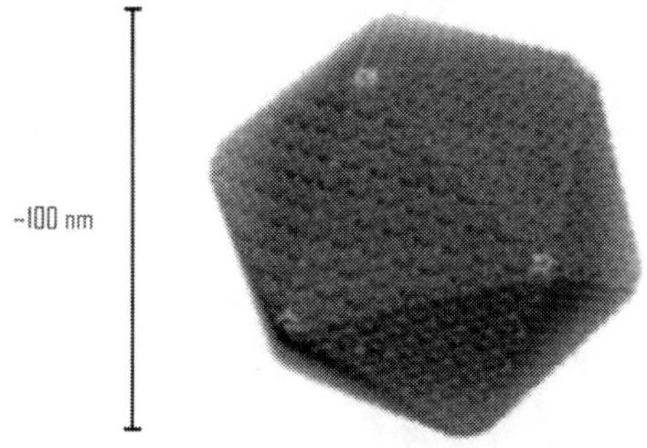

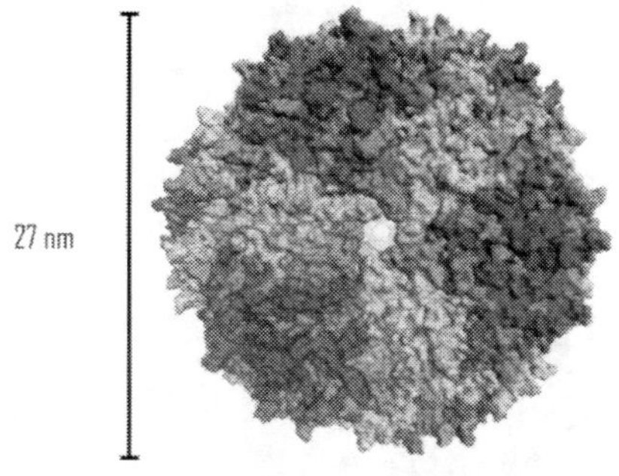

Pdu Bacterial Microcompartment

- More complex and harder to isolate and assay

MS2 Virus-like particles

- Coat proteins similarly form pores (about double the diameter)
- Easier to work with
- Useful for *in vitro* studies

Yeates et al. *Annu. Rev. Biophys.* 2010.
Grahn et al. *RNA* 1999.

We used alkaline phosphatase assay to probe the effect of pore changes on one step reaction

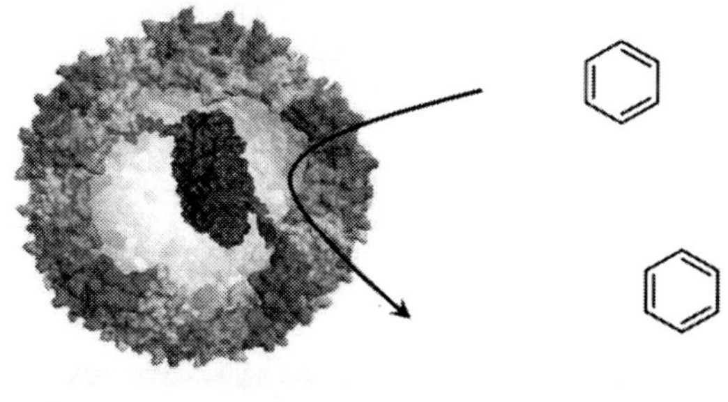

4-methylumbelliferyl phosphate

With Matt Francis

Glasgow et al. *ACS Nano* 2012.

We used alkaline phosphatase assay to probe the effect of pore changes on one step reaction

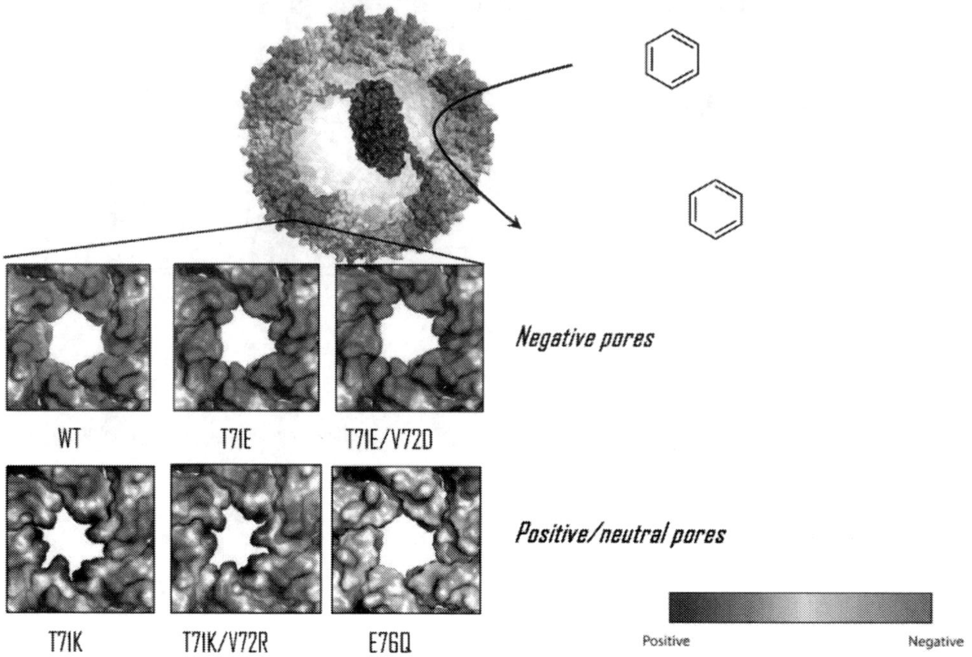

With Matt Francis

Glasgow et al. *ACS Nano* 2012.

For alkaline phosphatase substrates, negatively-charged pores greatly impact the enzyme Km

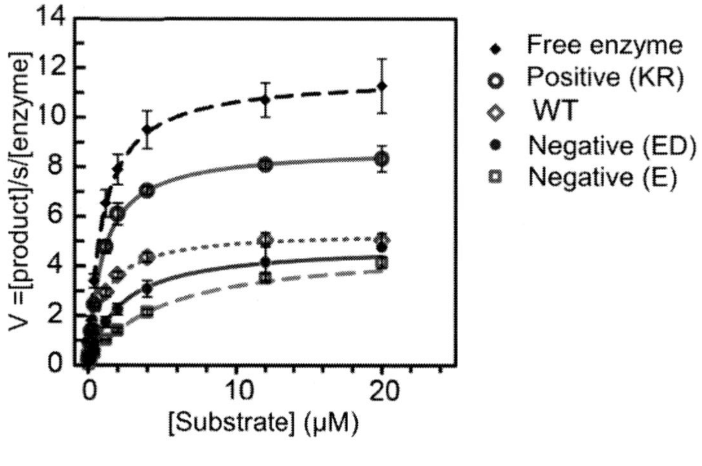

Glasgow et al. *ACS Nano* 2012.

For alkaline phosphatase substrates, negatively-charged pores greatly impact the enzyme Km

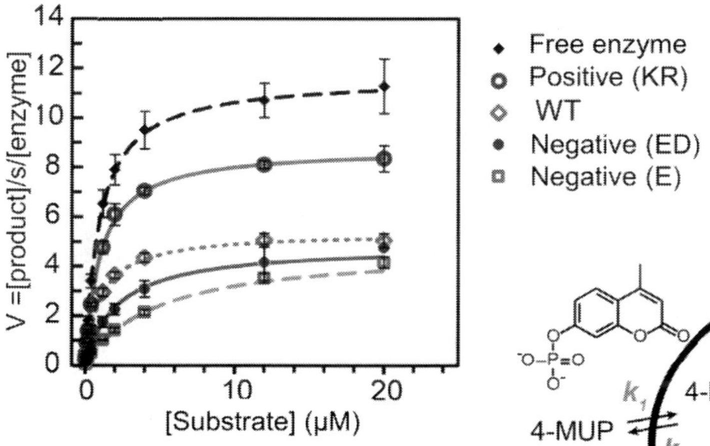

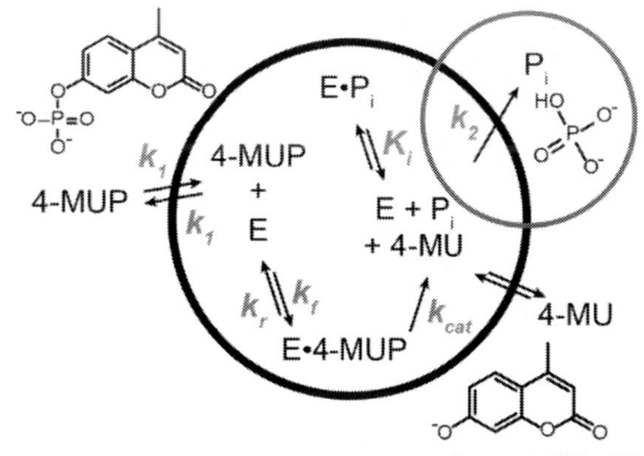

Glasgow et al. *ACS Nano* 2012.

We hypothesized that similar effects could occur in the Pdu MCP system

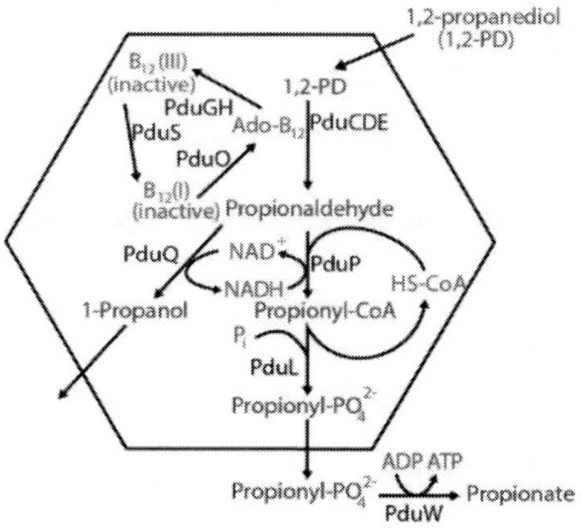

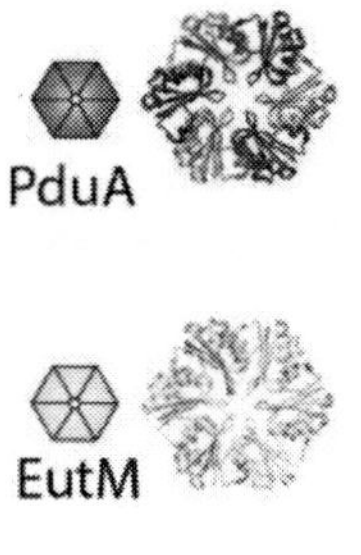

There are two ways to change shell properties

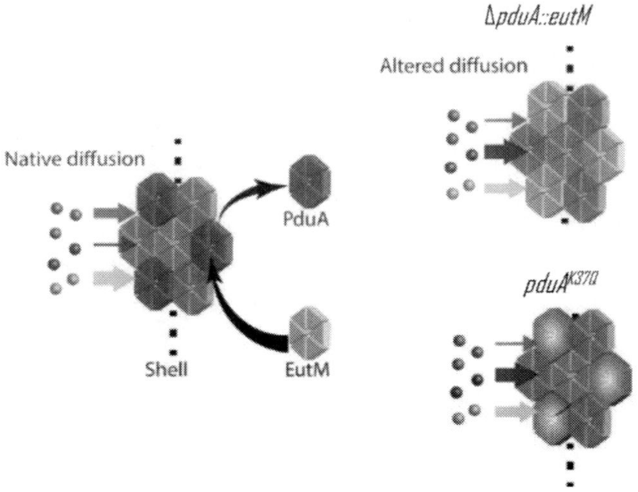

- Complete shell protein substitution
- Alteration of pore structure (Chimera)

There are two ways to change shell properties

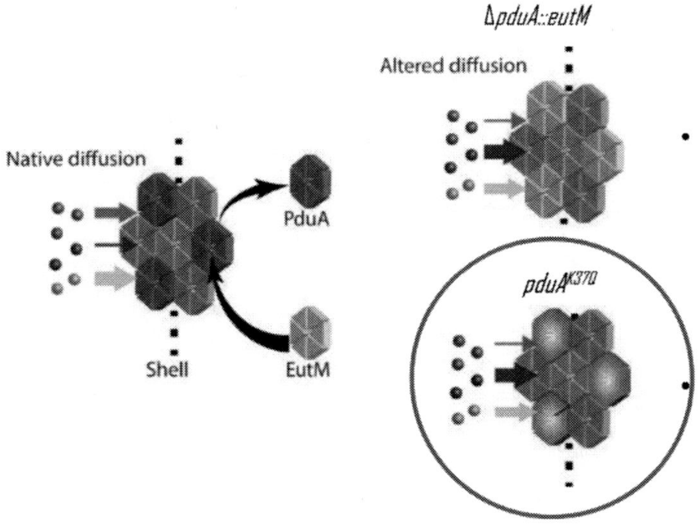

- Complete shell protein substitution
- Alteration of pore structure (Chimera)

Pdu MCPs may be visualized *in vivo* and *in vitro*

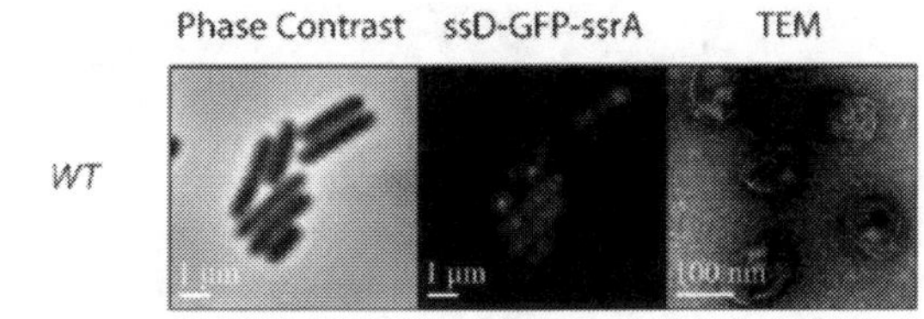

The *pduA*^{K37Q} mutant also forms MCPs

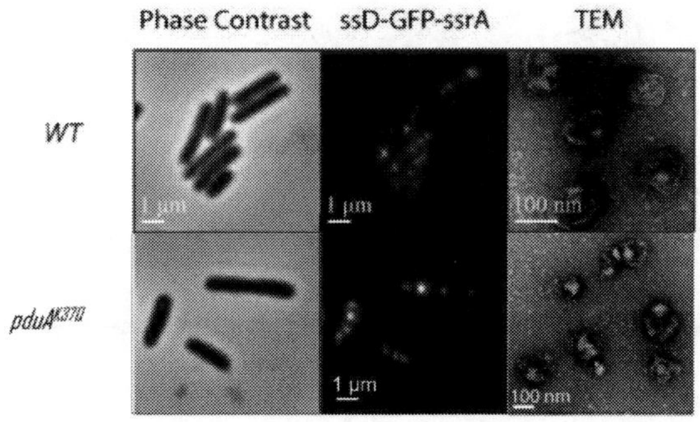

pduA^{K26A} is a non-MCP-forming control

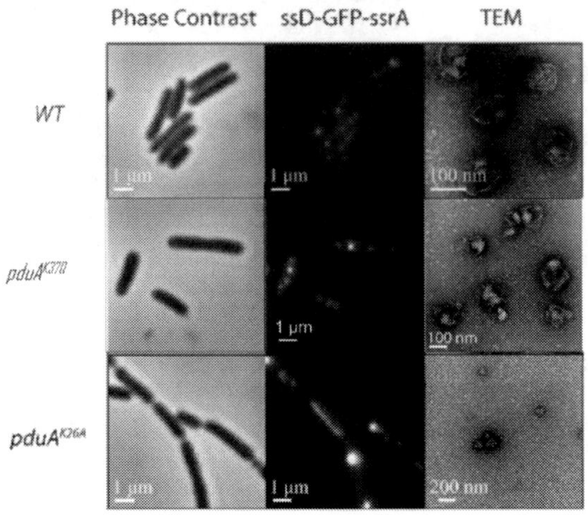

pduA^{K37Q} mutant has altered growth

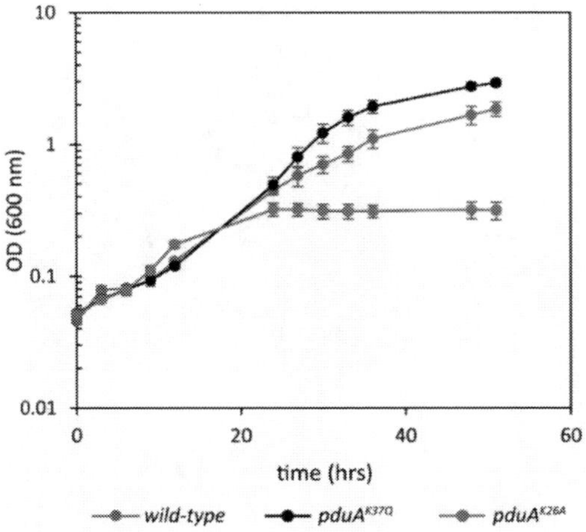

pduA^{K37Q} mutant consumes1,2-PD at the same rate as WT

Metabolite concentration in growth medium

mM 1,2-propanediol

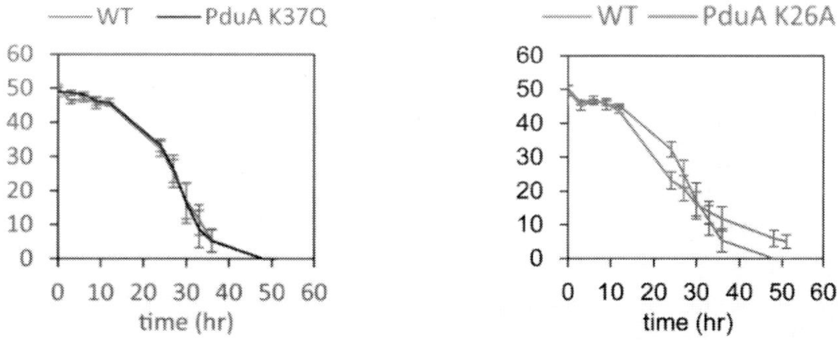

pduA^{K37Q} mutant has decreased propionate and propionaldehyde in media

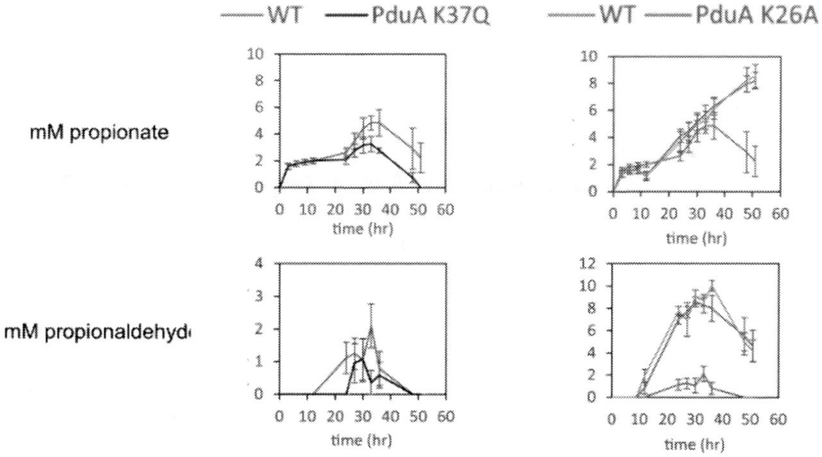

Two likely explanations stem from a change in shell permeability

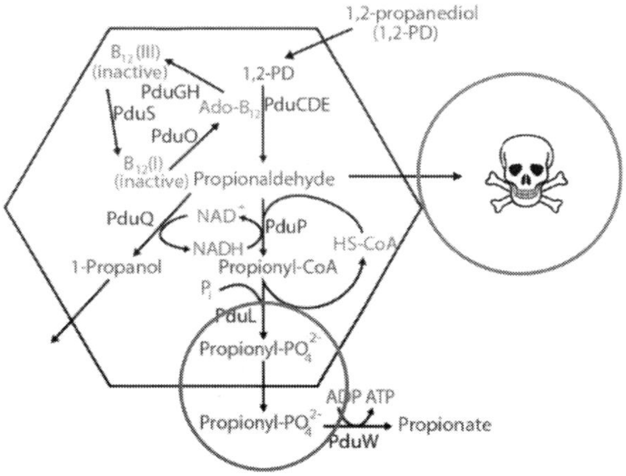

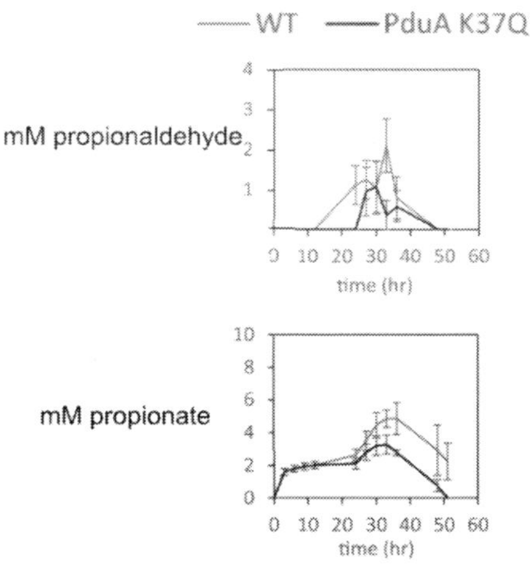

Mutagenesis of the pore shows hydrophobicity and charge affect pathway function

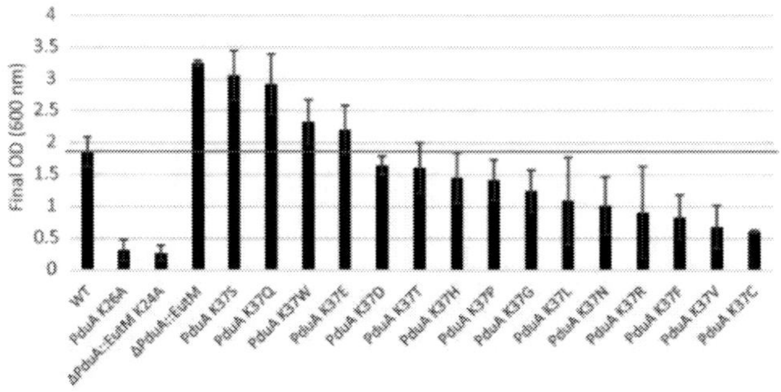

In summary: Pdu MCPs can alter behavior of encapsulated pathway by altering metabolite diffusion

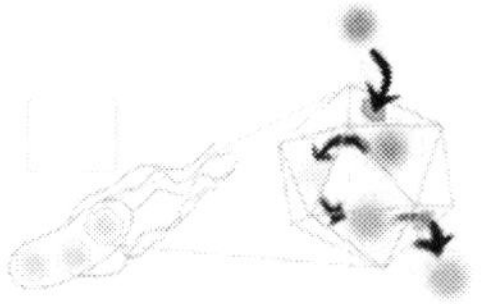

- Protein membranes serve as excellent starting points for bacterial organelle engineering

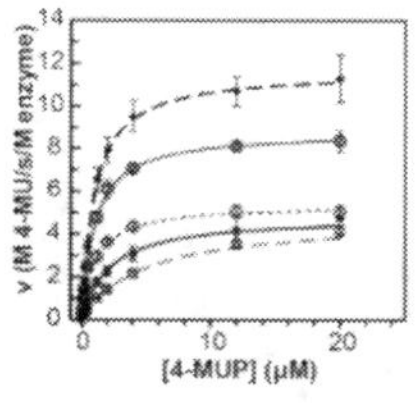

- MS2 capsid pores can impact encapsulated reaction kinetics by affecting small molecule transport rates *out of* the compartment

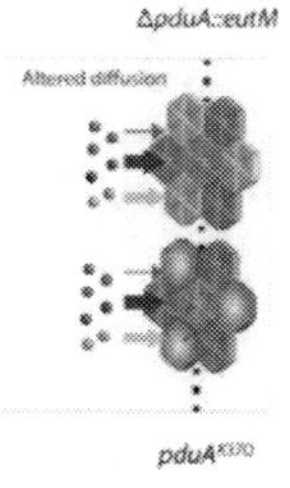

- Activity of Pdu enzymes encapsulated within the Pdu MCP can be affected by modifying shell protein pores.

Acknowledgements

Graduate students and postdocs:
Christopher Jakobson
Han Teng Wong
Lisa Burdette
Emily Hartman
Taylor Dickman
Douglas Griffith, PhD

Compartments Subgroup Alumni:
Eddy Kim, PhD
Jeff Glasgow, PhD
Mike Asensio

Project Collaborators:
Adam Arkin
Matthew Francis
Clay Radke

NSF
U.S. DEPARTMENT OF ENERGY Office of Science
Energy Biosciences Institute
CBGP Chemical Biology Graduate Program
jbei Joint BioEnergy Institute

EutM incorporates into the MCP and pore mutant forms normal compartment structure

Analysis of purified MCPs

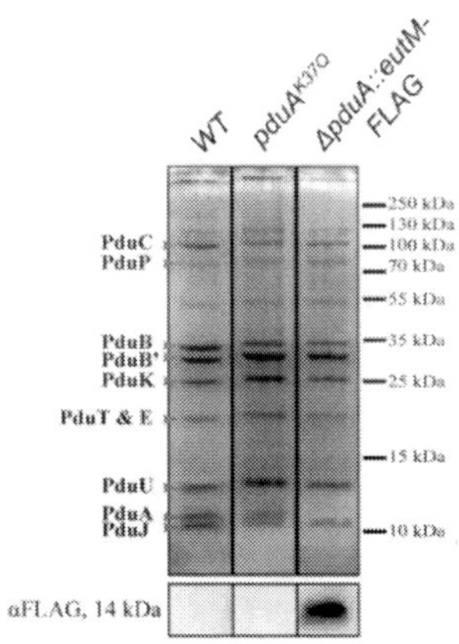

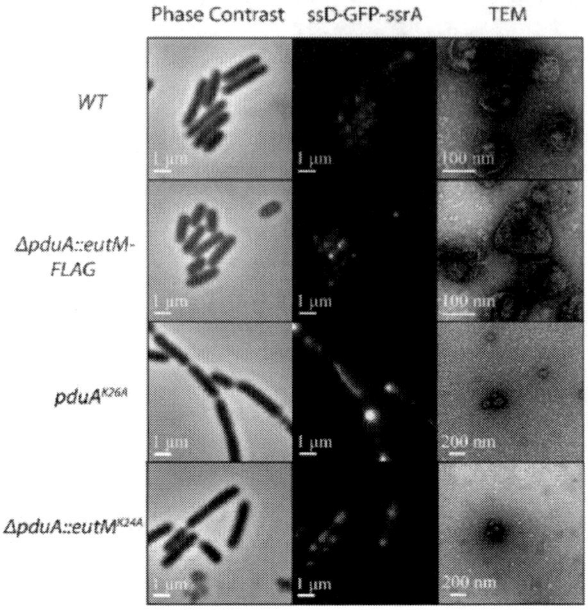

Making the analogous mutation to the substituted *eutM* gene causes the same phenotype.

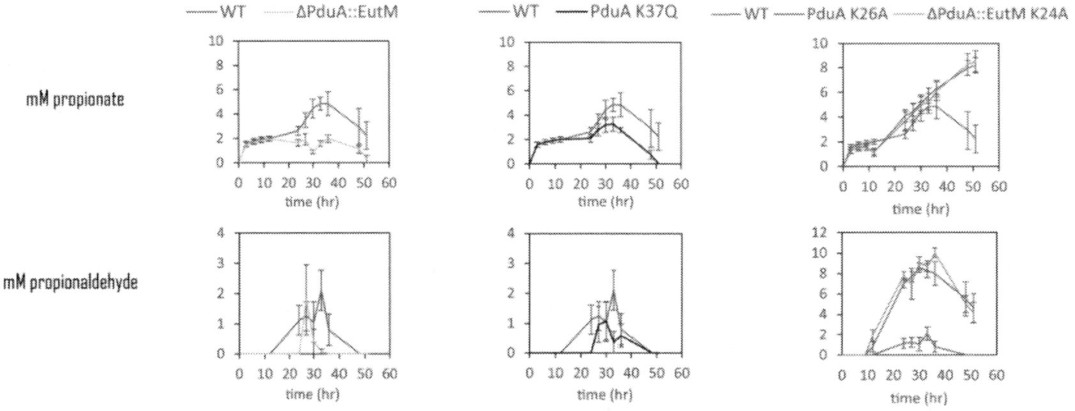

Addition of a degradation-tagged GFP enables an assay compatible with microscopy and flow cytometry

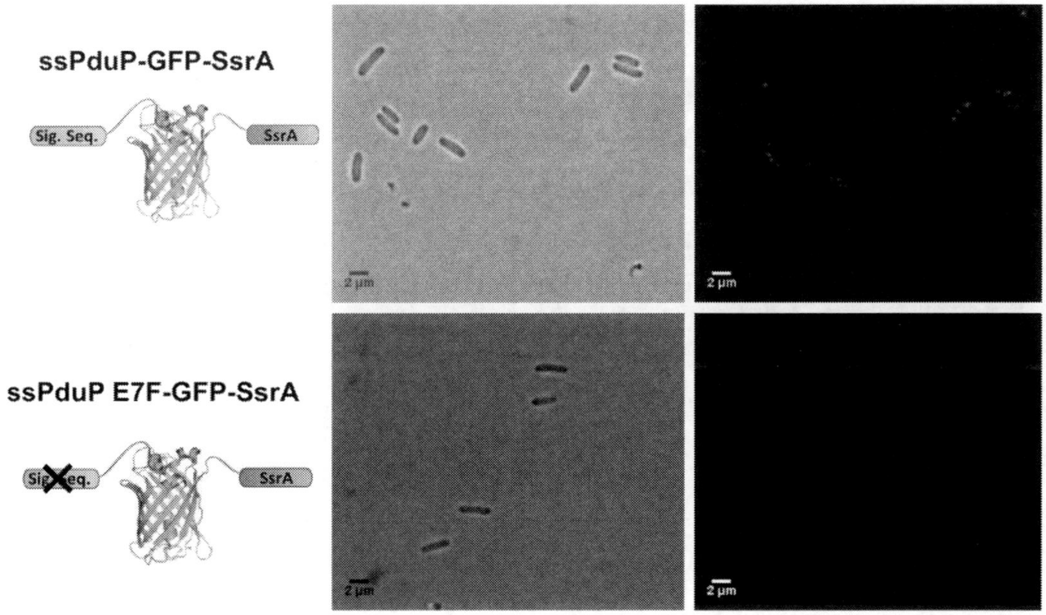

Kim and Tullman-Ercek, *Biotech J.* 2014.

Addition of a degradation-tagged GFP enables an assay compatible with microscopy and flow cytometry

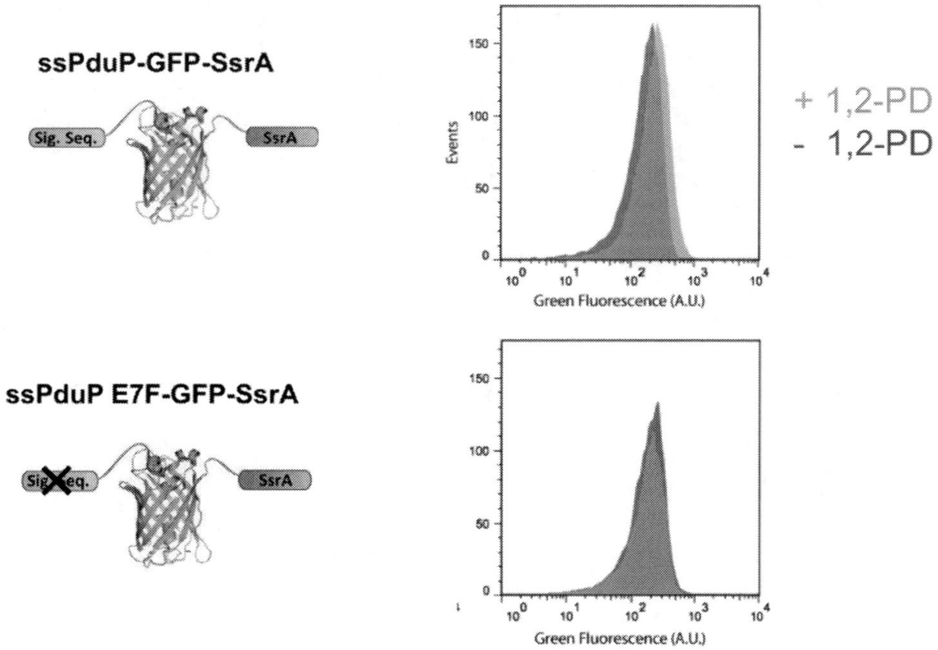

Native microcompartment formation is tightly controlled via a complex regulatory network

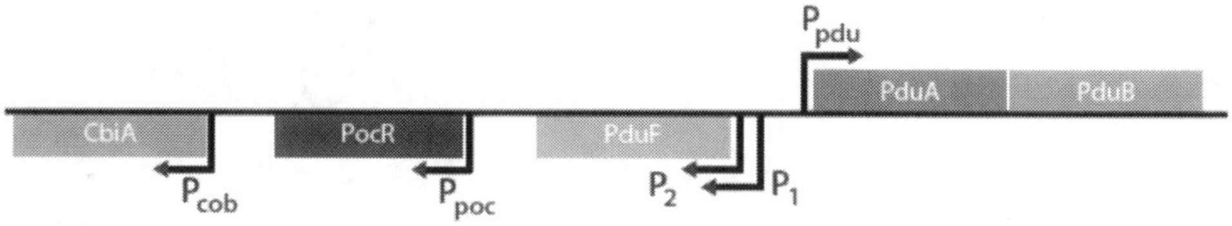

- 1,2-propanediol activates the Pdu operon
- Glucose represses the Pdu operon
- PocR is a known transcriptional regulator

We hypothesized that we could use PocR production in place of 1,2-propanediol addition and bypass the need for special media to create native microcompartments.

We can inducibly control native microcompartment formation in standard rich media...

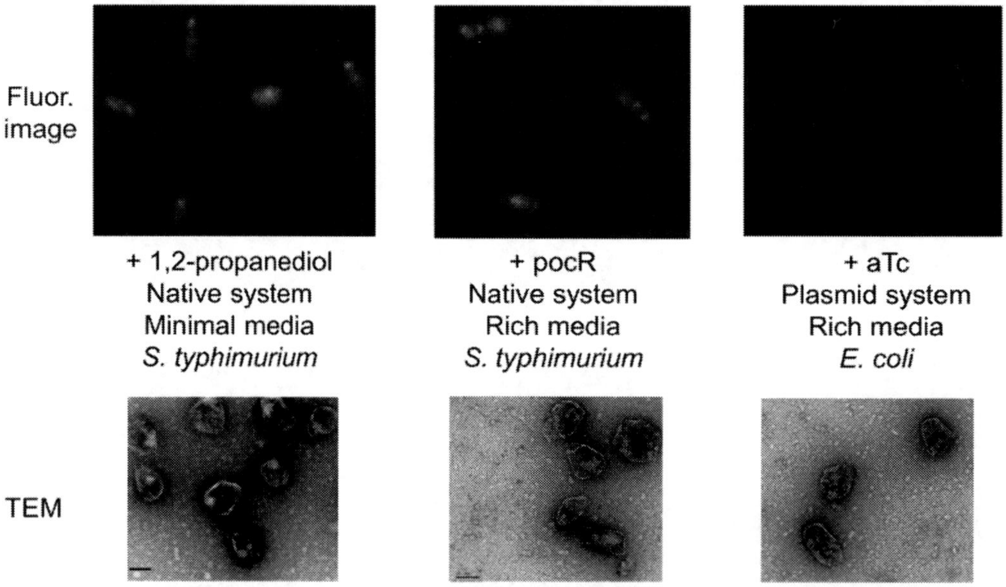

...and in *E. coli* using a minimal operon

Kim, Jakobson, and Tullman-Ercek, *Submitted.*

Engineering Trans-Acting Regulators to Dynamically Control Gene Expression in Bacteria

THE UNIVERSITY of WISCONSIN MADISON

Brian F. Pfleger

Metabolic Engineering 11

June 30th, 2016

Pfleger Lab?

Who are we?
- 9 grad students, 3 post-docs, ~10 undergrads
- Chemical Engineers, Biochemists, Chemists, Microbiologists

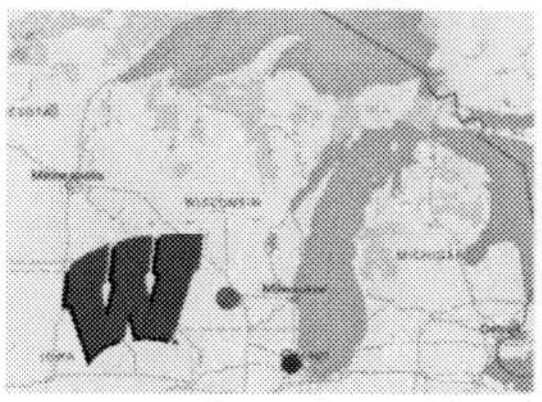

Chicago

THE UNIVERSITY of WISCONSIN MADISON

What do we study?
- Gene discovery and characterization
- **Synthetic biology tool development**
  - ***Cyanobacteria***
  - ***Trans-acting regulators***
- Metabolic engineering
  - Fatty acids
  - Fatty alcohols
  - Bioplastics
  - Fuels
  - Natural products
- Functional genomics and systems biology analysis
- Bioreactor studies and bioprocess design

Elements of Metabolic Engineering

1. Components of a Metabolic Pathway
   - What is the most efficient way from A→B?
   - Which enzymes are the most advantageous?
   - What does the host already do?
2. Regulation
   - What is the limiting step in the pathway?
   - Is it controlled at the transcript, protein, or activity level?
   - How will foreign genes impact native regulation and vise versa?
3. Controlling gene expression
   - How can [protein] be altered/controlled?
   - Is it quantitatively predictable?
   - Can expression be regulated in response to demand?

BioCyc Database Collection · BRENDA · PDB · KEGG · MetRxn 1.0 · KBase DOE Systems Biology Knowledgebase

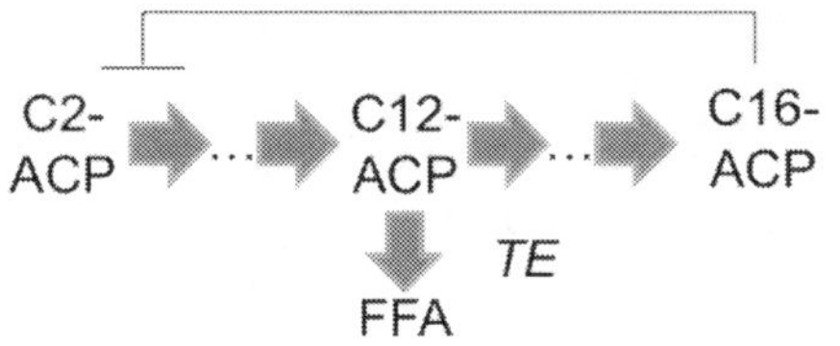

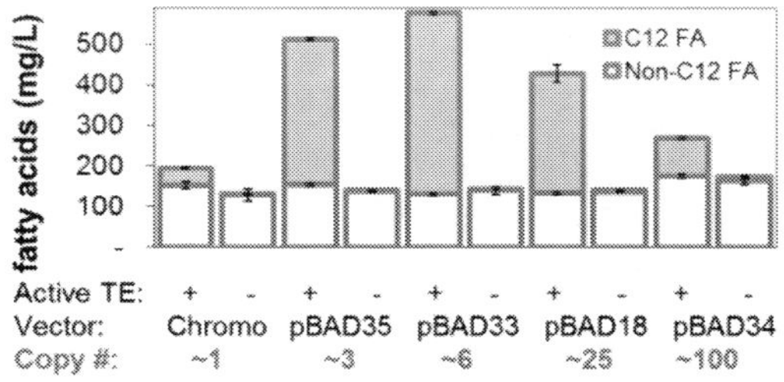

Lennen et al., B&B 2010

THE UNIVERSITY of WISCONSIN MADISON

Transcription factors recognize specific operator sequences

Van Hijum et al. MMB Reviews, 2009

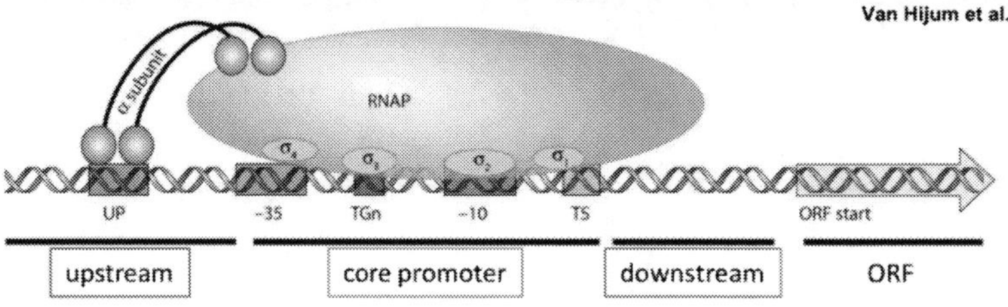

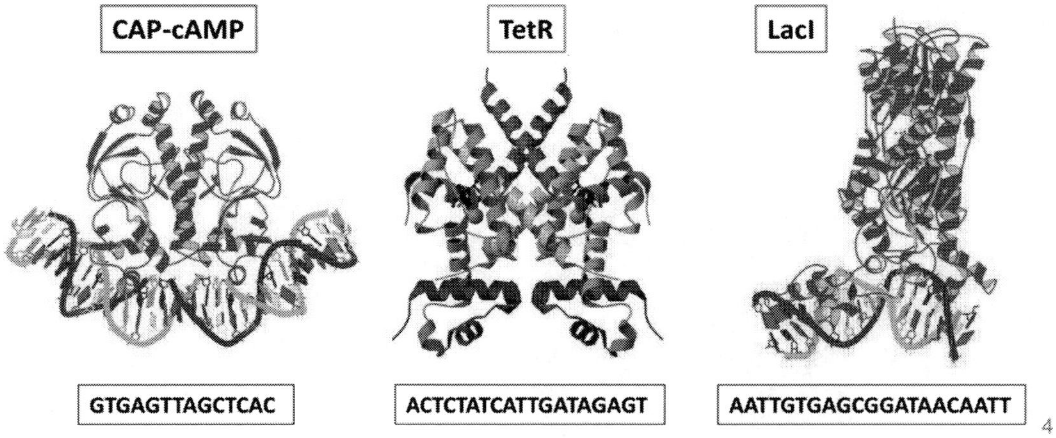

4

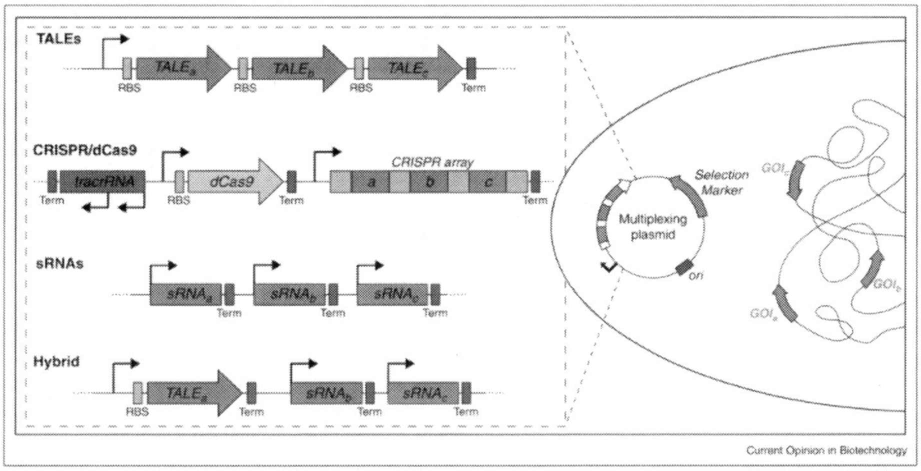

Copeland et al., 2014

SECTION 1: INDUCIBLE TRANS ACTING REGULATORS

THE UNIVERSITY of WISCONSIN MADISON

What are TALEs?

- TALEs = Transcription Activator-Like Effectors
- Injected by species of the plant pathogen *Xanthomonas* to promote host infection
- Native TALEs have three sections:
  - N-terminal translocation signal
  - Central tandem repeat domain
  - C-terminal region containing activation domains and nuclear localization signals
- Modular design
  - Repeat variable diresidues (RVDs) recognize/bind DNA bases

Mak *et al.*, Science **335**: 716-719

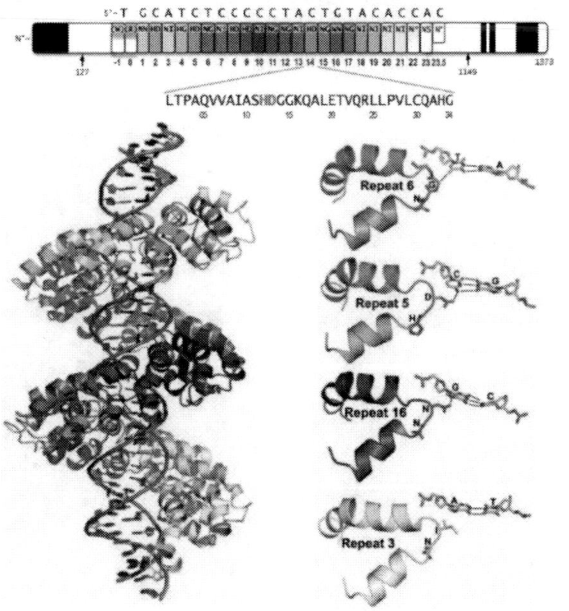

THE UNIVERSITY of WISCONSIN MADISON

TALE-based Repressors

Politz *et al.*, *ChemComm* (2013)

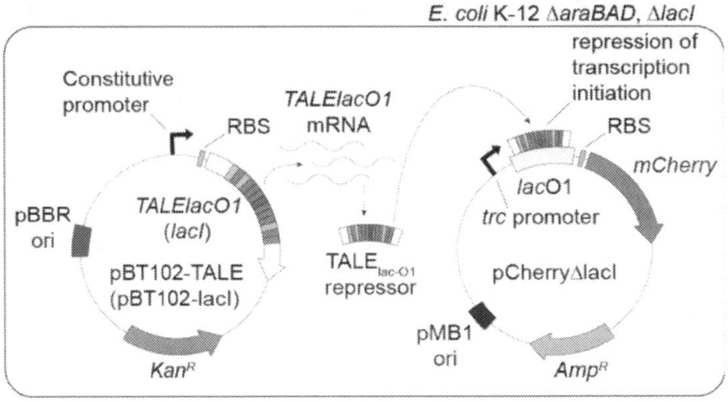

(b)

1 2 3 4 5 6 7 8 9 10 11 12 13 14 15 16 17 18 19 20 21

5'-A A T T G T G A G C G G A T A A C A A T T-3'

0 1 2 3 4 5 6 7 8 9 10 11 12 13 14 15 16 17 17.5

17.5-mer NG NN NG NN NI NN HD NH NN NI NG NI NI HD NI NI NG NG

THE UNIVERSITY of WISCONSIN MADISON

TALEs Repress Transcription

Politz *et al.*, *ChemComm* (2013)

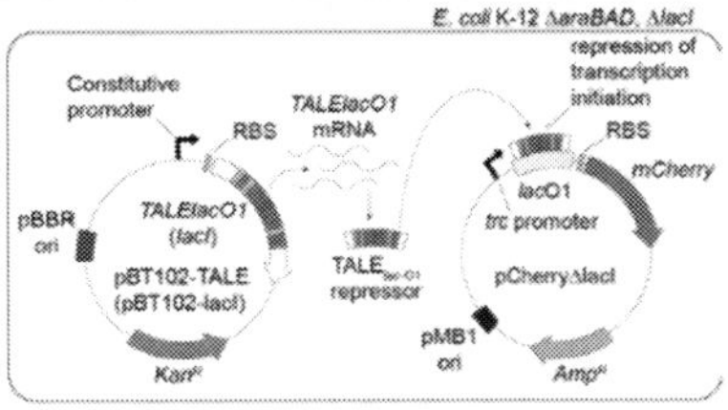

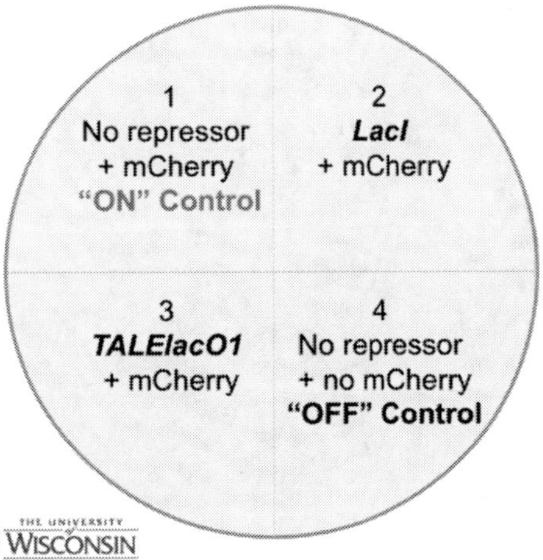

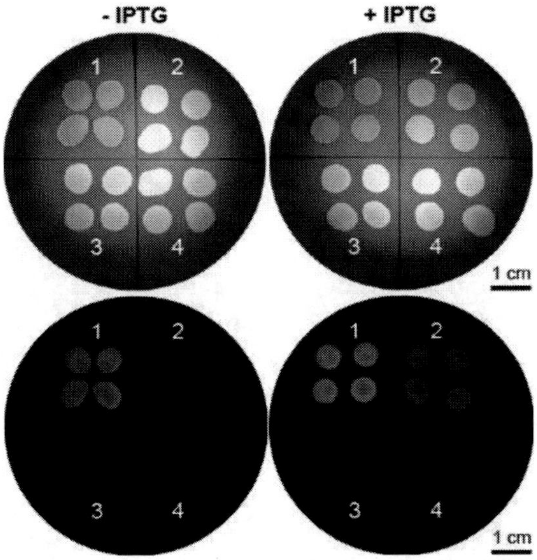

THE UNIVERSITY of WISCONSIN MADISON

of RVDs and [TALE] Impact Repression

Politz et al., Nature Chemical Biology (2016)

	0	1	2	3	4	5	6	7	8	9	10	11	12	13	14	15	16	17	17.5
Target lacO1 sequence A A	T	T	G	T	G	A	G	C	G	G	A	T	A	A	C	A	A	T	T
Final RVD sequence	NA	NG	NN	NG	NN	NI	NN	HD	NH	NN	NI	NG	NI	NI	HD	NI	NI	NG	NG
17.5	NA	NG	NN	NG	NN	NI	NN	HD	NH	NN	NI	NG	NI	NI	HD	NI	NI	NG	NG
14.5	NA	NG	NN	NG	NN	NI	NN	HD	NH	NN	NI	NG	NI	NI	HD	NI			
10.5	NA	NG	NN	NG	NN	NI	NN	HD	NH	NN	NI	NG							
9.5	NA	NG	NN	NG	NN	NI	NN	HD	NH	NN	NI								
8.5	NA	NG	NN	NG	NN	NI	NN	HD	NH	NN									
7.5	NA	NG	NN	NG	NN	NI	NN	HD	NH										
5.5	NA	NG	NN	NG	NN	NI	NN												
Empty	0																		

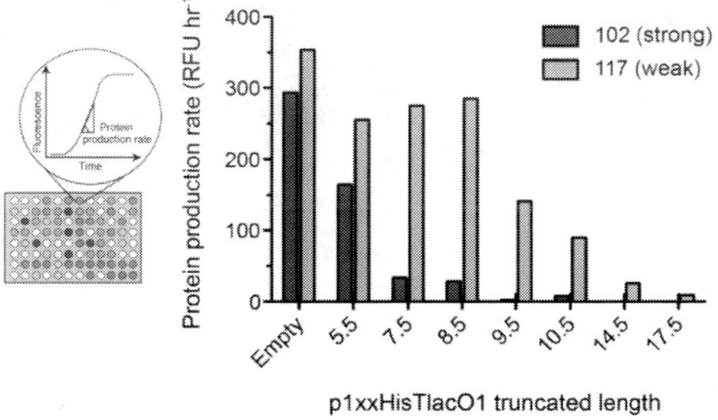

Protease Mediated TALE Induction System

Politz et al., Nature Chemical Biology (2016)

Repressed State

An engineered TALE binds to its operator and represses gene expression

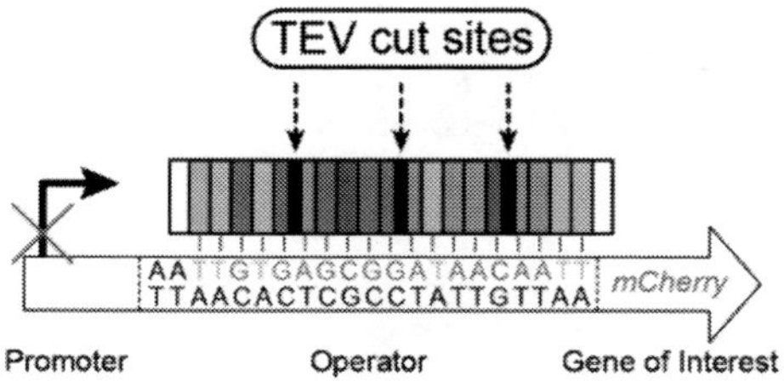

De-repressed State

An inducible TEV protease is expressed that will cleave the TALE and alleviate repression

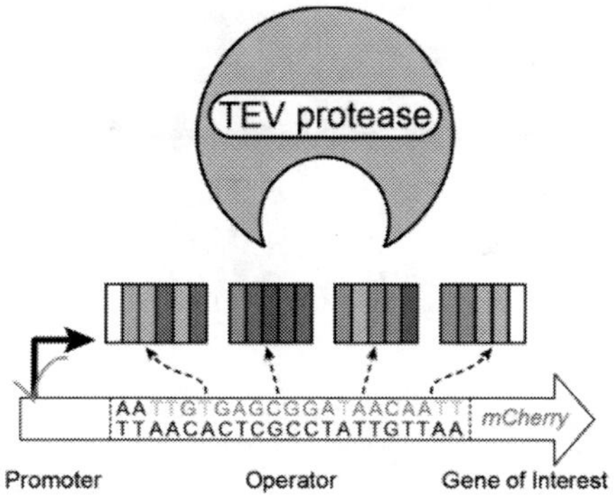

Where do TEV Sites Go?

Politz et al., Nature Chemical Biology (2016)

TALElacO1	TALE-Ins342
CQAHG-LTPEQ	CQAHGENLYFQ/GLTPEQ

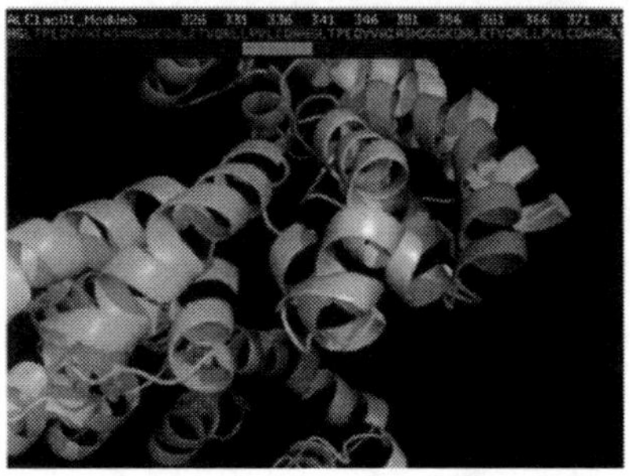

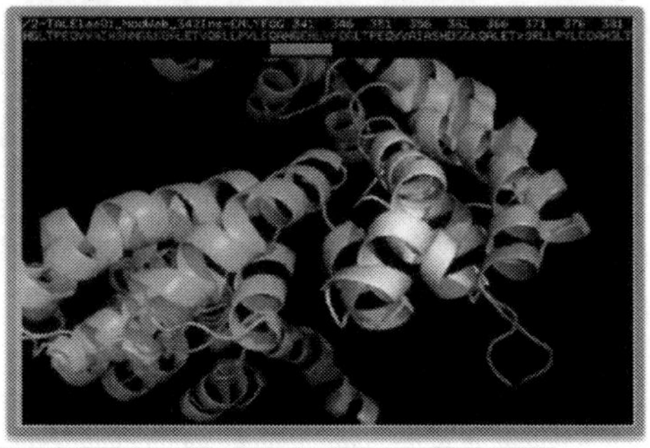

Locations for TEV site insertion – (ENLYFQ/G)

	0	1	2	3	4	5	6	7	8	9	10	11	12	13	14	15	16	17	17.5
Target lacO1 sequence A A	T	T	G	T	G	A	G	C	G	G	A	T	A	A	C	A	A	T	T
Final RVD sequence	NA	NG	NN	NG	NN	NI	NN	HD	NH	NN	NI	NG	NI	NI	HD	NI	NI	NG	NG
Piece 1	NA	NG	NN	NG	NN														
Piece 2						NI	NN	HD	NH	NN									
Piece 3											NI	NG	NI	NI	HD				
Piece 4																NI	NI	NG	NG

THE UNIVERSITY of WISCONSIN MADISON

11

Expression of TEV protease alleviates TALE-mediated repression

Politz et al., Nature Chemical Biology (2016)

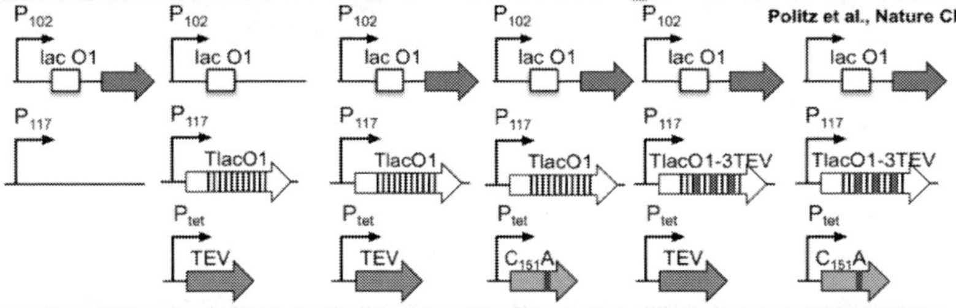

0 ng/mL atc

20 ng/mL atc

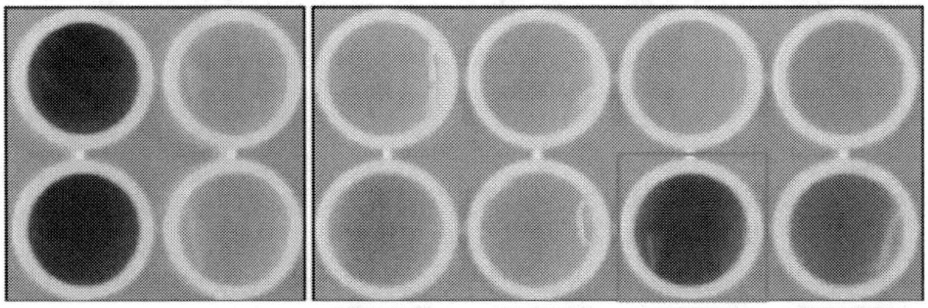

Photograph at 24 hours

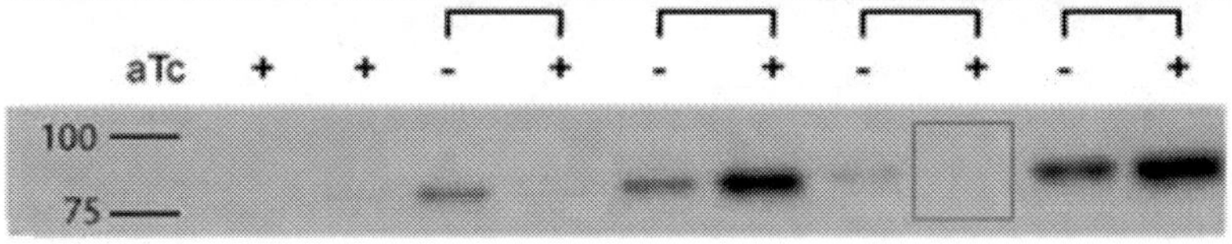

Probed for FLAG on TALE

% Full-Length TALE 0 14 100 8 140 453 39 0 331 587

TALE-TEV Outperforms LacI-IPTG

Politz et al., Nature Chemical Biology (2016)

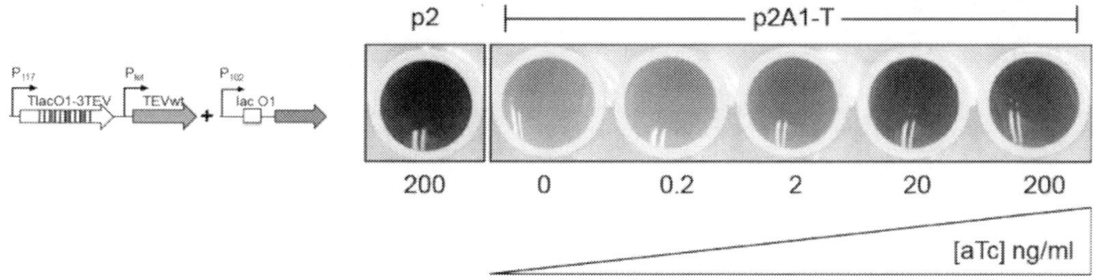

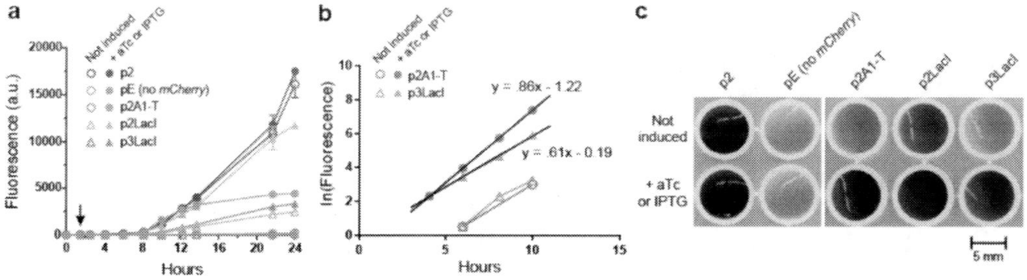

P3 stronger than P2

Improving the Scaffold

Politz et al., Nature Chemical Biology (2016)

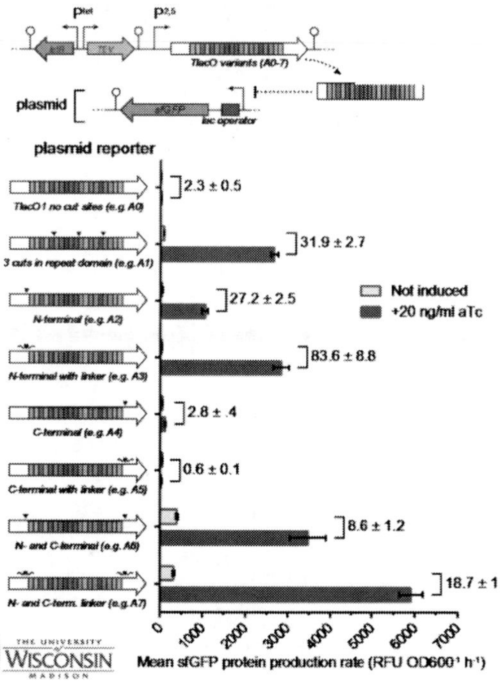

(Linker) Cut site (Linker)

...(GSG)ENLYFQ/S(GSG)...

Lysine Conditional Auxotrophy

Politz et al., Nature Chemical Biology (2016)

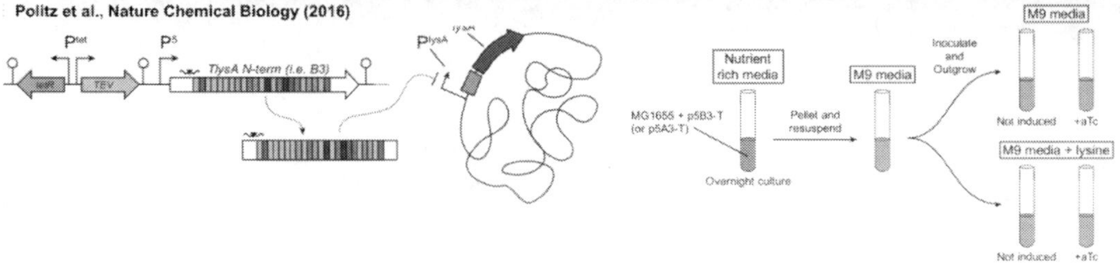

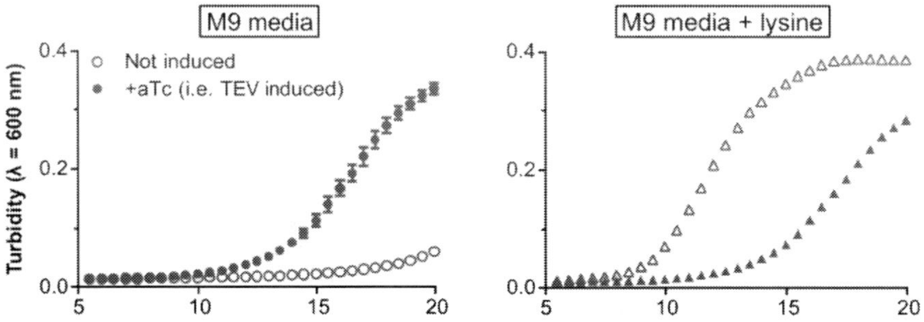

Targeting Multiple Sites

Politz et al., Nature Chemical Biology (2016)

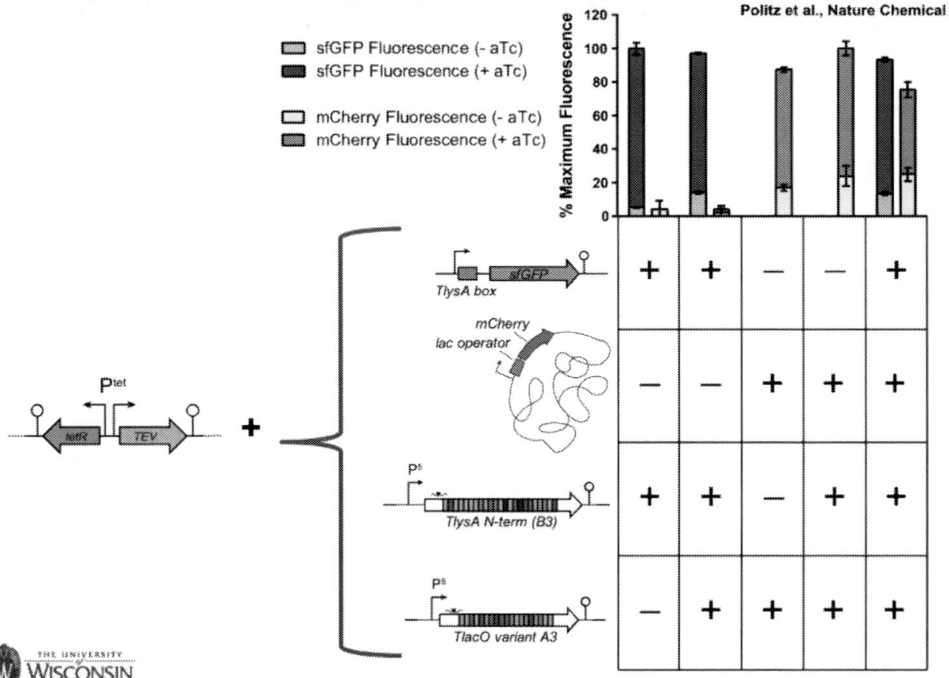

TlysA box – sfGFP	+	+	–	–	+
mCherry, lac operator	–	–	+	+	+
TlysA N-term (B3)	+	+	–	+	+
TlacO variant A3	–	+	+	+	+

Expression of TEV protease alleviates TALE-mediated repression

Politz et al., Nature Chemical Biology (2016)

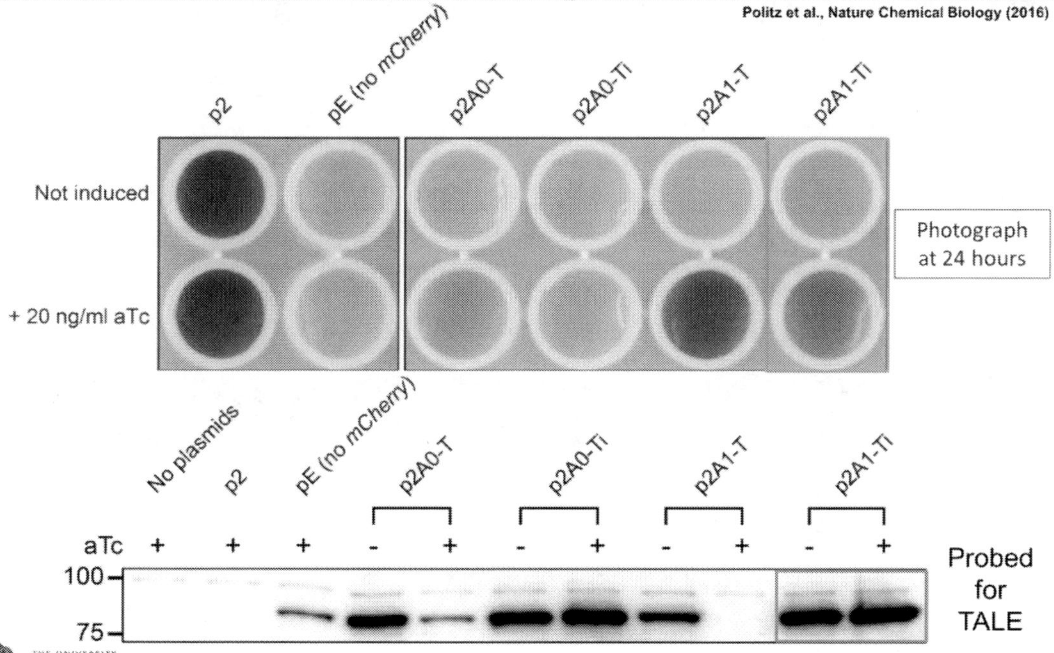

THE UNIVERSITY of WISCONSIN MADISON

De-repression by the inactive protease requires high affinity for the TALE substrate

Politz et al., Nature Chemical Biology (2016)

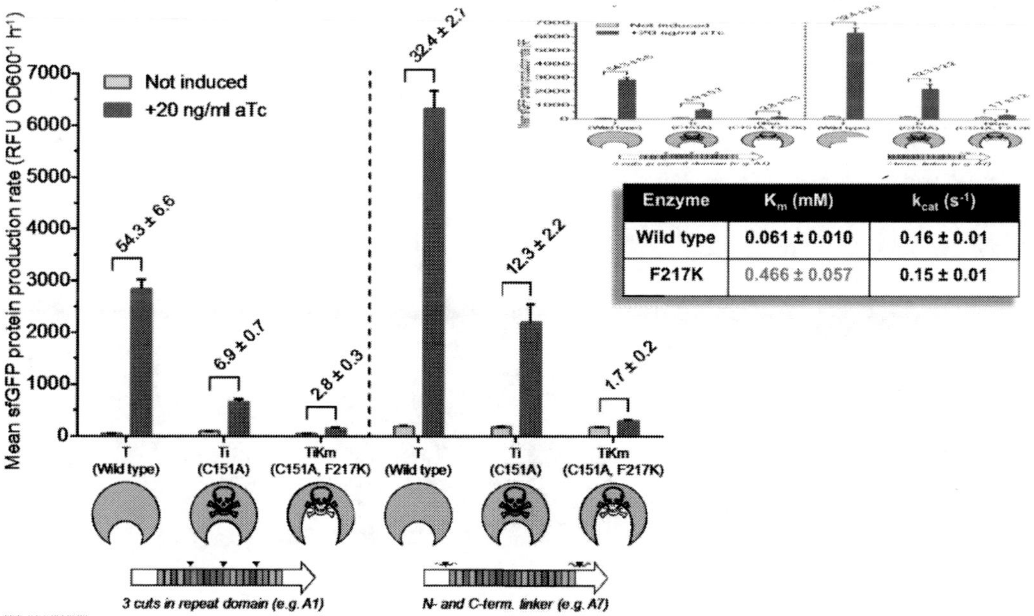

Enzyme	K_m (mM)	k_{cat} (s^{-1})
Wild type	0.061 ± 0.010	0.16 ± 0.01
F217K	0.466 ± 0.057	0.15 ± 0.01

THE UNIVERSITY of WISCONSIN MADISON

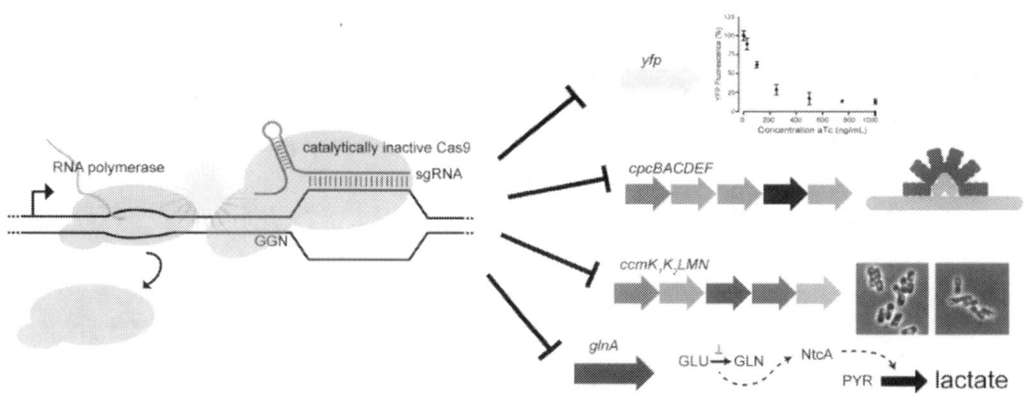

SECTION 2: ABC TALK "ANOTHER BLOODY CRISPR"

Model System

***Synechococcus* sp. PCC 7002**

- Isolated from Puerto Rico
- Uses CO_2 as a sole carbon source
- Marine/freshwater - halotolerant
- Fast doubling time
- Light tolerance
- Naturally transformable
- Sequenced genome
- Established selectable markers

Strain	Temp (°C)	CO_2 Supply	Light ($\mu E\ m^2\ s^{-1}$)	Growth rate (mg DCW $L^{-1}\ h^{-1}$)
PCC 7002[1]	37	2%	100	50
PCC 7942[2]	35	2%	65	18
PCC 6803[3]	30	40mM HCO_3^-	50	16

1. US Patent 8,465,954
2. Ducat et al. *AEM* (2012)
3. Angermayr et al. *AEM* (2012)

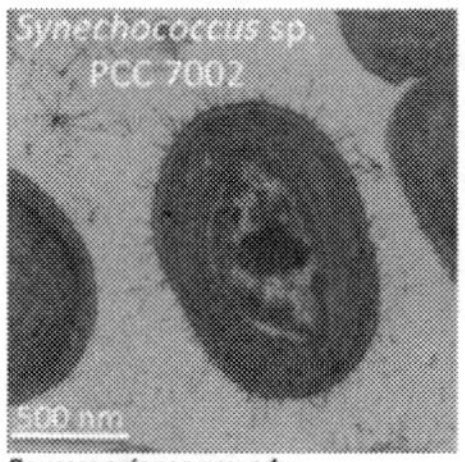

Source: science.psu.edu

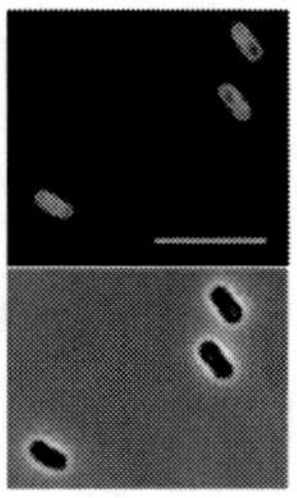

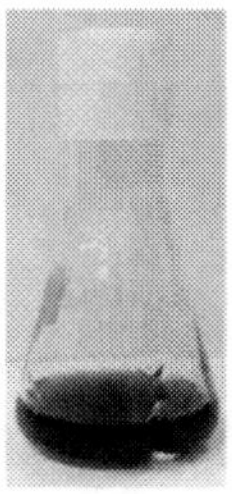

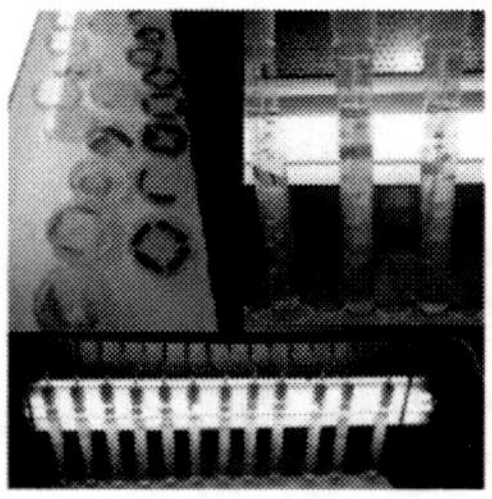

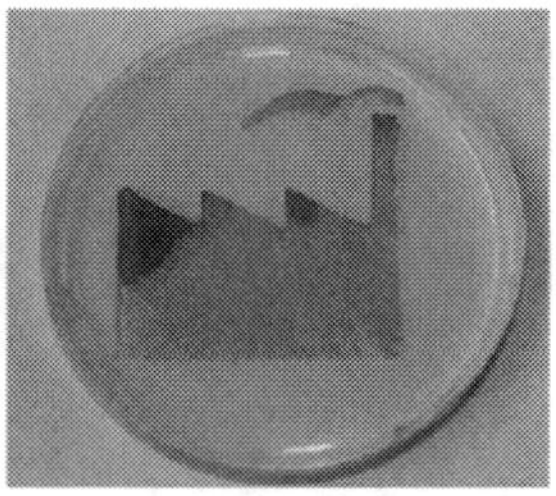

Anhydrotetracycline Induction

Zess et al., *Biotech & Bioeng (2016)*

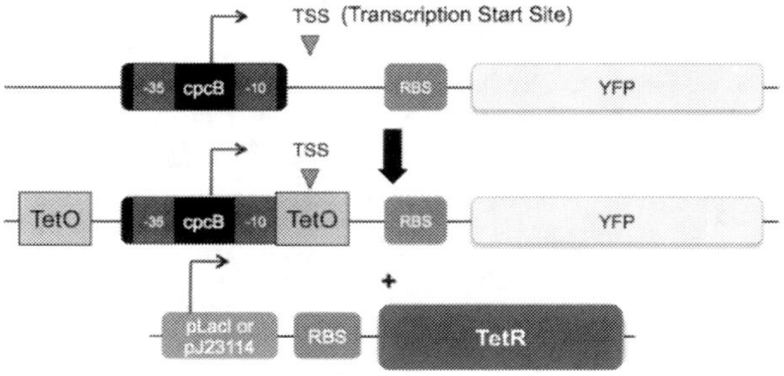

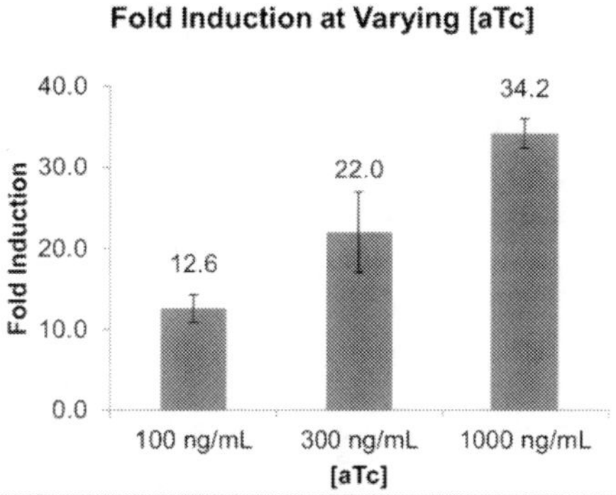

[aTc]	0 ng/mL	100 ng/mL	300 ng/mL	1000 ng/mL
% Max	3%	35%	61%	95%

THE UNIVERSITY of WISCONSIN MADISON

CRISPRi in PCC7002

Gordon et al., (Under Review)

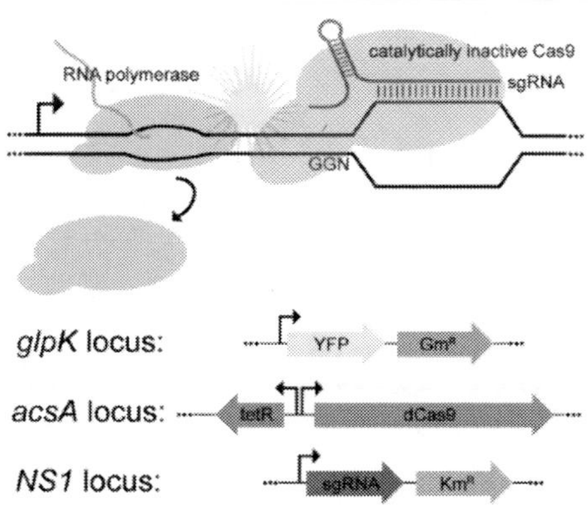

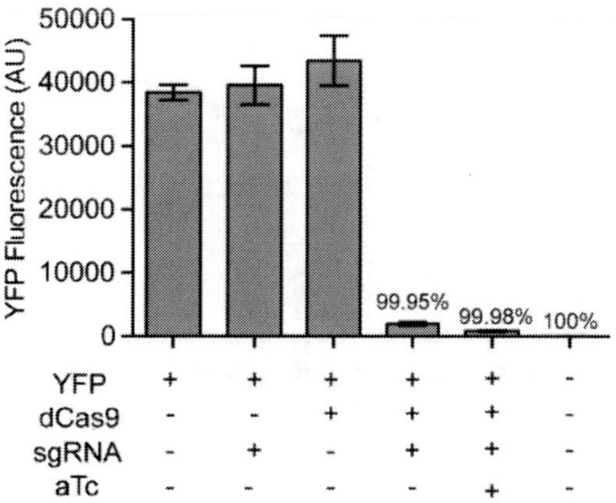

THE UNIVERSITY of WISCONSIN MADISON

Reducing sgRNA Expression

Gordon et al., (Under Review)

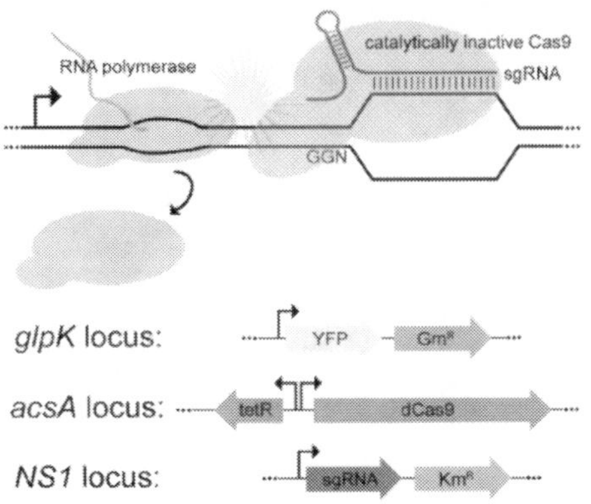

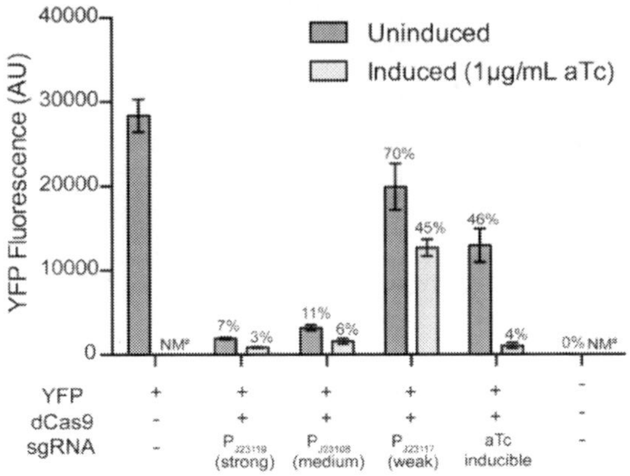

THE UNIVERSITY of WISCONSIN MADISON

Improving the System

Gordon et al., (Under Review)

Titrating System Components

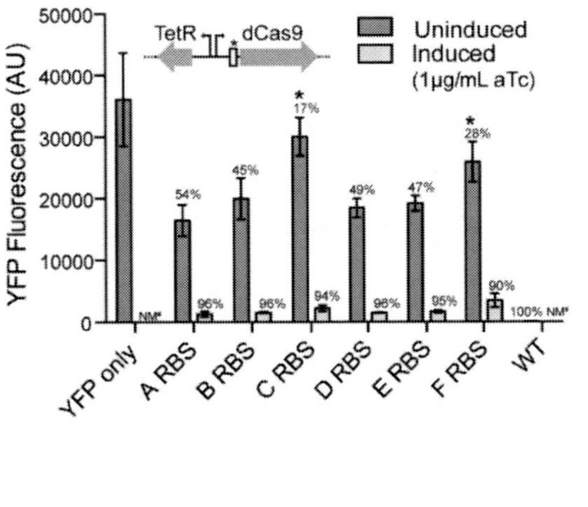

THE UNIVERSITY of WISCONSIN MADISON

Testing the System

Gordon et al., (Under Review)

Reversible Induction is Possible

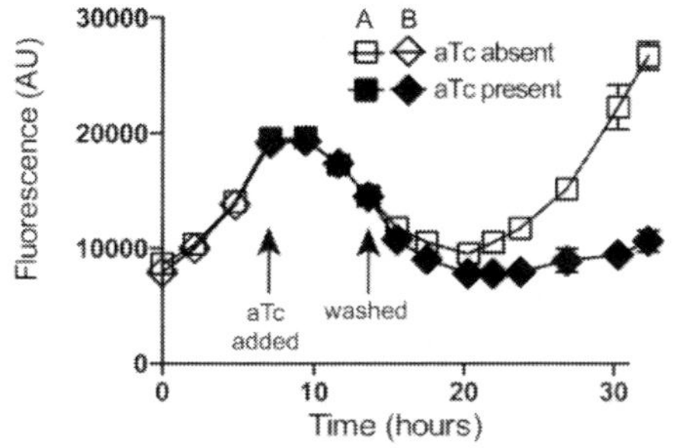

Targeting Carboxysomes

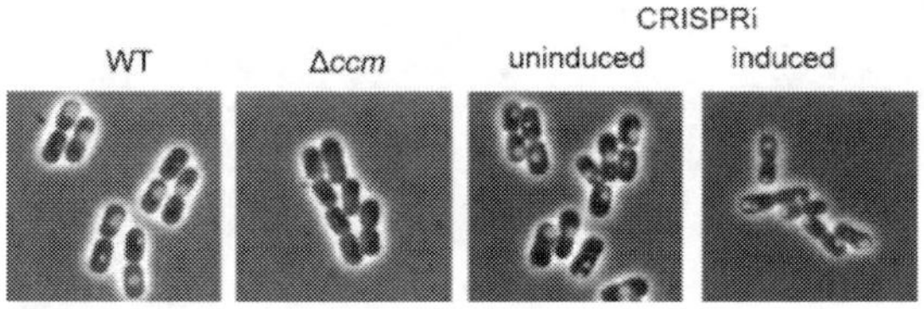

THE UNIVERSITY of WISCONSIN MADISON

Increasing Lactate Production

Gordon et al., (Under Review)

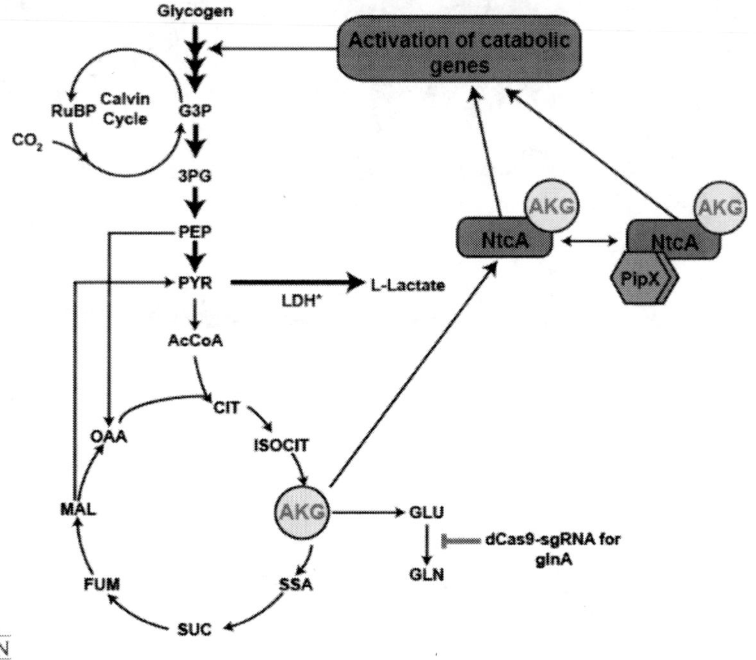

THE UNIVERSITY of WISCONSIN MADISON

Increasing Lactate Production

Gordon et al., (Under Review)

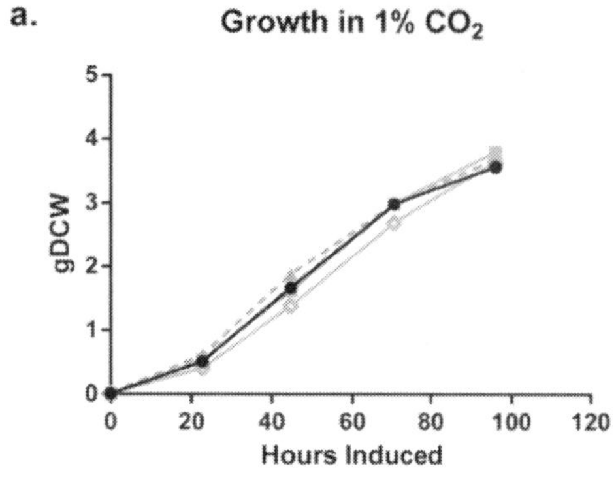

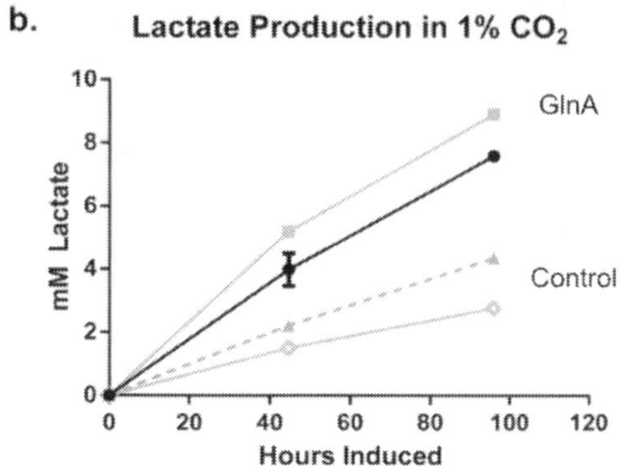

	CC130	CC131	CC133	CC142
LDH*	+	+	+	+
dCas9	-	+	-	-
sgRNA target	glnA	glnA	-	YFP

THE UNIVERSITY of WISCONSIN MADISON

Acknowledgements

Pfleger Lab
Taylor Cook
Austin Comer
Ryan Clark
Gina Gordon
Nestor Hernandez
Chris Jones, PhD.
Travis Korosh
Rung-Yi Lai, PhD.
Chris Mehrer
Andrew Markley, PhD.
Mark Politz
Jacqueline Rand

Alumni
Matt Begemann
Jeff Cameron, PhD
Matt Copeland, PhD.

Former Undergraduates
Erin Zess
Ryan Clarke
Charles Johnson

Funding

U.S. DEPARTMENT OF ENERGY
U.S. AIR FORCE
NSF
DOW
3M
WARF
BTP
NATIONAL INSTITUTES OF HEALTH
CBI-TP
THE UNIVERSITY of WISCONSIN MADISON
University of Wisconsin-Madison Genomic Sciences Training Program

THE UNIVERSITY of WISCONSIN MADISON

Engineering Nonphosphorylative Metabolism to Generate Lignocellulose-Derived Products

Zhang, K., University of Minnesota
Jambunathan, P., University of Minnesota
Tai, Y. S., University of Minnesota
Xiong, M., University of Minnesota
Wang, J., University of Minnesota
Wang, J., University of Minnesota
Stapleton, C., University of Minnesota
Originally presented on: 6/30/2016 11:35:00 - 11:50:00

Conversion of lignocellulosic biomass into value-added products provides significant environmental and economic benefits. Here we report the engineering of an unconventional metabolism for the production of TCA cycle derivatives from D-xylose, L-arabinose, and D-galacturonate. We designed a growth-based selection platform to identify several gene clusters functional in *Escherichia coli* that can perform this nonphosphorylative assimilation of sugars into the TCA cycle in less than 6 steps. To demonstrate the application of this new metabolic platform, we built artificial biosynthetic pathways to 1,4-butanediol (BDO) with a theoretical molar yield of 100%. By screening and engineering downstream pathway enzymes, 2-ketoacid decarboxylases and alcohol dehydrogenases, we constructed *E. coli* strains capable of producing BDO from D-xylose, L-arabinose, and D-galacturonate. The titers, rates, and yields are higher than those previously reported utilizing conventional pathways. This work demonstrates the potential of nonphosphorylative metabolism for biomanufacturing with improved biosynthetic efficiencies.

An Automated Adaptive Laboratory Evolution Platform for Industrial Biotechnology

Feist, A. M., University of California, San Diego

Originally presented on: 6/30/2016 12:00:00 - 12:20:00

Adaptive Laboratory Evolution (ALE) is a powerful experimental technique that exploits the inherent competition at play within a microbial culture and the natural accumulation of mutations from which to select desired phenotypic traits. Unlike rational engineering approaches, ALE is conducted without *a priori* knowledge on how such traits would arise and therefore can result in unintuitive biological mechanisms for adapting to desired selection pressures. By automating the ALE process, experiments can be run in high-throughput and with tight control over all environmental and process parameters. Omics characterization of evolved strains following an ALE experiment elucidates the genetic and metabolic changes underlying the new phenotype, with which genome engineering techniques can be used to guide the production or refinement of strains with specific biotechnological applications. We have generated and utilized an automated ALE technology in a number of applications, including maximizing strain growth rate in a given environment, increasing tolerance to inhibitory chemicals, and weaning cells onto new, previously unusable nutrient sources. As the use of ALE increases for both basic research and applied biotechnological goals, it will be important to develop a streamlined process for leveraging this ALE technology to facilitate optimal strain designs and production processes.

Synthetic Selection Devices Using Expression Regulators for Evolutionary Metabolic Engineering

Jung, G. Y., POSTECH

Originally presented on: 6/30/2016 12:20:00 - 12:40:00

Pathway optimization of microbial metabolism is essential for the production of commercially valuable chemicals such as biofuels, platform chemicals and biologically active compounds. To achieve the successful design or redesign of microbial metabolism, robustness of naturally occurring biological systems has to be relieved so that cells can be easily redesigned. Although extremely huge efforts have been made to find genetic target to improve metabolic function of the microorganisms, there still exists the additional room for the non-rational approach. Currently, typical approach for metabolic engineering uses both rational approach as well as non-rational methods such as combinatorial and evolutionary methods. One of the most critical problems of metabolic engineering is especially robustness of the biological systems. Bacterial cells are generally evolved at the various levels from DNA to protein for maintaining their robustness against the changing circumstances. Therefore, general strategy to modify cellular physiology depending the robustness or flexibility of the biological systems should be required. In this study, we developed synthetic selection devices using the expression regulators at transcription and translation levels. By combining the product-sensitive promoters and selection markers, a selection device turning the cellular phenotype into selective based on the intracellular concentration of the product. Additionally, intracellular metabolite sensor named "riboselector" to regulate metabolic distribution will be presented. The potentials of the platform technology developed in this study for the application to the production of biofuels and commodity chemicals.

Understanding and Exploiting Enzyme Promiscuity for Metabolic Engineering

Tyo, K. E. J., Northwestern University

Originally presented on: 6/30/2016 12:40:00 - 13:00:00

While the traditional view of biochemistry says one enzyme catalyzes one reaction, in reality enzymes are far from perfect catalysts. Many enzymes are known to be promiscuous, i.e., to react on chemically similar compounds. To date, only a modest understanding of the extent and nature of enzyme promiscuity has been incorporated into our view of metabolism and metabolic engineering. Enzyme promiscuity represents both an opportunity and a challenge in metabolic engineering. Enzyme promiscuity is an exciting opportunity as it provides a potential path for synthesizing chemicals that would not be possible with canonical enzymatic reactions. We can utilize a much larger set of biochemical reactions than has been documented. However, enzyme promiscuity is also a daunting challenge. An organism's native metabolic network may contain undocumented side reactions known as 'underground metabolism.' As we engineer organisms to produce chemicals, underground metabolism can frustrate efforts in unpredicted ways by introducing reactions that reduce yields or cause toxicity, especially in engineered systems, where substrate concentrations may be much higher than evolutionary optimal concentrations promoting promiscuous reactions.

To address these opportunities and challenges, the Tyo lab has actively developed cheminformatics algorithms to understand and predict enzyme substrate promiscuity. Our cheminformatics approach is unique and complementary to bioinformatics approaches, as we only analyze the compounds structure, not the enzyme. We rely on digital representations of compounds to make our predictions. Using a chemical fingerprinting representation, we have developed machine-learning algorithms that use existing enzyme/substrate information to predict reactions with new compounds. Our tools consistently perform with high accuracy. In parallel, we are using the Biochemical Network Integrated Computer Explorer (BNICE) to map underground metabolism. We have predicted the products of undocumented enzymatic side reactions on the genome scale. By knowing probable chemical structures of promiscuous products, we enable targeted metabolomics to experimentally verify underground reactions. Once underground reactions are identified, we can intervene to increase productivity and avoid toxicity.

These computational tools will significantly empower the future of metabolic engineering. By allowing the biosynthesis of a broad set of new compounds, we will generate new products and materials that were previously unthinkable. Likewise, by uncovering undocumented enzymatic reactions that affect metabolism will give us evolutionary insight into the arrangement of metabolic networks and critical information to guide troubleshooting efforts in metabolic engineering efforts.

Cyanobacterial Chemical Production

Shota Atsumi

To solve humanity's most pressing problems

Energy

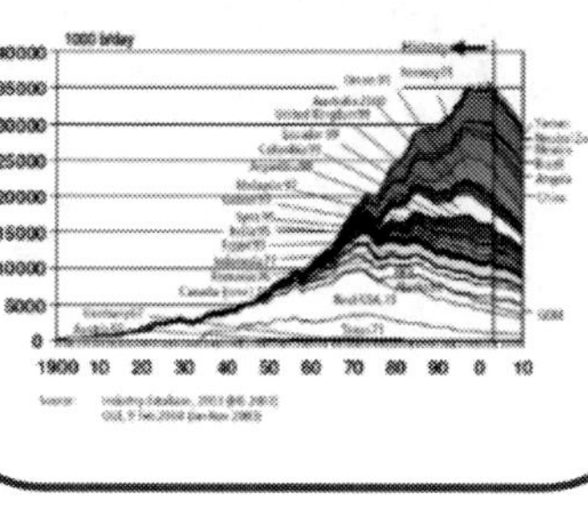

Environment

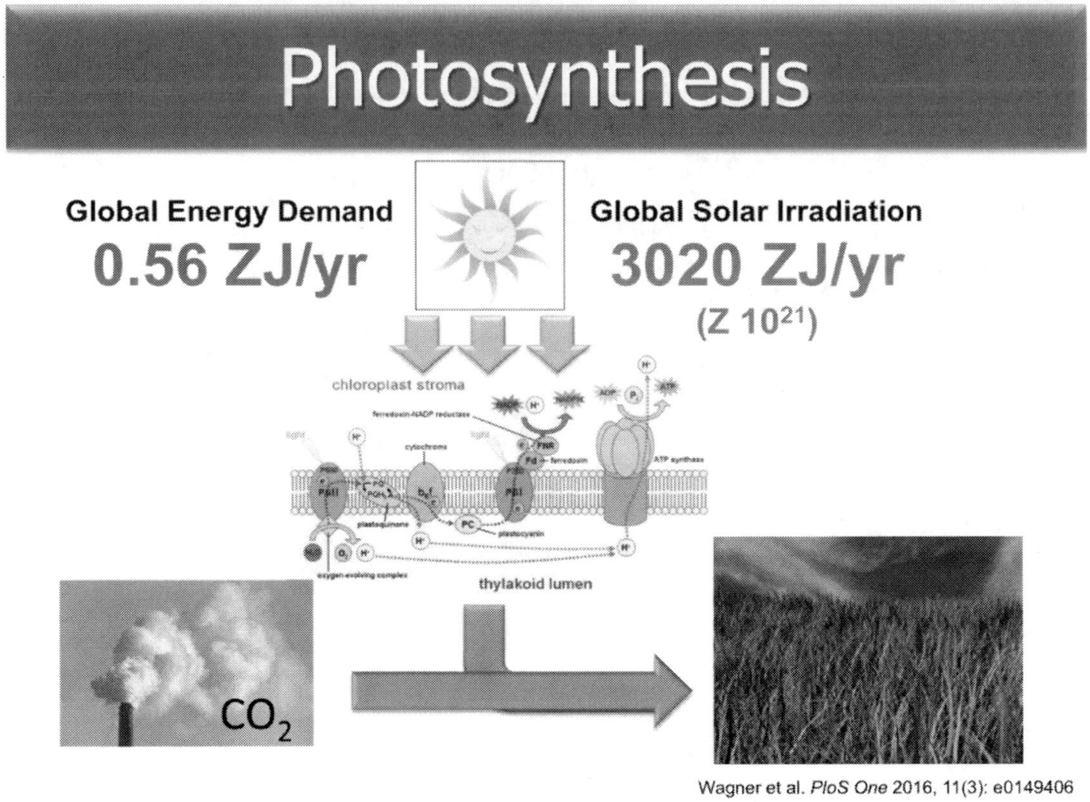

Photosynthesis
Global Energy Demand
0.56 ZJ/yr
Global Solar Irradiation
3020 ZJ/yr
(Z 10²¹)
chloroplast stroma
thylakoid lumen
CO₂
Wagner et al. PloS One 2016, 11(3): e0149406

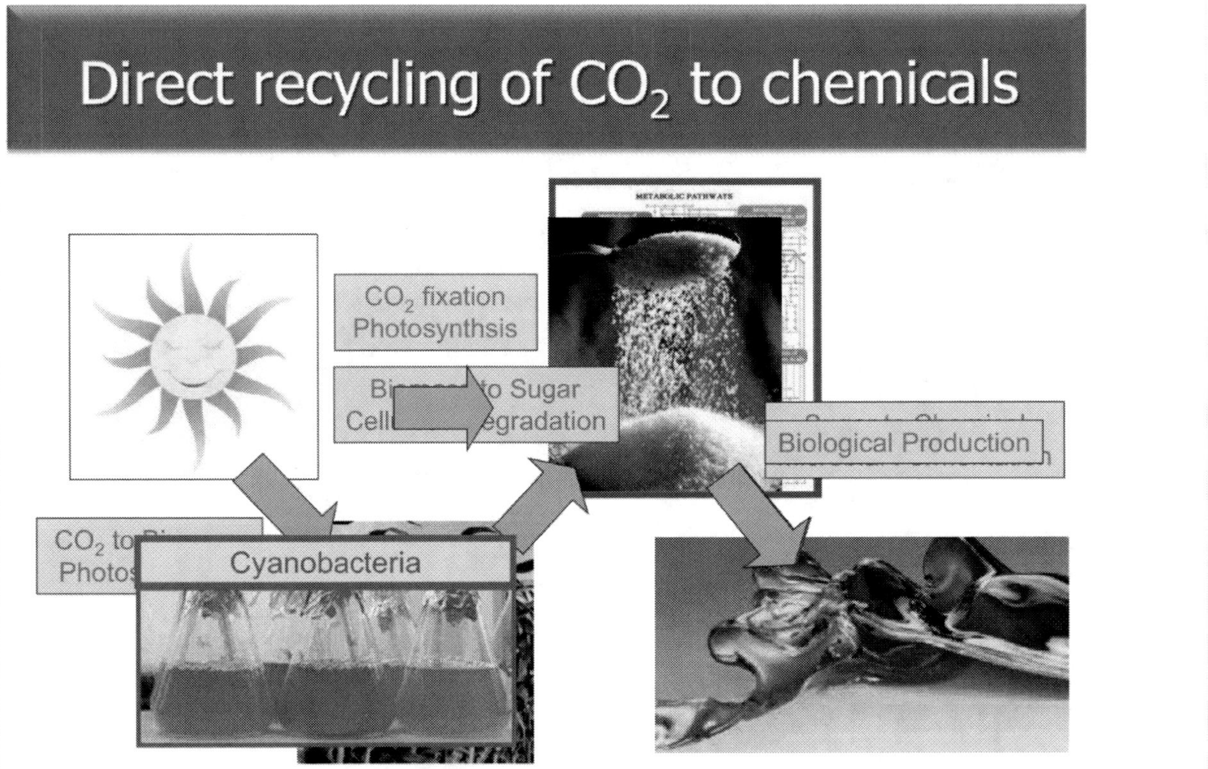

Direct recycling of CO₂ to chemicals
CO₂ fixation
Photosynthesis
Biological Production
Cyanobacteria

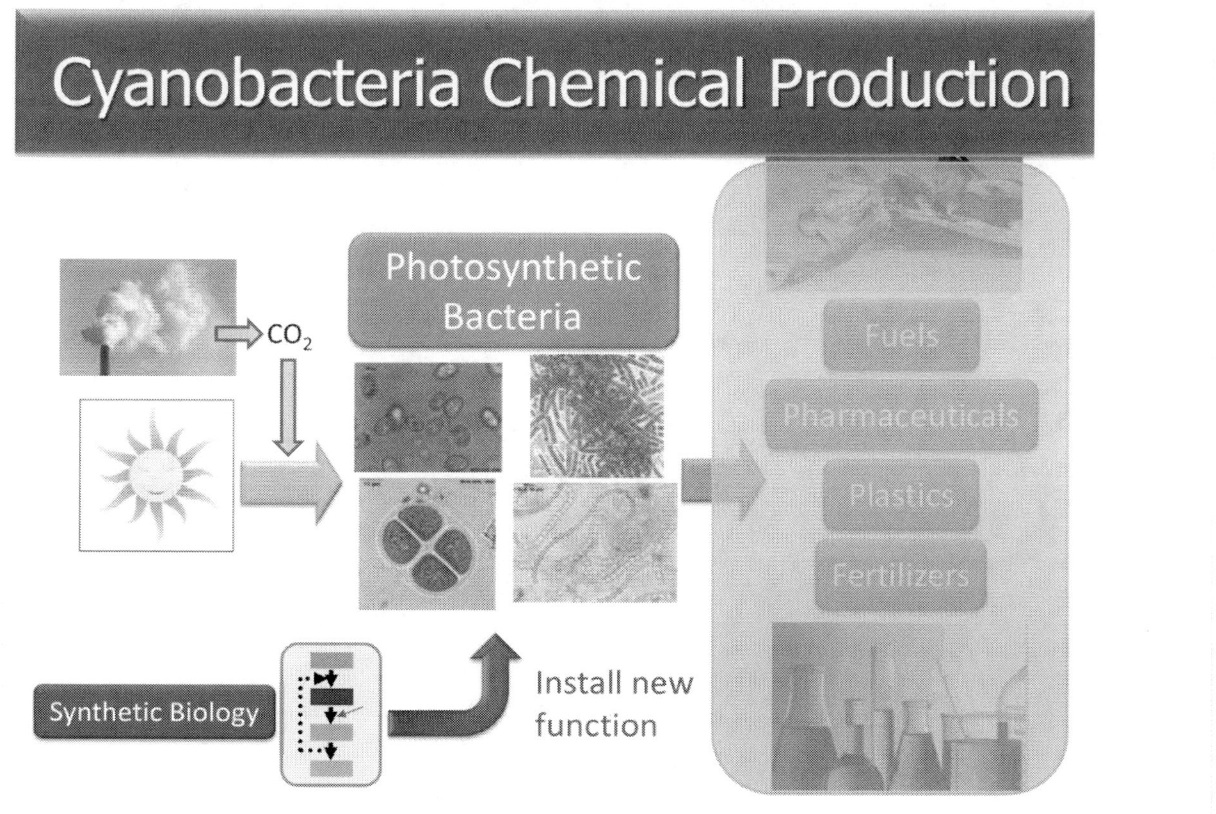
Cyanobacteria Chemical Production
CO2
Photosynthetic Bacteria
Fuels
Pharmaceuticals
Plastics
Fertilizers
Synthetic Biology
Install new function

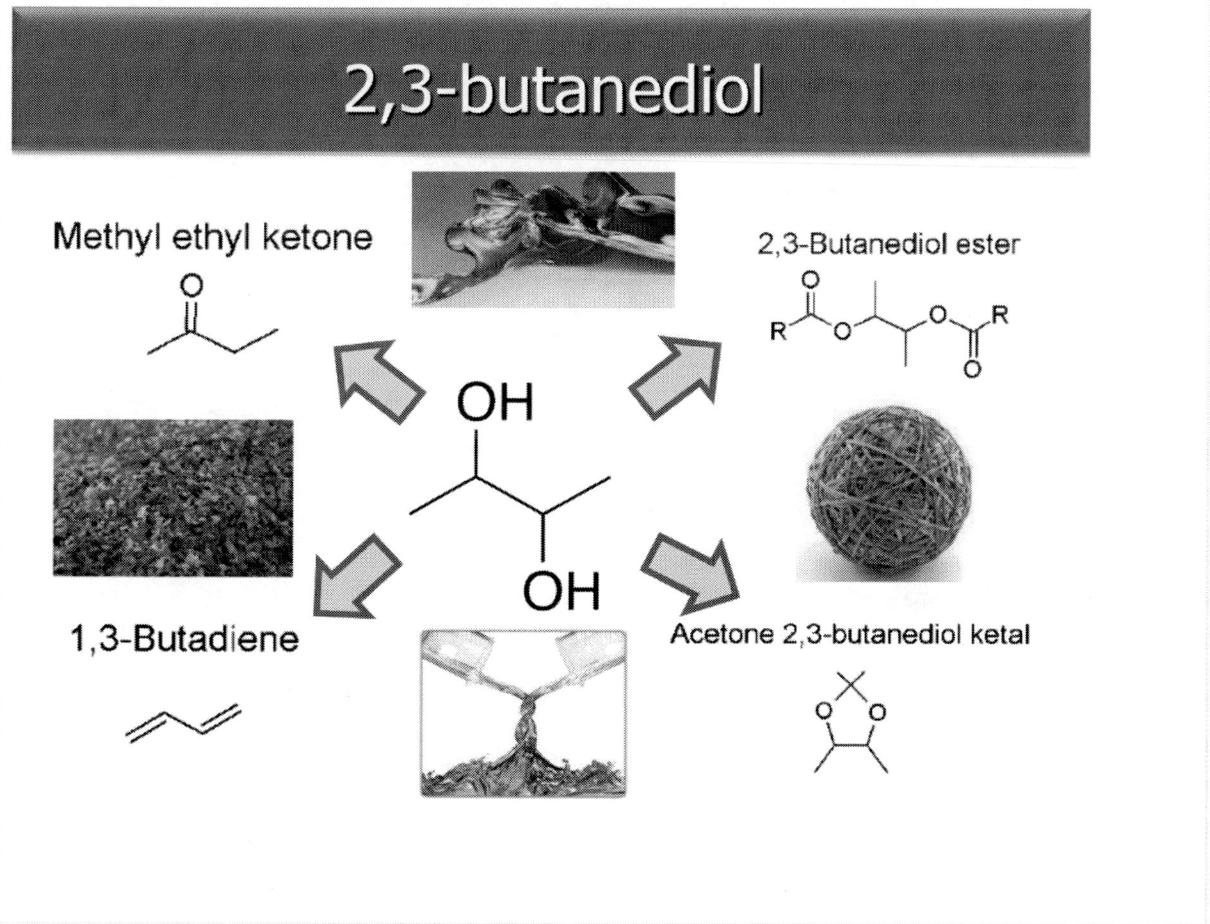
2,3-butanediol
Methyl ethyl ketone
2,3-Butanediol ester
OH
OH
1,3-Butadiene
Acetone 2,3-butanediol ketal

23BD biosynthetic pathway

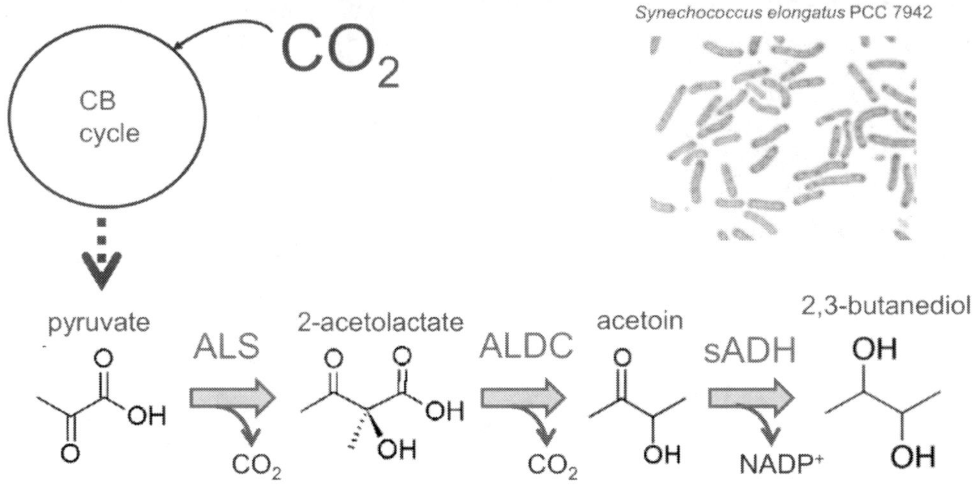

Oliver et al. *PNAS* 2013

A choice of target pathways

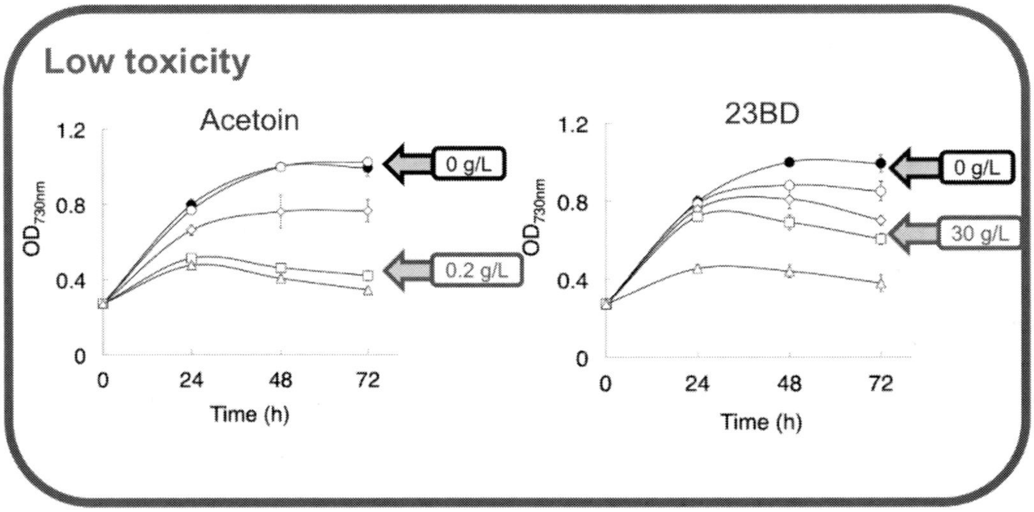

Oliver et al. *PNAS* 2013

A choice of target pathways

Irreversible reaction
To create a driving force to direct the flux toward the target

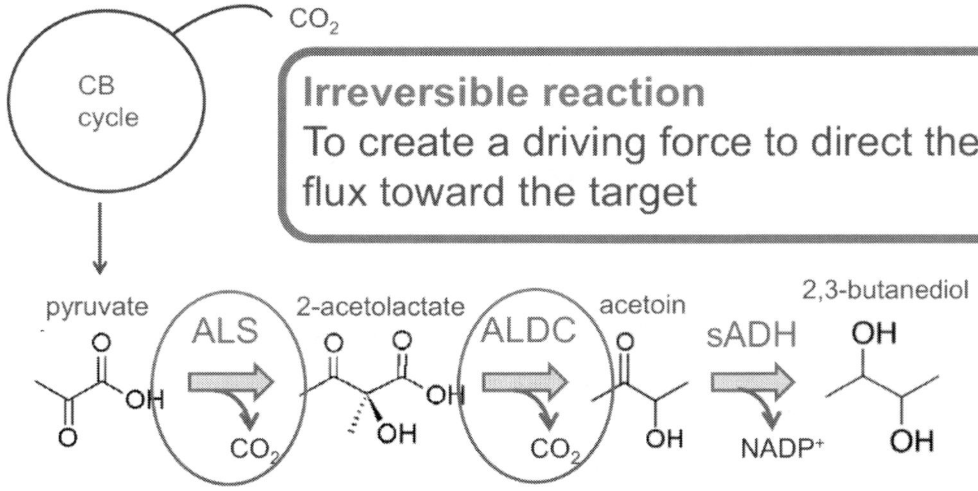

Bond-Watts et al *Nat Chem Biol* 2011
Shen and Liao *AEM* 2011

Oliver et al. *PNAS* 2013

A choice of target pathways

Chemicals are secreted from cells without additional modifications
To make the system simple and inexpensive

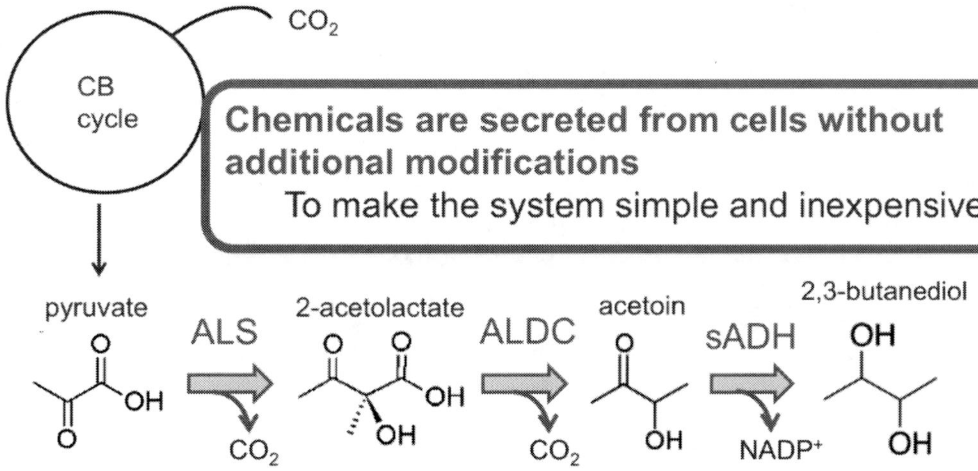

Oliver et al. *PNAS* 2013

23BD production from CO_2

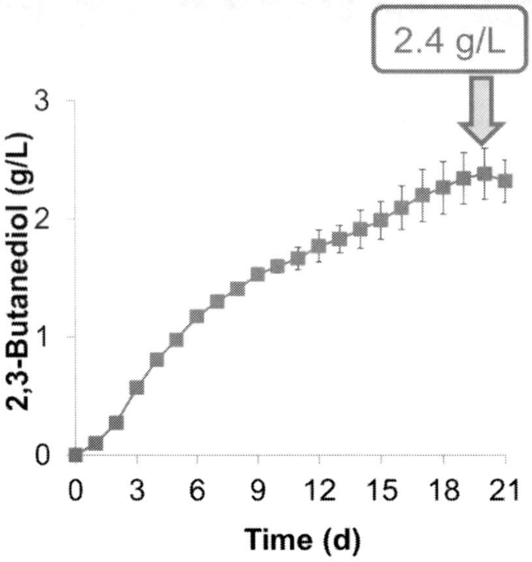

Oliver et al. *PNAS* 2013

Metabolic Engineering 22 (2014) 76–82

Contents lists available at ScienceDirect

Metabolic Engineering

journal homepage: www.elsevier.com/locate/ymben

Combinatorial optimization of cyanobacterial 2,3-butanediol production

John W.K. Oliver [a,1], Iara M.P. Machado [a,1], Hisanari Yoneda [a,b], Shota Atsumi [a,*]

[a] Department of Chemistry, University of California, Davis, One Shields Avenue, Davis, CA 95616, USA
[b] Asahi Kasei Corporation, 2-1 Samejima, Fuji, Shizuoka 416-8501, Japan

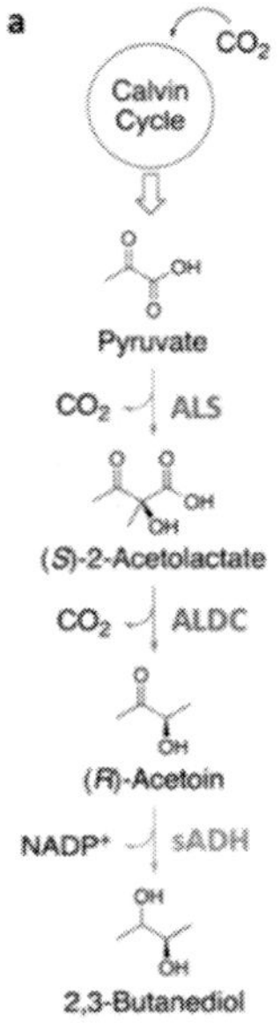

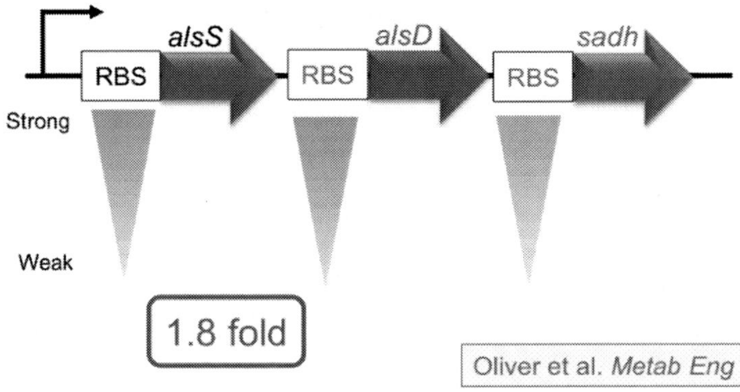

Oliver et al. *Metab Eng* 2014

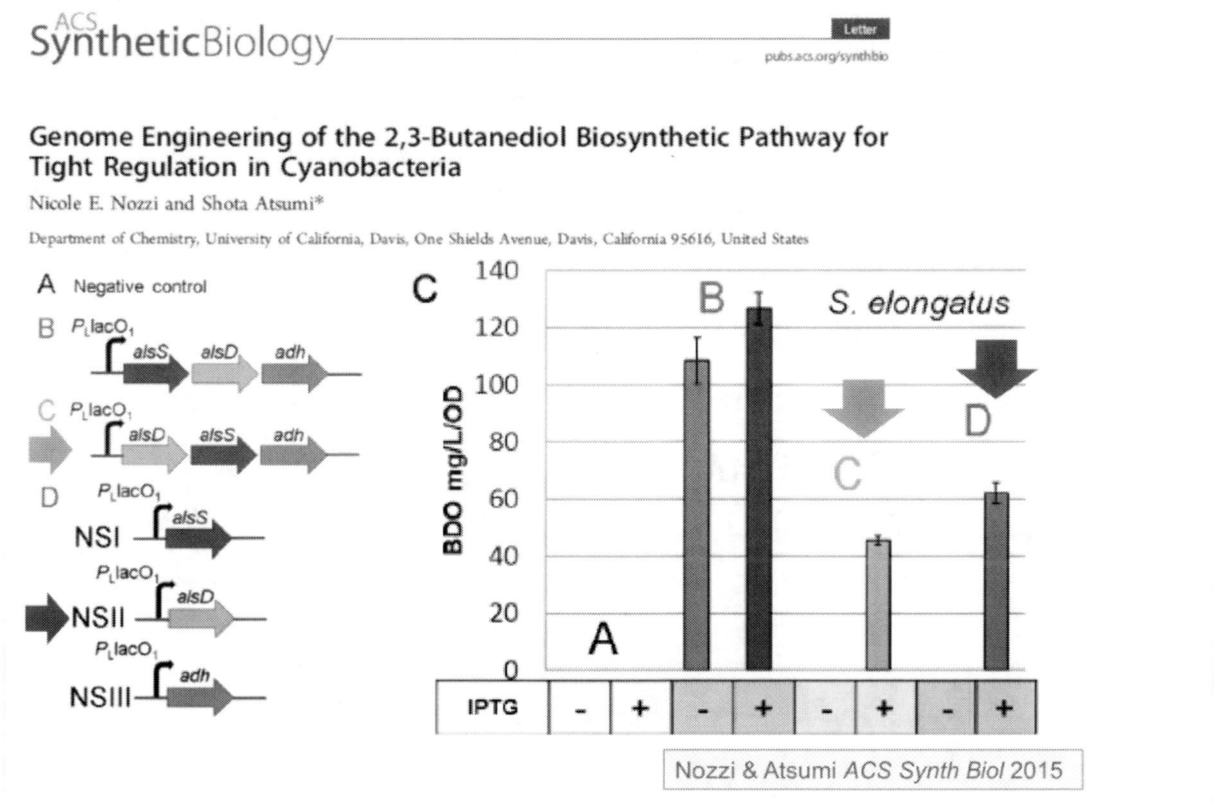
ACS Synthetic Biology
Letter
pubs.acs.org/synthbio
Genome Engineering of the 2,3-Butanediol Biosynthetic Pathway for Tight Regulation in Cyanobacteria
Nicole E. Nozzi and Shota Atsumi*
Department of Chemistry, University of California, Davis, One Shields Avenue, Davis, California 95616, United States
A Negative control
B $P_L$lacO$_1$
alsS alsD adh
C $P_L$lacO$_1$
alsD alsS adh
D $P_L$lacO$_1$
NSI alsS
NSII alsD
NSIII adh
C
BDO mg/L/OD
140
120
100
80
60
40
20
0
S. elongatus
A
B
C
D
IPTG
-
+
Nozzi & Atsumi ACS Synth Biol 2015

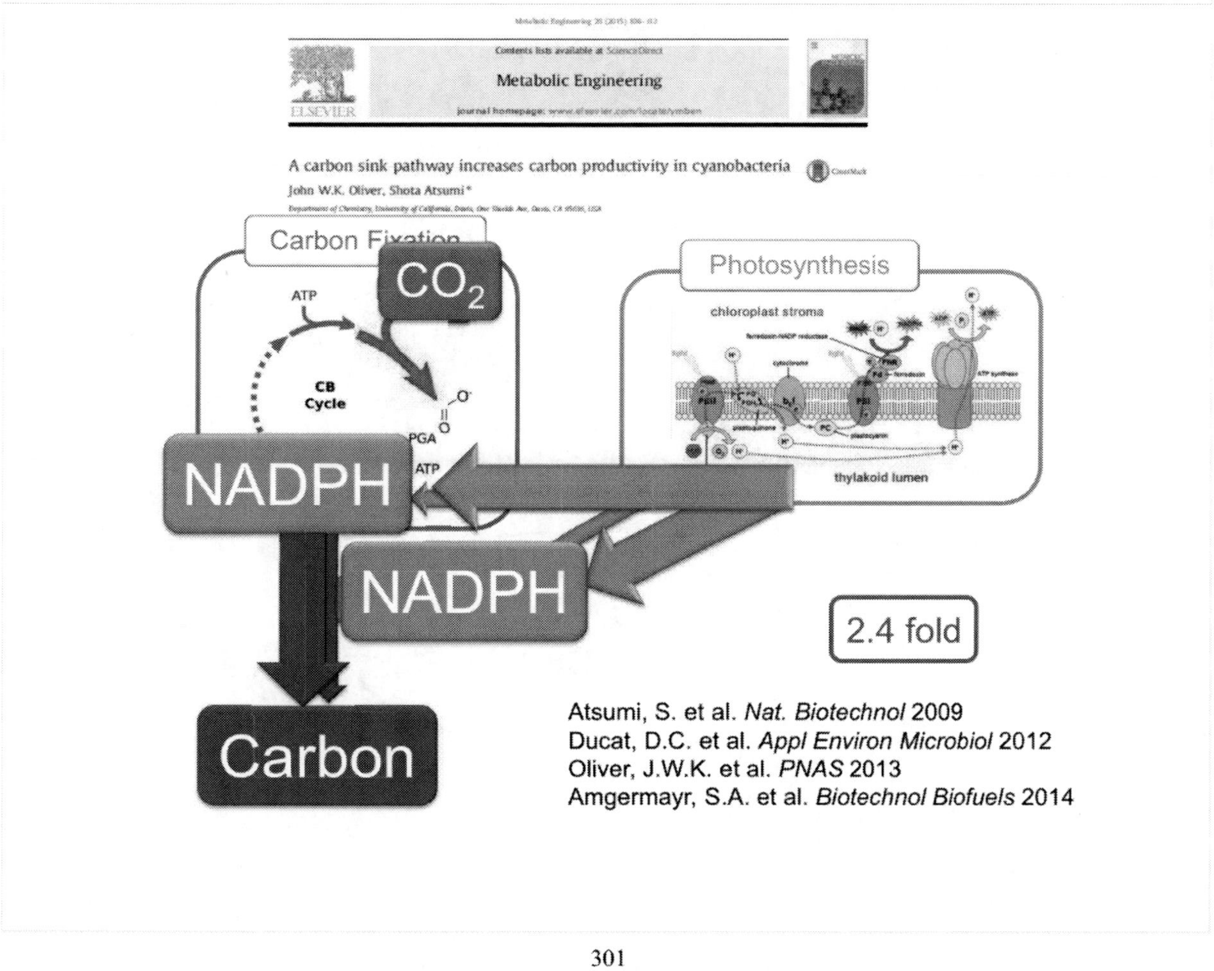
Contents lists available at ScienceDirect
Metabolic Engineering
ELSEVIER
A carbon sink pathway increases carbon productivity in cyanobacteria
John W.K. Oliver, Shota Atsumi*
Carbon Fixation
CO_2
ATP
CB Cycle
PGA
NADPH
Photosynthesis
chloroplast stroma
thylakoid lumen
NADPH
2.4 fold
Carbon
Atsumi, S. et al. Nat. Biotechnol 2009
Ducat, D.C. et al. Appl Environ Microbiol 2012
Oliver, J.W.K. et al. PNAS 2013
Amgermayr, S.A. et al. Biotechnol Biofuels 2014

Trophic conversion of autotrophic cyanobacteria

Energy from light

Carbon from CO_2

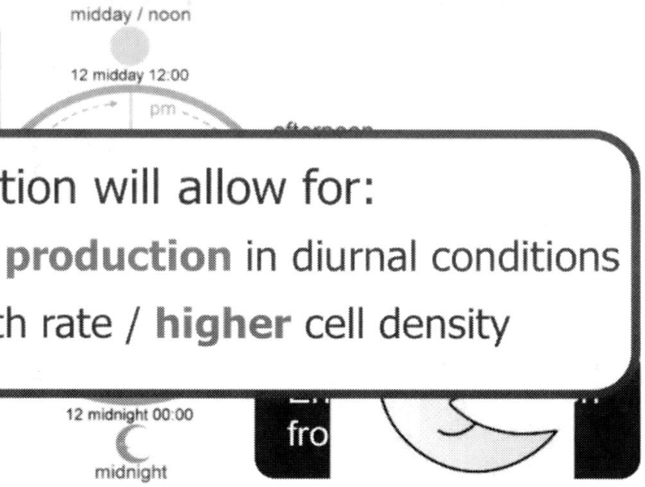

Sugar utilization will allow for:

- **Continuous production** in diurnal conditions
- **Faster** growth rate / **higher** cell density

To increase the overall productivity of the process

Hypothesis: Sugar Entry

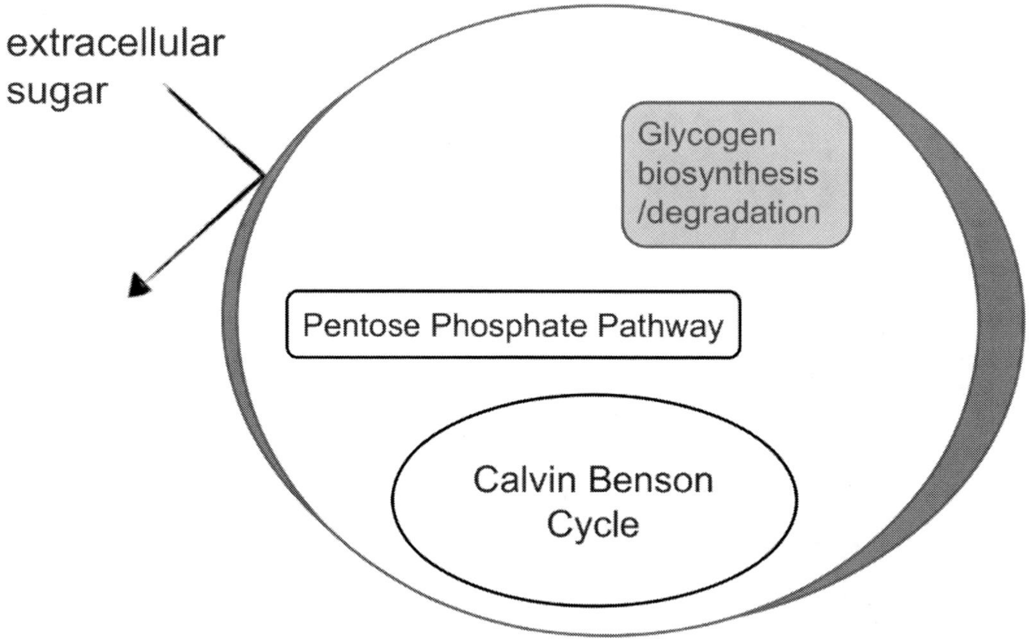

Hypothesis: Sugar Entry

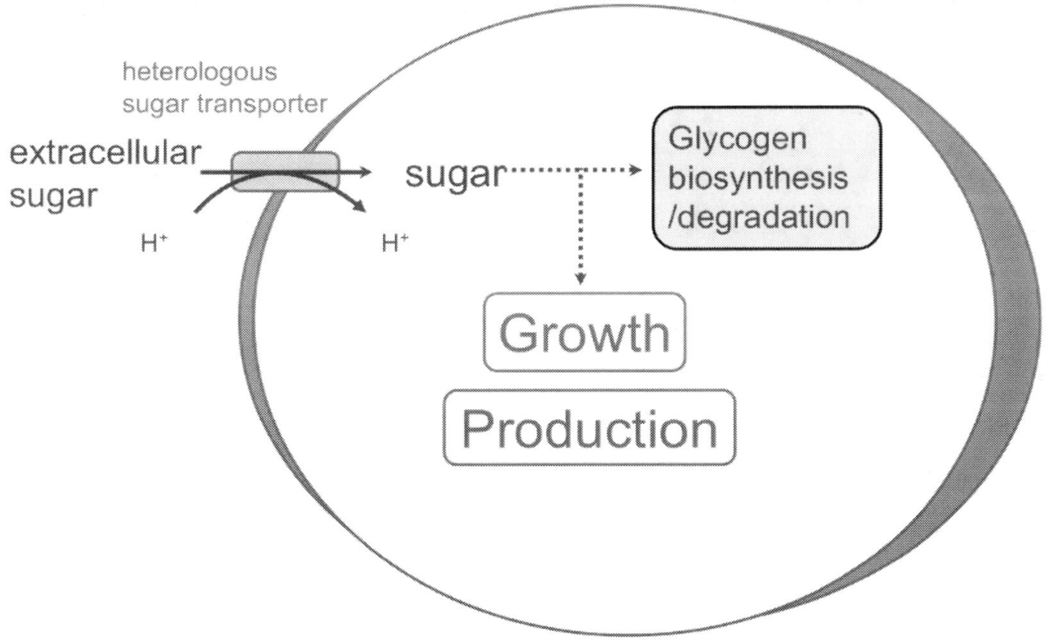

Glucose transporter mutants

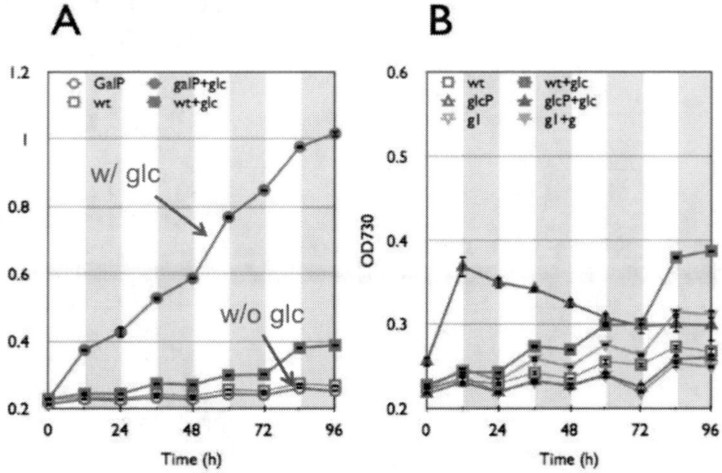

galP from *E. coli*

GalP: Galactose:H^+ sympoter
(inner membrane)

glcP from *Synechocystis* 6803
glut1 from human erythrocytes

McEwen et al. *AEM* 2013

Expanding synthetic transporter method

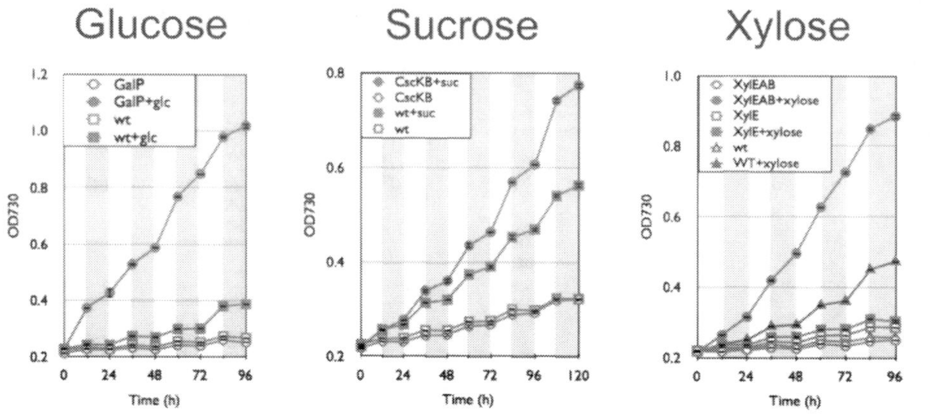

McEwen et al. *AEM* 2013

23BD production from CO_2 and glucose

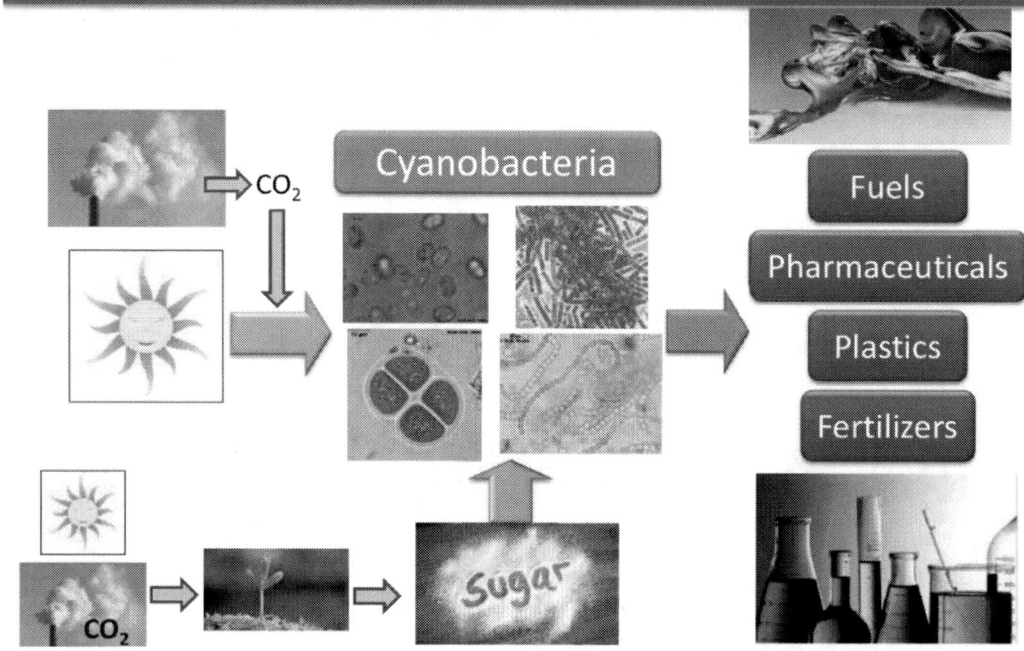

23BD production from CO_2 and glucose

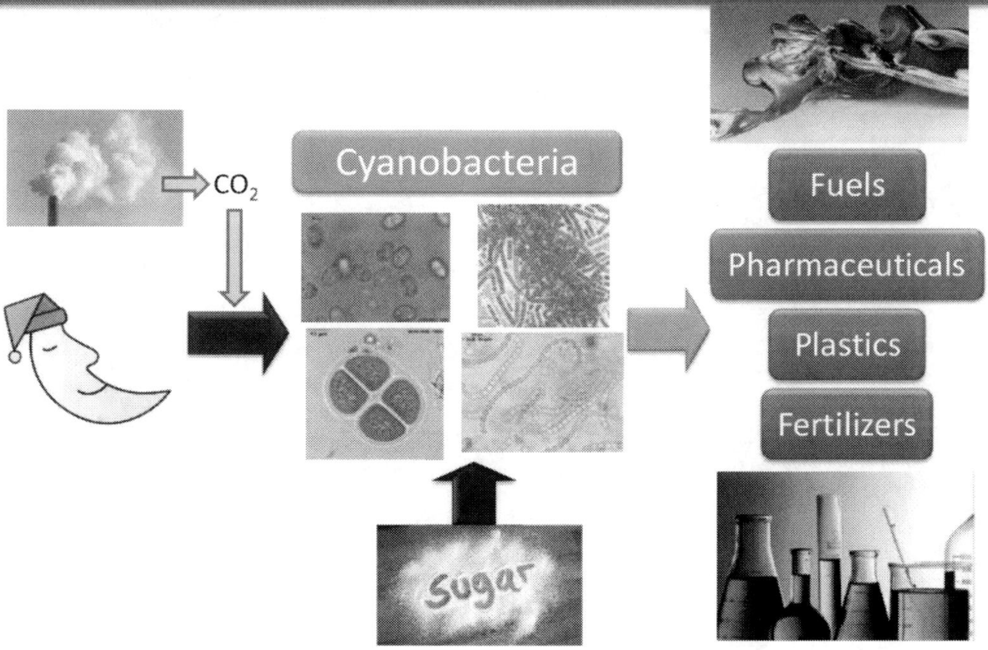

23BD production under diurnal conditions

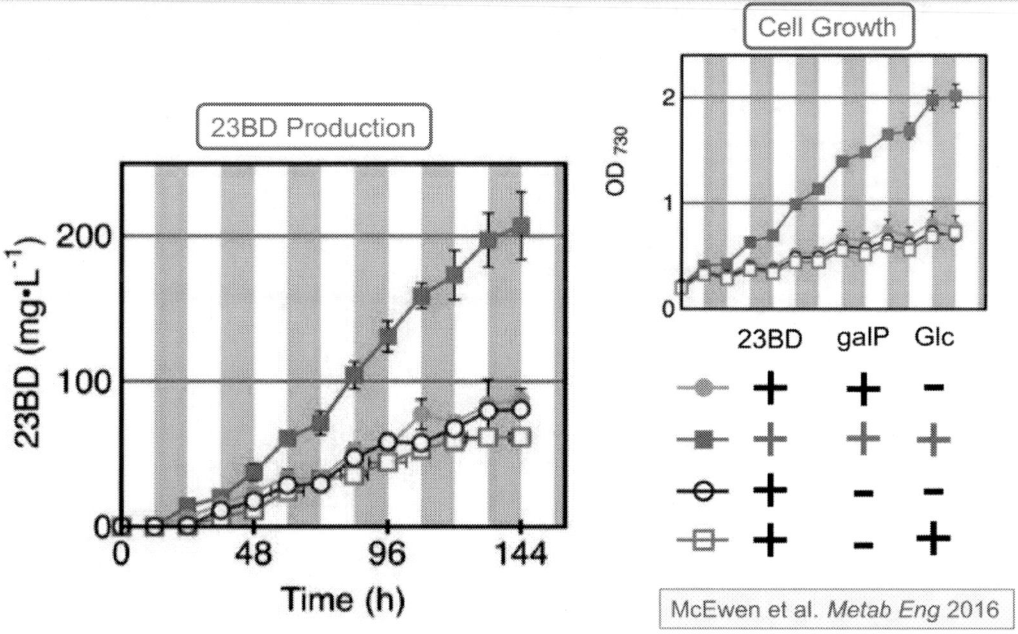

23BD production under dark conditions

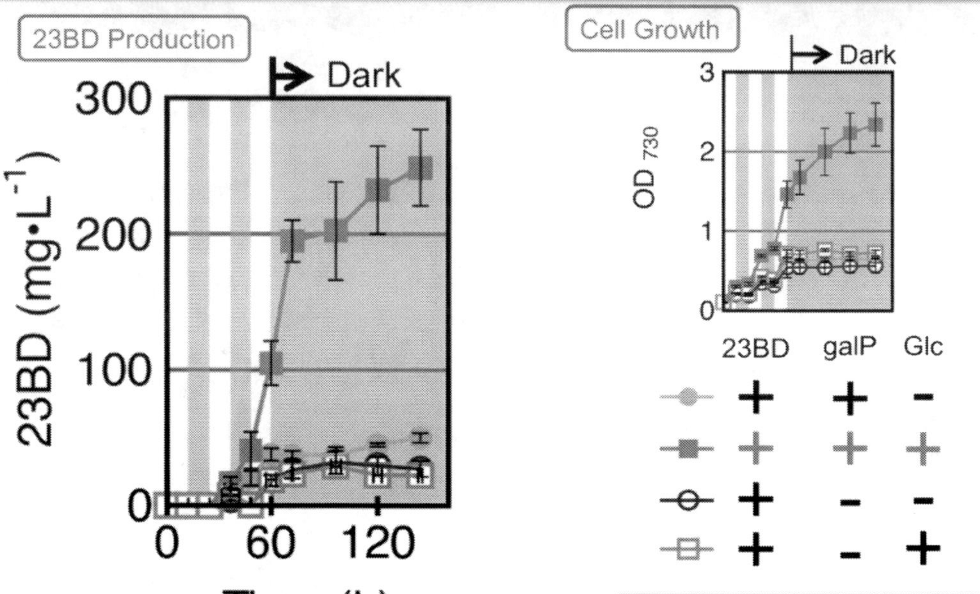

McEwen et al. *Metab Eng* 2016

Long term production under continuous light conditions

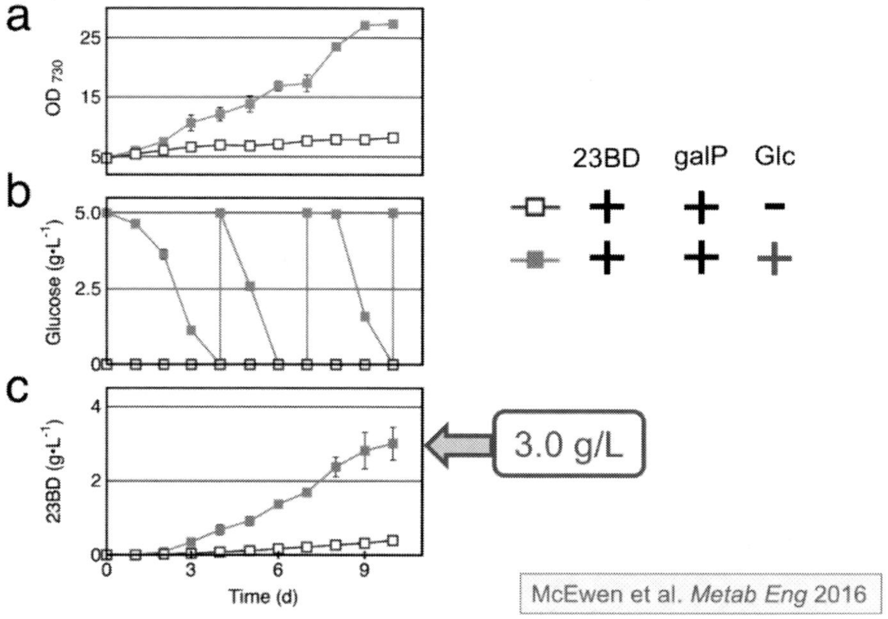

Trophic conversion of a photoautotrophic cyanobacterium

- Sugar transporter systems are the necessary genetic factors of *S. elongatus* to grow continuously in diurnal (light/dark) conditions using sugars.
- Glucose enhances cell growth and chemical production in light conditions.
- Glucose allows cell growth and chemical production in dark conditions.

McEwen et al. *AEM* (2013) 79(5): 1668-1675

McEwen et al. *Metab Eng* (2016) 36: 28-36

Acknowledgements

Atsumi Research Group

Postdocs
Christy Rabinovitch-Deere, Yohei Tashiro (12-15)
Michael Connor (10-11), Iara Machado (11-13)
Visiting Researcher
Masahiro Kanno, Hisanari Yoneda (11-13)
Graduate Students
Shuchi Desai, Nicole Nozzi, Anna Case, Austin Carroll, Neal Oliver, Morgan Matson
Jordan McEwen (PhD 15), Gabriel Rodriguez (PhD 14), John Oliver (PhD 14), Edna Lamsen (MS 12)
Undergraduate Students
Majed Eddin, Yousef Fararjeh, Kelly Nguyen

UC DAVIS
DEPARTMENT OF CHEMISTRY

Hellman Fellowship

AsahiKASEI

Asahi Kasei Corporation
DOE ARPA-E Electrofuel
DOE ARPA-E REMOTE
USDA-NIFA
NSF BBBE
NSF Energy for Sustainability
NSF CAREER

USDA

THE UNIVERSITY OF QUEENSLAND AUSTRALIA

Electro-fermentation as a Tool to Balance Redox in engineered *Pseudomonas putida*

me11
An IMES Conference

Jens O Krömer

Metabolic Engineering XI – Awaji, Japan
(30.06.2016)

CENTRE FOR MICROBIAL ELECTROCHEMICAL SYSTEMS

THE UNIVERSITY OF QUEENSLAND AUSTRALIA

CENTRE FOR MICROBIAL ELECTROCHEMICAL SYSTEMS

"Enable the development of electricity driven microbial processes scalable to industrial application in bio-production, bio-energy, bio-remediation and nutrient recovery "

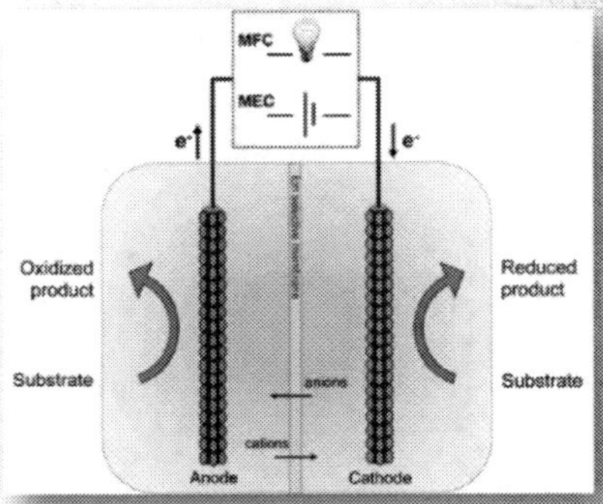

→ Improving bio-production processes by optimizing the cellular redox balance

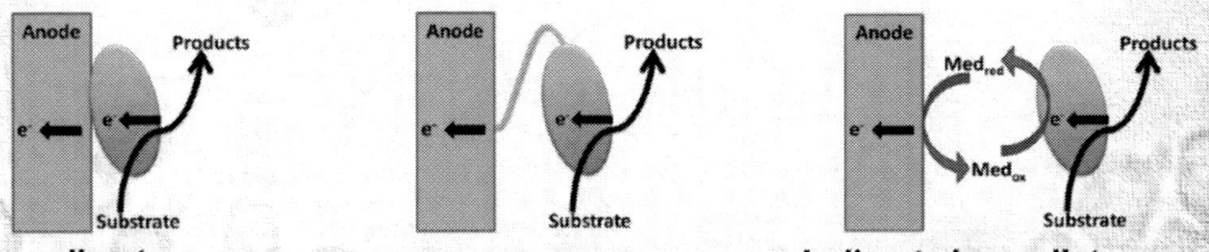

direct | **direct via pili** | **indirect via mediators**

CENTRE FOR MICROBIAL ELECTROCHEMICAL SYSTEMS

THE UNIVERSITY OF QUEENSLAND AUSTRALIA

Systems and Synthetic biology

Metabolic modelling; Metabolic engineering

Fundamentals of electron transport

Imaging; Electric biofilms; Mediator molecules

Material science

Electrode modifications; Advanced Electrochemistry

Process development and design

Bio cathodes and anodes; Reactor design

CENTRE FOR MICROBIAL ELECTROCHEMICAL SYSTEMS

THE UNIVERSITY OF QUEENSLAND AUSTRALIA

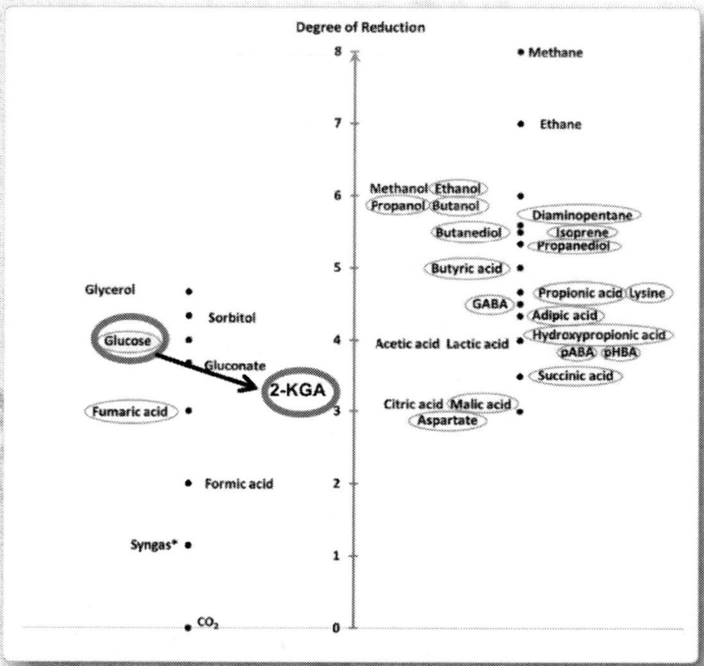

Redox limitations constrain many products

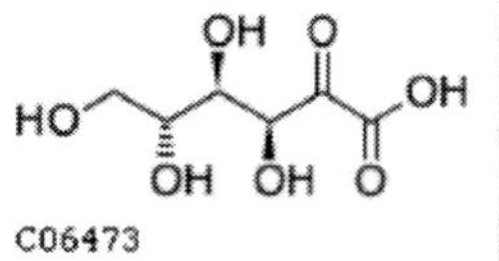

C06473

2-keto gluconic acid

Precursor for Erythorbic acid (E315)

CENTRE FOR MICROBIAL ELECTROCHEMICAL SYSTEMS

THE UNIVERSITY OF QUEENSLAND AUSTRALIA

Pseudomonas putida (F1):

- Motile flagella
- Gram-negative
- Obligate aerobe
- Diverse substrates
- High solvent tolerance
- Low by-products
- toxic chemicals production

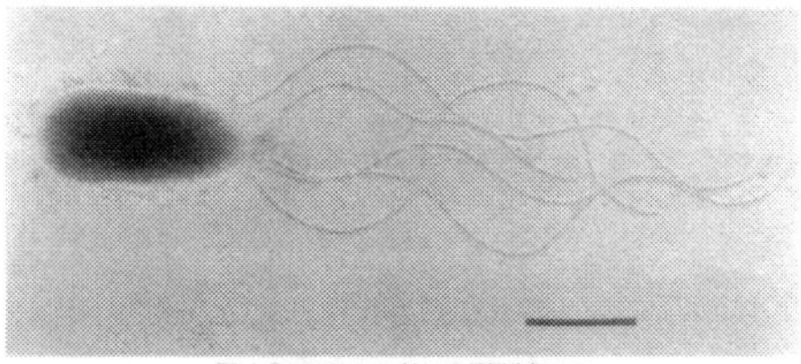

Harwood, C. S., et al, *J Bacteriol*, 1989

Q: Can *P. putida* operate anaerobically when an electrode is offered?

ELECTROCHEMICAL SYSTEMS

THE UNIVERSITY OF QUEENSLAND AUSTRALIA

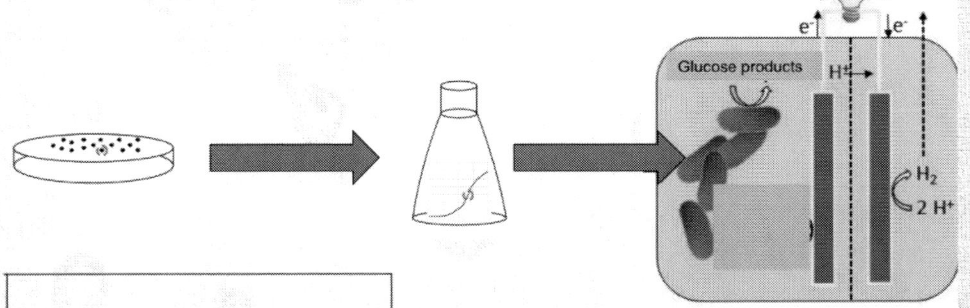

DM9 + glucose or LB medium;

-0.5~1.0V, vs Ag/AgCl/Sat.KCl;

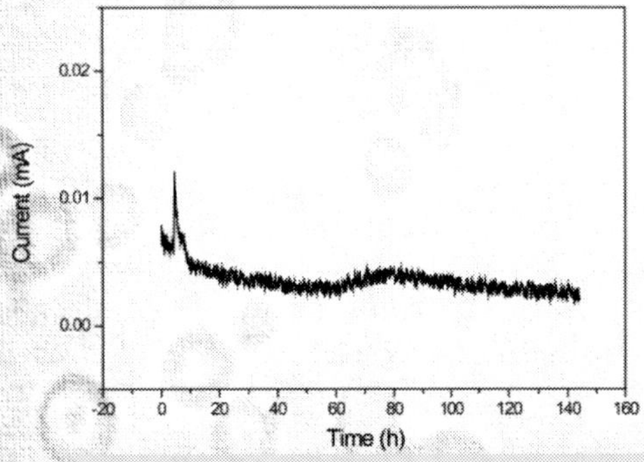

No significant current detected

No metabolism detected

Direct electron transfer not observed

Q: Can *P. putida* use mediators to achieve catalysis?

THE UNIVERSITY OF QUEENSLAND AUSTRALIA

Redox mediators:

- Broad redox potential
- Cover the whole membrane redox range

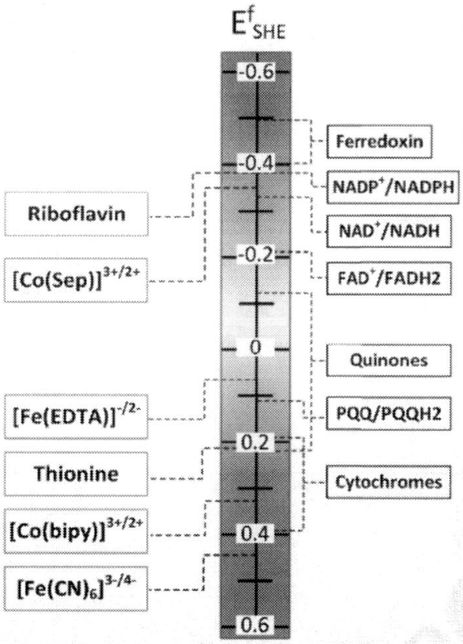

CENTRE FOR MICROBIAL ELECTROCHEMICAL SYSTEMS

THE UNIVERSITY OF QUEENSLAND AUSTRALIA

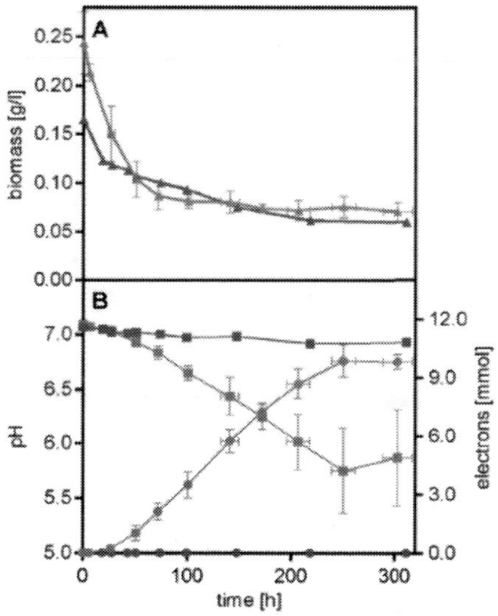

DM9 + glucose, 1mM $K_3[Fe(CN)_6]$
0.5V, vs Ag/AgCl/Sat.KCl;

No significant difference in biomass

pH drop detected

Electron transfer (current output) detected

A: Mediated Electron transfer possible.

Lai et al., 2016, Biotechnol Biofuels 9:39

CENTRE FOR MICROBIAL ELECTROCHEMICAL SYSTEMS

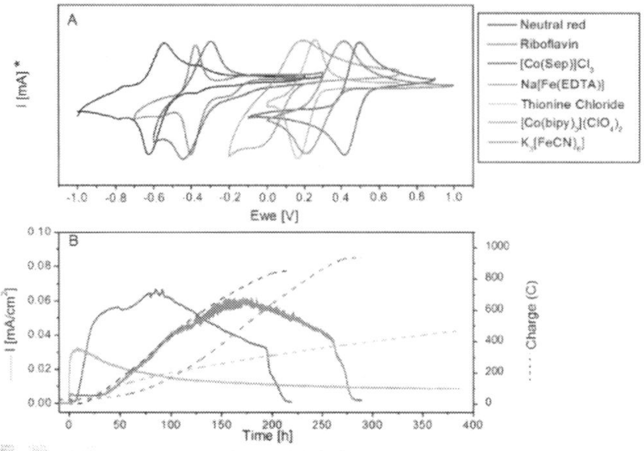

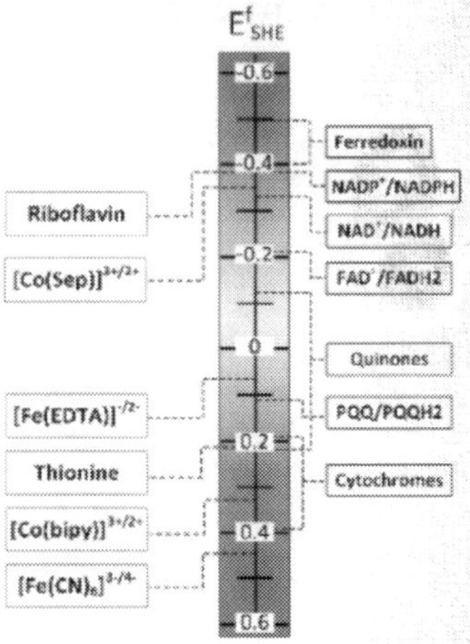

- DM9 + glucose, 1mM Med_{oxi}
- ~(E^f + 0.3)V, vs Ag/AgCl/Sat.KCl;

Effective: E^f >= 0.207V (Thionine)

Redox potential is key factor

Possibly gain electrons from PQQ or cytochrome

Q: Can we produce chemicals anaerobically with *P. putida*?

Anoxic catabolism of glucose

2-keto-gluconate major product (>90 % yield and CE).

Gluconate intermediate

Acetate minor by-product

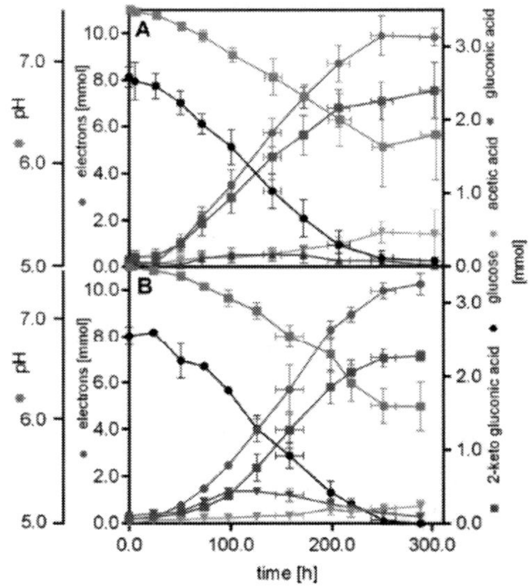

A: $[Fe(CN)_6]^{3-/4-}$; B: $[Co(bipy)]^{3+/2+}$

A: Anaerobic production from sugar with *P. putida* possible.

Q: Can *P. putida* generate ATP with the Anode?

Lai et al., 2016, Biotechnol Biofuels 9:39

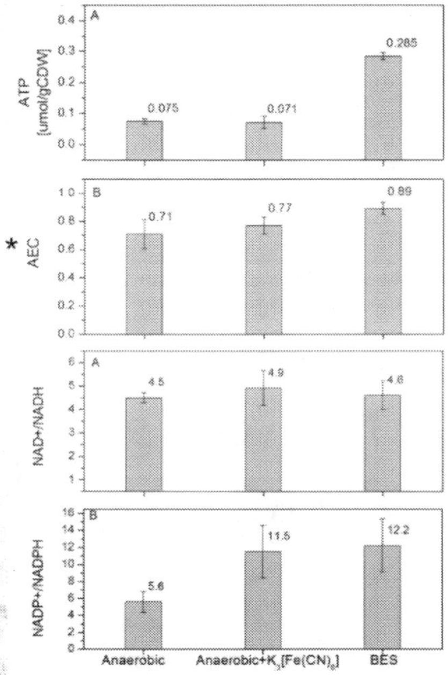

ATP regenerated by BES

AEC in range for growing cells

No effect on NADH pool

NADPH pool more oxidized

A: ATP generation possible when *P. putida* uses the Anode.

Lai et al., 2016, Biotechnol Biofuels 9:39

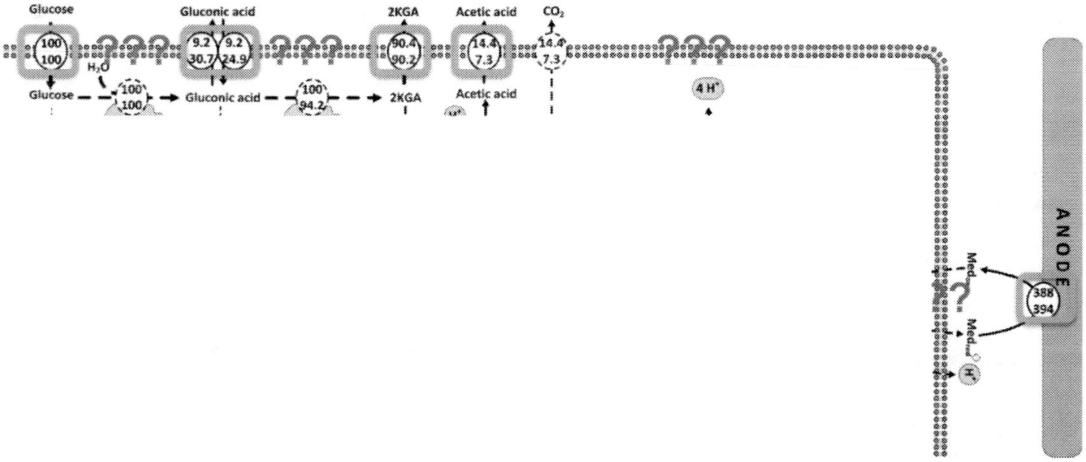

r [mmol ATP/(g$_{CDW}$ h)]	[Co(bipy)$_3$]$^{3+/2+}$	[Fe(CN)$_6$]$^{3-/4-}$
Glucose consumption	0.02	0.05
Redox power from BES	0.39	0.63

Reported NGAM

0.92 – 3.96

mmol ATP/(g$_{CDW}$ h)

Q: Can we engineer the cells to make it a faster process?

THE UNIVERSITY OF QUEENSLAND AUSTRALIA

Pseudomonas putida (KT2440):

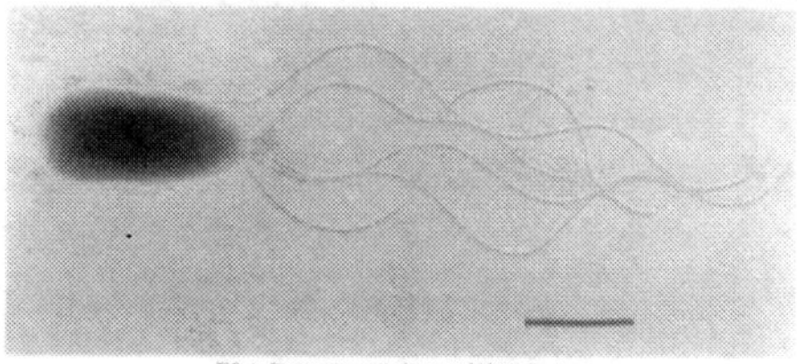

Glucose

↓ GCD (EC 1.1.5.2)

Gluconate

↓ KGDH (EC 1.1.99.3)

a-Ketogluconate

P. putida

P. putida GCD ↑

P. putida KGDH ↑

P. putida GCD ↑ KGDH↑

CENTRE FOR MICROBIAL ELECTROCHEMICAL SYSTEMS

THE UNIVERSITY OF QUEENSLAND AUSTRALIA

Rate is increasing for wildtype over time (biofilm? Expression changes?)

Overexpression leads to instant current output, higher peak-current.

Fermentation time reduced 70 %.

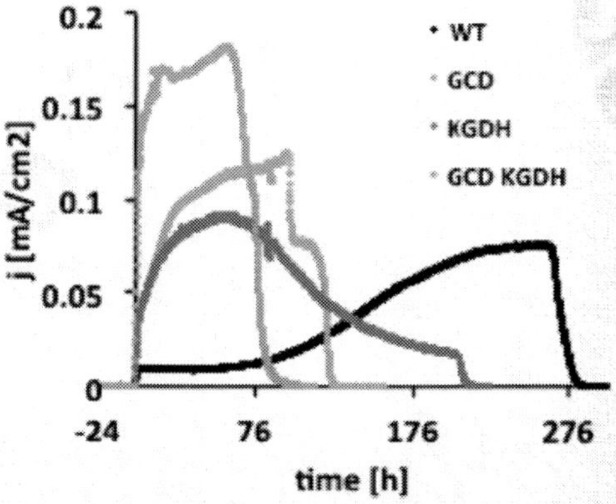

DM9 + glucose, 1mM $K_3[Fe(CN)_6]$

0.5V, vs Ag/AgCl/Sat.KCl;

A: Simple overexpression of key enzyme greatly enhances the process.

THE UNIVERSITY OF QUEENSLAND AUSTRALIA

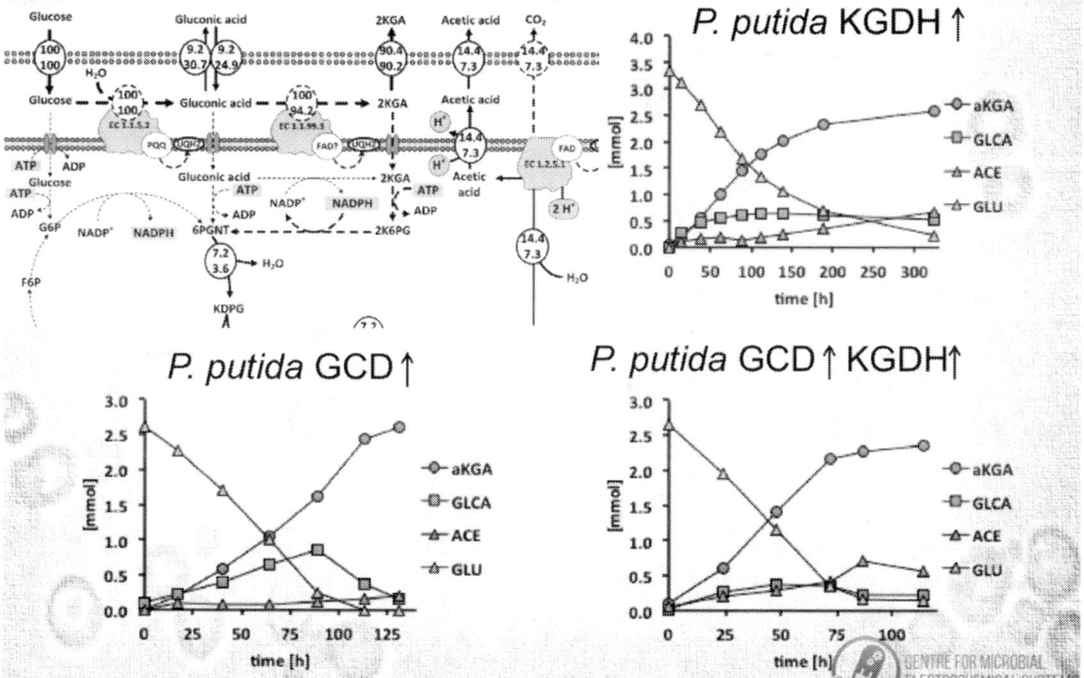

CENTRE FOR MICROBIAL ELECTROCHEMICAL SYSTEMS

THE UNIVERSITY OF QUEENSLAND AUSTRALIA

Summary

Bio-electrochemical systems can enable new processes (It works!...kind of)

Wild-type achieves commercial yields (> 90%, ok quite slow and diluted, but still....)

Mediator midpoint potential increases rate on electro-chemical level, metabolic engineering on a cellular level.

A new field for oxygen-free / but redox balanced peri-plasmatic redox biocatalysis is wide open.

CENTRE FOR MICROBIAL ELECTROCHEMICAL SYSTEMS

THE UNIVERSITY OF QUEENSLAND AUSTRALIA

Acknowledgements

Mr Bin Lai
Ms Shiqin Yu

Prof Paul Bernhardt
Dr Bernardino Virdis
Dr Gal Winter

Prof Korneel Rabaey (U Ghent)

Dr Manuel Plan (Metabolomics Australia)

Financial support

THE UNIVERSITY OF QUEENSLAND AUSTRALIA

CENTRE FOR MICROBIAL ELECTROCHEMICAL SYSTEMS

THE UNIVERSITY OF QUEENSLAND AUSTRALIA

Thank you for your attention…

… questions ?

CENTRE FOR MICROBIAL ELECTROCHEMICAL SYSTEMS

Video Presentation in
Original e-media

Random or Targeted Genetic Approaches for the Production of Isopropanol By Clostridium Beijerinckii DSM6423

Mate de Gerando, H.,
Wasels, F.,
López Contreras, A. M., Wageningen UR
Lopes Ferreira, N., IFP Energies nouvelles

Random and Targeted Genetic Approaches for the Production of Isopropanol by *Clostridium beijerinckii* DSM6423

Isopropanol (isopropyl alcohol) is a chemical compound widely used in cosmetics and for household use. It is mainly obtained by hydration of propylene but it can also be produced by microorganisms and used in a dehydrogenation process for biosourced propylene production. Despite their natural capacity to produce isopropanol, natural isopropanol producing strains such as *Clostridium beijerinckii*DSM6423 (NRRL B-593 ; George et al 1983) have been poorly studied. Recently, few publications have shown that this strain was able to produce relatively high levels of a mixture of isopropanol, butanol and ethanol either in continuous culture (Survase & Jurgens, 2011) or in an extractive fermentation system (Lopez-Contreras et al. 2013).

In order to improve the isopropanol production capacity of *C. beijerinckii*, two complementary approaches were investigated aiming at redirecting carbon fluxes toward this solvent. The first strategy consisted in combining random mutagenesis and genome shuffling. This technique is based on the possibility to mix entire genomes using protoplasts fusion of interesting mutated strains. It has already been successfully applied to *C. acetobutylicum* CICC 8012 (Gao et al., 2012) to improve yield and solvent titers of acetone-butanol fermentation. In our study (Mate de Gerando et al, 2016), we have been able to obtain 36 NTG mutants of *C. beijerinckii* DSM6423 on selective plates containing either high concentrations of isopropanol (up to 35 g/L), analogous substrates (ethyl-or methyl- bromobutyrate) or allyl alcohol, a well-known inhibitor of alcohol dehydrogenases. Considering stability and fermentation performances, the best NTG mutants were selected for the genome shuffling step. Three highly isopropanol tolerant strains able to withstand up to 50 g/L isopropanol were finally obtained. Moreover, the best strain produced up to 22% more isopropanol than the wild type. Genome analysis of the NTG and shuffled mutants revealed SNP's mutations located in several genes involved in the central metabolism or the transcriptional regulation network of the strain.

In the second approach, we used the genetic tools developed for engineering *Clostridia*. Targeted modifications of *C. beijerinckii* DSM6423 were tested but multiple attempts to transform this strain failed. To understand the recalcitrance of the strain to genetic modifications, its genome was sequenced and the physical map detailed. Genome reconstruction showed the presence of two extrachromosmic elements, a double strand linear DNA bacteriophage and a natural plasmid which could explain the difficulties to introduce another mobile genetic element for genetic engineering.This work on *C. beijerinckii* DSM6423 lead us to better understand the genetic content of the strain and its

IBE metabolic pathway, paving the way to future targeted engineering approaches.

References

George et al, 1983. Appl Environ Microbiol.45:1160-3

Survase et al, 2011. Appl Microbiol Biotechnol. 91:1305-13

Gao et al, 2012. Curr Microbiol. 2012 65:128-32

de Vrije et al, 2013. Bioresour Technol. 137:153-9.

Mate de Gerando et al 2016 Appl Microbiol Biotechnol DOI 10.1007/s00253-016-7302-5

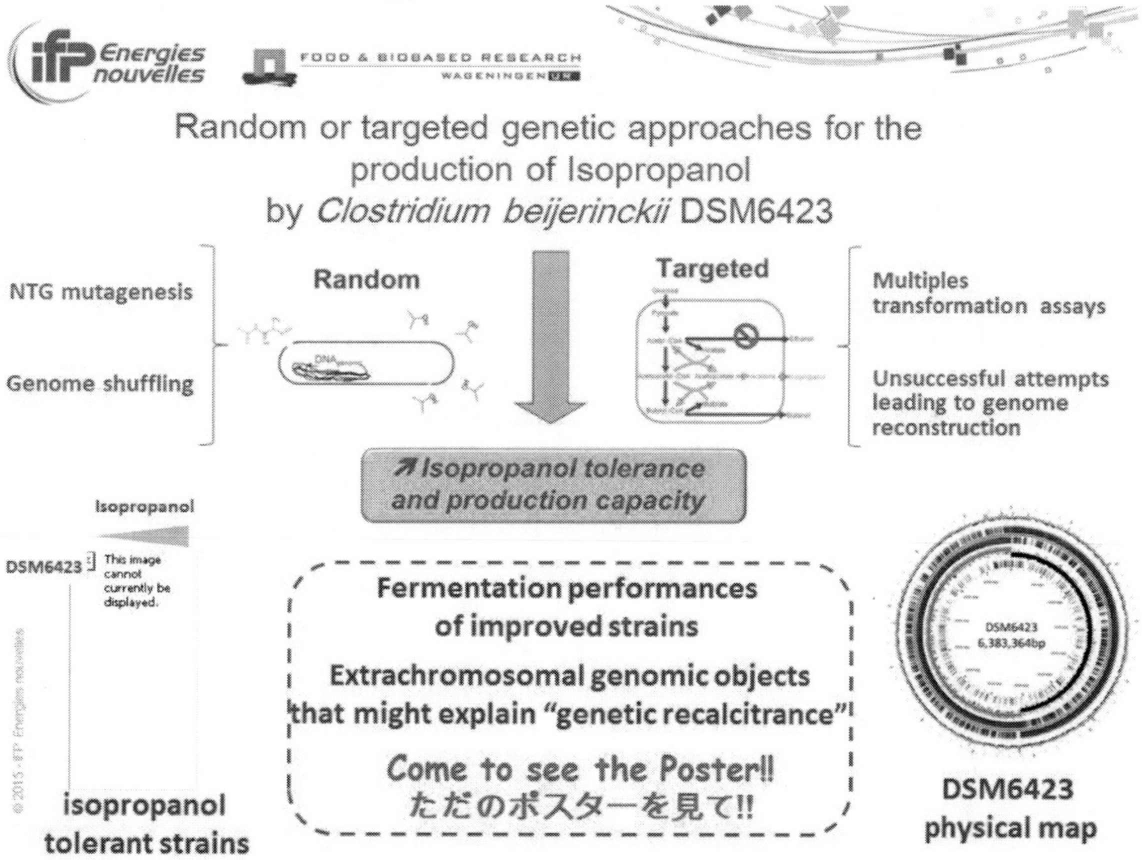

On-Chip Analysis, Indexing and Screening for Chemical Producing Bacteria in Microfluidic Static Droplet Array

Jang, S., POSTECH
Lee, B.,
Jeong, H. H.,
Jin, S. H.,
Jang, S.,
Kim, S. G.,
Jung, G. Y., POSTECH
Lee, C. S., Chungnam National University

Combinatorial engineering is known as one of the most powerful ways to create microbes for economic production of chemicals. The importance of efficient screening has been emphasized to investigate vast phenotypic diversity provided by advanced technologies. However, current methods place constraints on monitoring of intracellular concentration of diverse target products at the level of single cell. Here, we established a high-throughput screening platform for assessment of metabolites based on microfluidic static droplet array (SDA) and artificial riboswitch. In this system, a genetically encoded riboswitch visualizes intracellular concentration of a target molecule in a quantitative manner. Then, single cells from a library are individually entrapped in SDA to detect and collect highly productive variants. Using this approach, we analyzed intracellular L-tryptophan concentrations of single *Escherichia coli* cells. Fluorescence from an artificial L-tryptophan riboswitch enabled easy measurement of the metabolite at single cell resolution. Then, a mutant *E. coli* library was screened for improved L-tryptophan production and superior strains were obtained exhibiting up to 145% productivity in comparison with their parental strain. It is concluded that the combination of SDA and artificial riboswitch was successfully applied to high-throughput screening of microbial cell library for combinatorial metabolic engineering. This platform will be widely applied to a variety of screening projects by developing new artificial riboswitches for new target molecules.

On-chip analysis, indexing and screening for chemical-producing bacteria in microfluidic static droplet array

Sungho Jang / Rapid fire #3

Microfluidic device for static droplet array

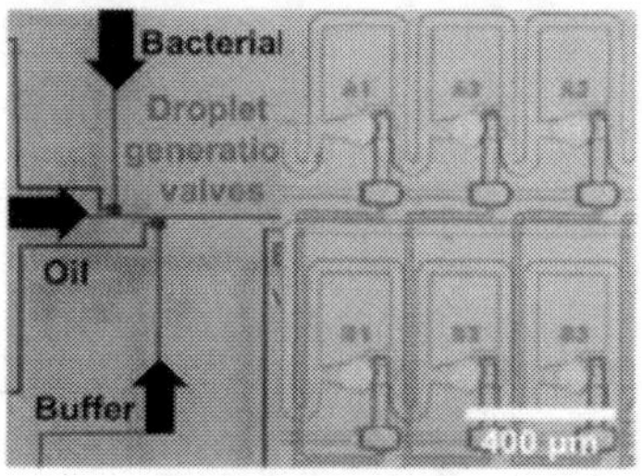

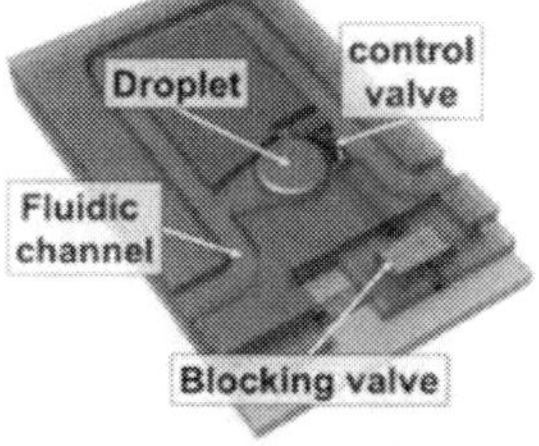

Analysis of single-cell for metabolite production & selective sorting

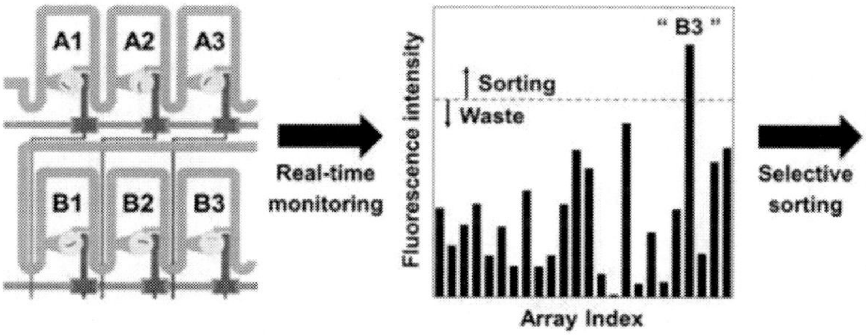

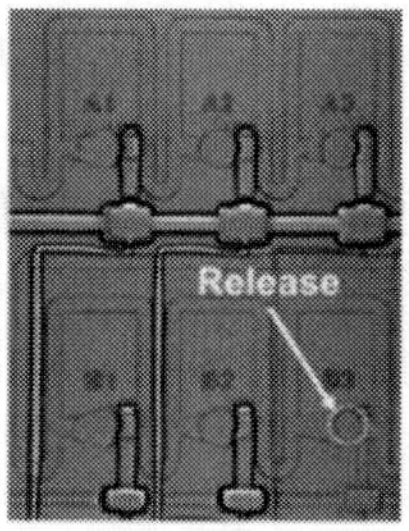

The Production of D-2,3-Butanediol and Meso-2,3-Butanediol with High Optical Purity in Engineered Bacillus Licheniformis WX02

Wang, Q.,
Qiu, Y.,
Chen, S.,

Abstract: Microbial production of 2,3-butanediol (2,3-BD) has received increasing interests because of its potential applications in cosmetics, foods, transport fuels, medicines, and polymers industries. *Bacillus licheniformis* WX-02 is a generally regarded as safe (GRAS) strain and is regarded as a potential 2,3-BD producer. However, this strain like other 2,3-BD producing microorganisms produces a mix of *D*-2,3-BD and *meso*-2,3-BD isomers. In order to reduce the cost of purifying a specific 2,3-BD isomer from fermentation broth, it is desirable for the microbial strain to produce the specific target isomer without synthesis of the other forms. In this study, the *budC* gene was identified responsible for the conversion of the precursor acetoin to *meso*-2,3-BD in *B. licheniformis*. A *budC* deletion strain of *B. licheniformis* WX-02 was successfully constructed to inhibit the flux of acetoin to *meso*-2,3-BD biosynthesis, and *D*-2,3-BD with high optical purity was successfully produced. Furthermore, glycerol dehydrogenase (Gdh) was identified as the catalyst in *D*-2,3-BD biosynthesis from acetoin in *B. licheniformis*. The *gdh* gene was therefore deleted from the wild-type strain to produce high purity *meso*-2,3-BD. In addition, the *acoR* gene involved in acetoin degradation was also deleted to increase the pool of the precurcor acetoin. The double deletion mutant WX-02Δ*gdh*Δ*acoR* produced 28.2 g/L of *meso*-2,3-BD isomer with > 99% purity in flask culture. Using fed-batch fermentation, the highest titer (98.0 g/L) ever reported for *meso*-2,3-BD was achieved by WX-02Δ*gdh*Δ*acoR*.

The production of D-2,3-butanediol and *meso*-2,3-butanediol with high optical purity using engineering *Bacillus licheniformis*

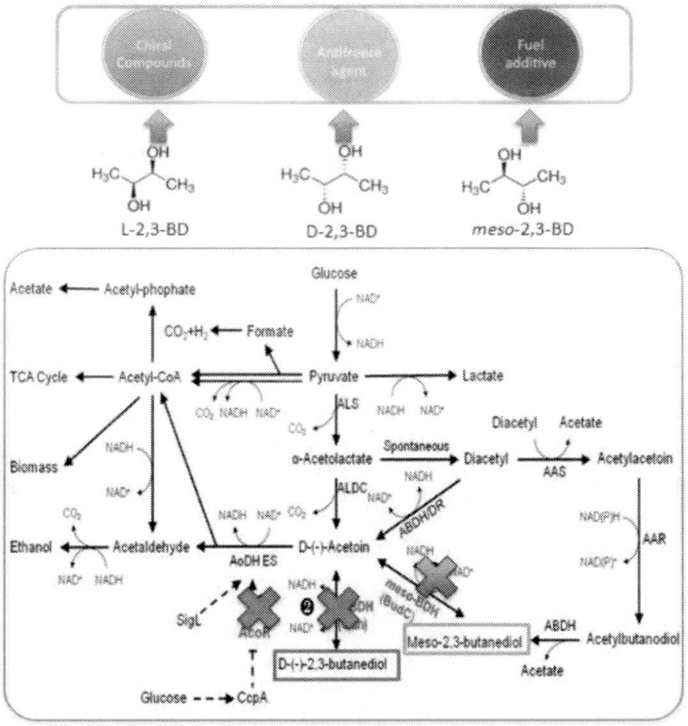

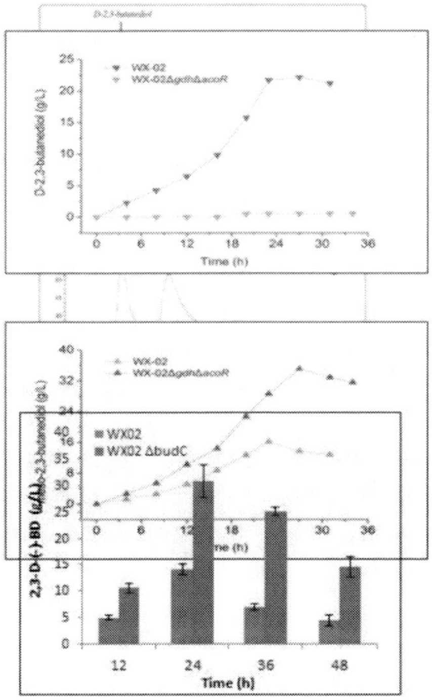

Sustainable Biorefining in Unsterilized Wastewater By Engineering of the Extreme Alkaliphile Bacillus Marmarensis

Wernick, D. G., The Weizmann Institute of Sciences, Rehovot, Israel
Pontrelli, S. P., UCLA
Pollock, A. W., UCLA
Liao, J. C., Department of Chemical Engineering, UCLA

Water usage, contamination susceptibility, and inability to co-utilize pentoses and disaccharides are some major challenges to industrialization of sustainable biorefining *in lieu* of fossil resource consumption. We have built tools for and engineered the unexplored, extreme alkaliphile *Bacillus marmarensis* as a platform for new bioprocesses which meet all these challenges. With development of a transformation technique, a genetic overexpression system, and RNA knockdown strategies, we engineered *B. marmarensis* to produce ethanol at 40 g/l titers at 65% the theoretical maximum yield in unsterilized glucose medium. Transitioning towards industrial substrates and conditions, we next achieved ethanol titers of 12 g/l at a 50% yield from xylose and cellobiose mixtures in unsterilized seawater and algal-contaminated wastewater media. *B. marmarensis* presents a promising approach for the contamination-resistant biorefining of a wide range of carbohydrates in unsterilized, non-potable seawater from this initial investigation.

Engineering Escherichia coli for the Production of Propionic Acid through the Wood-Werkman Cycle

Gonzalez-Garcia, R. A., The University of Queensland
Marcellin, E., The University of Queensland
Nielsen, L. K., The University of Queensland
McCubbin, T., University of Queensland

Biological production of propionic acid represents a viable alternative for the production of C3 chemicals. Propionic acid is natively produced through the *Propionibacteria*'s Wood-Werkman cycle. By coupling the electron transport chain during the reduction of fumarate to succinate in the dicarboxylic acid branch of the TCA cycle, the cycle offers an energetic advantage for cells. The unique functionality of one of the enzymes, the methylmalonyl-CoA carboxyltransferase, allows the direct exchange of a carboxyl group and a thiol group without ATP consumption, giving an even higher energy gain. These features make *Propionibacteria* an ideal candidate for the production of propionic acid. Unfortunately, engineering *Propionibacteria* has proven extremely challenging. Equipping *Escherichia coli* with the Wood-Werkman cycle is thus a viable alternative for heterologous propionic acid production. *In silico* analysis predicts that *E. coli* can use the cycle and take advantage of its energy benefits. The genome-scale model predicts that cells harbouring the cycle can sustain growth at a 43% higher growth rate and a yield of 1.135 $mol_{propionate}/mol_{glucose}$. Thus, seven genes from *P. acidipropionici* were transformed into *E. coli* K12, W and MG1655. Fermentations were performed in bioreactors, and cells were grown on minimal media. Initial results showed that *E. coli* was able to produce propionic acid through the Wood-Werkman cycle despite a stress related effect resulting in lactate overproduction. Intracellular metabolomics and proteomics identified limitations in the expression system as two key enzymes, namely methylmalonyl-CoA mutase and methylmalonyl-CoA carboxylase, were under-expressed.

The initial design, comprising a single synthetic operon under a *nar* inducible T7 promoter was too strong causing cell death after induction. This problem was addressed by testing different promoters and modifying the structure of the synthetic operon. Our results show for the first time that *E. coli* can express a functional Wood-Werkman cycle leading to propionic acid production. The next step is to evolve cells to adapt to the new metabolism. Cells are currently under adaptive evolution, and alternative by-products pathways are being knocked-out. We anticipate that the knowledge generated here in a heterologous host can help overcoming the current limitations in the biological production of C3 chemicals.

THE UNIVERSITY OF QUEENSLAND
AUSTRALIA

Engineering *Escherichia coli* for the production of propionic acid through the Wood-Werkman Cycle

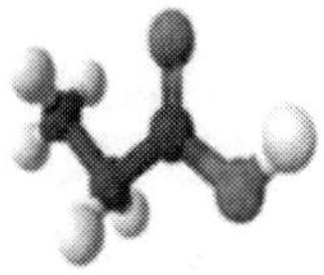

Axayacatl Gonzalez

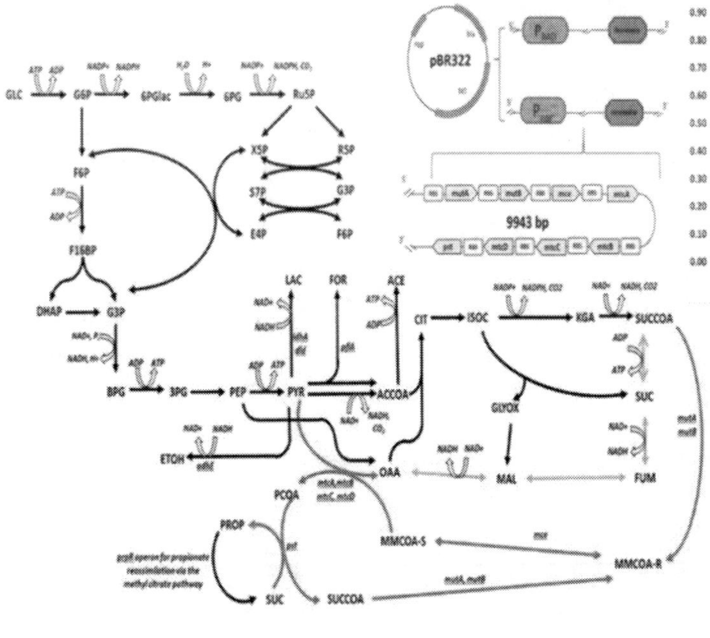

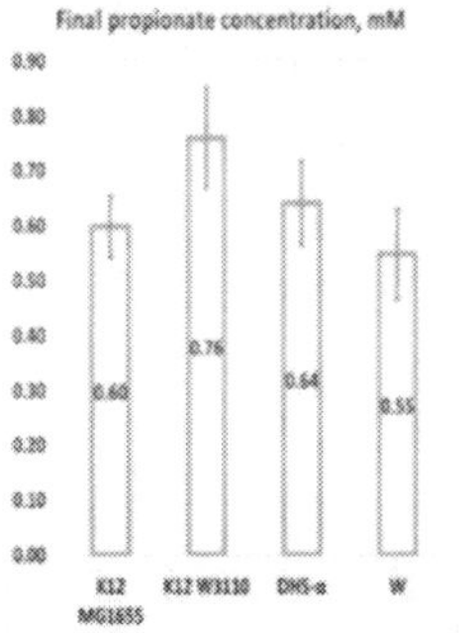

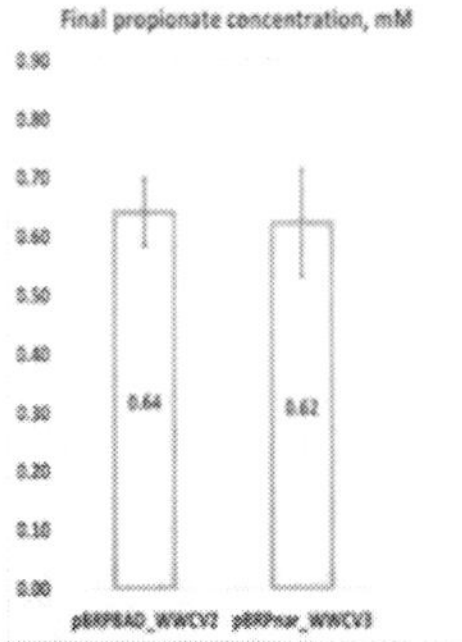

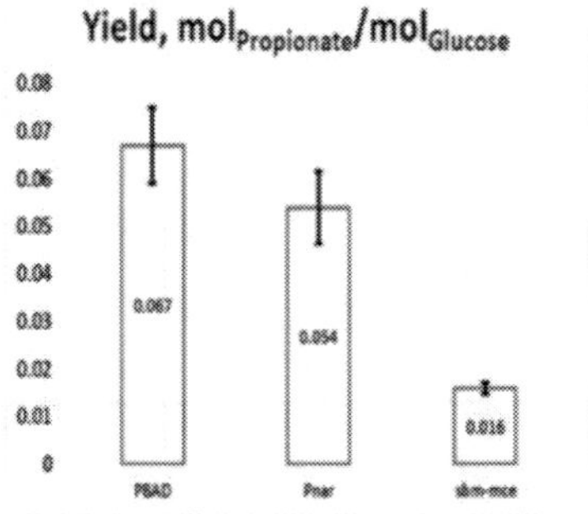

Engineering Yeast Cell Factories for Production of High-Value Isoprenoids

Krivoruchko, A., Chalmers University
Nielsen, J., Chalmers University of Technology

Isoprenoids comprise a large family of metabolites with various industrially-useful functions including biofuels, flavors, fragrances, nutraceuticals and pharmaceuticals. These metabolites are often produced in nature by organisms that are not optimal for industrial production. For example, many useful isoprenoids that are plant-derived are often found only in low levels in plants and are obtained by complex extraction procedures. Rising demands for such compounds, as well as the environmental impact associated with plant-based extraction creates a need for alternative production routes.

Reconstruction of plant biosynthetic pathways for valuable compounds in microbes provides a promising alternative to plant-based extraction. The yeast, *Saccharomyces cerevisiae*, makes for a particularly attractive host. It is a well-known and robust industrial organism, has a GRAS (Generally Regarded As Safe) status, is easily manipulated genetically and is able to functionally express eukaryotic cytochrome P450 enzymes. Our group makes use of different metabolic engineering approaches to construct yeast cell factories for production of industrially-relevant isoprenoid compounds including various sesquiterpenes, diterpenes and triterpenes. Our most recent advances in this field are presented.

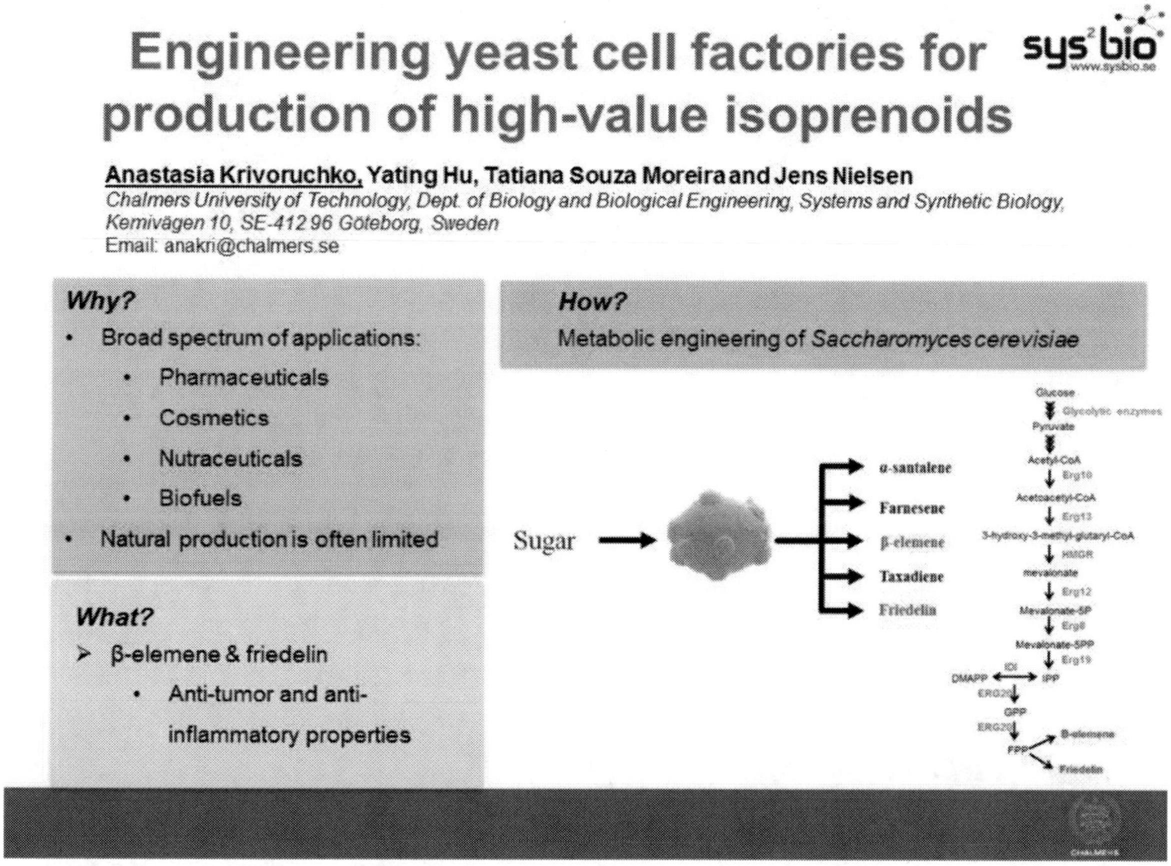

Engineering of a Novel Cellulose-Adherent Cellulolytic Saccharomyces Cerevisiae for Cellulosic Biofuel Production

Liu, Z., Kobe University
Inokuma, K., Kobe University
Hasunuma, T.,
Kondo, A., Graduate School of Engineering, Kobe University

Cellulosic biofuels are still impeded by the recalcitrance of lignocellulose that requires high dosages of cellulase to hydrolyze. Several cellulolytic yeast strains have been developed to reduce the need for enzyme addition, but exhibiting limited effect. This is because of the insufficient cellulolytic activity provided by yeast, and lacks of a novel strategy to achieve efficient cellulose degradation using cellulolytic strains. Here, we report the successful engineering of a high-efficiency cellulolytic *Saccharomyces cerevisiae* by displaying four different synergistic cellulases (β-glucosidase, endoglucanase, cellobiohydrolase I, and cellobiohydrolase II) on the cell surface. The cellulase-displaying yeast strain exhibited clear cell-to-cellulose adhesion and a "tearing" cellulose degradation pattern; its adhesion ability is correlated with enhanced surface area and roughness of the target cellulose fibers, resulting in higher hydrolysis efficiency. The engineered yeast could directly produce ethanol from rice straw, as well as cutting down more than 40% of the required enzyme dosage under high-dense fermentation process. We provided a novel strategy that increases cellulose hydrolysis via enhancing cell-to-cellulose interactions, and demonstrate a breakthrough in making the feasibility of cellulosic biofuel production.

Keyword: cell-surface display; cellulase; cellulose-adherent; cellulosic ethanol; *Saccharomyces cerevisiae*

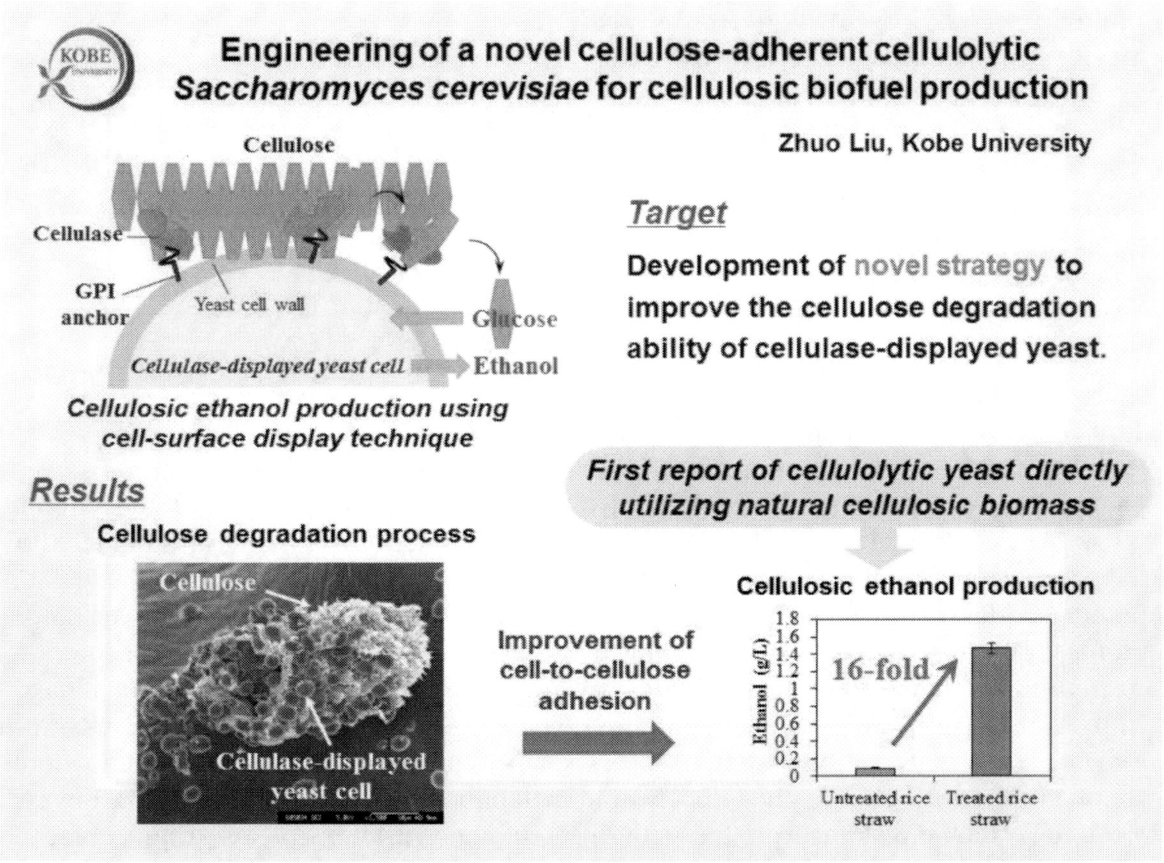
Engineering of a novel cellulose-adherent cellulolytic
Saccharomyces cerevisiae for cellulosic biofuel production
Zhuo Liu, Kobe University
Cellulose
Cellulase
GPI anchor
Yeast cell wall
Glucose
Cellulase-displayed yeast cell
Ethanol
Cellulosic ethanol production using cell-surface display technique
Target
Development of novel strategy to improve the cellulose degradation ability of cellulase-displayed yeast.
First report of cellulolytic yeast directly utilizing natural cellulosic biomass
Results
Cellulose degradation process
Cellulose
Cellulase-displayed yeast cell
Improvement of cell-to-cellulose adhesion
Cellulosic ethanol production
Ethanol (g/L)
1.8
1.6
1.4
1.2
1
0.8
0.6
0.4
0.2
0
16-fold
Untreated rice straw
Treated rice straw

Mathematical Model Based Understanding and Engineering of the Phenylpropanoid Pathway in Arabidopsis

Morgan, J. A., Purdue University
Chapple, C., Purdue University
Dudareva, N., Purdue University
Guo, L., Purdue University
Wang, P., Purdue University
Jaini, R., Purdue University
Ray, S., Purdue University
Lynch, J., Purdue University

2-Phenylethanol (2-PE) is a naturally occurring aromatic with advantageous properties compared to ethanol making it a potential biobased oxygenate in petroleum-derived gasoline. In plants, biosynthesis of 2-PE competes for the common precursor phenylalanine with phenylpropanoid pathway which directs about 30% of fixed carbon towards lignin biosynthesis. The efficient extraction of cellulosic biomass is hindered by the presence of lignin. Therefore genetic manipulation around the phenylalanine branch point provides a promising strategy to redirect carbon flux away from lignin to enhance biomass quality, and realize an economically valuable by-product 2-PE. Transgenic *Arabidopsis thaliana* were generated that overexpress the phenylacetaldehyde synthase from Arabidopsis (PAAS) in tandem with the phenylacetaldehyde reductase (PAR) from tomato. We selected primary stem as the experimental system for analyzing competition for phenylalanine between phenylpropanoid metabolism and engineered 2-PE pathway. Excised 5-week-old stems were fed with different concentrations of $^{13}C_6$-ring labeled phenylalanine, with both the amount and isotopic enrichment of downstream intermediates quantified with LC-MS/MS at multiple time points after feeding. A kinetic model of the phenylpropanoid network was constructed, and the parameters were identified through non-linear optimization with training datasets, and validated with data from an independent experiment. *In silico* analysis of the results from our model predicted that the cytosolic phenylalanine concentration limits 2-PE production in these transgenic plants. This prediction was tested by combining overexpression of PAR and PAAS with overexpressing a feedback-insensitive 3-deoxy-D-*arabino*-heptulosonate 7-phosphate (DAHP) synthase, the latter of which has previously been shown to have hyper-induced phenylalanine biosynthesis. More than an order of magnitude increase in 2-PE accumulation was successfully achieved using this strategy, compared to overexpression of the 2-PE pathway alone. We are currently investigating the effect on lignin accumulation and composition. Thus kinetic modeling combined with *in vivo* time-course metabolite profiling shows to be a promising approach to define a strategy for feedstock improvements.

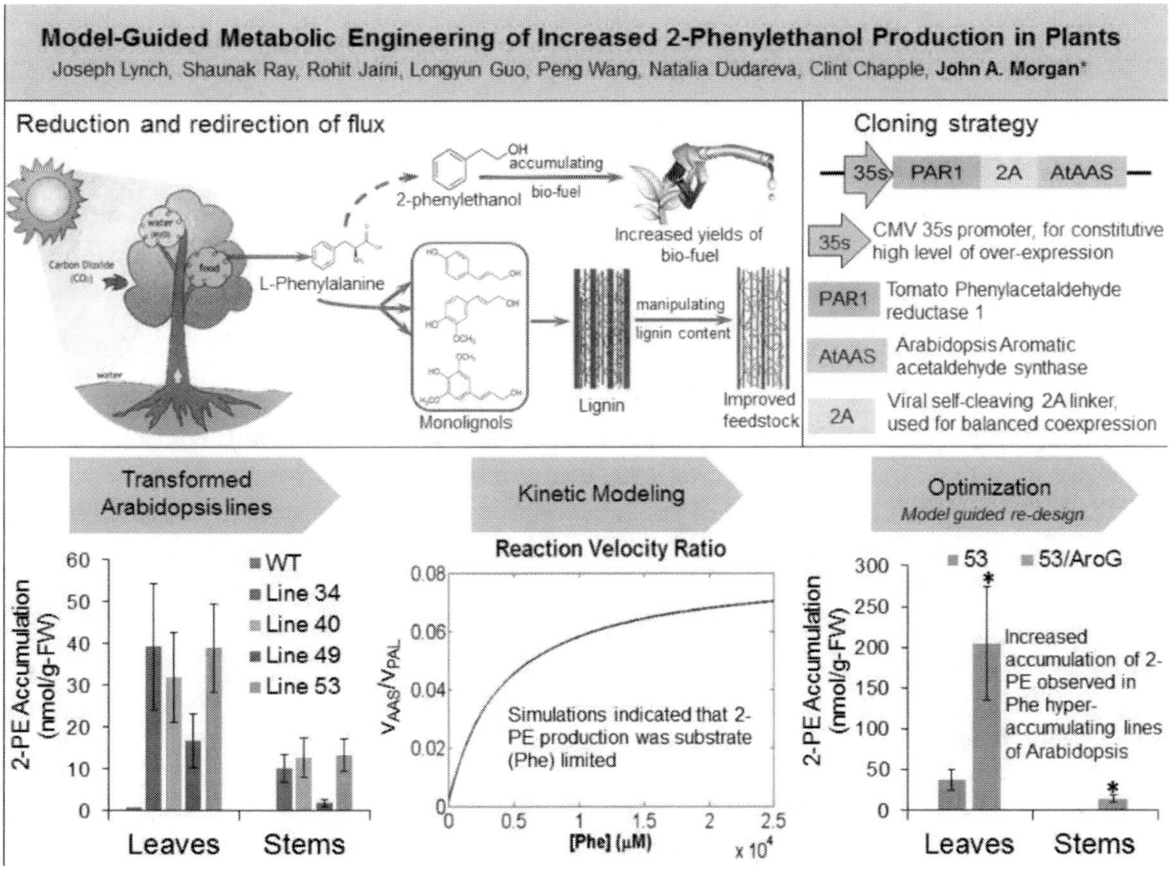
Model-Guided Metabolic Engineering of Increased 2-Phenylethanol Production in Plants
Joseph Lynch, Shaunak Ray, Rohit Jaini, Longyun Guo, Peng Wang, Natalia Dudareva, Clint Chapple, John A. Morgan*
Reduction and redirection of flux
2-phenylethanol
accumulating
bio-fuel
Increased yields of bio-fuel
L-Phenylalanine
Monolignols
Lignin
manipulating lignin content
Improved feedstock
Cloning strategy
35s
PAR1
2A
AtAAS
CMV 35s promoter, for constitutive high level of over-expression
Tomato Phenylacetaldehyde reductase 1
Arabidopsis Aromatic acetaldehyde synthase
Viral self-cleaving 2A linker, used for balanced coexpression
Transformed Arabidopsis lines
WT
Line 34
Line 40
Line 49
Line 53
2-PE Accumulation (nmol/g-FW)
Leaves
Stems
Kinetic Modeling
Reaction Velocity Ratio
V_AAS/V_PAL
Simulations indicated that 2-PE production was substrate (Phe) limited
[Phe] (μM)
x 10^4
Optimization
Model guided re-design
53
53/AroG
Increased accumulation of 2-PE observed in Phe hyper-accumulating lines of Arabidopsis

From Niche to Bulk - Glycolipids and Derivatives Synthesized from Sugar

Blank, L. M.,
Germer, A., RWTH Aachen University
Tiso, T.,

While rhamnolipids are discussed for decades as substitutes of oil-based molecules in such different applications as emulsifier in food or oil recovery, as foaming agent in shampoos, and surfactants in household detergents, very few applications exist in the market. Existing challenges are high production costs that arise besides other reasons from the use of plant oils as carbon source and the use of pathogenic bacteria as production host that immediately reduces application possibilities.

In this contribution we report on our rhamnolipid development platform with examples how we overcome the existing limitations. Using a GRAS organism as host[1] (*Pseudomonas putida* KT2440) and sugars (glucose, xylose) as carbon sources opens new possibilities for rhamnolipid production. The constructed production strain reaches an at least 6-times improved carbon yield on substrate when compared to the existing fermentations with *P. aeruginosa*[2]. Notably, a driven by demand metabolic engineering approach was used that reduces the number of genetic interventions[3]. The new challenge of hyper-foam formation due to the absent plant-oil organic phase is converted into an in-situ product removal strategy that allows simple product purification[4].

The developed technology enables the diversification of products by varying the number of rhamnose residues and the chain length of the OH-fatty acids. The different products (congeners) are quantified by a novel LC-MS technique[5]. Notably, while almost every publication on rhamnolipids or glycolipids in general starts with an argument on environmental friendliness and safe use, not much is reported on the latter, which we attended to change[6]. Without any rhamnosyltransferases, the engineered host produces OH-alkanoic acids (HAAs) at high rate, yield, and titer[7]. These HAAs compete besides other uses for bulk applications with neutral lipids from yeast and algae, however with the advantage of extracellular product accumulation. A novel application, the synthesis of 1,3-diols will be presented.

The results are discussed in the context of market needs for (glyco-)lipids.

References

1. Wittgens A, Tiso T, Arndt TT, …, Wichmann R,…, Rosenau F, Blank LM: Growth independent rhamnolipid production from glucose using the non-pathogenic *Pseudomonas putida* KT2440. Microb Cell Fact 2011, 10
2. Blank LM, Rosenau F, Wilhelm S, Wittgens A, Tiso T. (2013). Means and methods for rhamnolipid production, WO 2013/041670 A1
3. Tiso T, Sabelhaus P, Behrens B, Hayen H, Blank LM. Creating metabolic demand as engineering strategy in *Pseudomonas putida* - Rhamnolipd synthesis as

example. [submitted].
4. Blank LM, Küpper B, del Amor Villa EM, Wichmann R, Nowacki C. (2013). Foam adsorption, WO 2013/087674 A1
5. Behrens B, Engelen J, Tiso T, Blank LM, Hayen H. (2016). Characterization of rhamnolipids by liquid chromatography/mass spectrometry after solid-phase extraction. Anal Bioanal Chem [accepted].
6. Johann S, Seiler TB, Tiso T, Bluhm K, Blank LM, Hollert H. (2016). Mechanism-specific and whole-organism ecotoxicity of mono-rhamnolipids. Sci Total Environ 548-549: 155-163.
7. Blank LM, Tiso T, Germer A. (2015). Extracellular production of designer hydroxyalkanoyloxy alkanoic acids with recombinant bacteria, EP15175922

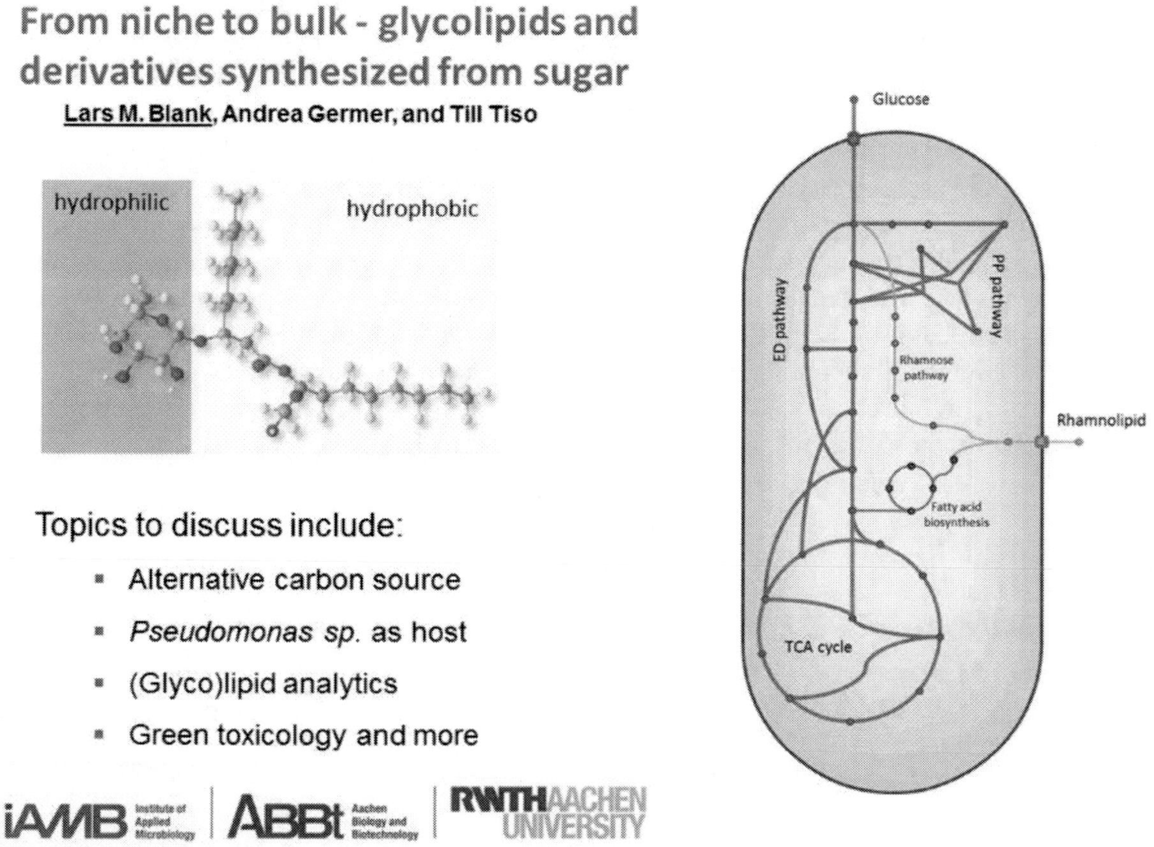

A Genetic Switch for Stable, Long-Term Fermentative Production of Anabolic Products in Yeast

Chua, P., Total New Energies USA
Jiang, H., Amyris, Inc.
Meadows, A.,

The strategic partnership between Total and Amyris is focused on producing biomolecules for renewable fuels and chemicals, and in particular for blending with petroleum-derived products to create diesel and jet fuel with reduced carbon footprint and pollution. Production and marketing of the flagship molecule farnesene (Biofene®) has already been commercialized with production scale in some markets. Farnesene has many potential applications as a renewable feedstock for the aforementioned diesel and jet fuels, as well as polymers, lubricants, solvents, and cosmetics. To reduce the production cost of farnesene, at Amyris we engineer strains using state-of-the art industrial synthetic biology platform to have high titer, yield, and productivity, and we perform fermentations in 200 m3 vessels over the course of many days or weeks. The challenge is that high-producer cells grow more slowly than spontaneous mutant low- or non-producer cells, especially in the nutrient-unlimited conditions of the seed train expansion, and yet must comprise the vast majority of the population. We have successfully addressed this challenge by developing an industrially-scalable genetic switch to successfully maintain high performance throughout lengthy fermentations. This genetic switch uses maltose (a cheap, non-toxic and metabolizable molecule) to control transcription such that when maltose is added in the seed train, product formation is shut off. This increased the growth of high-producer cells, resulting in higher inoculum purity and improved performance in bioreactors. The construction of the switch required the development of synthetic promoters that are tightly turned OFF in the absence of maltose.

Heterologous Production of n-Butanol at High Yield Using a Complete and Fully Functionnal Clostridial Pathway

Meynial-Salles, I., LISBP, INSA
Riviere, A., LISBP, Toulosue University
Tian, L., LISBP, Toulouse University
Foulquier, C.,
Soucaille, P.,

N-butanol, an important industrial intermediate chemical and a potential biofuel, is naturally produced in slow-growing native hosts that belong to genus *Clostridium*, as a mixture with acetone and ethanol in biphasic fermentation. Even if the acetone-butanol-ethanol (ABE) fermentation was one the first large scale industrial fermentation and despite the long research history on the fermentation process, obstacles still remain today. Butanol titer, yield, and productivity are still low mainly due to the co-production of acetone, ethanol and acetic and butyric acids. Advances in genetic manipulation of *clostridia* have led to large increases in yield and productivity. However, the regulation of solvent production is highly complex and cannot yet be achieved in a single-stage fermentation at high yields. To overcome some of the problems associated with clostridial n-butanol production processes, the clostridial CoA dependent n-butanol pathway was expressed in several heterologous organisms such as *E. coli B* or *K12* or *Saccharomyces cerevisiae* to convert acetyl-CoA to butanol. Reported results showed that the seven *C. acetobutylicum* known genes of the pathway are functionally expressed whatever the heterologous organism is, but the final butanol titer and yield are rather low. Neither species could produce more than 1g/l of butanol. A hypothesis could be that some key enzymes involved in butyryl-CoA synthesis in *C. acetobutylicum*, unknown up to now, were missing in the heterologous host.

To solve this problem and further improve final titer and yield of heterologous n-butanol production, our objectives were to i) identify all the enzymes involved in butyryl-CoA synthesis in *C. acetobutylicum* ii) demonstrate the neediness of these enzymes for *in vivo* butanol synthesis in *C. acetobutylicum* through knockout mutants construction and characterization iii) express the gene(s) encoding the novel enzyme (s) with genes encoding the complete already known metabolic butanol pathway from the *C. acetobutylicum* in a *E. coli* MG1655 strain. Results showed that the novel recombinant *E. coli* strain was able to produce n-butanol from glucose at high yield (73 % of the theoretical yield) never obtained before with any microorganisms (natural or non-natural n-butanol producers). Moreover, when the strain was grown in continuous culture, a stable production (for more than four months) of butanol with a low amount of other by-products was obtained.

Reference:

Soucaille P., Meynial-Salles I., Foulquier C., Rivière A. (2015) New polypeptide having ferredoxin $NADP^{+}$reductase activity polynucleotide encoding the same and uses thereof, European Patent application EP15306225.

Heterologous Production of n-Butanol at High Yield Using a Complete and Fully Functional Clostridial Pathway

Isabelle MEYNIAL-SALLES *et al.*

- ❖ Background : Heterologous expression of the n-butanol production pathway from *C. acetobutylicum* : ~ 1 g/l of n-butanol
- ❖ Limiting step : no enzyme to reoxidize reduced ferredoxin
- ❖ Strategy : identification of the Ferredoxin NADP$^+$ reductase of *C. acetobutylicum*

Creation of a new redox balanced anaerobic metabolism in *E. coli*

to link n-butanol production to glucose consumption

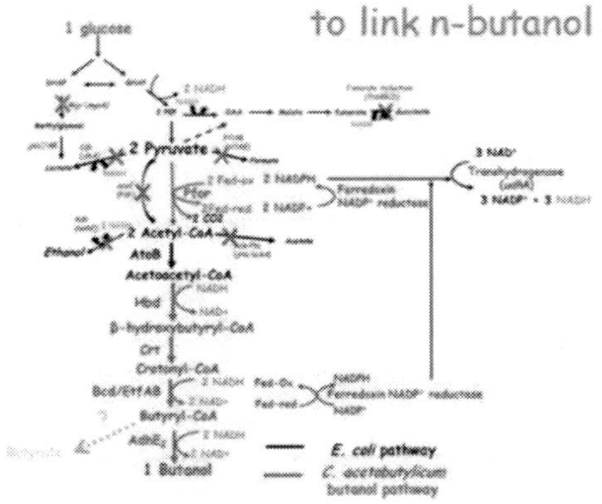

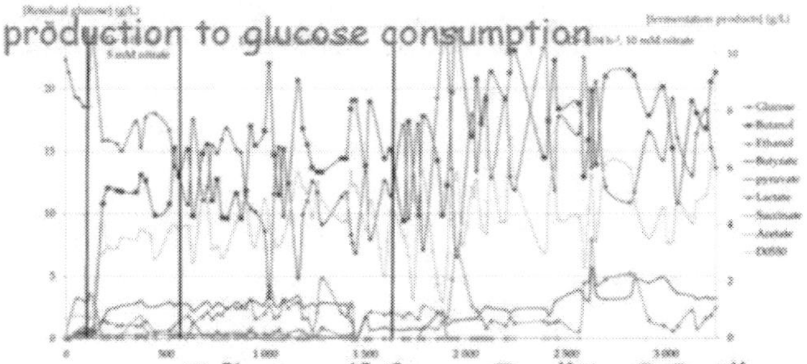

↗ [butanol] from 5 g/l to 8.2 g/l
Yield = 0.34 g butanol /g glucose consumed
Butanol/Ethanol = 11.2

Rational Metabolic Engineering of Escherichia coli for Ethylene Glycol Production from D-Xylose

Cabulong, R. B., Myongji University
Valdehuesa, K. N. G., Myongji University
Ramos, K. R. M., Myongji University
Maza, P. A. M. A., Myongji University
Pangan, J. O., Myongji University
Bañares, A. B., Myongji University
Nisola, G. M., Myongji University
Lee, S. P., Myongji University
Lee, W. K., Myongji University
Lee, C. R.,
Chung, W. J., Myongji University

The microbial production of renewable ethylene glycol (EG) has been gaining attention recently. This compound is mainly used as an antifreeze agent and precursor for polymer synthesis. EG has been successfully produced biosynthetically from D-xylose through several novel synthetic pathways. The first report on the biosynthesis of EG employs the Dahms pathway in *Escherichia coli*. Conversion of D-xylose is initiated by an oxidation reaction to produce D-xylonic acid catalyzed by a heterologous xylose dehydrogenase (Xdh). D-xylonic acid is further converted to glycolaldehyde and pyruvate through the native D-xylonic acid metabolism in *E. coli*. Pyruvate is then assimilated into the central carbon metabolism, while glycolaldehyde is either reduced to ethylene glycol or oxidized to glycolic acid. The yield achieved through this pathway is at 70 % of the theoretical. In this study, EG yield is further improved by implementing metabolic engineering strategies. D-xylonic acid accumulation was reduced when Xdh was placed under a weak promoter and tighter control of gene expression. This involves expression of Xdh under the *tac* promoter in a low-copy vector, and introduction of multiple copies of *lac* repressor gene. In addition, the suitable aldehyde reductase (ALR) for the EG pathway was identified in *E. coli*. EG yield improved after overexpression of the native YjgB. Finally, the EG yield was further increased by blocking the competing glycolic acid formation. The lactaldehyde dehydrogenase activity of *E. coli* was deactivated by disruption of the *aldA* gene, which encodes for AldA. The chromosomal *yjgB*, which encodes for YjgB, was also disrupted in the host genome. The combination of these genetic manipulations resulted in improved growth, D-xylose consumption, and EG yields of the final strain WTXB. The yield reached up to 99 % from D-xylose, which is the highest yield of EG from D-xylose so far. The main drawback in WTXB is the low productivity. Additional studies and pathway optimization are still necessary to enhance the overall performance of WTXB in order to reach industrial viability. *This work was supported by Basic Science Research Program through the National Research Foundation of Korea funded by the Ministry of Education (No. 2009-0093816), and Korea Research Fellowship Program through the NRF funded by the Ministry of Science, ICT and Future Planning (No. 2015H1D3A1062172).*

Advances in Engineering of Bacterial Methyl Ketone Synthesis for Biofuels

Beller, H. R., Lawrence Berkeley National Laboratory
Goh, E., Joint BioEnergy Institute (JBEI)
Baidoo, E., Joint Bioenergy Institute (JBEI)
Burd, H., Lawrence Berkeley National Laboratory
Lee, T. S., Joint BioEnergy Institute
Garcia Martin, H., Lawrence Berkeley National Laboratory
Keasling, J. D., Joint Bioenergy Institute

We have engineered *Escherichia coli* to overproduce saturated and monounsaturated aliphatic methyl ketones in the C_{11} to C_{17} (diesel) range; this group of methyl ketones includes 2-undecanone and 2-tridecanone, which have favorable cetane numbers and are also of importance to the flavor and fragrance industry. We have made specific improvements that resulted in more than 10,000-fold enhancement in methyl ketone titer relative to that of a fatty acid-overproducing *E. coli* strain, including the following: (a) overproduction of beta-ketoacyl-coenzyme A (CoA) thioesters achieved by modification of the beta-oxidation pathway (specifically, overexpression of a heterologous acyl-CoA oxidase and native FadB, and chromosomal deletion of *fadA*) and (b) overexpression of a native thioesterase (FadM). The first generation of engineered *E. coli* (Goh et al. 2012) produced ~380 mg/L of methyl ketones in rich medium. We have subsequently made additional genetic modifications that included balancing overexpression of *fadR* and *fadD* to increase fatty acid flux into the pathway, consolidation of the pathway from two plasmids into one, codon optimization, and knocking out key acetate production pathways (Goh et al. 2014). These modifications have led to a methyl ketone titer of 1.4 g/L with 1% glucose in shake flask experiments, which represents 40% of the maximum theoretical yield, and also attained titers of 3.4 g/L after ~45 h of fed-batch glucose fermentation (the best values reported to date). We have also conducted *in vitro* assays with purified pathway enzymes, which confirmed that a decarboxylase is not required to convert beta-keto acids into methyl ketones and that FadM is promiscuous and hydrolyzes not only beta-ketoacyl-CoAs but also other CoA-thioester pathway intermediates. These *in vitro* results have provided insight on how to fine-tune expression of pathway genes for further optimization of methyl ketone production. As part of efforts to further improve methyl ketone production, metabolic modeling was used to identify gene deletions that could improve flux through the methyl ketone pathway. One of the specified knockouts, Δ*scgE*, which was annotated as a homolog of Rpe (ribulose-5-phosphate epimerase) in the pentose phosphate pathway, improved methyl ketone production by >50% relative to the base strain (EGS1710). ^{13}C-glucose experiments have recently been performed on the knockout strain along with the DH1 wild-type and base strains to obtain more comprehensive metabolic flux profiles that will be leveraged to enable additional improvements in methyl ketone production.

Systems Metabolic Engineering of Corynebacterium Glutamicum for Production of Bio-Based 1,3-Propanediol from Lignocellulose

Chen, Z., Tsinghua University
Huang, J., Tsinghua University

1,3-Propanediol is an important chemical which is mainly used as monomer for the synthesis of valuable polymers. In this study, *Corynebacterium glutamicum* is systematically engineered for the high production of 1,3-propanediol from lignocellulosic sugars, including glucose, xylose and arabinose, etc. First, the heterologous modulars for glycerol and 1,3-propanediol synthesis were introduced and combinatorially optimized based on promoter engineering and enzyme screening. Next, the modulars for xylose and arabinose transport and degradation were imported. The fluxes to downstream pathway of glycolysis and competing products (acetate, lactate) were blocked or down-regulated. Furthermore, the sugar transprot system and the cofactor regeneration systems were optimized. The final engineered strain was able to produce more than 100 g/L 1.3-propanediol from glucose, xylose or their mixtures. This is an important step towards the enomical bio-production of 1,3-propanediol for polymer industry.

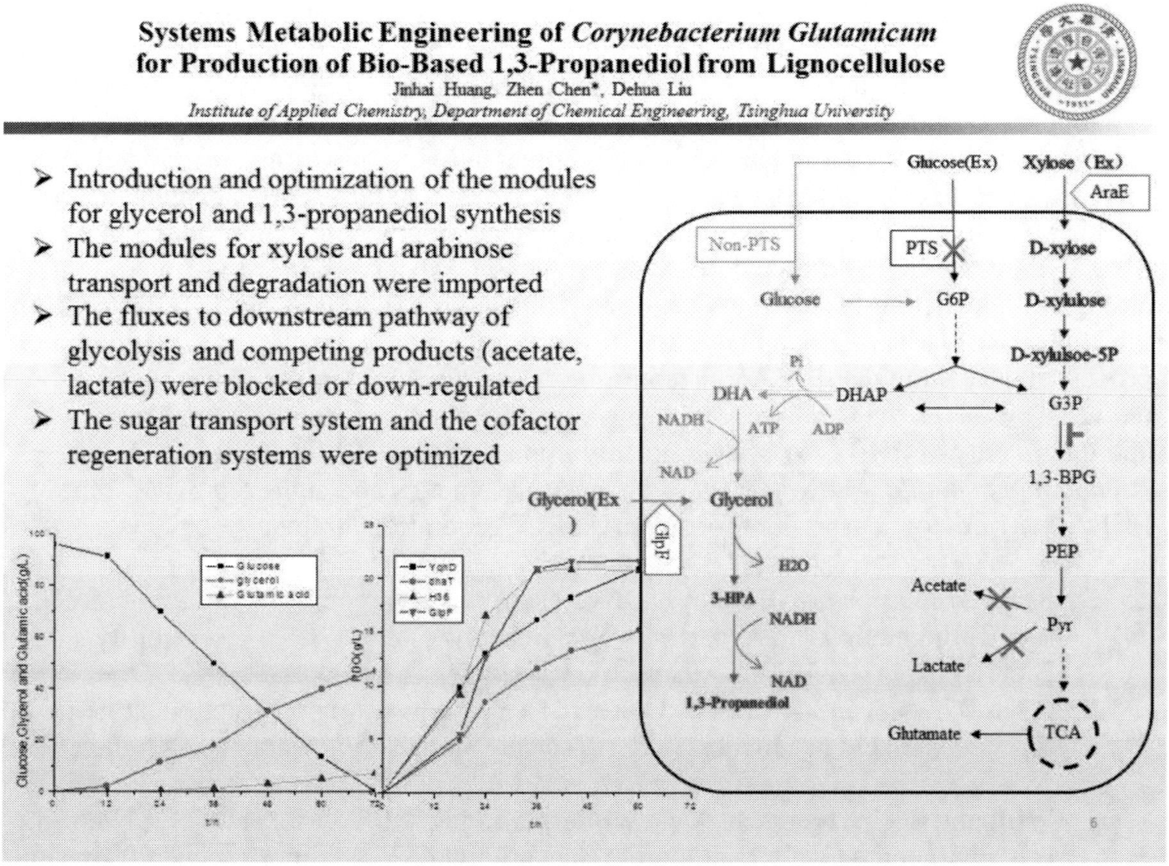

Production of Aromatics in Microbes

Yu, S., The University of Queensland
Averesch, N. J. H., Centre for Microbial Electrosynthesis (CEMES) within the Advanced Water Management Centre (AWMC)
Williams, T. C., The University of Queensland
Winter, G., Centre for Microbial Electrosynthesis (CEMES) within the Advanced Water Management Centre (AWMC)
Vickers, C. E., The University of Queensland
Nielsen, L. K., The University of Queensland
Krömer, J., The University of Queensland

Aromatics are amongst the most important bulk feedstocks for the chemical industry, however, no viable bioprocess exists today and production is still dependent on petro-chemistry. We have established toxicity levels and baseline production of two key molecules, p-hydroxybenzoic acid (PHBA) and p-amino benzoic acid (PABA), in *Saccharomyces cerevisiae* [1]. In order to further improve this baseline production we employed *in-silico* analysis for the prediction of suitable targets for strain improvement for products of the shikimate pathway [2]. A key reaction in metabolism was the knockout of pyruvate kinase, which is known to be non-viable on glucose. In addition, the knockout of competing pathways draining carbon from the shikimate pathway led to very poor growth and auxotrophic strains. To address these problems we have engineered dynamic regulation using a synthetic quorum sensing circuit in *Saccharomyces cerevisiae*. The circuit activates gene expression at a high population density, and is linked with an RNA interference (RNAi) module to enable targeted gene silencing. The circuit was used to control flux through the shikimate pathway for the production of PHBA, showing that wild-type equivalent growth was observed prior to activating the circuit while reaching PHBA titers of over 1.1 mM in production phase [3]. Building on this, the strain was further engineered, a feeding strategy was developed and a fed-batch process finally achieved PHBA concentrations over 20mM. Two current limitations of the yeast system is the Crabtree effect causing production of ethanol in batch processes and the toxicity of PHBA. We have recently implemented PHBA production in *Pseudomonas putida,* known for its resistance to aromatics and achieved over 12 mM PHBA and very low by-product formation in first generation strains.

1. Krömer, J.O.; Nunez-Bernal, D.; Averesch, N.J.; Hampe, J.; Varela, J.; Varela, C. (2013) Production of aromatics in *Saccharomyces cerevisiae*--a feasibility study. J Biotech, 163, 184-193.
2. Averesch, N.J. and Krömer, J.O. (2014) Tailoring strain construction strategies for muconic acid production in *S. cerevisiae* and *E. coli*. Metab Eng Comm, 1, 19-28.
3. Williams, T.C.; Averesch, N.J.; Winter, G.; Plan, M.R.; Vickers, C.E.; Nielsen, L.K.; Krömer, J.O. (2015) Quorum-sensing linked RNA interference for dynamic metabolic pathway control in *Saccharomyces cerevisiae*. Metab Eng, 29, 124-134.

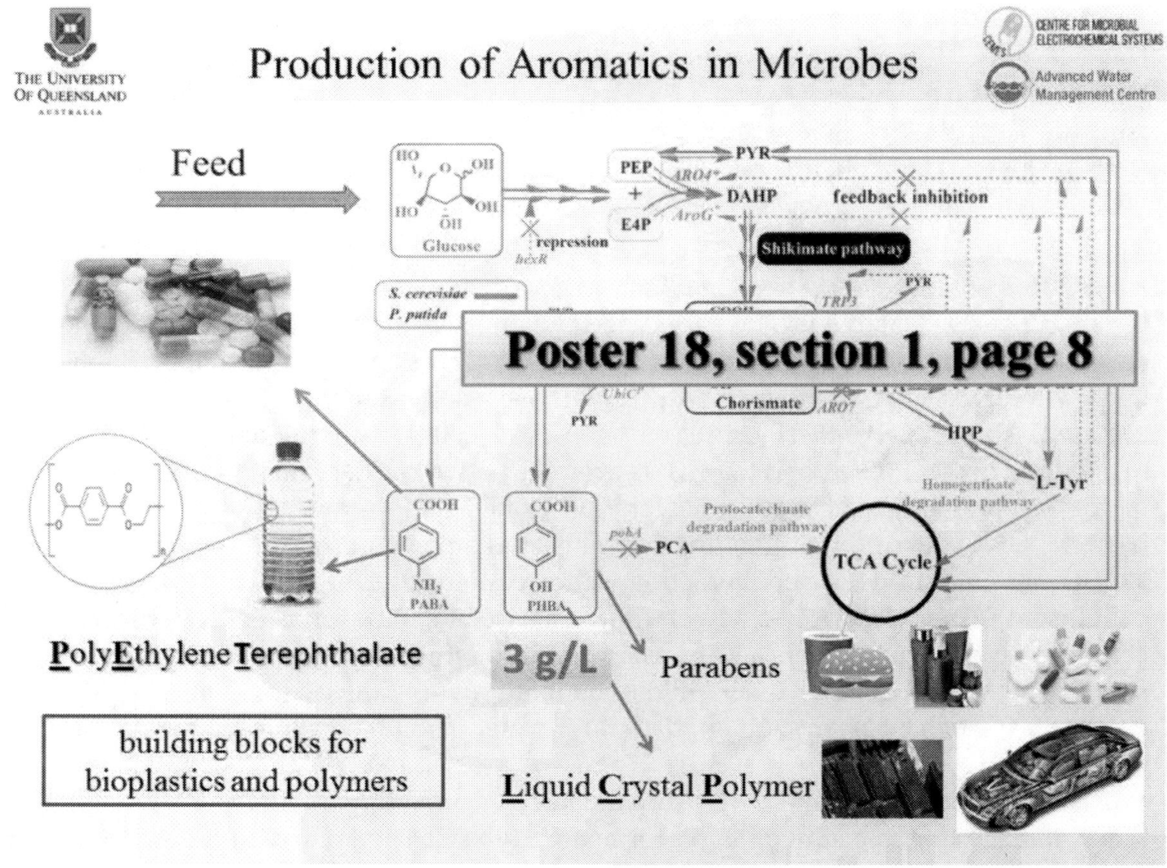

THE UNIVERSITY OF QUEENSLAND AUSTRALIA
Production of Aromatics in Microbes
CENTRE FOR MICROBIAL ELECTROCHEMICAL SYSTEMS
Advanced Water Management Centre
Feed
Glucose
repression
hexR
PEP
E4P
PYR
DAHP
feedback inhibition
Shikimate pathway
S. cerevisiae
P. putida
Poster 18, section 1, page 8
Chorismate
PYR
HPP
L-Tyr
Homogentisate degradation pathway
Protocatechuate degradation pathway
PCA
TCA Cycle
COOH
NH2
PABA
COOH
OH
PHBA
PolyEthylene Terephthalate
3 g/L
Parabens
building blocks for bioplastics and polymers
Liquid Crystal Polymer

Rapid Evolution for High Performance Nano-Cellulosome By Module Library Approach

Nakazawa, H., Tohoku University
Okada, I., Tohoku University
Ishigaki, Y., Tohoku University
Kobayashi, E., Tohoku University
Umetsu, M., Tohoku University

An economical conversion from inedible cellulosic biomass materials to water-soluble sugar units is expected to produce food, feed, chemicals and fuels. Microbial cellulolytic enzymes, called cellulase is an enzyme which hydrolyzes cellulose to water-soluble sugar units with low energy and low environmental loads, and more than 1000 cellulases are reported. About 33% of these cellulases have carbohydrate-binding domain (CBD), fused to catalytic domain (CD), and the CBDs promotes effective degradation of cellulose by increasing substantial substrate concentration around CD. Further, in some rumen bacteria, CDs with different functions are clustered on a giant scaffold protein containing a CBD, via cohesin-dockerin interaction, called cellulosome, to efficiently degrade cellulose. We have proposed a new design of artificial cellulosome, called hybrid nano-cellulosome (HyNaC), by clustering recombinant biotinylated CDs and CBDs on streptavidin-modified CdSe quantum dot via biotin-avidin interaction. This HyNaC shows drastic enhancement of degradation for water insoluble cellulose.

We selected optimal CD and CBD from the library containing 50 kinds of CDs and 20 kinds of CBDs for screening high performance HyNaC. The HyNaC structure format produced more reducing sugar than the corresponding native module structure, and a replacement of the kind of CBD contributed considerably to the enhancement of the degradation of amorphous cellulose on HyNaC, whereas, chimera module enzyme with each CBD did not show. In addition, mixture experiment of HyNaC with different CBD shows the synergistic effect between CBDs observed on HyNaC with most active CD of E2 from *Thermobifida fusca.*

In this study, to understand the mechanism of the enhanced degradation on HyNaC by variety of CBD, we investigated the binding affinity and the saturated binding amount of CBD library to amorphous and crystalline cellulose and extracted a rule to enhance the cellulolytic activity of HyNaC based on CBD.

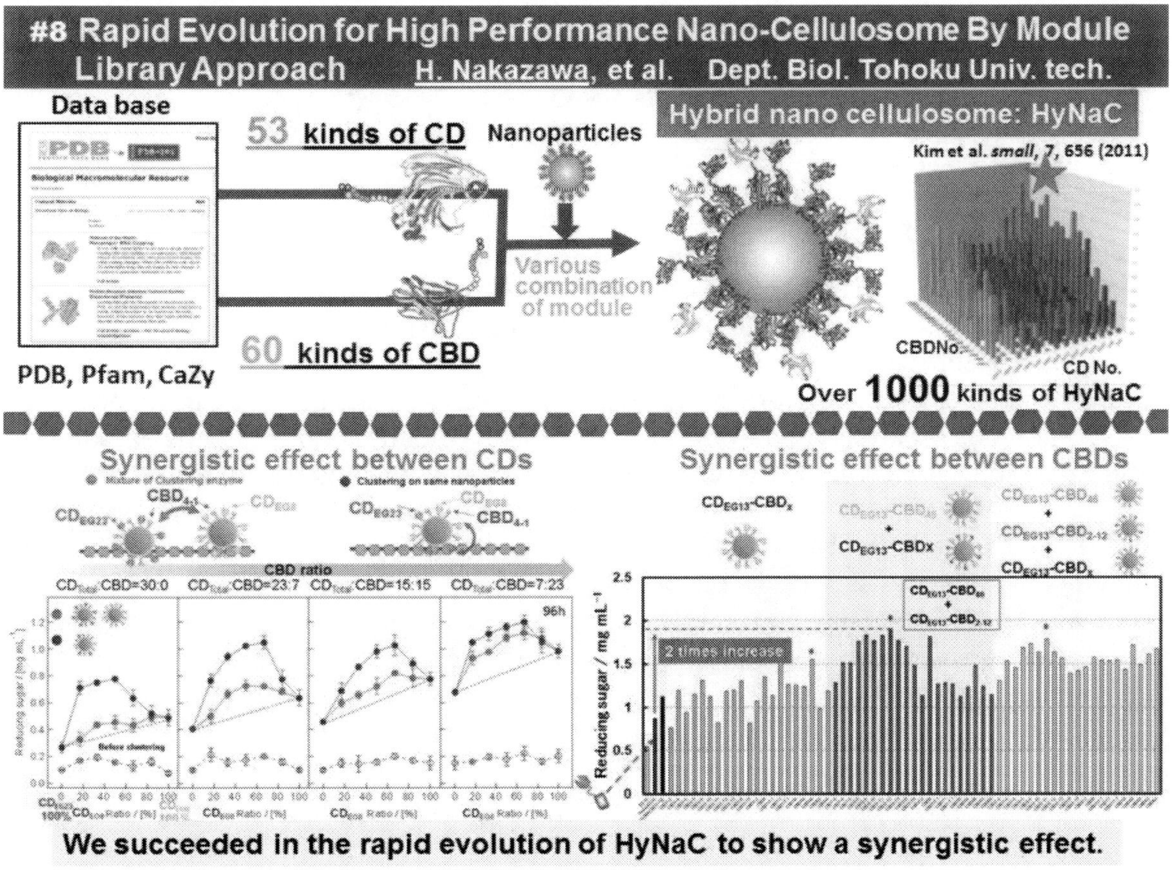
#8 Rapid Evolution for High Performance Nano-Cellulosome By Module Library Approach
H. Nakazawa, et al. Dept. Biol. Tohoku Univ. tech.
Data base
PDB, Pfam, CaZy
53 kinds of CD
60 kinds of CBD
Nanoparticles
Various combination of module
Hybrid nano cellulosome: HyNaC
Kim et al. small, 7, 656 (2011)
CBDNo.
CD No.
Over 1000 kinds of HyNaC
Synergistic effect between CDs
Synergistic effect between CBDs
CBD ratio
2 times increase
Reducing sugar / mg mL⁻¹
96h
Before clustering
We succeeded in the rapid evolution of HyNaC to show a synergistic effect.

Flux Control at the Malonyl-CoA Node through Hierarchical Dynamic Pathway Regulation in Saccharomyces Cerevisiae

David, F., Chalmers University of Technology
Nielsen, J., Chalmers University of Technology
Siewers, V., Chalmers University of Technology

The establishment of a heterologous pathway in a microbial host for the production of industrially relevant chemicals at high titers and yields requires efficient adjustment of the central carbon metabolism to ensure that flux is directed towards the product of interest. To achieve this, we established a ligand responsive transcription factor based sensor module in *Saccharomyces cerevisiae* and applied the concept for enzyme engineering and controlling flux at the malonyl-CoA node. We present a novel approach for dynamic modulation of pathway flux and enzyme expression levels. It is based on a hierarchical dynamic control system around the key pathway intermediate malonyl-CoA. The upper level of the control system ensures down-regulation of endogenous use of malonyl-CoA for fatty acid biosynthesis, which results in accumulation of this pathway intermediate. The lower level of the control system is based on use of a novel biosensor for malonyl-CoA to activate expression of a heterologous pathway using this metabolite for production of 3-hydroxypropionic acid (3-HP). The malonyl-CoA sensor was developed based on the FapR transcription factor of *Bacillus subtilis*, and it demonstrates one of the first applications of a bacterial metabolite sensor in yeast. Introduction of the dual pathway control increased the production of 3-HP by 10-fold. The approach can be seen as a prototype of a general strategy to engineer and dynamically control metabolic pathways for balancing precursor supply and can also be applied for production of other malonyl-CoA derived products.

Flux Control at the Malonyl-CoA Node through Hierarchical Dynamic Pathway Regulation in *Saccharomyces cerevisiae*

Florian David, Verena Siewers, Jens Nielsen
Systems and Synthetic Biology, Dept. of Biology and Biological Engineering
Chalmers University of Technology, Göteborg, Sweden, **davidfl@chalmers.se**

1. **Established an orthologous transcription factor based biosensor for Malonyl-CoA sensing in *S. cerevisiae***
2. **Dynamic 2-level control increased 3-HP production by 10-fold**

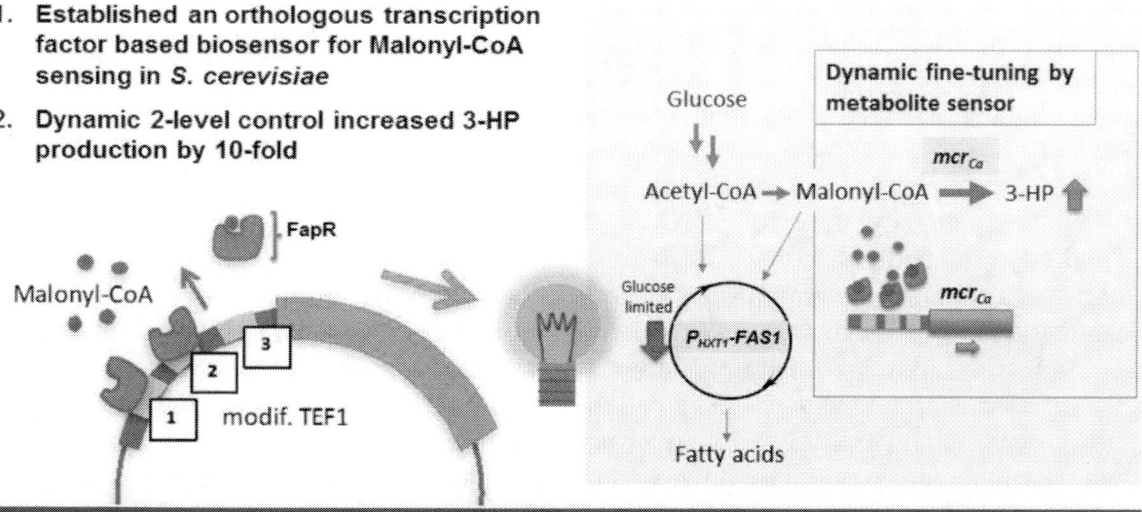

Engineering of Hydroxytyrosol Production in Escherichia coli

Yoshida, E., Joint BioEnergy Institute
Koma, D., Lawrence Berkeley National Laboratory
Keasling, J. D., Joint Bioenergy Institute
Lee, T. S., Joint BioEnergy Institute

Hydroxytyrosol (HTy) is one of the most powerful antioxidants with potential biological function as an anti-tumor, anti-atherogenic, anti-inflammatory and anti-platelet aggregation agent. It has a wide range of potential applications in industry, such as functional food, dietary supplement, cosmetics, and animal feed. Currently, HTy is produced from enriched olive extracts after chemical or enzymatic hydrolysis. There is an increasing demand for stable and sustainable production, and microbial fermentation can be a promising solution.

We have previously reported an engineered *E. coli* capable of producing HTy. Our synthetic HTy pathway consists of five heterologous genes for tyrosine hydroxylation and downstream conversion of L-DOPA to HTy. Although we have shown that this synthetic pathway is functional, the low titer from tyrosine or from glucose remained as an issue to be addressed. Here we report three metabolic engineering efforts to improve the efficiency of the synthetic HTy pathway. First, we have engineered the co-factor biosynthetic pathway for tyrosine hydroxylation. Next, we have identified inhibition of the tyrosine hydroxylation by downstream chemicals and applied co-culture strategy to overcome it. Lastly, the gene expression of pathway enzymes was optimized using metabolomics and proteomics data. With these engineering, we were able to improve the product yield from tyrosine about 3-fold. In addition, we have engineered the host strain to overproduce tyrosine. Using this engineered strain, we confirmed the production of L-DOPA from glucose without any external supplementation of tyrosine, and the resulting L-DOPA was further converted into HTy by co-culture strategy with an improved yield from the previous result.

This work demonstrates a potential industrial application of microbial HTy production and provides a good renewable microbial platform for the production of a wide range of chemicals that involve the hydroxylation of aromatic amino acid.

This work was supported by Ajinomoto (Study abroad program for research scientist) and the Joint BioEnergy Institute (http://www.jbei.org/), which is supported by the US Department of Energy, Office of Science, Office of Biological and Environmental Research, through contract DE-AC02-05CH11231 between Lawrence Berkeley National Laboratory and the US Department of Energy.

Membrane Engineering Strategies to Improve Production of Biorenewable Fuels and Chemicals

Jarboe, L., Iowa State University

Tan, Z., Iowa State University

Lian, J., Iowa State University

Jin, T., Iowa State University

Chen, Y., Iowa State University

Boggess, E., Iowa State University

Yoon, J. M., Iowa State University

Shanks, J. V., Iowa State University

Nielsen, D.,

Wen, Z., Iowa State University

Dickerson, J., Iowa State University

The production of fuels and chemicals from biomass in a manner that is economically competitive with petroleum-based processes is often hindered by microbial inhibition by the product and/or impurities in the biomass-derived sugar stream. In many of these cases, this inhibition of the microbial biocatalyst involves membrane damage and the threshold for this damage can differ for production vs exogenous challenge. We have demonstrated the effectiveness of various strategies for strengthening the *Escherichia coli*membrane.

The first engineering strategy is enabling homeoviscous adaptation in *E. coli* via expression of the *Pseudomonas aerugionase* cis-trans isomerase (Cti) enzyme, as described in Tan *et al*Metabolic Engineering 2016. The Cti enzyme converts existing cis-unsaturated fatty acids (CUFA) to the trans unsaturated fatty acid (TUFA) form, resulting in a significant increase in membrane fluidity and improved tolerance to a variety of inhibitors. TUFA production has a dual effect on the protection provided against exogenously-supplied octanoic acid: some amount of TUFA production is helpful, but too much TUFA production dampens the protective effect. Strains with the appropriately tuned TUFA/CUFA ratio not only showed improved tolerance to exogenously provided octanoic acid, but also demonstrated significantly increased octanoic acid titers. Specifically, the titer was increased 41% in minimal media and 29% in rich media. This association of improved tolerance and improved production was also observed for styrene. The tolerance of n-butanol, hexane, toluene, elevated temperature (42C) and low pH (5.5) were also significantly increased.

The second membrane engineering strategy involves altering the distribution of the native phospholipid head groups. One such engineering strategy was associated with increased tolerance of carboxylic acids, various biomass-associated inhibitors (such as furfural), aromatic monomers, low pH (6.0) and low temperature (20C). This increased tolerance of carboxylic acids was accompanied by a substantial increase in membrane integrity during carboxylic acid challenge. When this engineered strain was provided with appropriate fatty acid transporters, fatty acid production titers were increased nearly 40% relative to the corresponding control with the unaltered phospholipid distribution.

Finally, by implementing Orgel's Second Rule that "evolution is cleverer than you are", we can find inspiration for other membrane engineering strategies. Two such membrane engineering strategies inspired by reverse engineering of evolved *E. coli* strains will be discussed.

Membrane Engineering Strategies to Improve Production of Biorenewable Fuels and Chemicals

Zaigao Tan, Jacqueline V. Shanks, Laura R. Jarboe*

Product/substrate toxicity due to membrane damage is a common problem in biorenewables production.
Damage of the walls of a reaction vessel would be addressed by changing the vessel composition. Can we use this approach with microbial biocatalysts?

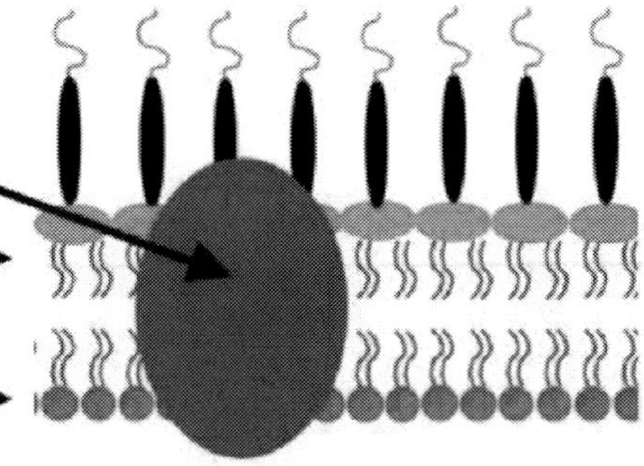

+50% fatty acid titer
+2.5-fold furfural IC_{50}
+50% phenol tolerance
+10% styrene titer

Chemical and Biological Engineering, Iowa State University
NSF Center for Biorenewable Chemicals Engineering Research Center (CBiRC)

Engineering Zymomonas Mobilis for the Production of Biofuels and Value-Added Chemicals

Zhang, M., National Renewable Energy Laboratory
Yang, S., National Renewable Energy Laboratory
Chou, Y. C., National Renewable Energy Laboratory
Franden, M. A., National Renewable Energy Laboratory
Himmel, M.,

Microbial conversion of abundant and sustainable lignocellulosic biomass to biofuels and chemicals offers the opportunity to reduce our current dependence on fossil fuels and chemicals. Known for its high specific glucose uptake rate, rapid catabolism and high ethanol yield, *Zymomonas mobilis* has been engineered to also efficiently convert the second and third most abundant plant derived sugars, xylose and arabinose, to ethanol at high yield. More recently, we have been applying systems biology and genomic tools to investigate and improve its tolerance to the specific inhibitors present in biomass hydrolysate obtained from dilute acid pretreatment. *Z. mobilis* is one of the top fermentation organisms currently developed for the cellulosic biomass to ethanol conversion process. With its capabilities for utilizing the vast majority of the biomass sugars in a toxic hydrolysate environment, we asked a question - Can we redirect carbon to make other fuel molecules or chemicals in addition to ethanol by applying metabolic engineering and synthetic biology tools? Our recent successes in engineering *Z. mobilis* for the production of value-added chemicals including 2,3 butanediol demonstrated a promising novel platform for future biorefineries. 2,3 Butanediol is a valuable bulk chemical building block which can be used for synthesis of polymers and can be also upgraded in high yields to gasoline, diesel, and jet fuel.

Engineering *Zymomonas mobilis* for the Production of Biofuels and Value-Added Chemicals

Min Zhang[1], Shihui Yang[1], Mary Ann Franden[1], Yat-Chen Chou[1], and Michael E. Himmel[2], [1]National Bioenergy Center, [2]Biosciences Center, National Renewable Energy Lab, Golden, CO

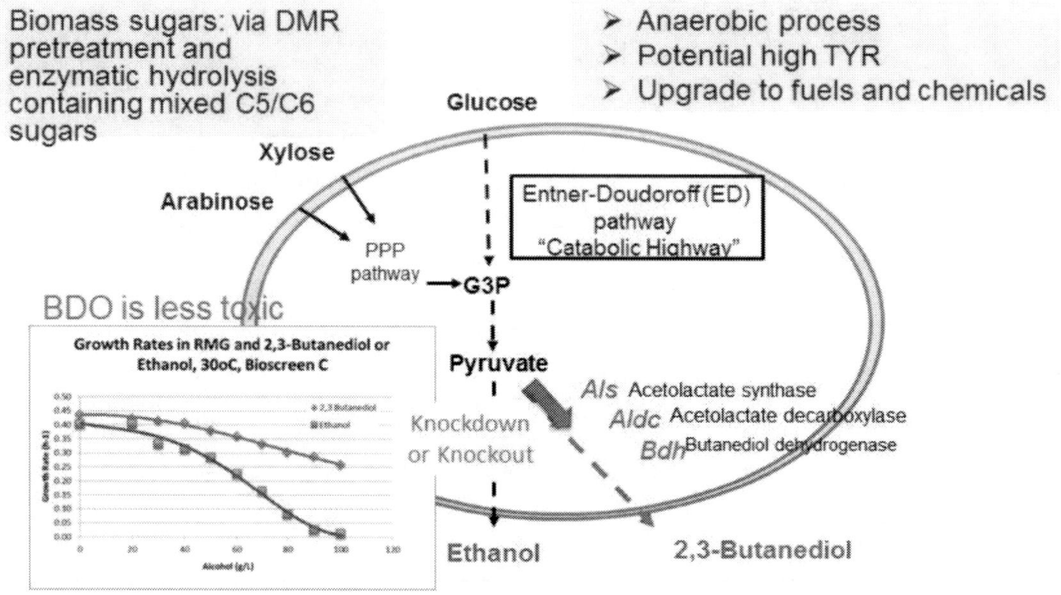

Ethanol Production from Alkaline Pretreated Miscanthus Using Respiration-Deficient Thermotolerant Yeast

Yang, J., Korea University Graduate School
Yoon, J. J., Korea Institute of Industrial Technology
Kim, K. H., Korea University Graduate School

The major obstacle of simultaneous saccharification and fermentation (SSF) is that enzymatic hydrolysis in SSF is not conducted at optimal temperature. To overcome this obstacle, thermotolerant *Saccharomyces cerevisiae* strains were screened from respiration-deficient mutants induced by ethidium bromide. Among the mutant strains, *S. cerevisiae* mbc 2 was selected as the most suitable strain for fermentation at 42°C. Meanwhile, *Miscanthus* was prepared as substrates after pretreatment with 1.5 M NaOH at 150°C for 30 min. Next, SSF was conducted with pretreated *Miscanthus* containing 3% glucan to estimate the fermentability of *S. cerevisiae* mbc 2 at 42°C. In results, ethanol concentration of 15.3 g/L was achieved by *S. cerevisiae* mbc 2 in 48 h, representing 90.1% of theoretical ethanol yield. Furthermore, SSF with increased substrate concentration up to 9% glucan resulted in 82.2% of theoretical ethanol yield. Thus, this approach for SSF with high solid loading using thermotolerant *S. cerevisiae* mbc 2 might contribute to processes and microbial strain development for ethanol production from lignocellulosic biomass.

An Engineered Saccharomyces Cerevisiae for Producing (S)-3-Hydroxybutyrate

Cho, K. M., Korea University Graduate School
Yun, E. J., Korea University Graduate School
Kim, K. H., Korea University Graduate School

(S)-3-Hydroxybutyrate (*S*-3HB) can be used as an initial material for the synthesis of biodegradable polymers such as polyhydroxyalkanoate and stereo-specific fine chemicals such as antibiotics, vitamins, perfumes, and pheromones. To produce *S*-3HB in yeast, the biosynthetic pathway for the production of *S*-3HB from acetyl-CoA was introduced into *Saccharomyces cerevisiae*. The synthetic pathway of *S*-3HB involves the three enzymes from different microbial sources, acetyl-CoA C-acetyltransferase (ACCT), acetoacetyl-CoA reductase (ACR), and 3-hydroxybutyryl-CoA thioesterase (HBT) originating from *S. cerevisiae*, *Clostridium acetobutylicum*, and *Escherichia coli*, respectively. An engineered yeast strain overexpressing *ERG10*, *hbd*, and *tesB* genes not only exhibited enzyme activities of ACCT, ACR, and HBT, but also produced *S*-3HB from ethanol. To get a higher titer of *S*-3HB, a fed-batch fermentation based on pulse feeding of C2 compound (*e.g.*, ethanol or acetate) as a carbon source was conducted, and a final *S*-3HB titer was 12.0g/L. To our knowledge, this is the first report on the production of 3HB by engineered yeast, utilizing ethanol as the carbon source, suggesting that the industrially preferred *S. cerevisiae* can be a promising host for producing S-3HB.

Characterization of a Keystone Enzyme in the Metabolism of 3,6-Anhydro-L-Galactose, a Main Sugar Component of Red Macroalgae

Yu, S., Korea University
Yun, E. J., Korea University Graduate School
Choi, I. G.,
Kim, K. H., Korea University Graduate School

Red macroalgae are emerging as sustainable sources for producing biofuels and chemicals. 3,6-Anhydro-L-galactose (AHG) is the major constituent of agar, which is the main carbohydrate of red macroalgae. Recently, the metabolic pathway of AHG and the enzymes involved in the AHG metabolism were discovered in a marine bacterium. In this study, a novel aldehyde dehydrogenase, AHG dehydrogenase (AHGD) which oxidizes AHG to 3,6-anhydrogalactonate (AHGA) as the first enzyme in the AHG metabolism was characterized. Both NAD^+ and $NADP^+$ were found to act as a cofactor for the enzymatic reaction of AHGD with AHG. The optimal reaction pH and temperature and substrate specificity of AHGD were investigated, and the kinetic parameters of AHGD were determined in this study. Since AHGD is involved in the first step of the AHG pathway, AHGD may be the keystone enzyme which enables AHG to enter its catabolic pathway of AHG in marine bacteria utilizing red macroalgal biomass.

Is Acid Pretreatment Beneficial for the Saccharification and Fermentation of Alginate from Brown Macroalgae?

Wang, D.,
Kim, D. H., Korea University
Yun, E. J.,
Kim, K. H., Korea University Graduate School

Brown macroalgae are actively considered for the production of biofuels and chemicals. In the overall process of utilizing brown macroalgae, the monomerization of alginate which is the main carbohydrate of brown macroalgae is critical and prerequisite. In general, the physicochemical pretreatment of biomass leads to increased enzymatic saccharification. To obtain a high reducing sugar yield from alginate, acidic pretreatment was performed at various temperatures such as 120–160°C for 15–45 min using 1% (w/v) sulfuric acid in this study. These acid pretreatment processes resulted in the formation of mannuronic and guluronic acids and oligosaccharides of alginate. The enzymatic saccharification of pretreated alginate by sequential treatments of endo- and exo-alginate lyases resulted in a 52.2% final reducing sugar yield, which was only 7.5% higher than when using unpretreated alginate as the substrate for the enzymatic saccharification. Rigorous analyses of reaction products of alginate by TLC, LC-MS, and MALDI/TOF-TOF MS showed that the depolymerization of alginate in acidic pretreatment occurs via hydrolysis and oligosaccharides saturated at their non-reducing ends are formed as a result of the acidic hydrolysis. The enzymatic saccharification of these saturated alginate oligosaccharides leads to the formation of mannuronic and guluronic acids which are not fermentable monomeric sugars. These result indicated that acidic pretreatment is not a desirable process for the saccharification and fermentation of alginate from brown macroalgae.

Fermentative Production of Enantiomerically Pure S-1,2-Propanediol from Glucose By Engineered E. coli Strain

Yu, B.,

The pure stereoisomers of 1,2-propanediol (1,2-PDO) could be used as starting materials to synthesize high value-added specialty chemicals and chiral pharmaceutical products. As the stereoisomers of 1,2-PDO cannot be obtained by traditional chemical synthesis processes, biotechnological processes have gained increasing attention. However, to our knowledge, the production of *S*-1,2-PDO directly from glucose has not been previously reported. In this study, we demonstrate a novel artificial pathway to convert L-lactic acid to *S*-1,2-PDO and its integration into the genome of *Escherichia coli* strain BW25113Δ*pox*B with synchronous deletion of genes responsible for branch metabolic pathways from glucose. L-lactate production was increased by replacing the native D-lactate dehydrogenase with the L-lactate dehydrogenase from *Bacillus coagulans*. The methylglyoxal bypass pathway was blocked to avoid synthesis of a racemic mixture of D- and L-lactate and prevent the accumulation of methylglyoxal, a toxic intermediate. To further improve the yield of *S*-1,2-PDO, a novel cofactor regeneration system was introduced by combining pyruvate decarboxylase and acetaldehyde-CoA dehydrogenase II to simultaneously regenerate NADH and the CoA donor of acetyl-CoA for the lactate conversion pathway. Finally, 13.7 mM *S*-1,2-PDO with >99% enantiomeric purity was directly produced from glucose by disrupting the major carbon-competing pathways and strengthening the lactate transformation pathway. This study demonstrates the first attempt to synthesize *S*-1,2-PDO by direct fermentation of glucose.

Productions of Fuels and Chemicals from Alginate and Mannitol, Promising Carbon Sources in Brown Macroalgae

Kawai, S., Kyoto University
Murata, K., Setsunan University

Marine brown macroalgae have attracted attention as a promising carbon source for biorefining due to their several advantages over terrestrial biomass. Alginate and mannitol are major carbohydrates in brown algae. Because of the huge potential availability of those carbohydrates, development of biotechnologies to produce fuels and chemicals from those carbohydrates has been awaited. We have focused on these technologies for several years. At this meeting we present our researches and current efforts to produce ethanol, pyruvate, and others from alginate and mannitol (1–6).

For the first time we reported a production of ethanol from alginate using a bacterium, *Sphingomonas* sp. strain A1 that greedily assimilates alginate (1) and analyzed this production system (2). We also found that this bacterium could secrete pyruvate during assimilation of alginate and established a production system for pyruvate from alginate (3).

We searched mannitol-assimilating ability of yeasts and demonstrated a production of ethanol using *Saccharomyces paradoxus* NBRC 0259-3 (4). However, budding yeast *Saccharomyces cerevisiae* that is a superior production system for fuels and chemicals does not assimilate mannitol as well as alginate. We found that *S. cerevisiae* could acquire an ability to assimilate mannitol through spontaneous mutation in Tup1-Cyc8 genes (5). Production of ethanol or pyruvate from mannitol was attempted using these *S. cerevisiae* strain (5, 6). Based on our studies on *Sphingomonas* sp. A1, we had started to confer the ability to assimilate a monomeric sugar of alginate, 4-deoxy-L-erythro-5-hexoseulose uronate (DEH) on *S. cerevisiae*. However, Enquist–Newman *et al.* were the first to successfully achieve this goal in 2014 [Nature, 505(7482):239-243]; they also succeeded in creating *S. cerevisiae* strain that can assimilate mannitol (Mtl+) through overexpression of the genes for a mannitol-2-dehydrogenase homolog (Ynr073c) and a putative transporter (Hxt17). Now we have concluded which method is better to create Mtl+ strain and successfully constructed *S. cerevisiae* strains that assimilate both DEH and mannitol; now we are conducting an adaptation experiments in DEH medium as Enquist–Newman *et al.* did. Moreover we comparatively analyzed exo-type alginate lyases that produce DEH from alginate. Systems to produce 2,3-butanediol and isobuthanol are under construction. However, an improvement of the abilities to assimilate DEH and mannitol is obviously needed. Our current candidate of brown macroalgae is Akamoku [*Sargassum horneri* (Turner) C. Agardh], a very promising macroalgae for the large-scale farming; a pilot test (□`2,000 $m^2$, □`2,000 kg wet wt.) has been conducted at Kyoto Prefecture.

1. H. Takeda, F. Yoneyama, S. Kawai, W. Hashimoto & K. Murata: **Energy Environ. Sci.,** 4(7):2575-2581 (2011).

2. M. Fujii, S. Yoshida, K. Murata & S. Kawai: **Bioengineered**, 5(1):38-44 (2014).

3. S. Kawai, K. Ohashi, S. Yoshida, M. Fujii, N. Sato & K. Murata: **J. Biosci. Bioeng.,** 117(3):269-274

(2014).

4. A. Ota, S. Kawai, H. Oda, K. Iohara & K. Murata: **J. Biosci. Bioeng.,** 116(3):327-332 (2013).

5. M. Chujo*, S. Yoshida, A. Ota, K. Murata & S. Kawai*: **Appl. Environ. Microbiol.,** 81(1):9-16 (2015). *Equally contributed.

6. S. Yoshida, H. Tanaka, M. Hirayama, K. Murata & S. Kawai: **Bioengineered,** 2;6(6):347-350 (2015).

Recombinant Saccharomyces Cerevisiae Construction through Rapid DNA Assembly and Evolutionary Engineering and High-Efficiency Lignocellulose Hydrolysates Fermentation

Geng, A., Ngee Ann Polytechnic

The bioethanol fermentation from lignocellulose hydrolysates requires efficient hexose and pentose co-utilizing and inhibitor-tolerant microbes. The objective of the present work is to develop a recombinant *Saccharomyces cerevisiae* strain which can make full use of mixed sugars in the presence of inhibitors in lignocellulose hydrolysates through metabolic engineering. In this study, a mixture of expression cassettes containing bacterial xylose and arabinose utilizing genes were assembled and integrated in the genome of a haploid S. cerevisiae strain Juk36a strain. The best functional strain 36aS1 was selected from consecutive adaptive evolution in synthetic medium containing of xylose and inhibitors. In the end, a high-efficiency hexose and pentose co-utilizing and inhibitor-tolerant strain 36aS1.10.4 was isolated from strain library of 36aS1.10 on lignocellulose hydrolysates plates. Strain 36aS1.10 exhibited a maximum specific growth rate of 0.176 1/h and 0.140 1/h in synthetic medium with 50 g/L D-xylose and 20 g/L L-arabinose as sole carbon source, respectively. Fermentation of industrial biomass hydrolysates obtained from two different companies using strain 36aS1.10.4 with a low yeast inoculum, maximum specific D-xylose consumption rates of 0.537 g/g DW/h and 0.541 g/g DW/h were obtained. The corresponding final ethanol titers were 54.11 g/L and 50.36 g/L with ethanol yields of 0.44 g/g and 0.41 g/g, respectively. After supplement of more sugars in one of the industrial hydrolysates, the final ethanol titer reached 78.37 g/L with about 94% total sugar conversion. The above results suggest that strain 36aS1.10.4 has great potential in industrial cellulosic ethanol production using lignocellulosic biomass sugars.

Re-Engineered β-Oxidation Reversal for the Synthesis of ω -Fuctionalized Carboxylic Acids in Escherichia coli

Cheong, S., Rice University
Clomburg, J. M., Rice University
Gonzalez, R., Rice University

β-oxidation reversal is a novel carbon-chain elongation platform with high carbon and energy efficiencies that has been exploited to synthesize fatty acids and alcohols with varying chain lengths. While these products include industrially important fuels and chemicals, product diversity has been limited by the use of alkyl-CoAs, typically acetyl-CoA, as the primer. In order to expand the product range outside of those with a methyl group at the omega carbon, this study focused on the use of primers with varying ω-functional groups. Product diversification was achieved through the utilization of ω-carboxylated succinyl-CoA and glutaryl-CoA as novel primers, in combination with iterative operation of carbon-chain elongation and usage of different pathway termination reactions at different β-functionalized intermediates. This expanded the product range of β-oxidation reversal to various ω-fuctionalized carboxylic acids such as α,ω-dicarboxylic acids, ω-1-keto fatty acids and ω-hydroxy fatty acids in *Escherichia coli*. Among all the products, adipic acid, a top industrial chemical as the monomer of nylon-6,6, showed 2.71 g/L of maximum titer. This re-designed reverse β-oxidation platform can be potentially used for the production of other important fuels and chemicals with diverse functionalities and applications.

Can Ag43 Mediate Surface Display Efficiently in Escherichia coli?

Ng, I. S., National Cheng Kung University
Guo, Y., Xiamen University
Kao, P. H., National Cheng Kung University

Cell surface display system enables the recombinant proteins being displayed on the surface of bacterial cells. The protein is convenient to access external substrates via surface display, and overcomes the difficulty caused by membrane penetration of substrates. This feature has many potential applications, which urgent in the high-throughput screening to whole-cell biocatalysts for biofuels production. Various microorganisms are used for surface displaying recombinant proteins, including yeast, lactic acid bacteria and *Escherichia coli*. Among all, *E. coli* is the most widely used microorganisms because of its history in genetic manipulation, intrinsic advantages of rapid growth, and high yield for producing heterologous proteins. In order to export the passenger to the cell surface, there were a variety of anchor proteins being developed for the surface display in *E. coli*, which are OmpT, PgsA, Blc, AIDA-I, ice nucleation protein (INP) and autotransporter proteins (ATs). ATs are a family of bacterial proteins composed of an N-terminal passenger domain and a C-terminal b-barrel-forming domain that is integrated into the outer membrane and thereby facilitates the surface display of the passenger domain. The ATs system we used to display recombinant protein is *E. coli* AT Antigen 43 (Ag43) via genetic replacement of the native passenger domain. Ag43-mediated display has been used in some studies of single-chain antibody (scFv) and recombinant chymotrypsin (rChyB) recently. However, these researches rarely show how to optimize the expression and performance of recombinant protein mediated by Ag43. The key factors may involve with promoters, target protein types and even the position at which passenger fused. In this study, we have successfully constructed a series of recombinant plasmids to display different proteins (i.e., red fluorescent protein from *Discosoma striata* (RFP), formate dehydrogenase from *Candida boidinii* (CbFDH) and endo-1,4-beta glucanase from *Bacillus subtilis* (BaCel5)). By adopted various backbones with three promoters, and fused at 138 aa and 700 aa of Ag43, we have different expression level for the construction selected. Firstly, no activity of CbFDH fused into Ag43 might cause by oxidoreductase would inactive due to the oxygen effect when being displayed on the cell surface. The recombinant RFP expressed in pET28a is better than pBAD99a, which can be observed by the color of cell turn red. All recombinant BaCel5 mediated by Ag43 produce a transparent hydrolytic zone in the congo red test. In our present result, surface display mediated by Ag43 is not a universal system but can be a potential tool in mutagenesis screening in the hydrolytic enzymes.

Non-Sterilized Fermentation for Ethanol or L-Lactic Acid from Waste Substrate By Metabolic Engineering of Thermoanaerobacterium Sp. strain

Zhu, M., South China University of Technology
Yang, X., South China University of Technology
Li, S., South China University of Technology

Thermoanaerobacterium aotearoense SCUT27 is one kind of thermophilic, obligate anaerobic bacterium, which could assimilate a variety of carbohydrates from lignocellulosic biomass, to produce primarily L-lactic acid, ethanol, acetic acid and hydrogen with little carbon catabolite repression. Since its unique culture condition, this strain can ferment mixed substrate with non-sterilized condition. In our previous study, we have redesigned and engineered the carbon flux of *T. aotearoense* SCUT27 to generate some mutated strains which can achieve higher yields of products. Herein, we will show the distinct improvements of ethanol and lactic acid production from lignocellulosic derived sugars by different mutants derived from *T. aotearoense* SCUT27.

In the first place, we blocked the lactate metabolic pathway (deleted the *ldh* gene) and redistributed total cellular NADH, to acquire an increase of 2.37 folds for ethanol yield compared to the wild strain, accompanied by 2 folds of hydrogen yield. By using 80 g/L glucose plus 40 g/L xylose as substrate in 5 L fermentor, 31.5 g/L of ethanol concentration was obtained with the yield of 1.5 mol/mol. Meanwhile, there was no lactic acid formation during all the fermentation which can dramatically reduce the cost of purification of the final product and facilitate fermentation to industrial production.

In the second place, the acetic acid synthetic pathway of *T. aotearoense* SCUT27 was deleted by genetic manipulation to enhance the optically pure L-lactic acid production, and the mutated strain of which is named as LA1002. With 50 g/L of substrate in bioreactor, the maximum L-lactic acid yield of LA1002 could reach 0.93 g/g glucose with an optical purity of 99.3%, which is very close to the theoretical value (1.0 g/g of glucose). Moreover, an advanced biorefinery technology was established to convert mixed bakery waste as a nutrient rich feedstock to optically pure L-lactic acid. After hydrolyzation by *A. awamori* and *A. oryzae*, a nutrient-rich hydrolysate was obtained from the mixed bakery waste collected from Starbucks, which contained 83.6 g/L glucose, 9.5 g/L fructose and 612 mg/L free amino nitrogen. By using an evolved *T. aotearoense* LA1002-G40, the overall lactic acid produced from mixed bakery waste was 0.18 g/g under non-sterile conditions, and the titer of which was 78.5 g/L. Here we proposal an innovative procedure involving a complete bioconversion process for L-lactic acid production from mixed bakery waste with non-steriliztion.

Metabolic Analysis of Solventogenic Clostridium Saccharoperbutylacetonicum N1-4 (HMT)

Saunders, E.,

The market for solvent production is predicted to reach $43.4 billion by 2018 with n-butanol having over 20% market share value. Its main use is for the production of biofuel, butyl-acetate, butyl-acrylate, glycol-ethers and plasticisers.

This project focuses on the metabolic and physiologic characterisation of the acetone-butanol producing model strain N1-4. We will use a systems biology approach involving the construction of a genome scale metabolic model of the microorganism, which will be experimentally validated and informed, through the calculation of parameters associated to growth, maintenance, and solvent biosynthesis. This project has created a working GSMN for *Clostridium saccharoperbutylacetonicum* N1-4 (HMT) with 1296 reactions informed by the genomes sequence sources from NCBI (reference: NC_020291.1) and data from the literature and is being validated via experimental data.

Chemostat cultures will be used to determine the effects of cell stress caused by acid production in solventogenic clostridia, on bacterial growth and energy metabolism. Organic acids are initial products formed during early stages of growth, while solvent production is produced later on, possibly triggered by changes in pH..

Solventogenesis in N1-4 is under control of the *sol* operon, which contains genes for solvent production, including *bld, ctfA, ctfB*, *adc*, and *adh*. However, the solventogenic genes contained within the *sol* operon vary between solvent producing species. The saccharolytic species *C. beijerinckii* and N1-4 have an operon structure identical to the amylolytic *C. acetobutylicum.* It was found that a N1-4 degenerate strain carrying the wild-type genotype of *sol* operon, failed to fully induce butanol production. This clearly indicates that other factors, regulatory proteins, 4RNA, and/or metabolites are important for the induction of solvent production. This highlights the need to understand how different operons and Induction mechanisms are connected in order to overcome such problems.

Linking metabolic network analysis, genome analysis and metabolic and physiological observations will help to elucidate the metabolic limitations in solvent production and to design metabolic engineering strategies to overcome those limitations.

Application of Escherichia coli Coculture System for n-Butanol Production

Chao, Y. P., Feng Chia University
Chiang, C. J., China Medical University

n-Butanol is a substituted fuel of poteintial and can be blended with gasoline at any concentrations for use. It has been historically produced by *Clostridium* species in a bioprocess known as the ABE fermentation. However, this fermentation process is difficult to operate efficiently due to a complicated regulation circuit. *Escherichia coli* is known as a biotechnology workhorse and is promising for n-butanol production. In this work, a new production platform of n-butanol in *E. coli* was presented. This production system consists of two strains, a butyrate-producing strain and a butyrate-conversion strain. The latter strain was developed by recruitment of endogenous *atoDA* and *Clostridium adhE2*. M9 mineral salt plus yeast extract (M9Y) was used and this strain produced 6.2 g/L n-butanol from supplemented butyrate. The molar conversion yield accounts for 92%. The butyrate-producing strain was equipped with a heterologous pathway leading to butyrate. The production of n-butanol was then developed by co-culturing these two strains. As a result, 5.5 g/L n-butanol from glucose was obtained in the coculture system in M9Y medium at 24 h. The production yield on glucose accounts for 69% of the theoretical yield. Overall, the result indicates that this new production platform of n-butanol is promising and may open a new avenue of bio-butanol production in *E. coli*.

Synthetic Yeast Cell Factories for High-Level Production of Oleo-Chemicals and Advanced Biofuels

Zhou, Y. J.,
Klaas, B.,
Zhu, Z., Chalmers University of Technology
Qin, J., Chalmers University of Technology
Siewers, V., Chalmers University of Technology
Nielsen, J., Technical University of Denmark

Volatile energy costs and pressure to conserve fossil fuel resources have ignited efforts to produce biofuels and renewable commodity chemicals via microbial fermentation of biomass. The fatty acid-based biofuels (fatty acids, alkanes, fatty alcohol, etc.) are considered as ideal alternative to fossil based chemicals and advanced biofuels. Though most of these molecules have been microbial synthesized,the titers remains to be largely improved for industrial process. As *Saccharomyces cerevisiae* is a well-studied industrial model microorganism and the overall metabolism including the lipid metabolism is well studied, it is feasible to engineer lipid metabolism for overproduction of oleo-molecules in *S. cerevisiae*. Here, we systemitically engineered the lipid metabolism by increasing the precursor supply, inhibition of by-product formation, and disrupting the reversal futile cycling, which enabled production of 10.4 g/L free fatty acid, the highest titer in any microorganism. Then this fatty acid overproducing strain was tailored for fatty alcohol biosynthesis by using a carboxylic acid reductase. After enhancing the fatty aldehyde reduction by engineering alcohol dehydrogenation and reduction, fatty alcohol production reached 1.5 g/L. These results provided many interesting targets for further enhancing the production of oleo-chemicals and advanced biofuels.

Pushing the Boundaries of Biofuels and Oleochemicals Production in Oleaginous Yeasts

Ledesma-Amaro, R., INRA
Nicaud, J. M., INRA

Pushing the boundaries of biofuels and oleochemicals production in oleaginous yeasts

Rodrigo Ledesma-Amaro[1,2] and Jean-Marc Nicaud[1,2]

rodrigo.ledesma@grignon.inra.fr

[1]INRA, UMR1319, MICALIS, Domaine de Vilvert, F-78352 Jouy-en-Josas, France
[2]AgroParisTech, UMR Micalis, Jouy-en-Josas, France

Microbial oils have emerged as promising alternatives to fossil oils, with substantial advantages over plant, animal and algae oils. Among the oleaginous microorganisms, *Yarrowia lipolytica* is by far the most studied and used by the industry. Additionally, in the last years, the natural capacity of this yeast to produced lipids has been maximized by sequential metabolic engineering approaches and culture optimization. Nonetheless, the cost of the process still limits its industrial production to high value compounds and therefore, more research is required to reduce these costs. Our lab is facing this issue by two independent and combinable approaches; expanding the usage of non-expensive substrates and improving lipid production by strain engineering. Both strategies are based on the use of metabolic engineering, synthetic biology, systems biology, metabolic modelling and process optimization techniques.

On the one hand, although *Y. lipolytica* can perfectly grow and produce lipids from glucose it is inefficient or incapable of using other preferred non-expensive carbon sources. We used metabolic engineering to expand the range of use of different hexoses, pentoses or polymeric sugars. The construction of this "Swiss army knife" is a combination of heterologous pathways expression and the awakening of endogenous and limiting sleeping genes. Some of these desired carbon sources are starch and lignocellulosic materials, which are inexpensive and widely spread in both nature and wastes. Therefore, we engineered *Y. lipolytica* to cope with these and other sugars. The results of these metabolic engineering approaches will be discussed.

On the other hand, we have combined previously known target with novel modifications in order to boost lipid accumulation. Additionally we identified the most favourable bioreactor conditions to maximize lipid production. We have recently achieved the highest lipid content described so far in a strain able to accumulate more than 90% of the DCW as lipids. The strategies leading to these strains will be presented and discussed.

The increasing researches in using microbial lipids to produce fuels and chemicals are spreading world wide in the last years, both in public and private sector, pointing out to a general interest and desire. This combined global effort brings us to a future independent of the undesired, limited, and pollutant fossil fuels.

Optimization of the Functional Expression of Plant Cytochrome P450s in Yeast

Lassen, L. M. M., The Novo Nordisk Foundation Center for Biosustainability
Zhang, J., Technical University of Denmark
Jensen, M. K., Technical University of Denmark
Keasling, J. D., Lawrence Berkeley National Laboratory

Development of microbial cell factories holds the potential to enable sustainable biobased production of a wide variety of valuable chemical compounds. Engineering microbial hosts for the biosynthesis of plant natural products often includes expression of cytochrome P450s. Heterologous expression of plant P450s in microbial hosts is known to be challenging, including the requirement for coexpression of redox partners. In order to support biobased production of plant natural products in yeast cell factories we have designed, built and tested a multi-parametric set-up to uncover principles related to optimal enzymatic conversion of a P450-mediated reaction. The parameters of interest include copy number, expression tuning, choice of redox partner and cytochrome B5. Here we present the results of our combinatorial approach using a candidate reaction of relevance to development of platform cell factories for plant natural products.

Glutamate Biosensor Based on Chimeric Two-Component Regulatory System Introduced into Escherichia coli

Ravikumar, S.,
CHOI, J. I., Chonnam National University

This study describes the construction of specific amino acid biosensors, which can be used as a high throughput screening system to screen microorganism producing glutamate. The biosensors are based on the chimeric two-component regulatory systems (TCRS). A chimeric DegS/EnvZ (DegSZ) TCRS was constructed by fusing the sensor kinase of DegS from *Planococcus sp.* PAMC21323 with the cytoplasmic signaling domain of the *E.coli* osmosensor EnvZ to control the expression of the *gfp* gene in response to glutamate. The *gfp* gene was mediated by the ompC gene promoter through the response regulator, OmpR. The biosensors showed a 4-fold increase of the Fluorescent signal after addition of glutamate. The linear correlation was observed between the Fluorescence intensity and exogenously added glutamate concentration. The chimeric TCRS based biosensor was successfully applied for detection of glutamate concentration at the single cell level by Fluorescence-activated cell sorting. Therefore We beleive this biosensors can be used to isolate novel gene products and optimize pathways involved in amino acid production.

Hyper-Exg1 Export Promotes the Biotransformation of Mogrosides in Saccharomyces Cerevisiae kre6 Mutants

Wang, R., Institute of Food Science and Technology
Lo, Y. C., National Taiwan University
Lu, T. J., Institute of Food Science and Technology
Huang, S. T.,
Lin, P. Y.,
Chiu, C. H., Institute of Food Science and Technology

Fungi are able to secrete extracellular enzymes to convert macromolecules into smaller units for growth and adaptation. Studies has been reported that hyper-production of extracellular enzymes and proteins is sometimes associated with cell wall alterations in fungus. Previously, we identified that *Saccharomyces cerevisiae kre6Δ* mutant can efficiently convert mogroside V into anti-diabetic bioactive compound mogroside IIIE. However, the underlying mechanism is unidentified, we therefore hypothesized the facilitative conversion of *kre6Δ* mutant might be cause by changes in cell wall structures and enzyme secretion behavior. In the current study, we examined the extracellular enzyme secretion in relation to mogrosides bioconversion efficiency of several cell wall mutants; while analyzing their cell walls differences by transmission electron microscopy (TEM), zymolyase sensitivity test, and mannoprotein release assays. We found only mutants, including *kre1Δ*, *las21Δ*, *gas1Δ*, and *kre6Δ*, defective in mannoprotein anchorage at outer cell wall layer, had an efficient mogrosides conversion property; such mannoprotein defects suggests with increased cell wall porosity in relation to increased Exg1 enzyme release responsible for mogrosides conversion. Wild-type cells, or *cwh41Δ* or *cwh43Δ* mutants without mannoprotein defects, however, had average enzyme release and mogrosides conversion rates. Moreover, the discovery of extra mannoprotein and enzyme present in the medium of *kre6Δ* mutants and other mutants may further used for increasing the color stability and aroma respectively in wine making industry and promote the bioactive compounds synthesis efficiency.

Genome Sequence of a Novel Marine Strain Alg07 Gives Insight into the Degradation Pathways of Algal Polysaccharides and Monosaccharide

Zhu, Y., Tianjin Institute of Industrial Biotechnology, Chinese Academy of Sciences
Men, Y.,
Yang, J.,
Chen, P.,
Li, J.,
Sun, Y.,

Algae are widely distributed in marine environments. Due to their special ingredients, algae can be applied in various fields, such as functional food, energy source, fertilizer, and so on. The brown alga is a large group of mostly marine algae, and a number of them are edible and good for health. The brown algae contain special polysaccharides, such as alginate, laminarin and fucoidan. The degradation products of these polysaccharides possess specific biological activities, which makes them a hot research topic. Moreover, the algal polysaccharides have been used as an energy source to produce ethanol.

The effective degradation of algal polysaccharides is very important to produce not only functional oligosaccharides but also ethanol and other energy substances. Recently, we isolated a strain Alg07 from marine area near the city of Weihai, China. It was shown that the strain was able to degrade kelp effectively and exhibited extraordinary activity to degrade alginate. According to the phylogenetic analysis based on 16S rRNA gene sequences, Alg07 was identified as the genus *Bacillus*, but exhibits lower similarities (< 95%) with several other well-studied *Bacillus* species. Furthermore, the genome of Alg07 was sequenced and annotated. We identified the enzymes associated with the degradation of polysaccharides in the genome of Alg07 by comparison with the carbohydrate-active enzymes database (CAZymes). The genes and gene cluster for degradation of alginate were predicted and the key enzymes were successfully heterologously expressed and characterized. Interestingly, two alginate lyases exist lonely in the genome, but not be included in the gene cluster. The signal peptides were found in the N terminal of both alginate lyases, which indicated that the alginate was degraded into oligoalginate in the extracellular. The genes responsible for transporting oligoalginate to intracellular, degrading oligoalginate into unsaturated monosaccharide, and reducing monosaccharide to 2-keto-3-deoxy-d-gluconate (KDG) were arranged in a gene cluster. The gene clusters for degradation of laminarin and mannitol were disclosed. Therefore, the degradation pathways of algal polysaccharides and monosaccharide were determined in the novel marine Alg07 strain.

Production of 3-Aminopropionic Acid, a Precursor for Nylon-3 Synthesis By a Metabolically Engineered Escherichia coli

Kim, J. W., Korea Advanced Institute of Science and Technology (KAIST)
Song, C. W., Korea Advanced Institute of Science and Technology (KAIST)
Lee, J., Korea Advanced Institute of Science and Technology (KAIST)
Ko, Y. S., Korea Advanced Institute of Science and Technology (KAIST)
Lee, S. Y., Korea Advanced Institute of Science and Technology (KAIST)

Metabolic engineering strategies were taken to produce 3-aminopropionic acid (3-AP), an important platform chemical for manufacturing acrylamide and acrylonitrile, in *Escherichia coli*. Using a fumaric acid producing *E. coli* strain as a host, the *C. glutamicum panD* gene (encoding L-aspartate-α -decarboxylase) was overexpressed and the native promoter of the *aspA* gene was replaced with the strong *trc* promoter to strengthen the aspartase-catalyzed reaction. The *aspA* and phosphoenolpyruvate carboxylase (*ppc*) genes were additionally overexpressed, and the ammonium sulfate was supplemented in the medium, which resulted in the production of 3.49 g/L 3-AP. Optimization of PPC expression level by using synthetic promoter and RBS sequences increased the titer upto 3.94 g/L. The amount of accumulated acetic acid was decreased by replacing the native promoter of the *acs* gene with strong *trc* promoter and the fed-batch culture of this final strain produced 32.3 g/L of 3-AP in 39h with an overall yield and productivity of 0.135 g 3-AP/g glucose and 0.828 g/L/h, respectively. (This work was supported by the Technology Development Program to Solve Climate Changes on Systems Metabolic Engineering for Biorefineries from the Ministry of Science, ICT and Future Planning (MSIP) through the National Research Foundation (NRF) of Korea (NRF-2012M1A2A2026556 and NRF-2012M1A2A2026557)).

Heterologous Expression of Cold-Adaptive Amaylase from Arthrobacter sp

Kim, S., Chonnam National University
CHOI, J. I., Chonnam National University

In this study, the gene encoding an α-amylase from a psychrophilic *Arthrobacter agilis* sp. (PAMC 27388) was cloned into pET-28a(+) vector and heterologously expressed in *Escherichia coli* BL21 (DE3). The recombinant α-amylase with a molecular mass of 80 kDa was purified by Ni^{2+}-NTA affinity chromatography and showed optimal activity at pH 3.0 and 30°C. This recombinant α-amylase were highly stable at varying temperature (30-60°C) for 24 h within the pH range of 3.0-5.0. The α-amylase activity was enhanced in the presence of $FeCl_2$ (1 mM), NaCl (1 mM), KCl (1 mM), β-mercaptoethanol (5 mM) and phenylmethylsulfonyl fluoride (1 mM), while $CoCl_2$ (1 mM), ammonium persulfate (5 mM), and urea (1%) inhibited the activity. The presence or absence of Ca^{2+} ion did not affect the enzyme activity. Thin layer chromatography (TLC) analysis showed that the cold-active α-amylase hydrolyzed starch, maltotetraose and maltotriose, and produced maltose as the major end product.

Metabolic Engineering of Escherichia coli for Production of Tyrosine and Cadaverine through Microbial Gene Repression System Using the Synthetic Small RNAs

Chung, H., Korea Advanced Institute of Science and Technology (KAIST)
Na, D., Chung Ang University
Yoo, S. M., Korea Advanced Institute of Science and Technology (KAIST)
Lee, S. Y., Korea Advanced Institute of Science and Technology (KAIST)

Small regulatory RNAs (sRNAs) are gene expression regulators which act on post-transcriptional phase of bacterial gene expression. Here, we developed a synthetic small regulatory RNA system to down-regulate gene expression. The target mRNA binding region of MicC, one of the well-studied sRNAs in *Escherichia coli*, was replaced to translation initiation sequence of our target genes. We found that MicC scaffold based synthetic sRNA properly repressed expression of DsRed2, a fluorescence protein. Synthetic sRNAs were applied to metabolic engineering to enhance the production of tyrosine and cadaverine. Through screening of 14 different strains, harboring one or combination of the synthetic sRNAs targeting *ppc*, *tyrR*, *csrA*, and *pgi*, we isolated a tyrosine producer producing 2 g per liter of tyrosine. Using a library of 130 synthetic sRNAs, we screened knockdown targets that increase cadaverine productivity substantially. Repression of *murE* led to 55% increase in cadaverine production compared to the base strain. [This work was supported by the Technology Development Program to Solve Climate Changes on Systems Metabolic Engineering for Biorefineries from the Ministry of Science, ICT and Future Planning (MSIP) through the National Research Foundation (NRF) of Korea (NRF-2012M1A2A2026556 and NRF-2012M1A2A2026557); the Intelligent Synthetic Biology Center through the Global Frontier Project (2011-0031963) of the Ministry of Education, Science and Technology (MEST) through the National Research Foundation of Korea]

Phenol Production By Metabolically Engineered Escherichia coli

Ha, J. Y., Korea Advanced Institute of Science and Technology (KAIST)
Kim, B., Korea Advanced Institute of Science and Technology (KAIST)
Park, H., Korea Advanced Institute of Science and Technology (KAIST)
Na, D., Chung Ang University
Lee, S. Y., Korea Advanced Institute of Science and Technology (KAIST)

Engineering *Escherichia coli* for biological production of phenol is considered infeasible due to its toxicity and complex biosynthetic network. To address these challenges, we took advantage of the synthetic sRNA-based gene knockdown technology to simultaneously engineer 18 *E. coli* strains and screen them for higher phenol tolerance and precursor tyrosine availability. The knock-down of *csrA* and *tyrR* genes were identified to be crucial for redirecting carbon flux toward tyrosine and applied to all the *E. coli* strains with over-expression of the key genes for the biosynthesis of tyrosine and phenol (tyrosine phenol-lyase). The engineered strains showed substantially different metabolic phenotypes. The engineered BL21 (DE3) strain produced 419 mg/L of phenol in flask culture and 1.69 g/L in fed-batch fermentation. A biphasic fed-batch fermentation using glycerol tributyrate was develope for *in situ* extraction while minimizing the phenol toxicity. The preferential partition of phenol from culture medium into glycerol tributyrate increased the titer and productivity to 3.79 g/L and 0.18 g/L/h, respectively.
[This work was supported by the Intelligent Synthetic Biology Center through the Global Frontier Project (2011-0031963) of the Ministry of Science, ICT & Future Planning through the National Research Foundation of Korea.]

Microbial Production of Poly(3-hydroxybutyrate-co-3-hydroxyvalerate) in Escherichia coli By a Rational Metabolic Engineering

Yang, J. E., Korea Advanced Institute of Science and Technology (KAIST)
Choi, Y. J., University of Seoul
Lee, S. H., Chonnam National University
Song, B. K., Korea Research Institute of Chemical Technology
Park, S. J., Myongji University
Lee, S. Y., Korea Advanced Institute of Science and Technology (KAIST)

Polyhydroxyalkanoates(PHAs) are bacterial polyesters accumulated in many bacteria. Bio-based PHAs production has great interest because of their unique characteristics, biodegradable and biocompatible thermoplastic. Among lots of PHA copolymers, poly(3-hydroxybutyrate-co-3-hydroxyvalerate) [P(3HB-co-3HV)] is one of the important copolymers because of lower melting point and much better flexible compared to current used in many application. So far, for the production of P(3HB-co-3HV), adding of second auxiliary carbon source was needed. However, due to the toxicity of auxiliary carbon source, it is very hard to maintain the balance between cell growth and P(3HB-co-3HV) production. Thus, we developed the *E. coli* can stably synthesize 3HB-CoA and 3HV-CoA in controlled ratio from glucose without feeding of exogenous auxiliary carbon source by metabolic engineering. Two different metabolic pathways for the production of propionyl-CoA from 2-ketobutyrate were constructed. The first one is composed of *Dickeya dadantii* 3937 2-ketobutyrate oxidase (2KO) or *E. coli* pyruvate oxidase mutant (PoxB V380A L253F) for the conversion of 2-ketobutyrate into propionate and *Ralstonia eutropha* propionyl-CoA synthetase (PrpE) for further activation of propionate into propionyl-CoA. Second one is consisted of *E. coli* 2-ketobutyate synthases (TdcE) or *Clostridium difficile* (PflB) for direct conversion of 2-ketobutyrate into propionyl-CoA. This engineered strain can efficiently synthesize P(3HB-co-3HV) independent with exogenous auxiliary carbon source. [This work was supported by the Technology Development Program to Solve Climate Changes on Systems Metabolic Engineering for Biorefineries from the Ministry of Science, ICT and Future Planning (MSIP) through the National Research Foundation (NRF) of Korea (NRF-2012M1A2A2026556 and NRF-2012M1A2A2026557).]

Heterologous Biosynthesis of Plant Secondary Metabolites Derived from Phenylpropanoid and Flavonoid Pathway in Escherichia coli

Ha, S. H., Korea Advanced Institute of Science and Technology (KAIST)
Ha, J. Y., Korea Advanced Institute of Science and Technology (KAIST)
Lee, S. Y., Korea Advanced Institute of Science and Technology (KAIST)

Plant secondary metabolites involve various groups of natural products which have diverse applications. Though these compounds are used widely for human healthcare or industrial processes, they are generated extremely low by plants naturally. For this reason, there have been many attempts to produce plant secondary metabolites by chemical synthesis or plant cell culture but they are not enough to meet the industrial needs. Together with the development of synthetic biotechnology, demands on sustainable resources of natural products make heterologous production of those compounds attractive. In this situation, we suggest heterologous biosynthesis of plant secondary metabolites using *Escherichia coli* as a platform strain. By introducing plant genes in phenylpropanoid and flavonoid pathway to *E. coli*, biosynthesis of secondary metabolites (i.e., coumaric acid, caffeic acid, ferulic acid, coniferyl alcohol, naringenin, taxifolin) could be achieved. [This work was supported by the Technology Development Program to Solve Climate Changes on Systems Metabolic Engineering for Biorefineries from the Ministry of Science, ICT and Future Planning (MSIP) through the National Research Foundation (NRF) of Korea (NRF-2012M1A2A2026556 and NRF-2012M1A2A2026557).]

Establishment of a Platform Strain for Production of Short-Chain Fatty Acids and Alkane through Metabolic Engineering

Ko, Y. S., Korea Advanced Institute of Science and Technology (KAIST)
Choi, Y. J., Korea Advanced Institute of Science and Technology (KAIST)
Lee, S. Y., Korea Advanced Institute of Science and Technology (KAIST)

Development of biofuels from renewable resources is important to replace limited fossil fuels. Here *Escherichia Coli* was metabolically engineered to produce short-chain alkanes, free fatty acids, fatty esters and fatty alcohols through the fatty acyl (acyl carrier protein (ACP)) to fatty acid to fatty acyl-CoA pathway. The *fadE* gene was deleted to block β-oxidation pathway and prevent the degradation of fatty acyl-CoAs generated *in vivo*. To increase the conversion of short-chain fatty acids to SCAs, the activity of 3-oxoacyl-ACP synthase (*FabH*) was enhanced by deleting the *fadR* gene and preventing upregulation of the *fabA* and *fabB* genes. A modified thioesterase was used to convert short-chain fatty acyl-ACPs to FFAs, which were then converted to SCAs. The final engineered strain produced up to 580.8□mg□l^{-1} of SCAs consisting of nonane, dodecane, tridecane, 2-methyl-dodecane and tetradecane with small amounts of other hydrocarbons. Furthermore, this platform strain could produce short-chain FFAs and short-chain fatty esters. (This work was supported by the Advanced Biomass Research and Development Center of Korea (NRF-2010-0029799) through the Global Frontier Research Program of the Ministry of Science, ICT and Future Planning (MSIP) through the National Research Foundation (NRF). And this work was supported by the Technology Development Program to Solve Climate Changes on Systems Metabolic Engineering for Biorefineries from the Ministry of Science, ICT and Future Planning (MSIP) through the National Research Foundation (NRF) of Korea (NRF-2012M1A2A2026556 and NRF-2012M1A2A2026557))

Developing Novel Pathways for Butanol Production in E. coli

Ferreira, S., University of Minho
Pereira, R., University of Minho
Rocha, I., University of Minho

In the biofuels market, n-butanol has been gaining attention due to its superior fuel properties when compared with ethanol. Although n-butanol is naturally produced by solventogenic bacteria through ABE fermentation, the low productivities obtained with this bioprocess discouraged its use. Thus, most n-butanol produced nowadays is chemically synthesized via petrochemical routes. One possible alternative is to express non-native pathways in microbial factories.

Our work focuses on the production of butanol through heterologous pathways, previously generated within our group [1] using a (hyper)graph-based algorithm, in *Escherichia coli*. First, we evaluated these pathways according to diverse criteria including the number of biochemical steps required, maximum yield, conservation of carbon atoms and novelty. A novel pathway generated by the algorithm, which is significantly different from the available in solventogenic bacteria, was then introduced in a genome-scale model of *E. coli* (iJO1366) and the production of butanol was optimized *in silico* using Optflux (http://www.optflux.org), a software platform for metabolic engineering. Several different environmental conditions and sets of gene knock-outs were evaluated with the aim of redirecting the cell metabolism to solvent production. The biological significance of the predicted phenotypes was analyzed to rationally design a strain that couples growth and butanol production.

Using biomolecular techniques, a novel strain of *E. coli* – able to produce butanol through the selected pathway in the same conditions simulated *in silico*– is currently being constructed.

[1] F. Liu, P. Vilaça, I. Rocha, and M. Rocha, "Development and application of efficient pathway enumeration algorithms for metabolic engineering applications," *Comput. Methods Programs Biomed.*, Dec. 2014.

Metabolic Engineering and Adaptive Evolution of Escherichia coli for 3-Hydroxypropionate Production Under Low pH Conditions

Wang, B., Key Laboratory of Systems Bioengineering, Ministry of Education, Tianjin University, Tianjin 300072, People's Republic of China
Wang, Z., Tianjin University
Chen, T., Tiamjin University
Zhao, X., School of Chemical Engineering and Technology, Tianjin University, Tianjin 300072, People's Republic of China

Metabolic engineering of *Escherichia coli* for the production of platform chemicals has become an emerging hotspot of industrial biotechnology. 3-hydroxypropionate, which has an active hydroxyl group and a carboxyl group on each terminal of its structure, can participate in many chemical reactions and be derived into many important chemicals. Biological production of 3-hydroxypropionate by metabolic engineering of *Escherichia coli* which utilized glycerol as a substrate have already reached a relatively high titer. Synthetic routes based on other intermediates like malonyl-CoA, beta-alanine and pyruvate have also been tested in model organisms like *Escherichia coli*, *Saccharomyces cerevisiae*. However, most of these processes were carried out under neutral pH conditions which would have an increased cost in fermentation broth neutralization and final product extraction in commercial scale production. In our recent work, 3-hydroxypropionate production was constructed based on lactate intermediate derived from glycolysis and fermentation followed by incorporation of the lactate utilization pathway from *Megasphaera elsidenii*. Protein engineering of CoA transferase was adapted to enable the transferring of CoA from 3-hydroxypropionic-CoA to lactate which facilitated the metabolic fluxing from lactate to 3-hydroxypropiniate. Also, adaptive evolution of *Escherichia coli* with minimal salt medium and glucose as sole carbon source was carried out to improve its productivity under low pH conditions.

Disruption of PHO13 Improves Ethanol Production Via the Xylose Isomerase Pathway

Bamba, T.,
Hasunuma, T.,
Kondo, A., Graduate School of Engineering, Kobe University

Xylose is the second most abundant sugar in lignocellulosic materials and can be converted to ethanol by recombinant *Saccharomyces cerevisiae* yeast strains expressing heterologous genes involved in xylose assimilation pathways. The oxidoreductase pathway involves the conversion of xylose to xylitol by xylose reductase (XR) and the conversion of xylitol to xylulose by xylitol dehydrogenase (XDH). In this pathway, the yield of ethanol is poor for the accumulation of xylitol. Another xylose assimilation pathway is catalyzed by xylose isomerase (XI), which directly converts xylose to xylulose. Therefore, XI-expressing strains do not produce excess xylitol, and suitable for high-yield ethanol production from xylose.

Recent research demonstrated that disruption of the alkaline phosphatase gene, *PHO13*, enhances ethanol production from xylose by a strain expressing the XR and XDH genes; however, the yield of ethanol is poor yet. In this study, *PHO13* was disrupted in a recombinant strain harboring multiple copies of the XI gene derived from *Orpinomyces* sp., coupled with overexpression of the endogenous xylulokinase (XK) gene and disruption of *GRE3*, which encodes aldose reductase. The resulting strain YΔGP/XK/XI demonstrated a xylose consumption rate of 2.08 g/L/h, 0.88 g/L/h volumetric ethanol productivity, and 86.8% theoretical ethanol yield (0.45 g-ethanol/g-xylose), whereas the xylose consumption rate, volumetric ethanol productivity, and ethanol yield of the control strain YΔG/XK/XI were 1.76 g/L/h, 0.57 g/L/h, and 80.3%, respectively, and only YΔGP/XK/XI demonstrated increase in cell concentration.

In the present study, the effect of *PHO13* disruption on intracellular metabolism of the XI-expressing strain was investigated by transcriptome and metabolome analysis using DNA microarray and LC-QqQ-MS methods, respectively. We revealed that the expression of pentose phosphate pathway (PPP) genes was upregulated. And, 125 genes involved in the cell cycle were dramatically altered by *PHO13* disruption. Metabolome analysis demonstrated a decrease in the level of S7P, an increase in the level of 6PG, and accumulation of ATP by *PHO13* deletion during xylose fermentation.

In previous reports, overexpression of PPP genes has improved ethanol yield in recombinant xylose-fermenting yeast strains. Especially, utilization of S7P is thought to be a rate-limiting step in the non-oxidative PPP.

In the present study, the non-oxidative PPP genes was upregulated by the deletion of *PHO13*, which should be the reason of the reduction in S7P pool size. And, the oxidative PPP genes (*SOL3* and *GND1*) was also upregulated by *PHO13* deletion, which might cause depletion of $NADP^+$ and reduction of the oxidative PPP flux. Since the oxidative PPP generates CO_2, reduced flux of the oxidative PPP might increase ethanol yield from xylose consumed. Previously, *PHO13* disruptants indicate high growth ability on xylose medium. In the present study, YΔGP/XK/XI demonstated higher cell concentration than the control strain on xylose fermentation.

We found that *PHO13* disruption upregulates PPP-gene expression in XI-harboling strain and that transcriptional level of 125 cell cycle genes was altered. NADPH supplied by PPP is required for biosynthetic processes and biomass yield. Thus, *PHO13* should be one of the factors affecting yeast cell growth.

High Level 3-Hydroxypropionate Production Via Malonyl-CoA Pathway

Zhao, G.,

3-Hydroxypropionate (3HP) is an important platform chemical, and malonyl-CoA reductase (MCR) of *Chloroflexus aurantiacus* can convert malonyl-CoA into 3HP. In our study, MCR was dissected into two functional distinct fragments, and malonyl-CoA was reduced with NADPH by MCR-C into malonate semialdehyde, which was further reduced into 3HP by MCR-N. It is discovered that the 3HP production via malonyl-CoA pathway was severely depressed by low activity of MCR-C and metabolic burden caused by excessive overexpression of MCR-N and MCR-C. To increase 3HP productivity, the enzymatic activity of MCR-C was improved by directed evolution, and a series of promoters was tested to achieve balance between enzyme activity required by 3HP synthesis and metabolic burden caused by heterologous protein expression. Finally, our engineered E. coli strain accumulated 40.6 g/L 3HP using glucose as sole carbon source under fed-batch fermentation conditions, about 270 times higher than that of our original strain. Compared with other 3HP biosynthetic routes, the malonyl-CoA pathway has some advantages, such as broad raw material spectrum and free of cofactor, and will be widely used in 3HP production in the future.

Direct Fermentation of L-Lactic Acid from Starch By Genetically Engineered Aspergillus Oryzae

Wakai, S., Kobe University
Sasakura, N., Department of Chemical Science and Engineering, Graduate School of Engineering, Kobe University
Asai-Nakashima, N., Kobe University
Tsutsumi, H.,
Hata, Y.,
Ogino, C., Graduate School of Engineering, Kobe University
Kondo, A., Graduate School of Engineering, Kobe University

Lactic acid can be produced biologically, and it is a source of polylactate. Polylactate has been a focus of recent studies for its use as a biodegradable alternative to petroleum-based plastic products. Nowadays, bio-based lactic acid is produced from starchy biomass, while bioproduction by lactic acid bacteria require for utilization of expensive amylolytic enzymes. A filamentous fungus *Aspergillus oryzae* has been used in Japan for more than 1,000 years for the production of fermented foods such as sake, miso, and soy sauce, and it produces various amylolytic enzymes. Therefore, we attempted to examine bioproduction of lactic acid from starch by *A. oryzae*.

Lactic acid-producing *A. oryzae* strains were constructed by genetic engineering. The bovine (*Bos taurus*) *L-LDH-A* gene sequence was modified based on the codon usage of *A. oryzae* and synthesized. The synthesized artificial gene fragment was cloned into a plasmid vector pIS1, which contains the *sodM* promoter and *glaB* terminator from *A. oryzae*. Using the resulting plasmid, *A. oryzae* btLDH which expresses the bovine L-lactate dehydrogenase gene was constructed. Similarly, L-lactate dehydrogenase genes from other organisms, a lactate-producing filamentous fungous *Rhizopus oryzae* and lactic acid bacteria *Lactobacillus casei*, were synthesized and then transformants were constructed. Furthermore, an authentic lactate dehydrogenase gene (*Ao-ldhA*) and pyruvate decarboxylase gene (*pdc*) on *A. oryzae* chromosome were disrupted by replacing with *pyrG* fragment.

A. oryzae btLDH was selected from the transformant pool because it showed highest titer in the transformants constructed. It produced 38 g/L of lactate from 100 g/L of glucose. To improve the lactate titer, *Ao-ldhA* disruption mutant *A. oryzae* btLDHΔ871 was constructed and it produced approximately 50 g/L of lactate from 100 g/L of glucose. Furthermore, *pdc* disruption mutant *A. oryzae* btLDHΔPDC was also constructed, because ethanol was simultaneously produced during the cultivation. This strain produced approximately 50 g/L of lactate from 100 g/L of glucose.

Next, direct fermentation of lactate from starch using these strains was examined, because *A. oryzae* strains innately secrete amylases. *A. oryzae* btLDHΔ871 and *A. oryzae* btLDHΔPDC produced approximately 30 g/L and 40 g/L of lactate from 100 g/L of soluble starch, respectively. Although the both strains showed titers at the same in the fermentation from glucose, *A. oryzae* btLDHΔPDC showed higher titer compared with *A.*

oryzae btLDHΔ871.

From these results, the disruption of *pdc* gene, that is disruption of ethanol production pathway, improved the lactate production in the genetically engineered *A. oryzae*. However, the titer and yield are lower than those of other lactic acid producers. Now we attempt to improve the titer and yield by constructing strains which a carbon flow into a TCA cycle is slowed down and the conversion of lactate from pyruvate is accelerated.

Direct and High-Productive Conversion from a Cyanobacterium Arthrospira Platensis to Ethanol

Aikawa, S., Graduate School of Engineering, Department of Chemical Science and Engineering, Kobe University
Inokuma, K., Kobe University
Hasunuma, T.,
Kondo, A., Graduate School of Engineering, Kobe University

In recent years, there is an urgent need to promote the industrial production of biofuels and biochemicals due to environmental issues. However, considering the increasing demand for edible food grains with population expansion, it becomes important to use inedible feedstocks for biorefinery.

Oxygenic photosynthetic microorganisms such as cyanobacteria or microalgae that accumulate polyglucan such as starch or glycogen, show promise as an alternative feedstock for biorefinery. They have benefits of high-photosynthetic activity and whole-year cultivation without utilizing arable lands. Based on these advantages, the bioethanol production from cyanobacteria or microalgae has been developed in the several previous reports. The microalgal or cyanobacterial polyglucan can be converted to ethanol through saccharification and microbial fermentation. However, the conventional ethanol production process from them composing of pretreatment step and hydrolysis step is high-energy demand and high-environmental load.

We aim to develop the conversion process with simple, low-energy demand, and low-environmental load from cyanobacteria or microalgae. In cyanobacteria and microalgae, the halophilic cyanobacterium *Arthrospira platensis*, commonly known as *Spirulina*, accumulates a large amount of glycogen. In addition, the intracellular glycogen is expected to be extracted relatively simple because of their brittle cell membrane architecture.

In the present study, to hydrolyze the glycogen in *A. platensis* without amylases addition, we used a recombinant *Saccharomyces cerevisiae* strain displaying α-amylase from *Streptococcus bovis* and glucoamylase from *Rhizopus oryzae* on the cell surface. As the result, *A. platensis* could be converted directly to ethanol through a consolidated bioprocess, which is combining glycogen extraction, glycogen saccharification, and ethanol fermentation by using the amylase expressing yeast. However, an ethanol titer (20 g L^{-1}), an ethanol yield (40% of theoretical yield), and an ethanol productivity (0.3 g L^{-1} h^{-1}) were insufficient to accommodate market demand. To improve the ethanol production from *A. platensis*, the intracellular glycogen has to be more effectively extracted.

The cell walls in *A. platensis* such as the peptidoglycan layer and the outer membrane covered with lipopolysaccharide would be barriers to obtain the intracellular glycogen. Lysozyme hydrolyzes cyanobacterial peptidoglycan layer. Additionally, cations may weaken outer membrane because the secretory amount of a recombinant protein from *E.*

coli is increased with the release of lipopolysaccharide by divalent cation such as Ca^{2+} or Mg^{2+}. Therefore, to enhance the glycogen extraction from *A. platensis*, we examined the effect of lysozyme and/or cations on the glycogen extraction. As the result, the ethanol titer and ethanol yield were enhanced by lysozyme addition with an increasing in the total amount of extracted glycogen. The ethanol productivity was accelerated by $CaCl_2$ addition with a releasing of lipopolysaccharide from outer membrane. Consequently, we successfully developed a direct ethanol conversion process by the amylase expressing yeast from *A. platensis* with a high titer (48 g L^{-1}), a high yield (94% of theoretical yield), and a high productivity (1.0 g L^{-1} h^{-1}) by the addition of lysozyme and $CaCl_2$. The present findings indicate that *A. platensis* is a remarkable carbohydrate feedstock, which is a promising material for the production of bioethanol and various other commercially valuable chemicals.

Selection-Based Molecular Breeding of Non-Mevalonate Pathway for Efficient Production of Terpenes

Tashiro-Iwasaki, M., Chiba University
Kawai-Noma, S., Chiba University
Saito, K., Chiba University
Umeno, D., Japan Science and Technology Agency (JST)

Terpenoids are one of the most diverse families of natural compounds that include valuable chemicals as pharmaceuticals, flavors, and fragrances. Dimethylallyl pyrophosphate (DMAPP) and isopentenyl pyrophosphate (IPP) are common building blocks for prenyl pyrophosphates, the precursors of all the terpenoids. To increase the production level of terpenoids in heterologous microbial hosts, numbers of gene targets that affect terpenoid synthesis have been identified using the color screening based on the carotenoid pigmentation. With the growth-coupled selection systems, throughput of the search for those targets could be much improved, enabling the rapid access to even more diverse set of gene targets that improves the supply capacity of the prenyl-pyrophosphates of the host.

It has been recognized that the overexpression of heterologous terpene synthases often exhibit the negative impact on the cell growth[1], due to the reduction of the biosynthesis of ubiquinones and lipid carriers. On the course of an attempt to hyper-produce geraniol in *Escherichia coli*, we accidentally found that the overexpression of the mutant of *Ocimum basilicum* geraniol synthase (GESM53), the variant previously obtained by the screening for improved precursor consumption[2], deprive some strains of *E. coli* almost entirely of their colony-forming capacities. Interestingly, the cell growth was completely restored by the additional overexpression of Idi, the known gene targets that elevate prenyl-pyrophosphate production upon overexpression. Therefore, we concluded that the depletion of prenyl pyrophosphates via GPP, not the product toxicity of geraniol, seemed to be responsible for the lethality of overexpression of GESM53. This also implies that, we could identify the genes having "booster" effect on precursor supply for terpernes, simply by searching for the clones that makes host cell grow in the presence of GESM53 overexpressed.

We developed an arabinose-inducible-GESM53-expression plasmid. Upon induction of GESM53 with 0.2% arabinose, the colony forming unit (c.f.u) of the cell harboring this plasmid dropped by 10^4-fold. We also tested that the additional overexpression of each and all genes involved in the endogenous MEP pathway and found that Idi and FPPS fully recovered the c.f.u. With this selection system set up, we conducted the directed evolution of IspH, which catalyze the final step of MEP pathway. While the wildtype IspH failed to restored the cell growth, four mutants of IspH were found to confer growth capacity where GESM53 being expressed. We found that the co-expression of these ispH mutants alone significantly elevate the production level of carotenoids and monoterpenes.

(1) Wang, C. *et al.*, *Metab. Eng. 13*, 648–655 (2011)

(2) Furubayashi, M. *et al.*, *PLoS One 9*, e93317 (2014)

Simultaneous Co-Fermentation of Glucose and Xylose By Single and Mixed Cultures of Engineered Yeast

Jin, Y. S., University of Illinois at Urbana-Champaign

Glucose and xylose are two prevalent sugars in plant cell wall hydrolysates. *Saccharomyces cerevisiae*, extensively used for producing ethanol from corn or sugarcane, cannot ferment xylose. In order to enable xylose fermentation in *S. cerevisiae*, heterologous xylose metabolic pathways and additional genetic perturbations eliciting enhanced xylose fermentation have been introduced. The resulting engineered yeast was able to ferment xylose, but xylose fermentation was severely repressed by glucose, which causes inefficient and slow fermentation of plant wall hydrolysates. Endogenous hexose transporters in *S. cerevisiae* can transport xylose, but their xylose transport is highly inhibited by glucose, resulting in sequential fermentation of glucose and xylose rather than simultaneous fermentation that is more advantageous. Recently two mutants of hexose transporter that are not subject to glucose inhibition have been isolated. Using these transporters, we constructed single and co-culture systems for rapid and simultaneous co-fermentation of glucose and xylose by engineered yeast. In order to eliminate the glucose inhibition on xylose transport by hexose transporters, we deleted eight hexose transporters (*HXT1-7* and *GAL2*) and overexpressed the mutant hexose sugar transporters in engineered yeast capable of fermenting xylose efficiently. Consistent to the reported kinetic properties of the mutant transporters, the engineered yeast expressing the mutant transporters individually was able to simultaneously co-ferment glucose and xylose. Overexpression of both mutant transporters in single engineered yeast showed faster simultaneous fermentation rates. Additionally, simultaneous co-fermentation of glucose and xylose can be achieved by co-culturing glucose-fermenting yeast and the xylose-fermenting yeast expressing the mutant transporter. Our results not only confirm that inhibition of xylose transport by glucose was a major reason why glucose and xylose are sequentially fermented in engineered yeast, but also demonstrate rapid simultaneous co-fermentation of glucose and xylose in plant cell wall hydrolysates by engineered yeast.

Correlation of the Nifh gene Expression and a Metabolic Model for the Hydrogen Production By a Phototropic Mixed Culture

Jurado Marbán, V. H., Unidad Profesional Interdisciplinaria de Biotecnología, Instituto Politecnico Nacional
Gonzalez García, A., Australian Institute for Bioengineering and Nanotechnology
Salgado Manjarrez, E., Unidad Profesional Interdisciplinaria de Biotecnología, Instituto Politecnico Nacional
García Pena, I., Unidad Profesional Interdisciplinaria de Biotecnología, Instituto Politecnico Nacional

Anaerobic Digestion (AD) as a process for the treatment of organic solid wastes and the production of energy is a promising technology. During this process, hydrogen (H2) and methane (CH4) are produced as by-products in addition to volatile fatty acid (VFA). Because H2 is considered a clean fuel and it contains a high energy content, its production is preferred. In recent years, the use of the volatile fatty acids (VFA) generated during AD for H2 production by photoautotrophic bacteria has been explored to increase the overall production yield. Unfortunately, the mechanisms and regulation systems for H□ production are not completely understood due to the complexity of the metabolic networks in photo-fermentation systems. Although several studies have been performed using single-strain cultures, they could present limitations during the scale-up of the process because the nature of the feedstock. The use of mixed cultures certainly provides a more robust system to be applied in larger scales. Here, we proposed a condensed metabolic model to describe the metabolic changes in an already characterized photo-autotrophic mixed culture during the H2 production. From previous results in our research group, the analyzed mixed culture is composed majorly by *Rhodopseudomonas spp* (85% of the bacterial distribution by the end of the fermentation). The information of the biochemical reactions involved in H_2 production during the photo-fermentation was gathered from KEGG and BRENDA databases, using *Rhodopseudomonas spp* and *Rhodobacter spp* global metabolic maps as the reference. Anaerobic fermentations were performed in a 5 L stirred tank photo-reactor. Acetate was used as carbon source and light intensity was set to be 1000 lux. Samples were taken regularly and analyzed by HPLC. By applying Flux Balance Analysis (FBA), the model predicts an increase in the reaction rates for the key enzymes (nitrogenase and hydrogenase) that participate in H2 production. The model predictions were validated by gene expression analysis. Previously, it was reported that H2 production in photo-bacteria happens as a way to keep the redox balance, and this mechanism is regulated by two enzymatic complex: nitrogenase and hydrogenase. Our results showed that for the mixed culture, the increase in hydrogen production is related to an increment in the expression of the nitrogenase complex from *Rhodopseudomonas* rather that the hydrogenase complex, as we observed an increase in the expression levels of the *nifH*. The model qualitatively mimics the nitrogenase flux as a response of the environmental conditions that favor nitrogenase transcription, which in turn increase the H2 production. Our results are a first approach to understand the mechanism involved in H2 production in a mixed culture. Even our results showed that *Rhodopseudomonas* it's the major contributor to H2 production, still the interactions with the other population should be considered in the context of whole metabolic interactive network.

Retron-Based Targeted Mutagenesis Enabling In Vivo Continuous Evolution in E. coli

Zheng, X., Tsinghua University
Wang, T., Tsinghua University
Lou, C.,
Zhang, C., Tsinghua University
Xing, X., Tsinghua University

By mimicking and accelerating Darwinian evolution, *in vivo* continuous evolution, when compared with the conventional *in vitro* methods, is supposed to be a highly effective approach to engineering tailor-made functions of enzymes or pathways with few human intervention. Generally used global mutagenesis methods, however, usually lead to high evolutionary escape ratio. Targeted mutation, which can largely improve the mutation rate of a target region against the relative steady genetic background, has raised much attention. However, existing targeted mutagenesis tools rely on mutagenic protein anchoring or bacteriophage assisting, which are limited in stringent targeting property and general applicability. Recently, two novel targeted mutagenesis tools based on orthogonal DNA replication system or error-prone reverse transcription process have been established in yeast. Effective targeted mutagenesis tools in prokaryote *E. coli* still remains underdeveloped.

In this work, we proposed a novel targeted mutagenesis tool based on *E. coli* native retron cassette. By inserting the target region sequence into the wild-type retron cassette and utilizing the error-prone T7 RNA polymerase, corresponding mutated ssDNAs can be generated via low-fidelity transcription and reverse transcription processes, and then integrated chromosomally into the target area through *bet*-guided single-stranded DNA recombination. As a proof of concept, the gene d*bla*, containing a premature stop codon in *bla* and encoding the inactivated β-lactamase, was integrated into *E. coli* DH1 genome as a selection marker for mutants. When the ssDNAs targeting the 100-bp stop-codon-containing region were generated in this strain, there was a chance of 10^{-7} to obtain mutants acquiring ampicillin resistance, corresponding to an increase in targeted mutation rate of around 100 fold. In the light of its strong ability to construct mutant library of an artificially designated region with high throughput and low workload, we believe that the retron-based targeted mutagenesis, when integrated into *in vivo* continuous evolution, will become a promising method in protein and metabolic engineering.

Metabolic Engineering for Upgrading Isopropyl Alcohol Production By Escherichia coli Based on 13c-Metabolic Flux Data

Okahashi, N., Osaka University
Matsuda, F., RIKEN CSRS
Yoshikawa, K., Osaka University
Shirai, T., Riken
Matsumoto, Y., Mitsui Chemicals Inc.
Wada, M., Mitsui Chemicals Inc.
Shimizu, H., Osaka University

[Introduction] Isopropyl alcohol (IPA) is a valuable precursor for the bioproduction of polypropylene. The metabolically engineered *Escherichia coli* strain producing IPA has been constructed by heterologous expression of IPA biosynthesis-related genes of *Clostridium*. Although the productivity has been improved by the optimization of culture conditions and the regulation of gene expressions, metabolic bottlenecks in the IPA production are remains unclear. In this study, ^{13}C-metabolic flux analysis (MFA) was performed for the IPA-producing *E. coli* strains to improve the IPA productivity.

[Materials and methods]

The engineered *E.coli* strains were constructed by introducing pUC119 plasmid harboring the GAPDH promoter and target genes. Cells were aerobically cultured at 30°C in Sakaguchi-flask with 100 mL of M9 medium containing [1-^{13}C]glucose : [U-^{13}C]glucose = 7 : 3 at growth phase. Non-growth associated IPA production was performed by inoculating grown cells to nitrogen-free medium containing [1-^{13}C]glucose : [U-^{13}C]glucose = 6 : 4. ^{13}C-MFA of growth and non-growth phase was performed by measuring ^{13}C-labeling of proteinogenic amino acids and intermediate metabolites, respectively. Flux distributions and 95% confidence intervals were calculated by "OpenMebius".

[Results and Discussion]

The *E.coli* strain expressing IPA biosynthesis pathway was constructed by the overexpression of endogenous *atoDAB* genes, *adc* gene from *C. acetobutylicum*, and *IPAdh* gene from *C. beijerinckii* (Ref strain). The fermentation test by the flask cultivation showed that acetate instead of IPA was produced at mid-log phase (OD_{600}~1). The metabolic bottlenecks was investigated by the ^{13}C-MFA. Proteinogenic amino acids were obtained from the *E. coli* cells cultured in the medium containing ^{13}C-glucose. A metabolic flux distribution was estimated from the ^{13}C-labeling patterns determined by GC-MS. The results indicated that NADPH regeneration rate was insufficient for IPA production since 1 mol of IPA is synthesized from 2 mol of acetyl-CoA (AcCoA) and 1 mol of NADPH. In this study, the NADPH supply was improved by two approaches. The first approach was the activation of NADPH regeneration by the glycolysis via the ED pathway. ED strain was constructed by the deletion of *pgi*, *gnd*, and *gntR* and overexpression of *zwf*. The ED strain produced IPA at mid-log phase with the yield of 0.13 mol/mol glucose. The result of ^{13}C-MFA confirmed that an over-regeneration of

NADPH was attained by the ED pathway-dependent glucose catabolism. The second approach was IPA production under the stationary phase to reduce NADPH demand for cell growth. The cultivation of Ref strain under a nitrogen-starving condition successfully increased the IPA yield to 0.20 mol/mol glucose. The metabolic flux distribution at the non-growing state revealed that NAPDH for IPA synthesis was supplied by the oxidative pentose phosphate pathway. Furthermore, IPA production with the yield of 0.55 mol/mol glucose was attained by the combination of two approaches. The level of flux through pyruvate to AcCoA was reached 95% of the theoretical maximal level. The analysis of intracellular NADPH levels showed that there is a positive correlation between specific IPA production rate and NADPH pool size, suggesting that the large NADPH pool size was a driving force for the IPA biosynthesis.

Bioconversion of Methane to Butanol By Metabolic Engineering of Methanosarcina Acetivorans and Escherichia coli

Gonzalez, J., University of Delaware
Whitaker, W. B.,
Bennett, R. K.,
Papoutsakis, E. T., University of Delaware
Antoniewicz, M. R., University of Delaware

Microbial bioconversion has been widely studied for the production of biofuels, an alternative to current fuel technologies. One target substrate for these processes is natural gas, a major energy resource primarily composed of methane, which can be converted to liquid alcohols, a more attractive fuel source. A scheme that converts methane to liquid fuel using bioconversion would combine two critical attributes for competitive fuel production: a higher energy density product and a highly specific and efficient process. A two-step process is being developed for the conversion of methane to butanol that involves two organisms, the anaerobic archaeon *Methanosarcina acetivorans* and the eubacterium *Escherichia coli*. Methane consumption and methanol production will be accomplished by an engineered *M. acetivorans* strain, while second conversion of methanol to butanol will be accomplished by an engineered *E. coli* strain. Strain development of both organisms is guided by modern metabolic engineering technologies including ^{13}C metabolic flux analysis (^{13}C-MFA). So far, a network model for *M. acetivorans* was developed and validated and the dynamics of methane metabolism were elucidated. Additionally, methanol consumption and utilization by *E. coli* was engineered and confirmed with ^{13}C-tracers. The use of a co-substrate with methanol was studied as well, which may be more economically viable than fermentation of methanol alone. To this end, flux analysis was performed on *E. coli* under various conditions using glucose and xylose as carbon sources to identify optimal conditions and co-substrates for methanol consumption. The simultaneous co-utilization of these sugars with methanol is also studied through ^{13}C-flux analysis. Further optimization of the process consists of using ^{13}C-MFA to determine fluxes of the engineered strains and drive carbon flow towards the desired product. Not only will the development of these engineered organisms produce an economical form of fuel production, but it will also demonstrate the capabilities of bioconversion and pave the way for further opportunities towards alternative fuels.

Systems-Level Analysis of Saccharomyces Cerevisiae Engineered for Production of 3-Hydroxypropionic Acid

Kildegaard, K. R., Technical University of Denmark
Jensen, N. B., Technical University of Denmark
Schneider, K., Technical University of Denmark
Czarnotta, E., RWTH Aachen University
Özdemir, E., Technical University of Denmark
Klein, T., Technical University of Denmark
Maury, J., Technical University of Denmark
Ebert, B. E., RWTH Aachen University
Christensen, H. B., Technical University of Denmark
Chen, Y.,
Kim, I. K., Chalmers University of Technology
Herrgård, M. J., Technical University of Denmark
Blank, L. M., RWTH Aachen University
Forster, J., Technical University of Denmark
Nielsen, J., Chalmers University of Technology
Borodina, I., Technical University of Denmark

Biologically produced 3-hydroxypropionic acid (3HP) is a potential source for sustainable acrylates and can also find direct use as monomer in the production of biodegradable polymers. For an economically viable process 3HP must be produced at high titer, rate and yield and preferably at low pH to minimize downstream processing costs. In this study, we engineered *S. cerevisiae* for high-level production of 3HP via a malonyl-CoA reductase (MCR)-dependent pathway by increasing the copy numbers of biosynthetic genes and improving flux towards precursors and redox cofactors. The final strain produced 9.8 $g \cdot L^{-1}$ 3HP with a yield of 13% $C\text{-}mol \cdot C\text{-}mol^{-1}$ glucose after 100 hours at pH 5 in carbon-limited fed-batch cultivation. The best 3HP-producing strain was further characterized by transcriptome analysis and ^{13}C metabolic flux analysis, which revealed several unexpected consequences of the undertaken metabolic engineering strategies. In conclusion, our study demonstrated that in strain engineering it is not sufficient to look at the resulting product titers, but one needs more in-depth characterization of the cellular physiology and cellular components and fluxes in order to follow and advance the strain development.

Screening and Characterization of Novel Xylose Dehydrogenase for the Production of D-Xylonic Acid from D-Xylose in Escherichia coli

Valdehuesa, K. N. G., Myongji University
Ramos, K. R. M., Myongji University
Cabulong, R. B., Myongji University
Maza, P. A. M. A., Myongji University
Pangan, J. O., Myongji University
Bañares, A. B., Myongji University
Nisola, G. M., Myongji University
Lee, S. P., Myongji University
Lee, W. K., Myongji University
Chung, W. J., Myongji University

D-Xylonic acid has several industrial applications as chelator, as concrete dispersion agent, and as precursor for synthesis of polymers and polyalcohols. This compound is derived from D-xylose though enzymatic catalysis of xylose dehydrogenases (Xdh), which are found in a few Bacteria (*Caulobacter* and *Gluconobacter*) and Archaea (*Haloarcula* and *Haloferax*). The microbial production of D-xylonic acid has been well studied in different kinds of yeasts and bacteria. The commonly used Xdh is the *xylB* (or *xdh*), which was recruited from *Caulobacter crescentus*. In this study, novel xylose dehydrogenase enzymes were identified and characterized for their potential application in the production of D-xylonic acid or other valuable compounds. Candidate Xdh from other microorganisms mined from available databases for annotated ORFs with high amino acid sequence homology to *C. crescentus* Xdh. Three candidate genes were selected from *Asticcacaulis*, *Brevundimonas*, and *Phenylobacterium* genus. The Xdh ORFs were then codon optimized, cloned, and expressed in *Escherichia coli*. The purified Xdh were characterized for substrate specificity, co-factor preference, and enzyme kinetics. All three Xdhs showed good activity towards D-xylose oxidation. The best performing Xdh was then tested for efficient conversion of D-xylose to D-xylonic acid in metabolically engineered *E. coli*. *This work was supported by Basic Science Research Program through the National Research Foundation of Korea funded by the Ministry of Education (No. 2009-0093816), and Korea Research Fellowship Program through the NRF funded by the Ministry of Science, ICT and Future Planning (No. 2015H1D3A1062172).*

Caffeic Acid Production By Simultaneous Saccharification and Fermentation of Kraft Pulp Using Recombinant Escherichia coli

Kawaguchi, H., Graduate School of Engineering, Kobe University
Katsuyama, Y., The University of Tokyo
Danyao, D., The University of Tokyo
Nakamura-Tsuruta, S., Kobe University
Teramura, H., Kobe University
Minami, H., Ishikawa prefectural university
Ogino, C., Graduate School of Engineering, Kobe University
Ohnishi, Y., The University of Tokyo
Kondo, A., Kobe University

Caffeic acid (3,4-dihydroxycinnamic acid) is a major hydroxycinnamic acids present in plant cells, and expected to serve as an important precursor for a variety of biologically active compounds due to the anticancer, anti-inflammatory, and antiviral activities. In the present study, *Escherichia coli* was metabolically engineered to produce caffeic acid from cellulosic feedstocks. To establish plant-specific phenylpropanoid pathway in a metabolically engineered *E. coli*, *hpaBC* genes encoding two component flavin dependent monooxygenase derived from *Pseudomonas aeruginosa* PAO1 was expressed in combination with *fevV* gene encoding tyrosine ammonia lyase derived from *Streptomyces* sp. WK-5344 under T7 promoter in a tyrosine overproducing *E. coli* AN219, which compensates a metabolic pathway from tyrosine to caffeic acid through a metabolic intermediate *p*-coumaric acid. A resulting recombinant strain YK01 produced caffeic acid from glucose as the sole carbon source. Kraft pulp was employed as a model cellulosic feedstock for caffeic acid production. After 84 h of cultivation, separate hydrolysis and fermentation (SHF) with kraft pulp hydrolysate yielded less caffeic acid than SHF with filter paper hydrolysate (3 and 6 mg/L, respectively). To avoid fermentation inhibition observed in SHF with kraft pulp hydrolysate, simultaneous saccharification and fermentation (SSF) of kraft pulp supplemented with cellulase cocktails was demonstrated. Notably, caffeic acid titer was markedly increased with reducing loads of cellulase, reaching at 240 mg/L in culture with the minimum cellulase load at 2.5 FPU/g kraft pulp after 310 h of cultivation. In the optimal cultivation conditions, glucose accumulation was no longer observed after 16 h of cultivation, indicating that the *E. coli* cells experienced glucose limitation under the conditions. In addition, the accumulations of a precursor *p*-coumaric acid and a variety of potential fermentation inhibitors were markedly lower and the formation of by-products of acetate, lactate, and succinate was significantly reduced as compared to both SHF with kraft pulp hydrolysate and SSF with higher loads of cellulase. The caffeic acid yield was negatively correlated with the initial glucose concentration, with a five-fold higher caffeic acid yield observed in culture medium containing 5 g/L glucose compared to 100 g/L. These results suggest that gradual enzymatic hydrolysis during SSF enhances caffeic acid production under glucose limitation and reduces the accumulation of fermentation inhibitors and metabolic overflow to by-products, collectively leading to increased caffeic acid yield.

Characterization and Engineering of a Formaldehyde-Responsive Promoter-Regulator System for Synthetic Methylotrophy and Beyond

Sandoval, N. R., University of Delaware
Rohlhill, J. R., University of Delaware
Bennett, R. K.,
Papoutsakis, E. T., University of Delaware

Understanding the quantitative sequence-function relationship in DNA is essential for furthering synthetic biology design principles. Transcription factors use protein-DNA interactions to regulate gene expression. Elucidation of these specific interactions is key for further engineering a range of inducible and tunable promoters.

Applications where engineered pathways produce toxic intermediates (e.g. formaldehyde) require dynamic internal pathway balancing to avoid growth and production inhibition. The formaldehyde-inducible promoter (Pfrm) transcribing the *frmRAB* operon is repressed in the presence of FrmR, the product of the first gene in the operon. FrmA and FrmB encode the glutathione-dependent formaldehyde dehydrogenase and S-formylglutathione hydrolase, respectively, which together oxidize formaldehyde to carbon dioxide. While the general locus of the Pfrm-FrmR interaction has been identified, the consensus sequence has not been previously elucidated.

Here, we describe the sensitivity, dynamic range, and noise of the response of the Pfrm promoter to varying levels of formaldehyde with construction of a GFP reporter plasmid and flow cytometry analysis. The system is sensitive to formaldehyde concentrations ranging from 10 to 1000 μM, where toxicity limits the upper end. We employed this formaldehyde reporter plasmid to successfully isolate functional heterologous methanol dehydrogenase genes in methanol-supplemented media using fluorescence-activated cell sorting (FACS).

To elucidate the sequence-strength and sequence-repression relationships of the promoter regulator pair, we screened a mutated library of synthetically-derived "mPfrm" promoters using FACS and analyzed the resulting activity-based binned populations with high throughput sequencing. Using information theory principles, we quantitatively describe the effect of mutations on promoter strength and efficacy of the repressor. This process enables a generalizable method for construction of custom tunable inducible promoter systems for refactoring pathways for precise metabolic flux control.

This work was supported by the US DOE ARPA-E agency through contract no. DE-AR0000432.

The DOE JGI DNA Synthesis Science Program

Yoshikuni, Y., University of California, Berkeley

The U.S. Department of Energy (DOE) Joint Genome Institute (JGI) started this program in 2012, as set forth in its 10-year strategic vision to serve community science as a user facility pioneering functional genomics to solve the most relevant bioenergy and environmental problems. The mission of the DNA Synthesis Science Program is to enable users to perform state-of-the-art functional genomics research and to help them translate genomic information into biological function. Our strategies are to provide users:

- Access to large-scale DNA synthesis and assembly capabilities (up to 400 K bases/proposal for independent PIs and 1.2 M bases/proposal for consortiums)
- Access to integrated genome technologies and informatics pipelines — from sequencing data generation and database mining including JGI genome databases (IMG, MycoCosm, and Phytozome), to construct design, DNA synthesis and assembly, microbial strain engineering, and transcriptome, proteome and metabolome analyses.

Here we also present several science projects supported by the program including "Discovery of novel TPP-dependent decarboxylase for enhanced medium-chain alcohol production (Nature Communications, 2015)", "Discovery of GH1 family enzymes compatible with JBEI pretreatment (high temp, low pH, and the presence of ionic liquid) (ACS Chem Bio, 2014)", and "Active site characterization of GH55 (JBC, 2015)".

Engineering for Efficient NADPH Cofactor Supply in Cytorchrome P450 Monoxygenase Reaction

Choi, K. Y.,

It is a general practice in metabolic engineering to amplify or disrupt enzymes to increase the productivity of target metabolite. Here it was rerouted the carbon metabolic flux from the EMP pathway to the PP pathway in *E. coli* to increase NADPH availability, and thereby increase the production in cytochrome P450 enzyme reaction which essentially requires NADPH reducing equivalent. In *E. coli*, the balance between NADH and NADPH can be maintained by two isoforms of pyridine nucleotide transhydrogenases, i.e. membrane-bound tranhydrogenase (*pntAB*) and soluble trnashydrogenasse (*udhA*). PntAB catalyses the transfer of reducing power from NADH to $NADP^+$ in an energy-dependent manner when intracellular NADPH level is low, while Udh mainly catalyzes NADPH to NAD^+ in an energy-independent manner when NADPH level is high. Here, NADPH concentration through metabolic engineering of the pentose phosphate pathway (PPP) was increased. The first strategy involved deleting the phosphoglucose isomerase (*pgi*) to increase flux through the (NADPH-producing) PPP resulting in improved redox potential for P450 enzymatic reaction. In the second strategy, NADPH level by expressing NADH kinase which can directly convert NADH to NADPH was also increased The third strategy is regeneration of NADP into NADPH using both glucose 6-phosphate dehydrogenase (GDH) and Transhydrogenase (STH) coupled with glycerol dehydrogenase (GLD).

The disruption of *pgi* and the over-exrpession of GDH/STH/GLH/NADK altered the intracellular redox equivalents levels, resulting in increased concentration of reduced form of NADPH. During cultivation at steady state, this mutant strain didn't change oxidized form of $NADP^+$, while greatly increased NADPH in cells more than five times resulting in the ratio shift to high redox potential (from 1.65 to 0.65). Using this *Δpgi*::GDH::STH::GLH::NADK strain as a biotransformation host, the selected CYP102D1 mutant L84A/F96V/A243S/L448A/T247R was investigated for their biotransformation efficiency. The resting cell assay result revealed that 500 μM of daidzein were converted into 350 μM of 3'-hydroxylated product with 70% of conversion which is 1.3 times higher than that by wild-type host system.

Breeding of Homo-Butanol-Fermentative Strain Using Clostridium Saccharoperbutylacetonicum ATCC27021

Mukoyama, M.,
Ichige, E.,
Nakanosho, M.,
Dobashi, Y., Nippon Shokubai
Nakayama, S.,
Tanaka, T., Kobe University
Kondo, A., Graduate School of Engineering, Kobe University

There are some problems in acetone-butanol-ethanol fermentation, because of low n-butanol yield owing to formation of significant amount of acetone and ethanol. We tried to breed a homo-butanol fermentative strain only to produce n-butanol as a solvent using *Clostridium saccharoperbutylacetonicum* ATCC27021. We constructed gene disruption plasmids pNS which can insert gene fragment including recyclable antibiotics-resistance marker based on GroupII intron function, and obtained strains that disrupted pta, ptb1, ctfB, adc, ldh1 gene which are involved in butanol fermentation using these plasmids. Disrupted strains were evaluated by cultivation. Δpta strain showed poor growth and low n-butanol yield. Δptb1 strain produced about half amount of acetone compared to wild type strain, its n-butanol yield was improved by 15%. Δptb1Δpta strain decreased in acetone formation drastically , n-butanol yield from glucose reached 85mol% of the theoretical yield. (Wild-type strain 53mol%). Δptb1ΔptaΔldh1 strain showed n-butanol yield 87mol%. Δptb1Δpta strain produced 52.5g/L(yield 84.1mol%) of n-butanol with high glucose concentration medium overlayed oleyl alcohol. (Wild type strain produced 23.6g/L (yield 41.2mol%) Metabolic analysis data of Δptb1Δpta strain indicate pyruvate and butyly-CoA concentration were significantly high and ratio of NAD/NADH was quite low. From these result, there were some redox unbalance, especially in dehydrogenation of pyruvate to acetyl-CoA and reduction of butylyl-CoA to n-butanol.

Enhanced Solvents Production in Consolidated Bioprocessing Consortia By Promotion of Metabolites Exchange Via Modules Metabolic Engineering

Wen, Z.,
Jiang, Y.,
Li, Q., Key Laboratory of Synthetic Biology, Institute of Plant Physiology and Ecology, Shanghai Institutes for Biological Sciences, Chinese Academy of Sciences, Shanghai, China
Liu, J., Key Laboratory of Synthetic Biology, Institute of Plant Physiology and Ecology, Shanghai Institutes for Biological Sciences, Chinese Academy of Sciences
Yang, S., Chinese Academy of Sciences

No natural microorganism or microbial community can produce butanol directly from lignocellulose. Assembly of modules involved in lignocellulose saccharification and butanol synthesis into one host strain failed to accomplish the long pathway construction and keep overall cellular health simultaneously. Therefore, artificial microbial consortium consisting of celluloytic and solventogenic clostridia was designed for division and coordination of metabolic labor.

Recently, we developed a synthetic CBP consortia for cellulosic butanol production, in which, *C. cellulovorans* decomposed AECC (alkali-extracted deshelled corn cobs) and provided fermentable sugars and organic acids for *C. beijerinckii* to produce butanol. Herein, we developed the genetic system of *C. cellulovorans* and then reconstructed the CBP consortia. The overall across species pathway was divided into four modules including butyrate synthesis and NADH redirection modules in *C. cellulovorans*, and butyrate reassimilation and pentose utilization modules in *C. beijerinckii*. The metabolites (mainly butyrate) exchange was promoted significantly by reinforcing butyrate supply by *C. cellulovorans* and butyrate reutilization in *C. beijerinckii* simultaneously via module metabolic engineering and consequently enhanced solvents production.

In 110 h, the optimal CBP consortia produced over 20 g/L solvents with AECC as sole carbon source, which approximated ABE output from starchy feedstock. Our work severe as a stepping stone for industrial production of ABE from lignocellulose.

Metabolic Engineering of Anaerobic Itaconate Production in E. coli

Weusthuis, R., Wageningen University
Vuoristo, K., Norwegian University of Life Sciences
Mars, A., Wageningen University and Research Centre
Sanders, J., Wageningen University and Research Centre
Eggink, G., Wageningen University and Research Centre

Itaconic acid is a potential new chemical building block. Industrial application is limited because of the high production costs. Currently itaconic acid is produced by *Aspergillus terreus*. The pathway yield of itaconic acid in this fungus is 1 mol/mol glucose, 25% lower than the maximum theoretical yield. This results in oxygen consumption, which has a negative effect on productivity.

To increase both yield and productivity, itaconate production should be performed without oxygen input. This requires the development of metabolic networks in which cofactor regeneration is possible without respiration. The mixed-acid fermentation pathway of *E. coli* was engineered to this end. Cis-aconitate decarboxylase (CadA), the key enzyme of itaconate production, of *Aspergillus terreus* was introduced in *E. coli*. To boost precursor availability, citrate synthase and aconitase from *Corynebacterium glutamicum* were introduced. To reduce byproduct formation both phosphotransacetylase and lactate dehydrogenase were knocked out. This resulted in anaerobic production of itaconic acid. Ethanol (+formate) and succinate production were used for cofactor regeneration.

It appeared that itaconate production was limited by CadA activity. The immunoresponse gene (Irg1) of mouse also catalyses the decarboxylation of cis-aconitate into aconitate *in vitro*. Replacing the CadA gene with the Irg1 gene resulted in itaconate production *in vivo*, at a similar rate as CadA. Using Irg1 genes from other origins may give the opportunity to increase itaconate production.

Although the mixed-acid-pathway does allow anaerobic itaconate production, the concurrent production of by products for cofactor regeneration may limit its commercial applicability. Two alternative pathways were designed showing a pathway yield identical to the maximum theoretical yield, combining the TCA cycle with either a reverse glyoxylate cycle or a reverse TCA cycle. Progress made with these pathways and their applicability for the production of other chemicals will be discussed.

Escherichia coli Chromosomal Engineering Toward High Titer Phenylalanine and Tyrosine Production

Koma, D., Lawrence Berkeley National Laboratory
Yoshida, E., Ajinomoto Co., Inc.
Yamanaka, H., Osaka Municipal
Moriyoshi, K., Osaka Municipal Technical Research Institute
Keasling, J. D., Joint Bioenergy Institute
Lee, T. S., Joint BioEnergy Institute
Ohmoto, T., Osaka Municipal Technical Research Institute

Synthetic pathways of phenylalanine and tyrosine are useful for conversion of renewable sugars to aromatic compounds. Expanding the pathways along with heterologous genes enable to produce variety of aromatic compounds, in *E. coli*, such as phenylethanol as rose flagrance, phenyllactate for potential biopolymer unit, tyrosol for medicinal resource, hydroxytyrosol for dietary supplement, and so on (Koma et al. Appl. Environ. Microbiol. 2012). To enhance the productivity of such aromatic compounds, engineering of the phenylalanine and tyrosine biosynthetic pathways is important, and most work so far has used plasmid systems to overexpress key enzymes in their biosynthetic pathways. In comparison to the plasmid systems, the metabolic flux in chromosomal engineered strains is generally difficult to control since expression levels of the key pathway genes from chromosome are usually quite lower than those from plasmid systems. In this study, we engineered the *E. coli* chromosome to generate superior phenylalanine and tyrosine producers using a strong promoter. We had two approaches for chromosomal engineering; the overexpression of several key genes for phenylalanine biosynthesis and the overexpression of whole pathway genes for tyrosine biosynthesis under P_{T7} promoters.

Two shikimate pathway genes, *aroG* and *pheA*, and two central metabolic pathway genes, *tktA* and *ppsA*, are known as key genes for phenylalanine synthesis. To generate chromosome-engineered phenylalanine producer, P_{T7}-*ppsA* and P_{T7}-*tktA* genes were integrated into the *E. coli* chromosome in addition to P_{T7}-*aroG*^{fbr} and P_{T7}-*pheA*^{fbr} genes. Overexpression of *tktA* gene caused metabolic burden and phenylalanine titer obviously decreased. When P_{T7}-*aroA* gene was further integrated into the chromosome, metabolic burden seemed to be relieved and the phenylalanine titer increased considerably. The yield reached to about 50% of theoretical maximum yield, which corresponded to 1.7- and 5-hold higher than the strains overexpressing only *aroG*^{fbr} + *pheA*^{fbr} genes and 4 key genes. Real-time PCR analysis revealed that overexpression of additional *aroA* gene along with overexpression of 4 key genes activated whole of the shikimate pathway.

JBEI has recently reported a tyrosine producer which contained two plasmids encoding 9 shikimate genes plus *ppsA* and *tetA* genes on 4 operons (Juminaga et al. Appl. Environ. Microbiol. 2012). In the current work, we integrated these 4 operons encoded in these plasmids into 4 loci of the *E. coli* chromosome. When the plasmid system was transferred into the chromosome without any modification of the promoters, titer and yield significantly decreased in comparison to those from the plasmid system. After changing the promoters of 4 operons from weaker ones (P_{lac}, P_{trc}, and P_{tet}) to stronger one (P_{T7}), titer increased up to 8-fold.

This work was supported by JSPS KAKENHI Grant Number 15K07402 and the Joint BioEnergy Institute (http://www.jbei.org/), which is supported by the US Department of Energy, Office of Science, Office of Biological and Environmental Research, through contract DE-AC02-05CH11231 between Lawrence Berkeley National Laboratory and the US Department of Energy.

Isotopically Nonstationary 13c-Metabolic Flux Analysis Under Photoautotropic Conditions for a Glycogen Mutant of Synechococcus Sp. PCC 7002

Hendry, J. I., Indian Institute of Technology Bombay
Prasannan, C.,
Ma, F., Donald Danforth Plant Science Center
Möllers, K. B.,
Digmurti, M.,
Allen, D. K., Donald Danforth Plant Science Center
Frigaard, N. U.,
Dasgupta, S.,
Wangikar, P. P., Indian Institute of Technology Bombay

Glycogen is the major carbon storage molecule in cyanobacteria and removal of this native carbon sink could increase the carbon flux towards other value-added products for biotechnological purposes. Comparative analyses of the wild type (WT) and a glycogen deficient mutant could provide insight into the re-routing of carbon through primary intermediates of metabolism and allow assessment of the flux ratios at critical branch points. Specifically, in this study, we have performed ^{13}C labeling studies on the model cyanobacterium *Synechococcus* sp. PCC 7002 using the WT strain and a mutant strain with inactivated glycogen synthesis made by deleting the genes encoding glycogen synthase (A1532/*glgA1* and A2125/*glgA2*). In both WT and mutant strains we observed slow labeling of the TCA cycle relative to glycolysis and Calvin cycle intermediates. In the WT strain, the ^{13}C-enrichment of different intermediates at steady state reveals the existence of inactive pools of metabolites that get labelled at reduced rates. Our results also indicate slow labeling of 2-phosphoglycolate (2PG) that is the first committed intermediate in photorespiratory metabolism, thus it is likely that the carboxysome-based carbon concentrating mechanism is fully active in these strains. As might be expected, the transient labeling of the glycogen precursor ADP-glucose was slow in the mutant strain compared to the WT. Currently, the transient labeling data for the mutant strain is being used to estimate the intracellular fluxes with an isotopically nonstationary metabolic flux analysis approach (INST-MFA). Comparison of flux maps for the WT and mutant strain will provide further insights which will be of value in metabolic engineering efforts in cyanobacteria.

Acknowledgement: Authors acknowledge financial assistance from Reliance Industries Ltd, India.

Construction of a Corynebacterium Glutamicum Platform Strain for the Production of High-Value Plant Secondary Metabolites

Kallscheuer, N.,
Vogt, M., Forschungszentrum Juelich
Marienhagen, J., Forschungszentrum Juelich

The soil-bacterium *Corynebacterium glutamicum* is an important microorganism in industrial biotechnology for the production of bulk chemicals, especially amino acids (van Ooyen *et al.*, 2012, Vogt *et al.*, 2014). However, until now, only a few *C. glutamicum* strains for the production of aromatic compounds are available, which can be mostly attributed to the presence of a complex (and partly unknown) catabolic network for a broad range of aromatic substances. We were able to uncover the complete catabolic route for phenylpropanoids, which allows growth of *C. glutamicum* using these compounds as sole carbon and energy source. Phenylpropanoids are channeled into the central carbon metabolism by the coupled action of an identified peripheral, CoA-dependent, β-oxidative degradation pathway and the well-known β-ketoadipate pathway (Kallscheuer *et al.*, 2016). By deleting four gene clusters comprising altogether 21 genes being essential for degradation of phenylpropanoids and other aromatic compounds, we obtained a *C. glutamicum* platform strain that was subsequently applied for the production of (2*S*)-flavanones and stilbenes (Marienhagen and Bott, 2013). Heterologous expression of codon-optimized genes encoding a 4-coumarate: CoA-ligase (4CL), a chalcone synthase and a chalcone isomerase led to the production of 30 - 35 mg/L of corresponding (2*S*)-flavanones from supplemented phenylpropanoids in the constructed strain. The strain also produced stilbenes when genes coding for a 4CL and a stilbene synthase (STS) were heterologously expressed. Several cultivation parameters influencing growth and gene expression were optimized with respect to improved product titers. Under optimal conditions, stilbene concentrations ranging from 60 to 160 mg/L could be obtained from supplemented phenylpropanoids. A strain converting the phenylpropanoid *p*-coumaric acid into the stilbene resveratrol was further modified with a focus on the increased supply of the precursor metabolite l-tyrosine. By introduction of an additional gene coding for a tyrosine ammonia lyase combining aromatic amino acid metabolism with phenylpropanoid synthesis, the strain was able to produce 60 mg/L directly from glucose in defined medium simultaneous to biomass formation. Beyond production of polyphenols, the strain will help to overcome current drawbacks of *C. glutamicum* resulting from degradation of aromatic compounds, offering a huge potential for production of other high-value aromatic compounds with this industrially relevant organism.

Kallscheuer, N., Vogt, M., Kappelmann, J., Krumbach, K., Noack, S., Bott, M., & Marienhagen, J. (2016). Identification of the *phd* gene cluster responsible for phenylpropanoid utilization in *Corynebacterium glutamicum. Applied microbiology and biotechnology*, 100(4), 1871-1881

Marienhagen, J., & Bott, M. (2013). Metabolic engineering of microorganisms for the synthesis of plant natural products. *Journal of biotechnology*, 163(2), 166-178.

van Ooyen, J., Noack, S., Bott, M., Reth, A., & Eggeling, L. (2012). Improved L□lysine production with *Corynebacterium glutamicum* and systemic insight into citrate synthase flux and activity. *Biotechnology and bioengineering*, 109(8), 2070-2081.

Vogt, M., Haas, S., Klaffl, S., Polen, T., Eggeling, L., van Ooyen, J., & Bott, M. (2014). Pushing product formation to its limit: metabolic engineering of *Corynebacterium glutamicum* for L-leucine overproduction. *Metabolic engineering*, 22, 40-52.

Identifying NAD-Dependent Methanol Dehydrogenases for Synthetic Methylotrophy

Bennett, R. K., University of Delaware
Sandoval, N. R., University of Delaware
Papoutsakis, E. T., University of Delaware

Methylotrophy describes the ability of microorganisms to utilize reduced one carbon compounds, such as methane and methanol, as growth and energy sources. The recent discovery of large natural gas reserves has prompted considerable interest in utilizing these compounds as substrates or co-substrates with sugars in industrial anaerobic fermentation of fuels and chemicals, as higher biomass and product yields are expected from these more reduced substrates. Native methylotrophs represent poor industrial microorganisms since many are strict aerobes, produce few metabolites and lack genetic engineering tools. Therefore, the development of synthetic methylotrophy is of considerable interest. Herein, we discuss the selection of suitable methanol dehydrogenase (MDH) candidates for engineering nonnative methanol metabolism in *Escherichia coli*, a well-developed host for anaerobic fermentation.

Methanol oxidoreductase enzymes are classified by the cofactor associated with methanol oxidation. Generally, methylotrophic yeast utilize a methanol oxidase, gram-negative methylotrophic bacteria utilize a pyrroloquinoline quinone-dependent MDH and gram-positive methylotrophic bacteria utilize a NAD-dependent MDH. NAD-dependent MDH enzymes offer the best selection towards synthetic methylotrophy since methanol oxidation supplies reducing equivalents in the form of NADH and PQQ biosynthesis is strictly aerobic and nonnative to *E. coli*. Of these NAD-dependent MDH enzymes, those from thermophilic *Bacillus methanolicus* and *B. stearothermophilus* have been well characterized *in vitro*. Of more importance, however, is heterologous *in vivo* activity of these enzymes in *E. coli*, which we demonstrate. We further discuss how to overcome the challenges involved in using these MDH enzymes, including unfavorable methanol oxidation and selectivity. Toward this, we identified a separate family of NAD-dependent alcohol dehydrogenase enzymes that exhibit high methanol selectivity.

We also report a high-throughput screening technique that was developed to identify improved MDH enzymes derived from protein engineering. This technique takes advantage of the native *E. coli* formaldehyde-responsive transcription factor/promoter system (*frmR-Pfrm*), derived from the formaldehyde detoxification *frmRAB* operon. We constructed a product-responsive reporter strain to isolate improved MDH variants via fluorescence-activated cell sorting (FACS). This reporter strain enables *in vivo* quantification of MDH activity via fluorescence detection upon methanol addition. We demonstrate that this technique successfully isolates MDH variants based on *in vivo* methanol oxidation activity.

SUPPORTED by the US DOE ARPA-E agency through contract no. DE-AR0000432.

Engineering a Short, Aldolase-Based Pathway for 1,3-Butanediol Production in Escherichia coli

Nemr, K., University of Toronto
Muller, J. E. N.,
Gawand, P., University of Toronto
Lu, S., University of Toronto
Venayak, N., University of Toronto
Mahadevan, R., University of Toronto

Aldolases have received wide interest as biocatalysts due to their ability to form new carbon-carbon bonds with high stereoselectivity, thereby forming long complex molecules from simple building blocks. They exhibit broad substrate specificity, which increases their utility in catalyzing the production of a wide range of compounds[1]. Of the Class I aldolases, deoxyribose-5-phosphate aldolase (DERA) naturally catalyzes the reversible aldol condensation of acetaldehyde and glyceraldehyde-3-phosphate to produce deoxyribose-5-phosphate. As of yet, efforts have been focused on increasing its activity and stability through enzyme engineering towards a wide range of aldehydes. Current DERA-based processes typically use the purified enzyme for biotransformation; however, using it to shorten synthetic pathways in whole-cell biotransformations has not been fully explored for the production of biochemicals directly from renewable resources. Shorter biosynthetic pathways reduce the complexity of strain and pathway optimization, including fine-tuning gene expression levels and metabolic burden on the cell. As such, we explored the possibility of producing chemicals with potentially high market value via shorter pathways using aldolases. We have used a pathway predictor algorithm tool (PPA), which employs generalized enzymatic reactions, and we included all known classes of aldolases. The addition of the aldolase class of reactions increased the number of unique chemicals by 5 fold, which can be produced from acetaldehyde using pathways composed of a maximum of three reaction steps. From these results, a short 2-step pathway from acetaldehyde to *(R)*-1,3-butanediol (BDO) was identified. *(R)*-BDO can be used to synthesize high-value compounds. Previous studies have demonstrated an acetyl-CoA based pathway for BDO production in *E. coli* via four enzymatic steps[2,3]. Our unique approach demonstrates the production of BDO via a shorter, 2-step pathway. To test our aldolase-based pathway, we used a systematic approach to engineer *E. coli* by: (1) screening and identifying the required enzymes as predicted by the pathway predictor algorithm, DERA and aldo-keto reductase (AKR); (2) expressing the pathway enzymes on a plasmid expression system; and (3) using a model-based strategy to delete native genes from the host strain to eliminate by-products, such as ethanol, and increase carbon flux from the acetaldehyde branch-point to pathway products by 97%. Our current strain produces 1.44 g/L of BDO with a yield of 0.09 g/g of glucose (17% of the maximum theoretical yield) by two-phase batch fermentation. Future work entails additional strain modifications to increase flux to BDO. Since DERA can accept a wide range of aldehydes as substrates, there is potential to design short, synthetic pathways for other long-chain chemicals.

1. Liu, J. & Wong, C.-H. Aldolase-catalyzed asymmetric synthesis of novel pyranose synthons as a new entry to heterocycles and epothilones. *Angew. Chem. Int. Ed.*

Engl. **41,** 1404–7 (2002).

2. Kataoka, N., Vangnai, A. S., Tajima, T., Nakashimada, Y. & Kato, J. Improvement of (R) -1,3-butanediol production by engineered Escherichia coli. *J. Biosci. Bioeng.* **115,** 475–480 (2013).

3. Kataoka, N. *et al.* Enhancement of (R)-1,3-butanediol production by engineered Escherichia coli using a bioreactor system with strict regulation of overall oxygen transfer coefficient and pH. *Biosci. Biotechnol. Biochem.* **78,** 695–700 (2014).

Bioalcohol Production By Metabolically Engineered Cyanobacteria Using Photobioreactor

Horiuchi, J. I., Kyoto Institute of Technology
Kumada, Y., Kyoto Institute of Technology
Hanai, T., Kyushu University
Hirokawa, Y., Kyushu University
Murakami, A., Kobe University, CREST
Konishi, M., Kitami Institute of Technology
Hashi, R.,
Kitamura, T.,

The 3rd generation biorefinery research using microalgae is ongoing in many countries because fuel/chemicals production from carbon dioxide by photogenic microorganisms is essentially sustainable. Cyanobacteria is recognized as a proper platform for chemicals production due to their high growth rate, but it requires genetic modification because the original strain never produce usefulchemicals. In this presentation, I would like to introduce the current status of our research concerning the production of iso-propanol (IPA) and 1,3 propandiol (1.3PDO) by genetically engineered cyanobacteria using photobioreactor system with LED irradiation equipment, which can selectively supply various wavelength of light with low power consumption.

To achieve IPA production in cyanobacteria, we constructed a metabolically engineered Synechococcus elongatus strain TA1741 which produces IPA from acetyl-CoA by four steps. Successful cell growth was observed for blue, and blue and red LED, however, IPA was not detected for all cases. On the other hand, higher intracellular glycerol accumulation was observed, which suggests that the carbon dioxide fixed in a Calvin cycle is converted primarily into Glycogen under light condition. So we thought that, by changing from light to dark condition, accumulated glycogen may be converted to IPA. To verify this idea, Cells grown in a jar fermentor was harvested, re-suspended in new medium and purged by nitrogen gas. Then the tubes were maintained under anaerobic and dark conditions for 5-10days and analyzed. After five or 10 days cultivation, the production of IPA was confirmed for all cases showing that the genes concerning IPA biosynthesis were working in TA1741. Maximum IPA concentration was approx. 23 mg/L under anaerobic and dark conditions, however the concentration is rather low and complicated operation is required for practical IPA production.

Then, we examined another product, 1.3-Propandiol (1.3-PDO) production by genetically engineered cyanobacteria. In cyanobacteria, metabolic flux passing through acetyle-CoA is very narrow under light condition. On the other hand, PDO can be produced from DHAP which is main intermediate product in Calvin cycle. To achieve PDO production in cyanobacteria, five genes were introduced into the host strain, which produces 1.3 PDO from DHAP through glycerol. We call this strain TA2984. Using this strain, batch experiments were performed in the same way. Cell growth was successful and maximum PDO concentration was approximately 0.7 mM/L. It should be noted that the higher production of glycerol was observed. Their production was clearly associated with cell growth. This characteristics allows us to try continuous process. So we are now examining the possibility of continuous process using immobilized cyanobacteria. As a bioreactor for continuous process, an airlift fermenter with an internal draft tube was used. Successful growth was observed using an airlift bioreactor, maximum PDO concentration increased to approximately 1.8mM.

TAR1 Is Required for Artemisinin Biosynthesis and Trichome Development in Artemisia Annua L

Tan, H., Second Military Medical University

Trichomes, small protrusions on the surface of many plants, can produce and store various secondary metabolic products. Artemisinin, a famous and potent medicine for malaria, is synthesized, stored and secreted by *Artemisia annua* trichomes. However the molecular basis regulating the biosynthesis of artemisinin and the development of trichomes remains poorly understood. Here, we revealed the crucial role of an AP2 transcription factor, TRICHOME AND ARTEMISININ REGULATOR 1 (TAR1), in controlling the development of trichomes and the biosynthesis of artemisinin in *A. annua*. *TAR1* is expressed mainly in young leaves, flower buds, and some trichomes, and its protein is specially located in cell nucleus. In *TAR1-RNAi* lines, the phenotype of trichomes and the composition of cuticular wax were altered. The content of artemisinin was dramatically reduced as well, which increased significantly when *TAR1* was overexpressed. Furthermore, the expression of many genes that participate in artemisinin biosynthesis was changed when *TAR1* was silenced or overexpressed, especially *ADS* and *CYP71AV1*, which were shown to be the likely direct targets of *TAR1* through electrophoretic mobility shift, yeast one-hybrid and transient transformation GUS assay. These results indicate that *TAR1* is a key component of the molecular network regulating *A. annua* trichome development and artemisinin biosynthesis.

Improved Co-Production of Hydrogen and Ethanol in Escherichia coli By the Deletion of Phosphofructokinase-1 and Activating Pentose-Phosphate Pathway

Sundara Sekar, B., Pusan National University
Seol, E., Pusan National University
Park, J. M., GS Caltex
Park, S., Pusan National University

Co-production of H_2 and ethanol has been suggested to overcome the low yield limitation in fermentative biohydrogen production and co-production of these valuable biofuels in a single process is beneficial [1]. We have suggested the activation of pentose phosphate pathway for improving the reduced co-factors needed for ethanol production. We tried to upregulate the carbon flux through PP pathway by down-regulation of Embden-Meyerhof-Parnas pathway and overexpressing Zwf and Gnd, the key enzymes in PP pathway. The overexpression of Zwf and Gnd could activate PP pathway and down-regulation of EMP pathway (Δ*pfkA*) additionally improved the flux through PP pathway. Further, acetate production pathway was deleted to divert more acetyl-CoA towards ethanol production. The deletion of acetate production pathway significantly reduced the co-production yield by excreting pyruvate. We also attempted to eliminate acetate/pyruvate production by increasing the expression levels of Zwf and Gnd. The co-production yield of 1.55 mol/mol H_2 and 1.41 mol/mol ethanol could be obtained with a slight accumulation of acetate. The production levels of NAD(P)H suggested that ~70% of carbon was consumed through PP pathway. The carbon distribution analysis of the mutant strain and enzyme activities of Zwf and Gnd were studied. The gene expression analysis was also performed and the effect of Zwf and Gnd overexpression was characterized. Our strategy should be very useful for the production of biochemicals requiring reduced co-factors.

References

1. Seol E, Ainala SK, Sekar BS, Park S. Metabolic engineering of *Escherichia coli* strains for co-production of hydrogen and ethanol from glucose. Int J Hydrogen Energ. 2014; 39:19323-19330.
2. Seol E, Sekar BS, Raj SM, Park S. Co□production of hydrogen and ethanol from glucose modification of glycolytic pathways in *Escherichia coli*–from Embden□Meyerhof□Parnas pathway to pentose phosphate pathway. Biotechnol J. 2015; 11:249-256.

Protein Engineering for the Production 5-Hydroxytrypophan and Serotonin in Escherichia coli

Mora-Villalobos, J. A.,
Zeng, A. P., Hamburg University of Technology

5-Hydroxytryptophan (5HTP) is a non-canonical amino acid with therapeutic effect over various symptoms (e.g. depression, insomnia, chronic headaches, etc.). It is also the precursor of serotonin which, unlike 5HTP, does not cross the blood-brain barrier. Both 5HTP and serotonin are interesting for the pharmaceutical industry and currently they are either chemically synthesized or extracted from seeds of the African plant *Griffonia simplicifolia*. Current efforts are focused on their biotechnological production, and in our group we aim to link their production with the glucose central metabolism in order to obtain valuable metabolites from inexpensive carbon sources.

So far, we have developed a strain that is capable to hydroxylate tryptophan and produce 5HTP by using rational design approaches. To this end, phenylalanine hydroxylase (PAH) from *Cupriavidus taiwanensis* (*Ct*PAH) was chosen as a starting point. Molecular modeling and docking analyses were conducted and prominent residues that may lead to an increase of the preference of tryptophan as a substrate were identified. Modeling analyses also suggested the need to eliminate part of the N-terminal region of the enzyme. Point mutations in five out of six residues showed higher activity toward tryptophan both *in vitro* and *in vivo*. The engineered *Ct*PAHs were then introduced into an *E. coli* strain that is capable of producing tryptophan with a high productivity. To the date, we are comparing the effects of different promoters that may control the expression of *Ct*PAH for a fine tuning.

To further increase the enzyme activity, we also endeavor to move from a rational design to a semi-rational evolutionary approach. For this regard, we are working in the generation of libraries, where we can combine different targeted sites with restriction enzyme and/or ligase independent methods. In order to screen the collection of mutants, a colorimetric method has been established that allows us to detect hydroxylated aromatic rings directly in agar plates. In parallel, we are working with *in vitro* cell free expression systems with the goal of developing multistep-single tube reactions that can facilitate the detection of the metabolites and therefore speeding up the screening of the library.

Regarding to the serotonin production, we are using a combination of three enzymes: one decarboxylase and a combination of P450 from *Oryza sativa* and its respective NADPH-cytochrome P450 reductase. So far, we can detect serotonin in cells supplemented with tryptophan, but low titers were detected when the corresponding genes were transformed into the tryptophan producer. To overcome this, we intent to follow the same approaches described above in order to increase the production either *in vivo* or *in vitro*.

Production of Organic Acids in Fungal Hosts

Toivari, M.,
Ruohonen, L.,
Penttilä, M., VTT Technical Research Centre of Finland

Biotechnological production of organic acids has gained significant interest in providing bio-based platform chemicals for various applications. Fungi, both yeast and filamentous fungi, offer a possibility for low pH biotechnological production, advantageous for downstream processing. Many filamentous fungi produce hydrolytic enzymes, allowing polymeric carbohydrates to be used as raw material in bioprocesses.

We have engineered various yeast and filamentous fungi for production of organic acids. Synthetic biology tools are essential for high through put strain development. For this purpose methods like CrispR, and synthetic promoters have been set up in various fungal hosts, in particular targeting for fungal production hosts with high industrial potential.

Nature biodiversity has enormous potential for creating novel industrial hosts. We have generated a pipeline to identify novel industrial fungal production hosts, including strain selection (tolerance to low pH, acids, hydrolysates etc.), construction of metabolic engineering tool boxes, and physiology studies - up to industrial implementation.

For metabolic engineering of a model pathway for production of xylonic acid, a number of steps were taken: enzyme candidates were identified by bioinformatic data mining, their properties characterized and structure determinations and site-directed mutagenesis were carried out. NMR methods were used to monitor the kinetics of selected pathway enzymes and in vivo reactions. Our studies also highlighted global transcriptomic responses and intracellular accumulation of acid.

Choice of host, gene, pathway and production process (typically batch or fed batch) have all been found to be important in achieving high levels of organic acid production.

Diversion of Fatty Acid Synthesis Pathway of Escherichia coli for Production of Long Chain Fatty Alcohol

Fatma, Z., International Centre for Genetic Engineering and Biotechnology
Jawed, K., International Centre for Genetic Engineering and Biotechnology
Mattam, A., International Centre for Genetic Engineering and Biotechnology
Yazdani, S., International Centre for Genetic Engineering and Biotechnology

Fatty alcohols are valuable chemical used in the industrial manufacturing of innumerable products as well as in transportation sector. Microbial fatty alcohol production can be achieved in *E. coli* through fatty acid synthesis pathway by diverting fatty acyl ACP towards alcohol in a two-step process where an acyl ACP reductase (AAR) converts fatty acyl ACP to fatty aldehyde and further an aldehyde reductase converts fatty aldehyde to fatty alcohol. The first step was achieved in this study via optimal expression of an AAR from *Synechococcus elongatus* in *E. coli*. For achieving second step an extensive *in silico* and *in vivo* analysis was undertaken to identify aldehyde reductase / aldehyde dehydrogenase that could play role in converting fatty aldehyde to fatty alcohol. We identified a long chain specific aldehyde reductase that converted long chain aldehyde to alcohol at much higher efficiency than other enzymes of similar category and characterized in terms of substrate specificity via enzyme kinetics. We further co-expressed cyanobacterial AAR along with newly characterized aldehyde reductase and produced significant amount of long chain fatty alcohols of C16-C18 chain length range. Importantly, we modulated endogenous pathway leading to membrane lipid formation and found 60% increase in fatty alcohol production. Our study provides the new insight that will accelerate the effort towards improvement and development of fatty alcohol production through microbial engineering.

Metabolic Engineering of Saccharomyces Cerevisiae for Cyclic Terpenoid Production

Ebert, B. E., RWTH Aachen University
Czarnotta, E., RWTH Aachen University
Walter, K.,
Knuf, C., DTU Denmark
Maury, J., Technical University of Denmark
Jacobsen, S. A., Technical University of Denmark
Lewandowski, A.,
Polakowski, T., Organobalance GmbH
Lang, C.,
Forster, J., Technical University of Denmark
Blank, L. M., RWTH Aachen University

Triterpenoids are terpenoids derived from squalene and consist of six isoprene units (C30). These compounds can be isolated from many different plant sources. They occur in countless variations and can be subclassified into several groups including lanostanes, dammaranes, lupanes, oleananes, ursanes, hopanes, cycloartanes, friedelanes, cucurbitacins, and miscellaneous compounds (Mullauer et al. 2010).

Many of them or their synthetic derivatives are currently being investigated as medicinal products for various diseases. The cyclic triterpenoid betulinic acid is of special interest for the pharmaceutical and nutritional industry as it has antiretroviral, antimalarial, and anti-inflammatory properties and has potential as an anticancer agent (Muffler et al. 2011). Despite their obvious industrial potential, the application is often hindered by their low abundance in natural sources. This poses challenges in a biosustainable production of such compounds since per gram active ingredient produced a high volume of solvent is needed in the purification process.

Here, we present a novel biotechnological process for the production of betulinic acid using tailored *Saccharomyces cerevisiae* strains. The multi-scale optimization of this microbial process included:

- pathway engineering by determination of optimal gene combination and dosage,
- compartment engineering to increase the reaction space of the betulinic acid pathway, and
- strain engineering by implementation of different push, pull and block strategies.

In parallel we developed the fermentation process and were able to boost the performance of the engineered yeast by optimization of medium composition, cultivation conditions, carbon source and mode of fermentation operation in lab scale bioreactors. Product purification was achieved by a one-step extraction with acetone.

The final process was evaluated in terms of economic and ecological efficiency and rated to be competitive with existing plant extraction procedures with potential for further performance improvement.

The engineered *S. cerevisiae* strain represents an ideal platform strain for the production of a variety of industrial compounds that can be derived from the betulinic acid precursors squalene and 2-oxidosqualene.

References

Muffler K, Leipold D, Scheller MC, Haas C, Steingroewer J, Bley T, Neuhaus HE, Mirata MA, Schrader J, Ulber R. 2011. Biotransformation of triterpenes. Process Biochemistry 46(1):1-15.

Mullauer FB, Kessler JH, Medema JP. 2010. Betulinic acid, a natural compound with potent anticancer effects. Anti-Cancer Drugs 21(3):215-227.

Discovery and Evaluation of Novel Pathways for Production of the Second Generation of Biofuels

Tokic, M., Swiss Federal Institute of Technology (EPFL)
Hadadi, N., Ecole Polytechnique Federale de Lausanne (EPFL)
Ataman, M., Ecole Polytechnique Federale de Lausanne (EPFL)
Miskovic, L., Ecole Polytechnique Fédérale de Lausanne (EPFL)
Neves, D., RWTH Aachen University
Ebert, B. E., RWTH Aachen University
Blank, L. M., RWTH Aachen University
Hatzimanikatis, V., Swiss Federal Institute of Technology of Lausanne (EPFL)

In an effort of overcoming the limited availability of fossil energy resources the focus of the research and development in the area of biofuels has moved towards developing the second generation of fuels that should be produced via microbial fermentation. The idea is to use as a feedstock inexpensive and abundant waste materials such as lignocellulosic biomass. The second generation biofuels should satisfy several criteria such as lower emission, higher energy density and should be less corrosive to engines. However, currently used industrial workhorses such as *E. coli* and *S. cerevisiae*do not produce many of these molecules naturally. Furthermore, for majority of these molecules there are no known biochemical pathways.

The need for discovery of novel biosynthetic pathways towards desired molecules sparked the development of computational tools that are capable of reconstructing feasible reaction steps between a given set of starting compounds (precursors) and a molecule of interest (a target compound). In this study, we performed the retrobiosynthesis analysis for all the candidates applying Biochemical Network Integrated Computational Explorer (BNICE.ch) and we reconstructed the metabolic network for 69 fuel compounds. We analyzed compounds (biological or chemical) and reactions one step away from these target molecules and the appearance of the potential substrates in their reconstructed metabolic network. Based on the results of this analysis, coupled with other criteria such as the Gibes free energy of formation, combustion potential and expert opinion, we chose the 5 highest ranked fuel candidates for further pathway reconstruction studies. For these 5 fuel candidates we have created 1.8 million *de novo*pathways with different number of reaction steps for further evaluation.

We have then embedded the reconstructed pathways in the genome-scale models of the two chassis organisms *E. coli* and *P. putida* and we performed the pathway evaluation with respect to their maximum theoretical yield using two methods (i) Flux Balance Analysis (FBA), and (ii) Thermodynamic-based Flux Analysis (TFA). All the pathways that did not satisfy the thermodynamic constraints were rejected. Our results indicated that in the evaluation and ranking of the pathways the thermodynamics is a necessary consideration. Specifically, we showed that the majority of the pathways that were determined as feasible based on FBA, they were not feasible based on TFA. Moreover, we also showed that some pathways that were determined as feasible in both organisms based on FBA, they were not feasible in both of them based on TFA. Also, the yield of the feasible pathways can differ depending on the organism. These results demonstrate how TFA can also guide the selection of a suitable chassis organism. For all discovered reactions in the feasible pathways we have also identified known enzymes from the KEGG database with the closest reaction similarity that could be engineered for the implementation of novel reactions.

This study shows the full potential of computational tools for discovery and design of novel synthetic pathways and their relevance for future developments in the area of metabolic engineering and synthetic biology.

Engineering Clostridium beijerinckii for Butanol Production from Corn Cob

Zhang, T., Tianjin Institute of Industrial Biotechnology, Chinese Academy of Sciences
Dong, H.,
Zhang, Y., Institute of Microbiology, Chinese Academy of Sciences
Li, Y., Institute of Microbiology, Chinese Academy of Sciences

Corn cob is one of the most potential renewable feedstocks for bio-chemicals production. Hydrolysis of corn cob with dilute acid generates fermentable sugars (xylose, a little glucose and arabinose) and some microbial inhibitors. Most bacteria prefer glucose to xylose due to the carbon catabolite repression, therefore, the utilization efficiency of xylose and the tolerance to inhibitors of the producing strain must be improved for efficient biorefinery of corn cob. In this study, strain *Clostridium beijerinckii* was engineered to produce butanol from corn cob hydrolysate. To improve the tolerance to inhibitors, a successive mutation and adaptation process was developed. As well, heat-shock proteins GroESL from *C. acetobutylicum, Thermoanaerobacter tencongenis,* and *Pseudomonas putida* were investigated to enhance the hydrolysate tolerance of the cells. As a result, mutant SMB626 was derived, which produced 14.6 g/L solvent ABE (9.9 g/L butanol, 67.8%) from corn cob hydrolysate (containing 44.2 g/L xylose, 4.8 g/L glucose and some inhibitors) in 72 h fermentation. The yield of ABE from corn cob hydrolysast reached 0.37 g/g, which was comparable to that from corn mash.

Assembly of New Minicellulosomes through Disulfide Bonds Using Yeast a-Agglutinin on Saccharomyces Cerevisiae Surface for Cellulosic Ethanol Production

Hou, J.,
Tang, H.,
Bao, X.,

Saccharomyces cerevisiae is an ideal candidate for combining cellulases production, cellulose hydrolysis and ethanol fermentation into a single step to achieve consolidated bioprocessing (CBP). Previous reports generally assembled minicellulosomes on yeast cell surface through cohesion and dockerin interaction and these heterologous minicellulosomes expression often leaded to low display and assembly efficiency. In this study, we designed a new assembled pattern of minicellulosome on *S. cerevisiae* surface through disulfide bonds using yeast a-agglutinin system. N-terminal 149 amino acids of yeast a-agglutinin Aga1p is used as the cohesin domain and repeated cohesin domains with a cellulose-binding domain (multi-aga1) are synthesized for the scaffodin construction. The display efficiency of multi-aga1 scaffodin was 4 times higher than the traditional CipA3 scaffodin from *Clostridium thermocellum*. Cellulases, including a cellobihydrolase, an endoglucanase and a β-glucosidase, are fused with N-terminal Aga2p as dockerin. The recombinant cellulases can functionally interact with N-terminal of yeast a-agglutinin Aga1p through disulfide bonds. Cell surface assembly of minicellulosome is confirmed by the immunofluorescence microscopy and flow cytometric analysis (FASC). The newly developed recombinant yeast strain with self-surface assembly of minicellulosome can utilize phosphoric acid-swollen cellulose (PASC) and produce 720 mg/L bioenthanol. The assembly of minicellulosomes through disulfide bonds using yeast a-agglutinin obtained much higher display efficiency and reduced the cellular burden to expression heterologous proteins. This system is also readily to be used for assembly of cellulosomes with two miniscaffoldins.

Enhanced Enzyme Stability through Fusion of Self-Assembled Cellulose Binding Domain

Kim, M., Chungnam National University
Yeom, S. J.,
Kim, H.,
Lee, D. H., University of Science and Technology
Lee, S. G., Korea Research Institute of Bioscience and Biotechnology

Enzyme stability has been a significant concern for protein engineers due to its industrial importance in biocatalytic processes. Here, we present the enhanced enzyme stability by fusing the enzyme to a nano-particle-like cellulose binding domain (CBD) expressed in engineered *Escherichia coli*. The CBD from *Cellulomonas fimi* fused with a β-glucuronidase (GusA) derived from *E. coli* was expressed, purified, and characterized. Whereas the fusion construct retained enzymatic activity, significant differences in thermostability and pH stability were observed. In particular, the CBD-GusA variant showed almost 2-fold increased half-lives compared to findings for the free GusA enzyme. In addition, the CBD-GusA protein exhibited increased methanol tolerance and decreased protease accessibility compared to those of the free GusA. The fusion strategy presented here reveals a surprising means for the stabilization of enzymes and stresses the importance for improving stability of enzymes used in industrial applications.

Molecular Toolbox for Co-Expression of Chaperones for Heterologous Protein Production in Lactic Acid Bacteriametabolic Engineering XI: Design, Synthesis and System Integration for Metabolic Engineering

Wong, F. T., Molecular Engineering Laboratory, A*STAR
Pek, H. B.,
Tay, S. V.,
Lim, P. Y., Bioprocessing Technology Institute, A*STAR
Tan, L. L., Molecular Engineering Laboratory, A*STAR
Ow, D. S. W., Bioprocessing Technology Institute, A*STAR

Lactic acid bacteria (LAB) are an important group of Gram positive genera traditionally used in the food industries due to its GRAS (generally recognized as safe) status. Members of LAB include *Lactobacillus*, *Bifidobacterium*, *Lactococcus*, *Aerococcus*, *Leuconostoc*, *Oenococcus* and *Pediococcus*. Considerable attention has been paid to them as commensal and probiotic bacteria in the gastrointestinal tract. To date, applications of LAB include acting as live vectors for targeted vaccine delivery and as microbial cell factories for production of recombinant proteins and bio-therapeutics. In its native environment, LAB are constantly exposed to environmental stresses such as heat, high salinity, organic acids and osmotic pressure. Consequently, they have evolved various adaptive mechanisms to response to these stresses. One of the classic response systems in LAB is the protein quality control system that consists of molecular chaperones and proteases that facilitate the quality control and folding of heterologous proteins.

Molecular chaperones are a group of proteins that mediates in the folding of other proteins. They are capable of binding to misfolded proteins with high affinity towards hydrophobic residues which are exposed in a non-native protein state. This causes a conformational change and subsequently releasing the correctly folded protein. Another group of chaperones namely the Clp family function as proteases that can help unfold protein aggregates. This adaptive mechanism adopted by LAB allows heterologous protein to be folded properly thus preventing aggregation. Currently, there have been little studies conducted on the co-expression of molecular chaperones in LAB to overcome protein aggregation during recombinant processes. This propose the need to examine chaperone co-expression to metabolically-engineer cells for production of difficult-to-express proteins. In this study, a strong and a moderate constituitive promoters (P59 and USP45) were cloned into a *Lactococcal* co-expression vector. Subsequenly, the chaperones (DnaK, GroES, GroEL, Trigger factor and ClpX) were cloned and transformed into *Lactococcus lactis* NZ900 strain. Nattokinase is chosen as the model protein. A comparision between the expression of Nattokinase in soluble fraction of recombinant *L. lactis* strains co-expressing various chaperones will be presented to detrmine their effect on Nattokinase production. The toolbox generated in this study will faciliate future applications to combat metabolic bottlenecks during heterologous protein production in LAB.

Diversifying the Chemical Structures of Polymer Precursors Produced in E. coli

Haushalter, R. W., Lawrence Berkeley National Lab
Keasling, J. D., Joint Bioenergy Institute

Worldwide demand for plastics and other polymers are on the rise, and sustainable alternatives to petroleum sources will help to meet this demand in the future. Therefore, much effort has recently focused on microbial production of hydroxy fatty acids and dicarboxylic acids from renewable sources. These classes of compounds can be polymerized to produce bio-based plastics, nylons and other useful materials.

Here we report comparison and evaluation of varying engineered pathways to produce both hydroxy fatty acids and dicarboxylic acids directly from glucose. We have explored variants of key enzymes in pathways to produce hydroxy fatty acids, and have optimized conditions for the production of various chain lengths and degrees of saturation. We have also developed a novel pathway specifically for the production of odd-carbon dicarboxylic acids, and we compare this pathway with previously reported systems. These results expand the portfolio of chemical products accessible through microbial fermentation and will allow metabolic engineers to tailor the chemical structures of hydroxy fatty acids and dicarboxylic fatty acids for specific industrial applications

Metabolic Engineering of Limonene Overproduction in the Oleaginous Yeast Yarrowia Lipolytica

Wei, L.,
Cao, X., East China University of Science and Technology
Chen, J., East China University of Science and Technology
Hua, Q., East China University of Science and Technology

Limonene is known for its use as an important precursor of many flavoring, pharmaceutical and biodiesel products. Here, we have reported the production of limonene in the non-limonene producing oleaginous yeast *Yarrowia lipolytica* and the improvement of strain productivity through the increasing of the endogenous supply of precursor metabolites to enhance pathway activity. The strain *Y.lipolytica* PO1f-LN-000, with the overexpression of two heterologous limonene synthesis genes encoding both *neryl diphosphate synthase 1* (*NDPS1*) and *limonene synthase* (*LS*), was used as initial limonene production strain. This initial strain yielded limonene production of 6 μg/g DCW. To maximize limonene production, several genes in the MVA pathway were overexpressed, sometimes in different copies of the same gene and sometimes in combination. We found that overexpressing *the HMG1* gene alone raised limonene content to 0.11±0.01 mg/g DCW, an 18-fold improvement. After the *ERG12* gene was co-expressed with the *HMG1* gene, the yield of limonene was further improved to 0.67±0.04 mg/g DCW, a 112-fold increase from the initial yield. Finally, the optimization of additive amounts of pyruvic acid and dodecane helps the limonene content reach 1.36 mg/g DCW which was 226-fold higher than the yield of the initial strain. The strategy using in this study lay a foundation for further producing value-added limonene derivatives, even other monoterpenes in oleaginous yeast.

Optimization of Displayed Cellulases in Yeast Suitable for Ionic Liquid Pretreated Biomass by Cocktail δ Integration Methodology

Ogino, C., Graduate School of Engineering, Kobe University
Ishizue, N., Kobe University
Kondo, A., Graduate School of Engineering, Kobe University

With problems such as exhaustion of fossil resources, the bioethanol production from abundance rich and renewable lignocellulosic biomass has recently become an attractive technology. The structure of lignocellulosic biomass is too rigid due to the high density of hydrogen bond within its structure, and thus the several pre-treatment methods had been proposed for deconstructing of biomass. Among them, biomass pre-treatment with ionic liquid has a lot of attention because this method is environmentally benign methodology with comparable low energy consumption. By using ionic liquid as a solvent for biomass suspension, we had achieved well the construction of bio-ethanol production system combined with our developed arming yeast previously [1]. In common sense, the degradation process of pre-treated biomass is usually required by enzymatic degradation with cellulase enzyme(s). Our developed system combined with arming yeast and ionic liquid biomass pretreatment could deduce the loading of cellulase regent for efficient degradation. In addition, recently, the biomass based ionic liquid for biomass pretreatment was discovered, and we also achieved the efficient pretreatment of Bagasse from sugarcane with this ionic liquid, cholinium ionic liquid [2].

For efficient degradation of biomass with cellulases, there are optimal combination ratio between these cellulases including Endoglucanase (EG), Cellobiohydrolase (CBH I and CBH II), and Î²-glucosidase (BGL). In our laboratory, we have been also developed the optimizing methodology of displaying several cellulases on the yeast cell surface by using cocktail Î´-sequence [3]. By using this method, we could be established the adequate yeast strain depended on target biomass.

In this study, the combination of cocktail Î´-integration method for cellulases displaying and cholinium ionic liquid biomass pretreatment was evaluated by simultaneous scarification and fermentation (SSF) with constructed arming yeast. As a result, we successfully stablished an engineered yeast, and could produce 0.8 g/L ethanol from 5 g/L bagasse without any addition of cellulose enzymes.

[1] K. Nakashima, K, Yamaguchi, N. Taniguchi, S. Arai, R. Yamada, S. Katahira, N. Ishida, H. Takahashi, C. Ogino, A. Kondo. Direct bioethanol production from cellulose by the combination of cellulase displaying yeast and ionic liquid pretreatment. *Green Chem.*, **13(10)**, 2948-2953 (2011)
[2] K. Ninomiya, S. Omote, C. Ogino, K. Kuroda, M. Noguchi, T. Endo, R. Kakuchi, N. Shimizu, K. Takahashi. Saccharification and ethanol fermentation from cholinium ionic liquid-pretreated bagasse with a different number of post-pretreatment washings. *Bioresour. Technol.*, **189**, 203-209 (2015)
[3] R. Yamada, N. Taniguchi,T. Tanaka, C. Ogino, H. Fukuda, A. Kondo. Cocktail Î´-integration: a novel method to construct cellulolytic enzyme expression ratio-optimized

yeast strains. *Microb. Cell Fact.*, **9**; 32 (2010)

Keywords
Celluase display; cocktail Î´Â integration; ionic liquid; lignocellulosic biomass pretreatment

Metabolic Engineering in Synechocystis PCC 6803 for the Production of Butanol

Miao, R.,
Liu, X.,
Lindblad, P., Uppsala University

Fossil fuel limitations have been a world wide problem from both energy resource perspective and global peace perspective. The consequence of this is an urgent demand of new sustainable energy resources. 1-butanol is a potential fuel substitute because of its high energy density, less explosive, low vapour pressure and high flash point. In addition to these advantages, 1-butanol can be blended into gasoline in any ratio, and no specific engine is required. As a photosynthetic microorganism, which lives on light, CO_2 and water, cyanobacteria can be used as an ideal platform to convert CO_2 into carbon based valuable compounds. Some pioneer study has been done by introducing various of enzymes from different organisms into *Synechococcus elongatus* PCC 7942 and *Synechocystis* PCC 6803 for the production of 1-butanol and isobutanol. In this study, for improving the productivity of butanol, selected exogenous and endogenous genes encoding enzymes which can constitute a metabolic pathway able to catalyze *Synechocystis* intermediate metabolites to 1-butanol, are expressed on the genome of wild type cells. A series of integrative vectors, which can generate stable chromosomal recombinations are used in order to introduce synthetic pathways and knock out competing pathways at the same time. Carbon and cofactor flux in the engineered strains will be examined as well as the comparison of final product titers.

A Glucose-Sensing Toggle Switch Enables Autonomous Control That Improves Production but Avoids Growth-Toxicity

Bothfeld, W., Northwestern University
Tyo, K. E. J., Northwestern University

Currently, most biosynthetic strategies are growth-coupled. This inherently limits product yield, as *Escherichia Coli* convert ~50% of feedstock carbon (e.g. glucose) to biomass during cell growth. Furthermore, the available gene targets for modification are fundamentally constrained by this growth compatibility requirement. Genes essential for growth, but detrimental to pathway flux, cannot be removed and production-enzyme levels are limited to prevent growth toxicity and system instability. Decoupling the growth and production phases enable separate biomass (catalyst) generation and product synthesis phases. This allows relaxed constraints during production phase to enable essential gene removal and pathway over-expression to reroute substrate flux for increase product yields. To successfully implement this bi-phasic strategy, we have adopted the bi-stable toggle for bioprocessing. Here we demonstrate the integration of native glucose-sensing into the toggle switch to precisely and autonomously active expression of the PHB operon using smart media design. The switch remains stably in the new state even after reintroduction of glucose. Our system enables PHB operon expression at levels that improve PHB yield compared to a constitutively expressing PHB strain. Promisingly, early activation of the PHB operon during growth results in a severe growth-rate reduction (~5 fold lower), indicating the system generates pathway levels incompatible with growth. This result demonstrates the utility of a two-phase production strategy. While timed induction is common in academic labs, the culture volumes at industrial scale make it infeasible due to expensive chemical inducers. Our switch is "programmed" to induce simply by media design. This glucose-sensitive switch allows precise control of genetic programs using only the ingredients necessary for cultivation in rich or minimal media. It uses native promoters only for initiating the switch, which decouples the activation mechanism from the design of each genetic program. Nutrient-sensing promoters as the activator of a product pathway are subject to native regulation, often have gradual induction and low activity, and may prevent desired pathway expression levels. This switch frees the production phase from a nutrient-limiting environment that is required for maintaining pathway activity, but is not ideal for production. The glucose toggle switch should be broadly useful to initiate any type of desired genetic program for metabolic engineering applications.

McCormick
Northwestern Engineering

A glucose-sensing toggle switch for autonomous, high productivity genetic control

William Bothfeld[1,2], Grace Kapov[1], Keith E.J. Tyo[1]
[1]Department of Chemical & Biological Engineering, [2]Biotechnology Training Program Northwestern University, Evanston, IL, USA

NORTHWESTERN UNIVERSITY

Motivation

Growth-coupled synthesis strategies are **inherently limited**:

1. feedstock (sugar) is converted to biomass, not product
2. essential genes must be maintained
3. growth toxicity must be managed.

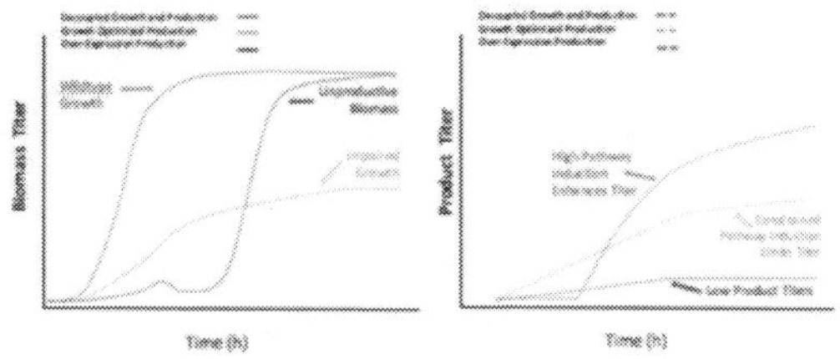

The First step needed is bioprocess friendly, **auto-inducible switch**

Two different strategies to integrate sensing into the Toggle Switch

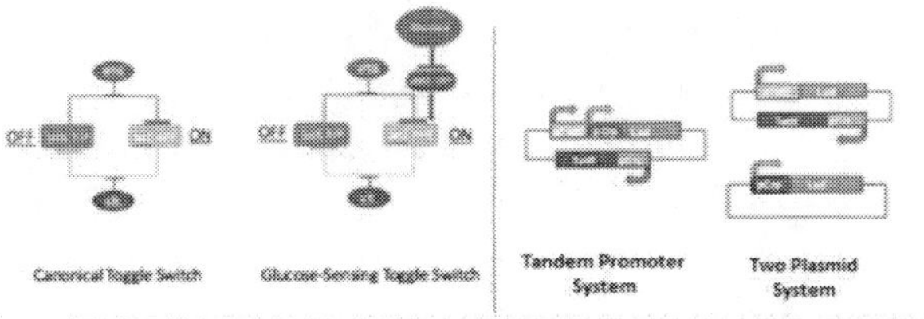

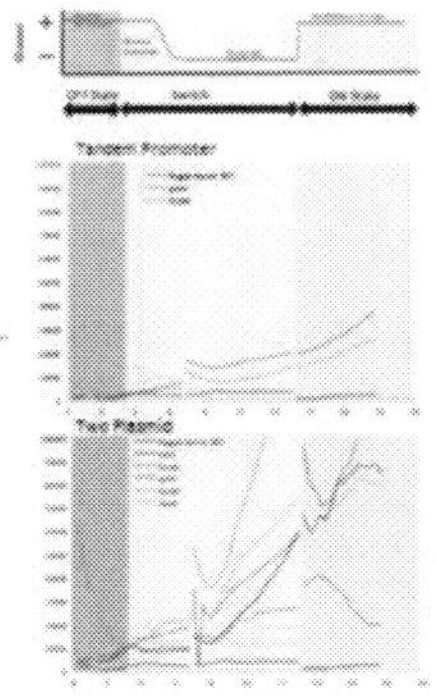

Strategy

1. Test Glucose Sensitive Promoters
2. Integrate Glucose sensing into the Toggle Switch in
3. Produce burdensome (PHB) pathway

PHB Production Results

1. Benchmarked vs. 29 copy constitutive Strain
2. Compared vs. varying chemical induction times
3. Tandem and Two plasmid systems both work

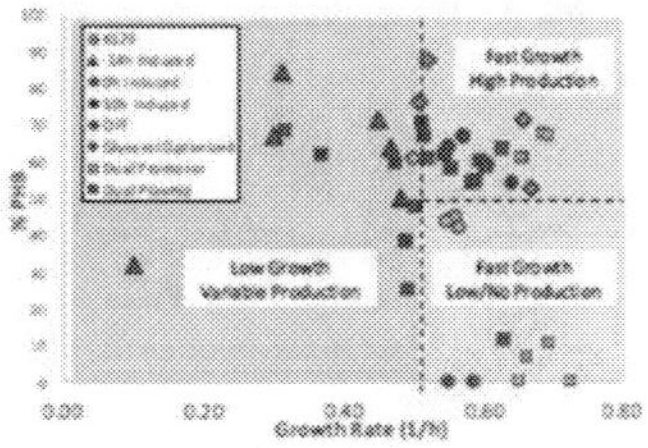

Key Results

- Developed an **auto-inducible genetic switch** that responds to **glucose availability** to precisely express PHB operon
- To our knowledge, first demonstration of using tandem promoters to encode logic
- Easily transferrable to any number of systems

Advances in Genome Engineering Techniques Enable Fast and Efficient Construction of Microbial Cell Factories

Jakociunas, T., The Novo Nordisk Foundation Center for Biosustainability
Jensen, M. K., Technical University of Denmark
Keasling, J. D., Joint Bioenergy Institute

One of the key components for successful construction of microbial cell factories is the speed and efficiency by which genome can be manipulated to produce desired compounds. In recent years, advances in synthetic biology and genome-engineering techniques, enabled much faster and efficient integration or removal of key components from the genome, hence allowing for rapid construction and optimization of metabolic pathways in host organisms. Here, we present various techniques based on CRISPR/Cas9 for yeast *Saccharomyces cerevisiae* genome engineering. Developed techniques, allow rapidly and in multiplexed manner integrate, delete, *in vivo* assemble/integrate components required for production of high value chemicals. Further, we developed and validated a set of vectors capable of carrying multiple gRNA expression cassettes and allowing easy exchange of them for targeting up to five desired genomic sites simultaneously. In addition, we employed CRISPR/Cas9 for protein engineering, which allows screening for more robust or higher specificity enzymes. All together, these genome manipulation techniques, allow for building and testing of various metabolic pathways in a faster, marker-free, multiplex, cost-effective manner and advances the design of yeast cell factories further.

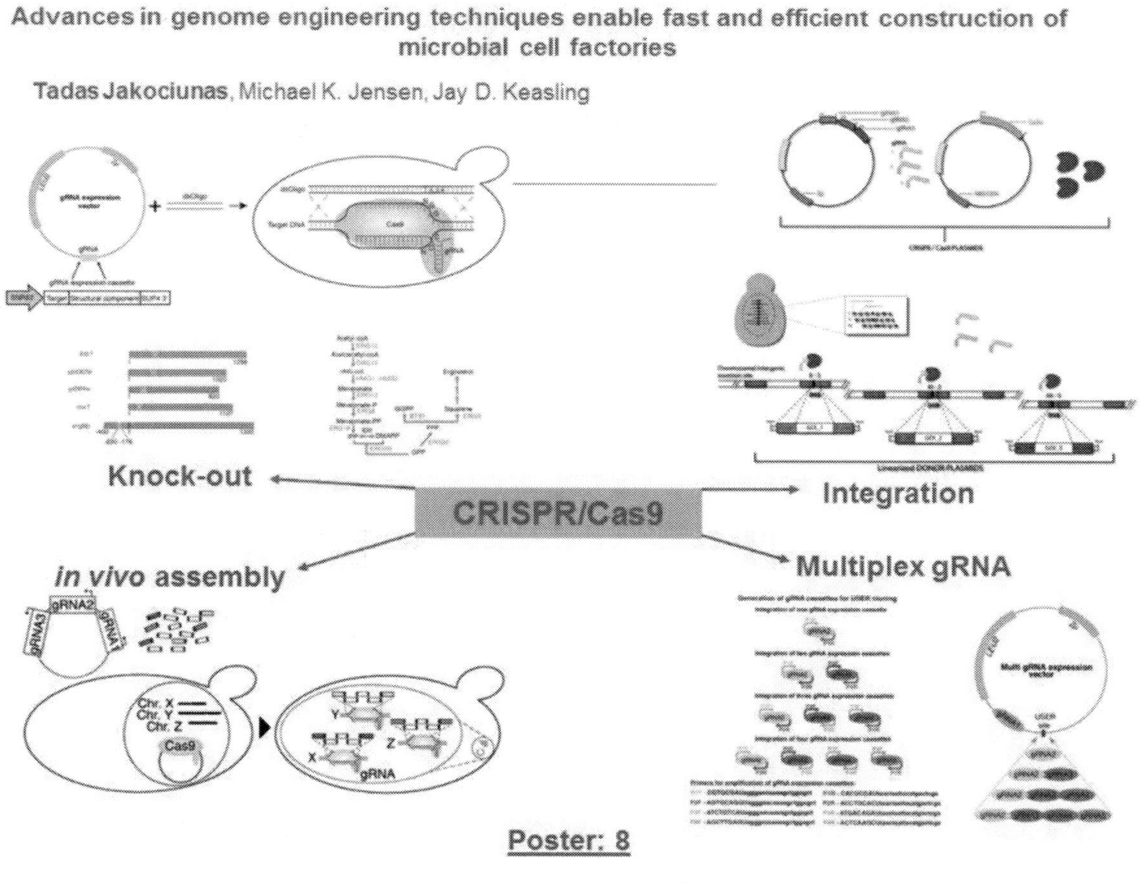

Poster: 8

Gamma-Aminobutyric Acid Production Via GABA Shunt By the Introduction of Synthetic Scaffolds in Recombinant Escherichia coli

Hong, S., University of Ulsan

Nylon 4 is a biodegradable polymer which can be produced from the monomer of pyrrolidone. Gamma-aminobutyric acid (GABA) is a precursor of pyrrolidone used for the production of bioplastics. In this study, Escherichia coli were engineered to produce gamma-aminobutyric acid from glucose via an alternative novel pathway by the introduction of synthetic scaffolds. The GABA pathway constructed contained succinate dehydrogenase, succinate-semialdehyde dehydrogenase and GABA aminotransferase to redirect the Krebs cycle flux to GABA production. By introduction of a synthetic scaffold, production of 0.64 g/L GABA was achieved at 30oC and pH 6.5. Final GABA concentration was increased by 11.3% via the inactivation of competing pathways, and higher initial glucose concentration led to the enhanced final GABA concentration of 1.01 g/L. This work was supported by a grant from the Next-Generation BioGreen 21 Program (SSAC, grant number: PJ01111601) by RDA, and Basic Science Research Program by the Ministry of Education (NRF-2014R1A1A2054726).

Iso-Propanol Production By Engineered Cyanobacteria, Synechococcus Elongatus pcc 7942

Hirokawa, Y., Kyushu University
Hanai, T., Kyushu University

Cyanobacteria converting carbon dioxide into organic compounds using solar energy is an attractive host for bio production. The introduction of synthetic metabolic pathway composed of multiple genes derived from other organisms enables some cyanobacteria to produce valuable chemicals and fuels. In this study, we used *Synechococcus elongatus* PCC 7942 as host strain and firstly accomplished iso-propanol production from carbon dioxide by engineered strain.

A synthetic metabolic pathway composed of acetyl-CoA acetyl transferase (encoded by *thl*), acetoacetyl-CoA transferase (encoded by *atoAD*), acetoacetate decarboxylase (encoded by *adc*), and secondary alcohol dehydrogenase (encoded by *sadh*) was introduced into *S. elongatus* PCC 7942. The extrinsic acetate, precursor of acetyl-CoA, enabled the constructed strain to produce iso-propanol, indicating that the introduced pathway is functional in *S. elongatus* PCC 7942. Although no iso-propanol was produced under light condition, iso-propanol and acetate were detected under dark and anaerobic conditions. Therefore, two-phase incubation was applied for iso-propanol production. At first, the constructed strain was grown under light condition (growth phase). Next, the grown culture was transferred to dark and anaerobic conditions (production phase). The culture at early stationary phase in growth phase was optimal for iso-propanol production with acetate accumulation. The shifting dark and anaerobic to light condition during production phase dramatically increased iso-propanol production by conversion to iso-propanol from acetate. After optimization of the conditions, 2.42 mM (146 mg/l) of iso-propanol was produced.

The transition to dark and anaerobic conditions is complicate process and not preferable for practical application. It was expected that the iso-propanol production under light condition is accomplished by self-production of acetate. The introduction of phosphate acetyltransferase (encoded by *pta*) catalyzing reversible conversion between acetyl-CoA and acetyl phosphate led *S. elongatus* PCC 7942 to produce acetate after stationary phase. To prevent produced iso-propanol evaporation, the strain introduced both a synthetic metabolic pathway and *pta* was incubated in tight-capped flask with addition of $NaHCO_3$. Under these conditions, the constructed strain successfully produced 0.55 mM (33.1 mg/l) of iso-propanol. The enforcement of metabolic flux to acetyl-CoA would be necessary for more improvement of productivity.

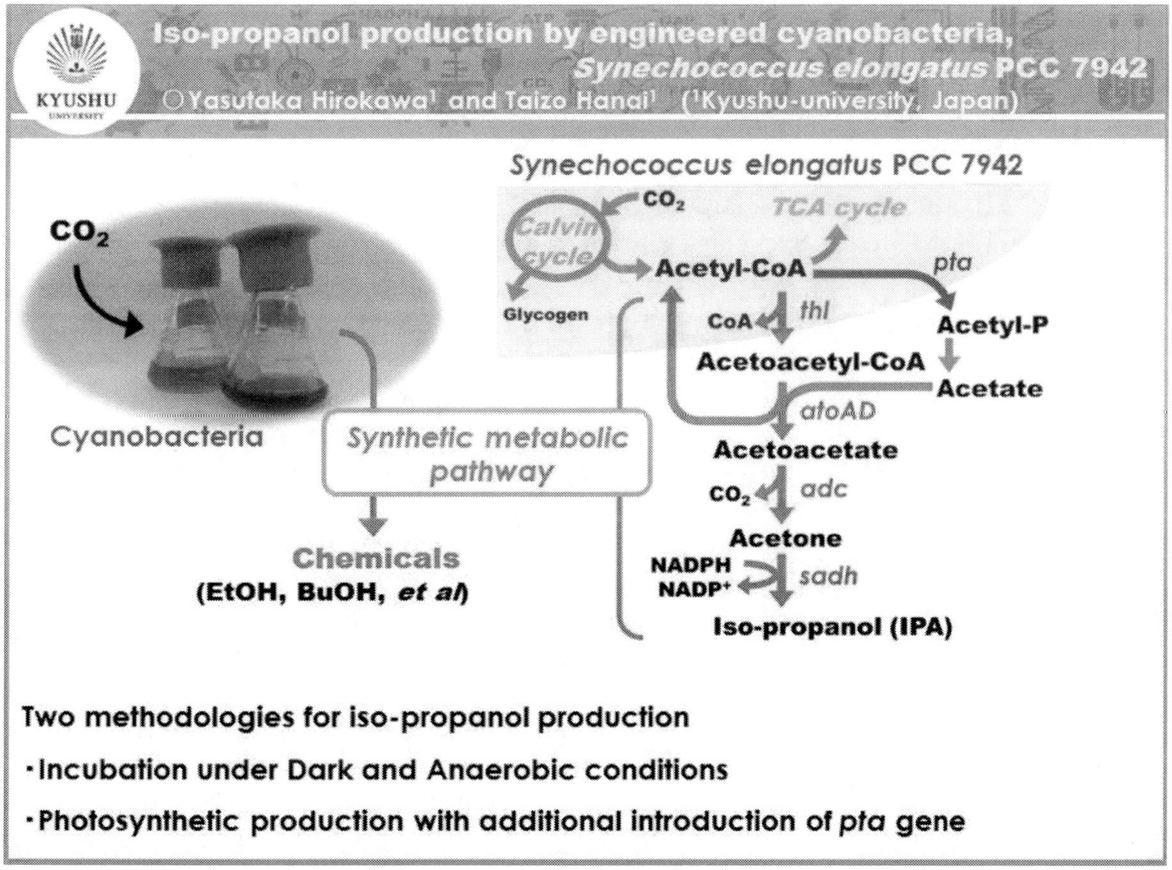
Iso-propanol production by engineered cyanobacteria,
Synechococcus elongatus PCC 7942
○Yasutaka Hirokawa[1] and Taizo Hanai[1] ([1]Kyushu-university, Japan)
KYUSHU UNIVERSITY
Synechococcus elongatus PCC 7942
CO2
Cyanobacteria
Synthetic metabolic pathway
Chemicals
(EtOH, BuOH, et al)
Calvin cycle
CO2
TCA cycle
Acetyl-CoA
pta
Glycogen
CoA
thl
Acetyl-P
Acetoacetyl-CoA
Acetate
atoAD
Acetoacetate
CO2
adc
Acetone
NADPH
NADP+
sadh
Iso-propanol (IPA)
Two methodologies for iso-propanol production
・Incubation under Dark and Anaerobic conditions
・Photosynthetic production with additional introduction of pta gene

Engineering of an NADPH/NADP+ Redox Sensor in Yeast

Zhang, J., Technical University of Denmark
Sonnenschein, N., Technical University of Denmark
Pihl, T. P. B., Technical University of Denmark
Pedersen, K. R., Technical University of Denmark
Herrgård, M. J., Technical University of Denmark
Jensen, M. K., Technical University of Denmark
Keasling, J. D., Lawrence Berkeley National Laboratory

Biosensors have emerged as powerful tools for timely and precise *in vivo* diagnosis and selection. In particular, biosensors that can couple intercellular cues with downstream signaling responses are currently attracting major attention within health science and biotechnology. In this study, we report the engineering and application of a genetically encoded NADPH/$NADP^+$ redox biosensor in the budding yeast *Saccharomyces cerevisiae*. Using the biosensor, we are able to monitor the NADPH redox state and experimentally validate genetic mutants predicted to have changed NADPH availability. We also show that the sensor can regulate the expression of an $NADP^+$-oxidoreductase to suppress NADPH deficiency observed in a mutant with severely impaired NADPH regeneration. Finally, we couple the biosensor with expression of novel native selector genes, which can be used for selection of cells with higher NADPH/$NADP^+$ ratios. Taken together, we show that hitchhiking and rational engineering of native signaling components can be applied for engineering of biosensors applicable to both monitoring and timely regulation of cell redox state. Such biosensors should be of interest to the metabolic engineering community for the facile identification of platform strains with high cofactor requirements.

Engineering of an NADPH/NADP$^+$ Redox Biosensor in Yeast

Jie Zhang,[1] Nikolaus Sonnenschein,[1] Thomas P. B. Pihl,[1] Kasper R. Pedersen,[1] Michael K. Jensen,[1,*] Jay D. Keasling[1,2,3,4,5]
[1] The Novo Nordisk Foundation Center for Biosustainability, Technical University of Denmark,
[2] Joint BioEnergy Institute, [3] Biological Systems & Engineering Division, Lawrence Berkeley National Laboratory, [4] Department of Chemical and Biomolecular Engineering & [5] Department of Bioengineering, University of California, Berkeley

Sensing Mechanism

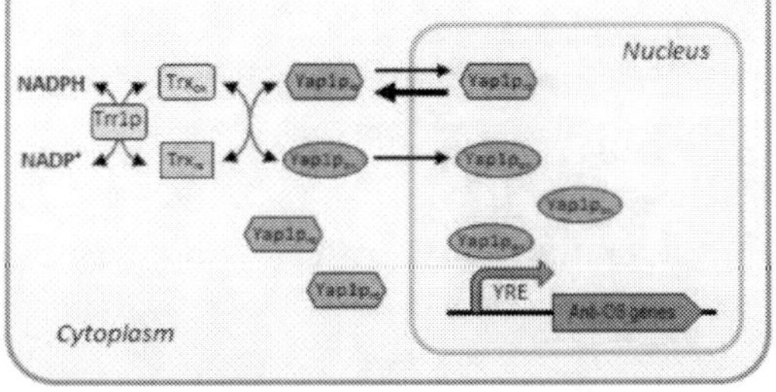

Sensor-Actuator

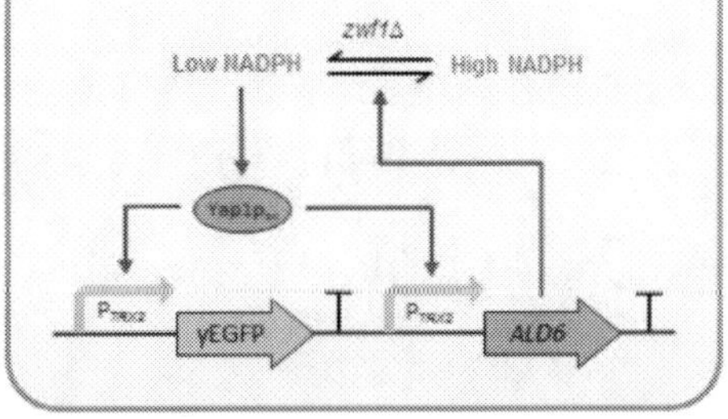

Sensor-Reporter

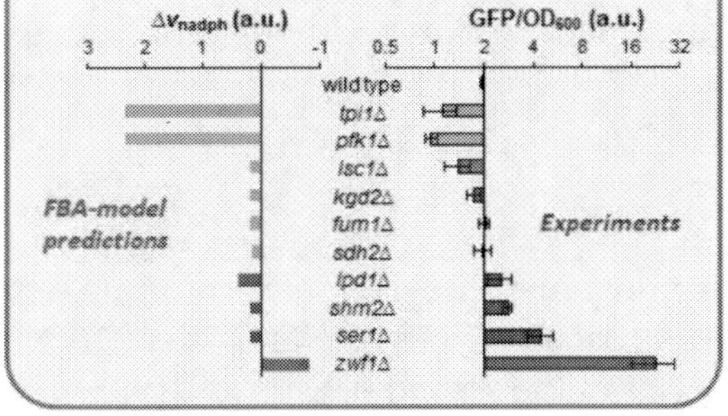

Sensor-Selector

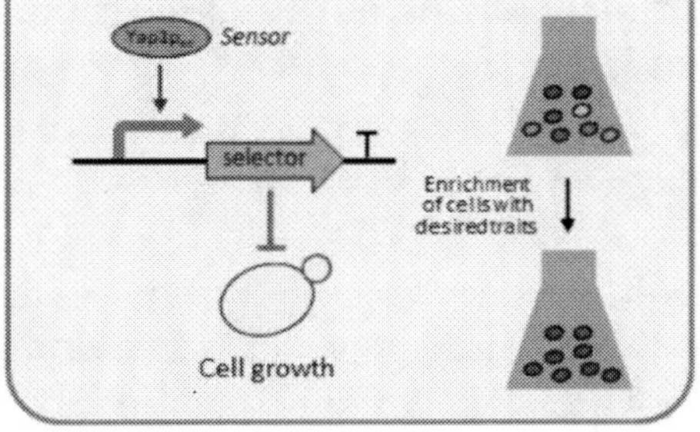

Enhanced Integration of Large DNA into E. coli Chromosome By Crispr/Cas9

Li, H., National Tsing Hua University
Sung, L. Y., Tsing Hua University
Wu, M. Y., National Tsing Hua University
Huang, C. H.,

Enhanced Integration of Large DNA into *E. coli* Chromosome by CRISPR/Cas9

Mu-En Chung[1§], I-Hsin Yeh[1§], Hung Li[1], Li-Yu Sung[1], Meng-Ying Wu[1],Chun-Hung Huang[1] Yun-Peng Chao[2], I-Son Ng[3], Yu-Chen Hu[1]*

[1] Department of Chemical Engineering, National Tsing Hua University, Hsinchu, Taiwan 300

[2] Department of Chemical Engineering, Feng Chia University, Taichung, Taiwan 407

[3] Department of Chemical Engineering, National Cheng Kung University, Tainan, Taiwan 701

Abstract

Metabolic engineering often necessitates chromosomal integration of multiple genes but integration of large genes into *E. coli* remains difficult. CRISPR/Cas9 is an RNA-guided system which enables site-specific induction of double strand break (DSB) and programmable genome editing. Here we hypothesized that CRISPR/Cas9-triggered DSB could enhance homologous recombination and augment integration of large DNA into *E. coli* chromosome. We demonstrated that CRISPR/Cas9 system was able to trigger DSB in >98% of cells, leading to subsequent cell death, and identified that mutagenic SOS response played roles in the cell survival. Acting in concert with the l-Red proteins and linear dsDNA, CRISPR/Cas9-induced DSB enabled homologous recombination of the donor DNA and replacement of *lac*Z gene in the MG1655 strain at efficiencies up to 99%, and allowed high fidelity, scarless integration of 2.4, 3.9, 5.4 and 7.0 kb DNA at efficiencies approaching 91%, 92%, 71% and 61%. The CRISPR/Cas9-faciliated gene integration also functioned in different *E. coli* strains including BL21 (DE3) and W. Our methodology facilitated precise integration of dsDNA as large as 7 kb into *E. coli* with efficiencies exceeding 60%, thus significantly ameliorating the editing efficiency and overcoming the size limit of integration using the commonly adopted recombineering approach.

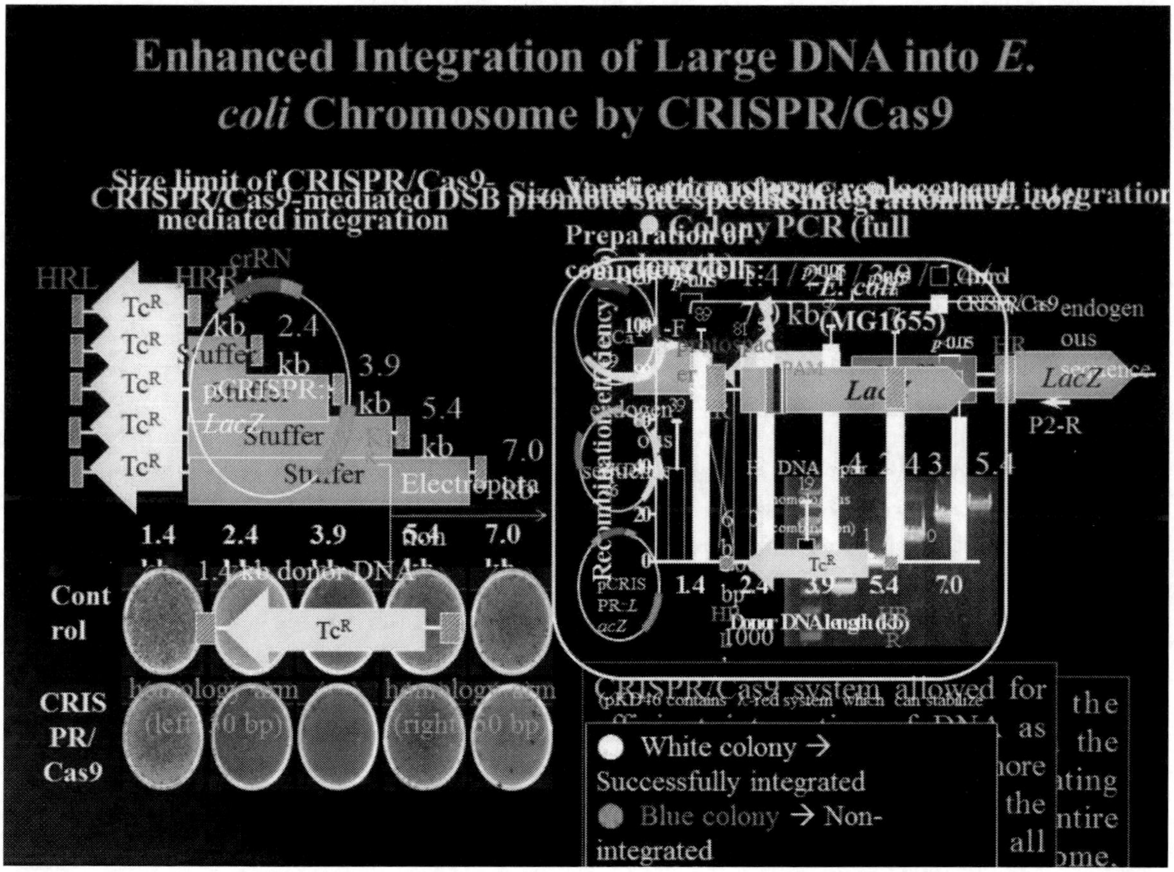

Enhanced Integration of Large DNA into E. coli Chromosome by CRISPR/Cas9
Size limit of CRISPR/Cas9-mediated integration
1.4
2.4
3.9
5.4
7.0
Control
CRISPR/Cas9
Stuffer
White colony → Successfully integrated
Blue colony → Non-integrated
Donor DNA length (kb)
LacZ
P2-R

Bioengineering Magnetic Nanoparticles in Magnetotactic Bacteria

Furubayashi, M., Chiba Univ.
Voigt, C. A., University of California, San Francisco

Background & Objective

Magnetic nanoparticles (MNPs) are in great demand in diverse modern technologies such as high-density data storage, electromagnetic shielding, and various applications for biomedicine including targeted drug delivery or hyperthermia. Compared to bulk magnets, MNPs shows unique and interesting properties, which largely dependent on their size, shape or crystal composition. The strong interest in the filed is to produce uniform-sized MNPs with desired properties required for each technological application. However, chemical synthesis of structurally perfect MNPs with narrow size distributions still remains a challenge [1]. In contrast, magnetic nanoparticles biomineralized by natural organisms are known to show narrow size distributions, high crystal perfections that could be biosynthesized under environment-friendly reaction conditions.

Our goal is to design and produce MNPs with various properties by genetically engineering *Magnetospirillum magneticum* AMB-1, a bacteria that can biosynthesize chains of magnetic nanoparticles (called magnetosomes). Magnetosomes are composed of magnetite (Fe_3O_4) nanoparticles with remarkably narrow size/shape distribution. By modifying bacterial MNPs from bottom-up and retaining its natural ability to produce them with narrow size distributions, we envision the production of "designer magnetic nanoparticles" which is otherwise difficult to achieve.

Results

Characterizing genetic parts for magnetotactic bacteria. Magnetospirillums are microaerobic alphaproteobacteria with 6-hour doubling time, and culturing this bacteria is known to be difficult due to several reasons, including the cell's genomic instability and their fastidious and microaerobic mode of growth. Several efforts have been reported to optimize and develop a genetic system for this strain [2]. We have engineered and characterized basic genetic parts for *M. magneticum* to further expand the genetic toolkit. First, we have characterized stable plasmids in this strain, which became useful to further characterize following genetic parts. Second, we have constructed and characterized several constitutive promoters which spanned 25-fold range of expression, and inducible promoters that showed high dynamic range, up to 260-fold.

Tuning the magnetic properties of magnetosome-MNPs. For many magnetic applications, maximized magnetic moment per particles is one of the ideal properties of MNPs. This could be achieved by several strategies such as: enlarging their size up to the single-domain limit, elongating their shape, or doping heavy metals such as cobalt or nickel. To explore these possibilities, several studies have reported their effort to change the size or composition of bacterial MNPs [3]. We used the genetic tools that we have constructed to tune the expression of magnetosome-related genes, such as *mamGFDC* or *mms6* cluster, to explore the landscape of expression level vs size/shape relationship. In the presentation, we will also discuss our attempts to decorate the surface or modify the chemical composition of MNPs.

References

1) Prozorov T et al. *Mater Sci Eng R* **74,** 133–172 (2013).
2) Schultheiss D & Schuler D. *Arch Microbiol* **179,** 89–94 (2003), Okamura Y *et al. Appl Environ Microbiol* **69,** 4274–4277 (2003).
3) Staniland S *et al. Nat Nanotech* **3,** 158–162 (2008), Tanaka M *et al. J Biol Chem* **286,** 6386–6392 (2011).

Computational Design Strategies for Synthesizing Fusion Proteins and Enhancing Their Expression

Lee, D. Y., National University of Singapore

We developed cost-effective computational platforms, SynLinker [1] and COOL [2,3], for rationally designing synthetic fusion proteins and enhancing their expression in hosts, respectively. These platforms, in conjunction with other computational tools, will facilitate synthetic fusion protein design and production for diverse biotechnological, biopharmaceutical and biomedical applications. In this talk, I will present a general workflow and working principles of *in silico* designing synthetic gene and fusion proteins [4] and demonstrate its applicability via several case studies. Current challenges and future direction are also discussed.

References:

[1] Liu, C., J. X. Chin, and D.-Y. Lee. 2015. SynLinker: an integrated system for designing linkers and synthetic fusion proteins. *Bioinformatics*, **31**(22): 3700-3702.

[2] Chung, B. K. S. and D.-Y. Lee. 2012. Computational codon optimization of synthetic gene for protein expression. *BMC Syst. Biol.*, **6**: 134.

[3] Chin, J. X., B. K.-S. Chung and D.-Y. Lee. 2014. Codon Optimization On-Line (COOL): a web-based multi-objective optimization platform for synthetic gene design. *Bioinformatics*, **30**(15): 2210-2212.

[4] Yu, K., C. Liu, B.-G Kim and D.-Y. Lee. 2015. Synthetic fusion protein design and applications. *Biotechnol. Adv.*, **33**: 155-164.

Computational Design Strategies for Synthesizing Fusion Proteins and Enhancing Their Expression

Dong-Yup Lee (NUS & BTI)

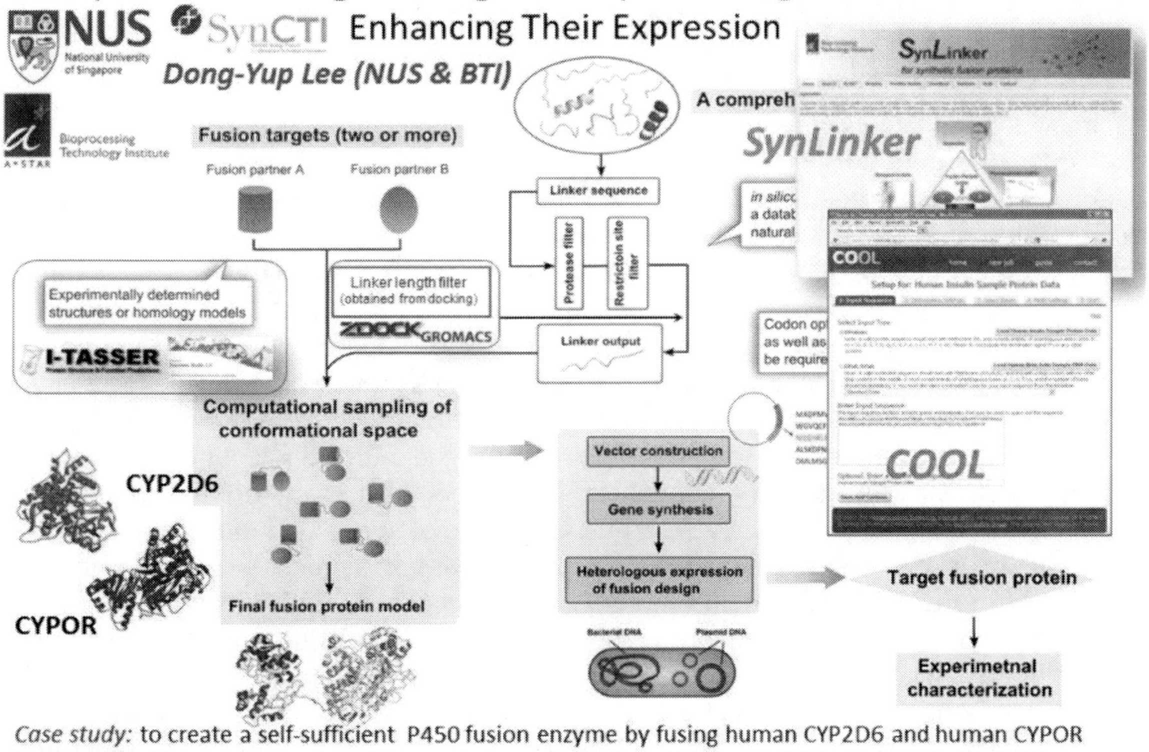

Case study: to create a self-sufficient P450 fusion enzyme by fusing human CYP2D6 and human CYPOR into one single polypeptide via a linker; to express the fusion enzyme in Pichia strain with efficient catalytic function as a drug screening tool.

Expanding the Substrate Range and Product Portfolio of Y. Lipolytica through Metabolic Pathway Engineering and Synthetic Biology Tool Development

Markham, K., The University of Texas at Austin
Alper, H., The University of Texas at Austin

Yarrowia lipolytica is an industrially attractive, non-conventional yeast due to advantageous traits such as its ability to bypass tight pathway regulation, grow on hydrocarbons and other diverse carbon sources, grow at lower temperatures, and tolerate natural inhibitors like salts. Several efforts of metabolic engineering have demonstrated the flexibility and potential of this platform. As an example, our lab has demonstrated the lipogenesis potential of this yeast by engineering a strain capable of accumulating greater than 90% dry cell weight lipid with titers exceeding 40 g/L. This high lipogenic potential, and more broadly, the high flux through CoA precursor molecules and key TCA cycle intermediates, makes this organism an attractive host for the production of a wide breadth of chemicals. However, several tools and pathways require improvements to further engineering capacity in this host. To further harness the full metabolic potential of Y. lipolytica, we have established a suite of synthetic biology tools to facilitate metabolic engineering of this host. We demonstrate here some recent successes in this effort to create synthetic promoters and terminators capable of tuning gene expression. Initial efforts from our lab have focused on a hybrid promoter approach capable of varying gene expression by tuning the number of repeated upstream activating sequences. Additionally, a collection of synthetic terminators expands this toolset to create tunable expression at levels that were previously unachievable in this yeast. More recent efforts from our lab, including creating synthetic transcription control elements, will be described. Finally, we demonstrate the utility of these parts and approaches through several applications. First, we developed a workflow to enhance utilization of alternative carbon substrates in Y. lipolytica. We demonstrate this through a case study of producing lipids (suitable for biodiesel) from sorghum syrup. Second, we demonstrate a diversified product portfolio demonstrating that high levels of CoA precursor molecules can be converted and rewired into valuable commodity and specialty chemicals including modified fatty acids, organic acids, and other industrially attractive small molecules at high titers (i.e. g/L levels). To this end, our expanded synthetic biology toolkit enables both an enhanced substrate range and expanded product portfolio for this industrially attractive host organism.

Expanding the Substrate Range and Product Portfolio of *Y. lipolytica* through Metabolic Pathway Engineering and Synthetic Biology Tool Development

Kelly A. Markham and Hal S. Alper

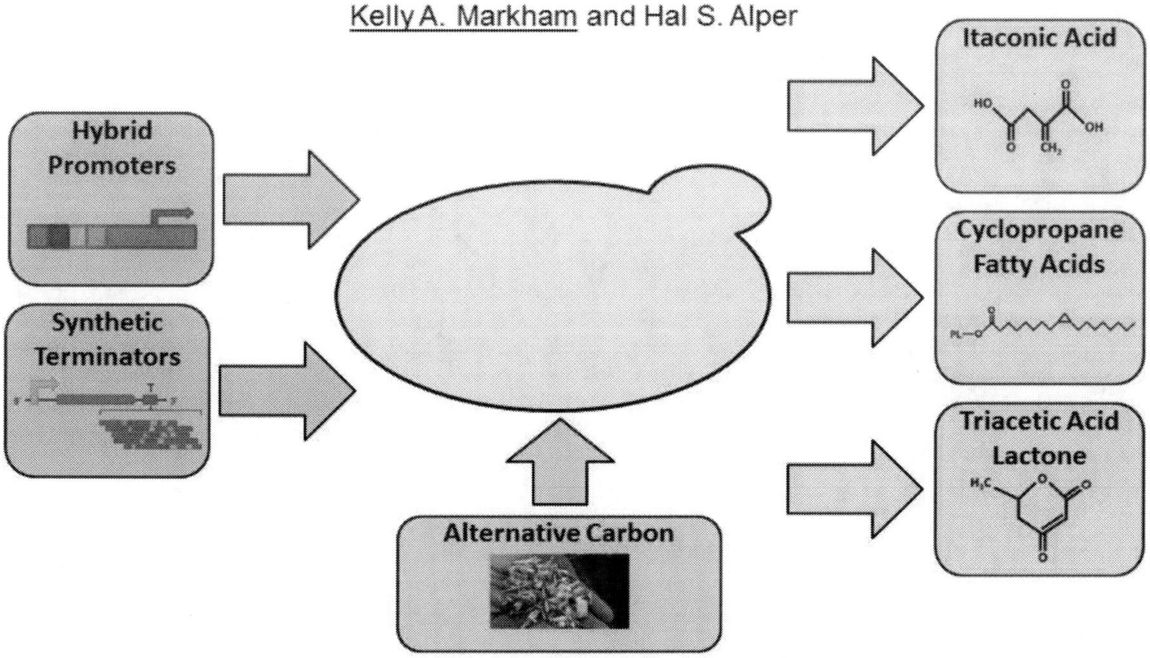

Fluorescent Biosensors for In Vivo Single-Cell Analysis of the Physiology of Saccharomyces Cerevisiae

Brink, D. P., Lund University

Tufvegren, C.,

Tueros, F. G.,

Gorwa-Grauslund, M. F., Department of Applied Microbiology, Lund University

Baker's yeast *Saccharomyces cerevisiae* is a main protagonist of modern applied and industrial biotechnology due to its robustness, well-established cultivation requirements and, last but not least, its highly plastic and engineering-friendly genome. Despite the vast knowledge of this model organism, many aspects regarding the physiology of this yeast still remain to be uncovered - especially in the context of the modern (lignocellulose) biorefinery. This discrepancy calls for novel and rapid methods for screening of beneficial phenotypes, with down-stream applications ranging from genetic library screening and novel metabolic target discovery to systems biology and modelling. As such, *in vivo* biosensors have become a promising step towards closing the gap between the possibilities of metabolic engineering and the currently limited methods of phenotypical assessment.

To dissect the intracellular response of this yeast to its many contemporary biotechnological applications, we have in our lab developed a set of integrative biosensor cassettes based on the yeast enhanced Green Fluorescent Protein (yEGFP) gene [1]. By coupling this fluorescent gene to different endogenous yeast promoters, we have established a panel of non-intrusive biosensor strains that allows us to rapidly monitor the changes in physiology of *S. cerevisiae* to a multitude of industrially relevant conditions on a single-cell level by assessing the changes in the fluorescent phenotype of the engineered yeast over time. An example of a recent application of this biosensor system is a set of *S. cerevisiae* redox sensor strains designed to monitor the intracellular NADH/NAD$^+$ ratio based on transcriptional regulation of the yeast GPD2 promoter [2]. In addition to this and to our recent advances in this field, we present a recently developed protocol for high-throughput, single-cell screening of fluorescent *S. cerevisiae* biosensor strains using flow cytometry that comes with an integrated *in silico* data analysis pipeline.

References:

[1] Cormack, B. P., Bertram, G., Egerton, M., Gow, N. A., Falkow, S., & Brown, A. J. (1997). Microbiology, 143(2), 303-311.

[2] Knudsen, J. D., Carlquist, M., & Gorwa-Grauslund, M. (2014). AMB Express, 4(1), 81.

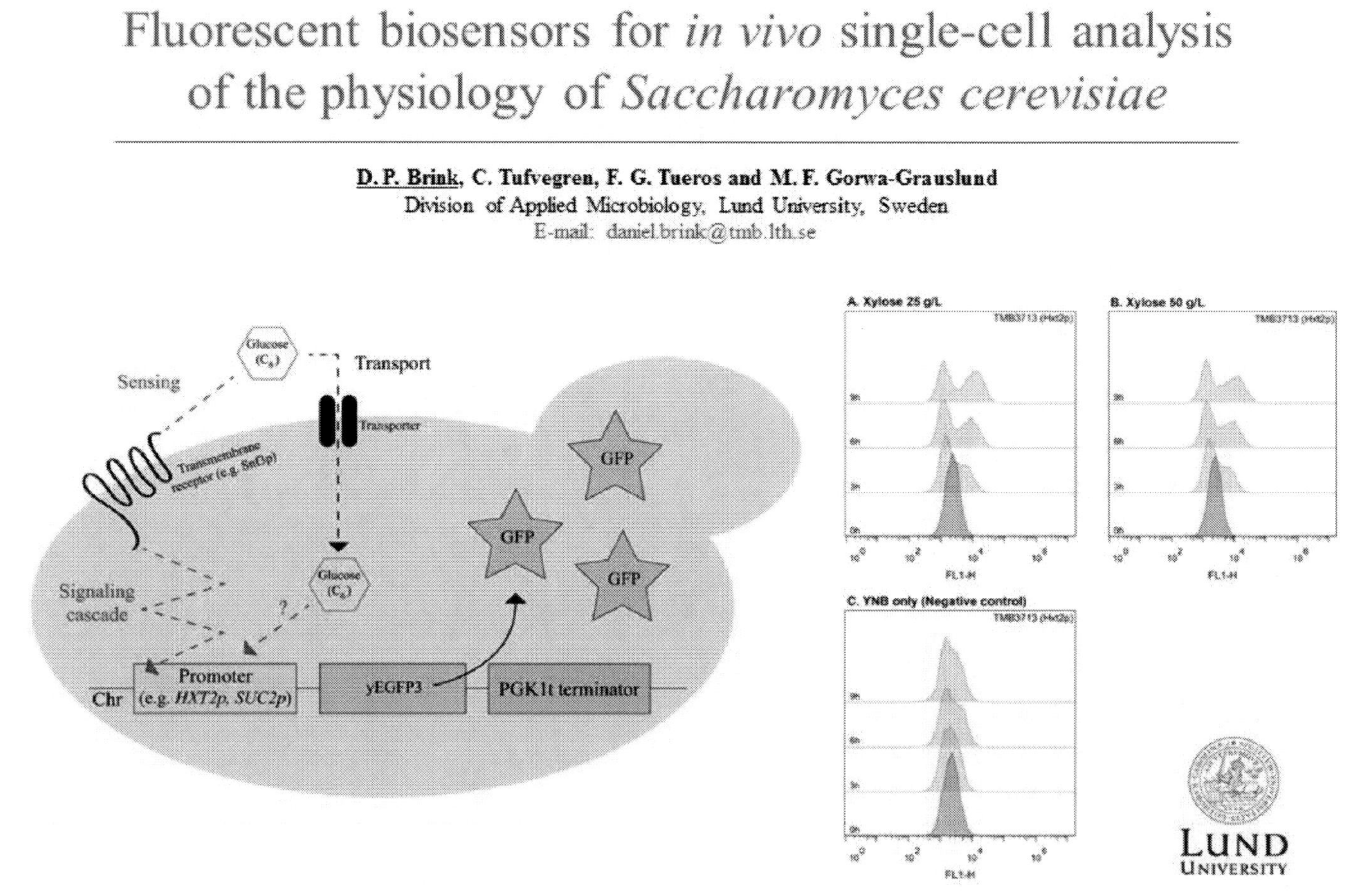

Synbio-Guided Metabolic Taming of Pseudomonas Putida

Nikel, P. I., Centro Nacional de Biotecnología (CNB-CSIC)
de Lorenzo, V., Centro Nacional de Biotecnología (CNB-CSIC)

The last few years have witnessed an exponential increase in the number of bacteria that can be used as alternative microbial cell platforms in several applications. Alas, the microorganisms which are the easiest to genetically manipulate (i.e., the so-called "model" bacteria, such as *Escherichia coli* or *Bacillus subtilis*) are often not adequate to perform given biotechnological applications (e.g., harsh oxidations or dehalogenation reactions). Much of the contemporary Synthetic Biology endeavors relies on the adoption of specific bacterial *chasses* for plugging-in and -out genetic circuits and engineer new-to-Nature functionalities. Attempts to design robust and predictable biological entities are reminiscent not only of the advances that accompanied the birth of modern engineering, but they also evoke some historical efforts to domesticate animal species for humans' sake. In this background, environmental bacteria, such as *Pseudomonas* strains, constitute ideal starting points to design flawless microbial platforms, since these microorganisms are pre-endowed with a large number of metabolic and stress-endurance traits which are optimal for biotechnological needs. Our recent developments on the taming of *P. putida* for biotechnological applications will be discussed in the context of Synthetic Biology strategies for [i] re-designing the metabolic architecture of central carbon catabolism and [ii] manipulating catalytic biofilms through Synthetic Morphology approaches.

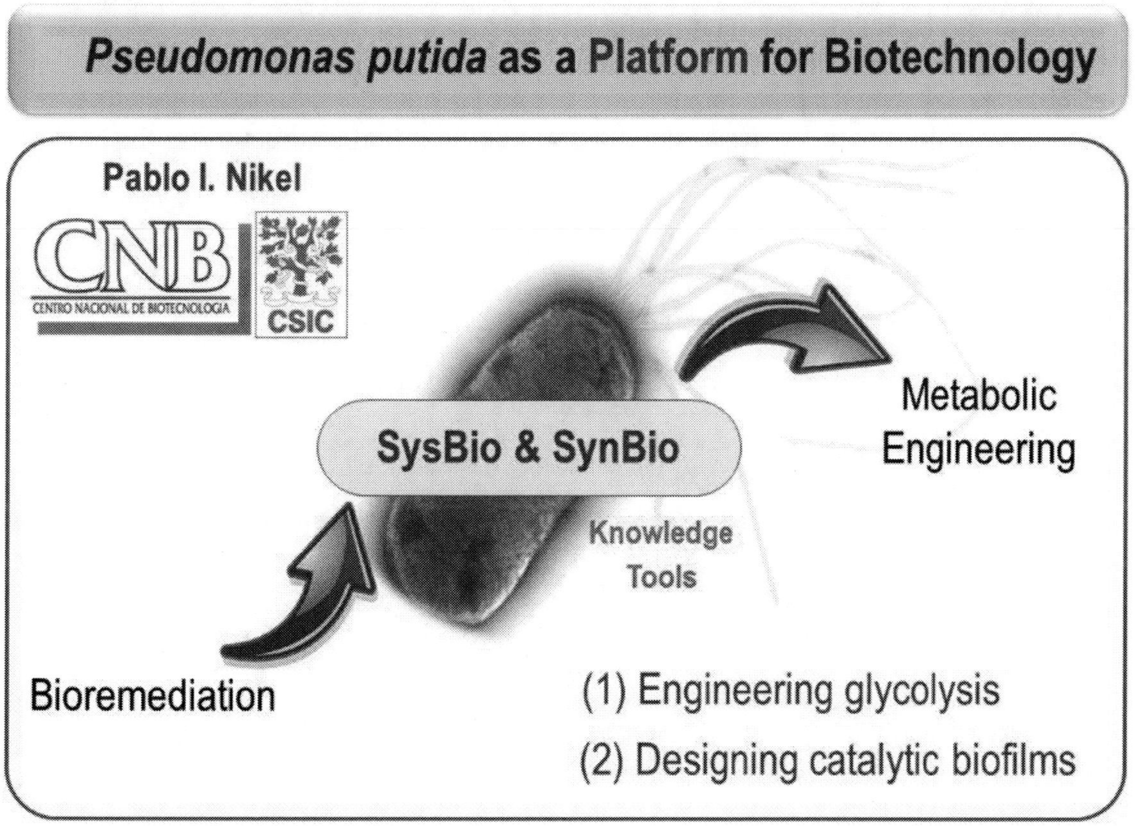

A Metabolic Valve of Glycolytic Flux for the Design of Optimal Cell Factory

Lim, J. H., POSTECH
Jung, G. Y., POSTECH

As an end metabolite of glycolysis in almost all organism, pyruvate is a critical intermediate from sugar uptake (i.e., glycolytic flux) to product formation and imbalance between glycolysis and product synthesis may accumulating pyruvate or reducing the productivity. Specifically, excess supply of pyruvate is inevitably secreted from the cell causing the significant reduction of yield whereas lack of pyruvate limits the product formation rate as the rate-limiting step. Therefore, balanced maximization of the pathway efficiency between glycolysis and product-forming pathway is required to construct the microbial cell factory with the maximum yield and productivity. Here, we propose synthetic UTR design of *ptsG* (encoding glucose transporter) is effectively controllable overall glycolytic flux to balance with the capacity of product formation pathway as a metabolic valve. Using the metabolic valve, optimal glycolytic flux was explored to maximize both yield and productivity of cell factory depending on the various efficiency of product-forming pathways, such as n-butanol, butyric acid, and 2,3-butanediol as model system. In result, tuning down of glycolytic flux was significantly reduced the leaking of pyruvate as major byproduct in n-butanol system, which increased 20% in yield while maximizing the parental productivity (84% of the theoretical maximum yield). Conversely, amplification of glycolytic flux transferred to the 7% and 12.5% higher productivity of butyric acid and 2, 3-butanediol, respectively, with the negligible yield change to the parental strain (around 83% and 60% of theoretical maximum, respectively). Our results demonstrated that fine control of the activity of PTS sugar transfer system allows a universal metabolic valve to accomplish balanced maximum metabolic performance for the maximum yield and productivity and consequently provides a breakthrough for designing the optimal microbial cell factory in the bio-based chemical industry.

A metabolic valve of glycolytic flux for the design of an optimal cell factory

Jae Hyung Lim[a] & Gyoo Yeol Jung[a,b]
[a] Department of Chemical Engineering, [b] I-Bio program, POSTECH, Korea

Motivation :
The discrepancy in pathway efficiency between glycolysis and product formation pathway

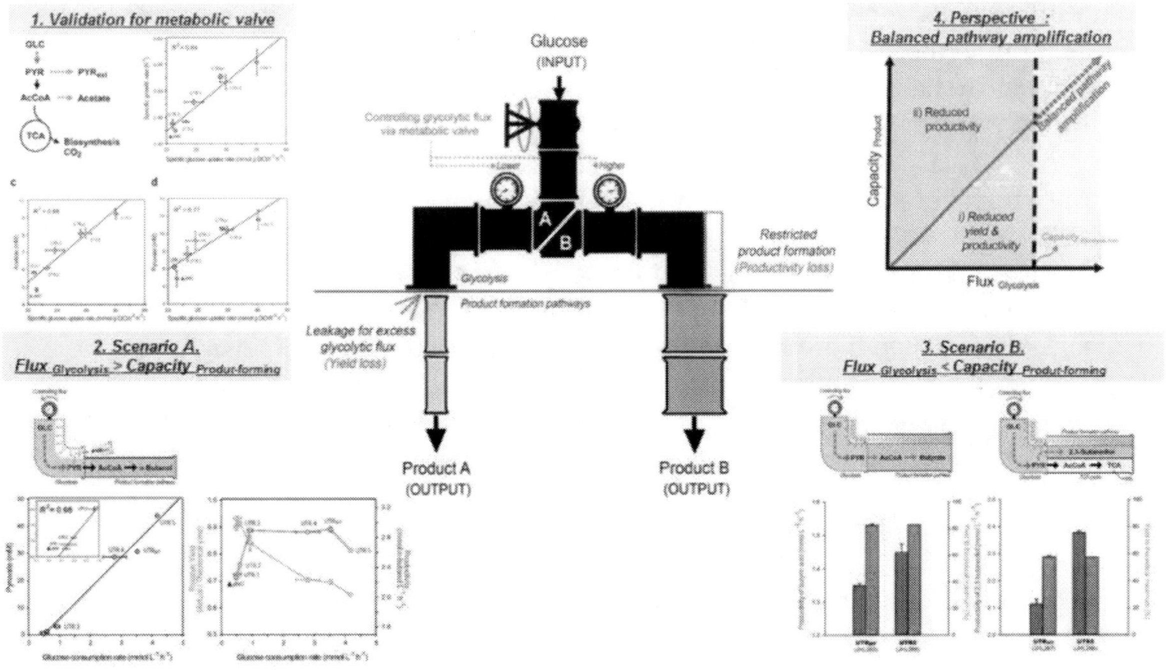

Expression of Native-Sized Spider Silk Protein in Escherichia coli through Metabolic Engineering and Synthetic Biology Approach

Chung, H., Korea Advanced Institute of Science and Technology (KAIST)
Xia, X. X., Jiao Tong University
Lee, S. Y., Korea Advanced Institute of Science and Technology (KAIST)

Naturally found spider silk and elastin protein attract attention due to its outstanding physical properties coming from the highly repeated structure and size. However, the exceptional structure and size limits expression in heterologous hosts, where the repetitive sequences in mRNA create extensive secondary structures. And these structures decrease ribosome processivity and assist mRNA degradation. Using the naturally found protein as a model we present techniques to solve biological problems that occurred: using metabolic engineering to supply aminoacyl-tRNA pool as well as decreasing the RNase activity to assist higher production of the spider silk protein. The newly synthesized native-size spider dragline silk protein produced highest titer in *Escherichia coli* than those reported previously. The results provide insight into approaches to control expression of useful recombinant proteins containing high molecular weight and highly repetitive sequence. [This work was supported by the Technology Development Program to Solve Climate Changes on Systems Metabolic Engineering for Biorefineries from the Ministry of Science, ICT and Future Planning (MSIP) through the National Research Foundation (NRF) of Korea (NRF-2012M1A2A2026556 and NRF-2012M1A2A2026557); the Intelligent Synthetic Biology Center through the Global Frontier Project (2011-0031963) of the Ministry of Education, Science and Technology (MEST) through the National Research Foundation of Korea]

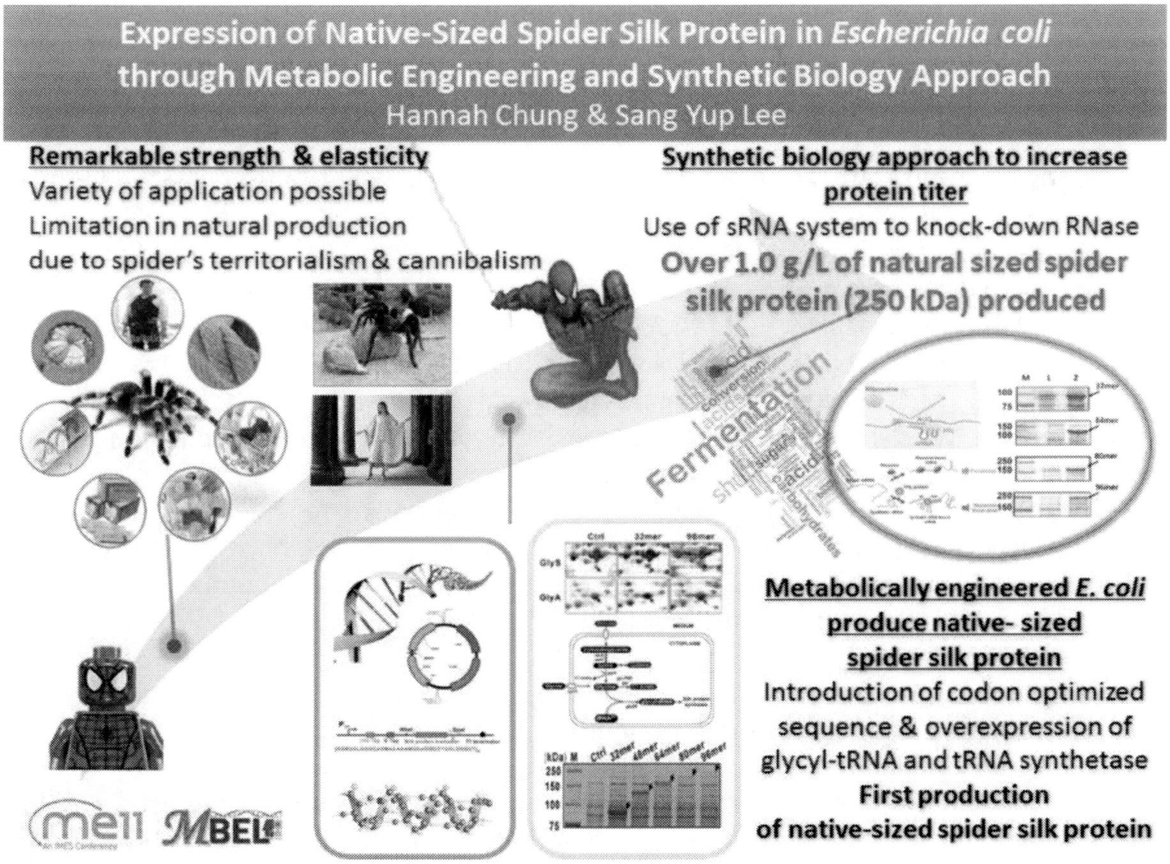

Expression of Native-Sized Spider Silk Protein in Escherichia coli
through Metabolic Engineering and Synthetic Biology Approach
Hannah Chung & Sang Yup Lee
Remarkable strength & elasticity
Variety of application possible
Limitation in natural production
due to spider's territorialism & cannibalism
Synthetic biology approach to increase protein titer
Use of sRNA system to knock-down RNase
Over 1.0 g/L of natural sized spider silk protein (250 kDa) produced
Fermentation
Metabolically engineered E. coli produce native- sized spider silk protein
Introduction of codon optimized sequence & overexpression of glycyl-tRNA and tRNA synthetase
First production of native-sized spider silk protein
GlyS
GlyA
Ctrl
32mer
96mer
kDa
250
150
100
75

Deriving Consistent Core Models from Genome-Scale Metabolic Networks and Their Application for Computing Metabolic Engineering Targets

Hädicke, O., Max Planck Institute for Dynamics of Complex Technical Systems
Klamt, S., Max Planck Institute for Dynamics of Complex Technical Systems

Stoichiometric and constraint-based modelling has become an essential tool for analysing properties and capabilities of metabolic networks and also for identifying suitable metabolic engineering targets. Up to now, more than 100 genome-scale metabolic networks have been reconstructed for different organisms and the process of compiling these models becomes faster and faster. However, with increasing size and complexity of genome-scale models, the computational effort for their analysis increased as well and some methods, such as elementary-modes analysis, ^{13}C flux analysis, or kinetic modelling cannot be applied to networks consisting of several thousand reactions. Furthermore, simulation results obtained with genome-scale models are sometimes difficult to interpret and studying basic principles of an organism's (central) metabolism might be easier with metabolic core models. An automated and unbiased reduction procedure delivering meaningful core networks from well-curated genome-scale reconstructions is therefore highly desirable.

For this purpose we have recently developed the *NetworkReducer* algorithm which derives core models (sub-networks) of genome-scale reconstructions by applying network pruning and network compression techniques [1]. *NetworkReducer* delivers reduced models that i) are consistent with the genome-scale network with respect to the parts maintained, ii) preserve defined key functionalities, and iii) are accessible for comprehensive and detailed analyses including full enumeration of elementary modes.

We demonstrate scope, predictive capabilities, and utility of a core model obtained by applying *NetworkReducer* to the latest genome-scale reconstruction of *E. coli* (iJO1366). We specified desired behaviours that have to be preserved by the reduced network. For example, the core model must allow for 99.9 % of the genome-scale optimal yields for biomass and selected metabolites and products.

The derived core model *EColiCore2.0* comprises 499 reactions and 519 metabolites. We present results of constrained-based analyses of *EColiCore2.0* (including flux variability, gene essentiality, and elementary-modes analysis) and relate them to results from the full model demonstrating the benefit of a reduced model being consistent with its genome-scale counterpart.

Further we analysed *EColiCore2.0* with respect to its capability to identify targets for metabolic engineering purposes. We used the concept of constrained minimal cut sets (MCSs) and enumerated in the genome-scale and in the core model reaction knockout sets for different engineering goals. These knockout sets were further analysed for commonalities and differences between the intervention targets found. Finally, we demonstrate how identified knockout sets of the core model can be used as seeds for computation of MCSs in genome-scale networks. This approach reduces the computational burden in large metabolic models considerably and enables the identification of intervention strategies with higher cardinalities in genome-scale networks with reasonable runtimes.

In conclusion, we demonstrate that consistent core models derived from genome-scale metabolic models may complement or/and simplify genome-scale analysis. *EcoliCore2.0* provides a valuable tool to analyse features of the central metabolism of *E. coli* and also supports the computer-aided search for intervention strategies for rational metabolic engineering.

[1] Erdrich P, Steuer R, Klamt S (2015) An algorithm for the reduction of genome-scale metabolic network models to meaningful core models. BMC Systems Biology 9:48.

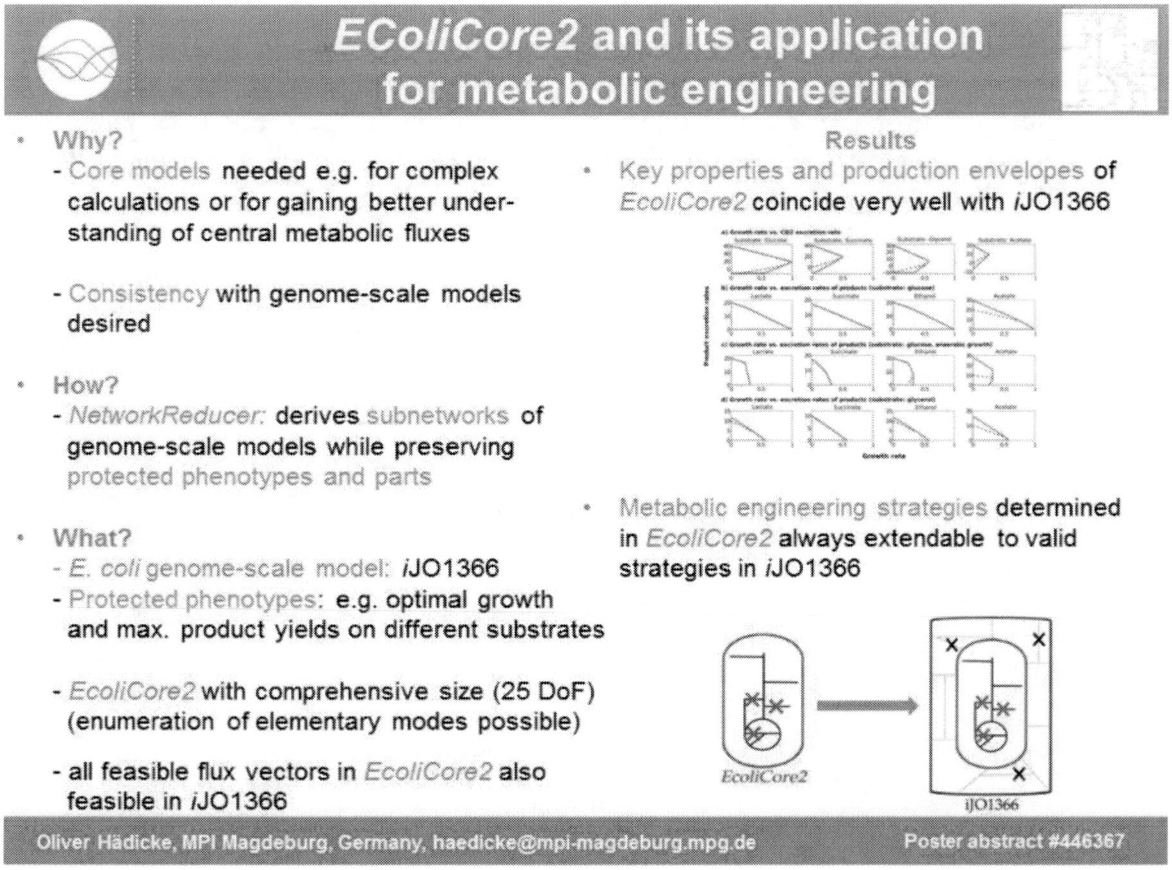

Rational and Evolutionary Engineering of Industrial Saccharomyces Cerevisiae Strains for Production of Chemicals from Xylose-Rich Feedstocks

Stovicek, V., Technical University of Denmark
Lis, A. V., Technical University of Denmark
Borodina, I., Technical University of Denmark
Förster, J., Technical University of Denmark

Biorefineries have been developed for industrial scale conversion of renewable biomass to gas, energy and various chemicals as an alternative to oil-based petrochemical industry. The biorefinery concept relies on robust and efficient microbial cell factories. Yeast *Saccharomyces cerevisiae* is one of the most used platform cell factories in the existing biorefineries. Second generation biorefineries process non-food agricultural or forest biomass and here efficient utilization of both C6 and C5 sugars is crucial. In this work we aim at developing of novel robust yeast cell factories for production of selected dicarboxylic acids that can be polymerised into bio-based polymers, from pentose-rich lignocellulose hydrolysates. Industrial yeast strains that are able to cope with the harsh environment in industrial settings have been selected as the production organism. We developed a molecular approach based on CRISPR-Cas9 (Stovicek et al., *Metab Eng Commun* 2015) for fast construction of gene disruptions and second generation of EasyClone vectors suitable for delivery and stable integration of genes into the genome of industrial yeast strains (Stovicek et al., *J Ind Microbiol* 2015). Such tools allow us to overcome genetic complexity of industrial strains and simplify the strain construction. Using the developed tools and adaptive evolution we constructed efficient xylose-fermenting industrial yeast strain performing in various media. Relevant metabolic and evolutionary engineering strategies will be discussed. This project is part of BioREFINE-2G (www.biorefine2g.eu), which is co-funded by the European Commission in the 7th Framework Programme (Project No. FP7-613771).

Rational and evolutionary engineering of industrial *Saccharomyces cerevisiae* strains for production of chemicals from xylose-rich feedstocks

Vratislav Stovicek, Alicia V. Lis, Irina Borodina, Jochen Förster

The Novo Nordisk Foundation Center for Biosustainability
Technical University of Denmark

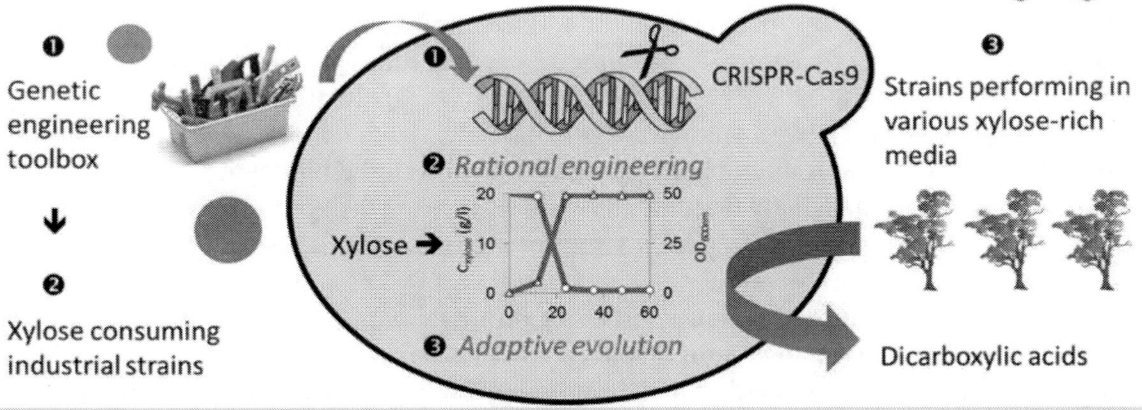

This project is co-funded by the European Union within the 7th Frame Programme. Grant Agreement n°613771.
The sole responsibility of this publication lies with the author. The European Union is not responsible for any use that may be made of the information contained therein.

Powering Nitrogen Fixation: the Metabolism of the Aerobic Diazotroph, Azotobacter Vinelandii, As Revealed By 13c-Metabolic Flux Analysis

Diaz, C. A. C., Stanford University
Antoniewicz, M. R., University of Delaware

Diazotrophs, or organisms capable of the enzymatic conversion of atmospheric nitrogen into ammonia, present a metabolic paradox. On the one hand, nitrogen fixation is an energy-intensive process, requiring 16 moles of ATP and 8 moles of reducing equivalents per mole of N_2 fixed. On the other hand, the nitrogenase enzyme is highly oxygen-labile, readily inactivated by oxidation of its metal cluster. Therefore, while an anaerobic lifestyle would circumvent the issue of the oxygen sensitivity of the nitrogenase, low ATP yields from fermentation present a significant limitation to the net ammonium output of anaerobic diazotrophs. Meanwhile, diazotrophs that are obligate aerobes (*e.g. Azotobacter vinelandii, Azorhizobium caulinodans*) are thought to cope instead by employing high respiration rates and protective membranes that limit the diffusion of oxygen. One of the most well studied diazotrophs, *Azotobacter vinelandii,* is an aerobic, fast-respiring bacterium found in soils globally. Although the biochemistry of this organism has been extensively documented, the features of its core metabolism that allow it to fix nitrogen and thrive in an aerobic environment have yet to be fully elucidated.

In this contribution, we have constructed a metabolic network model that captures the central carbon metabolism and amino acid biosynthesis pathways of *A. vinelandii*. The model was thoroughly validated by performing complementary parallel labeling experiments using all singly labeled glucose tracers, that is, [1-^{13}C]glucose through [6-^{13}C]glucose and measuring isotopic labeling of biomass by GC-MS. Subsequent ^{13}C-metabolic flux analysis (^{13}C-MFA) confirmed three key aspects of the metabolism in this organism, namely that *A. vinelandii*:

(1) exclusively employs the Entner-Doudoroff pathway for glucose catabolism.

(2) synthesizes alanine from an unusual pathway involving cysteine instead of pyruvate.

(3) diverts a significant portion of its carbon flux towards by-product formation and ATP generation rather than cell growth.

Additionally, for the first time, we present conclusive evidence that this diazotroph exhibits high flux through the TCA cycle during active nitrogen fixation, likely to supply reducing equivalents (and subsequently, ATP through the electron transport chain) for nitrogen fixation. Furthermore, we have observed fairly high growth rates for this microbe during active nitrogen fixation, even in minimal media (0.20 to 0.31 h^{-1}). Also, by comparing metabolic fluxes for wild-type *A. vinelandii* and an engineered *A. vinelandii* strain that overproduces ammonium we obtained additional insights into the metabolism that supports nitrogen fixation. Finally, by performing tracer simulations, we were able to identify [1,6-^{13}C]glucose paired with either [1,2-^{13}C]glucose or [3,4-^{13}C]glucose as the ideal tracers for routine studies of central carbon metabolism in *A.*

vinelandii.

The knowledge gained from studying the core metabolism that enables microbial nitrogen fixation in *A. vinelandii* offer valuable insight for future efforts to engineer diazotrophs as more environmentally and economically favorable alternatives to man-made sources of fixed nitrogen (*e.g.* the Haber-Bosch process) for industrial fermentation and agricultural applications.

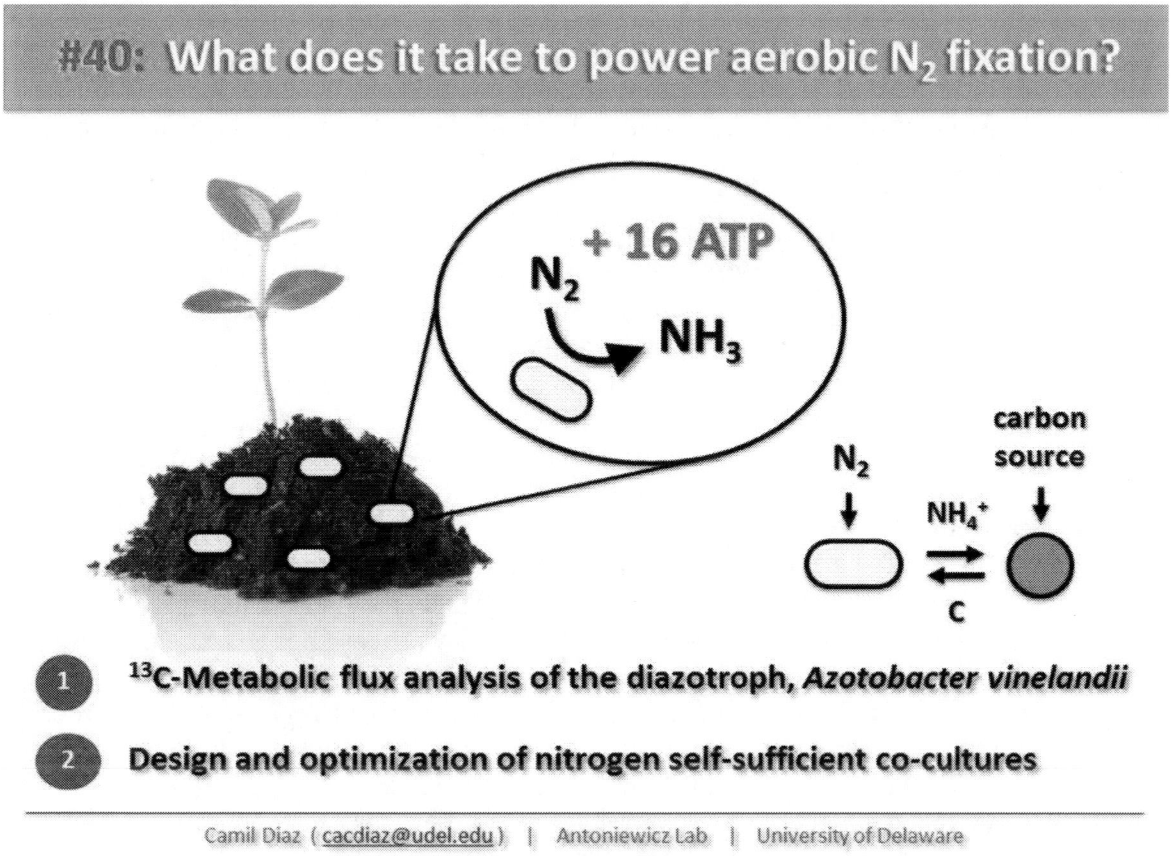

Identification and Engineering of n-Butanol Biosensors By Transcriptome Analysis in Yeast

Shi, S., Agency for Science, Technology and Research
Ang, E. L., Agency for Science, Technology and Research
Tan, M. H., Nanyang Technological University
Zhao, H., University of Illinois at Urbana-Champaign

Increasing concerns about depleting crude reserves have renewed interests in the biological production of n-butanol, due to its desirable fuel properties and its role as a platform chemical. There are advantages to utilize *Saccharomyces cerevisiae* as an n-butanol producer, due to its high n-butanol tolerance, phage resistance, well-established genetic tools and compatibility to current industrial infrastructures. However, current production level of n-butanol in *S. cerevisiae* is far less promising.

In vivo metabolite biosensors would significantly accelerate the construction and optimization of yeast producer via dynamic control of cell metabolism and serve as important high-throughput screening tools for the selection of the individuals carrying the desired phenotype. Thus, we propose to develop a systematic approach for the identification and engineering of biosensors to detect and regulate n-butanol in *S. cerevisiae* using transcriptomics analysis. The genome-wide gene expression profiling of *S. cerevisiae* exposed to n-butanol, n-propanol, and ethanol allowed us to identify novel genes responsive to the treatment of a specific alcohol. The gene hit list from each alcohol was cross-referenced with each other to identify genes that were only differentially expressed for n-butanol. We identified 435 genes that were differentially expressed only under the treatment of n-butanol rather than ethanol and n-propanol. Many genes were hypothetical, associated with ncRNA processing, DNA replication, response to DNA damage, steroid metabolism, and tRNA modification. Forty-eight promoters were selected and can be divided into two groups. One group of 24 promoters were selected because their genes had showed highest differential expression and stringent P value; the other group of 24 promoters were selected because they had showed conserved sequences in their promoter areas. For each group, promoters were selected within the region from 1 to 1000 bp upstream of translation start, and fused with the gene of mCherry fluorescent protein. The resultant reporter cassettes were introduced into *S. cerevisiae* to evaluate promoter strength and specificity for n-butanol pressure. Three promoters were shown to significantly drive the expression of mCherry upon exposure to n-butanol in a dose-dependent manner, with n-butanol concentration ranging from 0 to 1% (v/v). All these three promoters were from the group contained conserved sequences, highlighting the importance of data mining and processing. Upon further testing by truncating promoter areas, we have found sequences that contributed to the n-butanol response, indicating the possibility of re-design and re-construction of more efficient n-butanol sensors.

Our work demonstrates a general method for the discovery of endogenous promoters as biosensors to monitor alcohols, particularly n-butanol. This method should be generally applicable and highly useful for the development of biosensors in both high-throughput screening of large diversified libraries and implementation of synthetic circuits for dynamic control of cell metabolism.

Identification and engineering of n-butanol biosensors by transcriptome analysis in yeast

Shuobo Shi, Ee Lui Ang, Meng How Tan, Huimin Zhao

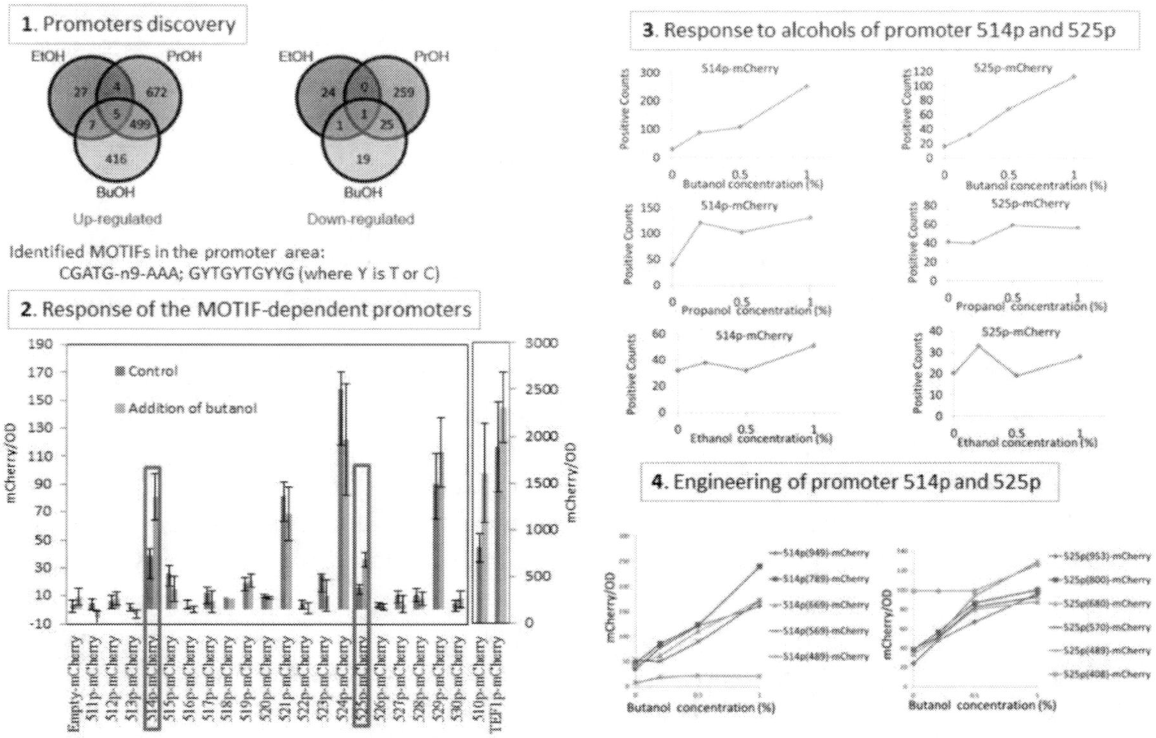

Multiple Gene Repression in Cyanobacteria Using Crispri

Yao, L., KTH-Royal Institute of Technology
Cengic, I., KTH-Royal Institute of Technology
Anfelt, J., KTH-Royal Institute of Technology
Hudson, P., KTH-Royal Institute of Technology

Cyanobacteria are promising hosts for production of biofuels and chemicals. However, the deficiency of robust genetic tools has been a barrier to metabolic engineering of cyanobacteria. The CRISPR/Cas system has emerged and been widely used for genome editing, gene expression regulation and imaging of genomic elements. However, CRISPR-based gene editing or regulation had not yet been demonstrated in cyanobacteria. Here, we used the nuclease-deficient Cas9 (dCas9) of *Streptrococcus pyogenes* for the repression of genes in the model cyanobacterium *Synechcocystis* sp. PCC 6803. This CRISPR-interference (CRISPRi) repressed green fluorescent protein (GFP) to negligible levels. The respression effect was inducible and reversible when using a synthetic aTc-inducible promoter. CRISPRi was also used to repress formation of carbon storage compounds polyhydroxybutryate (PHB) and glycogen during nitrogen starvation. By expressing multiple sgRNAs, we simultaneously knocked down 4 putative aldehyde reductases and dehydrogenases at 50–95% repression. Thus, CRISPRi is a robust tool and could greatly accelerate both basic research and metabolic engineering efforts in cyanobacteria. We also anticipate the development of other CRISPR-based synthetic biology tools in cyanobacteria.

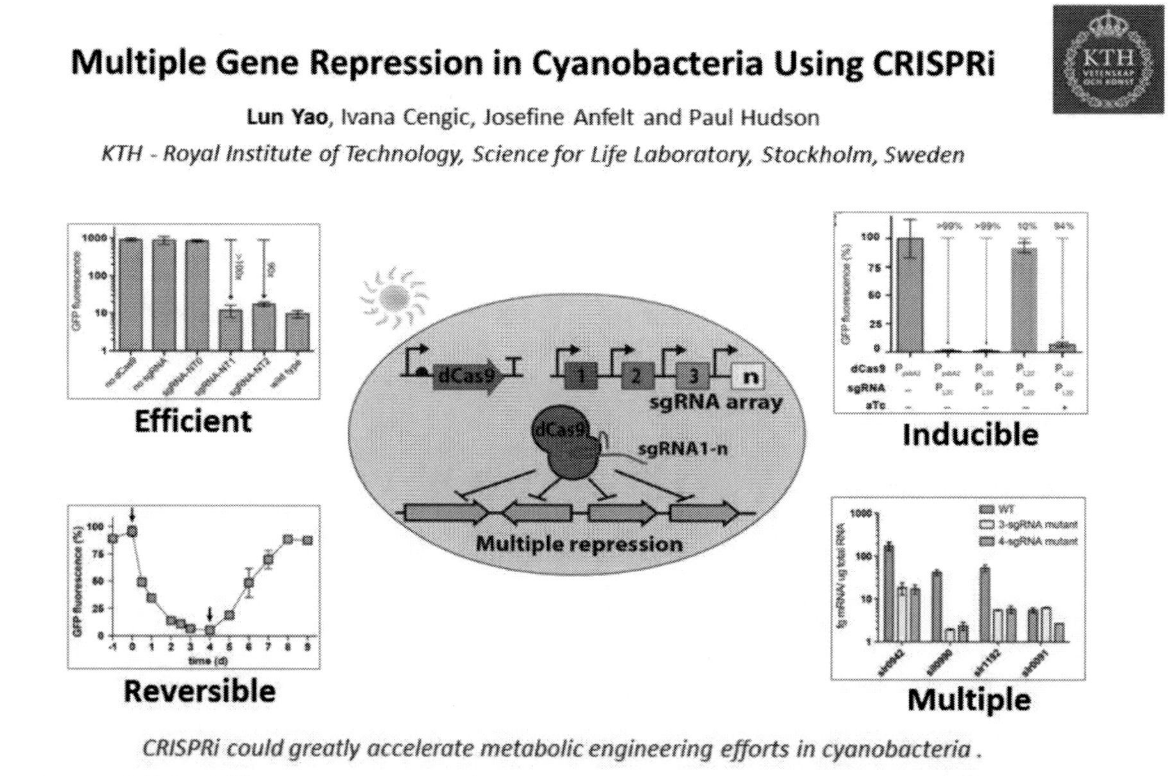

The Second Generation Ogab Method Makes It Feasible for Assembly More Than 50 DNA Fragments at One-Step

Tsuge, K., Keio University
Sato, Y., Keio University
Kobayashi, Y., Keio University
Gondo, M., Keio University
Hasebe, M., Keio University
Togashi, T., Keio University
Tomita, M., Keio University
Itaya, M., Keio University

Construction of an artificial gene cluster that is comprised of a series of genes responsible for one metabolic pathway has been expected to be one of the potent means to create high performance metabolic pathway in host cell. However, it is still not easy to prepare *de novo* designed-long DNA with several dozen kb in size, because DNA assembly methods devised so far is not good enough for at once assembly of small DNA blocks that would be prepared by chemical synthesis and following assembly of single stranded DNA. OGAB is multiple gene assembly method devised by us in 2003. This method adopts *Bacillus subtilis* plasmid transformation system, and is feasible for assembly as much as 15 DNA blocks at one-step. Different from other popular gene assembly methods, for example Golden Gate and Gibson Assembly, are designed to obtain circular form of ligation product *in vitro*, OGAB method intendedly makes long tandem repeat ligation products of material DNA blocks and assembly vector *in vitro*, which is achieved only after precise equimolar adjustment of the blocks. However, adjustment of DNA blocks in equimolar is quite difficult as long as there is size variety in those blocks, this is the reason why the maximum number of blocks at one assembly operation is limited to lower that 15. In order to overcome this circumstance, we have recently developed second generation OGAB method by altering original OGAB method. This method is specifically designed for construction of DNA up to 50 kb in size from as much as 50 of small DNA blocks (up to 1 kb) that can be obtained from synthetic DNA service. The key point of this method is preparing the small blocks as equal length by computer-aided manner without taking account for functional boundary of gene. These designed blocks are attached appropriate TypeIIS restriction enzyme site at both ends, and then cloned into conventional *E. coli* plasmid vector. Since length of the blocks is same and vector is common over the blocks, precise equimolar control of blocks can be easily achieved by just adjusting weight concentration of block-cloned plasmid DNA according to values obtained by a microvolume spectrometer. Resulted equimolar plasmid mixture can be treated as sole kind of plasmid as far as all the plasmids share same TypeIIS site. Thanks to this nature, plasmid vector separation by electrophoresis after relevant restriction enzyme digestion is conveniently finished only in one operation without distortion of equimolarity. Using small blocks thus obtained, assembly of unprecedentedly large number of DNA fragments at once becomes feasible.

June 27
#18

The Second Generation OGAB Method Makes It Feasible for Assembly More Than 50 DNA Fragments at One-Step

○Kenji TSUGE[1,3], Yukari Sato[1], Yuka Kobayashi[1], Maiko Gondo[1], Masako Hasebe[1], Takashi Togashi[1], Masaru Tomita[1,2], and Mitsuhiro Itaya[1,2]
1 Institute for Advanced Biosciences, Keio University, 2 Environment and Information Studies, Keio University, and
3 Science and Technology innovation, Kobe University

OGAB method is gene assembly method using *Bacillus subtilis* plasmid transformation.

Tsuge et al. Scientific Reports, 5 10655 (2015)

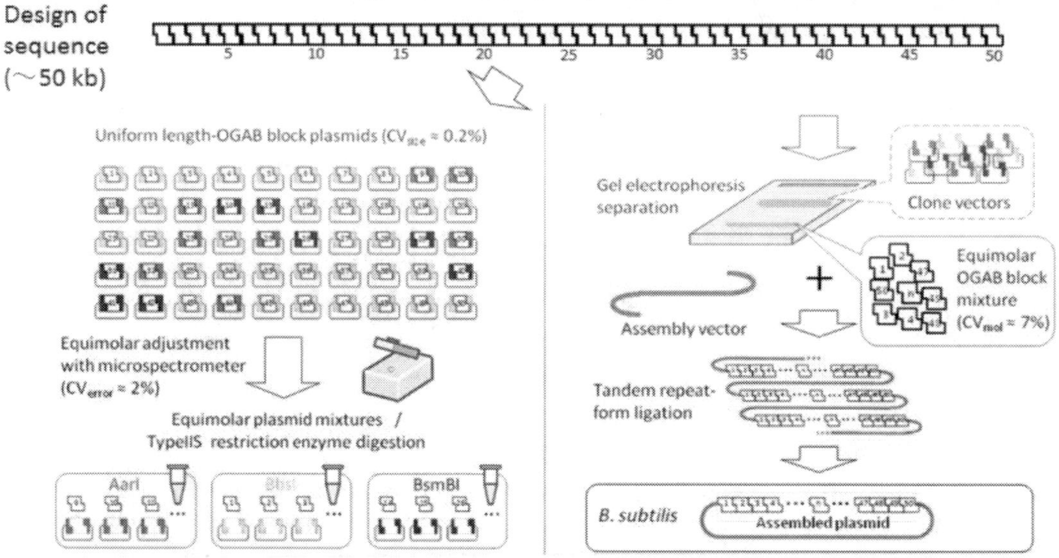

2nd generation OGAB method is feasible to make long designed sequence in one operation

How Uncertainty in Kinetic Parameters Affects Metabolic Control Analysis of Optimally Grown E. coli

Hameri, T., EPFL
Fengos, G., Swiss Institute of Bioinformatics
Ataman, M., Ecole Polytechnique Federale de Lausanne (EPFL)
Miskovic, L., Ecole Polytechnique Fédérale de Lausanne (EPFL)
Hatzimanikatis, V., Swiss Federal Institute of Technology of Lausanne (EPFL)

Large-scale kinetic models of metabolism are essential for understanding and predicting the behavior of cellular systems when subject to perturbations. Despite the advances in experimental measurement technologies, the numerous parameters that are required to build kinetic models remain scarce and involve uncertainty. Even after incorporating the partially available experimental data, models still have many degrees of freedom. Due to this parametric uncertainty, some reactions are able to operate in forward and reverse directions. An operational configuration consists of reactions that operate in a unique direction. This flexibility results in the existence of alternative operational configurations representing the same physiology, with very distinct regulatory properties. In this study, we focus on one operational configuration and we investigate how the underlying uncertainty in the kinetic parameters affects the robustness of the model predictions and regulatory capabilities.

To study this question, we used a large-scale non-linear kinetic model built using integrated fluxomics and metabolomics data describing the physiology for aerobically grown E. coli. Because of the under-determined nature of the system, there are multiple kinetic parameters that can render the model feasible within the selected operational configuration. To account for the variability of the kinetic parameters within the designated operational configuration, we selected a reference vector of fluxes and concentrations close to their nominal values. Then, we used the ORACLE (optimization and risk analysis of complex living entities) framework to build populations of kinetic models that are consistent with the given physiology, while satisfying the stoichiometric and thermodynamic constraints. From these, we built non-linear models to test how the system's response to perturbations changed with respect to the chosen kinetic parameters. This allowed us to quantify the effect of the uncertainty in the kinetic parameters on the robustness of the regulatory properties of the system.

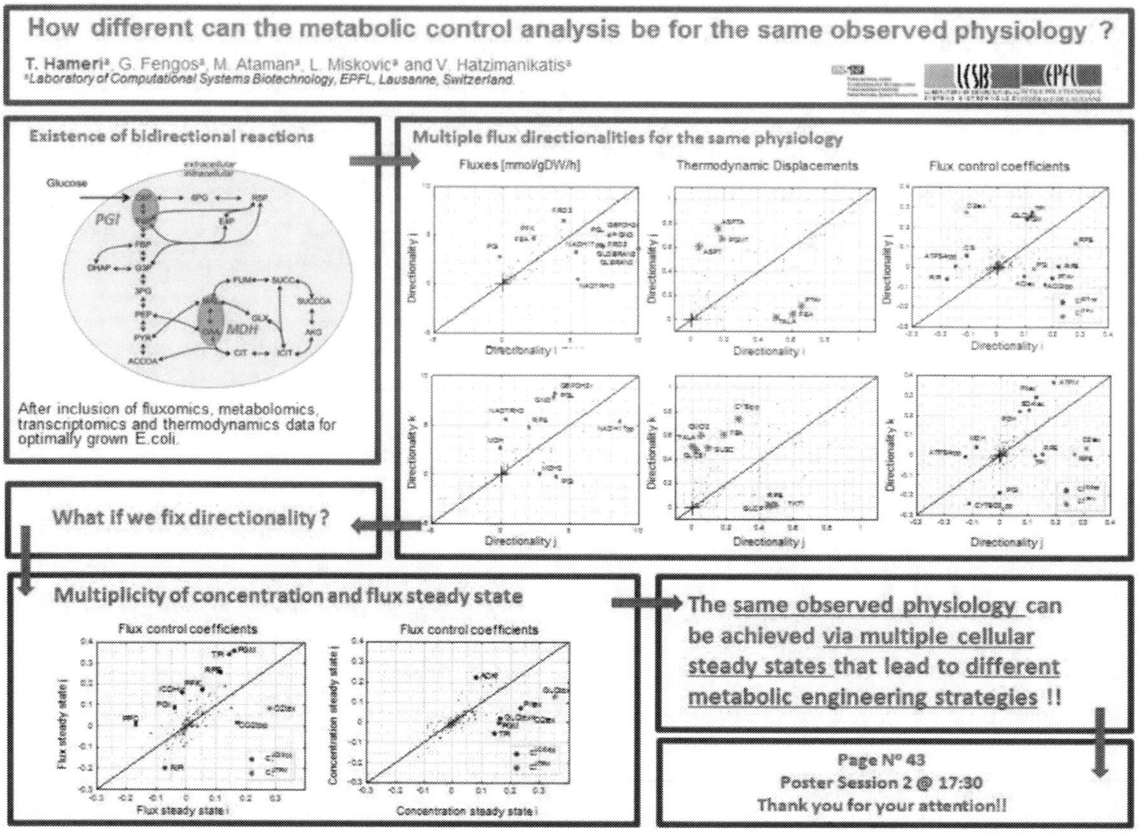
How different can the metabolic control analysis be for the same observed physiology ?
T. Hameri[a], G. Fengos[a], M. Ataman[a], L. Miskovic[a] and V. Hatzimanikatis[a]
[a]Laboratory of Computational Systems Biotechnology, EPFL, Lausanne, Switzerland.
Existence of bidirectional reactions
Glucose
PGI
MDH
After inclusion of fluxomics, metabolomics, transcriptomics and thermodynamics data for optimally grown E.coli.
Multiple flux directionalities for the same physiology
Fluxes [mmol/gDW/h]
Thermodynamic Displacements
Flux control coefficients
Directionality i
Directionality j
Directionality k
What if we fix directionality ?
Multiplicity of concentration and flux steady state
Flux control coefficients
Flux steady state i
Flux steady state j
Concentration steady state i
Concentration steady state j
The same observed physiology can be achieved via multiple cellular steady states that lead to different metabolic engineering strategies !!
Page N° 43
Poster Session 2 @ 17:30
Thank you for your attention!!

Growth-Coupled Fermentation Predicted By Next-Generation Genome-Scale Models

King, Z. A., University of California, San Diego
O'Brien, E. J.,
Feist, A. M., University of California, San Diego
Palsson, B. O., University of California, San Diego

Fermentation is an optimal growth strategy for microbes living in the absence of oxygen and other external electron acceptors. During fermentation, microbes naturally secrete large quantities of small molecules to balance their redox states, and these secretions have long been exploited for industrial production of ethanol, butanol, acetone, glycerol, and lactate. Recently, genetic engineering has been used to modify microbial fermentation and thereby produce a more diverse range of molecules. We generated a database of 69 engineered, growth-coupled strains of *Escherichia coli* from the literature and simulated these engineered strains in the next-generation genome-scale model of metabolism and expression (ME-model). In the ME-model, the tradeoff between pathway-enzyme cost and thermodynamics plays a major role in determining the optimal fermentation pathways. This tradeoff is absent from models strictly comprising of metabolic reactions (M-models), and we show that the ME-model predicts fermentation pathway activity more accurately than any existing M-model of *E. coli*. Furthermore, we use the ME-model to show that certain kinetic parameters can determine whether a target molecule will be growth-coupled. In particular, kinetics are necessary for understanding the production of succinate by *E. coli*. By predicting the effects of pathway-enzyme cost and kinetics on fermentation and growth coupling, ME-models can help guide the design of new growth-coupled production strains.

Growth-Coupled Fermentation Predicted By Next-Generation Genome-Scale Models

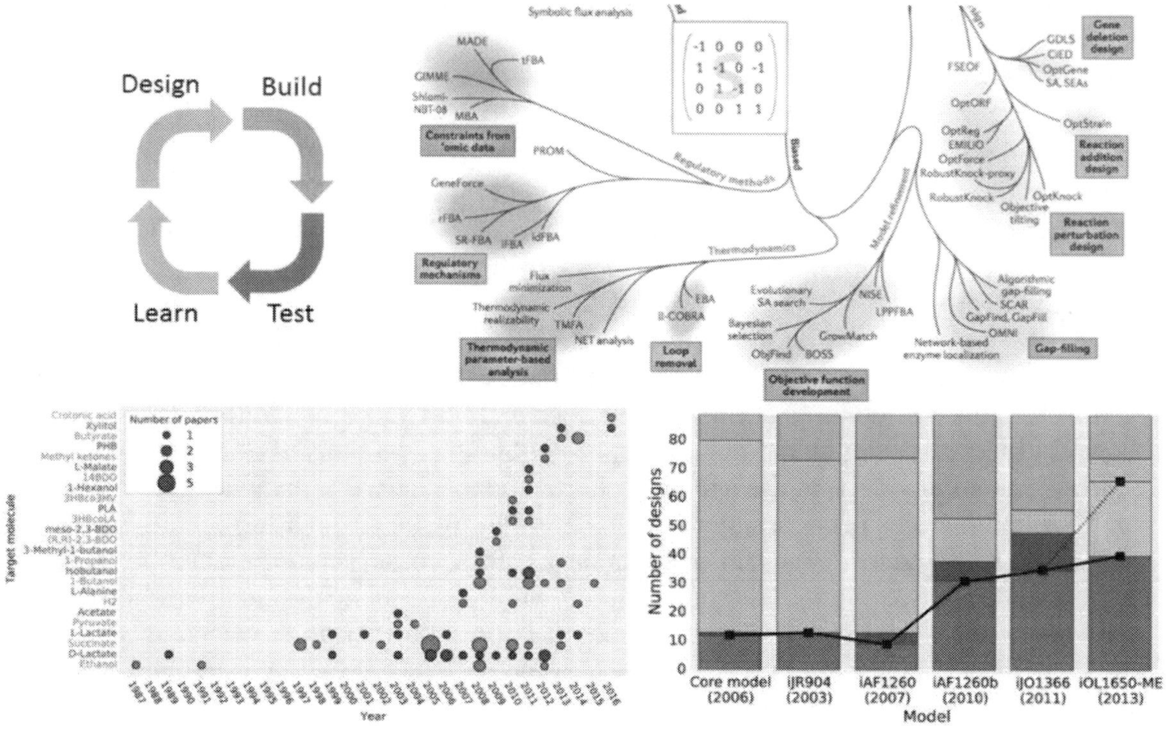

Ensemble Modeling of Cyanobacteria Metabolism Using Multi-Omics Data

Matsuda, F., RIKEN CSRS
Hiasa, N., Graduate School of Information Science and Technology, Osaka University
Uebayashi, K., Graduate School of Information Science a
Liao, J. C., University of California,
Shimizu, H., Osaka University

Introduction

Cyanobacteria is a promising host for a sustainable bio-production from carbon dioxide. The dynamic behavior of the central metabolism in cyanobacteria has been uncovered by the transcriptome, metabolome, and fluxose analyses of a glucose-tolerant strain of the *Synechocystis* sp. PCC 6803 under auto-, mixo- and heterotrophic conditions. However, its control mechanism remains unclear due to lack of the enzyme expression data and a kinetic metabolic model required for the metabolic control analysis. In this study, protein expression profile data were obtained from *Synechocystis sp.* PCC 6803 by the targeted proteome analysis. In addition to an integrated analysis with other omic data, kinetic models of the central carbon metabolism was constructed by the ensemble modeling technique with using the proteome, metabolome and fluxome dataset.

Method

Synechocystis sp. PCC 6803 cells were cultivated in 500 mL Erlenmeyer flasks with 100 mL modified BG11 medium (with 5 mM glucose and 20 μM atrazine) under continuous light (40 μmol of photons/m^2/sec). The peptide samples were analyzed using the nanoLC-MRM system (LC: LC-20ADnano (Shimadzu, Japan), MS: LCMS-8040 (Shimadzu, Japan)). A metabolic model including 62 reactions with the Michaelis-Menten and mass action type kinetics was used for the ensemble modeling. Flux control coefficients were determined by the metabolic control analysis (Fell et al. 1992). All *in silico* analyses were performed by in-house scripts written with Python2.7+Numpy1.9.

Results and discussions

Expression profiles of 93 central metabolism and photosystem related proteins were obtained from *Synechocystis sp.* PCC 6803 under the auto-, mixo- and heterotrophic conditions by the targeted proteome analysis using nanoLC-MS. Levels of rbcS protein in the mixotrophic and heterotrophic conditions were lower than that of the autotrophic condition. Comparisons with other omics data showed that the expression patterns of rbcS protein was correlated with that of metabolic flux levels in RuBisCo reaction. On other hands, there were poor similarities with *rbcS* gene expressions and ribulose-1,5-biophosphate accumulations.

An ensemble including 100,000 models of *Synechocystis sp.* PCC 6803 metabolism with randomly sampled parameters were produced using the proteome, metabolome and fluxsome data at the autotrophic condition as a reference. Photon flux was determined by minimizing ATP futile cycling. Following a modulation of the enzyme expression levels to that of heterotorophic condition according to the proteome data, the metabolic flux and metabolite concentration levels were predicted and compared with that of measured data of the heterotorophic condition. A subset of 100 metabolic models with better predictability were selected from the 100,000 models. The metabolic control analysis (MCA) using the 100 models showed that the expression levels of RuBisCo (rbcS) protein had low flux control coefficient (less than 0.01 in all models) for the RuBisCo flux level under the autotrophic condition studied here. The MCA also suggest a dispersed control of RuBisCo flux by several enzyme activities. These results suggested that the ensemble modeling using multi-omics data could be a strategy to investigate a control of the cyanobacteria metabolism. This work was supported by NSF/JST.

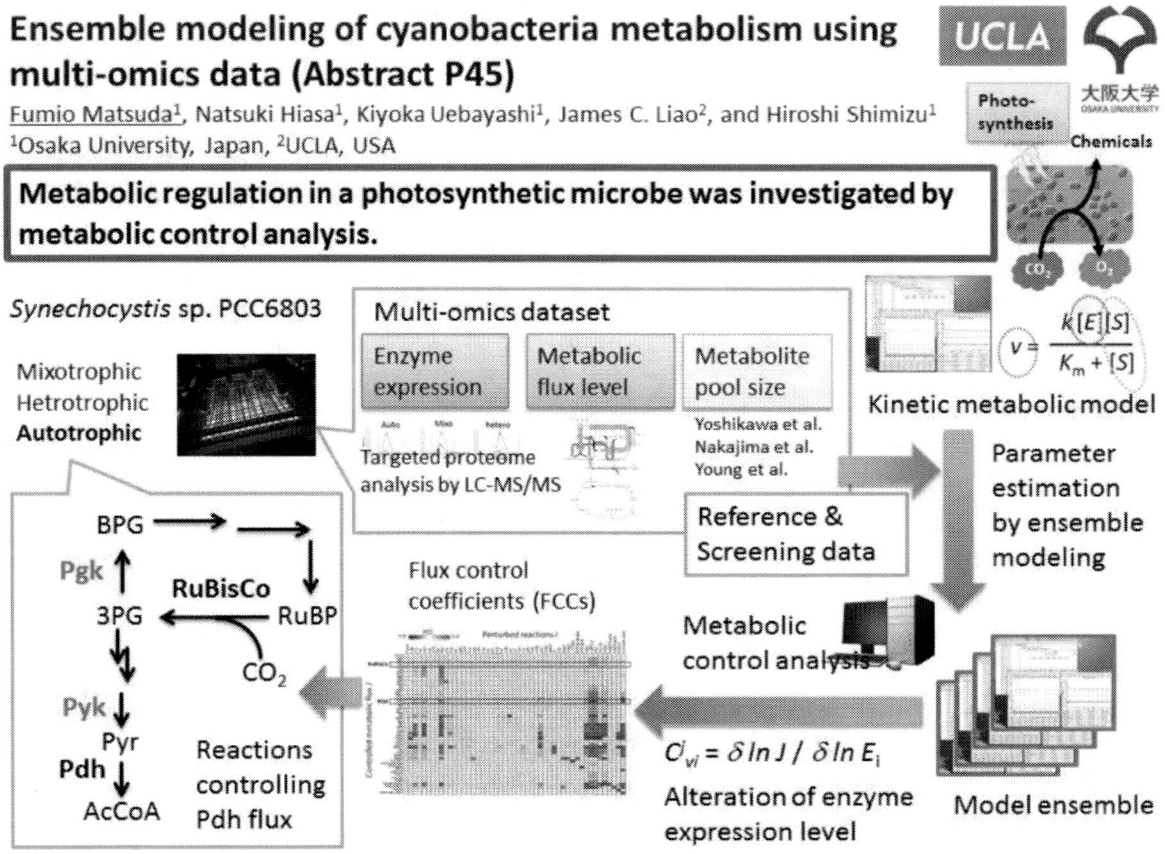

Crispri-Guided Balancing of a Biosynthetic Pathway Increases Terpenoids Production

Kim, S. K., Korea Research Institute of Bioscience and Biotechnology
Seong, W. J., Korea Research Institute of Bioscience and Biotechnology
Yoon, P., University of Science and Technology
Kim, M., Chungnam National University
Lee, S. G., Korea Research Institute of Bioscience and Biotechnology
Lee, D. H., University of Science and Technology

Simple and efficient regulation of gene expression is essential to the study of synthetic gene networks and to the biotechnological applications. In particular, well-modulated transcription of metabolic pathway genes plays a key role in maximizing product yield and minimizing metabolic burden caused by heterologous expression of multiple genes in operon context. CRISPR interference (CRISPRi) has shown tremendous promise as an efficient tool for transcriptional regulation, but there remain only a handful of applications described to date. Here, we report the systematic regulation of a biosynthetic mevalonate pathway that has been used for microbial production of the most numerous and diverse natural products, terpenoids. A tunable CRISPRi system was created for fine-tuning a metabolic pathway and implemented to efficiently channel carbon flux toward terpenoids synthesis in engineered *Escherichia coli* harboring exogenous mevalonate pathway and plant-derived terpene synthases through balancing the expression of multiple genes, resulting in enhanced production of (-)-α-bisabolol (C15), and lycopene (C40) along with alleviating the inhibition of cell growth. Coupling the CRISPRi to cell growth through regulation of endogenous essential gene further increased the production of isoprene (C5). Overall, CRISPRi is a platform for systematic modulation of synthetic gene expression to balance the synthetic metabolic pathways and subsequently for developing microbes as a smart cell factory to produce other industrially valuable terpenoids.

Mining Metabolism for Unannotated Enzymatic Functions and Serendipitous Metabolic Pathways

Jeffryes, J., Argonne National Laboratory
Lerma-Ortiz, C., University of Florida
Niehaus, T., University of Florida
Thamm, A., University of Florida
Elbadawi-Sidhu, M.,
Kind, T.,
Broadbelt, L. J., Northwestern University
Fiehn, O.,
Hanson, A., University of Florida
Tyo, K. E. J., Northwestern University
Henry, C. S., Argonne National Laboratory

The potential of metabolic engineering is greatly enhanced by enzymes' natural flexibility, which permits them to catalyze the transformation of not only their native substrate but also a range of related substrates. This trait enables cells to produce not only natural metabolic products, but also valuable compounds for which no natural pathway exists through rational design and directed evolution. However, unless physically sequestered, these enzymes and metabolic intermediates continue to interact with the chassis' preexisting metabolic network. Understanding potential off-pathway effects of overexpressed enzymes or the alternate fates of novel pathway intermediates permits the rationalization of observed phenotypes and informs the design of new pathways.

To address this challenge, we have constructed Metabolic In silico Network Expansions (MINEs) that expand existing metabolic models by using expert curated reaction rules to propose novel metabolites and reactions. The reaction rules include an enzymatic set, which has been demonstrated to reproduce a large fraction of known biochemical reactions and a set describing spontaneous (non-enzymatic) chemical transformations of metabolites under physiological conditions. We have applied these generalized reaction rules to compounds from various biochemical databases such as KEGG (www.kegg.jp) and the Model SEED (www.theseed.org/models) The resulting MINEs are freely accessible for noncommercial use at http://minedatabase.mcs.anl. The databases contain over 750,000 putative metabolites; over 90% of which are not found in PubChem, the largest freely available database of chemicals. MINEs have been used to annotate novel metabolites from 4 diverse organisms and propose potential sources for these compounds within known metabolism. MINE databases shine a light on unannotated enzymatic functions and serendipitous metabolic pathways, enabling more complete and predictive models of cellular metabolism.

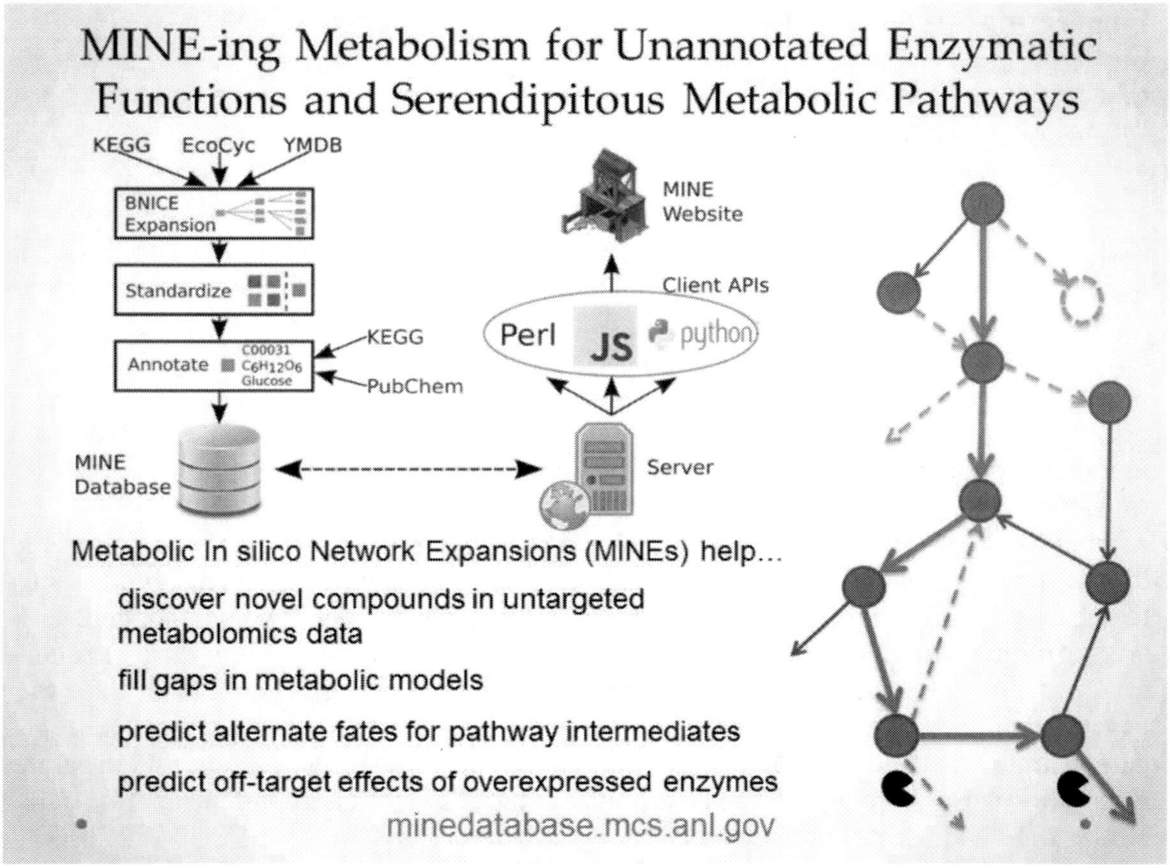
MINE-ing Metabolism for Unannotated Enzymatic Functions and Serendipitous Metabolic Pathways
KEGG
EcoCyc
YMDB
BNICE Expansion
Standardize
Annotate
C00031
Glucose
KEGG
PubChem
MINE Database
MINE Website
Client APIs
Perl
JS
python
Server
Metabolic In silico Network Expansions (MINEs) help...
discover novel compounds in untargeted metabolomics data
fill gaps in metabolic models
predict alternate fates for pathway intermediates
predict off-target effects of overexpressed enzymes
minedatabase.mcs.anl.gov

Harnessing the Crispr/dCas9 Crispri System for Rapid Assessment of Metabolic Engineering Strategies to Improve Natural Product Titers in E. coli

Cress, B. F., Rensselaer Polytechnic Institute
Linhardt, R. J., Rensselaer Polytechnic Institute
Koffas, M. A., Rensselaer Polytechnic Institute

The field of metabolic engineering is entering a new era in which readily addressable synthetic transcription factors like RNA-guided dCas9 are opening new possibilities for dynamic regulation and for rapid assessment of multiplexed genetic interventions to alter the metabolic landscape in both workhorse and previously intractable microbes. Toward this end, our group has focused on three major applications of CRISPRi-enabled metabolic engineering.

In a recent publication, we developed a modular assembly method for traditional restriction-ligation cloning of type II-A CRISPR array libraries for multiplex, combinatorial dCas9-mediated transcriptional repression. Importantly, we demonstrated for the first time that dCas9 can be utilized for metabolic engineering in *E. coli*, leading to enhancement of flavonoid titers through simultaneous repression of an endogenous transcription factor and several central carbon enzymes, including partial downregulation of a synthetic lethal pair. Furthermore, we showed that this system could be utilized to generate complex phenotypes (such as dual amino acid bradytrophs), and we also demonstrated that a single plasmid can be used to attenuate virulence genes in disparate *E. coli* lineages by repressing capsular polysaccharide secretion in both probiotic strain Nissle 1917 and pathogenic strain K5.

Second, we have recently expanded upon this work by adapting Golden Gate Shuffling for assembly of natural type II-A CRISPR arrays, enabling randomized one-pot assembly of array libraries simultaneously repressing all combinations of a user-defined set of target genes. We test this method for improving production of glycosylated natural products—including the pharmaceutically and nutraceutically valuable polysaccharides heparin and chondroitin and the colorful class of plant natural pigments known as anthocyanins—through repression of novel knockdown targets and partial downregulation of an essential glycolytic enzyme. Critically, we show that the plasmid libraries generated with this methodology can be transformed into a variety of host strains for rapid evaluation of target combinations in distinct production chassis. Direct comparison of CRISPRi strains with analogous deletion strains (obtained through lambda-red recombineering) shows that CRISPRi can lead to production improvements comparable to gene deletions.

Finally, we constructed orthogonal variants of the classic T7-lac promoter using site-directed mutagenesis, generating a panel of inducible hybrid promoters regulated by both LacI and dCas9 and covering a wide expression range. Remarkably, dCas9 orthogonality in our system is mediated by only 2-3 nucleotide mismatches in a narrow window of the RNA:DNA hybrid, neighboring the protospacer adjacent motif (PAM). We demonstrate that, contrary to many reports, one PAM-proximal mismatch is insufficient to abolish dCas9-mediated repression, and we show that mismatch tolerance and orthogonality is dependent upon target copy number. Finally, a subset of these refactored promoters were incorporated into the highly branched violacein biosynthetic pathway, where they act as orthogonal, dCas9-dependent valves capable of throttling and selectively redirecting carbon flux in *E. coli*.

Harnessing CRISPRi as a Tool to Improve Natural Product Titers in *E. coli*

Poster 51

Brady F. Cress, Robert J. Linhardt, and Mattheos A. G. Koffas

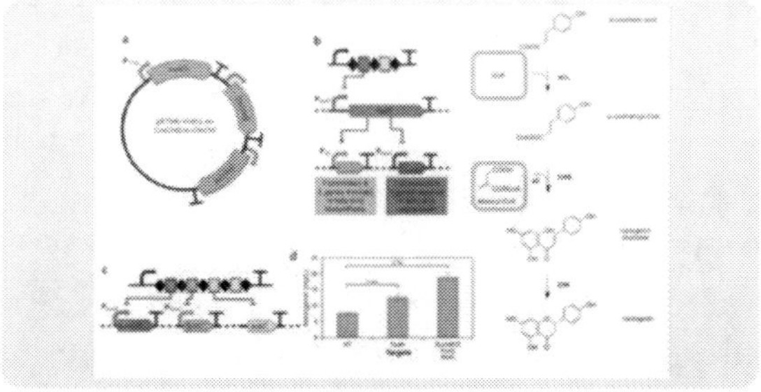

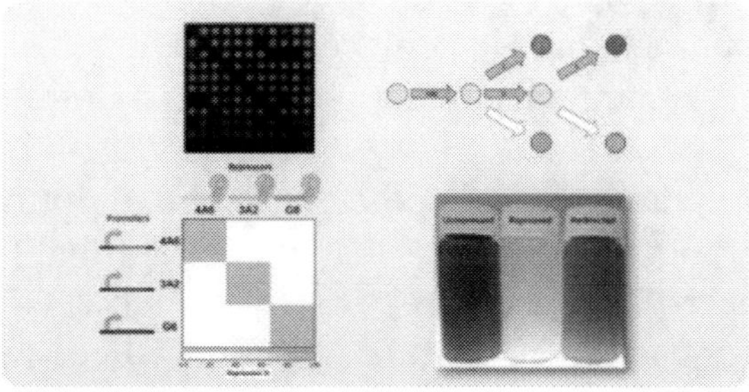

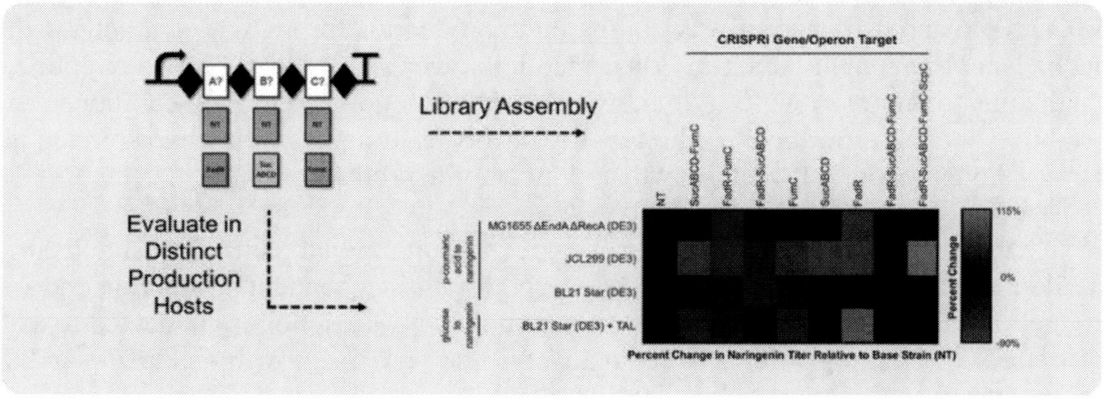

Orthogonal Design of Metabolic Pathways

Pandit, A. V., University of Toronto
Srinivasan, S., University of Toronto
Mahadevan, R., University of Toronto

Traditional metabolic engineering strives to achieve overproduction of target metabolites through growth coupling strategies that alter these structures and functions in such a way that neither cell growth nor metabolite production is optimal due to extensive network interactions. Hence, we are motivated to explore an alternative philosophy that prioritizes optimal network structures for metabolite production over cell growth. Our antithetical approach aims to achieve mutually exclusive operation of pathways responsible for cell growth and metabolite production by minimizing interactions between these pathways.

We show that it is possible to obtain optimal metabolite production through engineered orthogonal pathways that are not optimal for cell growth by reducing the number of shared enzymatic steps between pathways producing biomass precursors and those that produce the metabolite. We found succinate production through the Embden-Meyerhoff-Parnas, the Entner-Doudoroff and the methylglyoxal bypass to have orthogonality values of 0.42, 0.46 and 0.40 respectively compared to production through the synthetic pathway that had an orthogonality of 0.52. Consequentially, we establish that growth coupling, which aims to improve connectivity between the production and growth components of a metabolic network, lowers the orthogonality of the resulting network, characterized by increased interactions between growth and metabolite production.

We present an algorithmic approach to engineering orthogonal pathways using a cut sets based method. To facilitate switching between growth-independent production and production-independent growth phases, a metabolic valve reaction is connected to a node that acts as a branching point between pathways that serve as growth precursors and those that produce the target metabolite. Turning off the metabolic valve would result in true orthogonal pathways and could be an attractive target for two phase fermentation processes and dynamic metabolic engineering. We found isocitrate dehydrogenase to be this valve reaction for succinate production from glucose.

Constraining cellular pathways by the chosen substrate-product pair also constraints the quantity and quality of interactions witnessed within the production network. By making orthogonal pathways dependent on the chosen substrate-product pairing, we have enabled de novo engineering of pathways for growth independent optimal production from many non-conventional substrates that are not glucose. We show that using xylose instead of glucose, to produce succinic acid minimizes the interactions prevalent in natural pathways to obtain orthogonal production pathways. The orthogonality of the xylose-based production pathway (0.59) supports this statement.

Hence, our proposed orthogonal pathways could bring about a reduction in the number of interactions within the network without necessitating an equivalent number of gene deletions while also delivering greater independence between the production and growth components of metabolism. In doing so, it enables substrate-dependent design of cell factories to leverage the use of several substrates previously thought be unsuitable. Through our process, we have also developed a qualitative assessment of the ability of a given metabolic network to support the two different objectives of cell growth and metabolite production as a measure of connectivity within the metabolic network

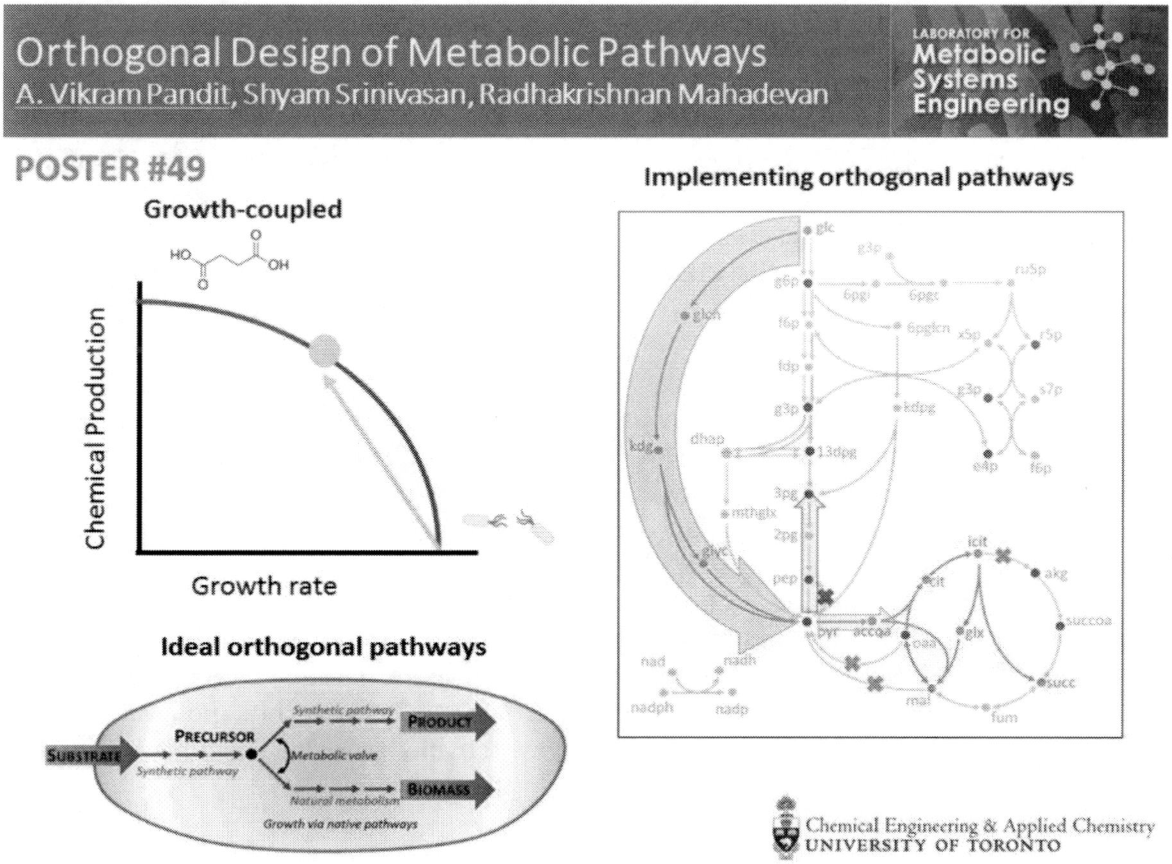

Creation of Novel Allosteric Regulation of Proteins for Synthetic Biology Based on a New Concept of Thermodynamic Model

Geng, F., Hamburg University of Technology
Ma, C. W., Hamburg University of Technology
Zeng, A. P., Hamburg University of Technology

A novel strategy to engineer ligand-induced allosteric regulation is proposed based on a new concept of thermodynamic model of protein dynamics with the aim to create novel allosteric regulation of proteins, especially regulated by non-natural ligands. In this model, allosteric process upon ligand binding is considered as the energetic development of a thermodynamic system proceeding from an initial ensemble to a final ensemble. The key feature of the thermodynamic model is that the allosteric process is divided into two sub-processes: conformational change and molecular binding and the allosteric process upon ligand binding is regarded as the overall effects of the two sub-processes. As a consequence of the separation of the allosteric process, the ligand-induced allosteric regulation can be examined from different perspectives. To prove the concept, aspartokinase III (AK3) from *E. coli* was used as a as a model system. After construction of the thermodynamic model for *E. coli* AK3, the thermodynamic model was first used to guide the design for enhancement of the allosteric effect of AK3 by its natural effector. Then the new concept was employed to establish new allosteric effects for both the natural and the non-natural ligands. For this purpose, an allosteric determinant was identified and followed by different strategies of focused protein design. Experimental results showed that the natural ligand was successfully altered from an inhibitor to an activator and both inhibition and activation effects have been established for the non-natural ligand guided by the same allosteric determinant. These results indicate that the new concept is powerful in constructing novel switchable biomolecules with potential applications in synthetic biology.

Creation of novel allosteric regulation of proteins for synthetic biology based on a new concept of thermodynamic model

Feng Geng, Cheng-Wei Ma, and An-Ping Zeng

BIO ENGINEERING

Goal: design novel allosteric proteins for dynamic metabolic pathway control

➢ Thermodynamic model of allostery

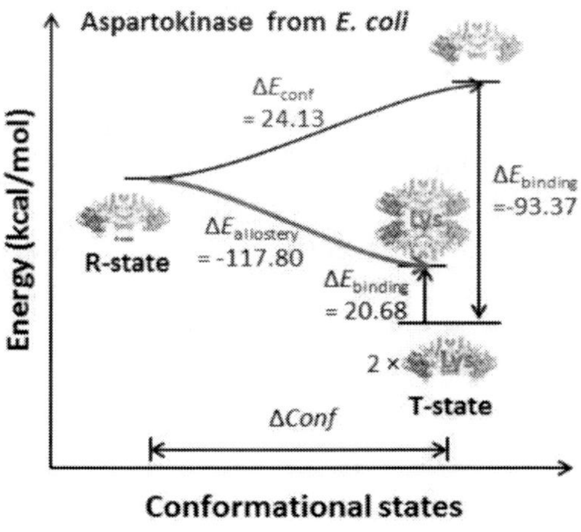

See our poster 46 for details.

➢ Model based design of allostery

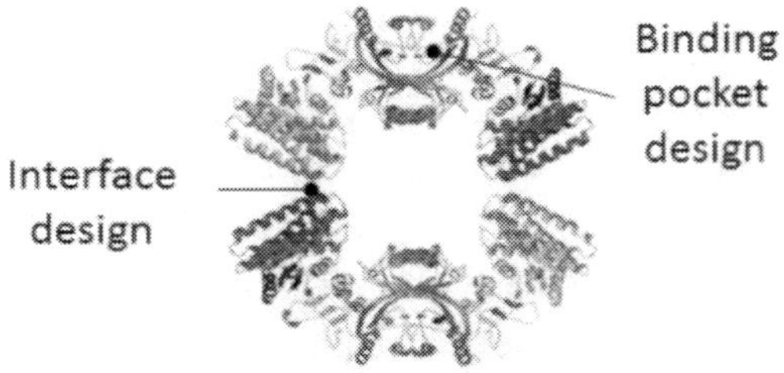

➢ Design of AK3 for a nonnatural ligand

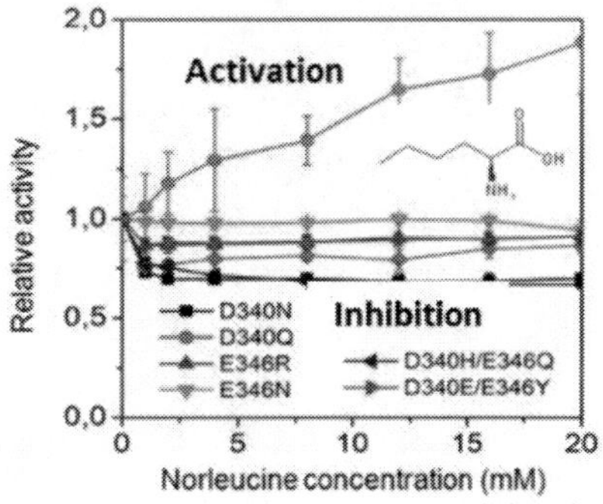

Development of Selection Method for Directed Evolution of Genetic Switches in Saccharomyces Cerevisiae

Tominaga, M., Kobe University
Kawai-Noma, S., Chiba University
Umeno, D., Japan Science and Technology Agency (JST)
Ishii, J., Kobe University
Kondo, A., Graduate School of Engineering, Kobe University

Genetic switches that turn gene expression "ON" and "OFF" in response to specific inputs are indispensable tools for metabolic engineering. Although there are hundreds of natural genetic switches in yeast, most of these systems cannot be directly used for metabolic engineering applications because of their inappropriate switching properties, such as insufficient switching sensitivity, expression intensity or basal leakiness. Directed evolution strategy is applicable to the improvement of any genetic switches in their properties. Desired switches can be obtained simply by coupling a rescuer and a conditional killer to their expression outputs. The rescuer leaves the variants to satisfy the sufficient output under "ON" states, while the conditional killer excludes the variants to exhibit the leaky output under "OFF" states. For the rapid construction of tailored genetic switches for various applications, all selection procedures should be completed by liquid handling that enables the parallel operation of multiple projects in a multi-well plate format. In this study, we report a rapid and liquid-based ON/OFF selection method in yeast *S. cerevisiae* to meet the wide demand for improving switching stringency, expression level, and sensitivity to the target metabolite.

In search for the robust OFF selection method that can be operated by liquid handling only, we chose the kinase activity of herpes simplex virus thymidine kinase (hsvTK) for non-natural nucleoside dP, initially used for the directed evolution of genetic switches/circuits in *Escherichia coli* (Y. Tashiro *et al.*, *Nucleic Acids Res.*, 39, e12 (2011)). However, yeast cells expressing *hsvtk* gene exhibited only a small loss of viability in the presence of 100 μM dP, which was over 100-fold higher concentration than the case in showing the complete loss of viability of *E. coli* cells expressing hsvTK. As the way to increase the dP sensitivity of yeast cells, we found that the co-expression of human equilibrative nucleoside transporter 1 (*hENT1*) gene significantly improved the dP sensitivity of the cells and caused the cell death with only sub-micromolar concentration of dP. Furthermore, *hsvtk* gene was fused to zeocin resistant gene from *Streptoalloteichus hindustanus* (*ble*) to conduct ON (zeocin) and OFF (dP) selection seamlessly. By using this fusion gene (*hsvtk::ble*), we completed both zeocin and dP selection with only 8-hour incubation in liquid media. Thus, we succeeded in significantly reducing the days of incubation steps using the conventional methods on solid media. After zeocin selection, the viability of hsvTK::Ble-expressing cells was 5×10^5 higher than the value of the control cells (not-expressing hsvTK::Ble). In contrast, after dP-selection the viability of hsvTK::Ble-expressing cells got 1×10^{-4} lower than the value of the control cells.

Finally, we further fused the *hsvtk::ble* gene with a fluorescent protein gene (monomeric Umikinoko-Green), establishing the novel selection system which enabled the ON/OFF selection and subsequent characterization of genetic switches by the rapid and seamless operations using only liquid handling. Having established the selection/characterization scheme, we will demonstrate the promising, high-quality selections to isolate new yeast genetic switches with varied switching properties.

#24: Development of Selection Method for Directed Evolution of Genetic Switches in *Saccharomyces cerevisiae*

Masahiro Tominaga (Kobe University)

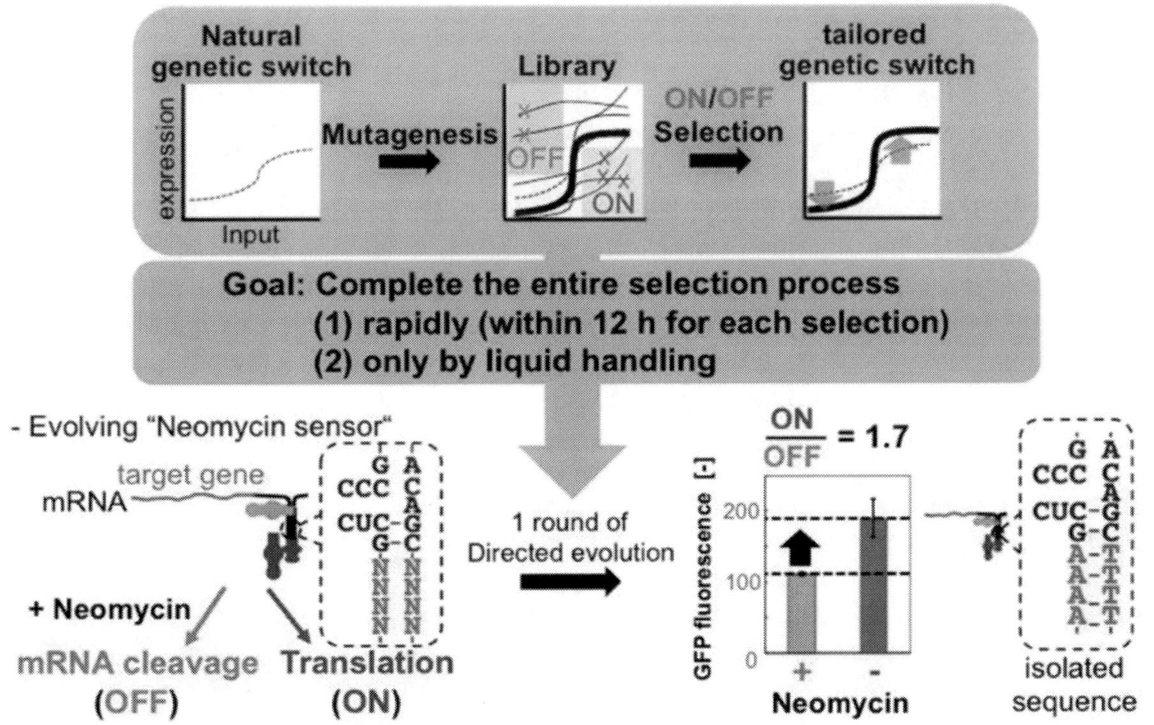

The Genome-Scale Metabolic Model for Chlorella vulgaris Utex 395 Reveals Quantitative Flux Distributions for Compartmental-Pathway Activity Under Polytrophic Growth

Zuniga P. C., University of California
Li, C. T., Johns Hopkins University
Huelsman, T., University of California
Levering, J., University of California, San Diego
Zielinski, D. C., University of California
McConnell, B. O., University of New Hampshire
Long, C. P., University of Delaware
Antoniewicz, M.,
Betenbaugh, M. J., Johns Hopkins University
Zengler, K., University of California, San Diego

Reconstructions are biochemically, genetically and genomically structured knowledge-bases that contain information of reaction stoichiometry, reaction reversibility, and the association between genes, proteins and reactions. Currently genome-scale metabolic reconstructions are available for a number of heterotrophic organisms capable of biofuel production. Here we report on the genome-scale reconstruction for the photoautotroph algae *Chlorella vulgaris* UTEX 395.

The reconstruction, *i*CZ843, consists of six compartments: the cytoplasm, mitochondrion, chloroplast, thylakoid, glyoxysome, and the extracellular space. It contains 843 out of 7,100 annotated genes (around 12%), delineating 1,770 metabolites and 2,294 reactions. *C. vulgaris* can grow under different trophic conditions (i.e. photoauto-, hetero-, and mixotrophic). Each of these growth conditions is represented mathematically through different biomass objective functions (BOFs). Every equation contains the stoichiometric coefficients for the most important metabolites being part of the biomass. Major components of the biomass, such as lipids, proteins, carbohydrates and ribose in RNA were measured experimentally and values were compared with model predictions. Also the model was validated against experimental growth results on 118 carbon and nitrogen sources.

The model accurately predicts growth rates under photoauto-, hetero-, and mixotrophic conditions. The simulated flux distributions revealed high correlation with the ^{13}C flux analysis data previously reported for *C. protothecoides*. Predicted flux distributions under different trophic conditions show that not only central carbon metabolism but also amino acid, nucleotide, and pigment biosynthetic pathways are impacted when the microalgae is under nitrogen starvation. Furthermore, prediction of growth rates under various medium compositions using *i*CZ843 suggested an increased growth rate when tryptophan or methionine were added. Experimental data verified these predictions and moderately exceed the expected results.

*i*CZ843 represents the most comprehensive model for any eukaryotic photosynthetic organism to date based on genome size and number of genes included in the reconstruction. The highly curated model was validated against experimental data and lays the foundation for advanced strain design and medium alteration to improve yield.

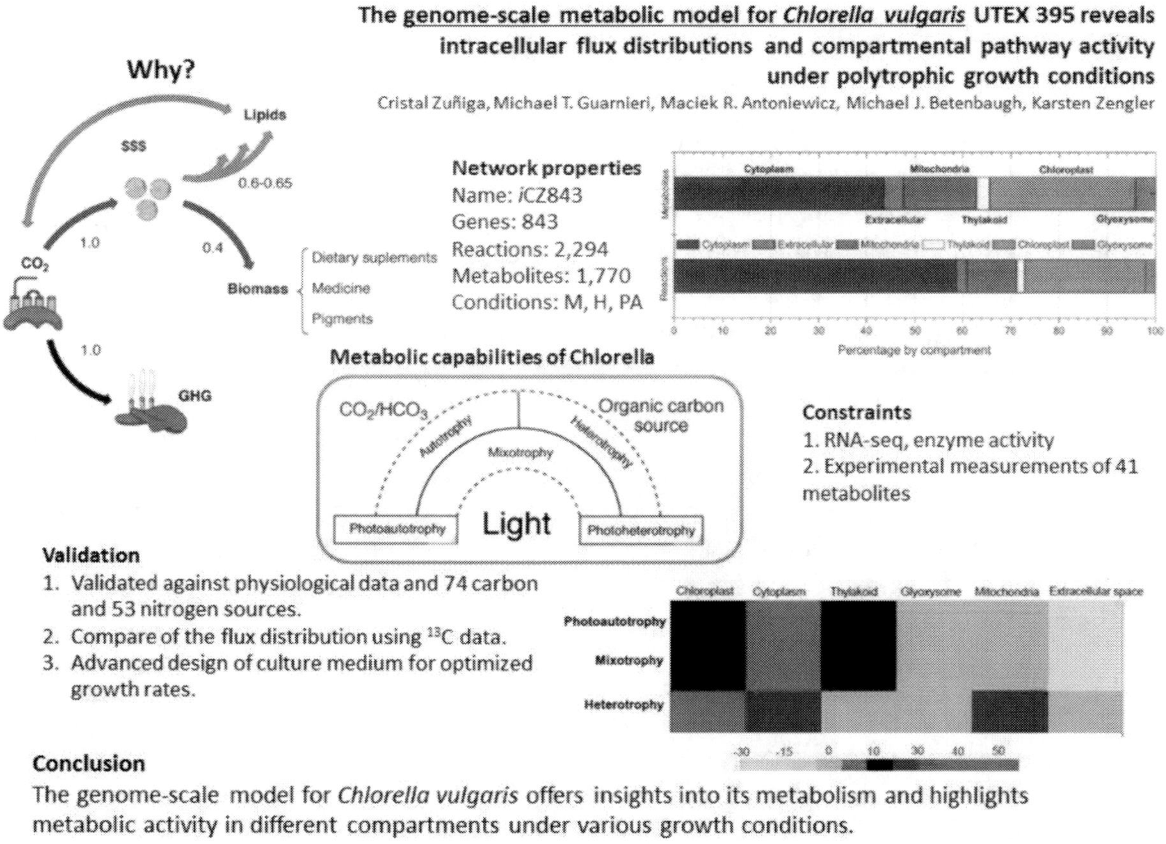

Synthetic Biology in the Design of Versatile Expression Systems in Synechocystis Sp. PCC 6803

Thiel, K., University of Turku
Mulaku, E., University of Turku

Photosynthetic cyanobacteria can be employed for producing diverse end-products directly from atmospheric CO_2 and water, using sunlight as energy. Some of the first applications are already in pilot-scale testing, but the main bottleneck is the efficiency which still remains below commercial feasibility. One of the limitations is the lack of efficient and modifiable tools for the generation of optimized production strains. Our research is focused on synthetic biology, aiming at engineering more efficient cyanobacterial production platforms with flexible means of modification and optimization. We are employing a modified Bio-Brick approach to effectively assemble operons in expression vectors designed for *Synechocystis sp.* PCC6803. This is used for the evaluation of different genetic elements (such as RBSs, promoters or chromosomal integration sites), as well as for the flexible construction, modification and optimization of multi-gene over-expression constructs. The ultimate aim is to combine (i) the tools for engineering optimized heterologous pathways for biofuel production with (ii) different modification strategies of the host metabolism for maximal photon conversion efficiency and metabolic flux from CO_2 towards the final products.

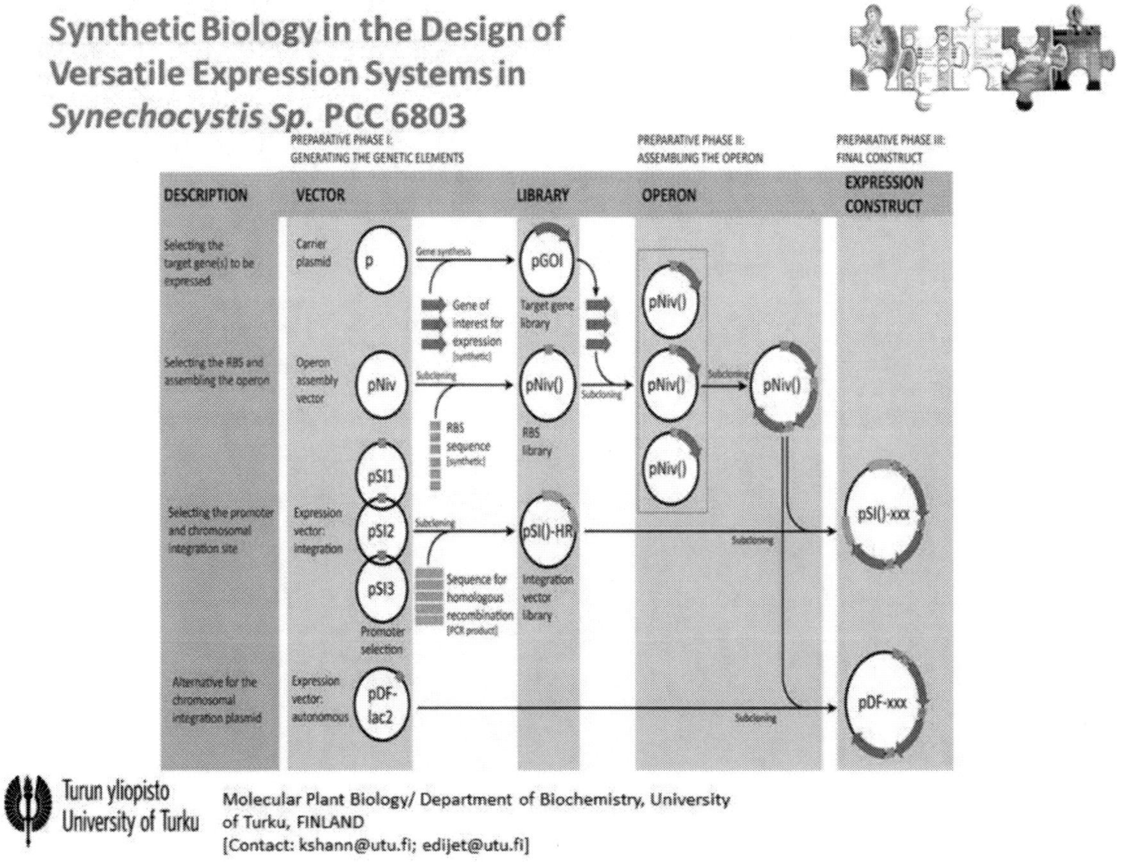

Approaches to Improve the Titers of Natural Products in E. coli

Jones, J. A., Rensselaer Polytechnic Institute
He, W., Rensselaer Polytechnic Institute
Linhardt, R. J., Rensselaer Polytechnic Institute
Koffas, M. A. G., University at Buffalo, the State University of New York

Approaches to Improve the Titers of Natural Products in *E. coli*

J. Andrew Jones, Wenqin He, Robert J. Linhardt, Mattheos A. G. Koffas

Rensselaer Polytechnic Institute, Department of Chemical and Biological Engineering, Center for Biotechnology and Interdisciplinary Studies, Troy, NY 12180

Metabolic engineering has enabled the intelligent design of microbial strains for the high-titer production of many chemicals ranging from biofuels to high-value pharmaceutical precursors. Here we present several recently published methods used to improve the titers of biologically interesting molecules in *E. coli*. First, we present our work on the development and optimization of polycultures (three or more strains in co-culture) for the extension of the flavonoid branch pathway *in vivo*. This technology has enabled, for the first time, the *de novo* production of flavan-3-ols in *E. coli*. Utilizing a computationally guided optimization approach, we were able to demonstrate up to a 970-fold improvement over previously published monoculture titers.

Second, we will present our method for combinatorial transcriptional optimization, ePathOptimize. Applying the traditional Design-Build-Test-Learn cycle we were able to fine-tune the genetic optimization of the violacein pathway. Using this approach for genetic optimization, coupled with traditional shake flask fermentation optimization, we were able to achieve titers of 1.8 g/L violacein, a 2.6-fold improvement over previous benchmarks in *E. coli*. Additionally, we will highlight the power of this technology through optimization of an exogenous methanol utilization pathway in *E. coli* enabling growth on methanol and, for the first time, flux from methanol to high-value specialty products *in vivo*.

Finally, we will demonstrate optimized production of the polysaccharide, chondroitin, through modification of transcriptional polycistronic operon configurations of the three-gene production pathway. After scale-up, 2.4 g/L of chondroitin was produced in a dissolved oxygen-stat fed batch bioreactor. Expanding upon this work, efforts to successfully express and purify human chondroitin 4-sulfotransferase (C4ST1) have been fruitful, resulting in high-level conversion of chondroitin to chondroitin sulfate type-A *in vitro*. These technologies are leading the way towards enabling industrially relevant titers, yields, and productivities for a variety of biologically interesting natural products.

Approaches to Improve the Titers of Natural Products in *E. coli*

Rapid Fire #25
Page 47

J. Andrew Jones and Mattheos A. G. Koffas

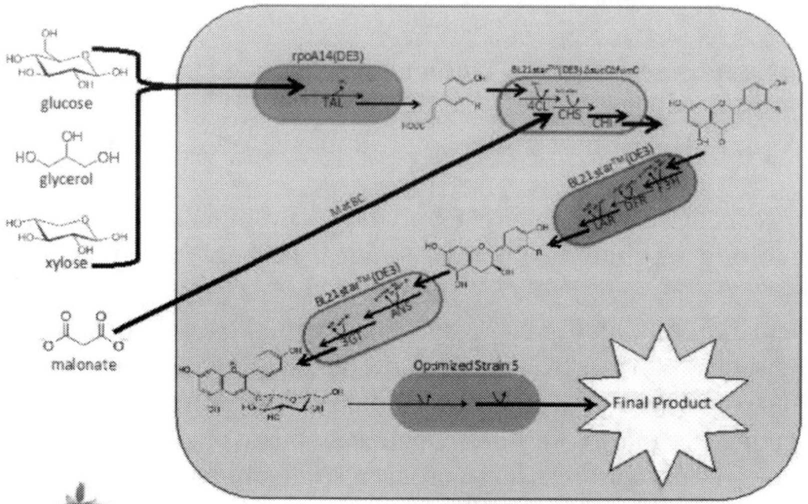

De novo production of Anthocyanins

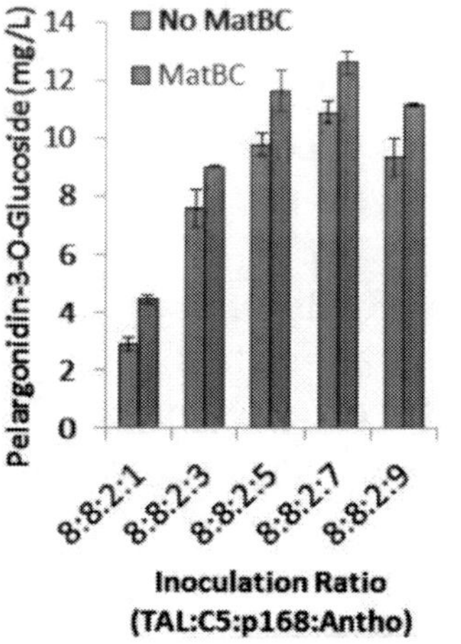

Strains:	4
Enzymes:	19

Rensselaer

How Much Is It? Predicting Enzyme Costs with a Genome Scale Model of Yeast: Applications to Metabolic Engineering

Sánchez, B. J., Chalmers University of Technology
Zhang, C., Science for Life Laboratory, KTH - Royal Institute of Technology
Kerkhoven, E., Chalmers University of Technology
Nielsen, J., Chalmers University of Technology

Genome scale modelling (GEM) is a recurrent and successful approach for modelling metabolic fluxes of *Saccharomyces cerevisiae* (budding yeast) and predicting genetic modifications for increased performance of mutant strains[1]. However, when considering the production of a metabolite of interest, these models typically make the major assumption that the uptake rate of the carbon source (e.g. glucose) is what limits production. This is hardly the case, since experimental yields are usually considerably lower than the maximum theoretical yields. Instead, what is likely happening in several cases is that there is a limitation of the amount of enzymes from the relevant pathways. GEMs of yeast so far are not able to encapsulate this phenomenon, because they do not include a direct way of connecting enzyme levels to metabolic fluxes. The latter should give insight into cellular capabilities at conditions of high enzymatic demand, such as yeast growing at a high growth rate or under stress, consuming non-typical carbon sources and/or overexpressing a given pathway.

Here we present the enhancement of a yeast GEM[2] to account for enzymatic constraints. It is done with the constraint-based approach[3] extended to include enzymes as part of reactions, using mass balances for both metabolites and enzymes[4]. With this formalism reaction fluxes are limited by the maximum flux the enzyme is able to catalyse (v_{max}), which can be estimated from the enzyme's abundance inside the cell together with the enzyme's turnover number (k_{cat})[5]. Using this approach with different levels of data, we show that we can predict metabolic strategies that regular constrained-based approaches cannot and significantly reduce variability of flux predictions. Most importantly, with different case studies we show that we gain insight into the distribution of enzyme costs inside any metabolic pathway and under different conditions. This way we can understand which could be the key steps that limit yeast growth/metabolite production in metabolic engineering applications.

References:

1 B. J. Sánchez and J. Nielsen, *Integr. Biol.*, 2015, **7**, 846–858.

2 H. W. Aung, S. A. Henry and L. P. Walker, *Ind. Biotechnol.*, 2013, **9**, 215–228.

3 N. E. Lewis, H. Nagarajan and B. Ø. Palsson, *Nat. Rev. Microbiol.*, 2012, **10**, 291–305.

4 E. J. O'Brien and B. Ø. Palsson, *Curr. Opin. Biotechnol.*, 2015, **34**, 125–134.

5 R. Adadi, B. Volkmer, R. Milo, M. Heinemann and T. Shlomi, *PLoS Comput.*

Biol., 2012, **8**.

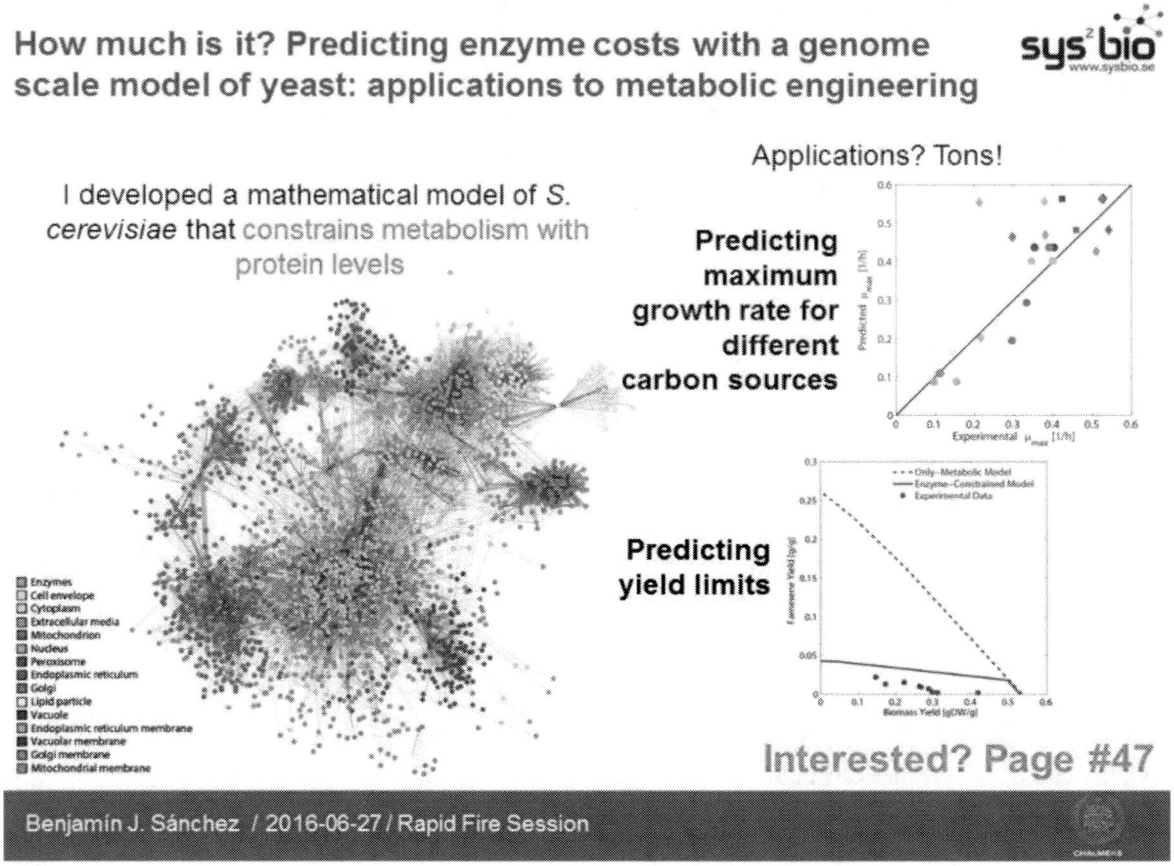
How much is it? Predicting enzyme costs with a genome scale model of yeast: applications to metabolic engineering
sys2bio
www.sysbio.se
I developed a mathematical model of S. cerevisiae that constrains metabolism with protein levels
Enzymes
Cell envelope
Cytoplasm
Extracellular media
Mitochondrion
Nucleus
Peroxisome
Endoplasmic reticulum
Golgi
Lipid particle
Vacuole
Endoplasmic reticulum membrane
Vacuolar membrane
Golgi membrane
Mitochondrial membrane
Applications? Tons!
Predicting maximum growth rate for different carbon sources
Predicting yield limits
Only-Metabolic Model
Enzyme-Constrained Model
Experimental Data
Interested? Page #47
Benjamín J. Sánchez / 2016-06-27 / Rapid Fire Session

Metagenomic Data to Molecules: Novel Metabolites Identified from the Human Gut Microbiome

Chang, F. Y.,
Voigt, C. A., University of California, San Francisco

Human microbiome-derived metabolites, including host modulators and therapeutic leads, have been isolated from a small fraction of bacteria that have been cultivated in lab. However, the majority of microbes in the human microbiome remains recalcitrant to conventional means of cultivation and has only been identified from metagenomic sequencing. As such, chemical diversity of the uncultured microbial majority remains largely unexplored. Here we developed a sequence-to-molecule pipeline for the discovery of novel compounds from the human metagenome, starting from sequence data as the only input source. Comparative bioinformatics analysis of metagenomic samples from the Metagenomics of the Human Intestinal Tract Project and the National Institutes of Health Human Microbiome Project led to the identification of an unprecedented family of nonribosomal peptide synthetase (NRPS) gene clusters that were widely and exclusively distributed in the human gut. Systematic reconstruction of these gene clusters designed for *E. coli* expression and heterologous production led to the isolation of novel compounds in two distinct structural classes, namely nitrogen-containing heterocycles and linear acylpeptides. Knockout studies of the NRPS machinery revealed that the production of two distinct compound structures was dictated by the functionality of the individual NRPS protein domains, where silent starting condensation domain and functional terminal reductase domain led to heterocycle production and the opposite led to acylpeptide production. Rational design of a synthetic NRPS hybrid with functional starting and terminal domains generated novel compounds with unique acylpeptidyl alcohol/aldehyde structures, suggesting a third structural class that this microbiome-derived gene cluster family encodes. With the improved DNA sequencing technology rapidly expanding the amount of genomic information, the capability to transform digital data into chemical entities makes this a powerful strategy for the systematic discovery of novel bacterial metabolites from the human metagenome and other genetic sources.

Cell-Free Biosynthesis of Complex Molecules Using Designed and Tailored Permeabilized Cells

Heinzle, E.,
Blass, L.,
Weyler, C.,
Krauser, S.,

Complex low molecular weight compounds comprising polyketides, polypeptides, carbohydrates as well as combinations and derivatives thereof are presently regaining importance as weapons against various diseases like cancer, infections or chronic illnesses. Metabolic engineering activities have been directed to the biosynthesis of such compounds using natural producers but increasingly also model organisms as *E. coli*. Though the biological diversity is enormous and increasingly accessible using modern genetic engineering methods, there are some crucial limitations: (i) Usually the obtained final titers are low, (ii) chemically similar byproducts produced complicate down-stream processing, (iii) the creation of new compound variants is generally possible by engineering the core synthesis machines, usually megasynthases, but still very difficult and time consuming, and (iv) *in vitro*studies of megasynthases are limited because of their inherent fragility.

We show that cell-free biosynthesis using permeabilized cells, in combination with computational tools for adequate pathway design, contributes significantly to solving problems sketched above and to develop organisms and processes for the production of complex low molecular weight compounds. Permeabilized cells still keep all their proteins in a state close to the original *in vivo* one. After removal of all original metabolites, a designed cocktail of substrates and cofactors is added to start biosynthesis.

First we developed computational tools to design metabolic networks supporting the cell-free biosynthesis. We start from a pan-organism metabolic network derived from databases as KEGG combined with thermodynamic data for reaction equilibria (eQuilibrator) and search for a biosynthesis path or network for a target molecule. A list of potential start metabolites and cofactors is used together with a set of constraints to identify pathway candidates using MILP. These candidates are then ranked using a set of criteria, e.g. length, thermodynamics or number of heterologous genes required. Promising candidates are finally evaluated by experts and implemented.

With some practical examples, where we applied engineered, permeabilized cells for production of uridinylated carbohydrates (UDP-glucose), polyketides (flaviolin and oxytetracycline) and non-ribosomal polypeptides (luminmides and new biosynthetic variants), we show that this methodology will significantly contribute to the production of complex low molecular weight compounds.

Cell-Free Biosynthesis of Complex Molecules Using Designed and Tailored Permeabilized Cells

E. Heinzle, L. Blass, S. Krauser, C. Weyler, Biochemical Engineering, Saarland University

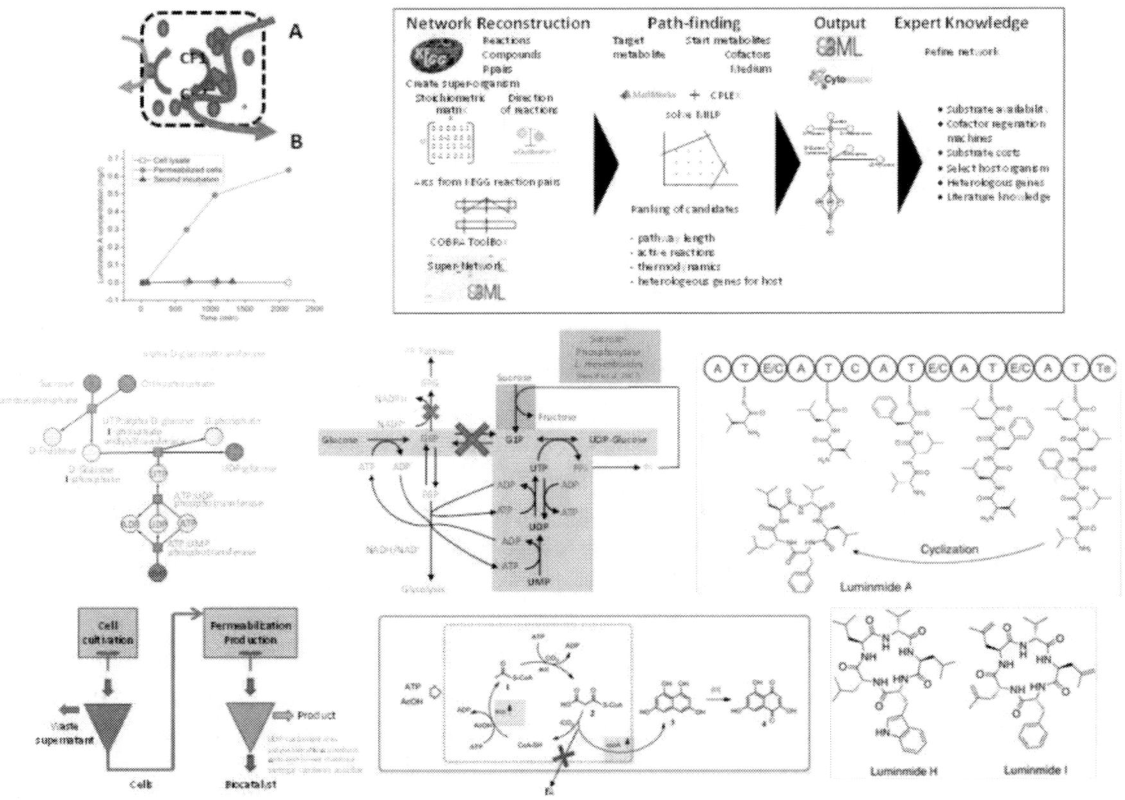

Crispr-Cas Assisted Metabolic Engineering of Lipid Metabolism in Saccharomyces Cerevisiae

Ferreira, R., Chalmers University of Technology
David, F., Chalmers University of Technology
Siewers, V., Chalmers University of Technology
Nielsen, J., Chalmers University of Technology

Engineering microbes for industrial-scale production of valuable chemicals involves extensive modulation of cellular metabolism for enhanced productivity. Genome engineering to introduce different types of genetic modifications including gene deletion, overexpression, and exact regulation is extremely important to improve pathway efficiency and product yield. In this context, CRISPR/Cas offers a powerful time efficient tool for targeting specific genomic loci of interest, which can conceivably be used to tune different gene expressions. As *Saccharomyces cerevisiae* is a well-studied industrially robust microorganism and the overall metabolism, including the lipid metabolism, is well known, it is feasible to engineer lipid metabolism for overproduction of oleochemicals reaching industrial standards. Here we use CRISPR/Cas to engineer the lipid metabolism of *S. cerevisiae* towards accumulation of triglycerides. Triacylglycerols (TAGs) are key molecules for the cell as they function as energy storage and also as feedstock for the production of oleochemicals and biodiesel. In order to achieve industrially relevant titers of TAGs, we regulated a series of genes involved in various pathways such as beta-oxidation, lipid biosynthesis, lipid droplet formation and autophagy to prevent these lipids from mobilisation/degradation by the cell. Ultimately, the knowledge obtained from the different strains generated will provide a better understanding on how lipid metabolism is regulated, and how it affects the cell´s phenotype, lipid composition, growth and size.

Metabolic Engineering of Lipid Metabolism in *Saccharomyces cerevisiae*

Raphael Ferreira, Paulo Teixeira, Florian David, Verena Siewers, Jens Nielsen
Systems and Synthetic Biology, Dept. of Biology and Biological Engineering
Chalmers University of Technology, Göteborg, Sweden, **rapfer@chalmers.se**

1. **Established Multiple Metabolic Engineering Platform in *S. cerevisiae***

2. **Towards a Minimal Lipid Metabolism**

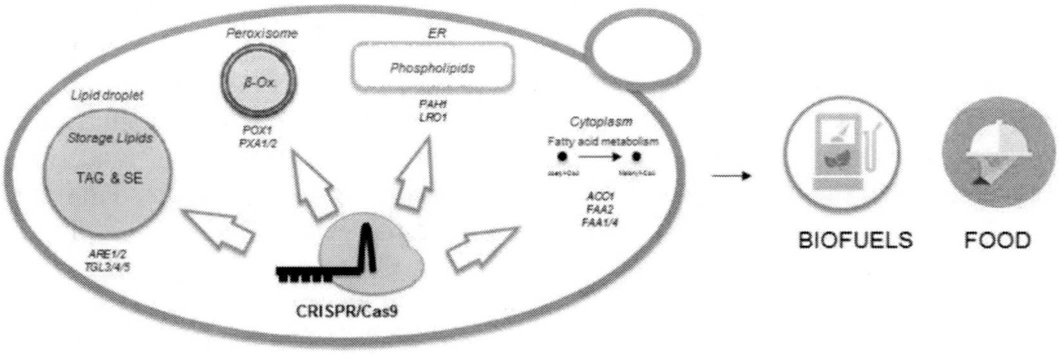

Engineering the Spliceosome of Saccharomyces Cerevisiae to Splice Heterologous Introns from Distant Fungi

DeNicola, A. B., University of California, Los Angeles
Tang, Y., University of California, Los Angeles

As filamentous fungi are prolific producers of valuable compounds in the fight against human disease, there has been renewed interest in leveraging the recent advances in genomics to discover biosynthetic pathways that yield new chemical scaffolds. The explosion of genomic sequence data has revealed a wealth of uncharacterized natural product gene clusters in microbial species and it is essential to develop new tools to rapidly decode and translate this genetic information in order to discover new bioactive compounds. Fungal gene clusters contain many non-coding introns, which are difficult to manually annotate or predict *in silico*. If the gene cluster is active in the native producer, intron-free cDNA can be obtained, but neither silent gene clusters nor gene clusters in uncultivable species can be interrogated by this approach. Further, the native spliceosome of the model eukaryote *Saccharomyces cerevisiae* cannot remove introns from distant fungi, thus hampering natural product characterization efforts. Two naturally-occurring introns were used in this study: the second *MATa1* intron from *S. cerevisiae* and the *PES1* intron from *Aspergillus fumigatus*. Using a prototrophic selection marker as a splicing reporter, intron sequence motifs from *PES1* were identified that abolish splicing in yeast. A chromosomally-integrated splicing reporter was developed for the rapid, cloning-free construction of intron variants and subsequent characterization of splicing activity via dosage-dependent copper selection. Through the mutation of specific splicing factors in *S. cerevisiae*, the intron sequence motifs from *A. fumigatus* are identified by the spliceosome and intron splicing is restored. A yeast strain capable of removing the heterologous introns in gene clusters from different fungal species will be a valuable synthetic biology tool for the discovery of new chemical compounds and potential therapeutics.

GNPN

Engineering the Spliceosome of *Saccharomyces cerevisiae* to Splice Heterologous Introns from Distant Fungi

Anthony DeNicola and Yi Tang

Motivation

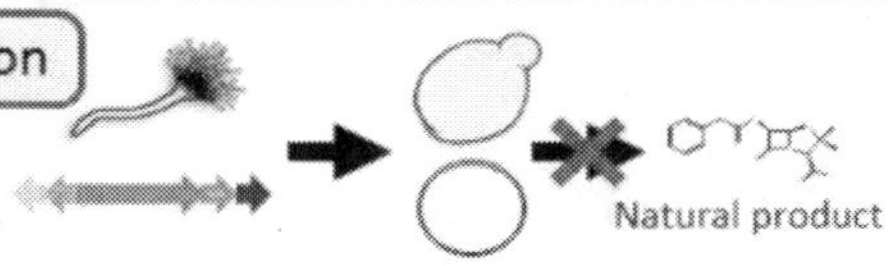

The Intron Problem

1. Incorrect *in silico* intron prediction
2. No cDNA available – silent biosynthetic pathways
3. Natural product genes contain an average of 5-9 introns

Aim 1. Engineer yeast with a copper-sensitive reporter to assay splicing efficiency

Aim 2. Rescue inefficiently spliced heterologous introns with mutant splicing factors

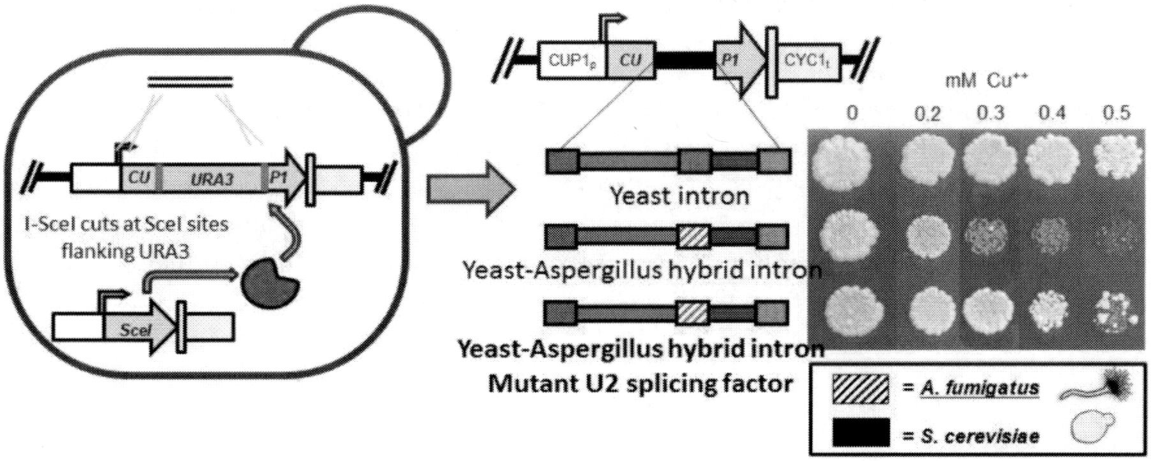

Combining Metabolic Engineering and Biocompatible Chemistry for High-Yield Production of Homo-Diacetyl and Homo-(S,S)-2,3-Butanediol

Liu, J., Technical University of Denmark
Lee, S. Y., Korea Advanced Institute of Science and Technology (KAIST)
Solem, C., Technical University of Denmark
Jensen, P. R., Technical University of Denmark

Biocompatible chemistry is gaining increasing attention because of its potential within biotechnology for expanding the repertoire of biological transformations carried out by enzymes. Here we demonstrate how biocompatible chemistry is used for synthesizing important value-added compounds as well as for linking metabolic pathways to achieve redox balance and rescued growth. By comprehensive rerouting of metabolism, activation of respiration, and finally metal ion catalysis, we successfully manage to convert the homolactic *Lactococcus lactis* into a homo-diacetyl producer with high titer (95 mM or 8.2 g/L) and high yield (87% of the theoretical maximum). Subsequently, the pathway is extended to (*S*,*S*)-2,3-butanediol (S-BDO) through efficiently linking two metabolic pathways by chemical catalysis. The outcome is efficient homo-S-BDO production with a titer of 74 mM (6.7 g/L) S-BDO and a yield of 82%. The diacetyl and S-BDO production rates and yields obtained are the highest reported, demonstrating the promising combination of metabolic engineering and biocompatible chemistry and the great potential of *L. lactis* as a new production platform.

Engineering Biofuel Tolerance Using Efflux Pumps

Basler, G., University of California, Berkeley
Tullman-Ercek, D., University of California, Berkeley

A major limitation for the efficient production of biofuels is that their intracellular accumulation leads to a disruption of membrane integrity, inhibition of growth, and cell death. Therefore, the engineering of efflux pumps to facilitate the cellular export of biofuels has proven a valuable strategy for improving production efficiency. We previously used directed evolution to develop pumps that secrete short-chain alcohols; the presence of these engineered pumps confer improved alcohol tolerance of *Escherichia coli*. Moreover, we have designed a stress-response system which dynamically optimizes the efficiency of efflux pumps by fine-tuning their expression levels.

Here, we describe our efforts to extend this approach to engineer the efflux pumps of *Pseudomonas putida*, which confer a high tolerance to several biofuels, including *n*-butanol. To exploit the use of these pumps for biofuels and achieve further enhanced tolerance in this promising production host, we apply targeted modifications to a native pump of *P. putida* that is homologous to our successful variant from *E. coli*. We further integrate the dynamic stress-response system for optimizing the expression levels and investigate the effect of the system on biofuel tolerance of *E. coli* and *P. putida*. The engineered strains represent highly promising host systems for enabling the large-scale production of biofuels.

Metabolic Engineering of Saccharomyces Cerevisiae for Production of Short / Medium Chain Fatty Acids

Zhu, Z., Chalmers University of Technology
Zhou, Y. J.,
Krivoruchko, A., Chalmers University
Siewers, V., Chalmers University of Technology
Nielsen, J., Chalmers University of Technology

Short / medium chain fatty acids (S/MCFAs, 6-12 carbons) are important industrial chemicals used to produce a range of commercial molecules, which serve as oleochemicals for coatings, plasticizers, soaps, detergents, lubricants and flavors, and transportation fuels alternative to current gasoline and jet fuel. Growing concerns over the environment and unsustainable petroleum resources have led to efforts to develop S/MCFA based fuels with improved quality rather than traditional bioethanol. However, limited annual output of coconut oil, the plant oil extracted from the fruits of tropical coconut palm and rich in medium-chain triglycerides, is insufficient for the bulk production of short / medium chain biofuels. Exploitation of the microbial biosynthetic pathways via metabolic engineering and synthetic biology tools provides a sustainable way to economically and steadily produce these kinds of fatty acids by transformation of lignocellulosic biomass based materials. And the yeast *Saccharomyces cerevisiae* is selected as the microbial cell factory for these products based on some superior features, such as easily accessible genetic manipulation tools, well-characterized cellular genetic, metabolic and regulatory processes, simple culture media, and inexpensive industrialization via the exist infrastructures of bioethanol. We therefore engineered yeast metabolism to produce S/MCFAs and this engineered strain represent an excellent platform cell factory for future development of bioprocesses for production of these valuable chemicals.

Construction of Platform Strains Producing Cosmeceutical Materials with Yarrowia Lipolytica

Yu, B. J., Korea Institute of Industrial Technology(KITECH)
Jang, J. Y., Korea Institute of Industrial Technology(KITECH)
Kim, S. K., Korea Institute of Industrial Technology(KITECH)

Yarrowia lipolytica which has been extensively used as the model oleaginous yeast for producing heterologous proteins has the capability to accumulate lipids (lipidbody, peroxisome *etc*) up to 50% of the DCW (dry cell weight). In addition, *Y. lipolytica* can be easily manipulated in gene deletion and insertion because it possesses a haploid genome and its genome sequence was completed and it is also known as GRAS (generally recognized as safe). Therefore, *Y. lipolytica* can be a high potential host strain in producing oleaginous (hydrophobic) and high value-added compounds used in cosmetic, medicine and nutraceutical materials which are hydrophobic and produced from plants by conventional extraction processes. Recently, these compounds have been produced in microorganism and it was well known that the DuPont company produced omega-3 using *Y. lipolytica* and industrialized it in 2010.

In this study, we used *Y. lipolytica* Po1g and constructed an ura3 auxotroph strain, *Y. lipolytica* (PolgDura3), in order to use URA3 blaster system which is a useful tool to carry out repeat genetic deletion and insertion on the genome. We also constructed a *Y. lipolytica* (PolgDku70Dura3) in which Ku70 involved in non-homologous recombination was deleted to increase recombination efficiency in this strain. Therefore, we confirm that *Y. lipolytica* (PolgDku70Dura3) showed higher homologous recombination efficiency (>5 folds) than *Y. lipolytica* (PolgDura3). In addition, we designed and constructed an expression vector pBJY_1 in which strong promoter, URA3 pop-out cassette and integrated target region were cloned for high expression of target genes and genome integration in *Y. lipolytica.* We synthesized target genes involved in biosynthesis of specific cosmetic and medicine materials and are cloning into the vector. We are sure that this strain and genetic tool are very useful to construct and develop various customized and industrial strains for producing various and valuable cosmeceuticals.

Total Biosynthesis of Opiates By Engineered Escherichia coli

Nakagawa, A., Ishikawa prefectural university
Sato, F.,
Minami, H., Ishikawa prefectural university

Opiates such as morphine, codeine, and thebaine have been used since ancient times and are the gold standard for pain relief. They are mainly obtained by extraction from opium poppies. To increase the production yields of specific alkaloids, extensive efforts have been made to characterize and subsequently modify the alkaloid biosynthetic pathway. However, the regulation of biosynthetic pathways in plants is complex; therefore, it is relatively difficult to increase production of desired products. Fermentative opiate production in microbes has also been investigated, and complete biosynthesis of opiates from a simple carbon source has recently been accomplished in the yeast system. Accordingly, microbial alkaloid production may facilitate to construct an efficient production system. The next challenge is to increase production. In this report, we demonstrate that *Escherichia coli* serves as an efficient, robust, and flexible platform for total opiate synthesis. Thebaine, the most important raw material in opioid preparations, was produced by stepwise culture of four engineered strains at yields of 2.1 mg/L from glycerol, which corresponds to a 300-fold increase from recently developed yeast systems. This large improvement was presumably due to strong activity of enzymes related to thebaine synthesis from (*R*)-reticuline in *E. coli*. Furthermore, by adding two genes to the thebaine production system, we were able to biosynthesize hydrocodone, a clinically important opioid. Large improvements in opiate production in this *E. coli* system represent a major step toward the development of a practical opiate production system.

Itaconic Acid and Beyond: New Insights into the Itaconic Acid Production Pathway of Ustilago Maydis

Wierckx, N., RWTH Aachen University
Geiser, E.,
Przybilla, S.,
Bölker, M.,
Blank, L. M., RWTH Aachen University

Itaconic acid is a bio-based platform chemical for the production of pharmaceuticals, adhesives and polymers. *Ustilago* is a promising fungal host for the production of itaconic acid [1]. Contrary to established filamentous itaconate producers, *Ustilago* grows unicellularly. This has distinct process advantages relating to morphology control, viscosity and aeration [2]. However, the itaconate production yield, titer and rate of *Ustilago* are thusfar below those of the commercial production hosts. In order to optimize itaconate production in *Ustilago* by metabolic engineering, detailed biochemical knowledge was needed.

We recently identified an itaconate gene cluster in *Ustilago*. The function of the cluster genes and their encoded proteins was characterized by knockout and overexpression. The activity of two essential catalytic proteins was determined *in vivo* by permeabilized cell assays and *in vitro* by *E. coli* expression and purification. In contrast to the known itaconate biosynthesis pathway of *Aspergillus terreus*, *Ustilago's* itaconate production proceeds first via an isomerization from *cis-* to *trans*-aconitate, followed by decarboxylation of the *trans*-aconitate [3].

Two additional genes in the cluster, *cyp3* (cytochrome P450 monooxygenase) and *rdo1* (ring-cleaving dioxygenase) are not directly involved in itaconate production. Instead, they appear to have a role in the further conversion of itaconate, likely into the more oxidized products 2-hydroxyparaconate and itatartarate. Although these products have been identified previously [2,4], the associated genes and pathway was until now unclear. We here present the application of this new biochemical knowledge to enhance the overall product titer and yield, but also to precisely control product specificity. The results are discussed in the context of the process advantages of smut fungi as novel hosts for industrial biotechnology.

References

[1] N. Maassen, M. Panakova, N. Wierckx, E. Geiser, M. Zimmermann, M. Bölker, U. Klinner, L.M. Blank. 2014. Influence of carbon and nitrogen concentration on itaconic acid production by the smut fungus *Ustilago maydis*. *Eng. Life Sci.* **14**:129-134.

[2] E. Geiser, V. Wiebach, N. Wierckx, L.M. Blank. 2014. Prospecting the biodiversity of the fungal family Ustilaginaceae for the production of value-added chemicals. *Fung. Biol. Biotechnol.* **1**:2

[3] E. Geiser, S.K. Przybilla, A. Friedrich, W. Buckel, N. Wierckx, L.M. Blank, M.

Bölker. 2016. *Ustilago maydis* produces itaconic acid via the unusual intermediate *trans*-aconitate. *Microb. Biotechnol.* **9**: 116-126

[4] Guevarra, E.D. and Tabuchi, T. (1990) Accumulation of itaconic, 2-hydroxyparaconic, itatartaric and malic acids by strain of the genus *Ustilago*. *Agric. Biol. Chem.* **54**:2353

Engineering Escherichia coli for High Selective Production of Butyric Acid

Chiang, C. J., China Medical University
Chao, Y. P., Feng Chia University

Butyric acid has many important applications in industry and is mainly produced by the petroleum-based process. The bio-based production method for butyrate is increasingly interesting because of a concern over global warming and fossil fuel shortage. *Clostridium* species have been long applied for the butyrate production. However, a mixture of butyrate and acetate is produced in the clostridial fermentation, which results in low selectivity of butyrate (ca. the butyrate/acetate (B/A) ratio within 10-14). To address this issue, *Escherichia coli* was engineered for the production of butyrate. This was carried out by pathway engineering of *E. coli* in a systematic way. First, the pathways leading to byproducts were removed to conserve NADH. Secondly, a heterologous pathway was constructed to direct acetyl-CoA to butyryl-CoA. Finally, endogenous *atoDA* was overexpressed for conversion of butyryl-CoA to butyrate at the expense of acetate. As a consequence, the resulting strain enabled production of 10 g/L butyrate from 20g/L glucose and 8 g/L acetate within 48 h. The result accounts for the B/A ratio of 143. It indicates that our approach is promising for production of butyrate in *E. coli*.

In Vitro Metabolic Pathway for the Salvage Synthesis of Nicotinamide Cofactor

Honda, K., Graduate School of Engineering, Osaka University

Excellent thermal and operational stabilities of thermophilic enzymes can greatly increase the applicability of biocatalysis in various industrial fields. On the other hand, since thermophilic enzymes are optimally active at high temperatures, there is an inherent limitation in the application of these enzymes, namely their incompatibility with thermally labile substrates, products, and cofactors. Particularly, the relatively low thermal stabilities of nicotinamide cofactors (NAD^+/NADH and $NADP^+$/NADPH) tend to be a major obstacle in the application of thermophilic enzymes owing to their pivotal roles in a wide range of enzymatic redox reaction. Recently, we developed a novel approach to engineer an in vitro artificial pathway by assembling multiple thermophilic enzymes and demonstrated one-pot bioconversions of glucose to lactate, malate, and 1-butanol through the in vitro pathways. Although they could be accomplished at sufficient production rates and yields, the conversions stopped within several hours due to the thermal decomposition of nicotinamide cofactors [1-3]. To overcome this limitation, in the present study, we constructed an in vitro metabolic pathway consisting of eight thermophilic enzymes for the salvage synthesis of NAD^+ from its thermal degradation products (nicotinamide and ADP-ribose). NAD^+ concentration could be kept almost constant for 15 h at 60ºC in the presence of the salvage enzymes whereas that in the control experiments without the enzymes decreased to less than 25% of the initial level after the incubation for the same period. Thus, the in vitro salvage synthetic system can markedly improve the apparent stability of NAD^+ at high temperatures and increase the feasibility of biocatalytic chemical manufacturing with thermophilic enzymes.

1) Ye X, et al. (2012) *Microb Cell Fact*, 11: 120

2) Ye X, et al. (2013) *J Biotechnol*, 164: 34-40

3) Krutsakorn B, et al. (2013) *Metab Eng*, 20: 84-91

Metabolic Engineering to Produce Rosmarinic Acid in Baker´s Yeast Saccharomyces Cereviciae

Borja Zamfir, G. M.,
Chen, X.,
Nielsen, J., Chalmers University of Technology
Borodina, I., Technical University of Denmark

Rosmarinic acid (RA) is a secondary plant metabolite with antioxidant, anti-inflammatory, antibacterial, and antiviral properties. Currently the rosmarinic acid is obtained by extraction from plants. Microbial production of RA is an attractive alternative, which can save time and cut the production costs. In this study we aimed to engineer the yeast Saccharomyces cereviciae to produce RA from a cheap carbon source, e.g., glucose. First, we reconstructed two variants of biosynthetic pathway from p-coumaric acid to RA, a plant-like pathway and a chimeric microbial-plant pathway, and obtained 20 ± 4 and 18 ± 3 µg·L-1 of RA respectively, when culturing the cells with p-coumaric acid. Second, we overexpressed tyrosine ammonia lyase Flavobacterium johnsoniae to obtain RA directly from glucose without supplementation of precursor. Both the plant-like pathway and the chimeric pathway resulted in similar titers of ~10 µg·L-1, when culturing the cells on minimal medium with glucose as the sole carbon source. Furthermore we engineered the strains to increase the supply of precursors, p-coumaric acid and 4-hydroxyphenyllactic acid, by overexpressing ARO4fbr & ARO7fbr and knocking out PDC5 and ARO10, leading to an increase of 2.8-fold of p-coumaric acid and 3.1-fold in 4-hydroxyphenyllactic acid compared to the non-engineered strains and with no significant effect on RA production. Subsequently, the copy number of rosmarinic acid synthase-encoding gene was increased using either a multiple integration into the genome or expression from a 2µ-vector to result in 3.3 and 6.6-fold increase of RA for plant and chimeric pathways respectively, compared to the parental strains. Finally, cultivating the best producing strain in feed-in-time medium resulted in the final concentration of 256±25 µg·L-1 of RA. To our knowledge this is the first time rosmarinic acid was produced in S. cerevisiae.

Model-Based Metabolic Engineering of Escherichia coli for High Yield Itaconic Acid Production

Harder, B. J., Max Planck Institute for Dynamics of Complex Technical Systems
Bettenbrock, K., Max Planck Institute for Dynamics of Complex Technical Systems
Klamt, S., Max Planck Institute for Dynamics of Complex Technical Systems

Itaconic acid is a high potential platform chemical which is currently industrially produced by *Aspergillus terreus*. It is primarily used for polymer synthesis and has the potential to replace petrochemically derived methacrylic acid. Heterologous production of itaconic acid with *Escherichia coli* could help to overcome limitations of *A. terreus* regarding slow growth and high sensitivity to oxygen supply. However, the performance achieved so far with *E. coli* strains is still low.

To enable heterologous itaconic acid production we introduced a plasmid carrying the codon-optimized genes of the cis-aconitic acid decarboxylase from *Aspergillus terreus* and the citrate synthase of *Corynebacterium glutamicum* into *E.coli* MG1655. We then applied a model-based approach to construct a high yield producer strain. Based on the concept of minimal cut sets, we identified intervention strategies that guarantee high itaconic acid yield while still allowing growth. One cut set was selected as starting point and some of the corresponding genes were iteratively knocked-out. As a conceptual novelty, we pursued an adaptive approach allowing changes in the model and initially calculated intervention strategy if a genetic modification induces changes in byproduct formation. Using this approach, we iteratively implemented 5 interventions leading to high yield itaconic acid production in Minimal Medium with glucose as substrate supplemented with small amounts of glutamic acid. In shake flask cultivations of the derived *E. coli* strain we achieved an itaconic acid titer of 2.2 g/l with an excellent yield of 0.77 mol/mol. In a fed-batch cultivation, this strain produced 32 g/l itaconic acid with an overall yield of 0.68 mol/mol and a peak productivity of 0.45 g/l/h. These values are by far the highest that have ever been achieved for heterologous itaconic acid production and indicate that realistic applications come into reach.

Our results demonstrate that itaconic acid can be produced with high yield and titer by a dedicated *E. coli* mutant strain. Furthermore it shows that constrained minimal cut sets, applied in an iterative manner, provide a suitable computational tool for *in silico* strain design and to identify intervention strategies for redirecting carbon flux to high product formation.

Heuristic Design and Engineering of pH-Inducible Yeast Promoters

Rajkumar, A. S.,
Arsovska, D., Technical University of Denmark
Kristensen, M., Technical University of Denmark
Jensen, M. K., Technical University of Denmark
Keasling, J. D., Lawrence Berkeley National Laboratory

While advances in synthetic biology have led to the development of synthetic promoters in eukaryotes and prokaryotes, few have been designed with environmental conditions imposed by fermentations in mind. Here, we present a set of strong, synthetic promoters for *Saccharomyces cerevisiae* that are inducible by low pH. By mining available transcriptome and transcription factor binding data and the relevant literature on transcriptional regulation and promoter architecture we identified transcription factor binding sites (TFBS) and other elements relevant to regulation by low pH. We then used these to modify the upstream activation sequence of a native promoter inducible by low pH, and improve its low-pH induction and overall regulatory output by a factor of six to ten. We subsequently used the same TFBS modifications to engineer a de novo response to low pH in a native promoter lacking one. After characterizing the regulatory response of the promoters at the RNA and protein level, we used the best promoter variants in industrial yeast to construct strains for the low-pH production of lactic acid. Ultimately, we achieved six-fold increased production of lactic acid with our promoters compared to titres obtained from a production strain with a popular constitutive yeast promoter. Our synthetic promoters are as strong as constitutive promoters typically used in yeast cell factories, and are a valuable addition to the toolbox of regulatory parts for metabolic engineering. Finally, our findings outline and validate a general strategy to iteratively design and engineer synthetic yeast promoters inducible to environmental conditions or stresses of interest.

Metabolic Pathways of Alginate-Derived Monomers in the Marine Bacterium Saccharophagus Degradans 2-40T

Kim, D. H., Korea University
Kim*, K. H., Korea University Graduate School

Marine plant biomass such as algae is a potential source for the sustainable production of biofuels and bio-based chemicals. Alginate, a major component of brown macroalgae, consists of two uronate monomers, mannuronic acid and guluronic acid, which are further non-enzymatically converted to 4-deoxy-L-erythro-5-hexoseulose uronic acid (DEH). DEH is known to be metabolized by a catabolic pathway consisting of DEH reductase, 2-keto-3-deoxy-d-gluconate (KDG) kinase, and 2-keto-3-deoxy-phosphogluconate (KDPG) aldolase, yielding two glycolytic intermediates, D-glyceraldehyde-3-phosphate and pyruvate. However, functional enzymes for the DEH pathway have rarely been identified. In the present study, a DEH catabolic pathway was investigated in *Saccharophagus degradans* 2-40^T, one of marine bacteria fermenting alginate. The genes encoding putative DEH reductase, KDG kinase, and KDPG aldolase were identified and individually overexpressed in *Escherichia coli*. Through *in vitro* tests assisted by GC-MS and GC/TOF MS, the purified enzymes were functionally confirmed and annotated as dehR, kdgK, and kdpgA, respectively. In a future study, the three genes will be used to build an engineered host expressing a complete DEH metabolic pathway for brown macroalgae-based bioconversion processes.

Chassis Optimization for Recombinant Rhamnolipid Synthesis

Tiso, T.,
Germer, A., RWTH Aachen University
Blank, L. M., RWTH Aachen University

Previously recombinant rhamnolipid production with *P. putida* KT2440 was established [1], delivering an alternative for the synthesis of this secondary metabolite with the pathogenic *P. aeruginosa*. While production rate and yield already compete with the conventional process based on *P. aeruginosa*, titers are far from being industrially interesting. Metabolic engineering of *P. putida* KT2440 to become a rhamnolipid synthesis chassis is presented here.

First, futile pathways were deleted. A strain deficient in the synthesis of the storage polymer polyhydroxyalkanoate (Δ*phaC*), and a knockout strain (Δ*fadB*) featuring a decelerated β-oxidation were tested. The novel chassis shows enhanced rhamnolipid production.

Second, precursor pathways were enhanced. For rhamnose synthesis phosphoglucomutase and the whole *rml* operon encoding the dTDP-rhamnose pathway were overexpressed. To facilitate the increased synthesis of the lipid precursor, one subunit of the pyruvate dehydrogenase complex was overexpressed, leading to triplication in surfactant titers compared to the control strain. Third, a mutant devoid of the energy intensive flagellar machinery [2] was tested, with the aim to reduce metabolic energy for maintenance.

Last, the export of rhamnolipids was improved using a putative transporter from *P. aeruginosa*.

The additive effects on rhamnolipid synthesis of the engineering targets identified here are investigated.

References

1. Wittgens A, Tiso T, Arndt TT, Wenk P, Hemmerich J, Müller C, Wichmann R, Küpper B, Zwick M, Wilhelm S, et al: **Growth independent rhamnolipid production from glucose using the non-pathogenic *Pseudomonas putida* KT2440.** *Microb Cell Fact* 2011, **10**.

2. Martinez-Garcia E, Nikel PI, Aparicio T, de Lorenzo V: **Pseudomonas 2.0: Genetic upgrading of *P. putida* KT2440 as an enhanced host for heterologous gene expression.** *Microb Cell Fact* 2014, **13**.

A Synthetic Biology Approach to Enhancing Metabolic Engineering of Clostridium Acetobutylicum

Mordaka, P.,
Heap, J.,

Transition from fossil carbon to sustainable fuels is crucial for the long-term energy security. *n*-Butanol is a natural product of the fermentation of carbohydrates by *Clostridium acetobutylicum* and can be used as biofuel and/or biofuel additive with superior properties over other bio-derived alcohols. Formerly, *C. acetobutylicum* was used in the industrial Acetone-Butanol-Ethanol (ABE) fermentation to produce solvents in a ratio of 3 parts acetone, 6 parts butanol to 1 part ethanol. However, further strain engineering and process optimization is needed to make production of bio-butanol economically feasible.

Metabolic engineering of *C. acetobutylicum* had been hampered for many years due to lack of suitable genetic tools, which some were developed very recently (Heap *et al.*, 2007; Heap *et al.*, 2007; Heap *et al.*, 2012). Libraries of expression control elements (promoters and RBSs) with known characteristics are important to allow construction of synthetic/heterologous metabolic pathways that perform predictably, and similarly to allow rational modification of existing pathways. To generate such libraries, reporter constructs are used to measure strengths of elements. However, classical fluorescent reporter proteins are oxygen-dependent and inactive in anaerobic bacteria. Therefore, we tested several oxygen-independent fluorescent reporter systems such as Light-Oxygen-Voltage proteins and SNAP/CLIP-tag protein fusions as well as enzymatic reporters phoZ and gusA. Among them gusA was found to give the highest sensitivity and the lowest background signal and was used to expand our existing toolkit of synthetic parts for controlling gene expression.

Subsequently, we attempted to create a high-yield strain of *C. acetobutylicum* by employing synthetic regulatory elements to develop and implement a system for metabolic flux redistribution. By using recently developed tools for genetic manipulation of Clostridia, we up-regulated and down-regulated expression of enzymes involved in the carbohydrate fermentation to improve production of solvents. This strategy will also be employed in the future to create high-yield strains of *C. acetobutylicum* and other industrially relevant strains for production of a wide range of commodity chemicals.

Heap, J.T., Pennington, O.J., Cartman, S.T., Carter, G.P. and Minton, N.P. (2007) The ClosTron: a universal gene knock-out system for the genus *Clostridium*. *J. Microbiol. Methods.* 70: 452-464

Heap, J.T., Pennington, O.J., Cartman, S.T. and Minton, N.P. (2009) A modular system for *Clostridium* shuttle plasmids. *J. Microbiol. Methods.* 78: 79-85

Heap, J.T., Ehsaan, M., Cooksley, C.M., Ng, Y.K., Cartman, S.T., Winzer, K. and Minton, N.P. (2012) Integration of DNA into bacterial chromosomes from plasmids without a counter-selection marker. *Nucleic Acid Res.* 40:e59

Marker-Free Crispr/Cas9 Based Genome Editing of Penicillium Chrysogenum

Pohl, C., University of Groningen
Kiel, J. A. K. W., University of Groningen
Driessen, A. J. M.,
Bovenberg, R. A. L., DSM Biotechnology Center
Nygård, Y., DSM Biotechnology Centre

CRISPR/Cas9 systems have emerged as versatile platforms for precise genome editing in a wide range of organisms. CRISPR/Cas9 tools offer high flexibility in target selection and the possibility to introduce genetic alterations without the introduction of epitopic sequences. These tools may in the future replace marker-based gene editing in filamentous fungi.

Here, we describe CRISPR/Cas9 tools for rapid genome engineering of *Penicillium chrysogenum*. We have developed powerful CRISPR/Cas9 tools for marker-based and marker-free gene modifications in *P. chrysogenum*, a model filamentous fungi but also an industrially relevant cell factory. The developed CRISPR/Cas9 toolbox is highly flexible and allows editing of new targets with minimal cloning efforts.

We have used the CRISPR/Cas9 tools for metabolic engineering of *P. chrysogenum*, for improved secondary metabolite (SM) production. We use a modular cloning approach for building novel fungal expression cassettes and complete pathways in *Escherichia coli* and we have created a library of genetic parts (including promoters, activators, transcription factors, terminators, fluorescent reporters and integration sites) for rapid and standardized assembly of genetic constructs and deletion cassettes.

CRISPR/Cas9 based technologies are expected to significantly advance the research and exploitation of novel natural products and the development of *P. chrysogenum* as an industrial chassis organism. A major obstacle in the exploration of the SM clusters in filamentous fungi has been the lack of efficient genome editing tools, therefore the CRISPR/Cas9 technology is expected to become a major driver in exploitation of fungal diversity.

Production of Sphingolipids By Metabolically Engineered Escherichia coli

Ha, J. Y., Korea Advanced Institute of Science and Technology (KAIST)
Lee, S. Y., Korea Advanced Institute of Science and Technology (KAIST)

Sphingolipids are active compounds that can be used as cosmetic ingredients. Previously, they are extracted from animals, but some issues has been raised for the animal derived cosmetic ingredients like bovine spongiform encephalopathy contamination. Thus, the alternative ways to produce sphingolipid compounds apart from animal based production are needed to be developed. Through this study, we suggest a sphingolipid biosynthetic pathway by introducing heterologous enzymes related to sphingolipid production into *Escherichia coli*. First, the expressions of heterologous enzymes were tested, and sphingolipids production was confirmed by flask cultivation supplementing *L*-serine and palmitic acid. After the confirmation, the serine and palmitoyl-CoA production pathway was metabolically engineered to produce sphingolipids from glucose as a sole carbon source. As a result, sphingolipids were successfully synthesized from glucose in *E. coli*.
[This work was supported by the Technology Development Program to Solve Climate Changes on Systems Metabolic Engineering for Biorefineries from the Ministry of Science, ICT and Future Planning (MSIP) through the National Research Foundation (NRF) of Korea (NRF-2012M1A2A2026556 and NRF-2012M1A2A2026557).].

Metabolic Engineering of Mannheimia Succiniciproducens for the Production of Gamma-Butyrolactone

Kim, W. J., Korea Advanced Institute of Science and Technology (KAIST)
Choi, S., KAIST
Kim, H. U., Korea Advanced Institute of Science and Technology (KAIST)
Lee, S. Y., Korea Advanced Institute of Science and Technology (KAIST)

γ-Butyrolactone (GBL) is one of the most important chemicals, because it has a wide range of industrial applications. Because GBL is easily converted by acid treatment of 4-hydroxybutyric acid (4-HB), microbial overproduction of 4-HB followed by its chemical conversion can be a useful strategy for the biosynthesis of GBL. 4-HB can be produced in *Mannheimia succiniciproducens* LPK7 (*ldhA, pflD, pta,* and *ackA* mutant) by the introduction of heterologous genes that encode succinyl-CoA synthetase, CoA-dependent succinate semialdehyde dehydrogenase, and either 4-hydroxybutyrate dehydrogenase in LPK7 (p3S4CD) or succinate semialdehyde reductase in LPK7 (p3SYCD). LPK7 (p3S4CD) and LPK7 (p3SYCD) successfully produced 6.37 and 6.34 g/L of 4-HB, respectively, and GBL was finally obtained from chemical conversion of 4-HB. This study presents that the engineered rumen bacterium *M. succiniciproducens* can produce industrially valuable chemicals. [This work was supported by the Technology Development Program to Solve Climate Changes on Systems Metabolic Engineering for Biorefineries from the Ministry of Science, ICT and Future Planning (MSIP) through the National Research Foundation (NRF) of Korea (NRF-2012M1A2A2026556 and NRF-2012M1A2A2026557)]

Elementary Mode Analysis for Efficient Succinic Acid Production in Mannheimia Succiniciproducens

Kim, W. J., Korea Advanced Institute of Science and Technology (KAIST)
Choi, S., KAIST
Kim, H. U., Korea Advanced Institute of Science and Technology (KAIST)
Lee, S. Y., Korea Advanced Institute of Science and Technology (KAIST)

Systems metabolic engineering using *in silico* tools has facilitated mass production of wide range of valuable chemicals. Elementary Mode (EM) analysis, one of the useful *in silico* tool, has the advantage of simple implementation, however, there are difficulties with analysis and application of thousands of EMs for metabolic engineering. Here, we designed method by selection of optimal EMs and subsequently combining of clustering method, thereby easily provides valuable information for metabolic engineering. EM analysis with clustering method applied to analyze optimal metabolic network for the production of succinic acid in *Mannheimia succiniciproducens*, and simulation results were validated with previous experiment results. Additionally, the yield of succinic acid production in *M. succiniciproducens* successfully improved through experiments from simulation results. [This work was supported by the Technology Development Program to Solve Climate Changes on Systems Metabolic Engineering for Biorefineries from the Ministry of Science, ICT and Future Planning (MSIP) through the National Research Foundation (NRF) of Korea (NRF-2012M1A2A2026556 and NRF-2012M1A2A2026557)]

Characterization of the Eucalyptus grandis Terpenome Using Engineered Yeast Strains

Li, R.,
Yoshikuni, Y., University of California, Berkeley
Keasling, J. D., Lawrence Berkeley National Laboratory

Sequenced in 2014, *Eucalyptus grandis* has more terpene synthases than any other plant studied to date. The essential oil of *E. grandis* is highly complex, making it difficult to purify and study the terpene constituents that may be useful as fuels, chemicals, and pharmaceuticals. To simplify the search for novel terpenes and to connect them with their corresponding synthases in *E. grandis*, we have engineered *S. cerevisiae* to be used as a chassis for terpene synthase expression. Each strain contains an engineered mevalonate pathway, including the expression of heterologous genes to circumvent the native feedback regulation at HMG1 and overexpression of the precursor prenyl diphosphate synthase. The remainder of the ergosterol pathway is downregulated by the introduction of a copper repressible promoter, diverting flux towards terpene production. In collaboration with the Joint Genome Institute, the 2014 *E. grandis* genome was scanned with stringent filters for terpene synthases, finding 73 terpene synthases and 25 of the triterpene-producing oxidosqualene cyclases. In total 98 enzymes were synthesized and are being expressed and characterized in the yeast strains engineered for high terpene flux.

Design of Novel Biosynthetic Pathways for the Production of Industrially Valuable Chemicals

Kim, D. I., Korea Advanced Institute of Science and Technology (KAIST)
Chae, T. U., Korea Advanced Institute of Science and Technology (KAIST)
Lee, S. Y., Korea Advanced Institute of Science and Technology (KAIST)

There have been many efforts in developing systematic platforms to predict novel metabolic pathways for the efficient production of industrially desirable chemicals. In this need, we developed a systematic framework which generates metabolic pathways along with their candidate enzymes in order to identify feasible and most efficient biosynthetic pathways for the target chemicals. This platform largely consists of two parts, route generation and prioritization process. Route generation process generates pathways based on reaction rule sets which were constructed based on the logics acquired from the analysis of reaction mechanisms of existing biochemical reactions observed in biological systems. After route generation, two prioritization factors, including chemical similarity, thermodynamic favorability were evaluated to estimate the feasibility of each generated reaction pathway from the framework. Using this framework, several novel pathways have been designed to produce valuable chemicals. To validate this framework, novel biosynthetic pathways for 3-hydroxypropionate (3HP), acrylic acid, toluene and other several chemicals were predicted using this system. This systematic framework should be an valuable resource useful for the advance of synthetic biology and metabolic engineering.[This work was supported by the Technology Development Program to Solve Climate Changes on Systems Metabolic Engineering for Biorefineries from the Ministry of Science, ICT and Future Planning (MSIP) through the National Research Foundation (NRF) of Korea (NRF-2012M1A2A2026556 and NRF-2012M1A2A2026557).]

Homo-Organic Acid Production with Minimal Byproducts through Multi-Objective Optimization of Microbial Metabolism

Kim, D. I., Korea Advanced Institute of Science and Technology (KAIST)
Kim, H. U., Korea Advanced Institute of Science and Technology (KAIST)
Kim, T. Y., Korea Advanced Institute of Science and Technology (KAIST)
Park, J. M., Korea Advanced Institute of Science and Technology (KAIST)
Lee, S. Y., Korea Advanced Institute of Science and Technology (KAIST)

Minimization of byproducts during strain development is a key to reduce operation cost by facilitating separation and purification of microbial bioproducts in the downstream process. In this context, we implement strain design not only to maximize the target chemical product at a sufficiently high cellular growth rate, but also to minimize byproducts using a concept of multi-objective optimization of the production host's metabolic model. We experimentally demonstrate this approach for productions of acetic, lactic and succinic acids in *E. coli* as proof-of-concept studies. *E. coli* strains producing homo-acetic and homo-lactic acids were constructed by knocking out a minimal number of gene targets predicted from the *E. coli* metabolic model. Meanwhile, gene knockout targets that enable the homo-succinic acid production were not predicted for *E. coli*. This prediction failure suggests that the multi-objective optimization of host's metabolic network can be used to evaluate whether a specific microorganism is a suitable host for the production of a homo-organic acid. This systems metabolic engineering approach can be considered for the production of other industrially important chemicals.[This work was supported by the Technology Development Program to Solve Climate Changes on Systems Metabolic Engineering for Biorefineries from the Ministry of Science, ICT and Future Planning (MSIP) through the National Research Foundation (NRF) of Korea (NRF-2012M1A2A2026556 and NRF-2012M1A2A2026557).]

Production of Astaxanthin Using Isoprenoid Pathways in Metabolically Engineered Escherichia coli

Park, S. Y., Korea Advanced Institute of Science and Technology (KAIST)
Lee, S. Y., Korea Advanced Institute of Science and Technology (KAIST)

Astaxanthin is a reddish keto-carotenoid classified as xanthophyll which could be found in microalgae, yeast, and in some types of seafood. It is 10 to 100 times more powerful antioxidant than other carotenoids like beta-carotene or lycopene. It can be used as food pigment, powerful antioxidant, anticancer applicant and cosmetic compound. Although it has many valuable uses, its yield from natural source is low for meet the increasing demand and because of the stereoisomers, chemical synthesis has low efficiency, too. Therefore the production of astaxanthin in *Escherichia coli* was conducted in this study. Firstly, the heterogenous genes from *Pantoea ananatis* were introduced into *E. coli* for production of beta-carotene, which is an important precursor of astaxanthin. After comparison of different gene expression degrees, downstream genes toward astaxanthin were introduced. After development of base strain, native DXP pathway in *E. coli* was overexpressed in chromosomal level and heterogenous mevalonate pathway from yeast was introduced for increasing production rate. The strageties for increasing precursor pools of carotenoids such as GGPP and beta-carotene could be applied for further research. [This work was supported by the Technology Development Program to Solve Climate Changes on Systems Metabolic Engineering for Biorefineries from the Ministry of Science, ICT and Future Planning (MSIP) through the National Research Foundation (NRF) of Korea (NRF-2012M1A2A2026556 and NRF-2012M1A2A2026557).]

Production of Fumaric Acid from Glucose By a Metabolically Engineered Escherichia coli

Kim, J. W., Korea Advanced Institute of Science and Technology (KAIST)
Song, C. W., Korea Advanced Institute of Science and Technology (KAIST)
Kim, D. I., Korea Advanced Institute of Science and Technology (KAIST)
Choi, S., KAIST
Jang, J. W., KAIST
Lee, S. Y., Korea Advanced Institute of Science and Technology (KAIST)

In this study, *Escherichia coli* was metabolically engineered to produce a fumaric acid, a naturally synthesized organic acid as an intermediate of the TCA cycle, under aerobic condition. To redirect carbon flux through the glyoxylate shunt, the *iclR* gene was deleted, and the *fumA, fumB,* and *fumC* genes were also deleted to enhance fumaric acid formation. The resulting strain was able to produce 1.45□g/L of fumaric acid from 15□g/L of glucose in flask culture. Additionally, *in silico* flux response anaylsis was performed and the native *ppc* gene (encoding phosphoenolpyruvate carboxylase (PPC)) was overexpressed on plasmid level from the strong *tac* promoter. This resulting strain produced 4.09□g/L of fumaric acid. Moreover, the *arcA* and *ptsG* genes were deleted to reinforce the oxidative TCA cycle flux, and the *aspA* gene was deleted to prevent the fumaric acid from converting into L-aspartic acid. For enhanced glucose uptake rate and fumaric acid productivity, the native promoter of the *galP* gene was replaced with the strong *trc* promoter and the *lacI* gene was also deleted to avoid the use of inducer. Fed-batch culture of the final strain CWF812 produced 28.2□g/L fumaric acid in 63□h with the overall yield and productivity of 0.389□g fumaric acid/g glucose and 0.448□g/L/h, respectively. (This work was supported by the Technology Development Program to Solve Climate Changes on Systems Metabolic Engineering for Biorefineries from the Ministry of Science, ICT and Future Planning (MSIP) through the National Research Foundation (NRF) of Korea (NRF-2012M1A2A2026556 and NRF-2012M1A2A2026557)).

Disruption of the Butyrate Kinase (buk) Gene Is Vital for the High Butyric Acid Selectivity in Clostridium Acetobutylicum

Ha, S. H., Korea Advanced Institute of Science and Technology (KAIST)
Jang, Y. S., Korea Advanced Institute of Science and Technology (KAIST)
Lee, S. Y., Korea Advanced Institute of Science and Technology (KAIST)

Butyric acid is a valuable industrial chemical which utilized in food, pharmaceutical, animal feed supplement, and chemical industries. Several *Clostridium* strains, including *Clostridium butyricum, Clostridium thermobutyricum, Clostridium tyrobutyricum, Clostridium beijerinckii* and *Clostridium acetobutylicum*, can naturally produce butyric acid. Generally butyric acid producing *Clostridium* strains are co-produce acetic acid with butyric acid and increasing the butyric acid selectivity is important to economical butyric acid production. However, complex metabolic pathways and physiologies of clostridia induce the difficulties of economical butyric acid production. In this work, the second butyrate kinase of *C. acetobutylicum* encoded by *bukII* gene was employed instead of butyrate kinase I encoded by *buk* gene for the high butyric acid selectivity in *C. acetobutylicum*. Moreover, further metabolic engineering was used for the enhancement of the NADH-driving force. The metabolically engineered *C. acetobutylicum* strain produced 32.5 g/L of butyric acid with a butyric-to-acetic acid ratio (BA/AA ratio) of 31.3 g/g from 83.3 g/L of glucose in batch fermentation at pH 6.0. These results suggested that the *buk* gene knockout was vital for the high butyric acid selectivity in *C. acetobutylicum*. [This work was supported by the Technology Development Program to Solve Climate Changes on Systems Metabolic Engineering for Biorefineries from the Ministry of Science, ICT and Future Planning (MSIP) through the National Research Foundation (NRF) of Korea (NRF-2012M1A2A2026556 and NRF-2012M1A2A2026557)]

Metabolic Engineering of Ralstonia Eutropha for the Biosynthesis of Polyhydroxyalkanoates from Sucrose

Yang, J. E., Korea Advanced Institute of Science and Technology (KAIST)
Park, S. J., Myongji University
Oh, Y. H., Korea Research Institute of Chemical Technology
Song, B. K., Korea Research Institute of Chemical Technology
Lee, S. H., Chonnam National University
Lee, S. Y., Korea Advanced Institute of Science and Technology (KAIST)

Sucrose is one of the most abundant and relatively inexpensive carbon sources extracted from sugarcane and sugar beet. To develop sucrose utilizing *Ralstonia*, which is polyhydroxyalkanoates (PHAs) producer, *Ralstonia eutropha* NCIMB11599 and *R. eutropha* 437-540 expressed *Mannheimia succiniciproducens* MBEL55E *sacC* gene that encodes β-fructofuranosidase. It was found that β-fructofuranosidase excreted into the culture medium could hydrolyze sucrose to glucose and fructose, which were efficiently used as carbon sources by recombinant *R. eutropha* strains. When *R. eutropha* NCIMB11599 expressing the *sacC* gene was cultured in nitrogen-free chemically defined medium containing 20 g/L of sucrose, a high P(3HB) content of 73.2 wt% could be obtained. In addition, *R. eutropha* 437-540 expressing the *Pseudomonas* sp. MBEL 6-19 *phaC1437* gene and the *Clostridium propionicum pct540* gene accumulated P(3HB-*co*-21.5mol%LA) to a polymer content of 19.5 wt% from sucrose by the expression of the *sacC* gene and the *Escherichia coli ldhA* gene. The engineered *R. eutropha* strains reported here will be useful for the production of polyhydroxyalkanoates from sucrose, one of the most abundant and relatively inexpensive carbon sources. [This work was supported by the Technology Development Program to Solve Climate Changes on Systems Metabolic Engineering for Biorefineries from the Ministry of Science, ICT and Future Planning (MSIP) through the National Research Foundation (NRF) of Korea (NRF-2012M1A2A2026556 and NRF-2012M1A2A2026557).]

Modulation of Carbon Preference in Enterobacter Aerogenes for Efficient Utilization of Sugarcane Molasses for 2,3-Butanediol Production

Jung, M. Y., Korea university
Oh, M., Korea university

Due to its cost-effectiveness and rich sugar composition, sugarcane molasses is considered to be a promising carbon source for biorefinery. However, the sugar mixture in sugarcane molasses is not consumed as efficiently as glucose in microbial fermentation due to complex interactions among their utilizing pathways, such as carbon catabolite repression (CCR). Therefore, *E. aerogenes* was engineered to alleviate CCR and improve sugar utilization by modulating its carbon preference. To improve sucrose utilization in *E. aerogenes*, a sucrose regulator (ScrR) was deleted because sucrose occupies more than half of the sugar in the molasses. The deletion mutation increased the sucrose consumption rate significantly when sucrose or sugarcane molasses was used as a carbon source, but the efficiency of fructose utilization in the *scrR* mutant was relatively reduced. Therefore, the gene encoding catabolite repressor/activator (Cra) was removed to increase the fructose consumption rate. The disruption of *cra* improved the consumption rate of fructose, while significantly decreasing sucrose utilization, resulting in the accumulation of sucrose in the fermentation medium. Cra regulation on expression of the *scrAB* operon involved in sucrose catabolism was verified by reverse transcription and real-time PCR, and the efficiency of sucrose utilization was restored by disrupting the *scrR* gene and overexpressing the *scrAB* operon. In addition, the overexpression of *ptsG* gene involved in glucose utilization enhanced the glucose preference among mixed sugars, which relieved glucose accumulation in fed-batch fermentation. These genetic modifications enabled the mutant strain to metabolize all sugars in sugarcane molasses efficiently and simultaneously, which increased fermentation duration and relieved the accumulation of certain carbon sources in fed-batch fermentation, and the maximum titer of 2,3-butanediol production by the mutant reached 140.0 g/L at 54 h.

Rational Design of Spider Silk-Based Biomaterials

Qian, Z. G.,
Song, W. W., Shanghai Jiao Tong University
Xia, X. X., Jiao Tong University

Nature has evolved a variety of complex yet elegant and functional biomaterials, providing a formidable source of inspiration for scientists. For example, the silk, elastin, resilin and collagen etc. are fantastic biomaterials with sophisticated micro- to macro-scale structures and properties for specific applications. However, natural biomaterials as harvested are often inconsistent in batch composition and across sources, resulting in an inability to precisely dictate the properties of most natural materials. Recent developments in synthetic biology and metabolic engineering offer an alternative to recapitulate and even tune the properties of native biomaterials. Here we show that biomimetic design and production of long, highly repetitive protein of spider silk that is ultimately spun into a strong fiber with native-like mechanical properties. We also show that the conserved C-terminal domain of spider silk self-assembles into dual thermosensitive hydrogels, thus offering a functional module for the design and construction of novel protein hydrogels. Furthermore, we demonstrate the formation of hydrogels with multiple sensitivity and tunable properties when the consensus repeats of spider silk and other functional modules of natural biomaterials are genetically fused to the C-terminal domain of spider silk. The diverse forms and tunable properties of newly designed spider silk-based biomaterials reflect the power of synthetic biology in modular design of protein-based polymers at the DNA, protein, and material levels (this project was supported by the National Natural Science Foundation of China Grant No. 31470216 and the Shanghai Pujiang Program Grant No. 14PJ1405200).

Bringing Ethyl Acetate Production into the Biobased Economy
Kruis, A.,

Aleksander Kruis[1], Astrid Mars[2], Mark Levisson[1], Servé Kengen[1], John van der Oost[1], Ruud Weusthuis[3]

[1]Laboratory of Microbiology, Wageningen UR, Dreijenplein 10, 6703 HB, Wageningen

[2]Biobased Products, Wageningen UR, Bornse Weilanden 9, 6708 WG Wageningen

[3]Bioprocess Engineering, Wageningen UR, Droevendaalsesteeg 1, 6708PB, Wageningen

Ethyl acetate is used extensively as a food additive or as a bulk industrial chemical. However, like most bulk chemicals, they are traditionally produced from petrochemical resources. The use of petrochemical resources has various negative environmental and social impacts, which is why sustainable, biobased alternatives are being researched. One such alternative is the production of ethyl acetate from biomass through microbial fermentations. The biggest challenge of microbial fermentations is often the design of an efficient production organism. The most commonly used production organisms are *E. coli* and *S. cerevisiae*, due to their well-researched metabolism and genetic accessibility. However, they naturally produce only trace amounts of ethyl acetate because they lack the necessary enzymes.

Several non-*Saccharomyces* yeasts are able to produce large amounts of ethyl acetate and could be used as enzyme sources to transfer this trait to other organisms. However, the metabolic pathways and enzymes involved in ethyl acetate synthesis are poorly understood, posing a major bottleneck in engineering more efficient ways of producing biobased ethyl acetate. To understand the biochemistry behind the ethyl acetate synthesis, we have studied the yeast *Wickerhamomyces anomalus*. We performed an RNAseq analysis in an attempt to identify the genes involved in ethyl acetate synthesis.

Production of Acrylic Acid By Metabolically Engineered Escherichia coli

Ko, Y. S., Korea Advanced Institute of Science and Technology (KAIST)
Song, C. W., Korea Advanced Institute of Science and Technology (KAIST)
Lee, S. Y., Korea Advanced Institute of Science and Technology (KAIST)

Acrylic acid is an important bulk chemical with an annual production of 4.2 million tons, and most of it is currently produced by oxidation of propylene. To replace petroleum-based process, production of acrylic acid from renewable resources is highly important. Herein, we report a novel biosynthetic route to produce acrylic acid from glucose *via* beta-alanine, beta-alanyl-CoA, and acrylyl-CoA. First, *Escherichia coli* was metabolically engineered to produce beta-alanine as a precursor, and deamination of beta-alanine was catalyzed by sequential enzymatic reactions. We screened the corresponding enzymes and optimized the production. Finally, the best strain produced 41 mg/ L acrylic acid in flask cultivation. (This work was supported by the Technology Development Program to Solve Climate Changes on Systems Metabolic Engineering for Biorefineries from the Ministry of Science, ICT and Future Planning (MSIP) through the National Research Foundation (NRF) of Korea (NRF-2012M1A2A2026556 and NRF-2012M1A2A2026557))

20 Years After Launch: 13C Metabolic Flux Analysis Becomes Visual

Nöh, K., Forschungszentrum Jülich
Droste, P., 2Omix Visualization GmbH & Co. KG
Wiechert, W., Forschungszentrum Jülich

The first papers on ^{13}C-Metabolic Flux Analysis integrating the full methodological triad of i) cultivation under well controlled conditions, ii) MS or NMR ^{13}C labeling measurements, and iii) mathematical modeling of the metabolic network were published 20 years ago [1]. Since then, ^{13}C-MFA experienced a fairly dynamic development and became an indispensable tool for directing flux to wanted targets [2]. The workflow of ^{13}C-MFA is highly organism and experiment specific and, thus, requires painstaking attention [3]. Apart from the composition of various experimental and analytical tasks, a major hurdle in generating high-quality flux maps is orchestration of several modeling and computational steps in an iterative and interactive workflow [4].

We here present the first fully graphical tool suite for ^{13}C-MFA based on the Omix visualization software [5] and featuring the high-performance simulator 13CFLUX2 [6]. Scientists are provided with an integrated visual modeling environment for network construction, atom transition specification, measurement modeling, input substrate composition, experimental design, or flux map presentation just to mention a few of the implemented tools. Almost 30 Omix plugins have been implemented for this purpose. Several advanced graphical workflows and ergonomic user interfaces support major domain-specific modeling and proofreading tasks. With these features, the novel suite is a productivity-boosting tool. Remarkably, part of the flux analysis workflow already found its way to Agilentâ??s VistaFlux, the first commercial software for qualitative MFA [7].

References:
[1] Marx A, et al. (1996) Determination of the fluxes in central metabolism of Corynebacterium glutamicum. BiotechnBioeng49:111-129.
[2] NiedenfÃ¼hr S, Wiechert W, NÃ¶h K (2015) How to measure metabolic fluxes: a taxonomic guide for ^{13}C fluxomics. CurrOpBiotechnol 34:82-90.
[3] Buescher JM, et al. (2015) A roadmap for interpreting ^{13}C metabolite labeling patterns from cells. CurrOpBiotechn 34:189-201.
[4] Wiechert W, Niedenführ S, Nöh K (2016) A primer to ^{13}C metabolic flux analysis, Wiley Encyclopedia Biotechnology, Vol. 1:97-142.
[5] Nöh K, Droste P, Wiechert W (2015) Visual workflows for ^{13}C-metabolic flux analysis. Bioinformatics, 31:346-354.
[6] M. Weitzel, et al. (2013) 13CFLUX2 - High-performance software suite for ^{13}C-metabolic flux analysis. Bioinformatics 29:143-145.
[7] Agilent Technologies (2016), VistaFlux Software, Santa Clara, USA 2016.

Convergent Rapid Evolution of Fructose Operon Reveals Its Important Role in Sugar Uptake and Its Effect on Lysine Production

Wang, Z., Technical University of Denmark
Chan, S. H. J., Technical University of Denmark
Marke, M. S.,
Jensen, P. R., Technical University of Denmark
Solem, C., Technical University of Denmark

The fructose operon is a conserved structure in many bacteria, which is involved in fructose assimilation. We have identified an evolutionary incentive to keep this operon intact in *C. glutamicum* by discovering that mutations in two of its members not only affect fructose metabolism, but also the metabolism of sucrose and glucose. The implicated genes are *ptsF*, encoding the fructose specific component of the sugar phosphotransferase system (PTS) and *pfkB*, encoding a fructose-1-phosphate kinase.

Inactivation of either of these genes affects growth in minimal medium on all three sugars negatively, but suppressor mutants, where both genes are inactive, appear with high frequency, leading to restored or improved growth on these sugars mentioned above. We adapt a genome scale model for *C. glutamicum* by overlaying transcriptomic data and are able to predict that the hampered growth of the ptsF knock-out strain can be attributed to perturbed expression of sugar PTS genes, which are found to be drastically down-regulated. We are able to validate this prediction that the hampered growth of KO-*ptsF* and KO-*pfkB* mutants are caused by different mechanisms.

Furthermore, when the *pfkB* and *ptsF* mutations are introduced in a lysine-producing derivative, the lysine yield is found to be enhanced by almost 200% compared to the control strain on fructose medium and even 100 % higher than the yield observed on glucose. By further over-expressing genes in pentose phosphate pathway and down-regulating the TCA cycle, even higher lysine yields are achieved. The results obtained reveal a novel strategy for producing lysine that takes advantage of the diauxic shift observed when growing *C. glutamicum* on either glucose/fructose or sucrose.

Rational Modular Design of Yeast Strains Towards Overproduction of Industrially Relevant Products

Lopes, H., University of Minho
Maia, P., - Computational Biology Solutions for the Life Sciences
Patil, K. R., EMBL
Rocha, I., University of Minho

Despite some progress over the last years, the development of optimized yeast strains for the production of chemicals and other materials is still a costly and time-consuming process. The availability of suitable chassis cells, pre-optimized for the overproduction of different compounds, can contribute to reduce this cost/time burden. Among several eukaryotic organisms that can be exploited in industrial biotechnology, *Saccharomyces cerevisiae* is uniquely positioned as a well-established, robust and scalable industrial production platform for a large portfolio of products, due to its high tolerance to harsh industrial conditions and available knowledge.

Following this rationale, we herein propose a conceptual framework for the design of platform *S. cerevisiae* strains towards enhanced production of families of industrially relevant compounds, through the identification of gene deletions to improve the production of the desired targets. The framework includes two main stages: i) Optimization stage, where an OptGene-based [1, 2] multi-objective metaheuristic approach is applied for all targets in a given family and ii) Chassis analysis, where the occurrence of each gene in each of the targeted metabolites' solution set is accounted for. Candidate genes for deletion are sorted based on a computed score that favors genes occurring for multiple targets. Top scoring genes are then selected for chassis generation. Afterwards, all the solutions are organized in groups with common chassis. Case studies include the production of metabolites that have found widespread applications in the food, chemical and pharmaceutical industries, namely C_4-dicarboxylic acids derived from the TCA cycle (succinate, fumarate and L-malate) and a set of heterologous products derived from the shikimate pathway.

Several chassis strains containing a common backbone of non-intuitive gene deletions to the selected targeted metabolites, along with additional genetic targets to guarantee the desired phenotype, were generated and characterized in terms of their biological feasibility. Candidate solutions were ranked based on computed metrics resembling productivity and number of associated gene knockouts. Although experimental validation is still lacking, there is support in the literature for some of the proposed candidate strategies.

This work establishes a proof-of-concept showing that it is possible to generate pre-optimized platform strains for enhanced production of different metabolites derived from the same metabolic pathways. Furthermore, we foresee that this model-driven modular design concept may constitute an important step towards realizing the full potential for economical and sustainable production of other families of industrially relevant chemicals.

[1] K. R. Patil, I. Rocha, J. Forster, J. Nielsen, BMC Bioinformatics 6(308) 2005.

[2] M. Rocha, P. Maia, R.Mendes, J. P. Pinto *et al. BMC Bioinformatics 9*(499). 2008

Metabolic Engineering of Escherichia coli for the Production of 2-Hydroxyacids Containing Polyhydroxyalkanoates

Choi, S. Y., Korea Advanced Institute of Science and Technology (KAIST)
Park, S. J., Myongji University
Kim, W. J., Korea Advanced Institute of Science and Technology (KAIST)
Yang, J. E., Korea Advanced Institute of Science and Technology (KAIST)
Lee, H., Korea Research Institute of Chemical Technology
Shin, J., Korea Research Institute of Chemical Technology
Lee, S. Y., Korea Advanced Institute of Science and Technology (KAIST)

Polyhydroxyalkanoates (PHAs) are bacterial polyesters having biodegradability, biocompatibility and general mechanical properties of plastics. In PHAs biosynthesis, the key enzyme is PHA synthase that polymerizes hydroxyacyl-CoAs into PHAs. PHA synthases show activity for broad range of substrates which are more than 150 kinds of 3-, 4-, 5- and 6-hydroxyacyl-CoAs. However, there is no report about PHAs containing 2-hydroxyacids in nature and PHA synthase showed negligible activity for lactate (2-hydroxypropionate). Here, we produced non-natural substrate, 2-hydroxyacids such as lactate, 2-hydroxybutyrate containing PHAs with engineered PHA synthase from *Pseudomonas* sp. MBEL 6-19 and engineered propionyl-CoA transferase from *Clostridium propionicum*. For production from unrelated carbon sources such as glucose, systems metabolic engineering was conducted to redirect and concentrate the metabolic flux toward the 2-hydroxyacids based on the *in silico* simulation with heterogeneous metabolic pathways harboring propionyl-CoA transferase and evolved PHA synthase for polymerization. It suggests the sustainable plastic production process which can replace the current chemical process. [This work was supported by the Technology Development Program to Solve Climate Changes on Systems Metabolic Engineering for Biorefineries from the Ministry of Science, ICT and Future Planning (MSIP) through the National Research Foundation (NRF) of Korea (NRF-2012M1A2A2026556 and NRF-2012M1A2A2026557)].

Cameo: A Python Library for Computer Aided Metabolic Engineering and Optimization of Cell Factories

Cardoso, J. G. R., The Novo Nordisk Foundation Center for Biosustainability
Jensen, K., The Novo Nordisk Foundation Center for Biosustainability
Özdemir, E., Technical University of Denmark
Herrgård, M. J., Technical University of Denmark
Sonnenschein, N., Technical University of Denmark

In this work we present Cameo, a platform independent software framework that enables in silico strain design of bacterial and eukaryotic cell factories on a genome-scale. It is written in Python and implements novel and state-of-the-art methods for enumerating and prioritizing knock-out, knock-in, over-expression, and down-regulation strategies and combinations thereof. Cameo targets both experienced modelers as well as experimental strain engineers: a modular and extendible library of classes and functions simplifies the implementation of new strain design algorithms, and a high-level 'design' function enables the generation of hundreds of strain designs for desired compounds. Long-lasting computations have been implemented with parallelization in mind and can take advantage of multicore systems and high-performance computing infrastructure. Since Cameo is based on the constraint-based modeling software cobrapy and thus part of a larger community and ecosystem of modeling software, users can easily take advantage of, for example, system-level data integration, interactive pathway visualizations, and programmatic access to a database of curated models. Furthermore, Cameo is tightly integrated with a number of widely used scientific Python tools, including the Jupyter notebook interface, pandas, and sympy. Cameo is an open source software project and welcomes contributions by the community on GitHub. It is freely available under the Apache License 2.0 and can be installed either from source or PyPI. A dedicated website including documentation, examples, and installation instructions can be found at http://cameo.bio.

Understanding Thermotolerance of Kluyveromyces Marxianus: A Constraint-Based Modelling Approach

Marcisauskas, S., Chalmers University of Technology
Ji, B., Chalmers University of Technology
Nielsen, J., Technical University of Denmark

Kluyveromyces marxianus is one of thermotolerant yeast, which can metabolise a broad range of carbon sources, including less conventional lactose, xylose, arabinose and inulin. Such phenotypic trait is sustained even up to 45 °C, what makes it relevant candidate for industrial biotechnology applications, such as ethanol production. It is thus imperative to unravel the key factors, contributing to such phenotype of *K. marxianus*. The recent yeast studies suggested, that thermotolerance is achieved by reducing the number of growth-determining proteins or suppressing oxidative phosphorylation. In our work we aimed to find the related factors, which contribute to thermotolerance of *K. marxianus*. We utilized publicly available *Kluyveromyces lactis* genome-scale metabolic model as a template model for *K. marxianus* model reconstruction. The model was then carefully refined to include the missing species-specific metabolic capabilities. Gene essentiality analysis was performed to evaluate the model quality. Thereafter we coupled the model with thermodynamic constraints and performed simulations in various growth temperatures. Such approach provided a detailed insight in the favourable flux distribution for particular temperature. We also tried to include the findings from other studies and evaluate the model behaviour.

Development of Pantoea species As a Platform for the Production of High Value Terpenoids

Williamson, J., University of Nottingham
Hill, P.,
Fray, R., University of Nottingham

Plant derived terpenoids represent a diverse class of chemicals with important roles as fragrance and flavour molecules (eg camphor, menthol and citral), dyes and colurants (eg astaxanthin) as well as high value medicinal compounds such as the antimalarial Artemisinin and the anti-cancer drug Paclitaxel (Taxol).

High volume production of these products can be difficult due reliance on plant production or economic difficulties in synthetic production. These high value terpenoids represent an interesting challenge to metabolic engineering and synthetic biology, not only in being able to produce these chemicals at scale, but also developing novel products via the intermediates of these pathways.

Research into methods of microbial production of Taxol and its derivatives have focused on the precursor pathways leading to geranylgeranyl diphosphate (GGPP), as these are the last non-dedicated steps. This method works well however the p450 mediated hydroxylation steps appear to have limited success in *E. coli*.

Instead of optimising the precursor pathways, we have attempted to utilise natural diversity in search of a better host for production of these high value terpenoids, by selecting an organism that already produces a large amount of GGPP and can potentially better mediate p450 hydroxylation reactions.

GGPP can also be converted into carotenoid pigments such as lycopene, β-carotene and canthaxanthin, therefore Carotenoids production can be used as a surrogate measure for predicting how well a species might perform for production of heterologous isoprenoids. Genes for carotenoid biosynthesis from *Pantoea* species have been used in many heterologous expression experiments but the organism itself has not been considered as a chassis. As a relatively close phylogenetic relation to *E.coli*, these Gram negative bacteria are easily transformable, genetically tractable and many tools developed for *E. coli* function in *Pantoea*.

We will present details of the progress made and pitfalls encountered during the engineering of wild type *Pantoea* to produce intermediates of Taxol.

Synthetic Transcription Regulation System for Orthogonal Control of Gene Expression

Rantasalo, A., VTT Technical Research Centre of Finland
Czeizler, E.,
Virtanen, R., VTT Technical Research Centre of Finland
Rousu, J.,
Lahdesmaki, H.,
Penttilä, M., VTT Technical Research Centre of Finland
Mojzita, D., VTT Technical Research Centre of Finland
Jantti, J., VTT Technical Research Centre of Finland

This work describes the development and experimental as well as mathematical characterization of a modular synthetic expression system that has compact bidirectional architecture and enables a broad range of adjustable and predictable expression outputs in *S. cerevisiae*. In the system, the input signal is transferred via a synthetic transcription factor (sTF) into a synthetic promoter and a defined core promoter to generate an output signal. The system activation is based on a set of bacterial DNA-binding domains, modular binding sites in the DNA and a selection of transcription activation domains. The system tuning is achieved through selection of three separate modules, each of which enables an adjustable output signal regulation: 1) the transcription-activation domain of the sTF, 2) the binding-site modules in the output promoter, and 3) the core promoter modules which define the transcription initiation site in the output promoter. The system contains also a collection of repressor modules that together with the activation modules enable diverse control of gene expression. In contrast to most existing modular gene expression regulation systems, the present system is independent from externally added compounds. Furthermore, the system provides robust functionality that is minimally affected by growth conditions. These features should make it useful also in large scale biotechnology applications. Recent results on the expansion of the catalogue of the various parts for more complex cellular control and description of the system's functionality beyond *S. cerevisiae* will be described.

Metabolic Flux Rewiring and Physiology in E. coli Upper Central Carbon Metabolism Knockout Strains

Long, C. P., University of Delaware
Antoniewicz, M. R., University of Delaware

Metabolic and physiological responses to genetic perturbations are of fundamental interest to systems biology and metabolic engineering. They reveal new information on the network structure, for example, uncovering previously "hidden" reactions, as well as metabolic regulation and kinetics. The accurate prediction of metabolic fluxes following a genetic perturbation, critically important for rational metabolic engineering, has been a persistent challenge due to the complexity of cellular systems. There has been much work done to develop *in silico* predictive models of metabolism, most notably the constraint-based models. Progress in this area has been limited by a lack of high-quality, reproducible experimental metabolic flux (fluxomic) data of genetically perturbed strains. In this work, we leverage recent advances in ^{13}C-metabolic flux analysis (^{13}C-MFA) and the availability of all viable *Escherichia coli*single gene knockouts from the Keio collection to perform a comprehensive physiological and fluxomic study of mutants associated with central carbon metabolism. The results will provide both novel biological insights and a significant new resource for advancing metabolic modeling and strain design tools.

The core metabolism of *E. coli* is studied by measuring metabolic fluxes and biomass composition for all single gene knockout mutants associated with central carbon metabolism (glucose transport, glycolysis, pentose phosphate pathway, ED pathway, TCA cycle, and anaplerotic reactions) during aerobic, exponential growth on glucose. Fluxes are estimated based on an optimized parallel experimental design using [1,2-^{13}C]glucose and [1,6-^{13}C]glucose tracers, coupled with mass spectrometry measurement of metabolite isotopic labeling and ^{13}C-MFA. Complimentary biomass composition measurements were carried out using a novel GC-MS based method for the quantification of amino acids, RNA, fatty acids, and glycogen. These measurements, along with growth and yield observations, constitute a thorough assessment of the metabolic states of each knockout.

Here we will present the described rigorous assessments of the wild-type *E. coli* and 23 knockout strains which comprise the complete "upper half" of central carbon metabolism: the key glucose transporters, the entirety of the pentose phosphate and ED pathways, and the upper portion of the EMP pathway (glycolysis). Global correlations and trends involving biomass composition and fluxes are reported. The fluxomic data is leveraged to develop new insights into specific kinetic and regulatory responses, as well as more general stress responses. Particularly notable are cases in which the knockout strains exhibit unusual pathway usage, and/or previously unannotated or "hidden" reactions. All fluxes are compared with predictions made by common constraint-based models such as FBA, MOMA, ROOM, and RELATCH. Ongoing efforts in this project will aim to extend this analysis to lower central carbon metabolism, and to apply the flux data to new model development efforts such as the ensemble modeling approach.

Genome-Scale Modeling Coupled with Crispri for Improving Biofuel Production in Cyanobacteria

Shabestary, K., KTH-Royal Institute of Technology
Yao, L., KTH-Royal Institute of Technology
Hudson, P., KTH-Royal Institute of Technology

Cyanobacteria are promising hosts for production of low-value products such as biofuels. However, competitive rates, yields and titers have not been reached to date. We are combining genome-scale modeling and CRISPR/Cas to create highly modified cyanobacteria cell factories. We used constraint-based modeling and optimization algorithms on a genome-scale model (GEM) of *Synechocystis* PCC6803 to predict ways to improve productivity of 1-butanol and other biofuels. We found that a combination of gene knockouts could couple biofuel production to cell growth; such strains are predicted to give high biofuel productivities and titers. High productivity of fermentation-based 1-butanol requires elimination of NADH sinks, while fatty-acid and terpene based products require creating imbalances in intracellular ATP and NADPH production and consumption. We are currently implementing these knockouts using CRISRP/Cas and characterizing mutant strains. Integration of transcriptomics and proteomics data from wild-type and mutant strains into the GEM will allow us to further refine our engineering strategies. Combining genome-scale modeling with CRISPR/Cas can lead a new era of metabolic engineering in cyanobacteria.

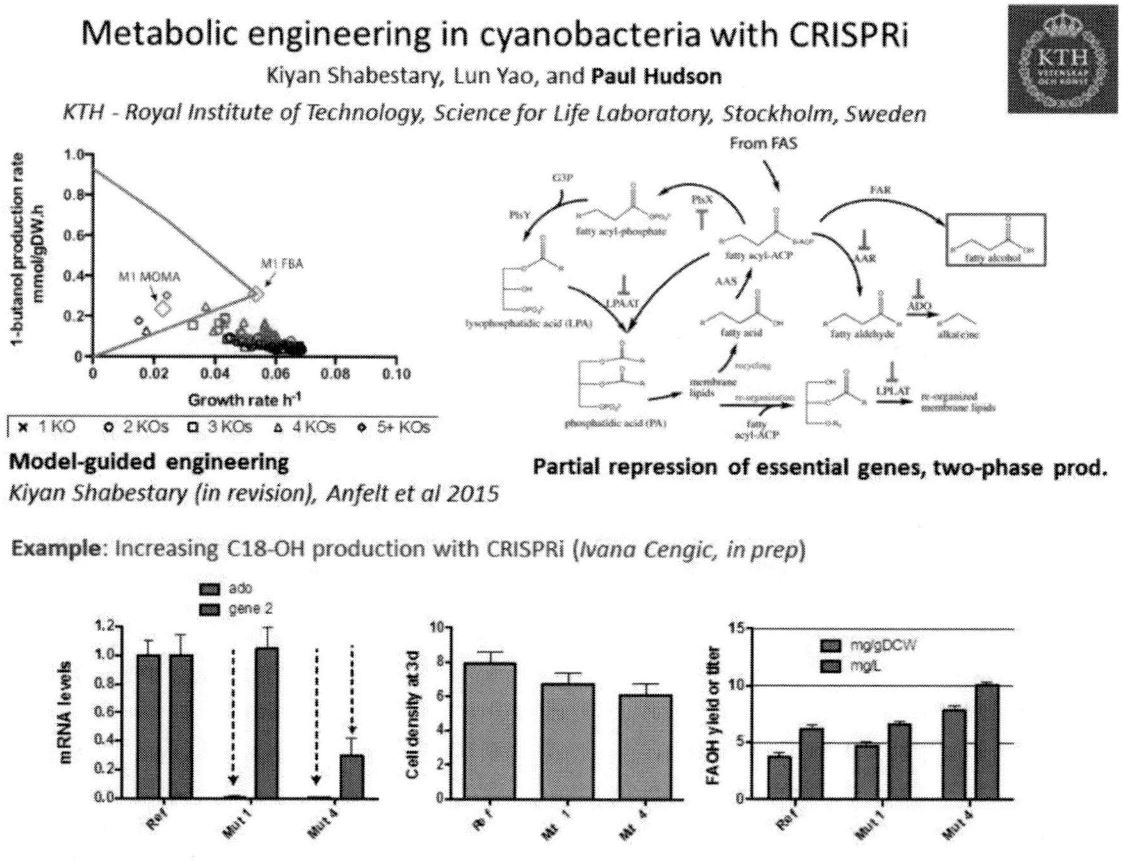

Elucidation of Clostridium Acetobutylicum metabolism Using Parallel Labeling Experiments and 13c Metabolic Flux Analysis

Au, J., University of Delaware
Antoniewicz, M. R., University of Delaware

Given their wide substrate range and high specific substrate utilization rates, solventogenic clostridia are seen as a promising class of organisms for biofuel and commodity chemical production. *Clostridium acetobutylicum* was historically used for industrial-scale fermentation and remains a potential candidate for butanol and butyric acid production today. However, although the biochemistry of *C. acetobutylicum* has been extensively reviewed, the central metabolic pathways have remained only partially resolved. Two recent reconstructions of genome-scale models have proposed different mechanisms for the biosynthesis of α-ketoglutarate, an important precursor for amino acid biosynthesis. Initial stable-isotope labeling experiments and qualitative ^{13}C-isotopomer analysis supported the idea of an incomplete tricarboxylic acid (TCA) cycle and suggested a *Re*-stereospecificity for the citrate synthase reaction.

In this work, we have rigorously validated the metabolic network model of *C. acetobutylicum* to guide future efforts aimed at the metabolic engineering of this organism for biofuels/chemicals production. Using parallel labeling experiments and ^{13}C-metabolic flux analysis (^{13}C-MFA), we quantitatively elucidated central carbon metabolism and amino acid metabolism. Contrary to previously proposed hypotheses, we found that while the TCA cycle runs in the oxidative direction, there is no notable flux between α-ketoglutarate and succinyl-CoA, succinate and fumarate, or malate and oxaloacetate, and that the conversion of succinyl-CoA to succinate proceeds independently. Additionally, using multiple ^{13}C-labeled amino acid tracers we identified a novel metabolic cycle that involves central carbon and amino acid metabolism where carbon flows from aspartate to threonine, serine, pyruvate, oxaloacetate, and back to aspartate. The physiological significance of this novel metabolic cycle in the evolution of *C. acetobutylicum* will be discussed.

The validated metabolic network model was used to study the metabolic stress response of *C. acetobutylicum* under butanol and butyric acid stresses. A major drawback to clostridia-based industrial production of these metabolites is their toxic effects on the cells, resulting in low product titers and inducing the metabolite stress response. While several studies have been carried out to understand the transcriptional and translational basis of this stress response, little is known about the impact of butanol and butyric acid stress on the metabolic pathways and fluxes of *C. acetobutylicum*. Here, we have used parallel labeling experiments and qualitative ^{13}C-isotopomer analysis to characterize the short term and long term effects of butanol and butyric acid stress on the metabolism of *C. acetobutylicum*. We show that while butanol and butyric acid result in lower absolute rates of glucose uptake and cell growth rate, the relative intracellular flux distributions remain relatively unchanged under stress. As there is growing evidence that the metabolite stress response is connected to tolerance, this work contributes to a systems understanding of the stress response and its regulation, which may lead to new strategies for achieving higher tolerance in *C. acetobutylicum* to these and other products.

Metabolic Engineering for the Production of Diverse Polyhydroxyalkanoates in Escherichia coli

Choi, S. Y., Korea Advanced Institute of Science and Technology (KAIST)
Park, S. J., Myongji University
Kim, W. J., Korea Advanced Institute of Science and Technology (KAIST)
Yang, J. E., Korea Advanced Institute of Science and Technology (KAIST)
Lee, H., Korea Research Institute of Chemical Technology
Shin, J., Korea Research Institute of Chemical Technology
Lee, S. Y., Korea Advanced Institute of Science and Technology (KAIST)

We previously reported production of polylactate (PLA) by engineered *Escherichia coli* harboring an evolved polyhydroxyalkanoate (PHA) synthase and propionyl-CoA transferase. The evolved PHA synthase from *Pseudomonas* sp. MBEL 6-19 is able to accept broad range of substrates from short chain length (SCL) hydroxyacyl-CoA to medium chain length (MCL) hydroxyacyl-CoA, therefore the PLA producing *E. coli* was further engineered to expand monomer spectrum. Various hydroxy acids including 3-hydroxybutyrate (3HB) and 4-hydroxybutyrate (4HB) were copolymerized with lactate as novel biomaterials. First, for *in vivo* synthesis of 3HB-CoA, *R. eutropha phaA* and *phaB* genes encoding β-ketothiolase and acetoacetyl-CoA reductase were introduced. Second, to produce 4-hydroxybutyrate, the *Clostridium kluyveri sucD* and *4hbD* genes encoding CoA-dependent succinate semialdehyde dehydrogenase and 4HB dehydrogenase were introduced. These copolymers showed enhanced polymer properties such as molecular weight, thermal properties as novel biopolymers for various applications. [This work was supported by the Technology Development Program to Solve Climate Changes on Systems Metabolic Engineering for Biorefineries from the Ministry of Science, ICT and Future Planning (MSIP) through the National Research Foundation (NRF) of Korea (NRF-2012M1A2A2026556 and NRF-2012M1A2A2026557)].

1-Propanol Production of S. Cerevisiae Engineering 2-Ketobutyrate Biosynthetic Pathway

Nishimura, Y., Kobe University
Ishii, J., Kobe University
Kondo, A., Graduate School of Engineering, Kobe University

As a means of handling global issues like warming impact and depletion of fossil resources, biofuels and products from sustainable biomass resource have received a lot of attention in recent years. Especially, since alcohols are available as next-generation fuels and building blocks of chemicals, they have been extensively studied. Therefore, we focused on 1-propanol. 1-propanol is generically used as solvent, food additive, paint, cosmetics and chemical intermediate like n-propylamine. However, a natural microbial producer of 1-propanol is yet to be identified.

The budding yeast *S. cerevisiae* is expected to become a producer strain of 1-propanol. As *S. cerevisiae* has been used to produce ethanol, it possesses tolerances to alcohols and other stresses during fermentation. The tolerance to high concentrations of alcohol should be useful for the industrial production of 1-propanol. Furthermore, *S. cerevisiae* utilizing not only glucose but also xylose as sugar have been developed to expand application range of biomass. For these reasons, *S. cerevisiae* would be more suitable for the production of 1-propanol than other microbial hosts.

Recently, it has been reported that the production of various alcohols from a-keto acids is enabled by using 2-Keto acid decarboxylase (KDC) and alcohol/aldehyde dehydrogenase (ADH) in metabolic engineering (Atsumi et al., 2008). Therefore, 1-propanol can be produced from 2-ketobutyrate (2KB). By using this method, 2.1 mM of 1-propanol was produced from 8 g/L of 2KB in *E. coli* (Atsumi et al., 2008). Therefore, it would be expected to convert 2KB produced from threonine into 1-propanol by reaction with KDC and ADH in *S. cerevisiae*.

Additionally, the pathway producing 2KB without passing through threonine was constructed in *E. coli*. Although *E. coli* primarily have no citramalate synthase (*cimA*), *E. coli* learn to produce 2KB from pyruvate through citramalate by introducing this heterogenous gene. With this method, 4.5 g/L of 1-propanol was produced in *E. coli* (Shen and Liao, 2013). Therefore, it would be expected to produce 2KB from pyruvate through citramalate by introducing *cimA* and methylmalate dehydratase (*leuC* and *leuD*) into yeast.

Furthermore, the yeast deleting acetolactate synthase (*ILV2*) in competing pathway produced 60 mg/L of 1-propanol (Eden et al., 2001). Against such a background, we tried to develop metabolically engineered *S. cerevisiae* suitable for the production of 1-propanol by the combination of overexpression of genes for 1-propanol biosynthesis pathway and deletion of competing pathway.

Efficient Fermentative Production of Polymer-Grade D-Lactate By an Engineered Alkaliphilic Bacillus

Xue, Y., Institute of Microbiology,CAS
Assavasirijinda, N.,
Ge, D.,
Yu, B.,
Ma, Y.,

Polylactic acid (PLA) is one important chemical building block that is well known as a biodegradable and a biocompatible plastic. The traditional lactate fermentation processes need CaCO3 as neutralizer to maintain the desired pH, which results in an amount of insoluble CaSO4 waste during the purification process. In this study, we developed a process for high-optical-purity D-lactate production using an engineered alkaliphilic Bacillus strain. The native L-lactate dehydrogenase gene (ldh) was knocked out, and the D-lactate dehydrogenase gene from Lactobacillus delbrueckii was introduced to construct a D-lactate producer. A fed-batch fermentation under non-sterile conditions was conducted using low-priced peanut meal as a nitrogen source and NaOH as a green neutralizer. The D-lactate titer reached 143.99 g/l, with a yield of 96.09%, an overall productivity of 1.674 g/l/h including with the highest productivity at 16h of 3.04 g/l/h. Moreover, high optical purities (approximately 99.85%) of D-lactate were obtained under both conditions. Given the use of a cheap nitrogen source and a non-sterile green fermentation process, this study provides a more valuable and favorable fermentation process for future polymer-grade D-lactate production.

A Hacked 3D Printer Towards the Printing of Materials through Biological Organisms

González, L. M., Carnegie Mellon University
Nielsen, A.,
Voigt, C. A., University of California, San Francisco

Over the last couple of years, 3D printing technology has received a lot of attention and this is not a surprise as the technology has an unlimited number of applications from a materials perspective. Addictive manufacturing surpasses subtractive manufacturing, as printing of intricate structures is now possible with a simple 3D CAD model. With this project, we are seeking to print 3D composite materials with multifunctional capabilities using synthetic gene networks to drive the printing process via encoding the materials to print at the DNA level in microbial organisms. To do the printing part, we modified a commercially available 3D printer, the MakerBot Replicator 2X. We first modified the printer's extruder head to be able to print a blend of agarose and cells from two individual reservoirs instead of the routinely used ABS or PLA plastics reel from a spool. Using an optical interrupter, we detect when the piston in the slider-crank configuration is moving up and fill in the chamber with the agarose and cells blend that eventually carves into the structures. A solenoid valve, normally closed, controls the agarose input and a DC liquid pump (and a second solenoid) controls the cell input via an Arduino microcontroller. When the piston is retracting (moving up), we applied a long enough time delay on the solenoid and the DC liquid pump to compensate for the lost in pressure in the mixing chamber. In addition, a set of check valves prevents the flow of these liquids back to their respective reservoirs. The piston displaces the agarose and cell blend through the nozzle as it is moving down. This design permits us to use an established skeinforge gcode generator with minor tweaks. In this case the sensing of the piston position occurs concomitantly with the machine-inherited 3D printing process. For the agarose feeding system, we have a hermetically seal chamber with one inlet for air and one outlet for the agarose to reach the printing head. To maintain the agarose in the molten state (upon exposing the agarose to 60°C) a magnet wire is wound around the tubing. The house air fills the chamber and because of the built-up pressure the agarose is propelled to this printing head. The agarose needs to be able to support itself without distortion as the patterned filaments are printed. Using a rheometer, we characterized the gelation time, the melting temperature and the shear thinning properties of this agarose. From a synthetic biology perspective, we are developing a broad host CRISPRi system (based on light sensing and recombinases tools) to control the production of cellulose in *Gluconacetobacter xylinus* during the printing process as the cells are exposed to ambient light.

Novel in silico Molecular Modeling for the Enhancement of Enzyme Activity

Yi, J. Y., Korea Research Institute of Bioscience and Biotechnology
Kang, D. K., Korea Research Institute of Bioscience and Biotechnology
Im, S., Korea Research Institute of Bioscience and Biotechnology
Lee, S. G., University of Science and Technology
Lee, J., Korea Research Institute of Bioscience and Biotechnology
Sohn, J. H., Korea Research Institute of Bioscience and Biotechnology
Sung, B. H.,

We proposed an effective molecular modeling concept for the enhancement of enzyme activity. The molecular dynamics of enzyme, substrate and cofactor was estimated with the CHARMM (Chemistry at HARvard Molecular Mechanics) and Generalized Born with a simple SWitching (GBSW). All variation of amino acid substitutions were introduced at the directly reacting residues with cofactor and the theoretical Delta Delta G values for each of the mutations was calculated. Also the correlation between changed surface area by the substitution and the Delta Delta G values was examined. Here we hypothesized that mutant enzymes revealed high Delta Delta G value at the same time does not show difference of surface area with wild type have higher activity than wild type enzyme. Therefore selected candidates were expressed in *Escherichia coli*, purified and its activity experimentally measured.

To verify the hypothesis, we used two enzymes, succinyl-CoA synthetase (SCS) from *Escherichia coli* and formate dehydrogenase (FDH) from *Thiobacillus sp.* SCS catalyzes a reversible reaction in the tricarboxylic acid (TCA) cycle. Succinate, CoA and NTP (N: adenosine or guanosine) were converted to succinyl-CoA, NDP and Pi by SCS and vice versa. Succinyl-CoA can be used as a precursor for the production of 5-aminolevulinate, porphyrins, heme, and chrorophyll. Therefore, increasing activity of SCS for succinyl-CoA production is needed for industrial usage. We discovered two SCS mutant enzymes that increased the production of succinyl-CoA two-fold than wild type. In addition, FDH is a NADH-dependent redox enzyme converting carbon dioxide and formate by ping-pong reaction. Oxidation activity of FDH to convert formate to carbon dioxide with NAD^+ is stronger than reduction activity to convert formate to carbon dioxide with NADH. However, if its reductivity can be increased by our *in silico* molecular modeling, economical benefit is expected such as reducing greenhouse gas and production of valuable organic chemical. As a result, we found a mutant FDH with 60 percent higher reductivity than wild type. This novel *in silico* molecular modeling can be applicable to design novel enzymes with enhanced activity.

Metabolite-Centric Flux Analysis for Rational Design and Engineering of Microbes

Lakshmanan, M., Bioprocessing Technology Institute
Lee, D. Y., National University of Singapore

The development of various constraint-based model-driven computational algorithms has immensely aided the rational redesign of metabolic networks for the targeted metabolite overproduction in microbial cell factories. While most such methods rely on a reaction-centric approach whereby multiple reactions are suggested to be overexpressed/knocked, we have recently presented a method which is metabolite-centric. In this work, we demonstrated how the previously developed concept of flux-sum, i.e. the turnover of metabolites in constraints-based flux analysis, can be utilized for identifying metabolic engineering targets by applying it to Escherichia coli, as a case study for enhancing ethanol and succinate production. Remarkably, while several metabolite targets identified well corresponded to the gene targets which have been identified by reaction-centric algorithms, certain unique targets such as the pyruvate attenuation for enhancing succinate production were also unraveled. Similarly, we also show how the concept of flux-sum can be utilized to analyze the intracellular redox balancing by estimating overall NAD(H) and NADP(H) turnover. For this purpose, we examine the NADPH regeneration potentials in E. coli, Saccharomyces cerevisiae, Bacillus subtilis, and Pichia pastoris across multiple environmental conditions and uncovered E. coli and glycerol as the best microbial host and most suitable carbon source, respectively, for producing NADP-dependent products. Moreover, we also identified the optimal cofactor specificity engineering (CSE) enzyme targets using our previously developed cofactor modification analysis (CMA), whose cofactors when switched from NAD(H) to NADP(H) augmented the overall NADP(H) flux-sum significantly. Overall, the metabolite-centric approach offer significant potential for rational design and engineering of industrial microbes in the field of systems metabolic engineering.

Metabolic Network of Synechococcus Sp PCC 7002: Genome Scale Metabolic Model, Flux Analysis and Network Modifications for Biofuel Production

Hendry, J. I.,
Prasannan, C. B.,
Joshi, A.,
Dasgupta, S., Reliance Industries Limited
Wangikar, P. P., Indian Institute of Technology Bombay

Synechococcus sp PCC 7002 is a model cyanobacterium with characteristics favorable for industrial scale bioprocess applications such as: shorter doubling time (~4h), ability to with stand high light intensities, tolerance to high salt concentrations, naturally transformable etc. In this study Genome Scale Metabolic model and Constraint Based Reconstruction Analysis (COBRA) are used to analyze the metabolism of this cyanobacterium. Gene essentiality and synthetic lethality analysis were used to analyse the robustness under gene perturbation. The number of essential and synthetic lethal genes revealed that the metabolic network of PCC 7002 is less robust than *Synechocystis* sp PCC 6803 under gene perturbation. Flux distributions under photoautotrophic, heterotrophic (in glycerol) and mixotrophic (in glycerol) conditions were predicted to gain insights into the metabolism of this organism. It was observed that the heterotrophic metabolism in glycerol is different from the glucose metabolism under dark. Heterotrophic growth in glycerol showed a complete TCA cycle and zero flux through the Oxidative Pentose Phosphate (OPP) Pathway. Minimization of Metabolic Adjustment (MOMA) algorithm, was used to identify candidate mutants that produce enhanced ethanol and butanol. An exhaustive analysis of Photoautotrophic flux distribution of all single and double gene reaction knock out mutants was done to identify mutants with high production rates of the target chemicals. The candidates with higher production rates were analysed in detail. It was found that double gene knock outs could divert only ~10% of the fixed carbon towards the target biofuel production.

Metabolic Engineering of an Acid-Tolerant Yeast Pichia Kudriavzevii NG7 for Gypsum-Free Production of D-Lactic Acid

Park, H. J., KAIST
Kim, M. J., KRIBB
Sohn, M. J., KRIBB
Lee, S. H., KRIBB
Bae, J. H., KRIBB
Sung, B. H., KRIBB
Han, J. I.,
Sohn, J. H., KRIBB

Lactic acid is a main building block of polylactic acid (PLA), a biodegradable bioplastic. Conventional lactic acid fermentation uses calcium hydroxide for the neutralization of acidic pH and eventually produces equal masses of insoluble by-product (gypsum), which decrease quality and value of the product in the product recovery process. To develop a robust yeast platform for gypsum-free lactic acid production, an acid-tolerant yeast strain was isolated from grape skins, and named as *Pichia kudriavzevii* NG7 by ribosomal RNA sequencing. This strain was able to grow at pH 2.0 and 42°C. For the commercial application of *P. kudriavzevii* NG7 as a lactic acid producer, the fermentation pathway from pyruvate to ethanol was redirected to lactic acid. Two copies of pyruvate decarboxylase gene in a diploid *P. kudriavzevii* NG7 were replaced with the D-lactate dehydrogenase gene derived from *Lactobacillus plantarum*. Then the engineered strain was sequentially exposed to the high concentration of lactic acid for adapted evolution. Final adapted strain produced 40 g/L of lactic acid without neutralization in flask cultivation. In a fed-batch fermentation, it produced 121 g/L of lactic acid with productivity over 4 g/L/h at a final pH of 3.5.

Production of 3-Hydroxypropionic Acid from Glycerol By Metabolically Engineered Klebsiella Pneumoniae L17 in Bioelectrochemical System

Kim, C., Pusan National University
Lee, C. R., Pusan National University
Jeon, B. H.,
Park, S.,
Kim, J. R.,

Conversions of glycerol to value-added chemicals are one of the interesting topics in bioenergy and biorefinery industry. 3-hydroxypropionic acid (3-HP) is commercially expensive materials due to their usefulness and difficulty of chemical synthesis. Interestingly, 3-HP could be produced biologically by microorganisms such as *Klebsiella pneumoniae* through glycerol fermentation. In *K. pneumoniae*, the fermentation of glycerol to 3-HP are proceeded by two enzymes and their cofactors; first glycerol to 3-hydroxypropionaldehyde (3-HPA) by glycerol dehydratase (DhaB) with Coenzyme B_{12} and then 3-HPA to 3-HP by aldehyde dehydrogenase (ALDH) with NAD^+. In the first step, anoxic condition is preferred because not only Coenzyme B_{12} could be produced in anaerobic condition but also DhaB enzyme could be inhibited by oxygen. However, NAD^+ regeneration are needed by oxidizing chemical/reagents (e.g. oxygen and nitrate etc.) for second step. To solve this inconsistency problem, we attempted to apply bioelectrochemical systems for regeneration of NAD^+ even in anaerobic condition. And also we metabolically engineered *K. pneumoniae* L17 for induction and overexpression of aldehyde dehydrogenase (KGSADH, pUC19 vector). Compared to conventional anaerobic fermentation, 3-HP production level was shown significantly high in bioelectrochecmical system. Low level of cellular $NADH/NAD^+$ ratio in *K. pneumoniae* L17 in bioelectrochecmical system was supported the results. As a result, 3-HP production yield was increased by recombinant *K. pneumonia*e L17 in bioelectrochemical system. This study is first attempt as the application study of bioelectrochemical system in 3-HP production, in our knowledge. Further study could be applied the recombinant *K. pnuemoniae* L17 in large scale fed-batch fermentation equipped with bioelectrochecmial system for the examination of commercially applicability.

Development of Shewanalla Oneidensis MR-1 for Bioelectricity Generation from Glycerol: Induction of Glycerol Uptake Module

Kim, M. Y.,
Kim, C.,
Munussami, G.,
Kim, J. R.,
Jeon, B. H.,

Glycerol is a waste byproduct from many biomass conversion processes; however it can be converted to valuable platform chemicals and biomaterials. Microbial fuel cells (MFCs) are the new type of bioprocess which can generate electricity and produce chemicals from biodegradable biomass using electrochemically active microorganism as biocatalyst. In the previous studies, the typical electrochemically active microbe, *Shewanella* sp. can metabolize lower carbon sources such as lactate and formate, but does not produce electricity from glycerol. In this study, we hypothesized that bioelectricity generation from glycerol could be obtained by introduction of glycerol oxidation modules through expression of the genes such as glpF, glpK, glpD. Voltage productions from lactate were obtained as 70 and 75 mV in MFCs using both wild type *S. oneidensis* MR-1 and recombinant *S. oneidensis* MR-1, respectively. In the case of glycerol substrate, significantly higher voltage productions were found in recombinant *S. oneidensis* MR-1 (80 mV) as compared to the wild-type *S. oneidensis* MR-1 (< 10 mV). The glycerol uptake activities were investigated by electrochemical analysis through cyclic voltammetry and power density curve. This study shows that recombinant *S. oneidensis* MR-1 oxidizes glycerol and increase cellular redox state by induction of glycerol metabolism. Using the genetically engineered glycerol metabolism, bioelectrochemical system could provide a platform to product value-added chemicals and bioconversion process in further study.

In Vivo Activation of Apo-Hydrogenases Using Synthetic Analogues of the Active Site

Khanna, N., Uppsala University
Meszaros, L., Uppsala University
Esmieu, C., Uppsala University
Berggren, G., Uppsala University
Lindblad, P., Uppsala University

Hydrogenases have attracted attention as catalysts for the processing of hydrogen. [FeFe] hydrogenases (HydA) consist of [4Fe4S] cluster directly tethered to an organometallic [2Fe] subsite. The [2Fe] subsite is dependent on a specific subset of maturation enzymes (HydEFG) for its biosynthesis and incorporation into HydA. Synthetic chemistry has made the synthesis of analogues of that subsite possible[1]. In this project, [FeFe] hydrogenase from *Chlamydomonas reinhardtii* was heterologously expressed in the absence of the Hyd-maturation machinery in *E. coli* BL21. *In vivo,* biomimetic hydrogen production was achieved upon incubation of the apo-hydA expressing cells with the synthetic mimics of the [2Fe] subsite. This is the first example of *in vivo* activation of apo-hydrogenases using synthetic analogues of the active site. The study demonstrates a powerful new tool for activating hydrogenases under *in vivo* conditions, in the absence of the still only partially characterized maturation machinery.

[1]Berggren, G., Adamska, A., Lambertz, C., Simmons, T. R., Esselborn, J., Atta, M., Gambarelli, S., Mouesca, J.M., Reijerse, E., Lubitz, W., Happe, T., Artero, V., Fontecave, M. (2014). Biomimetic assembly and activation of [FeFe] -hydrogenases. *Nature*, *499*(7456), 66–69.

Computational Pathway Design for Efficient Bio-Production during Stationary Phase

Toya, Y., Osaka University
Masuda, A.,
Shiraki, T., Osaka University
Shimizu, H., Osaka University

1. Backgrounds

Enhancing a target compound production is one of important challenges in bio-production using microorganism. Bio-production during the stationary phase, so called "non-growth-associated production", is expected to lead to a high target yield because the cells do not consume the substrate for growth. In the present study, the mevalonate production using *Escherichia coli*cells during the stationary phase was investigated. Metabolic pathway modification based on the stoichiometry has also been an effective approach for enhancing the target production. A method (solution space design [SSDesign]) that predicts the gene knockout candidates which are suitable for target production under the stationary phase has been developed (Toya et al., Biotech Bioeng, 2015).

2. Materials and Methods

A mevalonate-producing *E. coli* strain inducing *mvaES* gene derives from *Enterococcus faecalis*was used for the evaluations of bio-process during the stationary phase. The mevalonate-producing strain was aerobically grown on synthetic medium. After the growing, the cells were cultured under the synthetic medium without an essential nutrient such as nitrate, sulfur and magnesium. The network of optimal pathways for non-growth-associated production can be designed from the flux variability (solution space). Elementary mode analysis is a method to decompose the network into its smallest reaction sets, called elementary flux modes (EFMs). The proposed SSDesign determined the area over which EFMs should be removed from the solution space of the parent strain, and explored the gene knockouts that will eliminate these undesirable EFMs.

3. Results and Discussion

The mevalonate yields in the stationary phase increased by 1.5-1.9 times than that in the growth phase. The yield and the specific production rate of mevalonate were highest in magnesium starvation case. The mevalonate is synthesized from acetyl-CoA with NADPH as reductive cofactor in this strain. The metabolic and genetic analysis revealed that the NADPH for mevalonate synthesis was generated in a transhydrogenase reaction.

To evaluate the performance of SSDesign, the method was applied to non-growth-associated succinate production in *E. coli*. The predicted candidates for succinate production were the deletion mutants Δ*pntAB* Δ*sfcA* Δ*pykA,F* and Δ*sfcA* Δ*maeB* Δ*pykA,F* Δ*zwf*. According to the solution spaces, these strains allow high growth yield and inevitably produce succinate at zero biomass yield, since their metabolic pathways cannot sustain steady-state without discarding succinate from the cell.

A Hierarchy of Modeling Frameworks for Cyanobacterial Metabolism: From Gsms to ME Models

Mueller, T., Pennsylvania State University
Gopalakrishnan, S., Penn State
Simons, M., The Pennsylvania State University
Pakrasi, H. B., Washington University in St. Louis
Maranas, C. D., The Pennsylvania State University

Cyanobacteria serve as models for more complex photosynthetic organisms as well as promising platforms for bioproduction. Despite their ability to grow without complex carbon substrates, cyanobacteria have previously been limited in their applicability by their relatively slow growth rates. This has been mitigated by the recent discovery of the fast-growing strain *Synechococcus* UTEX 2973. To help expedite its adoption by the community we have developed stoichiometric, carbon mapping, and metabolism and expression (ME) models for *Synechococcus* 2973. Stoichiometric models form the foundation for the proposed hierarchy of modeling frameworks. They can deduce maximum theoretical yields for biomass formation, however growth rate prediction is unattainable as the rate of uptake of carbon substrates and cost of the catalytic machinery are fixed. By building upon this foundation, ME models can quantify growth rate as the cost of assembling/disassembling the catalytic resources is directly calculated. This provides the means of identifying the drivers that explain what confers fast growth for *Synechococcus* 2973 but not for *Synechococcus* 7942, a related organism that is 99.8% identical in its genome sequence. In addition, by integrating atom mapping information onto GSM models, improved and expanded flux predictions based on nonstationary MFA can be obtained.

Stoichiometric genome-scale metabolic (GSM) models were developed for both *Synechococcus* 2973 (*i*Syu627) and *Synechococcus* 7942 (*i*Syf686). These two models were developed using a semi-automated method that combined curated gene annotations with the previously developed *Synechocystis* GSM model *i*Syn731. The ability to grow on sugar varies across cyanobacteria and was analyzed through comparisons of both the metabolism (via GSM models) and genome of organisms. It has been experimentally shown that under similar conditions, *Synechococcus* 2973 grows 3.7 times faster than *Synechococcus* 7942. The inclusion of additional constraints has reconciled this difference in growth rates.

ME models are able to simulate in more detail the metabolic burden caused by enzymatic processes. An ME model of *Synechococcus* 7942 has also been constructed that expands upon the GSM model to include a complete description of the catalytic machinery. ME models integrate transcriptional and proteomic data and dynamically track metabolites and macromolecular products. The growth rate is calculated as the outcome of metabolism and cellular machinery's assembly/disassembly costs. An ME model simulation of *Synechococcus* 2973 was completed that minimized the differences between predicted relative gene expression levels and normalized transcriptomic data.

A genome-scale carbon mapping model in conjunction with 13C labeling data allows for more precise predictions of flux ranges using metabolic flux analysis (MFA) as compared to the ranges determined by a stoichiometric model. We found a 71% similarity between *i*Syn731 with the fully mapped iAF1260 *E. coli* model resulting in 221 unique cyanobacterial reactions. The first Genome-scale metabolic mapping (GSMM) models for cyanobacteria, *i*mSyn711 and *i*mSyf608 have been developed for *Synechocystis* and *Synechococcus*, respectively. These models reveal 67 novel carbon scrambling reactions unique to cyanobacteria arising from Calvin cycle, photorespiration, expanded glyoxylate metabolism, and corrinoid biosynthetic pathways. These models can be used to carry out non-stationary MFA to uncover cyanobacterial metabolism at a genome-wide level.

Engineering Bacterial Biocatalysts for the Production of Polyhydroxyalkanaote Biopolymers

Nomura, C., Hubei Collaborative Innovation Center for Green Transformation of Bio-Resources

Poly-3-hydroxyalkanoates (PHAs) are biodegradable polyesters produced by some bacteria as a carbon and energy from renewable resources. PHAs have attracted great interest as environmentally friendly replacements of petroleum-based plastics and have potential applications as bulk-commodity plastics and biomaterials for biomedical use. The physical properties of PHA polymers are dictated by their repeating unit composition and can be divided into three main classes: short-chain-length (SCL) PHA polymers are made up of repeating units of 3-5 carbons and exhibit thermoplastic properties, medium-chain-length (MCL) PHA polymers are made up of repeating units of 6-14 carbons and exhibit elastomeric properties and SCL-MCL PHA copolymers that exhibit a range of physical properties dependent on the mol ratio of SCL to MCL repeating units in the polymer. Our lab is interested in understanding what metabolic pathways in bacteria are used to generate PHA polymers. In this presentation, I will discuss some of the new pathways and methods we have designed to produce PHAs in bacteria and explore some new applications.

Metabolic Engineering of Clostridium Acetobutylicum with CoA Metabolite Analysis for Enhanced Butanol Production

Lee, S. H., Korea University Graduate School
Kim, S., Korea University Graduate School
Kim, J., Korea University Graduate School
Kim*, K. H., Korea University Graduate School
Shin, Y. A., R&D Center, GS Caltex Corporation

Clostridium acetobutylicum have been traditionally used for butanol production. However, economically feasible production is still limited by low solvent titer and productivity of traditional acetone-butanol-ethanol (ABE) fermentation. Here, we report the development of hyper solvent producing *C. acetobutylicum* having high volumetric productivity followed by integration of adsorptive continuous *ex situ* butanol recovery fermentation for high solvent titer and enhanced productivity. Metabolically engineered strain was constructed by simultaneous disruption of *pta* and *buk* genes, which encode phosphotransacetylase and butyrate kinase, and overexpression of *adhE1-ctfAB* operon, encoding aldehyde-alcohol dehydrogenase for high volumetric productivity. The strain produced ABE with the productivity of 1.4 g/L/h, which are 230% higher than those obtained from wild type in batch fermentation. Continuous *ex situ* butanol recovery fermentation was performed with the engineered strain to further improve ABE titer and productivity. The solvent productivity and desorbed ABE titer of 2.28 g/L/h and 46.3 g/L, respectively, was achieved. However, ethanol was accumulated and strain degeneration was occurred during continuous fermentation. Thus, target gene for metabolic engineering was examined with CoA metabolite analysis and the strain was further engineered by overexpressing thiolase (*atoB*) gene of *Escherichia coli*, which catalyzes condensation of two acetyl-CoA to an acetoacetyl-CoA molecule to reduce ethanol accumulation. The final recombinant strain was able to produce solvent with a high yield, productivity, and desorbed ABE titer of 0.38 g/g glucose, 2.64 g/L/h, and 55.7 g/L, respectively, with continuous *ex situ*recovery fermentation during 100 h. Ethanol titer was maintained less than 4 g/L in fermentor throughout the fermentation. Significant strain degeneration was not observed. The achieved ABE volumetric productivity is the highest among the reported adsorptive fermentations with Clostridia.

Keywords:

Clostridium acetobutylicum, metabolic engineering; CoA metabolite, high productivity, *ex situ* butanol recovery fermentation

This work was supported by the Center for Organic Wastes to Energy Business under the Environmental Technology Development Program funded by the Ministry of Environment, Korea (2013001580001).

Predictive Pathogen Biology: Genome-Based Prediction of Pathogenic Potential and Countermeasures Targets

Ray, D.,

Horizontal gene transfer (HGT) and recombination leads to the emergence of bacterial antibiotic resistance and pathogenic traits. Genetic changes range from acquisition of a large plasmid to insertion of transposon into a regulatory gene. HGT events can be identified by comparing a large number of fully sequenced genomes across a species or genus, define the phylogenetic range of HGT, and find potential sources of new resistance genes. In-depth comparative phylogenomics can also identify subtle genome or plasmid structural changes or mutations associated with phenotypic changes. Comparative phylogenomics requires that accurately sequenced, complete and properly annotated genomes of the organism. Due to dramatic advances in "short read" sequencing technology, the raw sequence coverage needed for sequencing a bacterial genome now can be obtained in a couple of days for a few dollars sequencing costs, starting with only a few nanograms of genomic DNA. Assembling closed genomes requires additional mate-pair reads or "long read" sequencing data to accompany short-read paired-end data. To bring down the cost and time required of producing assembled genomes and annotating genome features that inform drug resistance and pathogenicity, we are analyzing the performance for genome assembly of data from the Illumina NextSeq, which has faster throughput than the Illumina HiSeq (~1-2 days versus ~1 week), and shorter reads (150bp paired-end versus 300bp paired end) but higher capacity (150-400M reads per run versus ~5-15M) compared to the Illumina MiSeq. Bioinformatics improvements are also needed to make rapid, routine production of complete genomes a reality. Modern assemblers such as SPAdes 3.6.0 running on a standard Linux blade are capable in a few hours of converting mixes of reads from different library preps into high-quality assemblies with only a few gaps. Remaining breaks in scaffolds are generally due to repeats (e.g., rRNA genes) are addressed by our software for gap closure techniques, that avoid custom PCR or targeted sequencing.

Our goal is to improve the understanding of emergence of pathogenesis using sequencing, comparative genomics, and machine learning analysis of ~1000 pathogen genomes. Machine learning algorithms will be used to digest the diverse features (change in virulence genes, recombination, horizontal gene transfer, patient diagnostics). Temporal data and evolutionary models can thus determine whether the origin of a particular isolate is likely to have been from the environment (could it have evolved from previous isolates). It can be useful for comparing differences in virulence along or across the tree. More intriguing, it can test whether there is a direction to virulence strength. This would open new avenues in the prediction of uncharacterized clinical bugs and multidrug resistance evolution and pathogen emergence.

Engineering of a Cell Factory That Can Utilizing the Biomass for Biochemical Production

Qi, Q.,
Gao, D., Shandong University

A single-cell factory that can degrading large molecular biomass and can use various components in biomass or waste material is the key for biochemical and biofuels production. We **recently identified the catalytic domain of a cellulase (Cel-CD) from *Bacillus* sp. that can be secreted into the medium from recombinant *E. coli* in large quantities without its native signal peptide. This protein can degrading cellulose efficiently in the culture medium indicating a great potential. By subcellular location analysis, we proved that the secretion was a two-step process via the SecB-dependent pathway through the inner membrane.** Both the Cel-CD and its N-terminal sequence can serve as carriers for efficient extracellular production of select target proteins with a concentration from 101 to 691 mg/L in flask cultivation. A recombinant *E.coli* that can directly utilize cellulose as sole carbon source by fusion Cel-CD with a b-glucosidase was constructed.

By knocking out the genes involved in the phosphotransfer cascade of the Phosphoenolpyruvate-dependent glucose-specific phosphotransferase system (PTS^{Glc})^{c} in the host *E.coli*, the effect of each mutation on cell growth and substrate consumption in the presence of glucose was investigated. We found that the engineered strains have different capability for utilizing the mixture of glucose and xylose. Finally, we investigated the polyhydroxybutyrate (PHB) and succinate production using the different metabolic engineered mutants in the presence of both sugars with different ratios. The results suggested that this system has a potential application in lignocellulosic biomass degradation and biochemical biofuel production.

Construction of Highly Efficient Expression Systems in Halomonas Based on Orthogonal Transcriptional Machinery

Zhao, H.,
Zhang, H.,
Ouyang, Q., Peking-Tsinghua Joint Centre for Life Sciences
Chen, G. G. Q., Tsinghua University

The fact that genetic materials and design principles for synthetic biology have to be studied and formatted in an ad-hoc manner across different genetic context significantly hinders the transformation of achievements from model organisms into real bioengineering practice. One solution is to use orthogonal gene expression machinery to insulate the synthetic circuits and pathways from genetic context. Here we develop a set of orthogonal transcriptional machinery (OTM) that allows insulated transcriptional control in both *E.coli* and *Halomonas* TD01, a halophile of potent industrial value. Phage RNAP-promoter pairs as OTM candidates were first identified via genome mining and characterized in both genetic context; three pairs with best performance then engineered to allow genetically stable maintenance and strict inducible control (ON/OFF ratio >1000). Promoter library across a wide range of transcriptional activity was also built for each OTM. Results showed that the performance of OTMs and their libraries in *E.coli* and *Halomonas* TD01 are in strikingly linear correlation. OTMs were then used to control bacterial elongation cassette for industrial application. All OTMs enable *Halomonas* TD01 cells to elongate 100-fold in length compared with wild type, indicating the validity of OTMs

Metabolic Engineering for Chemicals Production By Solving Reducing Equivalent Problems

Zhang, X., Chinese Academy of Sciences

Redox balance is very important to realize efficient chemicals production in metabolic engineering. However, there are many redox imbalance problems during metabolic engineering for chemicals production. For example, the reducing equivalent type generated during anaerobic glucose metabolism is not consistent with that required for isobutanol production. The reducing equivalent quantities generated in the native host are not enough for those required for succinate and terpenoids production. We developed several strategies to solve these reducing equivalent problems. Three strategies were used for isobutanol production, including (1) activation of transhydrogenase PntAB and NAD kinase in combination for converting NADH to NADPH; (2) recruiting NADH-dependent keto acid reductoisomerase together with activating PntAB and NAD kinase; (3) recruiting Entner-Doudoroff pathway to change the reducing equivalent type generated during anaerobic glucose metabolism. Two strategies were used for succinate production, including (1) activation of pentose phosphate pathway to increase the reducing equivalent quantities generated during anaerobic glucose metabolism together with activating transhydrogenase SthA to convert NADPH to NADH for succinate production; (2) activation of pyruvate dehydrogenase under anaerobic condition in the transcriptional and protein levels. Pentose phosphate pathway and TCA cycle were activated to increase the reducing equivalent quantities generated during aerobic glucose metabolism for improving terpenoids production.

A Multi-Scale, Multi-Disciplinary Approach for Assessing the Technological, Economic, and Environmental Performance of Bio- Based Chemicals

Campodonico, M. A., Technical University of Denmark
Herrgård, M. J., Technical University of Denmark
Sukumara, S., Technical University of Denmark
Zhuang, K., Technical University of Denmark

In recent years, bio-based chemicals have gained interest as a renewable alternative to petrochemicals. However, there is a significant need to assess the technological, biological, economic, and environmental feasibility of bio-based chemicals, particularly during the early research phase. Recently, the Multi-scale framework for Sustainable Industrial Chemicals (MuSIC) was introduced to address this issue by integrating modeling approaches at different scales ranging from the cellular to the ecological scales. This framework can be further extended by incorporating modeling of the petrochemical value chain and the de novo prediction of metabolic pathways by using generic biochemical reaction operators in conjunction with genome-scale models.

Biosynthesis of a Novel Terpolymer P(LA-co-3HB-co-3HP) from Unrelated Carbon Source

Ren, Y., MOE Key Lab of Bioinformatics, Department of Biological Science and Biotechnology, School of Life Science, Tsinghua-Peking Center for Life Sciences, Tsinghua University, Beijing 100084, China

Polylactate (PLA) is currently commercially synthesized as a representative bio-based polyester by the chemo-bio process on the basis of metal catalyst mediated chemical polymerization of lactate (LA) supplied by microbial fermentation. PLA is turned into highly valuable materials for biomedical, food, and general purpose applications, but its poor ductility and heavy metal residue limited the application. Meanwhile, polyhydroxyalkanoates (PHA) are diverse polyesters synthesized by various bacteria as intracellular carbon and energy storage compounds. As bio-renewable and biodegradable materials, the diversity of PHA units confers to the different physical properties. So copolymerization of PLA with other hydroxyalkanoate (HA) units only using biological method is one of the effective methods to improve the physical properties of PLA and avoid the heavy metal contamination.

In this study, we established a mutant of PHA synthetase which could incorporate LA as well as other 3-HAs at the same time. We synthesized different PLA-PHA copolymers in recombinant *Escherichia coli*aiming at improving the properties of PLA and reducing the cost of copolymers by utilizing low cost carbon sources. The results obtained are summarized below:

The substrate specificity of the PHA synthases is the most important factor determining the monomer structures whether that could be incorporated into PHA or not. To introduce LA into the copolymer, we constructed mutated type II PHA synthetase derived from *Pseudomonas stutzeri* $phaC1_{Ps}$, which has a versatile substrate specificity, referring to the former reports about PLA polymerization. The mutant $phaC1_{Ps}$(E130D S325T S477G Q481K) could polymerize LA into PLA with precursor LA-CoA. Gene *pct* converting LA into LA-CoA and genes related to P3HB synthesis were co-expressed with the PHA synthetase mutant to produce P(3HB-*co*-LA). By optimizing the metabolite patheway and host strain, we finally produced P(3HB-*co*-23mol%LA) in recombinant *E. coli*. It is the most efficient PLA polymerizing enzyme in our system when we compared $phaC1_{Ps}$(E130D S325T S477G Q481K) with those phaCs reported before.

P(3HB-*co*-LA), as a classic PLA-PHA copolymer, has been widely studied. The physical properties of the homopolymers PLA and P3HB are both high strength and low ductility. P3HP has an opposite character that it possesses exccellent mechanical strength, tensile stength and elongation break. To synthsize a PLA-PHA coplymer with an enhanced ductility, we constructed an P3HP synthetic pathway and combined with PLA and P3HB synthetic pathway. P(LA-*co*-3HB-*co*-3HP) terpolymer was produced from unrelated carbon sources firstly. The nuclear magnetic resonance (NMR) analysis revealed that the random terpolyester P(3HB-*co*-3HP-*co*-LA) contained 90.41 mol% 3HB, 7.78 mol% 3HP and 1.81 mol% LA respectively. In addition, the PLBP terpolymer was produced from two types of unrelated carbon sources, and the monomer ratio of PLBP terpolymer was variable, by regulating the proportion of glycerol and glucose and modifying the metabolism of the host strain. By analyzing these terpolymers, we characterized a new PLA-PHA copolymer with better mechanical porperties.

In conclusion, we synthesized new PLA-PHA terpolymers with variable components from unrelated carbon source which will promote the developemnt of LA based copolymer and its application.

Key Words: PHA; PLA; PHA synthetase; P(3HB-*co*-3HP-*co*-LA); unrelated carbon source; *Escherichia coli*

Isoprene Production By Coupling Methylerythritol Phosphate and Mevalonate Pathway in Escherichia coli

Jiang, Y.,
Sun, B.,
Yang, C., Key Laboratory of Synthetic Biology, Institute of Plant Physiology and Ecology, Shanghai Institutes for Biological Sciences, Chinese Academy of Sciences, Shanghai 200032, China.
Gao, X., Key Laboratory of Synthetic Biology, Institute of Plant Physiology and Ecology, Shanghai Institutes for Biological Sciences, Chinese Academy of Sciences, Shanghai 200032, China.
Yang, S., Chinese Academy of Sciences

Isoprene, a key building block of synthetic rubber, is currently produced entirely from petrochemical sources. In this work, we observed 3□20-fold higher yield of isoprene production by coupling methylerythritol phosphate (MEP) and the mevalonate (MVA) pathway than individually overexpress single pathway in *E. coli*. Fed-batch cultivation of the engineered strain overexpressing the dual pathway resulted in production of 25.0 g/L isoprene with a yield of 0.277 g/g of glucose. The results strongly suggested that coupling of the complementary reducing equivalent demand and ATP requirement plays an important role in the synergy of the dual pathway, which was suggested to improve the production of a broad range of terpenoids in microorganisms.

A Simple Metabolic Flux Balance Analysis of Biomass and Bioethanol Production in Kluyveromyces Marxianus ATCC 26548 Batch Culture

Tompang, M. F., Universiti Sains Malaysia
Abd Karim, K., Universiti Malaysia Perlis
Harun @ Kamaruddin, A., Universiti Sains Malaysia
Ku Ismail, K. S.,

The yeast *Kluyveromyces marxianus* ATCC 26548 has been considered as a potential microorganism for many applications in industrial bioprocesses especially in bioethanol production due to its special traits as compared to others. Mathematical approach is used to analyze the flow of metabolites in biological system in order to improve desired product yield as well as the overall production process, thus development of a simple model could lead to sustainability and practicability of the process. In this study, the comparative analysis of a simple metabolic network and a black box description is carried out in order to evaluate the yeast growth and bioethanol production in batch fermentation. Metabolic flux balance methodology has shown to give more accurate estimation with complete analysis of the reaction rates. Furthermore, good interpretation of the yeast behavior in batch culture at different glucose concentrations was achieved based on its stoichiometric reaction analysis. No significant changes in biomass growth when glucose concentration is increased at a controlled aeration above 60% air saturation, but resulting in higher ethanol production in low dissolved oxygen concentration. The biomass and ethanol production are mostly depended on oxidative and reductive catabolism, respectively, which are controlled by glucose and oxygen uptake rates. Therefore, the analysis of reaction rates in minimal metabolic network provided substantial about the physiology of cultured yeast cells, and the production of bioethanol and biomass could be improved by changing the batch experimental conditions.

Heterologous Biosynthesis and Omics -Guided Metabolic Engineering for Overproduction of Spinosad in Streptomyces

Tan, G. Y., Shenyang Research Institute of Chemical Industry Ltd., Co., SINOCHEM Group
Liu, X., Key Laboratory of Combinatorial Biosynthesis and Drug Discovery (Wuhan University), Ministry of Education, and Wuhan University School of Pharmaceutical Sciences
Deng, K., Wuhan University
Tao, H., Wuhan University
Chen, K., Shenyang Research Institute of Chemical Industry Ltd., Co.
Li, X., Shenyang Research Institute of Chemical Industry Ltd., Co.
Deng, Z., Wuhan Institute of Biotechnology
Liu, T., E-mail: liutg@whu.edu.cn

Spinosad, produced by Saccharopolyspora spinosa, is an important macrolide antibiotics with potent insecticidal properties. As a safe and environment-friendly pesticide, spinosad has been widely used in agriculture. However, due to the low productivity, the industrial production of spinosad is quite limited in China. In this study, a large size of 110kb engineered BAC plasmid, which harbored biosynthetic genes of Spinosad, has been introduced into Streptomyces host, such as *Streptomyces albus* J1074 and *Streptomyces lividans*TK24. Then, by RT-PCR analysis, the transcription of all the biosynthetic genes can be detected. In the flask fermentation experiment, the heterologous production of spinosad in Streptomyces host has been detected by Q-Exactive high-resolution mass spectrometer. To furtherly improved heterologous production of targeted product, the "omics"-guided metabolic engineering of spinosad production has been applied. By using proteomics and metabolomics approaches, several of rate-limiting steps of spinosad heterologous production have been revealed. The successful heterologous production in this work lays the foundation for overproduction of antibiotics, and future metabolic engineering work will be applied for high efficient production of spinosad in Streptomyces.

Identifying Patterns in Kinetic Parameters That Determine the Impact of Rate-Limiting Enzymes

Miskovic, L., Ecole Polytechnique Fédérale de Lausanne (EPFL)
Hatzimanikatis, V., Swiss Federal Institute of Technology of Lausanne (EPFL)

Kinetic models are essential for studying complex behavior and properties of metabolism. However, a persistent hurdle for constructing these models is uncertainty in the kinetic properties of enzymes. Currently available methods for building kinetic models cope in an indirect way with uncertainties by integrating data from different biological levels and origins into models. In this context, we have recently proposed iSCHRUNK[1] a new approach for characterization and reduction of uncertainty that makes use of the ORACLE[2] (Optimization and Risk Analysis of Complex Living Entities) framework and machine learning classification techniques. ORACLE allows us to exploit synergies between different data sources and generate population of nonlinear kinetic models that are consistent with a reference physiological states and the available data. On the other hand, with the machine learning techniques we are able to data mine the integrated datasets *together* with the outputs of ORACLE in order to discern complex interactions and correlations between the parameters of kinetic models and available partial data such as metabolite concentrations, metabolic fluxes, thermodynamic data and information about kinetic properties of enzymes.

The proposed approach is versatile and can be used to infer more accurate ranges of kinetic parameters of all enzymes in large-scale and genome-scale metabolic networks that are consistent with the observed properties of the metabolic network.

In this work, we studied which enzymes determine the sign (positive or negative values) of important control coefficients, and then we identified what should be the configuration of the kinetic parameters of identified enzymes that will determine its role as a rate-limiting enzyme. Based on this analysis we categorized the enzymes in the network in groups that act on studied properties: (i) in a concerted fashion; (ii) antagonistically; or (iii) in a mutually independent way.

This work will allow us to investigate how to engineer the enzymes in order to achieve a pre-specified metabolic behavior and address the inverse metabolic engineering problems in a new way.

1. Andreozzi, S., Miskovic, L. & Hatzimanikatis, V. iSCHRUNK - In Silico Approach to Characterization and Reduction of Uncertainty in the Kinetic Models of Genome-scale Metabolic Networks. *Metabolic Engineering,* **33,**158-168 (2016).

2. Miskovic, L. & Hatzimanikatis, V. Production of biofuels and biochemicals: in need of an ORACLE. *Trends in biotechnology* **28,** 391–7 (2010).

Biosynthesis of Tetrahydrocannabinolic Acid - from Biotransformation Studies to Biotechnological Production

Kayser, O., Technical University Dortmund
Zirpel, B.,
Stehle, F., Technical University Dortmund

Tetrahydrocannabinolic acid (THCA) is a phenolic compound from the spectrum of cannabinoids in Cannabis sativa. THCA or its decarboxylated form THC are used for recreational Cannabis consumption, but both compounds are also of high interest in rational medicine. Due to legalization and acceptance of Cannabis sativa and the pure compound THC, there is a high demand by patients and a strong wish for stricter regulation and standardization of clinically used drugs.

To meet this demand a biotechnological and recombinant THC is of interest by physicians and regulation authorities world wide. In this work all relevant biosynthetic enzymes (CBGAS, THCAS) starting from olivetolic acid as feeding substance were cloned, functionally expressed and assembled in S. cerevisieae or P. pastoris. Biosynthetic primary pathways were in silico analyzed for the identification of metabolic bottle necks and metabolic engineering. Both enzymes were structurally mutated and engineered for improved efficacy and again assembled in metabolic altered production organisms. A full bioreactor design was worked out and implemented in a batch wise mode. By analysis of feeding conditions, temperature, pH, oxygen saturation an upstream process was developed to allow higher yield production by engineered microorganisms. In this talk concepts and strategies of a comprehensive bioprocess approach from biology to engineering is described in details.

References:

1. Zirpel, B., Stehle, F., Kayser, O. (2015) Production of Δ9-tetrahydrocannabinolic acid from cannabigerolic acid by whole cells of Pichia (Komagataella) pastoris expressing Δ9-tetrahydrocannabinolic acid synthase from Cannabis sativa L. Biotechnol. Lett., 37:1869-1875

2. Gruchattka, E., Hädicke, O., Klamt, S., Schütz, V., Kayser, O. (2013) In silico profiling of Escherichia coli and Saccharomyces cerevisiae as terpenoid factories. Microbial Cell Factories, 12: 84

3. Degenhardt, F., Stehle, F., Kayser, O. (2015) The biosynthesis of cannabinoids (Preedy, R. Editor) in The Handbook of Cannabis and Related Pathologies: Biology, Diagnosis, Treatment, and Pharmacology. Academic Press (Elsevier) in press

Phylogenetic and Molecular Dynamics Simulation of 3D-Structures of [Nifese] Hydrogenase: The Biological Relevance to Their Gas Cavity Development

Tamura, T.,

Oxygen tolerance of selenium-containing [NiFeSe] hydrogenases (Hases) is attributable to the high reducing power of selenocysteine (Sec) residues, which sustain the bimetallic Ni–Fe catalytic centre in the large subunit. Genes encoding [NiFeSe] Hases are inherited by few sulphate-reducing δ-proteobacteria globally distributed under various anoxic conditions. Ancestral sequences of [NiFeSe] Hases were elucidated and their three-dimensional structures were recreated *in silico* using homology modelling and molecular dynamic simulations, which suggested that deep gas channels gradually developed in [NiFeSe] Hases under absolute anaerobic conditions, whereas the enzyme remained as a sealed edifice under environmental conditions of a higher oxygen exposure risk. The development of a gas cavity appears to be driven by non-synonymous mutations, which cause subtle conformational changes locally and distantly, even including highly conserved sequence regions.

Microbial Block Copolymerization of 3-Hydroxyalkanoates

Wang, Y., Beijing Institute of Technology
Chen, X., Tsinghua University

Block polyhydroxyalkanoates (PHA) were reported to be resistant against polymer aging that negatively affects polymer properties. Recently, more and more attempts have been directed to make PHA block copolymers. It is possible to produce PHA block copolymers of various kinds using recombinant *Pseudomonas putida* KT2442. Polyhydroxyalkanoate (PHA) synthesis genes *phaPCJ*$_A$*.c* cloned from *Aeromonas caviae* were transformed into *Pseudomonas putida* KTOY06ΔC, a mutant of *P.putida* KT2442, resulting in the ability of the recombinant *P. putida* KTOY06ΔC (*phaPCJ*$_A$*.c*) to produce a short-chain-length and medium-chain-length PHA block copolymer consisting of poly-3-hydroxybutyrate (PHB) as one block and random copolymer of 3-hydroxyvalerate (3HV) and 3-hydroxyheptanoate (3HHp) as another block. PHA synthesis genes *phbC* and *orfZ* cloned from *Ralstonia eutropha* H16 were transformed into β-oxidation weakened *P. putida* KTOY08ΔGC, a mutant of *P. putida* KT2442. The recombinant *P. putida* strain termed KTHH06 was able to produce a short-chain-length PHA block copolymer consisting of poly(3-hydroxybutyrate) (P3HB) as one block and poly(4-hydroxybutyrate) (P4HB) as another block. Diblock copolymers PHB-*b*-PHHx consisting of poly-3-hydroxybutyrate (PHB) block covalently bonded with poly-3-hydroxyhexanoate (PHHx) block were for the first time produced successfully by a recombinant *P. putida* KT2442 with its β-oxidation cycle deleted to its maximum. These polymers were characterized by nuclear magnetic resonance(NMR), differential scanning calorimetric (DSC) , and thermo- and mechanical analysis. In comparison to a random copolymer with similar ratio and a blend of homopolymers PHB and PHVHHp , PHB and PHHx or P3HB and P4HB with similar ratio, the block copolymer showed improved structural related mechanical properties such as the highest tensile strength and Young's modulus. Compared with other commercially available PHA including PHB, PHBV, PHBHHx, and P3HB4HB, the block copolymers showed differences in terms of mechanical properties and should draw more attentions from the PHA research community.

Poly(3-hydroxybutyrate-co-4-hydroxybutyrate) Production By Halomonas TD01 in Continues and Open Fermentation Process with Low Sodium Concentration

Ye, J.,

Poly(3-hydroxybutyrate-*co*-4-hydroxybutyrate) [P(3HB-*co*-4B)], which is synthesized by a variety of bacterial strains, shows different mechanical properties with various 4HB ratios. So far, the industrial production of P(3HB-*co*-4HB) is generated from structurally related carbon sources, such as 1,4-butanediol or γ-butyrolactone. And the high cost of the fermentation process and raw materials are the disadvantages for the production and application of P(3HB-*co*-4HB). Now, a pathway for P(3HB-*co*-4HB) production from glucose was successfully constructed in *Halomonas* TD01, and further genes manipulation will be done to enhance the metabolic flux for higher 4HB ratio (10%-14%). On the other hand, *Halomonas* TD01 is a strong candidate for PHA production in continuous and open fermentation using seawater instead of fresh water for its high pH toleration. In our case, we can generate 60% PHA content of cell dry weight in shake flasks in MM medium with 30 g/L glucose only, and the 4HB ratio of P(3HB-*co*-4HB) copolymer production is around 1.8%. And the highest cell dry weight was more than 61 g/L and the 4HB ratio was about 1.3% in 1 litter continuous and open fermentation trial. At the same time, we are making effort to knock-out *gabD* gene to enhance the 4HB ratio, and three copies were successfully knocked out. Furthermore, we have brought down the NaCl concentration from 60 g/L to 7.5 g/L in MM medium for *Halomonas* TD01 to grow with some genes deletion, which can dramatically reduce the cost and corrosion of the bioreactor. All the above show the great effort and potential for the industrial low cost P(3HB-*co*-4B) production in the future.

Re-Assemblying the Peptidoglycan for Enhanced Production of Polyhydroxyalkanoates

Zhang, X.,

Morphology engineering for prokaryote cells involved in many genes related to bacterial shapes. And the manipulation of these genes made the enhancement of polyhydroxyalkanoates production possible. Due to the small size of microorganisms, especially prokaryote cells, it is usually very difficult to separate tiny cells from fermentation media. One of these possible solutions to overcome this obstacle is to manipulate bacterial cell-wall synthesis genes. Once the speed of bacterial cell-wall synthesis is consistent with the accumulation speed of polyhydroxyalkanoates granules, much more bigger cells can be obtained and separated from fermentation media. It is therefore desirable to find an alternative economic way to achieve the separation.

Characterization of UDP-Glycosyltransferase Bioparts and Metabolic Engineering of Saccharomyces Cerevisiae to Produce Bioactive Rare Ginsenosides

Wang, P., Institute of Plant Physiology and Ecology, Shanghai Institutes for Biological Sciences, Chinese Academy of Sciences

Wei, W., Institute of Plant Physiology and Ecology, Shanghai Institutes for Biological Sciences, Chinese Academy of Sciences

Wei, Y., Institute of Plant Physiology and Ecology, Shanghai Institutes for Biological Sciences, Chinese Academy of Sciences

Yang, C., Institute of Plant Physiology and Ecology, Shanghai Institutes for Biological Sciences, Chinese Academy of Sciences

Zhao, G., Institute of Plant Physiology and Ecology, Shanghai Institutes for Biological Sciences, Chinese Academy of Sciences

Yue, J.,

Yan, X., Institute of Plant Physiology and Ecology, Shanghai Institutes for Biological Sciences, Chinese Academy of Sciences

Zhou, Z., Institute of Plant Physiology and Ecology, Shanghai Institutes for Biological Sciences, Chinese Academy of Sciences

Ginsenosides, a group of triterpenoids, are the main pharmacological active constituents of ginseng. Most ginsenosides are glycosylated products of the triterpenoid aglycones protopanaxadiol (PPD) and protopanaxatriol (PPT). Rare ginsenosides, such as PPD-derived Compound K (CK), Rh2, Rg3 and PPT-derived F1 and Rh1, typically bear less glycosyl moiety but display a broad range of pharmacological properties, including anti-cancer, anti-inflammation, hepatoprotection and anti-diabetes bioactivities. Despite the available heterologous biosynthesis of their aglycones in yeast, de novo production of ginsenoside by a synthetic biology approach was hindered by the absence of UDP-glycosyltransferase (UGT) bioparts to glucosylate the C3, C6 or C20 hydroxyl of aglycones. We assembled 158 UGT contigs from *Panax* EST datasets, which were clustered into 59 operational taxonomic units (OTUs). 45 of them (UGTPg1-45) were successfully cloned from *Panax ginseng* and expressed in *Escherichia coli*. Through functional screening, UGTs responsible for the production of ginsenosides CK, Rh2, Rg3, F1 and Rh1 were identified and characterized. Based on the UGT bioparts and yeast chassis to produce PPD or PPT, yeast cell factories were built to produce these bioactive rare ginsenosides from glucose. Metabolic engineering and fed-batch fermentation of these strains were performed to further improve the titer. Our study provides a sound manufacturing approach for active rare ginsenosides in yeast via synthetic biology strategies, and reveals the possible biosynthetic pathways of ginsenosides in *Panax* plants.

Engineering Lactococcus Lactis for the Synthesis of Plant Polyphenols

Gaspar, P., Technical University of Denmark

Dudnik, A., Technical University of Denmark

Neves, A. R., Current address: Chr. Hansen A/S

Forster, J., Technical University of Denmark

Plant polyphenols are powerful antioxidants and have been shown promising effects in the combat of cardiovascular disease, certain types of cancer, neurodegenerative diseases, allergies, diabetes and inflammation. These health-promoting properties have led to the exploitation of this group of compounds for pharmaceutical and/or biotechnological applications. Nevertheless, polyphenols are produced in small amounts, difficult to purify from natural resources and due to their complexity chemical synthesis is a poor option. Hence, the utilization of microbes as platforms for effective production of plant bioactive metabolites emerges as a feasible alternative.

Lactococcus lactis is an industrially relevant microorganism used worldwide in food fermentations. In view of its economical value, *L. lactis* has been intensively studied and a large body of comprehensive data on their metabolism and genetics was generated. The technological breakthroughs in functional genomics, combined with a GRAS status, a relatively simple metabolism, a small genome, and rapid growth, makes this organism an ideal host for production of valuable food and pharmaceutical compounds [1]. Thus, *L. lactis* tailored for optimal production of nutraceutical plant-derived polyphenols is a promising tool to fortify fermented dairy products without the need for food supplements.

Here we describe our efforts to engineer *L. lactis* for the production of plant polyphenols. Synthetic pathways for resveratrol production from L-tyrosine were assembled and functional expressed in a LDH-deficient strain. Engineered strains accumulated resveratrol and the intermediate *p*-coumaric acid in different amounts depending on gene combination and pathway configuration (operon *vs* monocistronic). Additionally, cinnamic acid and the stilbene pinosylvin were produced from L-phenylalanine due to substrate unspecificity of enzymes in the pathway. During growth of the producer strains, the profiles of intermediates and stilbene pools indicated malonyl-CoA availability as a bottleneck in stilbene production. This hypothesis was substantiated by a remarkable increase of resveratrol and pinosylvin specific yields in the presence of the fatty acid synthesis inhibitor cerulenin. We are currently attempting to improve the malonyl-CoA levels through modulation of acetyl-CoA carboxylase activity in the engineered strains. In addition, we will report on the potential of *L. lactis* for the production of flavonoids, such as quercetin and fisetin [2]. Synthetic pathways for the production of these two flavonoids have been assembled and their functional expression in the LDH-deficient strain is being assessed.

[1] Gaspar et al., 2013. Biotechnol Adv, 31:764-88

[2] Stahlhut et al., 2015. Metab Eng 31:84–93

Combinatorial Design of Biosynthetic Pathways for the Generation of Beta-Arylamines

Warden-Rothman, R., MIT
Williams, B., University of California, San Francisco
Fischbach, M.,
Voigt, C. A., University of California, San Francisco

β-arylamines are a class of small-molecule pharmaceuticals with a wide array of effector functions, they act as stimulants, appetite suppressants, antidepressants, vasoconstrictors, and migraine prophylactics, among others. The class also includes a number of endocrine signaling molecules that are naturally derived from aromatic amino acids in eukaryotes and their associated microbiota. Because β-arylamines are ubiquitous, there are many enzymes that are known or thought to modify the aromatic amino acids and their derivatives. We have mined metagenomics databases to identify over 300 such enzymes, which were then sequence optimized for expression in *E. coli* and synthesized. The aim of this study is to generate libraries of pathways consisting of many classes of enzymes from all domains of life. It is our goal to create not only novel biosynthetic pathways, but also new β-arylamines.

In initial experiments, panels of decarboxylases and N-methyltransferases were assayed for their ability to modify phenylalanine, tyrosine, and tryptophan *in vivo*. While most of the enzymes converted the amino acids to products at levels detectable by LC-MS, 14% were able to robustly catalyze these transformations in the *E. coli* cytosol. By co-expressing two enzymes, the human indole-N-methyltransferase and a microbiome-derived decarboxylase, we synthesized N,N-dimethyltryptamine in *E. coli*, demonstrating the ability to combine enzymes from disparate species into a single biosynthetic pathway. Additionally, the enzymes have shown activity on non-canonical amino acids, such as meta-tyrosine and p-amino-phenylalanine, further supporting our ability to generate novel pathways.

In this study, we add another seven catalytic classes into our collection of enzymes: acetyl transferases, β-hydroxylases, monooxygenases, N-hydroxylases, O-methyltransferases, ring-halogenases, and ring-hydroxylases. Data obtained by individual enzyme assays will be used to inform the design of pathway libraries, through which we will vary both the included enzymes and their expression levels. LC-MS and NMR will then be used to determine the structure, titer, and novelty of the synthesized compounds.

3 UTR Engineering to Improve Functional Expression of Insoluble and/or Structurally Unstable Enzymes and Proteins in Escherichia coli

Park, J. B., Ewha Womans University
Song, J. W.,

3'-Untranslated region (3'UTR) engineering was investigated to improve solubility and functional expression of insoluble and/or structurally unstable enzymes and proteins in *Escherichia coli*. Insertion of the gene fragments containing putative RNase E recognition sites into the 3'UTR of the target genes (e.g., Baeyer-Villiger monooxygenase (BVMO) genes) led to reduction of the corresponding mRNA level in *E. coli*. However, the amount of BVMOs in soluble fraction was remarkably enhanced resulting in a proportional increase of *in vivo* catalytic activity. The increase of biocatalytic activity notably correlated to the number of putative RNase E endonucleolytic cleavage sites in the 3'UTR. For instance, biotransformation activity of the BmoF1 of *Pseudomonas fluorescens* DSM50106 in *E. coli* was linear to the number of RNase E cleavage sites in the 3'UTR. Thereby, it was assumed that 3'UTR engineering could be used to improve soluble expression of heterologous enzymes as well as to fine-tune the enzyme activity in microbial cells.

Microbial Cell Factory for C5 Platform Chemicals: Aminovalerate and Glutarate

Chae, T. U., Korea Advanced Institute of Science and Technology (KAIST)
Park, S. J., Myongji University
Kim, E. Y., Division of Convergence Chemistry, Korea Research Institute of Chemical Technology, Daejeon, South Korea
Noh, W., Korea Research Institute of Chemical Technology
Park, H. M., Korea Advanced Institute of Science and Technology (KAIST)
Oh, Y. H., Research Center for Biobased Chemistry, Division of Convergence Chemistry, Korea Research Institute of Chemical Technology, Daejeon, South Korea
Lee, S. H., Korea Research Institute of Chemical Technology
Song, B. K., Korea Research Institute of Chemical Technology
Jegal, J., Korea Research Institute of Chemical Technology
Lee, S. Y., Korea Advanced Institute of Science and Technology (KAIST)

5-Aminovalerate (5AVA) is an important C5 platform chemical. *Escherichia coli* was metabolically engineered for the production of 5-aminovalerate (5AVA) and glutarate. When the recombinant *E. coli* WL3110 strain expressing the *Pseudomonas putida davAB* genes encoding delta-aminovaleramidase and lysine 2-monooxygenase, respectively, were cultured and produced 3.6 g/L of 5AVA by converting 7 g/L of l-lysine. When the *davAB* genes were introduced into recombinant *E. coli* strain XQ56 allowing enhanced l-lysine synthesis, 0.27 and 0.5 g/L of 5AVA were produced directly from glucose by batch and fed-batch cultures, respectively. Further conversion of 5AVA into glutarate demonstrated by expression of the *P. putida gabTD* genes encoding 5AVA aminotransferase and glutarate semialdehyde dehydrogenase. When recombinant *E. coli* WL3110 strain expressing the *davAB* and *gabTD* genes was cultured in a medium containing 10 g/L l-lysine and 10 g/L α-ketoglutarate, 1.7 g/L of glutarate was produced.
[This work was supported by the Technology Development Program to Solve Climate Changes on Systems Metabolic Engineering for Biorefineries from the Ministry of Science, ICT and Future Planning (MSIP) through the National Research Foundation (NRF) of Korea (NRF-2012M1A2A2026556 and NRF-2012M1A2A2026557).]

13C-Metabolic Flux Analysis at Its Limits: Identifiability of Anaplerotic Reactions in Corynebacterium Glutamicum

Wiechert, W., Forschungszentrum Jülich
Noack, S., Forschungszentrum Jülich
Kappelmann, J., Forschungszentrum Jülich

^{13}C-Metabolic Flux Analysis has matured into a cornerstone of modern omics technologies [1], providing key insights in metabolic engineering applications as well as in basic pathway regulation [2]. Several studies performing ^{13}C-MFA on *Corynebacterium glutamicum*, an industrial workhorse exhibiting the simultaneous presence of five anaplerotic reactions, indicated non-identifiable fluxes for anaplerotic reactions [3, 4]. In light of these results we thoroughly investigated the question whether stationary isotopic data always allow inferring all fluxes for the best case of full isotopomer information.

Applying a suite of numerical methods and exact integer arithmetic we prove in full generality that for *C. glutamicum* the anaplerotic reactions of the wildtype are non-identifiable from stationary ^{13}C-isotopomer data, even if the labeling enrichment in each metabolite is completely known. We analyzed different scenarios of active anaplerotic reactions to conclude that assuming single reaction steps to be non-active does not necessarily render the remaining reactions identifiable. The omission of one reaction changes the topology of the network, induces a different label flow and thus alters the information content of the respective ^{13}C-enrichment. The results for different assumptions of active anaplerotic reactions are directly transferable to other compartmentation-free organisms [5].

On the one hand, our results show that fluxes in complex networks can remain non-observable regardless of the measurement information provided. Even parallel labeling experiments will not yield additional information. On the other hand, the gained structural information from our analysis allowed to deduce identifiable combinations of measurements and network topologies for *C. glutamicum* and, thus, represents further *a priori* knowledge for the future design of ^{13}C-MFA experiments.

[1] Wiechert W., Nöh K. (2013) Isotopically non-stationary metabolic flux analysis: complex yet highly informative. Curr Opin Biotechnol 24(6):979-986

[2] Niedenführ S., Wiechert W., Nöh K. (2015) How to measure metabolic fluxes: a taxonomic guide for ^{13}C fluxomics. Curr Opin Biotechnol 34:82-90

[3] Noack S., Nöh K., Moch M., Oldiges M., Wiechert W. (2012) Stationary versus non-stationary 13C-MFA: A comparison using a consistent dataset. J Biotech 154(2-3):179-190

[4] van Ooyen J, Noack S, Bott M, Reth A, Eggeling L (2012) Improved L-lysine production with Corynebacterium glutamicum and systemic insight into citrate synthase

flux and activity. Biotechnol Bioeng 109(8):2070-81

[5] Kappelmann J., Wiechert W., Noack S. (2015) Cutting the Gordian Knot: Identifiability of Anaplerotic reactions in *Corynebacterium glutamicum*. Biotechnol Bioeng 113(3):661-674

Production of Short Chain Butyl Esters By E. coli As a Strategy for Butanol Extraction

Chacón, M., University of Bath
Bolhuis, A.,
Leak, D. J., University of Bath

Interest in expanding our use of renewable energies has been gaining momentum over the last few decades. Biological production of transportation fuels, especially, has been of growing research interest, with ethanol fuel leading the way as a surrogate for gasoline. However, ethanol is not the most ideal bio-alcohol fuel because of its low energy density relative to gasoline, and water solubility. Compared to ethanol, butanol fuel is less hydroscopic, less volatile, less corrosive, and contains an energy density that is more similar to gasoline. Butanol is naturally produced by certain species of *Clostridia* through a metabolic pathway known as acetone, butanol and ethanol (ABE) fermentation, of which butanol constitutes the major product. This ABE pathway has been expressed in more tractable organisms such as *Escherichia coli* and *Saccharomyces cerevisiae*, and manipulation of these host microbes to maximize their butanol yield is a current stream of research, with one of its tributaries being the improvement of host butanol tolerance. Microbial butanol exposure is significantly toxic, with even native Clostridial producers only capable of accumulating it to approximately 2% (v/v) in culture, limiting final butanol titres.

In this research we investigate an alternative strategy to improve tolerance, and therefore yield, of butanol in microbial cultures. This strategy involves sequestering the endogenous butanol made by *E. coli* heterologously expressing the ABE fermentation pathway, into a more neutral ester molecule that can both accumulate to higher concentrations in culture with minimal toxic effect, as well as permeate into the media where it can be collected downstream for butanol recovery. Here we discuss the feasibility of such a strategy, from the initial selection of an appropriate candidate butyl ester through to engineering *E. coli* harbouring the ABE pathway for butyl ester synthesis.

Engineering of Glycolytic Pathway for Production of Isobutanol in Cofactor-Balanced Manner

Jung, H., Korea University
Oh, M., Korea University

Isobutanol is considered as one of the promising next-generation biofuels to replace petroleum based conventional fuels. We developed an isobutanol producing *Enterobacter aerogenes* through L-valine biosynthesis pathway. Naturally, *Enterobacter aerogenes* is well known as 2,3-butanediol producer using glucose as a carbons source. Therefore large amount of metabolic flux can be driven from glucose to pyruvate and alpha-acetolactate which are precursors of isobutanol. Firstly, we tried to increase intracellular pyruvate pool and reduce major by-products such as 2,3-butanediol and lactate by deleting *ldhA*, *budA* and *pflB* sequentially. Additionally, transformation of mutant strain was conducted to overexpress isobutanol synthesis pathway. The plasmids harboring isobutanol pathway genes, acetolactate synthase (*budB*), keto-acid reductoisomerase (*ilvC*), dihydroxyacid dehydratase (*ilvD*) of *K. pneumoniae*, alpha-ketoisovalerate decarboxylase (*kivD*), alcohol dehydrogenase (*adhA*) of *L. lactis* and alcohol dehydrogenase (*yqhD*) *of Escherichia coli* were constructed and introduced to mutant strains. We constructed engineered strain producing 12.5g/L of isobutanol (0.25g/g glucose) by metabolic pathway engineering.

When it comes to reducing equivalents generated in native glycolytic pathway, two molecules of NADH are regenerated by glyceraldehyde 3-phosphate dehydrogenase (GAPDH, GapA). However wild type keto-acid reductoisomerase (KARI, IlvC) oxidizes cofactor NADPH. In the aspects of enzymatic cofactor balance, additional NADPH should be supplied to approach theoretical maximum yield. We tried to increase intracellular NADPH pools by substituting NAD(H)-dependent GAPDH, GapA, to NADP(H)-dependent GAPDH, GapC, from *Clostridium acetobutylicum* together with deleting phosphofructokinase (PfkA) for further increasing NADPH generation through pentose phosphate pathway. By reference to the research implemented in S. Bastian et al., NAD(H)-dependent KARI was also constructed. Then four types of isobutanol production pathway were constructed such as 2 NADH dependent pathway ($ilvC^{WT}$, *adhA*), 1 NADH + 1 NADPH dependent pathway ($ilvC^{WT}$ *yqhD*, $ilvC^{Mut}$ *adhA*) and 2 NADPH dependent pathway ($ilvC^{WT}$ *yqhD*). These four pathways were introduced in the host microorganisms that have NADPH generating glycolytic pathway. The production yield increased about 32% in NADPH dependent pathway as the NADPH generating glycolytic pathway applied. This study suggests that engineering glycolytic pathway related to the generation of reducing equivalent can satisfy the cofactor requirement in isobutanol production pathway. Furthermore, 2,3-butanediol-producing *Enterobacteriaceae* have the potentiality as production host of isobutanol in the bio-fuel industry.

Optimization of Phenylpropanoid Production in Saccharomyces Cerevisiae through Disruption of Double Bound Reductase Activity

Lehka, B. J., Evolva
Eichenberger, M., Evolva
Garcia Vanegas, K., Evolva
Bjerg Jensen, N., Evolva
Dannow Dyekjær, J., Evolva
Jenssen, H., Roskilde University
Naesby, M., Evolva
Simon, E., Evolva

Flavonoids are secondary plant metabolites derived from the phenylpropanoid pathway. These bioactive compounds are of great commercial interest due to their varied properties, such as anti-oxidative, anti-tumor or antibacterial activities. However, mass production based on purification of flavonoids from plants can be problematic due to slow growth or limited access.

The objective of this study is to produce naringenin (as a model flavonoid) in *Saccharomyces cerevisiae,* without accumulating the undesired side product phloretic acid (dihydrocoumaric acid). Accumulation of phloretic acid results from the reduction of coumaroyl-CoA by an unknown endogenous yeast reductase activity.

In order to identify said reductases, 26 putative double-bond reductase knockout strains (KO) were transformed with plant genes coding for active 4-coumarate-CoA ligase (4CL) , then fed with p-coumaric acid (the strains were homozygous diploids for nonessential genes whereas in case of the essential genes the strains were heterologous diploids). Two of the KO strains consumed less p-coumaric acid, and produced less phloretic acid. To confirm their activity on coumaroyl-CoA, the two reductases were overexpressed in a naringenin producing strain: This revealed that one of the reductases had a high catalytic activity towards coumaroyl-CoA while the second had had only marginal activity. The identified gene encodes an essential enzyme. Therefore the next step was to find an alternative version of the enzyme that can perform its normal function in the cell but that cannot reduce coumaroyl-CoA. Different versions of the identified enzyme were obtained by constructing a site-saturated mutagenesis library focusing on the residues around the active site and the substrate binding pocket. The mutant library was introduced in a wild type yeast strain, also containing naringenin biosynthesis machinery, and afterwards the reductase gene in question was deleted. The resulting cells were analysed for their growth, production of coumaric acid, naringenin and phloretic acid. While some of the mutant reductase enzymes had significantly lower activity towards coumaroyl-CoA, their growth was negatively affected. We therefore looked at alternative strategies to the same end, some of which were successful – results and conclusions of these experiments will be discussed.

Sisbi: Software to Aid the Industrial Biotechnology Decision Process

Soares, S.,

The recent advances in the field of Metabolic Engineering (ME), regarding the optimization of microbes as cell factories, capable of producing compounds of high economic interest, have made this field one of the main providers of solutions for some of the main challenges faced by Industrial Biotechnology. Among other important contributions, some can be highlighted including strain optimization through the determination of suitable genetic changes, the optimization of culture medium conditions, the selection of suitable host microbes or the search for the most adequate pathways for the production of specific compounds through the insertion of heterologous pathways. In a previous project, the proponent consortium developed computational systems to address these and other ME tasks, creating an integrated platform for the selection and computational optimization of microbes seeking the production of added value compounds.

SISBI is a new project lead by SilicoLife seeking the development of an intelligent decision support system for Industrial Biotechnology (SISBI). This system promotes the development of innovative tools for the extraction and integration of knowledge, to provide an adequate support to reach important decisions in the context of Industrial Biotechnology, as it is the case of the selection of added value compounds, of adequate host microorganisms or of advantageous raw materials to be used in target bioprocesses.

SISBI will result in a knowledge hub including a set of knowledge extraction and integration modules, which will take into account diverse sources of available technical and economical data, boosted by powerful data and text mining tools capable of extracting knowledge from this raw information, useful knowledge for decision making.

Metabolic Engineering of Yarrowia Lipolytica for Value-Added Products Synthesis from Glycerol

Mironczuk, A. M., Wroclaw University of Environmental and Life Sciences
Rzechonek, D., Wroclaw University of Environmental and Life Sciences
Biegalska, A.,
Rakicka, M.,
Dobrowolski, A., Wroclaw University of Environmental and Life Sciences

Growing world population requires progressively more energy sources, therefore the production of biodiesel increased. Glycerol is an important renewable feedstock as it is the main side-product of the biodiesel production process, which is nowadays applied on a large commercial scale. Moreover, glycerol is produced by several others industries, such as fat saponification and alcoholic beverage production units. Despite the high contamination, crude glycerol might be easily utilized by yeast *Yarrowia lipolytica*. Oleaginous yeast *Y. lipolytica* is an interesting host for converting a diverse range of feedstocks, such as waste glycerol, into value-added products such as erythritol (sweetener) or citric acid. Glycerol in *Y. lipolytica* cells is assimilated by phosphorylation pathway, the substrate first is phosphorylated to 3-P-glycerol by a glycerol kinase (GK) and subsequently is dehydrogenated to dihydroxyacetone phosphate by glycerol-3-P dehydrogenase (GDH). To enhance glycerol assimilation we over-expressed the *GUT1* gene coding GK and *GUT2* gene encoding GDH. The modified strains have been tested for glycerol consumption rate and erythritol and citric acid synthesis. Erythritol is a natural sweetener, a four-carbon sugar alcohol, occurring in fruits, such as grapes or melons. Because of its low energy value and non- insulin stimulant properties, erythritol is used as a non-caloric sweetener and pharmaceutical excipient. Citric acid is mainly produced by the mycelial fungus *Aspergillus niger*, but in the past few years much consideration has been focused on the potential use of the yeast *Yarrowia lipolytica*.The engineered strains might be used as a starting platform for further modification for broad-range value-added products biosynthesis.

This work was financed by the polish National Centre for Research and Development under project LIDER/010/207/L-5/13/NCBR/2014

Engineering Yeast for Production of Glycolic Acid

Koivistoinen, O., VTT Technical Research Centre of Finland
Kuivanen, J., VTT Technical Research Centre of Finland
Pitkänen, J. P.,
Vartiainen, E.,
Penttilä, M., VTT Technical Research Centre of Finland
Richard, P., VTT Technical Research Centre of Finland

Glycolic acid is the smallest α-hydroxy acid and is used in a wide range of applications e.g. cleaning, industrial detergents, cosmetics and coatings. Importantly, it can be used as a precursor for biodegradable polyglycolic acid (PGA). PGA is a strong material with an exceptionally high barrier against oxygen and is, therefore, a potential material to be applied e.g. in packaging as a barrier layer. The value chain to economically viable, bio-based PGA can potentially start from lignocellulosic carbohydrates, which after hydrolysis are converted to glycolic acid by fermentation, and further polymerized to PGA – thus providing a high-value product concept for future biorefineries.

Our approach was to test several genetically engineered yeast species to develop a fermentation based process for glycolic acid production. Glycolic acid is produced in the glyoxylate cycle by converting glyoxylate into glycolic acid by glyoxylate reductase. Yeast has an endogenous gene coding for glyoxylate reductase but in order to make the conversion more efficient, we overexpressed an NADPH-specific glyoxylate reductase from *Arabidopsis thaliana*with a high affinity for glyoxylate.

For efficient conversion of glyoxylate into glycolic acid, malate synthase encoding gene was deleted to prevent glyoxylate to react to malate. Additionally, steps leading to glyoxylate formation were boosted by e.g. overexpressing isocitrate lyase encoding gene. Metabolic modelling was used for analysing different pathway options and for finding the targets for further modifications to enhance glycolic acid production. The resulting strains were cultivated in shake flasks in different media and production of glycolic acid was demonstrated. The engineered *Pichia kudriavzevii* produced over 30 gl^{-1} of glycolic acid from ethanol in a fed-batch bioreactor cultivation demonstrating the technical feasibility of the process.

Enhanced Production of Lipids from Glycerol By Engineered Yeast Yarrowia Lipolytica

Dobrowolski, A., Wroclaw University of Environmental and Life Sciences
Drzymala, K.,
Mitula, P.,
Mironczuk, A. M., Wroclaw University of Environmental and Life Sciences

Glycerol is an important renewable feedstock as it is the main side-product of the biodiesel production process, which is nowadays applied on a large commercial scale. The world biodiesel market is predicted to reach a 37 billion gallons by 2016. It was estimated that approximately 1 kg of crude glycerol is generated for every 10 kg of biodiesel produced. Furthermore, glycerol is produced by several other industries, such as fat saponification and alcoholic beverage production units. Taking into consideration the huge quantities of glycerol produced by above mentioned industries and its low costs, it is of urgency to find an alternative ways to convert this substrate into value added products. Unfortunately, crude glycerol may contains many impurities that significantly decrease its value and the purification process of crude glycerol is time and energy consuming. Despite the high contamination, crude glycerol might be easily utilized by yeast Yarrowia lipolytica, a well-known oleaginous yeast. *Y. lipolytica* is one of the most extensively studied "non-conventional" yeasts due to its biotechnological potential. This unconventional yeast, which has been classified as a GRAS organism, is able to metabolize this renewable feedstock and has a huge biotechnological potential to produce citric acid and other organic acids (pyruvic, α-ketoglutaric), single cell oils (SCO), erythritol, mannitol, coca-butter-like lipid or proteins. To enhance production of SCO from glycerol we over-expressed the *GUT1* (*YALI0F00484g*) gene coding GK and to direct carbon flow into lipid production additional we over-expressed the *SCT1* (*YALI0C00209g*) gene encoding G3P- acyltransferase. The modified strains have been tested for glycerol consumption rate, lipid content, fatty acid composition and biomass production. Subsequently, to improve the SCO production, we additionally over-expressed *DGA1* (*YALI0E32769g*) gene encoding DAG-acyltransferase involved in the last step of triglycerides synthesis. All genes were under control of a TEF promoter and enhanced TEF promoter (with 16 times repeated upstream activation sequence). Finally, we achieved strain of *Y. lipolytica* with enhanced production of lipids from glycerol.

This work was financially supported by the Ministry of Science and Higher Education of Poland – project no. IP2012 008972

Pyruvic Acid Production Using Thermotolerant Halomonas sp. KM-1

Tsuji, A.,
Nakazono, S.,
Kuramoto, K.,
Okamoto, N., BOST Kindai Univ.
Takei, Y., BOST Kindai Univ.
Niigawa, Y.,
Nishimura, T.,
Matsushita, I.,
Tsubota, J.,
Kawata, Y.,
Kiryu, T.,
Kiso, T.,
Murakami, H.,
Matsushita, K., Yamaguchi University
Azuma, Y., BOST Kindai Univ.

One of most important processes in industrial fermentation is temperature controlling of fermenters, actually huge electric energy is used to cool the fermenters to appropriate temperatures. While massive consumption of energy can become a factor causing climate change, it must be considered to decrease energy consumption in any fields, including even in the fields known as white-biotechnology, which requires less energy and creates less waste and environmental impact during the fermentative production. Thereupon we have attempted to develop a fermentation system without or with less temperature controlling using thermotolerant microbes. *Halomonas* sp. KM-1, isolated as a contaminant of *Spirulina platensis* culture, is an alkaliphilic, halophilic, aerobic, thermophilic, and heterotrophic bacterium requesting no organic nutrients including amino acids and vitamins except a carbon source for its cultivation. The KM-1 strain, which characteristics provide useful features, i.e. no requirement of sterilization and less cost for fermentation, accumulates poly-3-hydroxybutyrate (PHB) under aerobic fermentation and secret its monomer, 3-hydroxybutyrate (3-HB) under successive microaerobic one using varied carbon sources. Besides PHB and 3-HB, 2-oxopropanoic acid (pyruvic acid) was found to be accumulated in media of KM-1 culture using glucose as a carbon source. Pyruvic acid is known for having numerous benefits as a fine chemical compound to produce medicines and cosmetic products. It is also a relatively new supplement for human to decrease in body weight and fat mass and for athletes as an ergogenic compound to improve endurance during aerobic activity and increase ATP/energy production, but the roles are still controversial. To produce pyruvic acid at a commercial level using an environmentally benign process, we attempted to optimize the production of pyruvic acid by *Halomonas*sp. KM-1 and selected better mutants derived from the strain mutagenized using UV irradiation. Under a small-scale culturing of the KM-1 using 10% glucose at 33°C, pyruvic acid was accumulated up to 2% in the culture medium after 2-day incubation, and vanished within another 2-day culturing. First we tested a variety of carbon sources, their concentrations, and culturing temperatures. As a result, we could clarify a condition under which pyruvic acid was accumulated up to 5% in the culture at 45°C and no diminution of the product was observed even after longer incubation. We have selected mutants, which could not grow in media including glucose as a sole carbon source. Based on tests of growth on different carbon sources and productivities of pyruvic acid, it was shown that the mutants carried mutations on sugar transporters and enzymes for the glycolytic pathway. We also isolated several dozens of thermotolerant mutants, which could grow at 47°C. Those mutants might be useful to produce pyruvic acid or other chemicals in the glycolytic pathway at high temperatures, and capabilities of those mutants will be presented in this meeting.

This work was financially supported by the Advanced Low Carbon Technology Research and Development Program (ALCA), and a Grant-in-Aid for Scientific Research from the Ministry of Education, Culture, Sports, Science and Technology of Japan (KAKEN-HI: 22510222).

Efficient Bioproduction By Psychrophile-Based Simple Biocatalysts

Tajima, T., Hiroshima University
Hamada, M.,
Fuki, K., Hiroshima University
Nakashimada, Y.,
Kato, J.,

Bioproduction of valuable chemicals by conventional biocatalysts are some problem to improve the yields and productivities. Metabolic modifications or fermentation regulations are required for whole cell biocatalysts to enhance the metabolic fluxes. Extracted enzymes can directly reacted with the substrate, however, these approach are often extremely laborious and costly.

To conduct the effective bioproductions, psychrophile-based simple biocatalysts (PSCat) was developed to convert all substrate to products without byproduct formation. The approach is based on the expression of mesophilic enzymes in psychrophilic bacteria, which are then heat treated at moderate temperatures; as a result, intrinsic bacterial enzymes are inactivated, while recombinant mesophilic enzymes retain their catalytic activity. Thus, PSCat is expected to efficiently convert substrates into target compounds without the interference of by-products synthesized by natural bacterial enzymes.

PSCat has been successfully applied to produce high yield of 3-hydroxypropionaldehyde (3-HPA) from glycerol by heat treatment at 45 °C. 3-HPA was stoichiometrically produced with the complete consumption of glycerol at a high production rate of 8.85 mmol 3-HPA/g dry cell/h. PSCat could convert biodiesel-derived crude glycerol to 3-HPA. PSCat has significant advantages over conventional methods in that it is easy to prepare and implement since the method only requires heat treatment of bacterial cells [1].

PSCat could convert building-block chemicals such as aspartic acid which has been selected as a top value-added chemical from biomass by the US Department of Energy. Aspartic acid is produced by partially purified enzymes or whole bacteria including immobilized cells such as *Escherichia coli* bound to polyurethane carriers. However, in intact cells, a fraction of the fumaric acid was lost because of the activity of constitutively expressed bacterial fumarase that converts fumaric acid to L-malate. To generate an efficient biocatalyst by a simple procedure, PSCat was constructed by expressing *E. coli* aspartase in a psychrophilic bacterium, *Shewanella livingstonensis*Ac10. PSCat treated with heat at 50°C for 15 min was produced aspartic acid without L-malic acid as a major byproduct. Furthermore, the immobilized PSCat with sodium alginate was repeatedly used for the conversion. The yield for nine reactions maintained over 97.5% as well as the immobilized extracted aspartase [2].

PSCat also applied for fructose production from mannitol, the major component in marine biomass resource. PSCat expressed mannitol dehydrogenase and NADH oxidase was convert 50 mM mannitol to fructose with 97% yield, and fructose was not converted to further metabolites.

We are trying to produce valuable chemicals with high yields by PSCat catalyzing multiple enzymatic reactions by constructing the strong gene expression tool.

References

1. T. Tajima et al., AMB Express, 3(1), 69 (2013)
2. T. Tajima et al., Journal of Industrial Microbiology & Biotechnology, 42(10), 1319-1324, (2015)

Substrate Channeling for the Production of 2,3-Butanediol in Saccharomyces Cerevisiae

Kim, S., Seoul National University
Bae, S. J., Seoul National University
Hahn, J. S., Seoul National University

Substrate channeling is a process of transferring an intermediate from one enzyme to the next enzyme without diffusion into the bulk phase, thereby leading to an enhanced reaction rate. In the first part, we newly designed substrate channeling modules in *Saccharomyces cerevisiae* based on a high affinity interaction between dockerin and cohesin domains, which is a key process in the formation of cellulosome structure. Synthetic scaffolds containing two, three, or seven cohesin domains were constructed, and the assembly of dockerin-tagged proteins onto the scaffolds was confirmed by pull-down assay and bimolecular fluorescent complementation (BiFC) assay in vivo. This system was applied to produce 2,3-butanediol in *S. cerevisiae* by using dockerin-tagged AlsS, AlsD, and Bdh1 enzymes, resulting in a gradual increase in 2,3-butanediol production depending on the number of cohesin domains in the scaffold. In the second part, we investigated the effect of substrate channeling on the flux partitioning at a metabolic branch point, focusing on pyruvate metabolism. As a platform strain for the channeling of pyruvate flux, PYK1-Coh-Myc strain was constructed in which *PYK1* gene encoding pyruvate kinase is tagged with cohesin domain. By tethering the pyruvate-forming enzyme Pyk1 to heterologous pyruvate-converting enzymes, lactate dehydrogenase and α-acetolactate synthase, using high-affinity cohesin-dockerin interaction, pyruvate flux was successfully redirected toward desired pathways producing lactic acid and 2,3-butanediol, respectively, with a concomitant decrease in ethanol production even without genetic attenuation of the ethanol-producing pathway.

Transcriptional and Translational Regulation of Enzymes Involved in 3-Hydroxypropionic Acid Production Using Synthetic Cassette Architecture

Sankaranarayanan, M., Pusan National University
Somasundar, A., Noroo Holdings Co., Ltd
Seol, E.,
Kwon, S. J., Pusan National University
Nguyen-Vo, T. P.,
Park, S., Pusan National University

Commercial production of 3-hydroxypropionic acid (3-HP) from glycerol encounters several challenges such as 3-hydroxypropionaldehyde (3-HPA) accumulation, cofactor availability and 3-HP toxicity. Among them, 3-HPA accumulation is considered as a serious issue as the 3-HPA inactivate glycerol dehydratase (DhaB) and aldehyde dehydrogenase (KGSADH), the key enzymes for 3-HP production, and reduce the cell growth. Therefore, balancing the activity of DhaB and KGSADH is very important. To this end, we tried to control the expression levels of DhaB and KGSADH by range of transcription levels of *dhaB* with higher transcription level of *KGSADH*. The translation efficiency was normalized for both the genes. We used different strength synthetic constitutive promoters to alter the transcription levels and a bicistronic ribosome binding site that nullify the effect of mRNA secondary structures on translation efficiency. *dhaB* and *kgsadh* were cloned under the control of low and medium strength promoters, respectively and expressed in *E. coli* W. The controlled expression of DhaB and KGSADH were validated in both transcriptional and translational level, using RT-PCR and SDS-PAGE analysis, respectively. The recombinant *E. coli* did not accumulate even small amount of 3-HPA throughout the production and produced ~52 mM 3-HP which is ~20% higher than the reference strain (~41 mM). The activity of DhaB and KGSADH were also present till the end of fermentation. Further tuning of this system will increase the productivity and titer of 3-HP production.

Enhancement of Protocatechuate Decarboxylation for the Effective Production of Muconic Acid from Renewable Feedstocks

Sonoki, T., Hirosaki University
Kikuchi, A.,
Ogata, T.,
Takahashi, K.,
Kamimura, N.,
Masai, E.,

Protocatechuic acid (PCA, 3,4-dihydroxybenzoic acid) is an intermediate to produce a valuable platform chemical, *cis,cis*-muconic acid (ccMA), from renewable feedstocks such as sugar. For the effective production of ccMA, PCA decarboxylation (Pdc) step has been described as a bottleneck, and it has also been required to find out some methods to enhance the enzymatic activity for developing the efficient ccMA production. The Pdc from *Klebsiella pneumoniae* subsp. *pneumoniae*A170-40 (AroY) had been identified, and it has been mainly used to produce ccMA from sugar, however, no reports on the enhancement of Pdc activity and biochemical characteristics of AroY have been published. In addition to improving the Pdc activity, our current work focuses on ccMA production in particular, from lignin-related aromatic compounds because sugars are already in high demand in the development of many bio-based products.

We identified a way to increase Pdc activity in *Escherichia coli* through the expression of both AroY and the B subunit (KpdB) of 4-hydroxybenzoate decarboxylase complex. The Pdc activity increased approximately 20-fold higher than that without the KpdB expression. The effects of pH and temperature on Pdc activity under the condition that AroY was expressed either with or without KpdB. Regardless of the KpdB expression, the optimal pH and temperature for Pdc activity were 5.5 and 35°C, respectively. AroY could decarboxylate not only PCA but also Gallic acid (3,4,5-trihydrroxybenzoic acid) and 3-*O*-methylgallic acid (4,5-dihydroxy-3-methoxybenzoic acid) although the decarboxylated products from *p*-hydroxybenzoic acid, vanillic acid (4-hydroxy-3-methoxybenzoic acid), syringic acid (3,5-dimethoxy-4-hydroxybenzoic acid) were not detected. We are now developing a new transformed bacterial strain using *aroY* , *kpdB* and other required genes to produce ccMA from variety of aromatic compounds, vanillic acid, *p*-hydroxybenzoic acid, ferullic acid (trans-ferullic acid, 4-hydroxy-3-methoxycinnamic acid), and *p*-coumaric acid (trans-*p*-coumaric acid, 4-hydroxycinnnamic acid) etc, with high yields.

This work is supported in part by a Grant-in-Aid for Scientific Research (JSPS KAKENHI 26850041), and Advanced Low Carbon Research and Development Program, Japan Agency for Science and Technology (JST-ALCA).

Metabolic Engineering for Co-Production of Fuels and Chemicals Using a Single Microorganism

Seol, E., Pusan National University
Ko, Y., Pusan National University
Sundara Sekar, B., Pusan National University
Park, S., Pusan National University
Song, H., GS Caltex
Seung, D., GS Caltex

Microorganisms have evolved to optimize their growth, not for the formation of metabolite(s) useful to human. Microbial growth occurs with or through generation of multiple metabolites, but metabolic engineering usually focuses on the production of a 'single' compound by re-designing and re-directing metabolic pathways [1]. This often seriously interferes with cellular metabolisms, hurts cell growth and viability, and makes the microorganism to function inefficiently as microbial cell factory. Production of properly-selected, multiple metabolites, instead of a single metabolite, can be a useful strategy to improve the performance of microbial cell factory by alleviating cellular stresses caused by metabolic engineering. Co-production can offer flexible balancing of the *in vivo* metabolic flux and easier maintenance of cellular homeostasis including redox state. To demonstrate the potential of the co-production strategy, we have studied *Klebsiella pneumoniae* for co-production of 3-hydroxypropionic acid (3-HP) and 1,3-propanediol (1,3-PDO) or 1,3-PDO and 2,3-butanediol (2,3-BDO) from glycerol, and *Escherichia coli* for co-production of H_2 and ethanol or 3-HP from glucose. Co-production of 3-HP and 1,3-PDO made the generation and consumption of NAD(H) well-balanced, reduced the accumulation of the toxic intermediate, 3-hydroxypropionaldehyde, and resulted in high glycerol-to-products yield (>0.8) at high titer (>70 g/L) [2]. Co-production of 1,3-PDO and 2,3-BDO by *K. pneumoniae* alleviated metabolic traffic at the pyruvate node, reduced acetate accumulation, and enabled the maintenance of redox balance. With glycerol as sole carbon source or along with glucose, >100 g/L products (1,3-PDO plus 2,3-BDO) could be obtained. For co-production of H_2 and ethanol from glucose, carbon flux was diverted to the pentose phosphate (PP) pathway from the Embden-Meyerhof-Parnas (EMP) pathway by over-expressing the two major enzymes of PP pathway, Zwf and Gnd, and by disrupting or down-regulating Pfk or Pgi of EMP pathway [3]. Increase in NAD(P)H supply enhanced ethanol production, and H_2 and ethanol could be obtained with the theoretical maximum yields and negligible acetate. During the development of various biocatalysts, to meet the specific requirement for each biocatalyst and/or carbon substrate used, deletion and amplification of many pathways were made by employing carefully-selected enzymes and gene expression systems (dynamic promoters, 5'UTR engineering, etc). This lecture also discusses the challenges of the co-production strategy.

References

1. Kumar, Vinod, Somasundar Ashok, and Sunghoon Park. "Recent advances in biological production of 3-hydroxypropionic acid." Biotechnology advances 31 (2013): 945-961.

2. Kumar, Vinod, et al. "Simultaneous production of 3-hydroxypropionic acid and 1, 3-propanediol from glycerol using resting cells of the lactate dehydrogenase-deficient recombinant Klebsiella pneumoniae overexpressing an aldehyde dehydrogenase." Bioresource technology 135 (2013): 555-563.
3. Seol, E., Sekar, B. S., Raj, S. M. and Park, S. "Co-production of hydrogen and ethanol from glucose by modification of glycolytic pathways in *Escherichia coli* – from Embden-Meyerhof-Parnas pathway to pentose phosphate pathway." Biotechnology Journal 11 (2016): 249–256

Acknowledgement

This work was supported by the Advanced Biomass R&D Center (ABC) of Global Frontier Project funded by the Ministry of Science, ICT and Future Planning (ABC-2011-0031361). The authors are grateful also to the BK21 Plus program at Pusan National University.

Co-Production of 3-Hydroxypropionic Acid and 1,3-Propanediol Using Glycerol By Klebsiella Pneumoniae J2B: Reduction of Carbon Traffic at Pyruvate Node

Ko, Y., Pusan National University
Seol, E., Pusan National University
Sundara Sekar, B., Pusan National University
Kwon, S., Pusan National University
Park, J. M., GS Caltex
Park, S., Pusan National University

Microbial production of 3-hydroxypropionic acid (3-HP) using glycerol is a two-step reaction involving glycerol dehydratase (GDHt) and aldehyde dehydrogenase (ALDH). GDHt requires coenzyme B_{12} and ALDH requires NAD^+ as cofactors for the production of 3-HP from glycerol. Regeneration of NAD^+ requires high aeration which will in turn inhibit GDHt stability and coenzyme B_{12} production. Also, the imbalance between GDHt and ALDH activities cause the accumulation of toxic intermediate, 3-hydroxypropionaldehyde (3-HPA). As a solution to these problems, co-production of 1,3-propanediol (1,3-PDO) along with 3-HP has been suggested. 1,3-PDO production could regenerate NAD^+ and utilize 3-HPA, thus avoiding 3-HPA accumulation and also facilitate the process to be performed under microaerobic conditions. To this end, co-production of 3-HP and 1,3-PDO was attempted in *Klebsiella pneumoniae*, the natural producer of 1,3-PDO. *K. pneumoniae*could coproduce 3-HP and 1,3-PDO successfully but still the process was inefficient in terms of titer and productivity in the later stages. The strain also accumulated huge amounts of acetate and lactate which could reduce the co-production. Especially, acetate accumulation was reported to inhibit the cell performance even at 0.5 g/L by decreasing biomass formation, inhibiting recombinant protein production, etc. Therefore, in this study, to reduce the carbon traffic at pyruvate node and eliminate acetate production, we attempted the down-regulation of glycolytic pathway and disruption of byproducts formation pathway. We also attempted to operate TCA cycle/glyoxylate shunt to reduce the carbon traffic at pyruvate node. The performance of mutant strains were analyzed under different aeration conditions and the efficiency of each approaches in reducing acetate accumulation and improving co-production were addressed.

References

[1] Ashok, Somasundar, et al. "Development of recombinant *Klebsiella pneumoniae* D*dhaT*strain for the co-production of 3-hydroxypropionic acid and 1, 3-propanediol from glycerol." Applied microbiology and biotechnology 90.4 (2011): 1253-1265.

[2] Kumar, Vinod, et al. "Co-production of 3-hydroxypropionic acid and 1, 3-propanediol from glycerol using resting cells of recombinant *Klebsiella pneumoniae* J2B strain overexpressing aldehyde dehydrogenase." Applied microbiology and biotechnology 96.2 (2012): 373-383.

Engineering Lipid Homeostasis Mechanisms for an Efficient Production of Fatty Acid-Derived Chemicals in S. Cerevisiae

Teixeira, P. G.,
Ferreira, R., Chalmers University of Technology
Zhou, Y. J.,
Siewers, V., Chalmers University of Technology
Nielsen, J., Chalmers University of Technology

Saccharomyces cerevisiae has been widely studied and applied as a platform for production of fatty-acid derived chemicals. Although engineering of these cells is today a somewhat straightforward process, overcoming the natural limitations of this organism for accumulation and production of lipid molecules in high amounts brings forward many barriers that deny the progress of metabolic engineering as a tool for industry. A root factor that negatively reacts towards engineering attempts is the tight and sometimes not well understood regulation mechanisms surrounding lipid metabolism. Furthermore, recent progress in cell and molecular biology have been revealing either new concepts of lipid pool regulation and associations between lipid metabolism and many other cell homeostasis processes.

In this study, we explore strategies that target factors related to lipid metabolism regulation which could turn out as key points to unlock control and fine-tuning of fatty acid and acyl-CoA-related pathways for production of fatty acid-derived molecules, such as fatty alcohols, in an efficient way. Due to the unknown characteristics of such processes, we use a combination of rational and blind engineering approaches to both achieve desirable phenotypes and understand the process underlying it after analysis.

Unraveling the Metabolic Interactions in Plasmodium Falciparum Using a Novel Genome-Scale Model Reconstruction

Chiappino Pepe, A., Swiss Federal Institute of Technology (EPFL)
Tymoshenko, S., Swiss Federal Institute of Technology (EPFL)
Ataman, M., Ecole Polytechnique Federale de Lausanne (EPFL)
Soldati-Favre, D., University of Geneva
Hatzimanikatis, V., Swiss Federal Institute of Technology of Lausanne (EPFL)

The identification of novel and more efficient antimalarial therapies is a highly pressing need to fight against multi-drug-resistant parasites. Computational methods that analyze the metabolism of the pathogen can provide testable hypotheses and guide the experimental efforts for the identification of drug targets in metabolic networks. In this study, we aim at answering the following questions: (i) which of the substrates available at the blood stage are essential for growth simulation? (ii) what metabolic enzymes are essential *in silico* for replication? (iii) what are the intracellular metabolites that determine the directionality and function of metabolic enzymes? For this purpose, we have first developed iPfa, a newly reconstructed genome-scale metabolic model of *P. falciparum*, which extends the scope of the existing metabolic models of this parasite. We have then performed advanced computational analyses of iPfa, which involve the integration in iPfa of metabolomics data measured in *P. falciparum* at the blood stage and its thermodynamics-based flux analysis (TFA).

In our computational studies, we identify the minimal nutritional requirements of *P. falciparum* at the blood stage and we identify the *in silico* minimal media, which is composed of only 23 substrates that are required for growth. We also identify the genes and enzymes that may represent drug targets for the blood-stage *P. falciparum* infection. With TFA and metabolomics data integrated, we predict 63 genes and 25 pairs of genes to be essential for the intraerythrocytic growth of the parasite. In total, we found supporting evidence for 35 of these predictions and no information was found for 28 genes that remain to be tested. We also identify metabolites, like glutamine or CDP-ethanolamine, whose intracellular concentration determines the directionality and function of key reactions in the metabolic network. Our computational results provide novel insights about the metabolic capabilities and interactions of *P. falciparum* that can guide experimental efforts towards better understanding of the pathogen's physiology and, ultimately, towards the identification of novel antimalarial drug targets.

Construction of Saccharomyces Cerevisiae Cell Factory for Î²-Amyrin Production

Zhu, M., Tianjin University
Zhang, G., Beijing Institute of Technology
Li, C., Beijing Institute of Technology

The production of terpenoids in microorganisms especially yeast can be a consistent and sustainable way compared with other methods such as extraction from plants and chemical synthesis. Here, an engineered *S. cerevisiae* capable of producing β-amyrin, a commercially valuable compound and precursor of triterpenoids, was constructed via various refactoring approaches beginning with the heterogenous expression of β-amyrin synthase from *Glycyrrhiza glabra*. The titer of β-amyrin was increased 33 folds by introducing a heterogenous squalene monooxygenase from *Candida albicans* with high squalene affinity, which efficiently dragged intermediate metabolite squalene toward β-amyrin. Squalene biosynthesis module was further strengthened via introducing isopentenyl pyrophosphate isomerase from *Escherichia coli* along with overexpressing farnesyl diphosphate synthase and squalene synthase from *S. cerevisiae,* which consequently increased 49 folds of β-amyrin titer. Through reconstructing the promoters with the binding site of transcription factor UPC2, directed transcriptional regulation on the engineered pathway was availably achieved, resulting in β-amyrin titer increased by 65-folds. Ethanol fed-batch fermentation was also employed to enhance β-amyrin synthesis. After optimization, β-amyrin production was finally up to 138.80 mg/L with a yield of 16.30 mg/g DCW in the fermenter experiments, which was increased by 185 and 232 times compared with the initially engineered strain, respectively. Our refactored yeast strain can serve as a cell factory for β-amyrin production as well as the biosynthesis of various β-amyrin derived triterpenoids such as glycyrrhizin and soysaponin.

Keywords: β-Amyrin, Triterpenoids, Transcriptional regulation, *Saccharomyces cerevisiae*, cell factory

A Synthetic Biology Approach for the Design of Flexible and Efficient Production Platforms in Synechocystis Sp PCC6803

Dandapani, H., University of Turku
Thiel, K., University of Turku
Kallio, P., University of Turku
Mulaku, E., University of Turku
Aro, E. M., University of Turku

One of the challenges in the current biofuel research is to effectively harness the native photosynthetic capacity of cyanobacteria for the production of desired metabolites. Despite the potential, the obtained photon conversion efficiencies remain low and yields non-competitive in regards to viable industrial-scale applications.

Our current research is focused on synthetic biology, aiming at engineering more efficient production platforms with flexible means of modification and optimization. We are employing a modified Bio-Brick approach [1] to effectively assemble operons in expression vectors designed for *Synechocystis sp.* PCC6803. The systems can be used, for example, for RBS optimization of individual genes or entire operons, and evaluation of different integration sites used for protein over-expression from the host genome.

The ultimate aim is to combine (i) the tools for engineering optimized heterologous pathways for biofuel production with (ii) different modification strategies of the host metabolism for maximal photon conversion efficiency and metabolic flux from CO2 towards the final products.

Redgem and Lumpgem: Systematic Approaches for Reduction of Genome-Scale Metabolic Models

Ataman, M., Ecole Polytechnique Federale de Lausanne (EPFL)
Hernandez, D., Swiss Institute of Bioinformatics
Fengos, G., Swiss Institute of Bioinformatics
Hatzimanikatis, V., Swiss Federal Institute of Technology of Lausanne (EPFL)

Genome-scale metabolic reconstructions (GEMs) are invaluable resources for understanding and redesigning cellular networks as they encapsulate all the known biochemical processes occurring in organisms. Complexity of these large metabolic networks often hinders their utility in various practical applications. This difficulty sparked the interest for models of reduced sizes that are consistent with the genome-scale models. Currently available core models are reduced *ad hoc* with different aims and criteria, and up to date, there are no systematic reduction methods in the literature.

In this work, we have developed two methods, **redGEM** and **lumpGEM**, for constructing reduced metabolic models from compartmentalized or non-compartmentalized GEMs. redGEM uses as inputs: (i) definition of the metabolic subsystems/pathways that are of interest for a selected physiology; (ii) information about the extracellular nutrients and possible by-products; and (iii) any available physiological data from metabolomics, fluxomics, proteomics and transcriptomics. Firstly, redGEM employs a directed graph search method to find the pairwise connections (reactions) between the selected subsystems/pathways to account for all possible carbon flow routes. These subsystems (reactions and metabolites), along with the identified connections define the core network for the reduced model. Then, we apply lumpGEM to construct subnetwork(s) and corresponding lumped reaction(s) that are capable of synthesizing all biomass building blocks from this core network. redGEM/lumpGEM allows the generation of different sized models with the following criteria: (i) different degree of connections between subsystems/pathways of interest, and (ii) alternative lumped reactions for the same biomass building block. The result of this procedure is reduced core model that are consistent with the original GEM in terms of flux profiles and essential reactions and genes. We used these two methods to generate reduced models for *E. coli, S. cerevisiae* and human metabolic model Recon 2 under aerobic/anaerobic conditions and generated *GEM-compatible* core models. We also compared the 3 organisms and highlighted the differences and similarities of their core carbon metabolism.

"ATLAS of Biochemistry", a Repository of All Possible Biochemical Reactions for Synthetic Biology and Metabolic Engineering Studies

Hafner, J., Swiss Federal Institute of Technology (EPFL)
Hadadi, N., Ecole Polytechnique Federale de Lausanne (EPFL)
Hatzimanikatis, V., Swiss Federal Institute of Technology of Lausanne (EPFL)

Knowledge gaps in metabolic models and continuously growing biochemical databases indicate that our understanding of metabolism is far from being complete. Recent technical and analytical progress in the field of metabolomics has lead to vast amounts of new compounds being identified in living organisms, but how these metabolites integrate in metabolic networks remains largely unknown. To address these problems of incomplete knowledge, we propose a computational approach that identifies novel hypothetical reactions between known metabolites, it further integrates experimentally measured molecular structures into existing metabolic networks, and it finally predicts chemical compounds that are probable to exist in metabolism.

The computational framework BNICE.ch is used to exploit the known biochemistry contained in the Kyoto Encyclopedia of Genes and Genomes (KEGG). We summarize the vastly diverse functionalities of enzymatic reactions in a few hundred expert-curated reaction rules, each generalizing multiple biochemical reactions. We then apply these rules to all metabolites known to KEGG in order to create a database of all the biochemically plausible reactions between compounds reported to occur in living organism. This extrapolation of the known metabolism results in a network of more than 130'000 known and novel reactions, each connecting two or more KEGG compounds. For KEGG compounds that cannot be connected to any metabolic pathway through single known or novel reactions, we propose multi-step reactions involving not only KEGG compounds as intermediates, but also compounds appearing in purely chemical databases like PubChem. These hypothetical intermediate metabolites are highly probable to play a role in metabolism, even though they are not reported as biological compounds in KEGG.

For all the known and novel reactions generated by our approach we further estimate the Gibbs free energy using the Group Contribution Method (GCM), and we also assess the structural similarity of the hypothetical reactions to the known KEGG reactions. To validate the consistency of our results with known biochemistry, we show that several reactions that have been added in the last KEGG update can be predicted as *novel reactions* from compounds in the earlier versions of the KEGG database.

The generated information is organized in the "Atlas of biochemistry", an online database that allows the user to search for all the possible routes from any substrate compound to any product. These pathways involve known and novel enzymatic steps and they provide potential targets for protein engineering to alter substrate specificity. Our approach of introducing novel biochemistry into pathway design can be of great interest in metabolic engineering and synthetic biology projects.

Development and Application of Metabolic Network Models

Zhang, C.,
Gu, D.,
Jian, X.,
Hua, Q.,

In recent years, computer aided redesigning methods based on genome-scale metabolic network models (GEMs) have become a popular tool in the design of genetically engineered strains and helped biologists to decipher metabolism. Most of these methods, however, are hindered by intractable computing times, such as the prediction of high level knockout strategies for the overproduction of desired metabolite targets. In this study, a framework called IdealKnock has been developed to efficiently break through the limitation of maximum knockout number in reasonable time and suggest knockout strategies with better performance. Moreover, the complex gene-reaction relationships in most GEMs have resulted in limited capabilities with respect to directly predicting accurate genetic design for metabolic engineering. We have therefore proposed a couple of methods to simplify the gene-reaction associations by introducing intermediate pseudo reactions or pseudo gene controlling reactions, making it possible to generate genetic design and allowing more metabolic engineering applications such as direct integration of expression data into GEMs.

Kinetic Modeling of Metabolism in Obligatory Anaerobes

Dash, S.,
Khodayari, A., The Pennsylvania State University
Holwerda, E. K., Dartmouth College
Olson, D., Dartmouth College
Theisen, M., UCLA
Liao, J. C., University of California, Los Angeles
Stephanopoulos, G., Massachusetts Institute of Technology
Lynd, L. R., Dartmouth College
Maranas, C., Penn State University

Industrially relevant anaerobes such as *C. thermocellum* can metabolize cellulose and *C. ljungdahlii* can metabolize syngas to produce biofuels, but they remain poorly characterized with significant uncertainty in their metabolic repertoire. To this end, we develop cell-wide dynamic metabolic models for the two organisms using the Ensemble Modeling (EM) paradigm which requires curated genome-scale metabolic (GSM) model of the organism as its foundation. We constructed a second-generation GSM model (*i*Cth446) for *C. thermocellum* with 446 genes, 598 metabolites and 660 reactions, along with gene-protein-reaction associations by correcting cofactor dependencies, restoring elemental and charge balances and updating GAM and NGAM values. The GSM model was subsequently used to construct a core kinetic metabolic model of the *C. thermocellum*'s central metabolism containing 119 reactions and 93 metabolites with cellobiose as the carbon source under anaerobic growth condition. It encompasses the cellobiose degradation pathway, glycolysis/gluconeogenesis, the pentose phosphate (PP) pathway, the TCA cycle, pyruvate metabolism, anaplerotic reactions, alternative carbon metabolism, nucleotide salvage pathway, along with all biomass precursors and 22 substrate level regulatory interactions extracted from BRENDA. Model parameterization was carried out by simultaneously imposing the mutant library data recently measured and provided by the Lynd group. This dataset includes *C. thermocellum* variants with mutations in lactate, malate, acetate, and hydrogen production pathways and combinations thereof defining 22 specific mutants with measured concentrations of various fermentation products such as acetate, lactate, formate, hydrogen, pyruvate, ethanol, and cellobiose (19 measured concentrations per mutant). The kinetic model accurately predicted metabolic phenotypes in multiple mutant strains not included in the model parametrization. Examples include cytosolic concentrations for fourteen out of eighteen metabolites in the *Dldh* mutant. The kinetic model also alludes to a systemic level effect of limiting nitrogen source resulting in increased yields for lactate, pyruvate and amino acids and an increase in ammonia and sugar phosphates concentrations (~ 1.5 fold) due to down-regulation of fermentation pathways under ethanol stress. In addition, a leave-one-out cross validation analysis revealed that the robustness of the estimated parameters remained limited to the mutants located in the same vicinity of the parameterization mutants. The constructed model will also be analyzed using the ensemble modeling robustness analysis (EMRA) to identify the robustness of enzymes and the metabolic conditions which render them unstable. A similar effort is also underway for *C. ljungdahlii*. A core metabolic model composed of 77 reactions and 63 metabolites was constructed with 41 substrate level regulatory interactions from

BRENDA based on other Clostridia species. The network spans glycolysis, the Wood-Ljungdahl pathway, and TCA cycle. Experimental flux data for a wild-type and mutants will be used to estimate the core model parameters using the same proposed framework. The constructed kinetic models will be ultimately used to explore metabolic drivers that underpin the over-production of iso-butanol in *C.thermocellum* and acetate in *C. ljungdahlii.*

Effective Estimation of Dynamic Metabolic Fluxes Using 13c Labeling and Piecewise Affine Approximation: From Theory to Practical Applicability

Schumacher, R., Delft University of Technology
Wahl, S. A., Max Planck Institute

The design of microbial production processes relies on rational choices for metabolic engineering of the production host and the process conditions. These require a systematic and quantitative understanding of cellular regulation. As an extension of the Dynamic metabolic flux analysis (DMFA) framework to underdetermined metabolic network, a method for dynamic flux identification using quantitative metabolomics and ^{13}C labeling to identify piecewise-affine (PWA) flux functions has been described recently. Obtaining flux estimates nevertheless still required frequent manual reinitalization to obtain a good reproduction of the experimental data and, moreover, did not optimize on all observables simultaneously (metabolites and isotopomer concentrations).
In our contribution we focus on measures to achieve faster and robust dynamic flux estimation which leads to a high dimensional parameter estimation problem. Specifically, we address the following challenges within the PWA problem formulation: (1) Fast selection of sufficient domains for the PWA flux functions, (2) Control of over-fitting in the concentration space using shape-prescriptive modeling and (3) robust and efficient implementation of the parameter estimation using the hybrid implicit filtering algorithm. With the improvements we significantly speed up the convergence by efficiently exploiting that the optimization problem is partly linear. This allows application to larger-scale metabolic networks and demonstrates that the proposed approach is not purely theoretical, but also applicable in practice.

Analysis and Design of Metabolic Engineering and Synthetic Biology Strategies Using Large Scale Dynamic Models of Metabolism

Fengos, G., Swiss Institute of Bioinformatics
Miskovic, L., Ecole Polytechnique Fédérale de Lausanne (EPFL)
Hatzimanikatis, V., Swiss Federal Institute of Technology of Lausanne (EPFL)

Systematic analysis for the redirection of carbon flux in metabolite-producing microorganisms requires the comprehensive understanding of their complex metabolic processes. The use of large-scale dynamic models of metabolism plays key role in the understanding of these processes and the study of possible metabolic engineering interventions. However, the generation of such models is hampered by the intrinsic nonlinearities of enzymatic reactions, and the uncertainties at different levels of model building and data integration. In particular there is limited knowledge about the exact kinetic mechanisms, and many of the parameters involved in these mechanisms remain largely unknown.

In this study we propose a systematic methodology to generate large populations of dynamic non-linear models of metabolism using the ORACLE (Optimization and Risk Analysis of Complex Living Entities) framework. Instead of seeking for an optimal value of the estimated kinetic model parameter values, we integrate thermodynamics, available omics, and kinetic data to construct populations of models that are locally stable, and consistent with the observed physiology.

To demonstrate the utility of this methodology we constructed a population of large-scale dynamical models of optimally grown *E. coli* that involves 174 metabolites and 350 reactions. We used these models to (i) study the response of *E. coli* metabolism upon large-scale perturbations, such as gene knockouts; (ii) identify and analyze multiple steady states; and (iii) characterize basins of attraction around the identified steady states. The aforementioned analyses provide valuable insight for the design of metabolic engineering strategies towards amplification of desired product-forming pathways.

Synthetic Promoter Systems for Controlling Saccharomyces Cerevisiae Gene Expression

Cox, R. S. III, Graduate School of Engineering, Kobe University,Kobe University
Nambu, Y.,
Hasunuma, T.,
Kondo, A., Graduate School of Engineering, Kobe University

We characterized the previously reported artificial transcription factor Z3EV, which activates synthetic yeast promoters in response to the inducer beta-estradiol. We designed and constructed a series of synthetic promoters responsive to Z3EV and controlling expression of the reporter LacZ. We characterized their expression in different growth conditions, including aerobic glucose fermentation and growth on ethanol, and compared their expression levels to natural promoters from Saccharomyces cerevisiae.

We found that when Z3EV is expressed at moderate levels from the chromosome, the response to estradiol is very tight, and saturates at low levels of inducer. When Z3EV is expressed from a stronger natural promoter, its response to estradiol becomes more pronounced and the synthetic promoter expression increases with estradiol concentration. Higher expression is generally useful for genetic engineering applications, though for Z3EV this increased expression occurred at the cost of increased basal expression (leakiness) in the absence of estradiol inducer.

We examined the performance of the Z3EV system during the diauxic shift from glucose fermentation to growth on ethanol in aerobic batch culture. Many genes related to glycolysis are strongly expressed during glucose fermentation, but then decrease expression after glucose depletion and diauxic shift. We measured glucose consumption and ethanol production along with regulation by Z3EV of a LacZ reporter controlled by a synthetic promoter. We found that the Z3EV system increases gene expression after glucose has been depleted. This provides a tool for expression of heterologous genes at high levels after diauxic shift in batch culture.

Production of Itaconic Acid By Enhancing Functional Expression of Cada in Escherichia coli

Choi, J. H., Korea Research Institute of Bioscience and Biotechnology (KRIBB)
Cheong, D. E., Korea Research Institute of Bioscience and Biotechnology (KRIBB)
Jeon, H. G., Korea Research Institute of Bioscience and Biotechnology
Lee, B., Korea Research Institute of Bioscience and Biotechnology (KRIBB)
Song, J. J., Korea Research Institute of Bioscience and Biotechnology (KRIBB)

To date, although the microbial fermentation process has been considered as the sustainable technology to produce valuable chemicals, it is economically unfavorable than the petrochemical process in the production costs and yields. Accordingly, the increase of the productivity achieved by the biological process is to be the first priority problem. Itaconic acid (IA) is an unsaturated dicarboxylic acid which utilized as biochemical building block for production of resins, plastics, paints and synthetic fiber. IA is considered as future bio-based platform chemical. However, IA hasn't been improved during the past 40 years owing to physiological and genetical disadvantages of *Aspergillus terreus*. In these contexts, we previously verified that both the functional expression of cis aconitate decarboxylase (CadA), which converts the citric acid cycle intermediate cis-aconitate to IA, and nitrogen limitation were important factors in IA production using *Escherichia coli*. The latter by which contents of direct and indirect precursors in the cytosol was increased could be further optimized through metabolic engineering. To this goal, genes such as citrate synthase (*glt*A) and phosphate acetyltransferase (*pta*) were overexpressed and/or deleted. We observed that the overexpression of genes to additionally supply precursors was more effective on the productivity of IA than deletion.

Mmsr-Mediated 3-Hydroxypropionic Acid-Inducible Gene Expression System of Pseudomonas Denitrificans

Nguyen, N. H.,
Zhou, S.,
Ainala, S. K., Pusan National University
Somasundar, A., Noroo Holdings Co., Ltd
Shim, J.,
Park, S., Pusan National University

3-Hydroxypropionic acid (3-HP), an important platform chemical that boats a variety of industrial applications, can be biologically produced from glycerol by two enzymes consisting of coenzyme B_{12}-dependent glycerol dehydratase involving dehydration of glycerol to 3-hydroxypropionaldehyde (3-HPA) and aldehyde dehydrogenase for oxidation of 3-HPA to 3-HP. To increase the yield of 3-HP production, it is essential to optimize this pathway through a gene expression system which can be inducible by 3-HP and control the expression of these two enzymes in high and balanced manner. Genome and transcriptome analyses showed that *Pseudomonas denitrificans* can natively synthesize coenzyme B_{12}, an essential for activation of glycerol dehydratase and has a 3-HP inducible system for genes related to valine degradation, *mmsA* and *hbdH*-4, which encode methylmalonylsemialdehyde dehydrogenase and 3-hydroxyisobutyl dehydrogenase, respectively. As a result, it showed 3-HP functions as an inducer relying on the action of transcriptional regulatory protein, MmsR. The underlying mechanism of MmsR-mediated 3-HP inducible system was studied *in vivo* and *in vitro*. In the presence of 3-HP, MmsR regulated the transcription of *mmsA* and *hbdH*-4 genes by higher than 140-fold, indicating that two promoters controlling these genes, P_{cm} and P_{c4}, are strong for gene expression in *P. denitrificans*. The transcription activation was depleted when *mmsR* was deleted from the chromosome, but was recovered by complementation of *mmsR* from a plasmid. The *mmsR* was cloned in *Escherichia coli* and the recombinant MmsR was purified. The result exhibited MmsR was monomeric at the concentrations below 65 nM whereas oligomeric at higher concentrations. Electrophoretic mobility shift assay indicated that MmsR could bind to two adjacent operator sites located in the divergently overlapping promoter region present between *mmsR* and *mmsA*. The two operators, designated O1 and O2, contained sequences of dyad symmetry centered at position –58 and –9, respectively, upstream of the *mmsA* transcription start site. When both O1 and O2 were present, binding affinity of MmsR to the operator regions was higher than either O1 or O2 alone, the result suggests that the binding is cooperative. MmsR could bind to the operator regions in the absence of 3-HP, indicating that 3-HP is not compulsory for binding of MmsR to the operator regions; however, the transcription of downstream genes requires 3-HP as an inducer. This inducible system is novel and robust for expressing important enzymes in 3-HP production, excretion pathways and enhances the 3-HP titer in *P. denitrificans*.

Engineering Escherichia coli for Terminal (Ï‰) - Functionalized Long Chain Dicarboxylic Acids Production from Renewable Fatty Acids and Plant Oils

Chandran, S. P., Ulsan National Institute of Science and Technology (UNIST)
Lee, S. K., Ulsan National Institute of Science and Technology

Dicarboxylic acids particularly, long-chain- ω - functionalized dicarboxylic acids (LDCAs, ≥ C12) are widely used as a raw material for preparing various commodities and polymers, such as nylon, polyamides, polyesters, perfumes, adhesives, and high quality lubricants. Industrial production of LDCAs relies mainly on chemical conversion processes from petrochemical sources that have a number of disadvantages and limitations including limitations in the range of products produced, use of multi-step conversion processes, dependence on non-renewable petrochemical feedstock, and generation of unwanted and hazardous byproducts. Biotechnology offers an innovative solution for overcoming these limitations. With increasing attention towards on reducing petroleum footprint, recent researches have been focused on the sustainable production of industrial important bulk chemicals from low-cost and renewable resources. Fatty acids (FAs) are considered as one of the most abundant renewable resources found in nature. Several studies have previously proposed the production of renewable fuels and industrial chemicals from fatty acids by *Escherichia coli* and suggesting that fatty acids could become a sustainable feedstock for industrial production.

The production of α,ω-dicarboxylic acids involves the following three steps: (i) terminal (ω) oxidation of fatty acids by a hydroxylase complex composed of cytochrome P450 monooxygenase (CYP450) and an NADPH:cytochrome P450 oxidoreductase (NCP) to form ω-hydroxy fatty acids; (ii) subsequent oxidation of hydroxy fatty acids into oxo fatty acids or fatty acid aldehydes by alcohol dehydrogenase and (iii) finally, the fatty aldehyde is oxidized by aldehyde dehydrogenase to the corresponding dicarboxylic acid (DCA). Hydroxylation of fatty acids by CYP450 and NCP is the rate limiting step and is therefore a critical stage in the production of α,ω-DCA, as the CYP enzymes show broad substrate specificity as well as high regioselectivity.

E. coli is a highly versatile organism that has been widely used for heterologous protein production and for genetic recombination studies because of its rapid growth and potential for high-density cultivation on inexpensive substrates. Additionally, its well characterized genetics and the availability of an increasingly large number of cloning vectors and mutant host strains make *E. coli* an attractive model for genetic studies. In the present study, an *E. coli* strain was engineered to produce α,ω-dicarboxylic acids (C12 and C14) directly from fatty acid substrates, through heterologously expressed ω-oxidation pathway. By whole cell biotransformation, the resulting engineered *E. coli* produced a maximum of 41 mg/L and 163 mg/L of C12 DCA and C14 DCA, respectively, from respective fatty acid substrates. In addition, the production of DCAs was increased (159 mg/L of C12 DCA and 410 mg/L of C14 DCA) by the addition of a heme precursor and the hydroxyl radical scavenger in a shorter culture duration than that of the corresponding controls. The constructed biocatalytic system was used to synthesize DCAs of various chain lengths from coconut oil hydrolysate. Therefore, our system could effectively be used for the production of DCAs from any plant oils high in fatty acids, including the non-edible vegetable oils. This novel synthetic biocatalytic system could be served as an efficient platform for the industrial production of α,ω-dicarboxylic acids.

Metabolic Engineering of Cupriavidus Necator for Heterotrophic and Autotrophic Alka(e)Ne Production

Crépin, L.,
Guillouet, S.,

Alkanes of defined carbon chain lengths can serve as alternatives to petroleum-based fuels. Recently, microbial pathways of alkane biosynthesis have been identified and enabled the production of alkanes in non-native producing microorganisms using metabolic engineering strategies. The chemoautotrophic bacterium *Cupriavidus necator* has great potential for producing chemicals from CO_2: it is known to have one of the highest growth rate among natural autotrophic bacteria and under nutrient imbalance it directs most of its carbon flux to the synthesis of the acetyl-coA derived polymer, polyhydroxybutyrate, (up to 80% of intracellular content). Alkane synthesis pathway from *Synechococcus elongatus (*2 genes coding an acyl-ACP reductase and an aldehyde deformylating oxygenase) was heterologously expressed in a *C. necator* mutant strain deficient in the PHB synthesis pathway. Under heterotrophic condition on fructose we showed that under nitrogen limitation, in presence of decane overlay, the strain produced up to 670 mg/L total hydrocarbons containing 434.8mg/l of alkanes consisting in 286.2mg/l of pentadecane, 130.6mg/l of heptadecene, 18.0 mg/l of heptadecane, and 235.6mg/l of hexadecanal. We report here the highest level of alka(e)nes production by an engineering *C. necator* to date. We also demonstrated the first reported alka(e)nes production by a non native alkane producer from CO_2 as the sole carbon source.

Understanding the MEP Pathway for the Production of Terpenoids

Volke, D.,
Jennewein, S., Frau

Facing an increasing demand of sustainable energy and chemicals, new strategies have to be explored to synthesize these in low-cost and environmental friendly processes. Isoprenoids are a promising source for a broad spectrum of valuable molecules, including bulk chemicals, fuels, and pharmaceuticals. Isoprenoids are one of the largest as well as most diverse class of chemical molecules synthesized by the natural world. All isoprenoids are derived from two common precursors, isopentenyl phosphate and dimethylallyl phosphate, which are synthesized via two distinct pathways. The long known mevalonate (MVA) pathway uses Acetyl-CoA as substrate, while the recently discovered methylerythritol 4-phosphate (MEP) pathway uses glycerinaldehyd-3-phosphat and pyruvate.

The economical production of many terpenoids is hindered by either the inefficiency of the terpene synthase used or the limited supply of precursor from the upstream pathway. Most economic applications use the MVA pathway, because it is well understood, despite its theoretical lower yield than the MEP pathway. This is due to our lack of knowledge about the MEP pathway and its regulation.

The aim of this study is to elucidate the regulation and the kinetics of the MEP pathway. We investigate the pathway on several levels, in order to identify limiting factors in the MEP pathway, Through the use of recombineering we introduced variation in the expression of the enzymes involved in the MEP pathway. Different strains with different expression characteristics were investigated on their behavior through targeted proteomics and metabolomics as well as non-stationary ^{13}C metabolic flux analysis. These methods revealed intrinsic regulation in the pathway and the data is further used to establish an *in silico* kinetic model of the pathway and will allow us to engineer the pathway more efficient.

Efficient 3-Hydroxypropionic Acid Production from Glycerol Using Metabolically Engineered Klebsiella Pneumonia

Jiang, J.,
Huang, B.,
Wu, H., East China University of Science and Technology
Li, Z., East China University of Science and Technology
Ye, Q.,

3-Hydroxypropionic acid (3-HP), an important platform compound, has great potential application in producing chemical derivatives, including acrylamide, acrylic acid, 1,3-propanediol, malonic acid, etc. In this study, metabolically engineered *Klebsiella pneumonia* strains were constructed for efficient 3-HP production from glycerol. The wildtype strain overexpressing aldehyde dehydrogenase (*aldH*) from *E. coli* produced reasonable amount of 3-HP (18.3 g/L), however, it still accumulated lot of 1,3-propanediol and lactic acid. Since lactate dehydrogenase (encoding by *ldhA*) and 1,3-propanediol oxidoreductase (encoding by *dhaT*) were the key enzymes in the lactic acid and1,3-PDO biosynthesis pathways of *K. pneumonia*, respectively, the strains with *ldhA* single deletion (JJQ01) or *ldhA* and *dhaT* double deletion (JJQ02) were constructed by using the method of the λ Red recombination system. The aldehyde dehydrogenase from *E.coli* were also overexpressed in these two strains, respectively. In the fed-batch experiments of 5-L bioreactor, the metabolically engineered *K. pneumonia* strain JJQ01(pUC18K-*aldH*) produced 45.8 g/L of 3-HP with a yield of 0.31 mol/mol glycerol. The 3-HP production was further improved when both of *ldhA* and *dhaT* were knockout. About 63 g/L of 3-HP was accumulated in JJQ02(pUC18K-*aldH*), and the yield of 3-HP was increased to 0.58 mol/mol glycerol.

Bacterial Fermentation Platform for Producing Artificial Aromatic Amines and Its Utilization for High-Performance Aromatic Polyamide Synthesis

Masuo, S., University of Tsukuba
Tateyama, S., Japan Advanced Institute of Science and Technology
Kaneko, T., Japan Advanced Institute of Science and Technology
Takaya, N., University of Tsukuba

Aromatic amines containing an aminobenzene or an aniline moiety comprise versatile natural and artificial compounds including bioactive molecules and resources for advanced materials. However, a bio-production platform has not been implemented. Here we constructed a bacterial platform for *para*-substituted aminobenzene relatives of aromatic amines. We found 4-aminophenylalanine synthesis gene cluster in a *Pseudomonad* bacterium. Recombinant *Escherichia coli* expressing the corresponding genes produced 4-aminophenylalanine from glucose. Optimization of the metabolic pathway in the recombinant *E. coli* cells converted biomass glucose to 4-aminophenylalanine with high efficiency (4.4 g L^{-1} in fed-batch cultivation). We designed and produced artificial pathways that mimicked the fungal Ehrlich pathway in *E. coli* and converted 4-aminophenylalanine into 4-aminophenylethanol and 4-aminophenylacetate at 90% molar yields. We also succeeded to produce 4-aminophenylethylamine and 4-aminocinnamic acid from 4-aminophenylalanine by taking advantage of phenylalanine decarboxylase and phenylalanine ammonia-lyase, respectively. Combining these conversion systems, the 4-aminophenylalanine-producing platform fermented glucose to 4-aminophenylethanol, 4-aminophenylacetate, and 4-phenylethylamine. Furthermore, we obtained 200 g of 4-aminophenylalanine by a large-scale fermentation, and converted them into 4-aminocinnamic acid by using phenylalanine ammonia-lyase from *Rhodotorula glutinis*. From the resulting 4-aminocinnamic acid, we synthesized high–performance bioplastics of aromatic polyamide by means of photodimerization and polycondensation reactions. These results denotes that the platform for synthesizing bio-derived aromatic amines constructed in the present study increases the potential for producing a range of bio-based aromatic amine materials for bio-polymers and other applications.

Design of a Cost-Effective Strategy for Secretory Production of Recombinant CCN Proteins By Mammalian Cells

Leu, S. J. J., Taipei City Hospital

CCN proteins (CCNs), a family of secreted extracellular matrix (ECM)-associated proteins, constitutes six members (CCN1-CCN6). CCN protein family plays the important role in various pathophysiological conditions such as human cancers and cardiovascular diseases. In human tumorigenesis, clinical expression levels of CCNs have been shown to be correlated with malignancies of several human cancer types. In cellular levels, CCNs can modulate multiple cellular activities, including cancer cell migration, invasion, stem-like properties that drive cancer initiation, and metastasis processes. In addition, studies indicate that members of CCNs, particularly the CCN1 and CCN5, can act as the conditional tumor suppressors in certain cancers. This suggests that recombinant CCNs-based protein products may have potential of being developed as therapeutic agents. Our aim is to utilize mammalian cell expression system to establish a cost-effective protocol for production of recombinant CCN proteins, followed by testing the anti-cancer effects of the CCNs-derived proteins.

In the present study, we utilized the lentivirus-based transgene vector to express CCN1 and CCN5 proteins by human HEK293 cells. Following viral infection and antibiotic selection, the stable CCNs-expressing cells were further enriched by cell line selection and isolation. High-level expression and secretion of recombinant proteins were optimized by the processes relevant to metabolic engineering of mammalian host cells, including addition of sodium butyrate, change of medium osmolarity, as well as hypoxia-reoxygenation cultivation procedure. Finally, the recombinant CCN proteins were successfully collected from serum-free media, and then purified by affinity chromatography. The recombinant CCN proteins were proved to be bioactive by the effective intervention in inhibiting cancer cell malignancy such as reduction of cellular mobility and impaired cancer stemness.

Together, we have successfully established a mammalian cell expression system to produce CCNs-derived protein products of therapeutic potential. The CCNs-related proteins may be applied to target relevant human diseases in the near future.

Metabolic Engineering and Synthetic Biological Approach for Enhancing and Structurally Modulating Nargenicin A1 from Nocardia Sp. CS682

Dhakal, D., Institute of Biomolecules Reconstruction
Chaudhary, A. K., University of Nebraska
Pokhrel, A. R.,
Shrestha, A.,
Shrestha, B.,
Sohng, J. K.,

The infectious diseases caused by bacteria, particularly those gaining the multi-drug resistance, are major causes of mortality. Hence, the discovery of new antimicrobials or expanding the utility of existing antibiotics by overproduction and targeted modification resulting in better activities or pharmacological properties is crucial to combat such havoc [1, 2]. Recently, nargenicin A1 was rediscovered as potential antimicrobial candidate through whole-cell anti-sense assay and established as the first natural product having unique capability to inhibit DnaE, which is protein actively involved in DNA replication [3].Hence, in this study we attempted to enhance the production of nargenicin A1 by rational metabolic engineering approach and synthetic biological tools coupled with sensible precursor supplementation, assisted by response surface methodology to attain ~24 fold higher production in recombinant strain than the wild type strain, Nocardia sp. CS682. For this purpose a userdefined multimonocistronic vector was developed, evaluated and utilized for cloning genes responsible for precursor refactoring *viz. ngnN2*, *ngnN3*, *ngnN4* and *ngnN5* for pyrrole biogenesis, and *glf,glk* and *ACC*ase complex genes for aglycon synthesis. Furthermore, by incorporating pikC, a substrate flexible cytochrome P450 [4,5] in the nargenicin A1 biosynthetic pathway, novel derivative of nargenicin was isolated from the recombinant strain. To the other end, the nargenicin A1 was structurally diversified utilizing efficient single vessel reaction system in which the syntheses of uridine diphosphate (UDP)-α-D-glucose and UDP-α-D-2-deoxyglucose were modified and combined flexible glycosyltransferase (GT) from *Bacillus licheniformis*DSM 13[2]. Novel nargenicin derivatives were characterized by UPLC-PDA, HR-QTOF ESI-MS/MS analyses and NMR analyses as par needed. The overall conversion studies indicated both the *in vivo* and *in vitro* approaches were suitable approach for structural modification of nargenicin. Furthermore, assessment of solubility indicated that there was enhanced water solubility by almost ~1.8 fold in the case nargenicin A1 11-O-β-D-glucopyranoside than nargenicin, which is beneficial with pharmacokinetics and pharmacodynamic point of view. But, the evaluation of antibacterial potency of the derivatives of nargenicin showed that there was substantial reduction in biological potency of the novel derivatives in comparision to parental compound.

1. **Dhakal, D.**, Le, T. T., Pandey, R. P., Jha, A. K., Gurung, R., Parajuli, P., Pokhrel, A.R., Yoo, J.C. & Sohng, J. K. (2015). Enhanced Production of Nargenicin A1 and Generation of Novel Glycosylated Derivatives. Applied biochemistry and biotechnology, 175: 1-16.
2. **Dhakal, D.** & Sohng, J. K.(2015) Commentary: Toward a new focus in antibiotic and drug discovery from Streptomyces arsenal. Fronteirs in Microbiology,

2015:6.

3. Painter, R.E., Adam, G.C., Arocho, M., DiNunzio, E., Donald, R.G., Dorso, K., Genilloud, O., Gill, C., Goetz, M., Hairston, N.N., Murgolo, N.(2015) Elucidation of DnaE as the antibacterial target of the natural product, Nargenicin. Chemistry & biology. 22:1362-73.
4. Jha, A.K., **Dhakal, D.**, Van, P.T.T., Pokhrel, A.R., Yamaguchi, T., Jung, H.J., Yoon, Y.J. & Sohng, J.K. (2015). Structural modification of herboxidiene by substrate-flexible cytochrome P450 and glycosyltransferase. Applied microbiology and biotechnology, 99:3421-3431.
5. Pokhrel, A.R., **Dhakal, D.**, Jha, A.K. and Sohng, J.K. (2015). Herboxidiene biosynthesis, production, and structural modifications: prospect for hybrids with related polyketide. Applied microbiology and biotechnology, 99:8351-8362.

Application of Synthetic Biology for the Production of Flavonol Glycosides in Escherichia coli

Parajuli, P., Sun Moon University
Pandey, R. P., Sun Moon University
Sohng, J. K., Sun Moon University

Previously characterized multi-monocistronic synthetic vector was used to assemble multiple genes of a nucleotide diphosphate (NDP)-sugar biosynthetic pathway to construct robust genetic circuits for the production of valuable flavonoid glycosides in *Escherichia coli*. Characterized functional genes involved in the biosynthesis of uridine diphosphate (UDP)-glucose and thymidine diphosphate (TDP)-rhamnose from various microbial sources along with glucose facilitator diffusion protein (*glf*) and glucokinase (*glk*) from *Zymomonas mobilis* were assembled and overexpressed in a single synthetic multi-monocistronic operon. Independently overexpression of these genes in multi-monocistronic fashion increased the pool of each NDP-sugar in microbial host. The newly generated NDP-sugars biosynthesis circuits along with regiospecific glycosyltransferases from plants were introduced in *E. coli* BL21 (DE3) to probe the bioconversion of fisetin, a medicinally important polyphenol produced by various plants. As a result, significant amount of fisetin 3-*O*-glucoside and fisetin 3-*O*-rhamnoside were produced in UDP-glucose and TDP-rhamnose biosynthesis systems after 48 h of incubation in 3 L fermentor while supplementing fisetin exogeneously. These yields of fisetin glycosides represent ~99% of bioconversion in constructed microbial cell factories. The systems were also found to be highly effective in bio-transforming other flavonols (quercetin, kaempferol, myricetin) into their respective glycosides, achieving over 95% substrate conversion. The construction of a synthetic expression vector for bacterial cell factory followed by subsequent re-direction of metabolic flux towards desirable products have always been revolutionized the biotechnological processes and technologies. This multi-monocistronic synthetic vector in a microbial platform is customizable to defined task and would certainly be useful for applications in producing and modifying such therapeutically valued plant secondary metabolites.

Optimization of the Production of Butanol in Recombinant Pseudomonas Putida Using Large-Scale Kinetic Models

Tokic, M., Swiss Federal Institute of Technology (EPFL)
Ataman, M., Ecole Polytechnique Federale de Lausanne (EPFL)
Miskovic, L., Ecole Polytechnique Fédérale de Lausanne (EPFL)
Hatzimanikatis, V., Swiss Federal Institute of Technology of Lausanne (EPFL)

The production of valuable biofuels and biochemicals by conversion of renewable resources is a basis of sustainable economy. Heterologous expression of biosynthetic pathways taken from natural producers or expression of *de novo* synthetic pathways into microbial work horses such as *E. coli* or *S. cerevisiae* allows for production of a wide spectra of biofuels and biochemicals. *P. putida* has emerged as one of the amenable production hosts with a number of advantages over natural producers. *P. putida* is a non-pathogenic soil bacterium known for its versatile metabolism and its tolerance to high toxicity compounds. For instance, it is reported that *P. putida* can grow in the presence of high concentrations of butanol[1]. This highly adaptive bacterium has been found to survive and grow on a wide range of substrates from pure caffeine to toxic industrial waste.

In this contribution, we devised metabolic engineering strategies for improving butanol production in recombinant *P. putida* strain using large-scale kinetic models. Large-scale kinetic models allowed us to simultaneously optimize several factors such as specific productivity and yield. We started by embedding the butanol biosynthetic pathway[2] in the genome-scale model of *P. putida* and then we derived a consistently reduced, core stoichiometric model that was consistent with its genome-scale counterpart. We then used the ORACLE[3,4,5] (Optimization and Risk Analysis of Complex Living Entities) framework to integrate available experimental information and information from literature and databases to build a population of large-scale kinetic models of recombinant *P. putida* producing butanol. We used these models to identify the enzymes with the highest impact on the production of butanol and we proposed metabolic engineering strategies for improved specific productivity and yield.

This work demonstrates the potential and usefullness of ORACLE as a framework for optimizing production of biofuels and biochemicals.

1. Rühl, J., Schmid, A. & Blank, L. M. Selected Pseudomonas putida strains able to grow in the presence of high butanol concentrations. *Appl. Environ. Microbiol.* **75,** 4653–6 (2009).

2. Nielsen, D. R. *et al.* Engineering alternative butanol production platforms in heterologous bacteria. *Metab. Eng.* **11,** 262–273 (2009).

3. Miskovic, L. & Hatzimanikatis, V. Production of biofuels and biochemicals: in need of an ORACLE. *Trends in biotechnology* **28,** 391–7 (2010).

4. Chakrabarti, A., Miskovic, L., Soh, K. C. & Hatzimanikatis, V. Towards kinetic

modeling of genome-scale metabolic networks without sacrificing stoichiometric, thermodynamic and physiological constraints. *Biotechnol. J.* **8,** 1043–57 (2013).

5. Andreozzi, S. *et al.* Identification of metabolic engineering targets for the enhancement of 1,4-butanediol production in recombinant E. coli using large-scale kinetic models. *Metab. Eng.* 1–12 (2016). doi:10.1016/j.ymben.2016.01.009

Overexpression Self-Repressing Regulator Dnro Enhances Production of Daunorubicin in Blda Deficient Streptomyces Peucetius ATCC 27952

Raj Pokhrel, A.,
Kumar Chaudhary, A.,
Thi Nguyen, H.,
Dhakal, D., Institute of Biomolecules Reconstruction
Shrestha, A., Sun Moon University
Sohng, J. K., Sun Moon University

In daunorubicin (DNR) biosynthesis pathway of *Streptomyces peucetius* ATCC 27952, *dnrO* is the first regulatory gene to be activated. DnrO is self-repressing gene, which activates rest of the DNR biosynthesis pathway through cascades of regulatory events. *S. peucetius* was found to contain no functional copy of *bldA* -tRNA while closer look at *dnrO* codons reveals the presence of TTA codon, which is rarely encoded by *bldA*. The *dnrO* gene contains 341 codons, where 86th codon was found to be the rare TTA codon. Therefore, for evaluating the role of *dnrO* in DNR production, we generated multiple engineered strains of *S. peucetius* by heterologously expressing *bldA, dnrO* and combination of *bldA* and *dnrO*. Using these strains, we evaluated the effects of heterologously expressed *bldA* and overexpressed *dnrO* on pathway specific regulators, cell densities and production of DNR. The increase in transcription level of *dnrO* and *dnrI* along with enhanced DRN production in strain with *bldA* as compared to wild type strain was found.The DNR biosynthesis was found to be ~1.7 mg/l higher than that of *dnrO* overexpression strain. *S. peucetius* and its variants had optimum mycelial density in NDYE medium from 48 h to 60 h and maximum anthracycline compounds production at 60 h. *S. peucetius* ATCC 27952 produces 3.24 mg/l of DRN while the strains with *bldA* and *dnrO* produces 4.72 mg/l and 4.15 mg/l of DRN, respectively. Combining the results of transcriptional level of pathway specific regulators and biosynthesis of DNR from strain with *bldA* (heterologous expression), *dnrO* (overexpression), and combination of both, we found the translation of UUA codon on *dnrO* mRNA is partially *bldA* dependent.

Principal Studies on the Scopolamine Biosynthesis in Duboisia Spp. for Heterologous Reconstruction of Tropane Alkaloid Biosynthesis

Kohnen, L., Technical University Dortmund
Ullrich, F., Technical University Dortmund
Averesch, N. J. H., Centre for Microbial Electrosynthesis (CEMES) within the Advanced Water Management Centre (AWMC)
Kayser, O., Technical University Dortmund

The alkaloid scopolamine is an important bulk compound in the semi-synthesis of drugs for clinical medicines like Buscopan or Spiriva. To date scopolamine is still a plant-based product; the majority is obtained via extraction from *Duboisia* species. As demand is expected to increase in the future and to compensate for fluctuations in crop yield of the medicinal plants, development of a biotechnological process is becoming of great interest. In order to assemble a heterologous pathway fundamental understanding of the merging (acetate- and shikimate-) pathways, including biosynthetic genes, their transcription and regulation, is crucial. Therefore several fundamental studies were conducted: i) Analysis of the principal pathway and most important enzymes for their expression and biochemical profile, ii) Construction of a cDNA library from cytosolic root cells of *Duboisia* species, iii) Analysis of metabolic and metabolomic profiles of *Duboisia* species, iv) *In silico* investigation of a heterologous production system utilising metabolic network modelling.

Principal pathway analysis. Based on database annotation data of Solanaceae species primers were designed for all relevant genes involved in the scopolamine biosynthesis, starting with the first precursors in the acetate- and shikimate-pathway through to tropanol and phenyllactic acid, respectively. Despite the unavailability of comprehensive information on the *Duboisia* genome, we were able to identify and detect the genes for H6H, Cyp80F1, tropine reductase and putrescine *N*-methyltransferase using degenerated primer pairs. In addition, the key enzymes H6H and Cyp80F1 were quantified using RT-PCR.

cDNA library construction. Using parenchymatic root cells from *Duboisia myoporoides*, a cDNA library was constructed and sequenced (Ilumina 454). Contigs were assembled and full gene ORFs reconstructed and compared with database published sequences from other Solanacea plants in the genus of *Atropa*, *Hyoscyamus* and *Burgmansia*. Obtained genes, like H6H and Cyp80F1, were compared and aligned for bio-functional characterisation. Based on sequencing data significant base differences were determined in the upstream region of the genes.

Metabolic and metabolomic profiling. From various organs of root-tissue the metabolic profile of eight tropan alkaloids including putrescine, tropanol, hyoscyamine, 6-OH-hyoscyamine, and scopolamine was characterised. Quantitative analysis was performed by means of LC-MS to allow a correlation with RT-PCR results. The main compounds hyoscyamine and scopolamine were found to be in the range of 5 to 22 μg/mg, respectively. Interestingly, scopolamine glucoside was found in stem tissue, which is likely the form of transported scopolamine. In-detail metabolome analysis was conducted based on NMR studies: recorded NMR resonances were subject to Principal Component Analysis. A statistical analysis revealed that various amino acids as well as inositol and myo-inositol were significantly increased indicating potentially important roles of these as precursors to supply flux to the product pathways.

Metabolic network analysis. Using a novel tool for network-embedded thermodynamic analysis of elementary flux modes, a heterologous tropane-alkaloid pathway was studied in the cell envelope. This identified thermodynamic bottlenecks and major stoichiometric constraints. Ongoing studies focus on identifying key reactions and engineering targets that influence the viability of a bio-based process.

Integration of Quality By Design, Programming Languages, and Cross-Platform Automation for Closing the Design, Build, Test Cycle in Metabolic Engineering

Grant, C., University College London

Building robust manufacturing processes by metabolic engineering is a task that is highly complex and requires sophisticated tools to design and describe processes, inputs and measurements; as well as to administrate management of knowledge, data, and materials.

We argue that for bioengineering to succeed in the most difficult cases, it will require easier and more routine application of engineering and quality approaches such as design for manufacturing (DfM) and quality by design (QbD) as well as the application of statistically designed experiments to derive detailed empirical models of underlying genetic systems in context with changing environmental conditions. This requires execution of large-scale structured experimentation for which laboratory automation is necessary.

This will be greatly facilitated by the development of expressive, high-level languages that allow reusability of protocols, characterization of their reliability, breaking the artificial separation between design and execution and abstraction of implementation details in favor of functional properties.

To directly address these issues, we have developed Antha, a high-level language (antha-lang) and operating system (anthaOS) for describing, executing and transferring bio-R&D and biomanufacturing processes.

With a particular emphasis on automation, workflows are written in a high-level form avoiding device-specific low-level instructions. The structure is designed to enforce good quality practices, reusability and faciliation of mathematical approaches. This eases transferability between equivalent but distinct laboratory and manufacturing infrastructures and presents an opportunity to increase insight by facilitating more structured and sophisticated experiments; generating clean well-defined datasets. Reproducibility and transferability are enhanced as well as higher productivity. Furthermore, execution by non-lab scientists is enabled as well as the capacity to execute more ambitious experiments necessary to address metabolic engineering problems which would be impossible to reliably execute manually.

We present a series of case studies using Antha to exemplify what we believe is an exciting trend that could help further drive the reliability, utility and success of metabolic engineering approaches.

1. www.antha-lang.org

2. Sadowski, Michael I., Chris Grant, and Tim S. Fell. "Harnessing QbD, Programming Languages, and Automation for Reproducible Biology." Trends in Biotechnology. doi:10.1016/j.tibtech.2015.11.006.

Multiscale Design of Cell Factories

Herrgård, M. J., Technical University of Denmark

Developing commercially viable cell factories for production of chemicals and proteins requires design at multiple levels from DNA, to proteins, pathways, organisms, processes, and industries. At each level computational tools can be used to guide and constrain design choices. The metabolic engineering community is most familiar with organism level design to identify genetic manipulation targets, but it is equally important to make correct design choices at other levels. Unfortunately, different sets of tools and methodologies are used at each level making it difficult to perform multiscale design. We are developing a comprehensive software and conceptual framework for designing pathways, organisms, processes and industries simultaneously. This framework can be used to assess biological, technical, economic and ecological performance of possible target molecules for metabolic engineering. The framework will be demonstrated with case studies focusing on commodity chemical production in microbial host organisms.

Î'-Keto Acids Platform for Ketones and Alcohols Production in Saccharomyces Cerevisiae

Dai, Z., Chalmers University of Technology
Siewers, V., Chalmers University of Technology
Nielsen, J., Chalmers University of Technology

Depleting oil reserves and growing environmental concerns have aroused general interest for developing sustainable processes to fuels and chemicals. β-Keto acids, or 3-oxoacids, such as acetoacetate, have the ketone group at the second carbon from the carboxylic acid. They are easily decarboxylated to produce ketones and ketones could be reduced to secondary alcohols. Ketones and alcohols are important platform chemicals with widespread applications in the fragrance, flavor and pharmacological industries. In nature, β-keto acids could be formed by the hydrolysis of β-ketoacyl-CoA which is an intermediate of fatty acid β-oxidation. *Saccharomyces cerevisiae* is a very important cell factory and has been widely used for many biomolecules production. In *S. cerevisiae*. In yeast, fatty acid β-oxidation process was performed in peroxisome which is a natural workshop for our target chemicals production. Hereby, (a) POT1 gene was deleted to interrupt fatty acid β-oxidation process and increase the precursor β-ketoacyl-CoA, (b) native acyl-CoA oxidase, enoyl-CoA hydratase and 3-hydroxyacyl-CoA dehydrogenase were overexpressed to increase the accumulation of β-ketoacyl-CoA, (c) fatty acids transporters *PXA1* and *PXA2* were also expressed to improve the provision of fatty acids from cytosol to peroxisome, (d) a thioesterase FadM from *E. coli* was expressed with peroxisome signal peptide to convert β-ketoacyl-CoA to β-keto acid. To prove the concept, different chain length fatty acids were added into the culture medium to test the production of methyl ketones. Results showed methyl ketones (2-Undecanone, 2-Tridecanone, 2-Heptadecanone) could be detected when adding dodecanoic acid.

"Metabolic Engineering of Bacillus Subtilis for the Production of 3-Hydroxypropionic Acid"
Kalantari, A., Chalmers University of Technology
Ji, B., Chalmers University of Technology
Mijakovic, I., Chalmers University of Technology

"METABOLIC ENGINEERING OF BACILLUS SUBTILIS FOR THE PRODUCTION OF 3-HYDROXYPROPIONIC ACID"

Aida Kalantari, Boyang Ji, Ivan Mijakovic

Over the past few years a major interest is focused on the substitution of petrochemical products by processing bio-products from renewable raw materials. In this context, our project focuses on the production of 3-Hydroxypropionic acid (3-HP) from glycerol. 3-HP is one of the top-valued platform chemicals used for the production of many industrially important chemicals such as acrylic acid, and glycerol is a relatively cheap byproduct of biodiesel industry. 3-HP can be produced by several microorganisms which possess glycerol dehydratase (*dhaB*) and aldehyde dehydrogenase (*aldH*) genes, however, bio-production of 3-HP has never reached a commercially exploitable level. In this project, we explore the possibility of synthesizing 3-HP for the first time in *Bacillus subtilis*, by inserting the heterologous production pathway. The key advantages of *B. subtilis* are its ability to grow efficiently on glycerol and reach very high cell densities in bioreactors. Our genetically engineered *B. subtilis* contains *dhaB* and *aldH* genes from *Lactobacillus reuteri* and *Klebsiella pneumonia.* Metabolic network of *B. subtilis* was optimized *in silico* to guide our metabolic engineering strategy *in vivo* which resulted in 3-HP producing *B. subtilis* strains.

A Widget-Based Data-Analytic Metabolomics Pipeline

Porokhin, V., Tufts University
Alden, N., Alexion
Colebrook-Soucie, J., Tufts University
Cokova, E., Tufts University
Lee, K., Tufts University
Hassoun, S., Tufts University

Owing to recent technological advances in measurement platforms, it is now possible to simultaneously detect and characterize a very large number of metabolites covering a substantial fraction of the small molecules present in a biological sample. This presents an exciting opportunity to develop potentially transformative approaches to study cells and organisms. One major challenge in realizing this potential lies in efficiently processing and analyzing the data. A typical dataset from an untargeted experiment contains thousands of "features," each of which could correspond to a unique metabolite. While several tools have emerged to pre-process the data, annotating and analyzing the data to extract meaningful biological information remains a challenge.

We present in this work a software-based customizable workflow, which allows users to visually create their own metabolomics workflow for interpreting metabolomics data within a relevant biological context. We describe a suite of widgets (software gadgets that plug into and interact with a larger system) organized into a 'metabolomics toolbox', within the plug-and-play Orange framework (http://orange.biolab.si). Orange has over 100 widgets, organized in various toolboxes centered on data curation, visualization, and classification. To create a metabolomics analysis pipeline, users select and connect various widgets from our metabolomics toolbox and amongst other Orange widgets. Our plug-and-play visual workflow differs from current web-based approaches, offering several advantages. The ability to combine small processing steps into complex analysis pipelines allows the user to create customized and flexible data-processing workflows. This flexibility is an important advantage in efficiently analyzing data from different types of metabolomics studies. For example, a comparative study seeking to identify differences in metabolite profiles between two sets of samples requires a different workflow from a study that seeks to characterize all of the metabolites present in a sample. Another advantage is that the widgets can integrate existing data analysis tools (e.g. machine learning) already implemented in Orange to analyze data at differing points in the pipeline. Third, the integrated data visualization toolbox within Orange can be used to view relationships in the data in many different ways, thus departing from the traditional method of visualizing metabolite data in the context of metabolic pathways or atlases. Fourth, a custom workflow, once created, can be saved, published and shared with collaborators, thus facilitating not only data sharing but also method dissemination, which should be of broad benefit to the metabolomics research community.

We illustrate this workflow by describing the results obtained from two representative case studies. The datasets of these case studies describe the metabolite profiles of 1) Chinese hamster ovary (CHO) cells in fed-batch culture during monoclonal antibody production; 2) an *in vitro* culture of isolates collected from murine cecum. The first dataset involves samples from a culture of a single cell type from an organism (hamster) with an annotated genome in the KEGG database, while the second dataset is more complex, involving samples from a culture of many different microbial species. Using the metabolomics toolbox within Orange, we identify growth inhibitory metabolites accumulating in CHO cell culture, and discover novel microbiota

Computational Tools for the Synthetic Biology of Natural Products

Breitling, R., University of Manchester

The engineering of microbial production systems for high-value natural products relies heavily on the availability of computational tools to assist all steps of the design–build–test cycle of synthetic biology. Recent developments include:

- New methods for the detection and analysis of biosynthetic gene clusters (antiSMASH, multiGeneBlast) and the identification of novel compounds (Pep2Path).
- Bespoke computational modelling approaches for metabolic networks to aid host cell optimization and pathway engineering (MultiMetEval, Uncertainty Modelling).
- Metabolomics analysis software for the evaluation and debugging of engineered microbes (MetAssign, mzMatch, mzMatch-ISO).

These tools are currently implemented as part of the design and engineering pipeline of the Manchester SYNBIOCHEM centre, and the talk will illustrate some of these applications.

Tracing Optimal Conversion Stoichiometries Using Native and Novel Biotransformation

Ng, C. Y., The Pennsylvania State University
Chowdhury, A., Pennsylvania State University
Kumar, A., Penn State
Maranas, C. D., The Pennsylvania State University

De novo metabolic pathways can be carefully designed to outperform native pathways with higher carbon and energy efficiencies as well as expand the product range of a host organism. Scanning through the huge repertoire of microbial enzymes to identify potential components of a novel pathway is a challenging task as one has to consider all the possible factors affecting the pathway efficiency. They include the stoichiometric and cofactor balance, thermodynamic feasibility, co-substrate and co-product choices and compatibility of the enzymes with the host. To this end, we have developed a two-stage MILP-based (optStoic and minRxn/minFlux) formalism. The first step, optStoic, explores the maximum extent of converting carbon substrate(s) to desired product(s) through non-intuitive combination of co-reactants and co-products while remaining within the limits of mass, energy and thermodynamic balances. Subsequently, minRxn/minFlux identifies a minimal reaction network to achieve the overall conversion. Identified designs consisted of improvement on existing strategies in terms of carbon and energy efficiency. Specifically, multiple alternatives for non-oxidative glycolysis and methanotrophy were identified that showed non-intuitive combinations of a core set of known pathways can lead to higher carbon and energy efficiency. Pareto analysis of metabolic pathways revealed interesting tradeoffs between pathway energy efficiency and enzyme expression. Native central metabolic pathways were found to be optimized through natural evolution for energy efficiency. Upon further analysis, improved synthetic alternatives with lower enzyme expression costs were also identified after redesigning native pathways with closely related orthologs. In addition, we developed a reaction-rule based optRule algorithm for designing *de novo* biotransformations absent in databases to extend the optStoic formalism towards novel metabolites. In this study, a combined (mass-balance and reaction-rule) approach was used to identify non-intuitive biological designs for optimally converting carbon substrates from renewable sources, gases and feedstock to economically valuable molecules. In particular, we identified various pharmaceutical and platform molecule bioproduction strategies from inexpensive industrial feedstocks such as toluene and naphthalene.

Metabolic Engineering of Pseudomonas Putida KT2440 for Complete Mineralization of Methyl Parathion and Gamma-Hexachlorocyclohexane

Gong, T., Nankai University

Agricultural soils are often cocontaminated with multiple pesticides. Unfortunately, microorganisms isolated from natural environments do not possess the ability to simultaneously degrade different classes of pesticides. Currently, we can use the approaches of synthetic biology to create a strain endowed with various catabolic pathways that do not exist in a natural microorganism. Here, we describe the metabolic engineering of a biosafety *Pseudomonas putida* strain KT2440 for complete mineralization of methyl parathion (MP) and γ-hexachlorocyclohexane (γ-HCH) by functional assembly of the MP and γ-HCH mineralization pathways. The engineered strain was genetically stable, and no growth inhibition was observed. Such a strain not only would reduce the toxicity of MP and γ-HCH but also would prevent the accumulation of potentially toxic intermediates in the environment. Furthermore, expression of *Vitreoscilla* hemoglobin improved the ability of the engineered strain to sequester O_2. The inoculation of the engineered strain to soils treated with MP and γ-HCH resulted in a higher degradation rate than in noninoculated soils. Moreover, introduced GFP may be used to monitor the activity of the engineered strain during bioremediation. The engineered strain may be a promising candidate for in situ bioremediation of soil cocontaminated with MP and γ-HCH.

Production of Sulfated Organic Compounds in Cell Factories

Jendresen, C. B., Technical University of Denmark
Nielsen, A. T.,

Sulfation of organic molecules is a reaction that has a significant potential value for biological synthesis of organic chemicals in cell factories.

Some sulfated compounds may be of direct interest for biological production. One example is zosteric acid, a sulfate ester of *p*-coumaric acid, which has potential uses as an anti-biofoulant and as a precursor for functional polymer membranes. Sulfation may also give an added functionality to chemical precursors, such as those used in polymers, due to the charged nature of the sulfate group. Thirdly, sulfation may be used to change the chemical nature of a chemical of interest, namely, by increasing solubility or decreasing toxicity towards the production organism. As example, the antioxidant of red wine, resveratrol, has very limited solubility that hampers biological production, however, the sulfated molecule has significantly increased solubility and bioavailability. The change in toxicity may also find an alternative use in bioproduction. The preferred biomass substrate for biological production of low to intermediate valued compounds may be derived from lignocellulose and contains a number of inhibitory phenolic compounds. When sulfated the inhibitory compounds become less toxic, and sulfation is this way a means to limiting the inhibitory effect during growth in lignocellulosic biomass.

We show that sulfation of a range of compounds is possible in a biological system, and that sulfation may be used to detoxify inhibitory compounds, such as those found in biomass hydrolysate. We present metabolic engineering approaches for optimizing the sulfation activity.

Harnessing Sigma Factor Gene Overexpression for Production By Corynebacterium Glutamicum

Taniguchi, H., Bielefeld Univeristy
Wendisch, V. F., Bielefeld University

Sigma factor is one of the component of RNA polymerase holoenzyme and responsible for promoter recognition and transcription initiation. Bacteria usually have multiple genes of sigma factor and strictly control their activities at the transcription, translation and post-translation level. Activation of a specific sigma factor alters the promoter recognition of RNA polymerase holoenzyme and enhances the transcription of a specific group of genes with the similar promoter sequence. Sigma factor can be a good candidate to activate a set of genes at the same time and may be helpful for engineering bacterium.

In this study, the potential of sigma factor gene expression for production was examined in *Corynebacterium glutamicum*. This bacterium was first identified as a natural glutamate producer and has been used in industry for production of amino acid, especially glutamate and lysine. Because of its genetic amenability, various strains has been developed to produce industrial relevant compounds using metabolic engineering. *C. glutamicum* possesses 7 sigma factor genes, *sigA*, *sigB*, *sigC*, *sigD*, *sigE*, *sigH* and *sigM*.

Each of these sigma factors was overexpressed in two different background strains, the wild-type strain (WT) and the recombinant strain from lycopene production (Lyc5). Sigma factor genes were overexpressed in an IPTG-inducible system with different IPTG concentrations. Overexpression of each sigma factor gene influenced the growth rate differently and the effect was dependent on the IPTG concentration, thus, on expression strength. The strongest effect was observed for overexpression of *sigD* and *sigH*, respectively.

Notably, *sigD* overexpression in *C. glutamicum* WT also reduced foaming possibly due to secretion of anti-foaming compounds. Transcriptome analysis revealed that genes related to the cell wall integrity such as mycolate synthesis showed increased RNA levels as consequence of *sigD* overexpression.

Overexpression of *sigH* in *C. glutamicum* WT resulted in a phenotypic change, i.e. accumulation of a yellow compound in the supernatant, which was identified as riboflavin by HPLC analysis. Transcriptome analysis revealed increased expression of genes of riboflavin biosynthesis, the pentose phosphate pathway and of enzymes requiring FAD, FMN or NADPH as cofactor (Taniguchi and Wendisch, 2015). Since riboflavin is a precursor of FMN, *sigH* overexpression was harnessed as strategy for production of FMN. Balanced expression of *sigH* and the bifunctional riboflavin kinase/FMN adenyltransferase gene *ribF* improved accumulation of riboflavin (20 ± 1 μM) and allowed for its conversion to FMN (33 ± 2 μM).

Overexpression of *sigA* in the lycopene producing strain Lyc5 increased lycopene production. In addition, *sigA* overexpression in WT also increased biosynthesis and accumulation of the endogenous C50 carotenoid decaprenoxanthin. As consequence of *sigA* overexpression accumulation of lycopene and decaprenoxanthin continued into the stationary phase.

In this study, we showed that sigma factor gene overexpression is not only relevant to study the gene regulatory network, but pertains to *C. glutamicum* strain development with respect to production of value-added chemicals.

Taniguchi H & Wendisch VF (2015) Exploring the role of sigma factor gene expression on production by *Corynebacterium glutamicum*: sigma factor H and FMN as example. Front Microbiol 6: 740

Metabolic Engineering 11, Rapid Fire Poster Session 3 (Tuesday, June 28, 2016, 6:15-7:15PM)

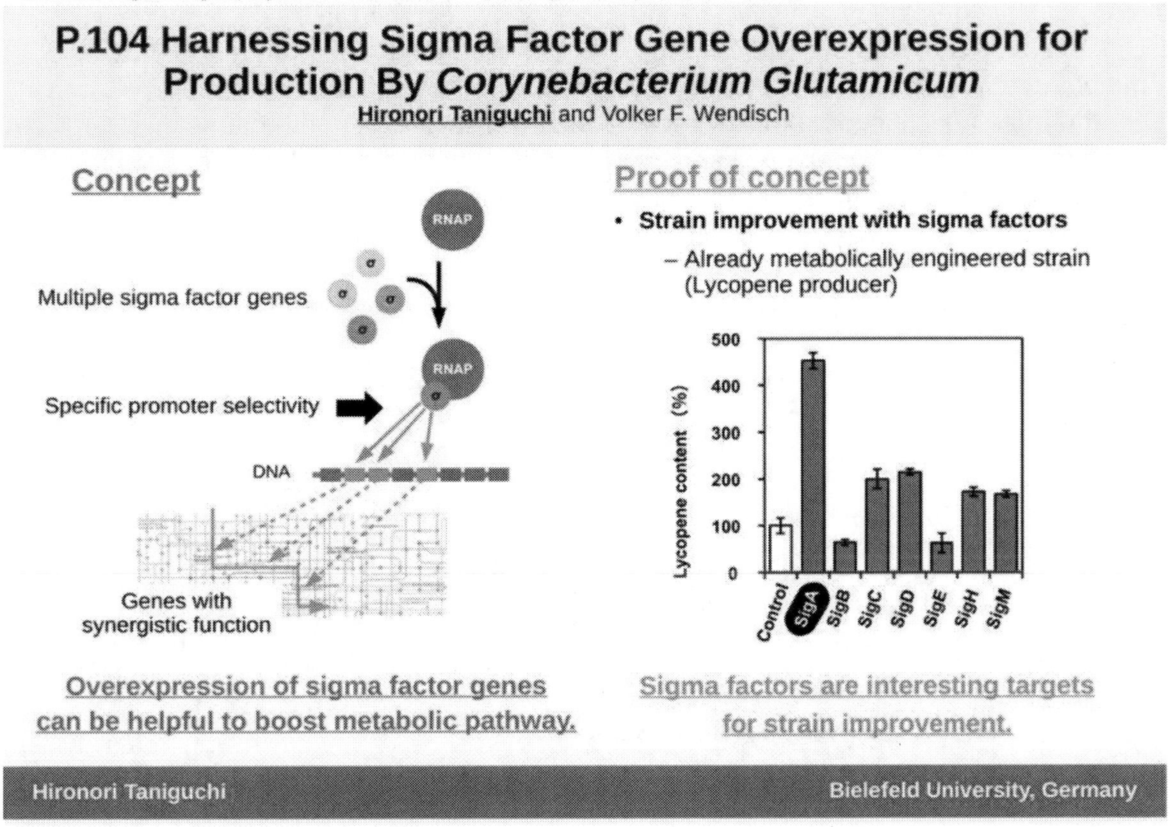

Industrial Strain Improvement for Food Functionality

Neves, A. R., Current address: Chr. Hansen A/S

Lactic acid bacteria (LAB) are a functionally related group of Gram-positive bacteria known essentially for their roles in food bioprocessing. The successful biotechnological application of LAB depends to a great extent on their unique phenotypic traits, which among others include fast acidification of the medium, texture and flavour forming abilities, bioprotection, and health promoting properties. However, a highly dynamic food market and ever-changing preferences from health-conscious consumers demands constant focus on product development. Fostered by this market push, Chr. Hansen places continuous efforts into the development of new LAB strains and cultures with novel properties and superior performance for the dairy industry. In the 21st century, genetic engineering could make such tasks easier to accomplish. However, the tight requirements of regulatory agencies and the negative perception by consumers of genetically modified foods impose the exclusive use of natural strain improvement methodologies. Thus, at Chr. Hansen innovative product development includes combining classical strain improvement techniques such as random mutagenesis, dominant selection, adaptive laboratorial evolution, with the insights gained from the molecular understanding of our model systems. This approach will be illustrated with relevant example cases of strain improvement for the dairy industry.

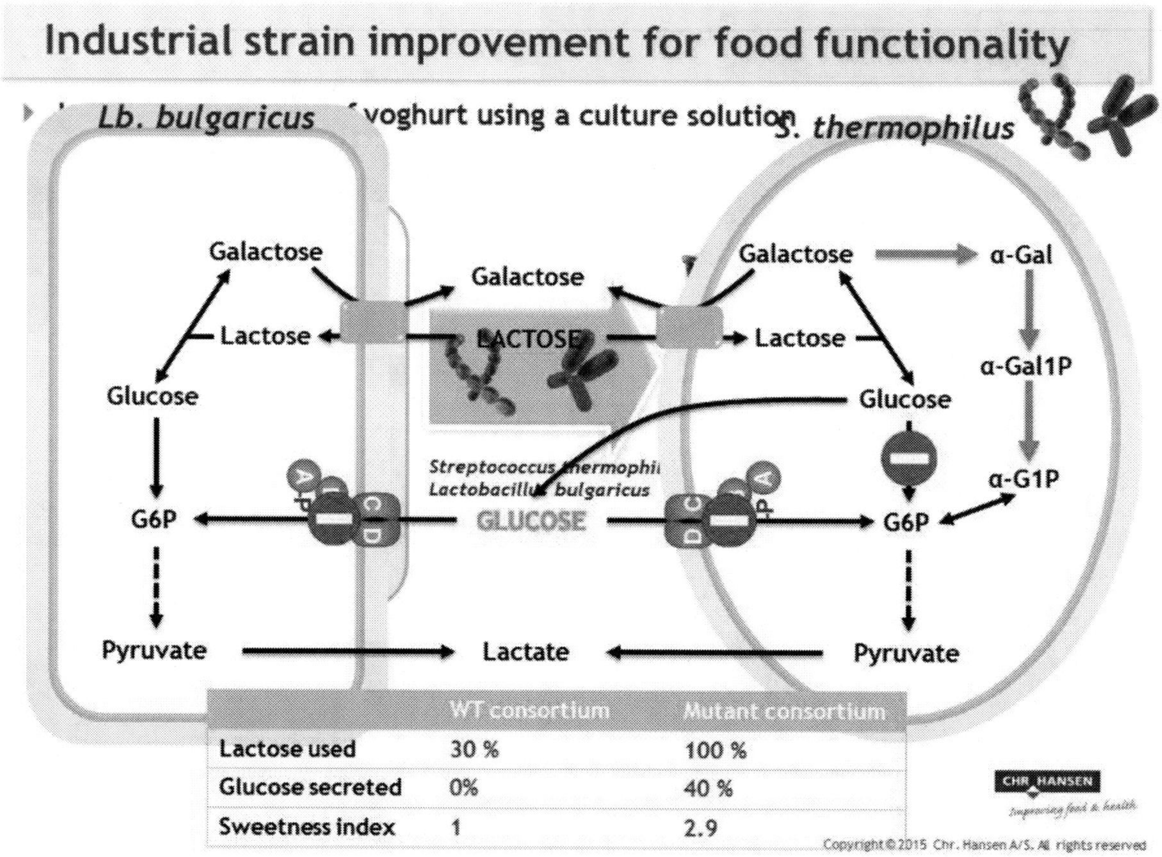

(13)C-Metabolic Flux Analysis of Ethanol-Assimilating Saccharomyces Cerevisiae for S-Adenosyl-L-Methionine Production

Hayakawa, K., Osaka University
Matsuda, F., Osaka University
Shimizu, H., Osaka University

Background:

Saccharomyces cerevisiae is a host for the industrial production of *S*-Adenosyl-L-methionine (SAM). The ^{13}C-metabolic flux analysis (^{13}C-MFA) of glucose-assimilating *S. cerevisiae* Kyokai 6 strain revealed that the larger metabolic flux level in the TCA cycle contributed the SAM production via an enhanced ATP regeneration (Hayakawa et al., 2015). It has been also reported that the feed of ethanol in cultivation improved SAM content (Shiozaki et al., 1989). However, a metabolic state in ethanol-assimilating *S. cerevisiae* remained unclear. In this study, the ^{13}C-MFA was conducted to compare metabolic flux distributions in the energy metabolism between ethanol-assimilating Sake yeast Kyokai no. 6 (high SAM-producing) and laboratory yeast S288C (control) strains.

Method:

S. cerevisiae Kyokai 6 and S288C strains were cultivated in aerobic carbon-limited chemostat culture for ^{13}C-MFA. The synthetic medium used for cultivation contained 3.8 g L^{-1} ethanol, 5.0 g L^{-1} $(NH_4)_2SO_4$, 0.50 g L^{-1} K_2HPO_4, and 3.4 g L^{-1} yeast nitrogen base without amino acids and ammonium sulfate with the moderate addition of vitamins and minerals. After reaching steady-state cell growth, the medium containing ^{13}C-labeled ethanol was fed to the chemostat cultures. The biomass samples were harvested at 5 residential times after feeding of ^{13}C-labeled ethanol. Following the acid hydrolysis and derivatization, ^{13}C labelling pattern of proteinogenic amino acids was measured using GC-MS. Two compositions of the carbon tracers were employed for ^{13}C-MFA (The ratios of (U-^{12}C):(1-^{13}C):(U-^{13}C) ethanol in the media were 6:4:0 and 6:2:2). The metabolic flux distributions were estimated by using OpenMebius (Kajihata et al., 2014).

Results and Discussion:

S. cerevisiae Kyokai 6 and S288C strains were cultured in the medium containing 3.8 g L^{-1} of ethanol as solo carbon source. Intracellular SAM level in Kyokai 6 strain [1.8 mg (g cell dry weight)$^{-1}$] was higher than those in S288C strain (0.86 mg g CDW^{-1}). The two strains were cultivated in the carbon-limited chemostat culture using the medium containing ^{13}C-labeled ethanol. Metabolic flux distributions were estimated from the ^{13}C-enrichment of proteinogenic amino acids and the specific rates of growth, ethanol consumption, and acetate production. The metabolic flux distributions were determined by using the two composition of ^{13}C-labeled ethanol [(U-^{12}C):(1-^{13}C):(U-^{13}C) were 6:4:0 and 6:2:2]. The comparison of 90% confidence intervals showed that the mixture at 6:2:2 was better for more precise metabolic flux estimation. The estimated metabolic flux distributions showed that the flux level of TCA cycle in Kyokai 6 strain was larger than that of S288C strain whereas the flux to SAM precursors (oxaloacetate and ribose-5P) was almost the same level. Analysis of ATP balance showed that more excess ATP was produced in Kyokai 6 strain due to the active oxidative phosphorylation. These results suggest that the high SAM accumulation should be also derived from the enhanced ATP regeneration in the ethanol assimilating Kyokai 6 strain.

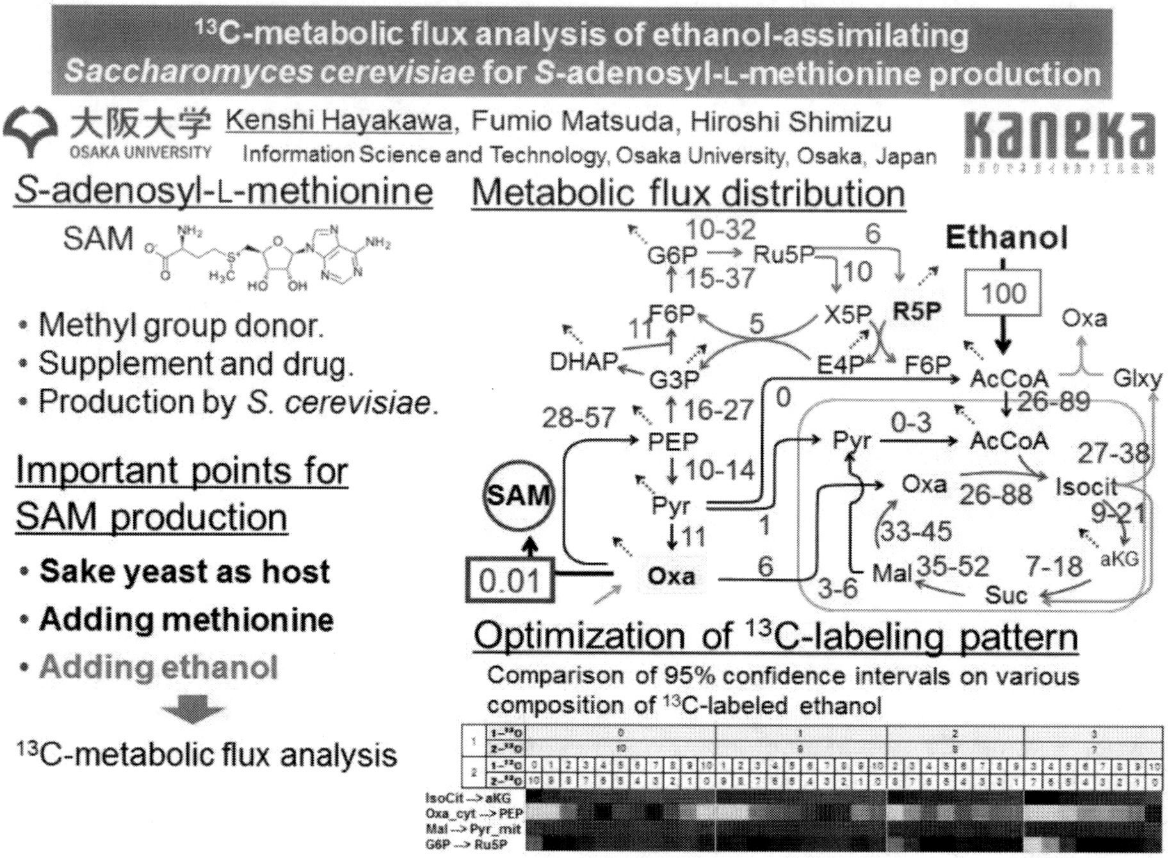

Advancing Dual-Phase Production Strategies with Corynebacterium Glutamicum through Systems Biology

Lange, J., Institute of Biochemical Engineering
Busche, T., Center for Biotechnology (CeBiTec)
Kalinowski, J., Center for Biotechnology (CeBiTec)
Takors, R., Institute of Biochemical Engineering
Blombach, B., University of Stuttgart

Background. Bacteria encounter varying oxygen concentrations in manifold situations e.g. in their natural habitat and especially in large scale industrial processes generating viability and production deficiencies. In particular, dual-phase approaches, e.g. for isobutanol or succinic acid production, evoked yield problems due to an abrupt shift from aerobic cultivation to anaerobic production (1, 2). In order to adapt to the changing environment bacteria have to remodel their entire metabolism (3, 4). Despite its overarching relevance for pathogenicity and pharmaceutical and bio-based production processes, the molecular events during these transitions are poorly understood. To address this question, we systematically investigate the adaptation of the industrially relevant *Corynebacterium glutamicum* to altering conditions that range from aerobiosis via microaerobiosis to anaerobiosis.

Methods. A "triple-phase" batch bioprocess with *C. glutamicum* was established that depicts the three successive phases (aerobiosis, microaerobic interface and anaerobiosis) in a single bioreactor. Throughout the process, samples were collected and analyzed for substrate consumption and organic acid production and additionally used for whole transcriptome analysis by RNA-sequencing.

Results. A definition of the three phases was directly feasible by the bacterium's physiological changes, i.e. a decreasing growth rate with increasing oxygen limitation. Furthermore, L-lactic acid, succinic acid and acetic acid were the main fermentation products secreted to the culture supernatant, interestingly in a manner, that their respective differential product yields clearly bordered each process phase. A closed carbon balance indicated that all significant products were analyzed. RNA-sequencing analysis revealed differential expression of an abundance of genes of the central and peripheral metabolism matching the expectations of previous work (5) but also delivered novel insights into the regulatory regime required for adaptation.

Conclusion. The established bioprocess is an elegant approach for the systematic understanding of *C. glutamicum*'s adaptation to a progressive oxygen deprivation. Deeper analysis of the RNA-sequencing data especially focusing on regulatory networks and their hierarchy could reveal novel targets for strain optimization to increase productivities in a dual-phase production process. A comprehensive understanding of the molecular events during the transitions might be also projected to other organisms and applications.

1. **Blombach B, Riester T, Wieschalka S, Ziert C, Youn J-W, Wendisch VF, Eikmanns BJ**. 2011. *Appl. Environ. Microbiol.* **77**:3300–10.

2. **Vemuri GN, Eiteman MA, Altman E**. 2002. *J. Ind. Microbiol. Biotechnol*. 28:325–32.

3. **Patschkowski T, Bates DM, Kiley PJ**. 2000. p. 61–78. *In* Storz, G, Hengge-Aronis, R (eds.), Bacterial stress responses. ASM Press, Washington, D.C.

4. **Bunn HF, Poyton RO**. 1996. *Physiol. Rev*. **76**:839–885.

5. **Inui M, Suda M, Okino S, Nonaka H, Puskás LG, Vertès AA, Yukawa H**. 2007. *Microbiology* **153**:2491–504.

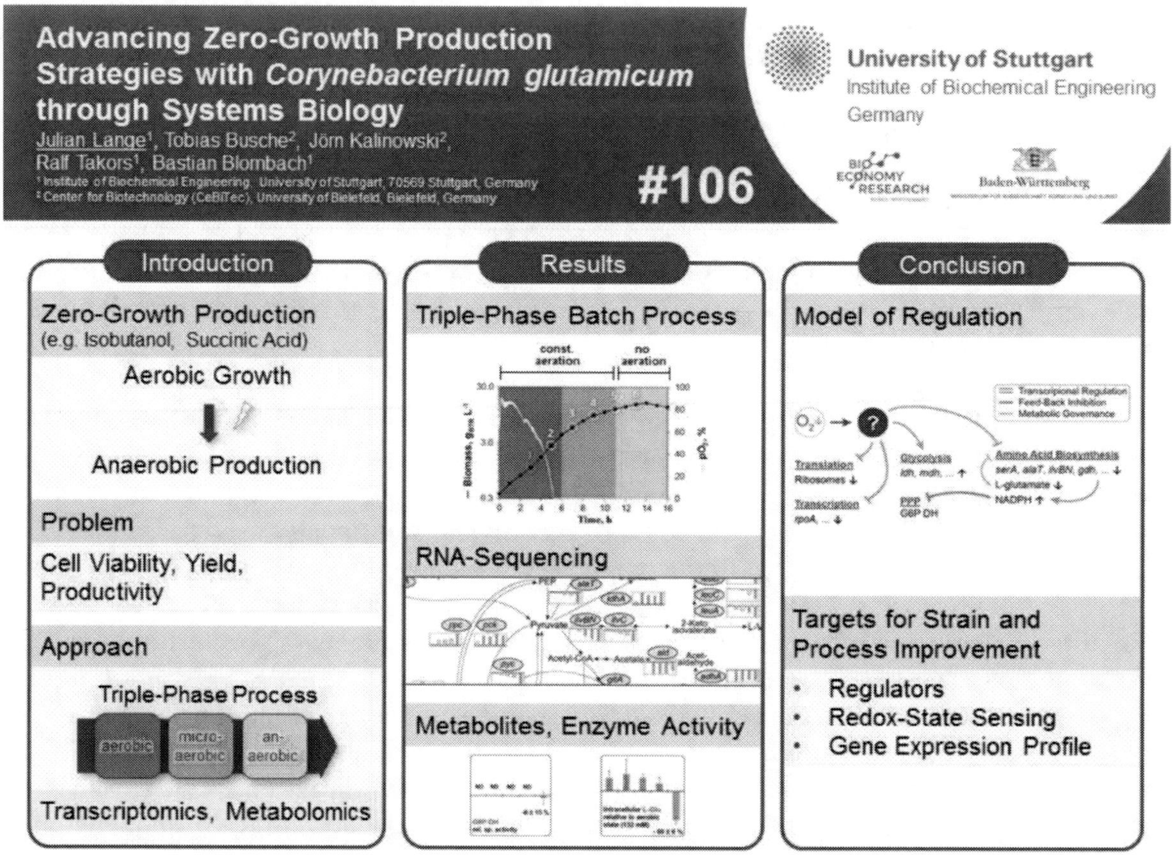

Characterization of Physiological and Transcriptional Properties of Improved Protein Secretion Yeast Strains

Huang, M., Chalmers University of Technology
Bao, J., Chalmers University of Technology
Hallström, B. M.,
Petranovic, D., Chalmers University of Technology
Nielsen, J., Technical University of Denmark

The yeast *Saccharomyces cerevisiae*is one of the most used organisms for protein production and developing an efficient secretion yeast system is favorable for industrial applications. In our previous study, several improved protein (α-amylase) secretion yeast strains were selected from UV mutagenesis libraries by microfluidic droplet screening and mutations of mutant strains have been identified by whole genome DNA sequencing. These identified mutations can be used as potential targets for strain optimization for protein secretion. However, except enhancement of protein production, the impacts of these mutations on mutant strains remain unknown.

In this work, mutant strains together with wildtype strain are cultured in SD-2×SCAA medium for batch fermentation to characterize their physiological and transcriptional properties. Compared with the wildtype strain, the mutant strains show higher α-amylase production rate, higher maximum specific growth rate on glucose, higher specific glucose uptake rate, higher ethanol production rate, lower yield of biomass from glucose and higher yield of ethanol from glucose. Lower percentage of intracellular α-amylase of mutant strains indicates that less α-amylase is retained by the cell, reflecting efficient secretory pathways of mutant strains.

To further explore the molecular mechanisms in the improved α-amylase secretion mutant strains, genome-wide transcriptional responses at exponential phase of mutant strains and the wildtype strain are analyzed. More genes are significant up- or down-regulation in higher α-amylase production strains, indicating wide global transcriptional adjustments are important to α-amylase production increasing. *PHO89*is the common down-regulation gene in all mutant strains. Up-regulation of most hexose transporter genes and down-regulation of genes in the glycolysis and TCA cycle are observed in mutant strains. Reporter GO term analysis reveals that GO term related to mitochondrial function, generation of precursor metabolites and energy, cellular respiration, cellular amino acid metabolic process, etc. are down-regulated. In contract, GO term related to ribosome function, cytoplasmic translation, regulation of organelle organization, Golgi vesicle transport, etc. are up-regulated. Genes controlled by transcriptional factors Mss11p, Rox1p, Tup1p, Upc2p and Yox1p are up-regulated, and genes controlled by transcriptional factors Adr1p, Bas1p, Gcr1p, Hap1p, Hap2p/3p/4p/5p CCAAT-binding complex, Opi1p and Sut1p are down-regulated. These results suggest anaerobic behaviors of yeast are favorable for protein production, which is in agreement with other studies from our lab.

The findings in this study help to understand underlying mechanisms of improved protein production strains and will contribute to design yeast cell factories for efficient protein production.

Robust Identification of Metabolic Engineering Targets Via Pool Efflux Capacities (PECs)

Takors, R., Institute of Biochemical Engineering
Teleki, A., Institute of Biochemical Engineering

Motivation:

The identification of promising genetic targets for the straightforward optimization of production strains is the epitome of metabolic engineering. The framework of metabolic control analysis provides the mindset for likewise approaches. In dynamic studies the metabolism of strains is properly perturbed for deriving characteristic parameters such as elasticities and flux control coefficients from the dynamic metabolic response. It has been shown that linlog models are a convenient tool for modelling the observed dynamics and for deducing the desired metabolic engineering targets. However likewise approaches demand for profound modelling expertise which may not always be present by wet-lab experts who are supposed to be the main applicants of the results.

Results:

Here we present the thorough application of the pool efflux capacity (*PEC*) criterion. *PEC* analyzes the observed metabolic dynamics after perturbation via mass balances for the individual pools. Accordingly a series of net efflux rates is calculated for each pool after the stimulus was set. Resulting efflux rates are ranked thus indicating the maximum capacity of the respective enzymes. Consequently high *PEC* values refer to low flux control and vice versa.

The approach was thoroughly applied for optimizing L-methionine production with a recombinant *E. coli* strain. It can be shown that *PEC* identifies the same dominating metabolic engineering targets as conventional methods. Noteworthy no tedious modelling is needed. Target identification is straightforward and purely data driven.

Furthermore an extended *PEC* approach will be presented integrating ^{13}C-labeling information. It will be shown that experimentalists only have to follow the ^{13}C-to-^{12}C ratio of the respective pools for identifying the same metabolic engineering targets. This finding offers great opportunities for application because it excludes the well-known problems of measuring proper intracellular pool sizes. The latter are often biased by non-wanted cell leakage, a phenomenon that does not affect the ^{13}C-to-^{12}C ratio.

Summarizing *PEC* analysis offers the straightforward identification of metabolic engineering targets without tedious modelling and without the bias of non-accurate intracellular pool sizes.

Metabolically Engineered Escherichia coli for Renewable Production of a Three Carbon Diamine, 1,3-Diaminopropane

Chae, T. U., Korea Advanced Institute of Science and Technology (KAIST)
Kim, W. J., Korea Advanced Institute of Science and Technology (KAIST)
Choi, S., KAIST
Park, S. J., Myongji University
Lee, S. Y., Korea Advanced Institute of Science and Technology (KAIST)

Bio-based production of chemicals is important for sustainable chemical industry. Here, *Escherichia coli* is metabolically engineered to produce 1,3-diaminopropane (1,3-DAP), a monomer for polyamide. Comparison of heterologous C_4 and C_5 pathways for 1,3-DAP production by *in silico* flux analysis revealed that the C_4 pathway employing *Acinetobacter baumannii dat* and *ddc* genes, encoding 2-ketoglutarate 4-aminotransferase and L-2,4-diaminobutanolate decarboxylase, respectively, was more efficient. In a strain having feedback resistant aspartokinases, the *ppc* and *aspC* genes were overexpressed to increase flux towards 1,3-DAP synthesis. Also, knocking out *pfkA* was found to increase 1,3-DAP production by applying 128 synthetic small RNAs. Overexpression of the *ppc* and *aspC* genes in the *pfkA* deleted strain resulted in even higher production of 1,3-DAP. Fed-batch fermentation of the final engineered *E. coli* strain allowed production of 13 g/L of 1,3-DAP in a glucose minimal medium. [This work was supported by the Technology Development Program to Solve Climate Changes on Systems Metabolic Engineering for Biorefineries from the Ministry of Science, ICT and Future Planning (MSIP) through the National Research Foundation (NRF) of Korea (NRF-2012M1A2A2026556 and NRF-2012M1A2A2026557).]

Metabolically Engineered *Escherichia coli* for Renewable Production of a Three Carbon Diamine, 1,3-Diaminopropane

Tong Un Chae, Won Jun Kim, Sol Choi, Si Jae Park and Sang Yup Lee

1. Target

2. Pathway

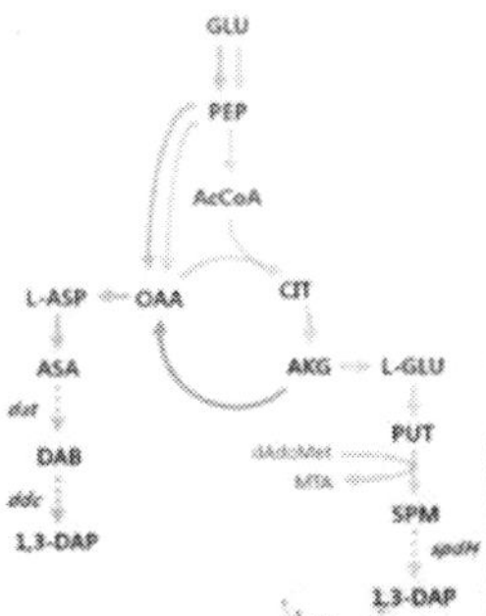

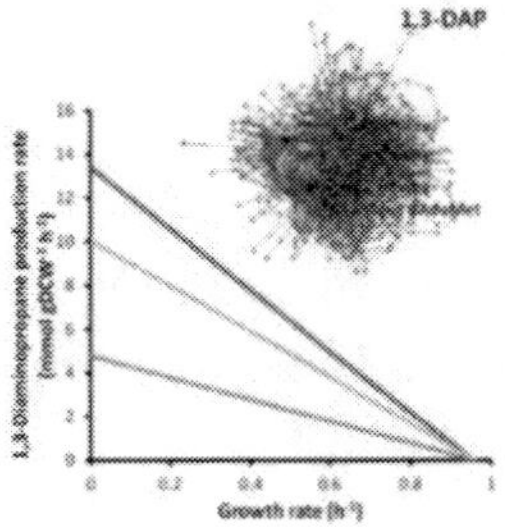

3. Metabolic Engineering

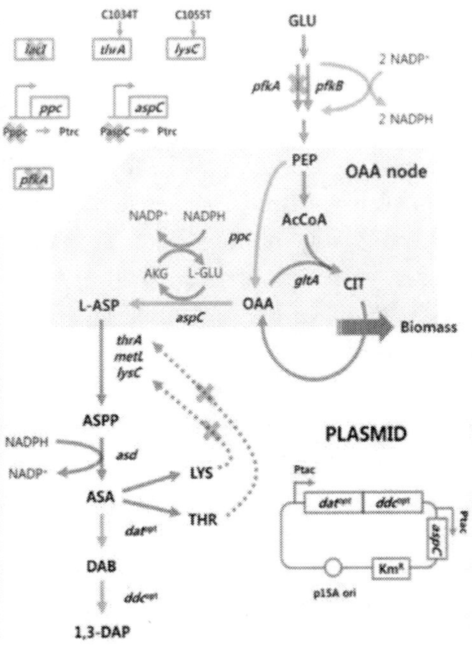

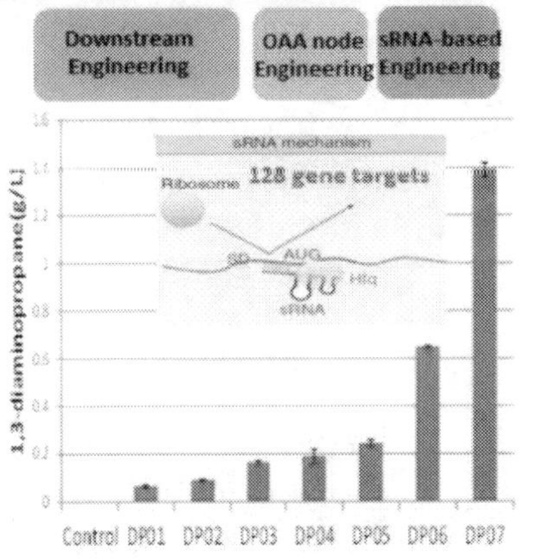

4. Fed-Batch Fermentation

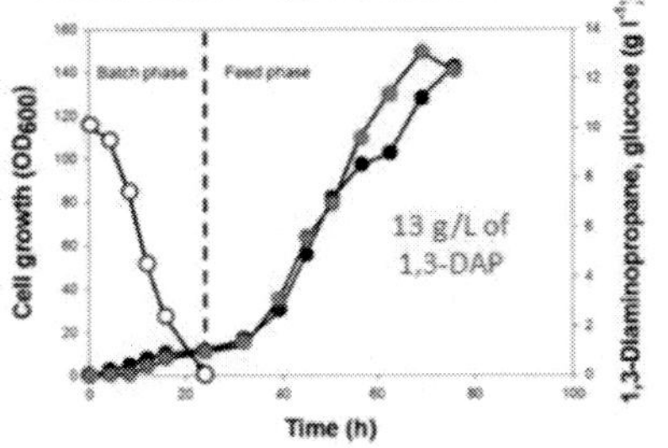

Prediction of Stress Resistance By Gene Expression Profiles

Furusawa, C., Osaka University
Suzuki, S., Quantitative Biology Center, RIKEN
Horinouchi, T., Quantitative Biology Center, RIKEN

Advances of technologies now make it possible to reveal the phenotypic and genetic changes responsible for adaptive evolution. Such detailed information of evolutionary traits provides a basis for understanding the nature of adaptive evolution, for example, what phenotypic changes are due to genetic mutation and which are not. In this study, we performed parallel evolution experiments of *Escherichia coli* under various stress environments, and as the results, we obtained several independently evolved stress resistant strains. Then, for these resistant strains, we performed gene expression analysis by microarrays and resequencing analysis by next-gen sequencer. Furthermore, cross-resistance/sensitivity to other stresses were quantified for these resistance strains. The results revealed that, although the patterns of cross-resistance/sensitivity and expression profiles were similar among the resistant strains for same stress. Based on these large-scale data, we demonstrated that the resistances could be quantitatively predicted by the expression changes of a small number of genes. Several candidate mutations contributing to the resistances were identified, while phenotype-genotype mapping was suggested to be complex and included various mutations that caused similar phenotypic changes. The integration of transcriptome and genome data enables us to extract essential phenotypic changes for stress resistances, and based on these results, we will discuss possible applications of these phenotype-genotype analyses of evolutionary dynamics on bioengineering.

Prediction of Stress Resistance By Gene Expression Profiles

Chikara Furusawa, RIKEN/Univ. Tokyo

Experimental evolution of *E. coli* under antibiotics

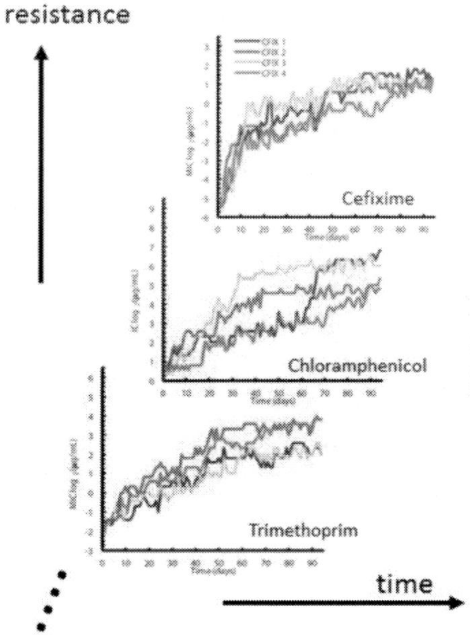

$$MIC_j^k = \sum_{i=1}^{N} \alpha_i^k X_{ij} + \beta^k$$

Extraction of essential expression changes from transcriptome data (expressions of 7～8 genes)

Quantitative prediction and control of resistance acquisitions

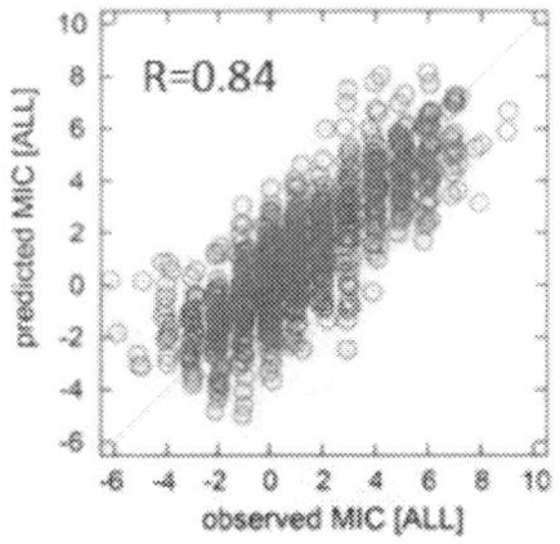

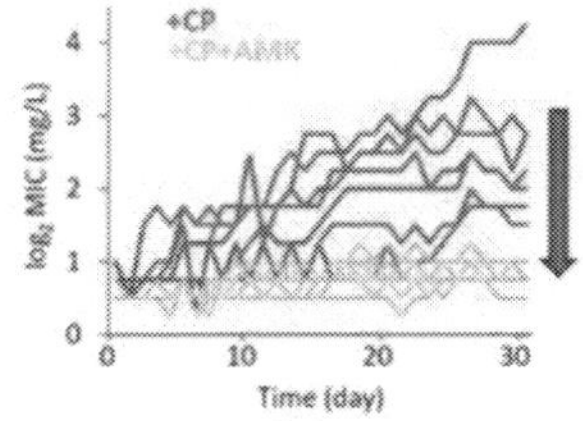

Suzuki, Horinouchi, Furusawa, **Nature Comm.** (2014)

Production of Terpenoids in Escherichia coli: Optimizing the Metabolic Network and the Production Process

Valderrama-Gomez, M. A.,
Wagner, S. G., Technical University of Munich
Kremling, A.,

Terpenoids are functionalized terpenes, which are naturally synthetized compounds formed by the ligation of isoprene molecules. They have a wide range of applications in the food, cosmetic, biotechnology and pharmaceutical industry. Lycopene, taxoids and artemisinin are some prominent examples of terpenoid molecules with remarkable applications. The microbiological terpenoid production is a relatively new field of research. Consequently, much work has yet to be done to improve current experimental product yields and productivities. Here, we report some achievements on the strain and process optimization of the *E. coli*-based taxoids production.

Firstly, we developed a protocol for Metabolic Engineering which guides the development and improvement of both the production strain and its subsequent usage in a bioreactor [Springer Protocols Handbooks, Chapter: "Computer-guided Metabolic Engineering", http://dx.doi.org/10.1007/8623_2015_118]. The protocol provides clear advices on the proper order of applying many standard computational methods widely used nowadays for rational Metabolic Engineering. By applying the tools described in the protocol, we could identify different carbon sources maximizing various performance criteria, e.g., carbon yield and profit. Additionally, we could reduce carbon leakage via addition of a synthetic non-oxidative glycolysis, as well as identify optimal process parameters. The *in silico* process optimization was performed by using experimental batch data at different production temperatures and time point of induction to construct a dynamic flux balance analysis model capable of predicting system behavior at new, untested conditions. By doing this, it was possible to find optimal process parameters that increased the highest experimental productivity by 25%. Further improvements of current host strain require a deep understanding of the metabolic stress induced by the expression of large enzyme cascades, typically required for the synthesis of complex taxoids. For that purpose, we introduced a reporter plasmid expressing the fluorescent protein mCherry under the control of a housekeeping promoter. Additionally, we introduced a second plasmid that provides a controllable load. This system enables us to systematically analyse genes regarding their replicative, transcriptional, and translational burden. A mathematical model which will not only consider metabolic fluxes but also transcriptional and translational processes is being developed and will be validated based on experimental data. By using this model, we will be able to calculate and predict energy and resource requirements for a heterologous burden and consequently, estimate to what extent inserted genes can be expressed. The developed model should also be useful at proposing genetic modifications as well as process parameters for resource rearrangement and energy delivery to accomplish the burden.

Production of Terpenoids in *E. coli* -

Optimizing the Metabolic Network and Production Process

TUM

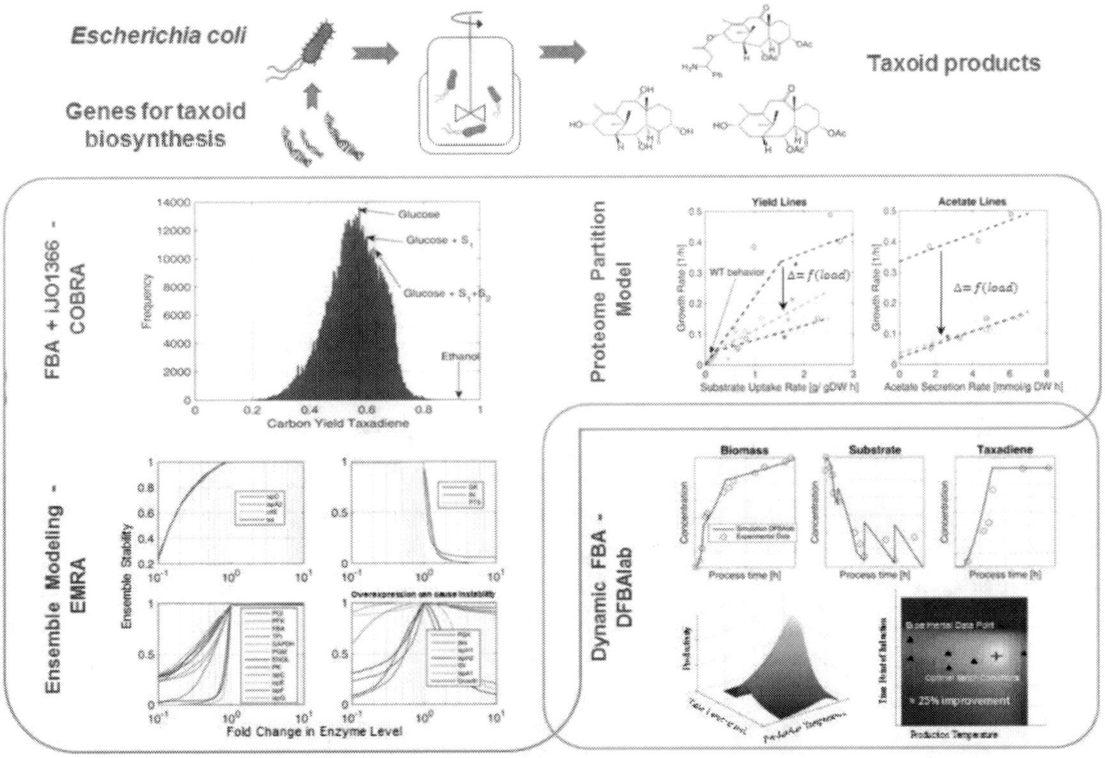

3-Amino-4-Hydroxy Benzoic Acid Production Via 3,4-AcAHBA in Escherichia coli

Krylov, A.,
Nishio, Y., Institute for innovation, Ajinomoto Co., Inc.,
Khrushcheva, A., Ajinomoto-Genetika Research Institute
Shozui, F., Research Institute for Bioscience Products & Fine Chemicals, Ajinomoto Co., Inc.
Slivinskaya, E., Ajinomoto-Genetika Research Institute
Tajima, Y., Institute for innovation, Ajinomoto Co., Inc.,
Fedorova, E.,
Yokoyama, K., Ajinomoto Co., Inc.

The bio-based production of chemicals is presently of great importance in the field of biotechnology. Production of aromatic compounds from renewable bio resources is recognized as one of the key aspects of biorefinery research. 3-Amino-4-hydroxy benzoic acid (3,4-AHBA) is a precursor of bioplastics as it increases the thermostability of the polymer when used as a monomer. Most aromatic compounds are produced via the shikimate pathway; however, in *Streptomyces griseus,* 3,4-AHBA is produced through a two-step catalyzed reaction from L-aspartate-4-semialdehyde (ASA) and dihydroxyacetone phosphate (DHAP). 3,4-AHBA was found to be a metabolic intermediate in grixazone biosynthesis in *S. griseus* and it is synthesized by the activities of the enzymes GriI and GriH. GriI catalyzes an aldol condensation reaction between ASA and DHAP, while GriH converts the resulting C7 metabolite into 3,4-AHBA.

However, there are two problems with biological production of 3,4-AHBA in *E. coli*. The first problem is its rapid conversion to other chemical compounds. 3,4-AHBA is non-enzymatically converted to 2-aminophenoxazin-3-one-8-carboxylic acid (APOC) under aerobic/anaerobic condition, while enzymatically, it is converted to 3-acetylamino-4-hydroxybenzoic acid (3,4-AcAHBA). Approximately 39% of the exogenously added initial amount of 3,4-AHBA is converted into 3,4-AcAHBA in the presence of *E. coli* cells. The second problem is an inhibitory effect of 3,4-AHBA on the growth of *E. coli* cells. Addition of 60 mM exogenous 3,4-AHBA led to significant growth retardation of MG1655 cells.

In contrast to the instability of 3,4-AHBA in culture medium, exogenously added 3,4-AcAHBA maintains a constant concentration in culture medium during aerobic/anaerobic cultivation conditions. Furthermore, 3,4-AcAHBA is non-toxic for *E. coli* cells. Taking these observations into account, biotechnological production of 3,4-AcAHBA and its chemical deacetylation were considered to be feasible processes.

In *S. griseus*, N-acetyltransferase (NAT) encoded by *natA* is solely responsible for the N-acetylation of 3,4-AHBA. Homology analysis showed that an *S. griseus* NAT-like protein, NhoA, is conserved in *E. coli*. NhoA functions as an arylamine acetyltransferase that catalyzes the acetyl-CoA dependent N-acetylation of a variety of arylamine substrates. Our study demonstrated that the disruption of *nhoA* led to a loss of function and consequently the loss of 3,4-AcAHBA production, thus indicating that NhoA is solely responsible for the N-acetylation of 3,4-AHBA in *E. coli.*

For the synthesis of 3,4-AHBA, NspI and NspH from *Streptomyces murayamaensis* (the GriIH homologs) were expressed in *E. coli.* The *E. coli* MG1655 strain harboring the *nspIH* operon expression plasmid produced around 2.7 %(mol/mol) of 3,4-AHBA from glucose and by-produced AHBA derivatives. By combining genetic modification to improve ASA supply in the host strain and expression of *nspI*, *nspH* and *nhoA*, 3,4-AcAHBA production increased up to 13.3 %(mol/mol) without any by-production of AHBA derivative compounds. These results suggest a technological advantage of 3,4-AcAHBA fermentation.

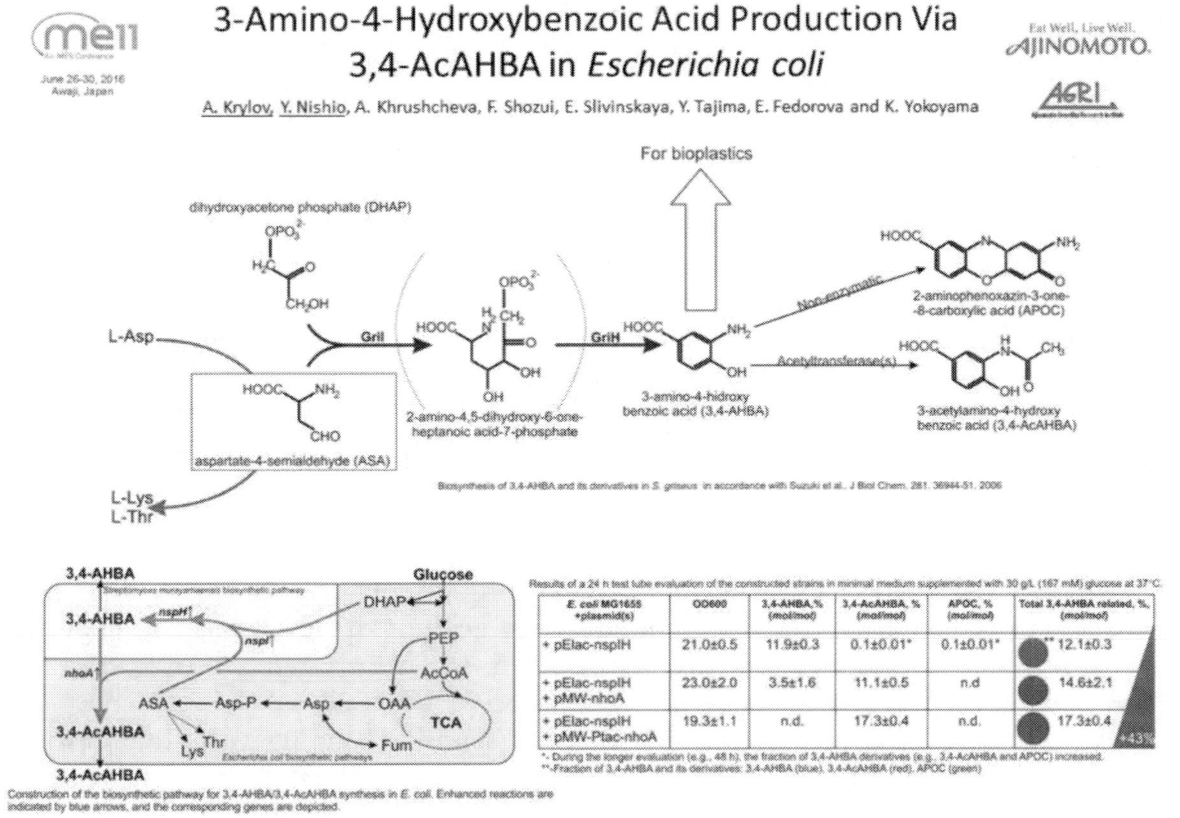

Biosynthesis of 3,4-AHBA and its derivatives in *S. griseus* in accordance with Suzuki et al., J Biol Chem, 281, 36944-51, 2006

Construction of the biosynthetic pathway for 3,4-AHBA/3,4-AcAHBA synthesis in *E. coli*. Enhanced reactions are indicated by blue arrows, and the corresponding genes are depicted.

Results of a 24 h test tube evaluation of the constructed strains in minimal medium supplemented with 30 g/L (167 mM) glucose at 37°C.

E. coli MG1655 +plasmid(s)	OD600	3,4-AHBA,% (mol/mol)	3,4-AcAHBA, % (mol/mol)	APOC, % (mol/mol)	Total 3,4-AHBA related, %, (mol/mol)
+ pElac-nspIH	21.0±0.5	11.9±0.3	0.1±0.01*	0.1±0.01*	** 12.1±0.3
+ pElac-nspIH + pMW-nhoA	23.0±2.0	3.5±1.6	11.1±0.5	n.d	14.6±2.1
+ pElac-nspIH + pMW-Ptac-nhoA	19.3±1.1	n.d.	17.3±0.4	n.d.	17.3±0.4

*- During the longer evaluation (e.g., 48 h), the fraction of 3,4-AHBA derivatives (e.g., 3,4-AcAHBA and APOC) increased.
**-Fraction of 3,4-AHBA and its derivatives: 3,4-AHBA (blue), 3,4-AcAHBA (red), APOC (green)

Metabolic Engineering of Cyanobacteria Using Genome-Scale Modeling and Crispri

Hudson, P., KTH-Royal Institute of Technology

Photosynthetic cyanobacteria are attractive hosts for low-value fuels and chemicals as $CO_2$and sunlight are cheap and abundant feedstocks. However, the photoautotrophic metabolism has been difficult to manipulate for high specific productivities, due to a lack of synthetic-biology tools and systems-level analyses. Low specific productivities are particularly problematic if high-density culture is not possible due to light shading.

I will discuss my group's approach and recent results on metabolic engineering of cyanobacteria for biofuel production[1, 2]. We use genome-scale modeling to find gene knockout strategies to enhance biofuel productivity from photoautotrophic metabolism. Two strategies can be applied for a range of biofuel pathways. One strategy exploits cofactor usage and another creates an imbalance in ATP and NADPH supply and demand. By considering expression data, we can reduce the number of require knockouts to achieve product-growth coupling[3]. To execute engineering strategies, we have implemented inducible, multiplex CRISPRi gene knockdown, which has not been used previously in cyanobacteria[4]. This tool will make it much easier to test metabolic engineering strategies as well as perform genetic logic in cyanobacteria. The cycle of prediction, execution, and analysis that we are following is expected to provide new cyanobacteria strains with specific productivities much higher than those reported so far. Finally, we are creating mutation libraries for cyanobacteria directed evolution and I will present our work on a microfluidics-based platform that can screen and sort cyanobacteria based on productivity and cell growth[5].

References

1. Anfelt J, Kaczmarzyk D, Shabestary K, Rockberg J, Renberg B, Uhlen M, Nielsen J, Hudson EP: **Genetic and nutrient modulation of acetyl-CoA levels in Synechocystis for n-butanol production.** *Microb Cell Fact* 2015, **14**:1–12.

2. Anfelt J, Hallström B, Nielsen J, Uhlén M, Hudson EP: **Using transcriptomics to improve butanol tolerance of Synechocystis sp. strain PCC 6803.** *Appl Environ Microbiol* 2013, **79**:7419–27.

3. Shabestary K and Hudson EP:**Computational metabolic engineering strategies for growth-coupled biofuel production by *Synechocystis*.** *Manuscript*

4. Yao L, Cengic I, Anfelt J, Hudson EP: **Multiple gene repression in cyanobacteria using CRISPRi**. *ACS Synth Biol*2015, 10.1021/acssynbio.5b00264

5. Hammar P, Angermayr S, Sjöström S, van der Meer J, Hellingwerf K, Hudson EP, Jönsson H: **Single cell screening of photosynthetic growth and lactate production by cyanobacteria**. *Biotechnol Biofuels* 2015, **8**:1–8.

Production of Specialized Fatty Acids in Saccharomyces Cerevisiae

de Carvalho, Â.,
Heider, H.,
Dalgaard Mikkelsen, M.,
Halkjær Hansen, E., Evolva

The primary aim of this project is to design *Saccharomyces cerevisiae* strains for the production of specialized fatty acids that are precursors in the biosynthesis of new ingredients that have high commercial value. To this, metabolic engineering will be employed using key genes to establish new routes or improve endogenous routes for the production of the desired fatty acid precursors.

As *S. cerevisiae* does not typically feed on fatty acids its production relies on endogenous biosynthesis. *De novo* synthesis of fatty acids takes place in the cytosol and in mitochondria followed by elongation and desaturation in the endoplasmic reticulum. Almost all subcellular organelles are involved in fatty acid metabolism, so maintenance of fatty acid homeostasis requires regulation at multiple levels.

In this project, metabolic enzymes that are required to redirect fatty acid metabolism will be identified and evaluated. Rational metabolic engineering of *S. cerevisiae* using key enzymes will be applied, to tailor make novel GM strains that produce relevant fatty acids.

Engineering Key Functions Conferring Cellular Robustness: Detoxification, Membrane Permeability and Protective Metabolites in Yeast

Bettiga, M., Chalmers University of Technology
Olsson, L., Chalmers University of Technology

The establishment of the so-called bioeconomy, where biomass will represent an important source of hydrocarbons, relies on the successful development of bio-processes for the production of fuels and chemicals. Lignocellulosic material will be adopted as the primary source of sugars to be converted in a biorefinery concept not to ethanol or fuels only, but to a portfolio of chemicals. Regardless of the product, robust microorganisms are a prerequisite for the feasibility of lignocellulose bioconversion.

Current research carried out by our group focuses on the yeast *Saccharomyces cerevisiae* and aims at investigating the molecular bases of microbial robustness. We have identified strain engineering strategies to improve robustness based on the concept of structural engineering key functions, rather than pursuing the quest of "magic bullets", i.e. single gene-encoded properties.

In this contribution we will present our results demonstrating how detoxification, membrane permeability and protective metabolites are key functions to improve *S. cerevisiae*'s robustness towards lignocellulosic inhibitors. In particular, specific examples on engineering detoxification, membrane properties and protective metabolites, applied to respectively the metabolic conversion of lignin derived phenolic compounds, acetic acid tolerance and accumulation of the cellular protectant glutathione will be presented.

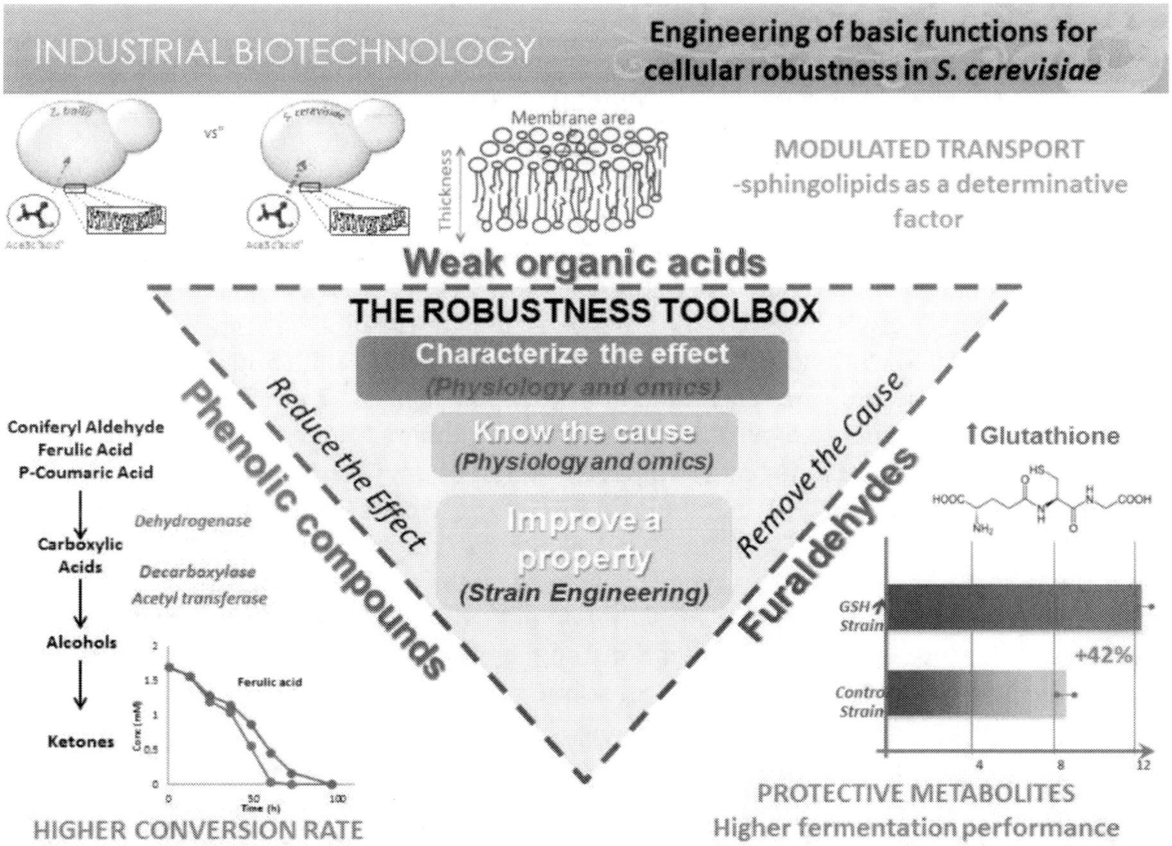
INDUSTRIAL BIOTECHNOLOGY
Engineering of basic functions for cellular robustness in S. cerevisiae
Membrane area
Thickness
MODULATED TRANSPORT
-sphingolipids as a determinative factor
Weak organic acids
THE ROBUSTNESS TOOLBOX
Characterize the effect
(Physiology and omics)
Know the cause
(Physiology and omics)
Improve a property
(Strain Engineering)
Reduce the Effect
Remove the Cause
Phenolic compounds
Furaldehydes
Coniferyl Aldehyde
Ferulic Acid
P-Coumaric Acid
Dehydrogenase
Carboxylic Acids
Decarboxylase
Acetyl transferase
Alcohols
Ketones
Ferulic acid
HIGHER CONVERSION RATE
Glutathione
GSH Strain
Control Strain
+42%
PROTECTIVE METABOLITES
Higher fermentation performance

Design of a Bio-Based Adipic Acid Microbial Cell Factory: The Challenging Choice of Microbial Host and Metabolic Pathway

Karlsson, E., Chalmers University of Technology
Olsson, L., Chalmers University of Technology
Mapelli, V., Chalmers University of Technology

Adipic acid is a six carbon- dicarboxylic acid used mainly as a building block in the nylon industry for the synthesis of nylon 6,6. According to the report (Bioenergy Task 42, Biorefinery) released by the International Energy Agency (IEA), adipic acid is the dicarboxylic acid derived from fossil raw materials with the highest annual production volumes. The aim of this study is to engineer a microbial host that would be able to convert sugars deriving from renewable bio-based materials, such as forest residues, into adipic acid. In this work, we present the metabolic pathway of our choice, using lysine as substrate. Preliminary results on the characterization of enzymes we identified as good candidates to catalyze specific metabolic steps are presented. Furthermore, we show the results of a study aimed to define the tolerance to adipic acid of an array of potential microbial hosts, including three yeast species, the filamentous fungus *Aspergillus niger* and bacterial species, such as *Escherichia coli* and *Corynebacterium glutamicum*. The results of this study give crucial indications on the choice of the most suitable microbial host to be considered for the design of an adipic acid microbial cell factory.

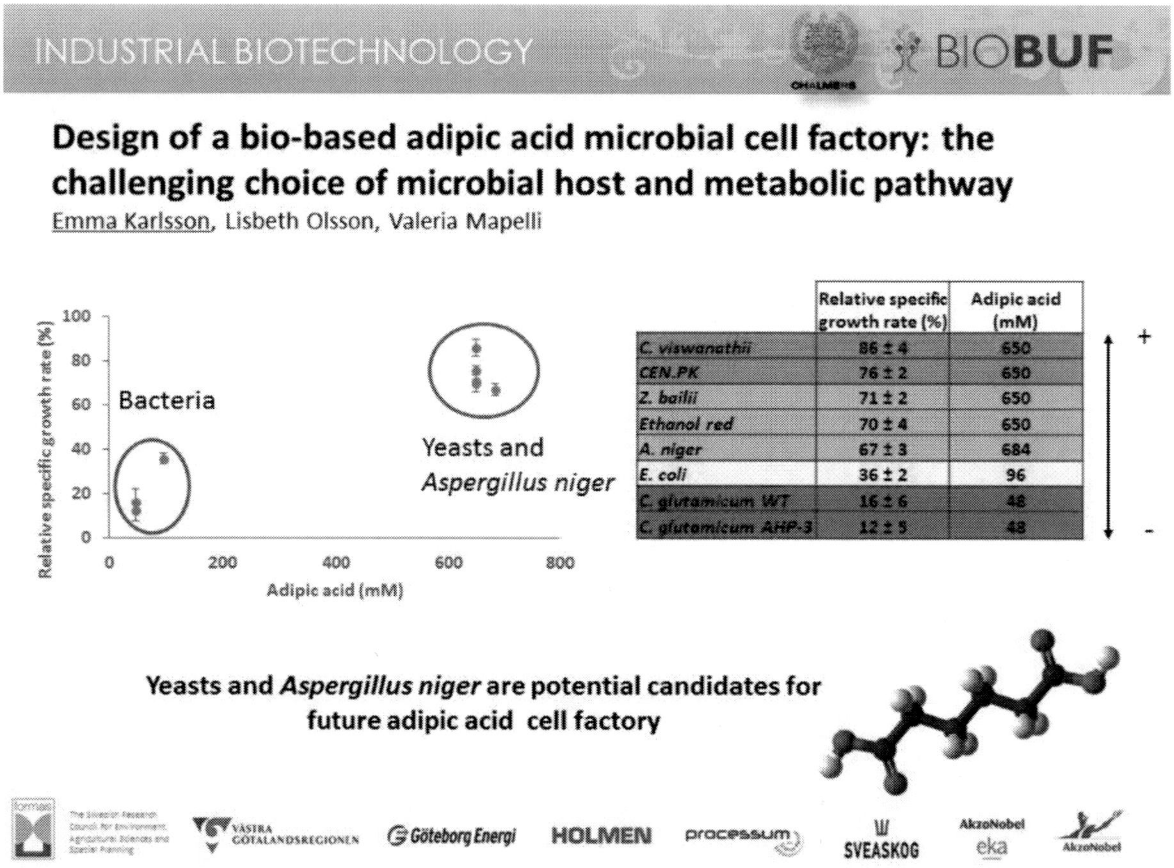

Design, Construction, and Validation of Modular Chassis Cells for Efficient Combinatorial Biosynthesis of Chemicals

Trinh, C. T., University of Tennessee, Knoxville

Transforming biology into engineering practice has a great potential to shape the industrialization of biology that will drive rapid development of novel microbial manufacturing platforms. These platforms will be capable of producing a vast number of industrially relevant chemicals at scale from alternative renewable feedstocks or wastes (e.g., biomass residues, biogas methane, syngas, CO2) without harming the environment. The challenge is to develop novel microbial manufacturing platforms to produce targeted chemicals with high efficiency in a rapid, predictable, and reproducible fashion. The conventional approach involves engineering each desirable cell in many trial-and-error optimization cycles. To engineer new desirable cells to produce other target chemicals, the entire process must be repeated, which is laborious and expensive. To address these bottlenecks, we have developed the MODCELL (Modular Cell) tool to design modular cells that can metabolically and genetically couple with a diverse class of exchangeable production modules for combinatorial biosynthesis of novel chemicals in a plug-and-play fashion requiring minimum iterative strain optimization cycles. We will present the design, construction, and validation of an *E. coli* modular cell for a combinatorial biosynthesis of novel bioesters that can be used as fragrances, flavors, solvents, and biodiesels from biomass-derived sugars and waste streams.

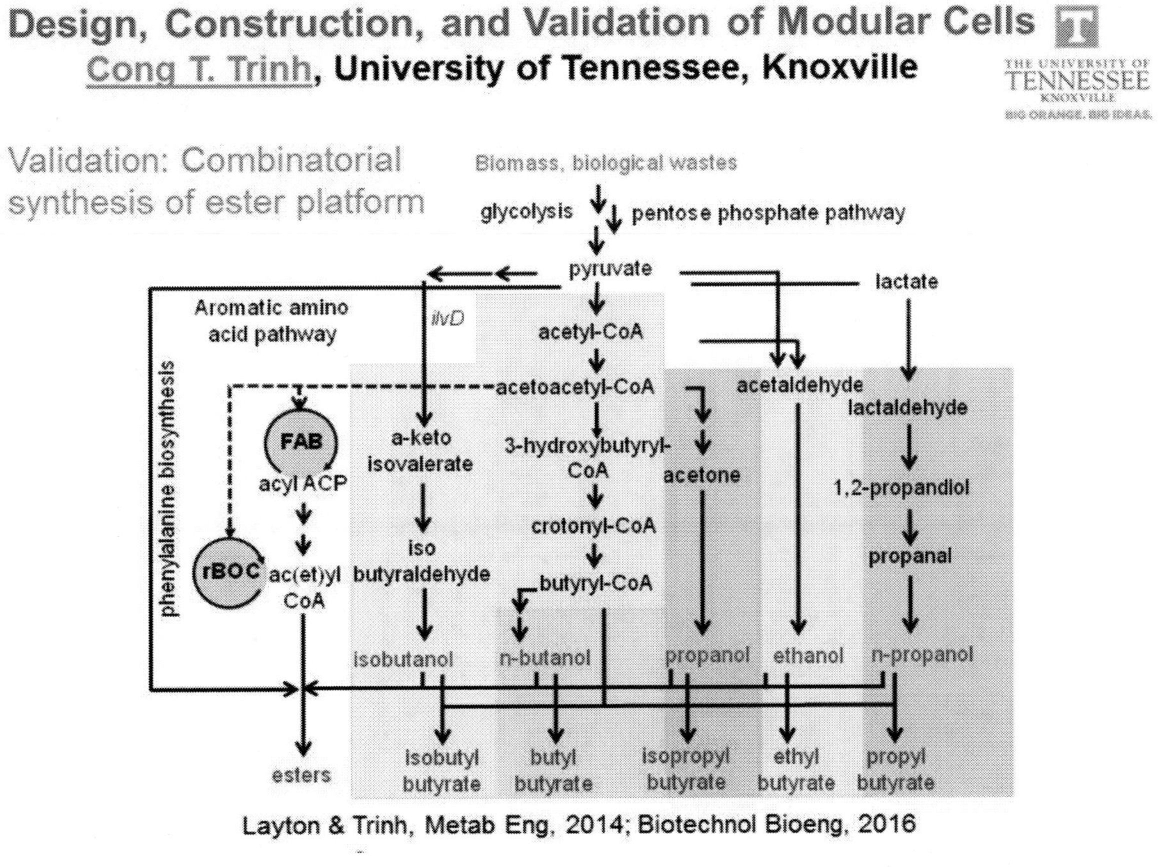

An Evolutionary Approach for Development of Novel Riboswitches with Desired Dose-Response Curve

Jang, S., POSTECH
Jang, S., POSTECH
Jung, G. Y., POSTECH

Various riboswitch-based genetic devices have been developed and applied to diverse fields such as live-cell imaging and screening for metabolite production. Since riboswitches should have distinct dose-response curves to meet the requirements for different applications, they have been redesigned by rational approaches based on sequence and structural information. However, rational approach is applicable only to few riboswitches whose structure and regulatory mechanism are well understood. Therefore, we developed an evolutionary approach for riboswitch engineering that is independent of preliminary knowledge about the riboswitch. First, we obtained naringenin (NRN) aptamer library by systematic evolution of ligands by exponential enrichment (SELEX), and constructed NRN riboswitch library composed of the aptamer library, random linker sequences, selection marker gene (*tetA*), and reporter gene (*sgfp*). Then we performed enrichments by dual selection in low, medium and high NRN concentrations. We chose enrichment conditions by investigation of the relationship between expression level of *tetA* and resistance to selection pressures (tetracycline or nickel ion concentration). We measured the dose-response curves of the enriched NRN riboswitches, and half maximal effective concentration (EC_{50}) of riboswitches gradually increased as the NRN concentration of enrichments. In conclusion, we established a novel approach for construction of riboswitches that have desired dose-response curves from scratch.

Engineering Genome-Reduced Corynebacterium Glutamicum for Isoprenoid Production

Henke, N. A., Bielefeld University
Peters-Wendisch, P., Bielefeld University
Heider, S. E. A., Bielefeld University
Wendisch, V., Bielefeld University

Terpenoids are one of the most abundant metabolite classes in nature with over 10000 different structures and exhibit a wide variety of fragrance, flavoring and coloring properties. The yellow-orange-red to violet carotenoids, a subfamily of terpenoids, can be found as pigments in plants, fungi, algae and bacteria. Due to their color and anti-oxidant effects this class of terpenoids becomes more and more attractive to the pharmaceutical and cosmetic industries. The orange-red colored ß-carotene-derived pigment astaxanthin and its precursor zeaxanthin and canthaxanthin are high-value ketocarotenoids with an increasing market. The biotechnologically important bacterium *Corynebacterium glutamicum* is naturally yellow pigmented due to the synthesis of the C_{50}-carotenoid decaprenoxanthin[1]. It is used for million ton-scale amino acid production and its biotechnological potential has been widened to include other high-value[2]. Genome reduction has led to *C. glutamicum* chassis strains[3].

The genomic background of a prophage-cured, genome-reduced *C. glutamicum* strain MB001[4] was engineered for enhanced production of rare carotenoids. Optimization of the supply of the precursor isopenthenyl pyrophosphate (IPP) through chromosomal promoter exchange of *dxs*, deletion of the endogenous carotenoid genes *crtEbY* and plasmid encoded expression of homologous genes *crtE*, *crtB* and *crtI* in this *C. glutamicum* strain led to efficient production of lycopene and the addition of a heterologous *crtY* gene from *Pantoea ananatis* resulted in β-carotene production[5].

Furthermore, the production of other industrially relevant carotenoids such as canthaxanthin, zeaxanthin and astaxanthin through heterologous gene expression with a two vector system was performed. Variations of *crtZ* and *crtW* genes from marine and non-marine bacteria encoding for β-carotene hydroxylases and ketolases, respectively, were compared and with this approach production of canthaxanthin and astaxanthin in the mg/g CDW range was achieved.

[1] Heider *et al.* (2014), *Appl Microbiol Biotechnol* 98:10, 4355-4368

[2] Peters-Wendisch & Wendisch (2014) In: Industrial Biocatalysis. Grunwald (Ed); London: LLC Taylor & Francis: 373–416

[3] Unthan *et al.* (2015) *Biotechnol J*, 10(2):290-301

[4] Baumgart *et al.* (2013) *Appl Environ Microbiol*. 79(19):6006-15.

[5] Heider *et al.* (2014), *Frontiers in Bioengineering and Biotechnology*, doi: 10.3389/fbioe.2014.00028

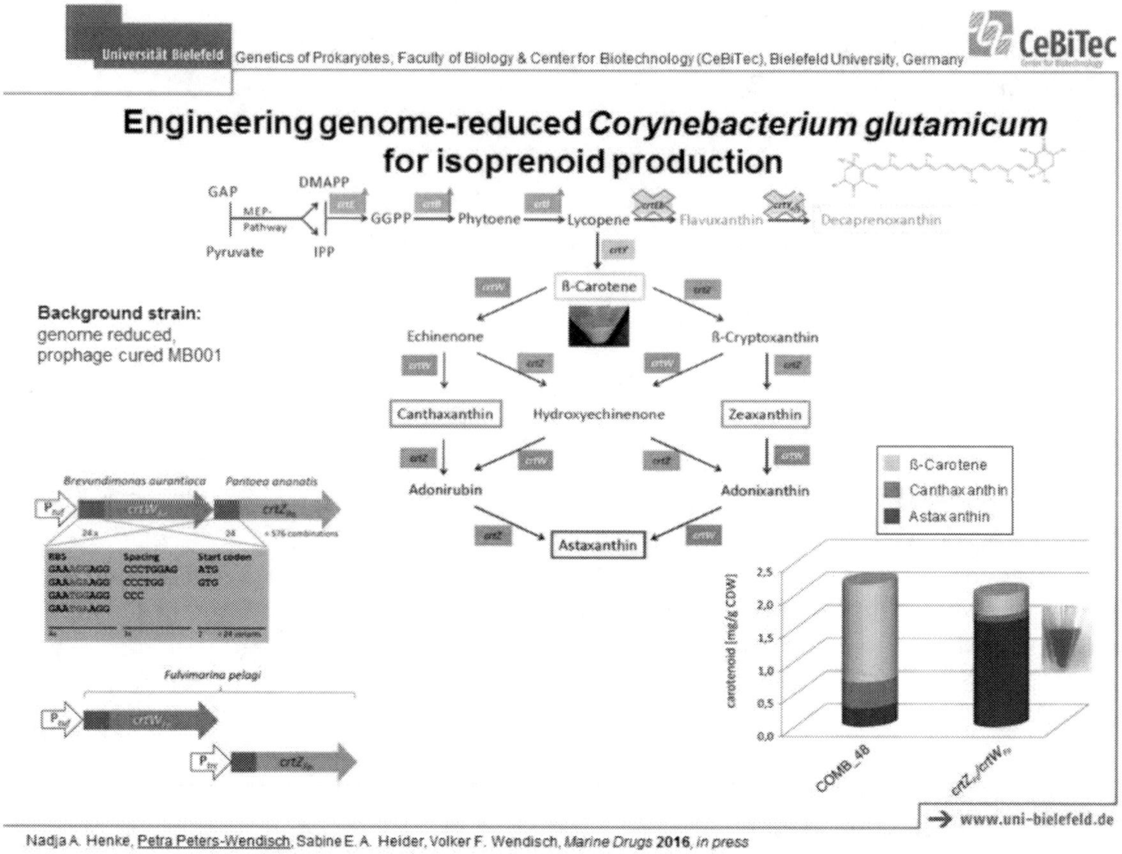
Universität Bielefeld
Genetics of Prokaryotes, Faculty of Biology & Center for Biotechnology (CeBiTec), Bielefeld University, Germany
CeBiTec
Engineering genome-reduced Corynebacterium glutamicum for isoprenoid production
GAP
MEP-Pathway
Pyruvate
DMAPP
IPP
GGPP
Phytoene
Lycopene
Flavuxanthin
Decaprenoxanthin
ß-Carotene
Echinenone
ß-Cryptoxanthin
Canthaxanthin
Hydroxyechinenone
Zeaxanthin
Adonirubin
Adonixanthin
Astaxanthin
Background strain:
genome reduced,
prophage cured MB001
Brevundimonas aurantiaca
Pantoea ananatis
Fulvimarina pelagi
ß-Carotene
Canthaxanthin
Astaxanthin
carotenoid [mg/g CDW]
COMB_4B
www.uni-bielefeld.de
Nadja A. Henke, Petra Peters-Wendisch, Sabine E. A. Heider, Volker F. Wendisch, Marine Drugs 2016, in press

Genome-Wide Identification of Genes Involved in NADPH Supply

Ye, L., Technical University of Denmark
Cardinale, S., Technical University of Denmark
Siedler, S., Technical University of Denmark
Ellabaan, M., Technical University of Denmark
Kim, S. H., Danmarks Tekniske Universitet
Sommer, M.,

NADPH is a crucial cofactor in reductive enzymatic reactions to synthesize valuable bioactive compounds. Previous designs to increase intracellular NADPH availability mainly rely on regeneration of NADPH by overexpressing $NADP^+$-dependent dehydrogenases, or strengthening of the oxidative branch of pentose phosphate pathway by abolishing the function of certain enzymes. However, modification in central metabolic pathway may cause unexpected growth defect, such as the deletion of phosphoglucose isomerase (*pgi*) gene would slow down more than 50% of the growth rate compares to wildtype. In this study, we prove the possibility to enrich the NADPH pool via engineering the *E. coli* K-12 strain based on genome wide screens. To identify mutants with both higher NADPH level and improved fitness, we combined the fast fitness quantification method by sequencing random barcoded transposon library (RB-TnSeq) and fluorescence-activated cell sorting (FACS) based on a NADPH biosensor. The mutant library (approximately 6×10^5 mutants each with unique barcode distributed in 3,550 genes) was constructed and characterized, and then transformed with plasmids containing the NADPH biosensor and different NADPH regenerating enzymes under an inducible promoter. Libraries were selected in NADPH consuming and regenerating scenarios, fitness/fluorescence oscillation between the two opposite redox conditions allows us to profile the intracellular NADPH level of each mutant. Thirty mutants were selected for better performance in NADPH consuming condition. Fifteen of these mutants were confirmed to have increased $NADPH/NADP^+$ ratio from 12% to 62% compared to the wildtype. These mutants have transposon insertion particularly in oxidoreductases, flagella assembly components, transcriptional regulators, or secretion proteins.

Genome-wide identification of genes involved in NADPH supply

L. Ye, S. Cardinale, S. Siedler, M. Ellabaan, S. H. Kim, M. Sommer lumye@biosustain.dtu.dk

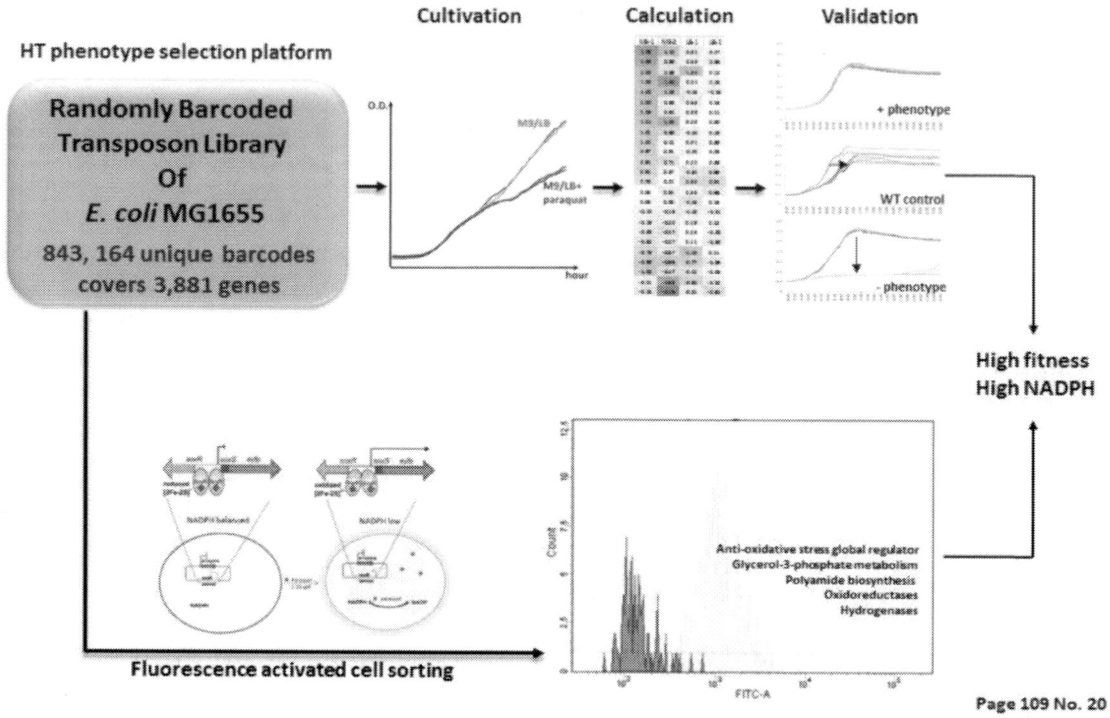

Page 109 No. 20

Engineering and Analyzing Streptomyces strain for the Recombinant Protein Production

Kashiwagi, N., 1Org. Adv. Sci. Tech., Kobe Univ.
Niimi-Nakamura, S.,
Nishioka, M.,
Hirose, S.,
Sota, M.,
Ogino, C.,
Kondo, A., Grad. Sch. Eng., Kobe Univ.

Streptomyces strains are aerobic, gram-positive and mycelia-forming soil bacteria that produce useful compounds such as secondary metabolites and secreted hydrolytic enzymes. *Streptomyces* is an attractive host for various products, for example antibiotics, antifungals, and secreted native or recombinant proteins. Genetic tools for *Streptomyces* hosts have recently been developed in strain engineering. Previously, we have reported that some recombinant proteins were overexpressed in *Streptomyces lividans* host by using the engineering vector under the control of *Streptoverticillium cinnamoneum* Phospholipase D promoter.

For the purpose to expand the cell factory system for the recombinant protein production, we investigated the following two things; 1) searching for useful enzymes from sequenced *Streptomyces*genomes and 2) improving and mechanistic understanding of the protein production.

To obtain novel useful enzymes, putative enzyme genes annotated from several *Streptomyces* genomes were expressed in the *Streptomyces*host strain. We found some novel enzymes which showed unique activity and oligosaccharide production after the characterization of these expressed enzymes.

To improve and understand the recombinant protein production, we examined the engineered and analytical approach of the host strain. We focused on the transcription factors for strain engineering. The mutation of the transcription factors is expected to change the expression levels of a wide range of the internal genes and to have an effect on the production yield. Several mutant strains were actually found to change the expression level of the recombinant protein compared to the wild-type strain. At the same time, we took the analytical approach to understand important factors during the protein production in some aspects, such as the metabolic aspect. Here, we summarized our research and findings in the presentation.

Engineering and analyzing *Streptomyces* strain for the recombinant protein production

Poster P116

Norimasa Kashiwagi (Kobe University)

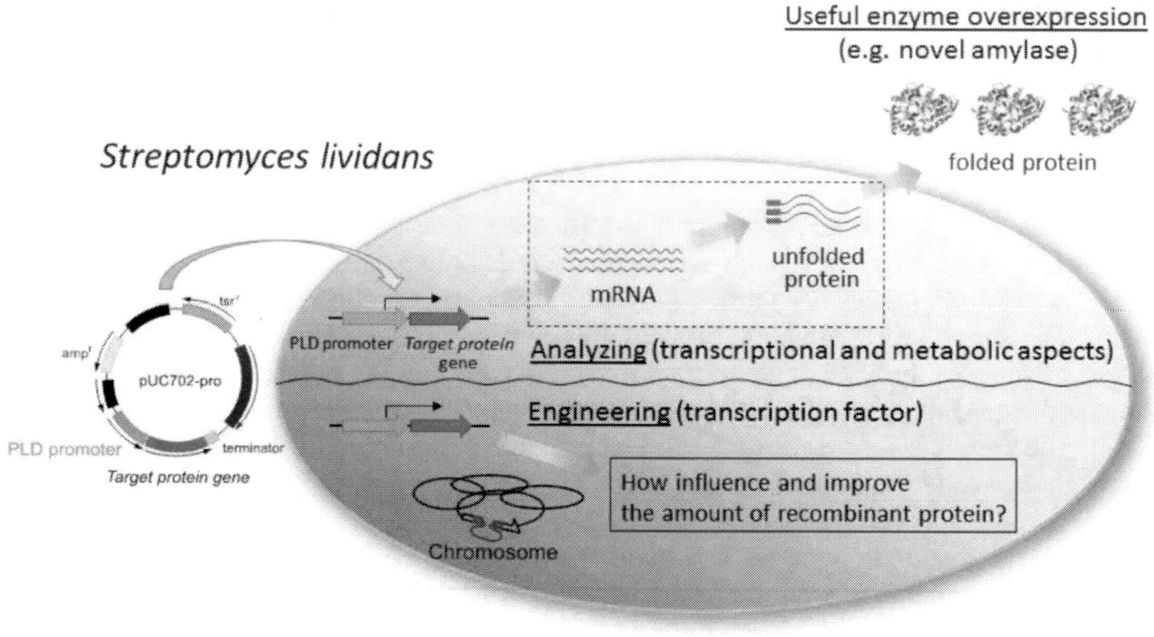

Biosensor-Driven Adaptive Laboratory Evolution of Amino Acid Production in Corynebacterium Glutamicum

Mahr, R., Forschungszentrum Juelich
Frunzke, J., Forschungszentrum Jülich

Rational engineering of microbial cell factories requires a detailed knowledge of the underlying metabolic networks. However, the high complexity of microbial physiology reduces rational targets to a low number of well-described gene functions. Random mutagenesis as well as adaptive laboratory evolution expand the toolbox of engineering approaches, which allow for a system-wide improvement of microbial production strains. So far, adaptive laboratory evolution was only applied for the improvement of fitness-linked production phenotypes, e.g. by exposure to thermal, solvent or chemical stressors (1).

In a recent study, we established a biosensor-driven adaptive laboratory evolution strategy to improve growth and production of inconspicuous, small metabolites. Sensor cells with the highest fluorescent output and hence, increased metabolite production, were iteratively isolated by fluorescence-activated cell sorting (FACS) and re-cultivated. Thereby an artificial selective pressure was imposed on the fluorescent output of the transcription factor-based biosensor. This strategy was successfully applied to the pyruvate-dehydrogenase-deficient L-valine production strain *Corynebacterium glutamicum* Δ*aceE* using the Lrp biosensor (2), which visualizes the intracellular concentration of branched-chain amino acids and L-methionine (3). Evolved strains featured improved growth, on average about 25% increased production and 3-4-fold reduced by-product (L-alanine) formation. Out of seven identified mutations, four mutations were reintroduced into the cured Δ*aceE* background. One mutation in the global regulator GlxR was attributed to the reduced by-product formation. Remarkably, a loss-of-function mutation within the urease-accessory protein UreD was found to improve L-valine production by about 100%. Altogether, these results emphasize biosensor-driven adaptive laboratory evolution as a straightforward approach to improve metabolic cell factories.

(1) Abatemarco, J. A. Hill and H.S. Alper (2013) Expanding the metabolic engineering toolbox with directed evolution. *Biotechnol J* 8, 1397-1410.

(2) Mahr, R., C. Gätgens, J. Gätgens, T. Polen, J. Kalinowski, and J. Frunzke*, (2015) Biosensor-driven adaptive laboratory evolution of l-valine production in *Corynebacterium glutamicum*. *Metab Eng* 32:184-194. doi: 10.1016/j.ymben.2015.09.017

(3) Mustafi, N., A. Grünberger, D. Kohlheyer, M. Bott & J. Frunzke*, (2012) The development and application of a single-cell biosensor for the detection of L-methionine and branched-chain amino acids. *Metab Eng* 14: 449-457.

Page 110

JÜLICH
FORSCHUNGSZENTRUM

Biosensor-driven evolution of L-valine production

Regina Mahr & Julia Frunzke

Challenge: Identification of non-intuitive mutations for metabolic engineering

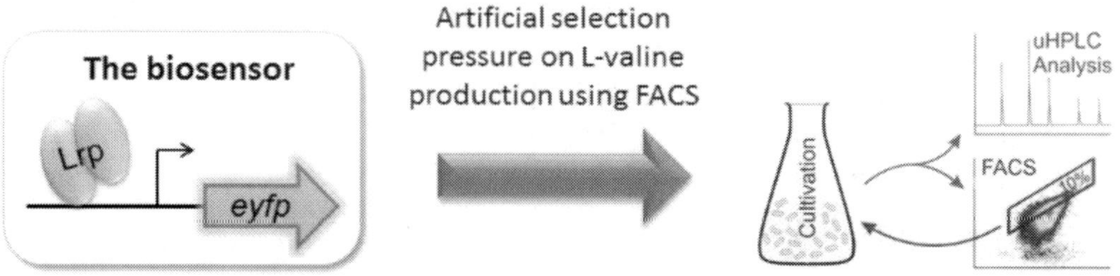

→ Improved growth and L-valine production of *Corynebacterium glutamicum*

Page 137 (session 3)

Metabolic Engineering of *Corynebacterium glutamicum* for the utilization of biomass-derived carbon and energy sources

Jan Marienhagen

Matching the Demand of Protein Production By Metabolic Engineering of the Central Metabolism: Lessons Learned from Cofactor and Amino Acid Metabolism in Yeasts

Mattanovich, D., Austrian Centre of Industrial Biotechnology (ACIB GmbH)
Russmayer, H., University of Natural Resources and Life Sciences
Nocon, J.,
Hohlweg, J., University of Natural Resources and Life Sciences
Steiger, M. G., University Leiden
Sauer, M., Austrian Centre of Industrial Biotechnology (ACIB GmbH)
Gasser, B., University of Natural Resources and Life Sciences

Metabolic engineering has been successfully applied to redirect metabolic fluxes towards desired products of the primary and secondary metabolism. Complex polymeric products like heterologous proteins also demand redistributions of primary metabolic fluxes, their rational design however is far less obvious. Therefore cell engineering for protein overproduction has concentrated mainly on transcription, codon usage, protein folding and secretion.

Amino acid synthesis has been proposed to be a potential bottleneck during the synthesis of an extra load of heterologous protein, especially when the target protein differs largely in its amino acid composition from the host proteome. We have evaluated such potential limitations in the yeast *Pichia pastoris* by a combined transcriptomics, proteomics and metabolomics approach. By focusing on genes or proteins which were upregulated and metabolites which strongly decreased in their intracellular concentration upon overexpression of several human and fungal proteins, we pinpointed 6 amino acid synthesis pathways. The identified rate-limiting genes in these pathways were overexpressed, leading not only to larger pools of free amino acid but also to a more than two-fold increase in heterologous protein production. This success could, however, not be transferred to traditional fed batch cultures. Reasons for this failure will be presented, illustrating the added complexity of cultivation to such metabolic engineering procedures.

In a complementary approach we integrated the synthesis of heterologous protein to a genome scale metabolic model and predicted metabolic engineering targets for enhanced productivity with FSEOF for gene overexpression and MOMA for gene deletion. Enhanced productivity of the target protein was verified in more than 50 % of these genetic interventions. The most beneficial mutations were related to reduction of the NADP/H pool, leading to four fold increased product formation in the best case.

Taken together, we demonstrate that genome scale metabolic modeling and metabolomics are suitable to design genetic interventions to the primary metabolism to increase recombinant protein production.

Matching the demand of recombinant protein production by metabolic engineering of the central metabolism

acib

Hannes Rußmayer[1,2], Justyna Nocon[1], Jonas Hohlweg[1,2], Matthias G. Steiger[1,2], Michael Sauer[1,2], Brigitte Gasser[1,2], Diethard Mattanovich[1,2]

[1] *Department of Biotechnology, University of Natural Resources and Life Sciences Vienna, Austria*

[2] *Austrian Centre of Industrial Biotechnology, Vienna, Austria*

Systems biology based analysis of the yeast *Pichia pastoris*

1. *Metabolic model-based predictions point to the pentose phosphate pathway as a bottleneck (NADPH supply?)*
2. *Amino acid synthesis is a major bottleneck of recombinant protein production*

Up to 4-fold increase of product titers

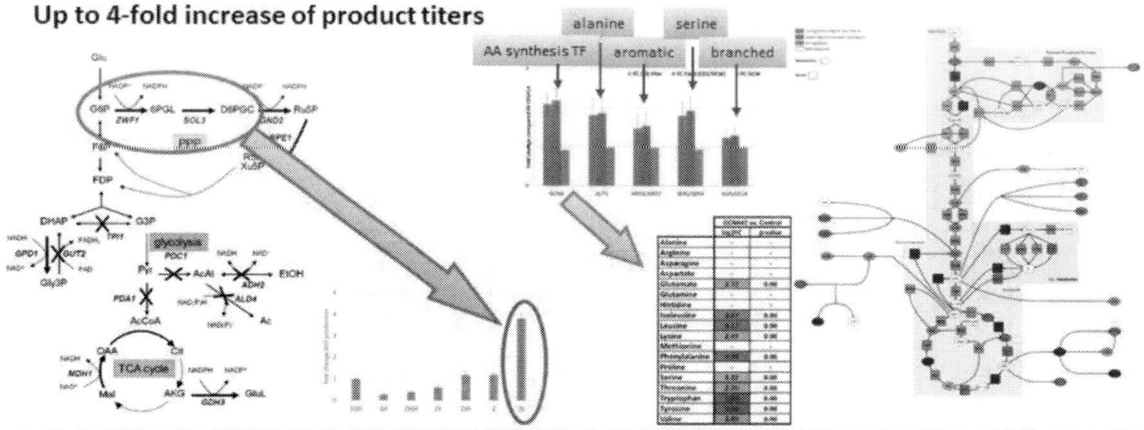

Microbial Biotechnology

Supported by the Austrian BMWFW, BMVIT, SFG, Standortagentur Tirol, Government of Lower Austria and Business Agency Vienna through the Austrian FFG-COMET Funding Programme, and by the Austrian Science Fund

FWF BioToP Biomolecular Technology of Proteins

A Toolbox for Accelerating Strain Engineering of Oleaginous Yeast Yarrowia Lipolytica

Holkenbrink, C.,
Beder, J.,
Dam, M. I.,
Borodina, I., Technical University of Denmark

The non-conventional oleaginous yeast *Y. lipolytica* is valued for its capability to accumulate high amounts of storage lipids and to degrade hydrophobic substrates. It is an attractive host for the production of lipids and fatty acid-derived compounds and for bioremediation, however the strain development is hampered by the lack of genetic tools. We seek to develop a genome editing toolbox, which will allow rapid iterative combinatorial strain engineering, i.e., repeatedly introducing combinations of multiple edits, such as gene knock-ins/knock-outs. First, we developed a set of standardized integration vectors for gene overexpression from high expression genome spots. Second, we adopted CRISPR-Cas9 technology for *Y. lipolytica* to enable genome editing with high efficiency without the need of a selection marker. We demonstrate the utility of the toolbox by engineering a strain for production of a potential platform chemical, a-ketoglutarate, from renewable feedstocks.

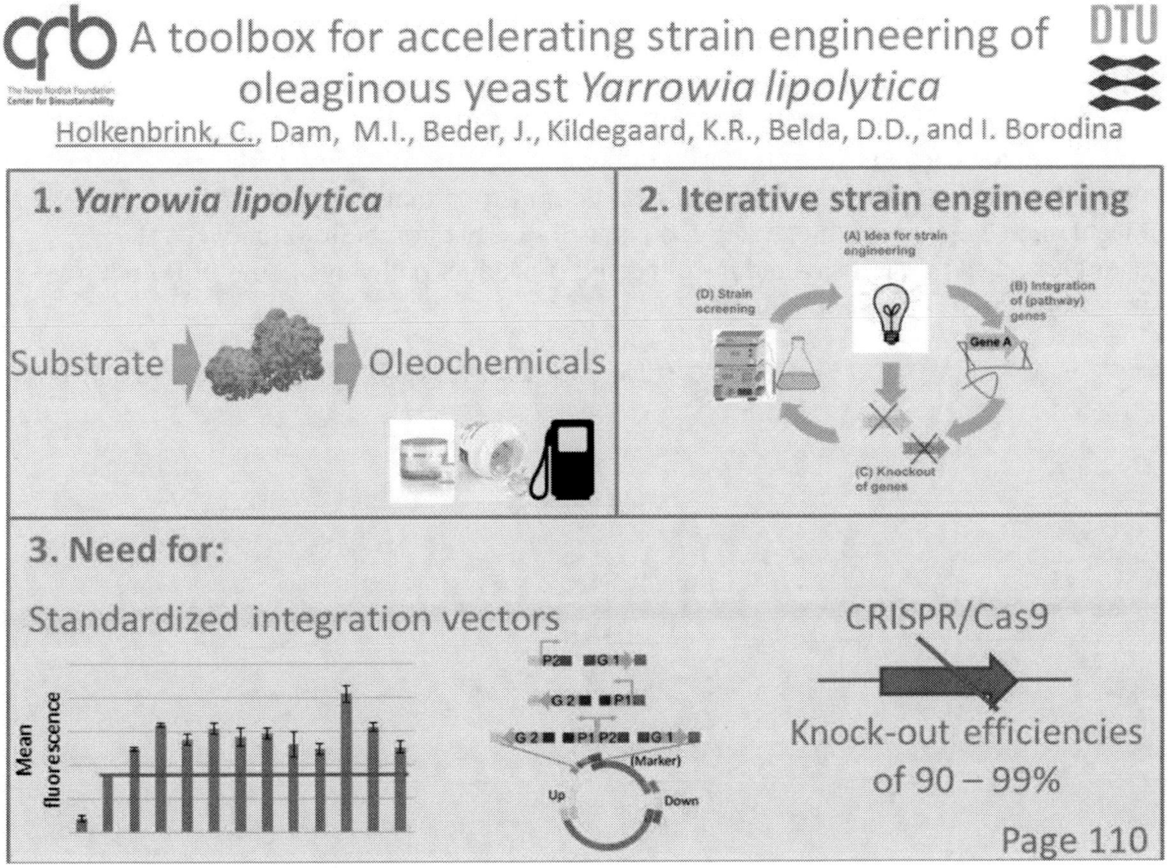

Modulating Flux-Competing Enzyme Squalene Synthase By Protein Degradation Mechanism to Improve Sesquiterpene Production in Saccharomyces Cerevisiae

Peng, B., The University of Queensland
Nielsen, L. K., The University of Queensland
Vickers, C. E., The University of Queensland

Sesquiterpenes have functions as fragrances, flavours, and semiochemicals, and as active molecules in pharmaceuticals and traditional herbs. Because of their high energy content, sesquiterpenes can also be applied as future alternative aviation fuels. Microbial fermentation is being examined as a competitive approach for bulk production of these compounds. Methods to avoid competition between production of endogenous isoprenoids and the introduced sesquiterpene are required to achieve economical yields. In the current study, the yeast *Saccharomyces cerevisiae* was used as the platform organism to produce an acyclic sequeterpene alcohol, trans-nerolidol. Nerolidol production was firstly improved by enhancing the upstream mevalonate pathway for the synthesis of the precursor farnesyl pyrophosphate (FPP). However, excess FPP produced was instead directed towards squalene by squalene synthase (Erg9p), resulting in squalene accumulation to 1 % biomass. It was shown that Erg9p is located on endoplasmic reticulum (ER) membrane through a C-terminal ER-targeted transmembrane peptide. An endoplasmic reticulum-associated protein degradation mechanism was developed to decrease cellular levels of Erg9p. This improved nerolidol titre by 86 % to ~100 mg L^{-1}. To further improve titres, the strong inducible *CUP1* promoter was applied for overexpression of the mevalonate pathway and nerolidol synthase. This resulted in a nerolidol titre of 248 mg L^{-1} in the pathway-enhanced, Erg9p-destabilized strain. In this strain, squalene levels were similar to the wild-type control strain, and there was no negative effect on cell growth. These data demonstrate that the protein degradation mechanism can be a useful strategy to balance the flux competition between the endogenous sterol pathway and introduced bio-product pathways at the FPP node for metabolic engineering applications.

Modulating flux-competing enzyme squalene synthase by protein degradation mechanism to improve sesquiterpene production in *Saccharomyces cerevisiae*

Bingyin Peng, Lars K. Nielsen, Claudia E. Vickers*

* Corresponding author c.vickers@uq.edu.au
Australian Institute for Bioengineering and Nanotechnology (AIBN), The University of Queensland, St. Lucia, QLD 4072, Australia

The University of Queensland Australia | AIBN Australian Institute for Bioengineering and Nanotechnology

Problem

Squalene synthase Erg9p drains nerolidol precursor (Farnesyl-PP) pool.

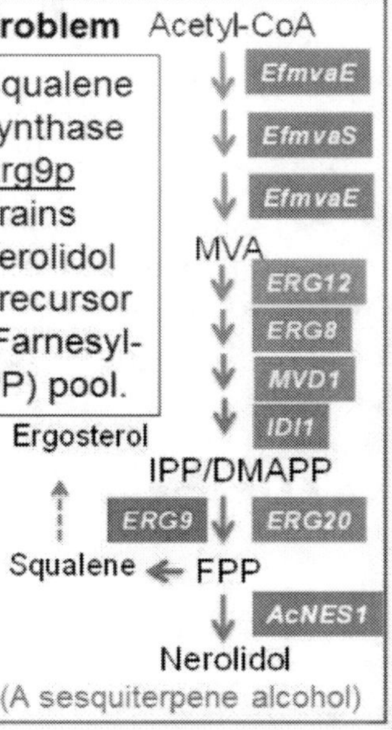

Solution

- Erg9p is an ER protein

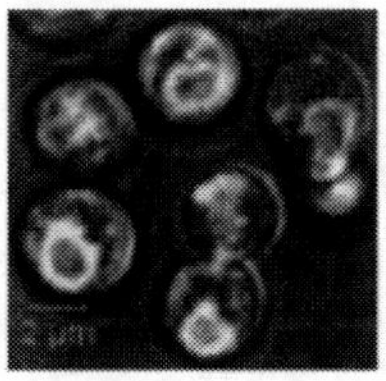

- Engineered degradation

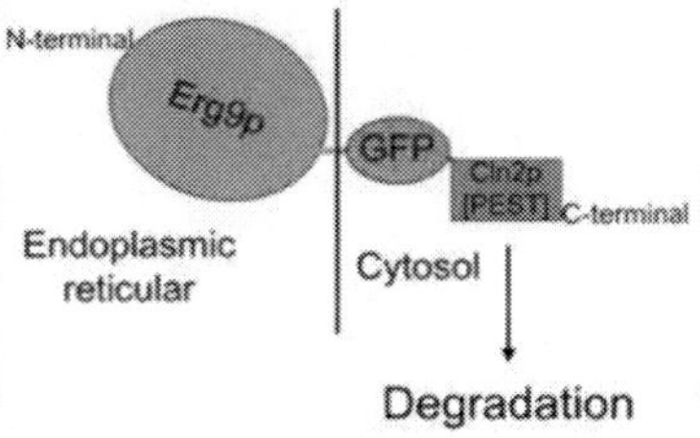

Results

- Reduced squalene contents
- Increased nerolidol production by 86%

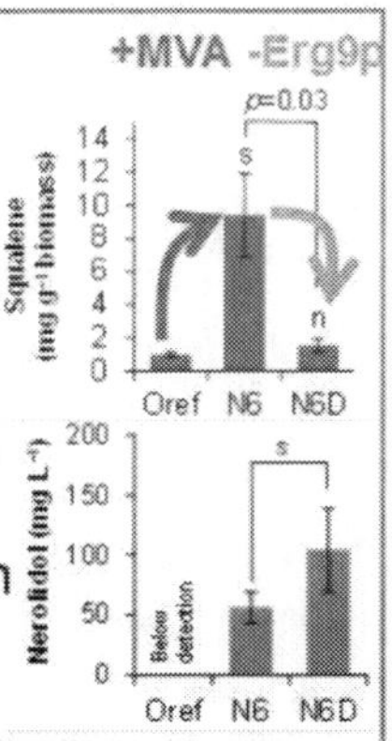

- Next-generation strain with improved transcriptional regulation yielded 5 g/L nerolidol in fed-batch cultivations

Microbial Production of Natural and Unnatural Betalain Pigments

Yamamoto, Y., MITSUBISHI CHEMICAL GROUP SCIENCE AND TECHNOLOGY RESEARCH CENTER, INC.
Miyaoku, K., Mitsubishi Chemical Group Science and Technology Research Center, Inc.

Betalain pigments are one of the most famous pigments in plant kingdom. They are comprised of red-purple colored betacyanins and yellow-orange colored betaxanthins. For their high antioxidant properties, sunlight absorbing ability, they have been isolated from plants and commercially utilized as food additives, cosmetic ingredients and dye-sensitized solar cells. However, the cost of current betalain pigments production process depending on extraction from plants is very high due to their extremely low content in plant. Therefore, development of other economically efficient industrial production process has been strongly expected. One of the solutions is considered to be microbial production process. In this study, we successfully engineered metabolic pathway of *Escherichia coli* and *Saccharomyces cerevisiae* to produce betalain pigments and confirmed that both betacyanins and betaxanthins were produced by these engineered microorganism in an appropriate culture condition. This is the first report of integrated microbial betalain pigments production from simple carbon source. This result may contribute to the development of cost efficient and sustainable industrial production process of betalain pigments.

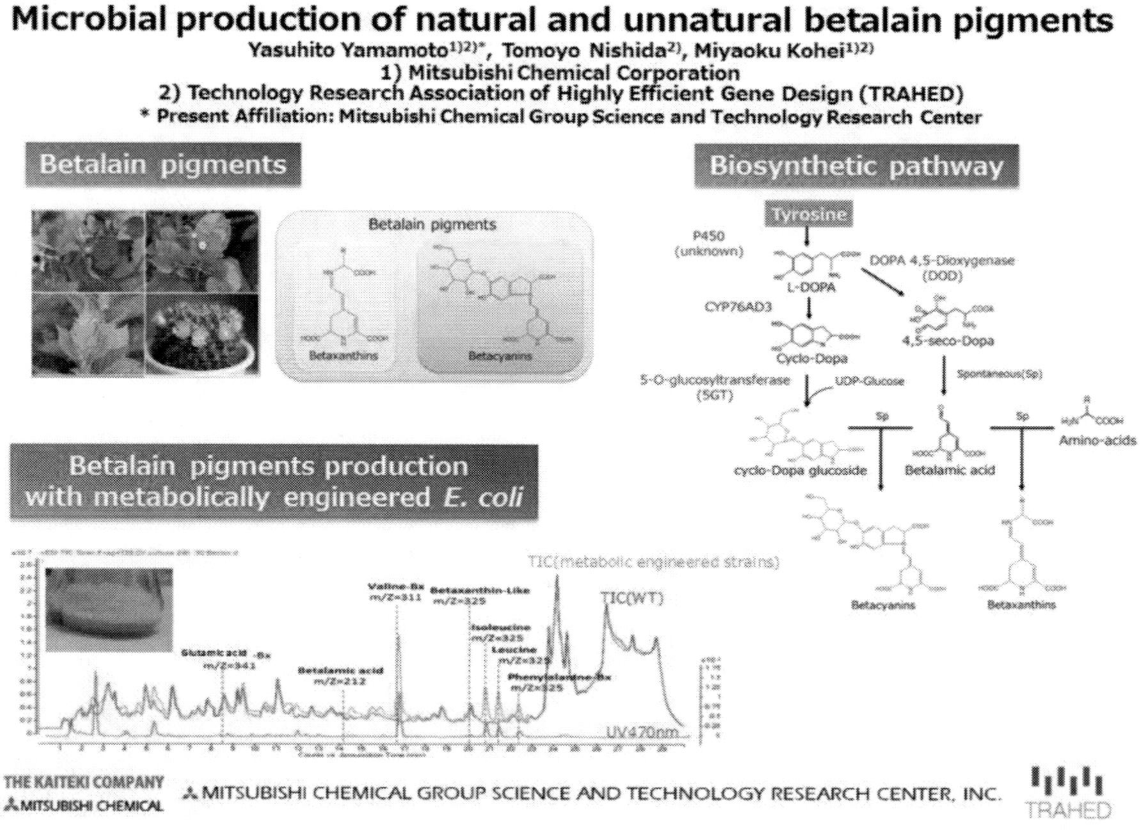

A Novel, Ultra-Sensitive Multiple Input-Output System for Target Identification in Systems Metabolic Engineering of E. coli

Song, L., Institute of Bioprocess and Biosystems Engineering, Hamburg University of Technology
Zeng, A. P., Hamburg University of Technology

Identification of relevant engineering targets is one of the key challenges in metabolic engineering. To avoid laborious try-and-error experiments, many studies have been focused on building genome-scale models to make predictions. One of the most popular systems-based tools is the genome-scale metabolic reconstruction and flux analysis. The recent studies in kinetic modeling show many difficulties and are limited to small scale networks. Metabolic and kinetic models lack the crucial information of regulation and interactions which are essential for prediction and reprograming of cellular functions. Although cells are composed of molecules and their viability relies on extracting and using energy to maintain them, they are not 'just' matter and energy. Information processing, also called "cellular computing", is essential for cellular function. Previous studies have proved that such computational abilities could be used rationally. It's interesting to ask the question "Can we rein the computation abilities of cells for systems-level prediction and optimization of microorganisms?" If so, we then avoid of building laborious mathematic models since the perfect "model" is the cell itself. The key issue is how to let the cells "to compute" the processes we are interested in and "to output" the right results corresponding to the different inputs?

Here, we aimed at reining the computation abilities of *E. coli* cells by designing multiple input and output (IO) systems. The input system is implemented by using various M13 phage derivatives which can carry out up or down regulations targeting various genes. Using a rationally designed synthetic circle, the signal changes within the cells after gene operations introduced by phage infection are linked to the phage reproduction process, which in turn is linked to the total phage population. Thus, the various signals are 'recorded' in form of the population of corresponding phage derivatives. Biological systems are complex but highly adaptive, meaning that the cells always try to reduce the perturbations introduced. For the traditional methods, we always do genetic modification and pick the expected colonies before detecting the signal changes. During this process, the cells have already made adaption and the final signal strength is lower than that while cells are making adaption. With the current IO system, the recording process happens immediately after the perturbations introduced. Thus, the signal changes during the whole adaption process could be captured, making current system more sensitive than the traditional methods principally.

For proof of concept, various gene operations related or not related to lysine biosynthesis in *E. coli* were used as inputs and the intracellular lysine concentration changes were used as output signals. Correct predictions of beneficial genetic manipulations for enhanced lysine production in *E. coli* were obtained. The IO system shows ultra-sensitivity in capturing the signal changes caused by the perturbations introduced. For example, while using operations that reduce TCA cycle activity as inputs, slight increase of intracellular lysine concentration could be detected although the strict allosteric regulations of lysine production pathways by lysine still work. This system developed in this work opens up new possibilities for target identification in metabolic engineering.

An Ultrasensitive Multiple I/O System for Target Identification, Evaluation and Evolution of Molecular Designs in Systems Metabolic Engineering

Lifu Song and **An-Ping Zeng** TUHH

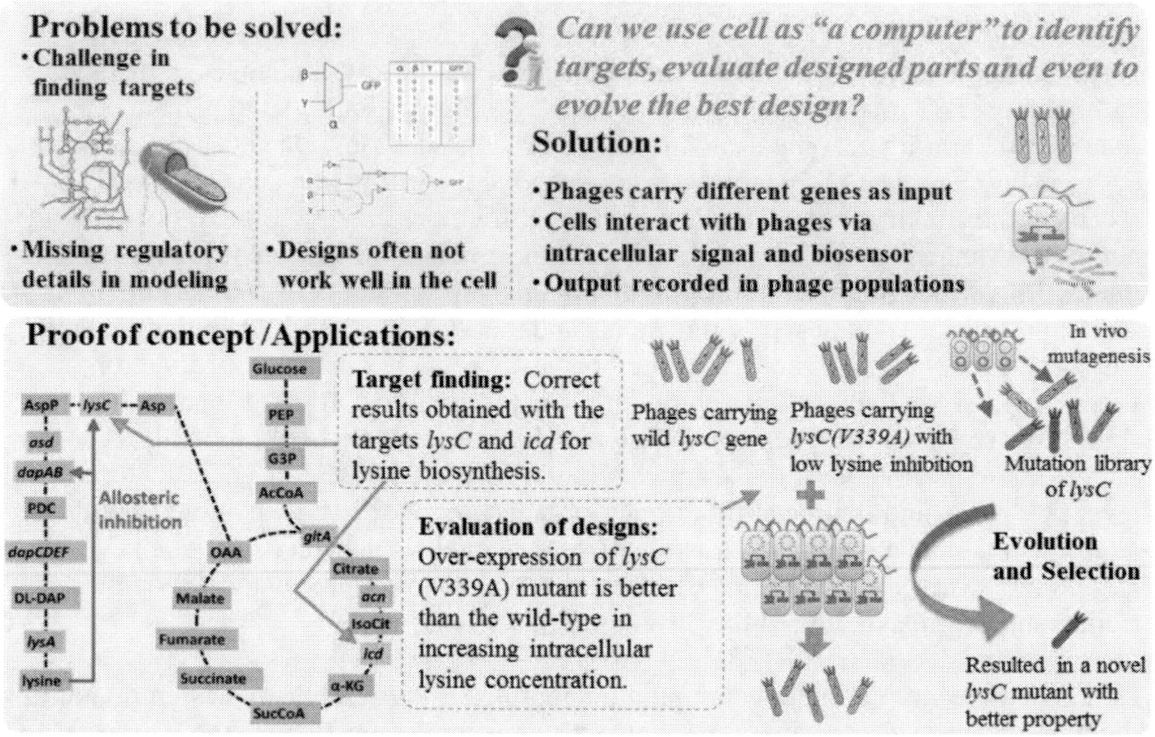

Engineering Sensors to Facilitate Multiplex Tuning of Biosynthetic Pathways

Meyer, A. J.,
Voigt, C. A., University of California, San Francisco

Metabolic pathways comprise several enzymes that progressively modify precursor molecules, resulting in a final product. Varying the expression level of each component enzyme results in tradeoffs of metabolic load, enzymatic flux, and buildup of intermediates. Therefore, in order to maximize the yield of a pathway, the optimal level of each enzyme must be determined. In this work we present a novel method for rapid exploration of such an expression landscape. Each component gene is placed under the control of a different, orthogonal small-molecule inducible system. Adjusting the concentration of each inducer alters the expression of each gene, and the entire expression landscape can be probed using a single genetic construct. In order to facilitate this approach, we continue to expand the number of inducible systems with the orthogonality and dynamic range necessary for such a system to function.

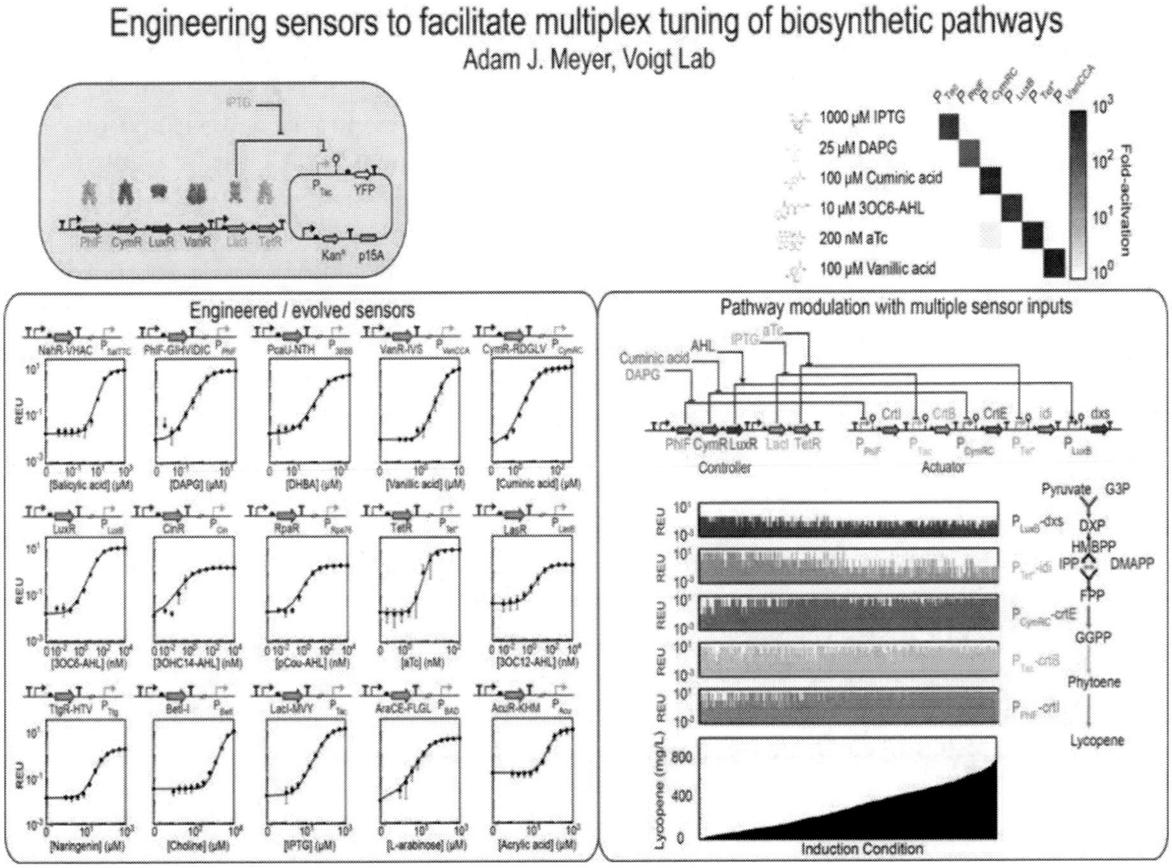

Development of Mini-Bioreactors for Evolution of Thermotolerance

Lehnen, M.,
Ebert, B. E., RWTH Aachen University
Blank, L. M., RWTH Aachen University

In biotechnological processes thermotolerance is often a crucial issue. Especially in applications of yeast energy intensive cooling is required to maintain a constant temperature level when a process produces heat. While *Saccharomyces cerevisiae* is a rather thermosensitive organism, there are also tolerant and even well-established strains like *Hansenula polymorpha* und *Kluyveromyces marxianus*, which have been shown to grow at temperatures up to 50° C [1, 2]. Although a theoretical maximum temperature for eukaryotic growth is widely accepted to be 60° C, improvement of thermotolerance is desirable in any industrial organism. As the molecular mechanisms of thermotolerance are highly complex, a direct genetic manipulation is a very difficult task and has shown only limited success in previous approaches. Adaptive laboratory evolution seems to be more promising and might also be helpful for further understanding of the molecular mechanisms. Towards this goal we developed an array of mini-scale bioreactors, which are based on a previously described morbidostat design and assembled from low-cost customized materials and commonly available electric parts [3]. The focus on thermotolerance required the development of individually controllable heating units, which was realized with perforated ceramic elements connected to separate electric circuits with controllable voltage. Each reactor comprises a standard 20 mL glass vial (12 mL working volume), equipped with in-/outlets for feed, waste and gas alongside a probe for temperature measurement. Medium and air are pumped by piezo actuated micro membrane pumps. The vial is placed in a shell, in which an LED and a detector diode are integrated to approximate biomass formation inside the reactor. Stirring is achieved by a magnetic stirrer driven by a controllable motor connected to a magnet and embedded in a bottom construction beneath the reactor. A set of four reactors is operated via an Arduino® microcontroller board and controlled by a LabView based software, equipped with an intuitive user interface.

The morbidostat operation is based on maintaining a certain level of stress induced growth inhibition by raising the stressor once the evolving culture has recovered from the previous step of stressor increase. Apart from stressor increases, reactors are operated in a sequential batch mode in a narrow range of biomass concentration, where dilution steps are triggered by a maximum threshold of the biomass signal and terminated at a minimum threshold. The time interval required to trigger the next dilution step is used to estimate growth rate and thus growth inhibition.

Besides temperature stress, alternative modes of operation allow adaptive evolution to rising chemical stressor concentrations or evolution of growth rate at constant conditions. Several security loops have been integrated into the software to provide a fully automated operation of the device between manual inoculation and harvest of the evolved culture.

References:

[1] Saraya, R. *et al.*, FEMS Yeast Res., 12:271-8, 2012.

[2] Fonseca, G. G. *et al.*, Appl. Microbiol. Biotechnol., 79:339-354, 2008.

[3] Toprak, E. *et al.*, Nat. Protocols, vol. 8:555-567, 2013.

ABBt Aachen Biology and Biotechnology RWTHAACHEN iAMB Institute of Applied Microbiology

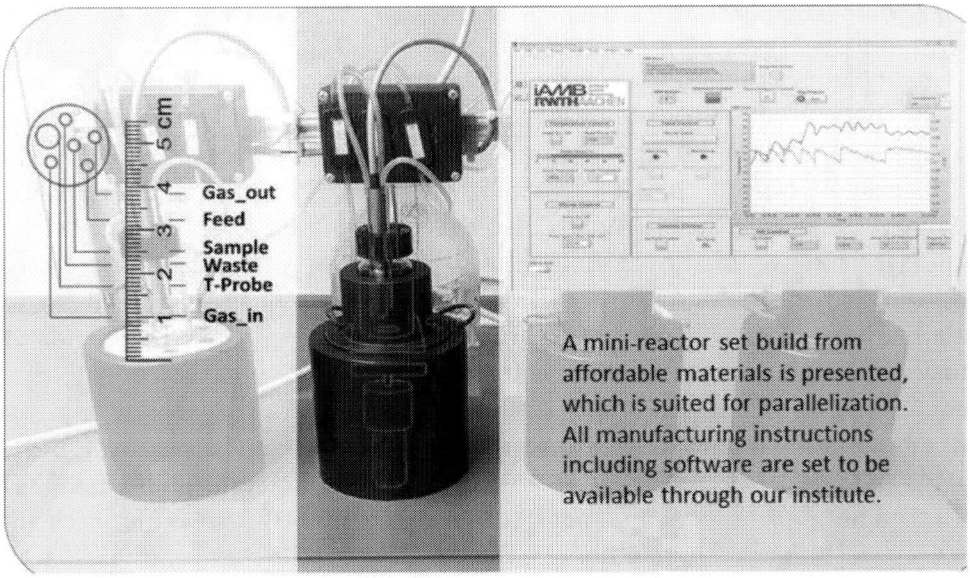

Development of Mini-Bioreactors for Evolution of Thermotolerance

Mathias Lehnen

28.06.2016

Reducing Exogenous Enzyme Requirements in Biofuels Production By Metabolic Engineering

Kumar, D.,
Singh, V., University of Illinois at Urbana-Champaign

Conventional corn dry grind ethanol process requires exogenous alpha and glucoamylases enzymes to breakdown starch into glucose, which is fermented to ethanol by yeast. Currently these alpha and glucoamylases enzymes are added externally in liquid form during the liquefaction and saccharification process respectively. This study investigates the use of genetically engineered corn and yeast, which can eliminate or minimize the use of these external enzymes and potentially reduce the operating cost and simplify the ethanol production process. The amylase corn used in the study has been genetically engineered to produce α-amylase in endosperm, which eliminates the need of external enzyme addition during liquefaction. Due to high expression levels of enzymes, only a small amount of the corn is mixed with the conventional dent corn. Similarly, the GA yeast investigated in this study has also been engineered to produce endogenous glucoamylases and provide novel metabolic pathways for high ethanol yields. The fermentation characteristics of amylase corn mix (15% amylase corn) using GA yeast (evaluated at 0% and 50% loadings of glucoamylases enzyme) were evaluated using laboratory dry grind process and the results were compared with the control treatments (100% dent corn, conventional yeast). The ethanol yield of amylase corn mix was 4.16% higher than that of dent corn using conventional yeast. For dent corn, no significant differences were observed in the ethanol yields between the control and using GA yeast without any addition of glucoamylases. Addition of only 50% glucoamylases along with GA yeast use resulted in a high ethanol concentration (18.96%) for amylase mixed corn. The ethanol yields for dent corn and amylase corn were found 3.08 and 3.04 L/kg corn respectively, which were 7.83 and 2.08% higher than that of conventional yeast at 100% glucoamylases loading. The findings from this study demonstrated the potential of getting high ethanol yields and reducing exogenous enzyme requirements using metabolic engineered corn and yeast.

From Gene to Activity - Exploring the Potential of the Nestlé Culture Collection

Bolten, C. J., Nestec S.A.
Duboux, S.,
Sybesma, W.,
Mercenier, A.,
Moine, D.,
Ngom Bru, C.,

Providing nutritious, healthy and sustainably produced food is one of the main objectives of food companies such as Nestlé. A culture collection of more than 3000 food grade strains (Nestlé Culture Collection, NCC), has been integrated into an R&D network covering microbiology, fermentation, food processing, nutrition and health, as well as clinical trials. Thanks to this approach and its research and technological capabilities, Nestlé has been able to develop and produce a large range of functional foods containing beneficial microbes.

Today the genomes of almost all NCC strains have been sequenced, assembled and annotated by using the complementary Illumina or Pacific Biosciences sequencing technologies, followed by assembly and annotation pipelines. Web based bioinformatic software enabling the storage and the analysis of the genomes have been implemented. The resulting genome databank allows direct evaluation and exploration of the NCC microbes for the development of fermented foods with enhanced taste and texture, functional benefits, and/or probiotics. The sequenced NCC can also be used to identify and source specific enzymes. In addition, the gene collection enables an easy screen for the presence/absence of undesired metabolic pathways or antibiotic resistance genes.

In this presentation we will illustrate how the newly developed bacterial genomic platform was used to go from gene to activity, bridging between *in silico*analysis and preclinical or proof of concept human trials. Examples relate to the use of lactic acid bacteria to deliver functional molecules or enzymes that could alleviate the consequences of specific food adverse reactions; identification and application of lactic acid bacteria as delivery systems for micro nutrients like iron; the use of specific lactic acid bacterial enzymes for the conversion of sugars into fibers; and the identification of metabolic routes for the production of desired flavor molecules.

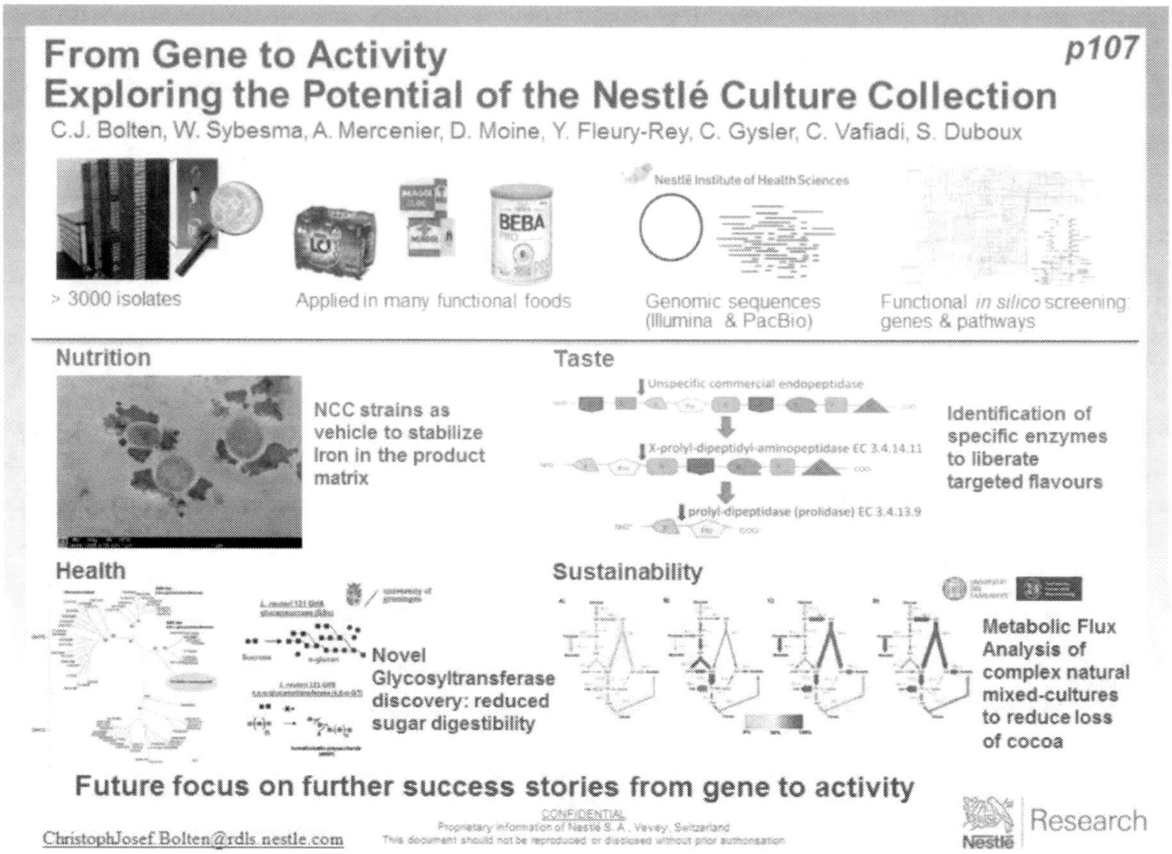
From Gene to Activity
Exploring the Potential of the Nestlé Culture Collection
p107
C.J. Bolten, W. Sybesma, A. Mercenier, D. Moine, Y. Fleury-Rey, C. Gysler, C. Vafiadi, S. Duboux
> 3000 isolates
Applied in many functional foods
Nestlé Institute of Health Sciences
Genomic sequences (Illumina & PacBio)
Functional in silico screening: genes & pathways
Nutrition
NCC strains as vehicle to stabilize iron in the product matrix
Taste
Unspecific commercial endopeptidase
X-prolyl-dipeptidyl-aminopeptidase EC 3.4.14.11
prolyl-dipeptidase (prolidase) EC 3.4.13.9
Identification of specific enzymes to liberate targeted flavours
Health
Novel Glycosyltransferase discovery: reduced sugar digestibility
Sustainability
Metabolic Flux Analysis of complex natural mixed-cultures to reduce loss of cocoa
Future focus on further success stories from gene to activity
ChristophJosef.Bolten@rdls.nestle.com
CONFIDENTIAL
Proprietary information of Nestlé S.A., Vevey, Switzerland
This document should not be reproduced or disclosed without prior authorisation
Nestlé Research

Construction of a Single-batch Fermentation System to Simulate Human Intestinal Microbiota

Sasaki, K.,
Kondo, A., Graduate School of Engineering, Kobe University
Osawa, R.,

Recently, various functional food components called prebiotics, probiotics, and biogenics have attracted attention as potential ways to manipulate the gut microbiota and to improve human health. Functionality of such food components usually has been evaluated by human intervention trials or animal feeding trials. However, human trials often are constrained by ethical considerations, while the animal feeding trials often yield results that are not reproducible in humans. Therefore, it was necessary to develop a relatively simple and short-term *in vitro* evaluation system. The aim of the present study thus was to construct a single-batch fermentation system that was as simple as possible, to prove a human intestinal model.

The anaerobic conditions in our system were carefully constructed and maintained by using a previously defined medium for the growth of anaerobes, Gifu anaerobic medium (GAM), and constantly flushing the medium with N_2/CO_2 gas before and during the fermentation. Operation of the batch fermentation system was initiated by inoculating with small amounts of human volunteer fecal sample. The total number of the eubacteria was calculated to range from 60.9×10^{11} to 128×10^{11} copies/wet-g of feces. After 24 h of fermentation, eubacterial copy numbers reached 4.86×10^{11} to 6.93×10^{11} copies/mL. 16S rRNA gene sequence analysis of bacteria grown in the system revealed that occupancies of the major phyla, including Bacteroidetes, Firmicutes, and Actinobacteria, as well as overall species diversity, were consistent with those of the original feces. At the earlier stages of fermentation (up to 9h), trace mixtures of acetate, lactate, and succinate were detectable; at the later stages (after 24 h), larger amounts of acetate accumulated along with low levels of propionate and butyrate. These patterns were similar to those observed in the original feces. Thus, this system could serve as a simple model to simulate the diversity as well as the metabolism of human intestinal microbiota.

Finally, in order to substantiate the practicality using the mode, we evaluated the functionality of prebiotics, in this case using oligosaccharides. Supplementation of the system with several prebiotic oligosaccharides (including fructo-, galacto-, isomalto-, and xylo-oligosaccharides; lactulose; and lactosucrose) yielded elevation of the population of bacteria belonging to the genus Bifidobacterium, consistent with significant increases in acetate production. Thus, the system was able to detect the known bifidogenic effects of prebiotic oligosaccharides. Therefore, our single-batch fermentation system is expected to serve as a simple but versatile tool for evaluating the functionality of various food components.

Keywords: single batch fermentation system, human intestinal model, microbiota, prebiotics

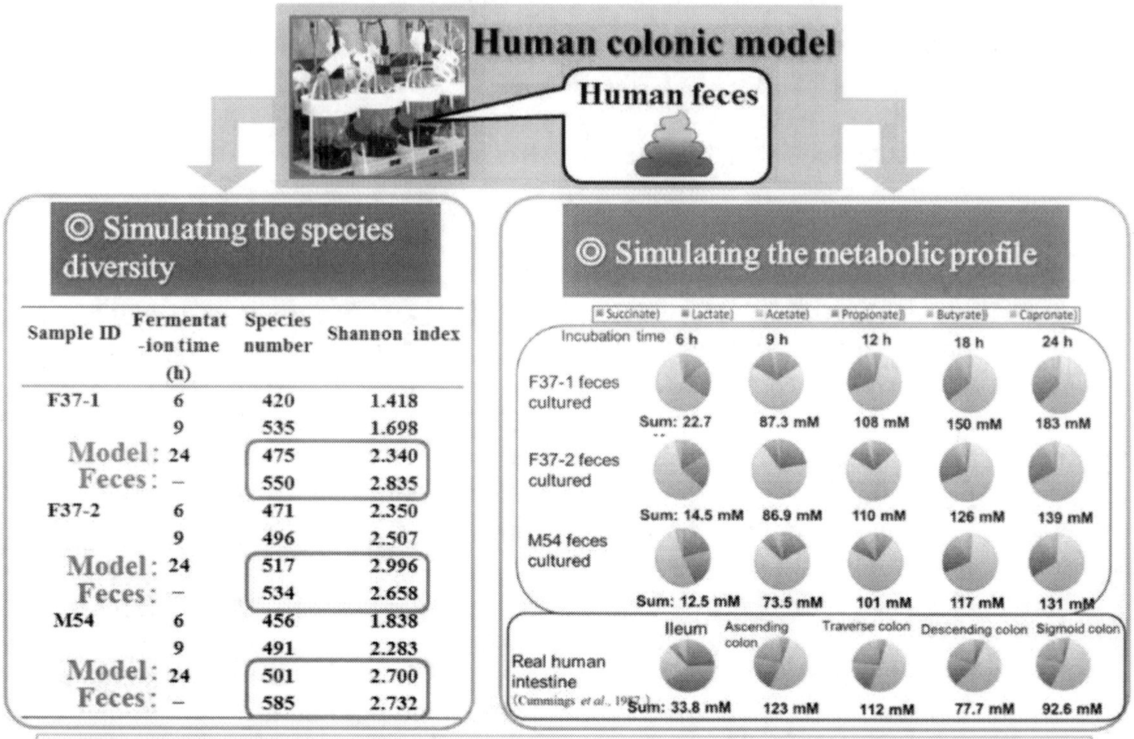

Our model serves as a tool for evaluating functionality of food components prior to *in vivo* human trials

Comparison of Metabolite Profiles of Escherichia coli and Saccharomyces Cerevisiae Depending on Minimal Vs. Complex Media

Kim, J., Korea University Graduate School
Kim, K. H., Korea University Graduate School

Metabolomics is one of the omics studies, which reflects global changes of the entire metabolites in various organisms. It directly shows the abundance of various metabolic intermediates of many biochemical pathways in response to genetic or environmental changes. In metabolomics studies, culture medium is one of the most important experimental factors because uptake of different nutrients dramatically alters the flux of metabolism. However, there is no studies of evaluating the effect of media on global metabolism of microorganisms. For this reason, we used gas chromatography/time-of-flight mass spectrometry (GC/TOF MS) to evaluate how intracellular metabolites are differently expressed depending minimal and complex media. We identified a total of 119 intracellular metabolites and performed statistical analyses including principal component and hierachical clustering analyses. The metabolite profiles of *Escherichia coli* and *Saccharomyces cerevisiae* clearly showed differences between the two groups such as cells grown in minimal vs. complex media. These results suggest that the medium complexity significantly affects the overall metabolism and metabolite profiles of the cells.

Sample Preparation Procedures for the Global Metabolite Profiling of Yeast

Song, J. H., Korea University
Kim, S., Korea University Graduate School
Kim, K. H., Korea University Graduate School

Saccharomyces cerevisiae is the most frequently used yeast strain. Metabolomics is the study of global changes of metabolome in a living organism and can be used for metabolic and physiological studies of the yeast. To obtain reliable and reproducible metabolomic data, the optimization of sample preparation methods is important. In this study, we evaluated various quenching methods and extraction solvents for the gas chromatography/time-of-flight mass spectrometry (GC/TOF MS) analysis. Among them, the cold methanol quenching methods showed significant losses of intracellular metabolites caused by severe cell membrane damage of yeast. Fast filtration method with acetonitrile/water mixture (1:1, v/v) at -20°C was the most suitable metabolome sampling method for the yeast in this study.

Microbial Platform for Glycosylation and Polysaccharide Biosynthesis

Yan, Y., the University of Georgia

Microorganisms utilize simple carbon sources such as glucose and xylose to propagate and generate a variety of small molecules, which forms the foundation of the biotechnology industry. With metabolic engineering efforts, large-scale production of these small molecules such as amino acids, antibiotics, organic acids and alcohols has been achieved through microbial fermentation. While the productivities, titers and yields of these molecules have been continuously improved through various metabolic engineering approaches, the low efficiency associated with microbial synthesis of the compounds that use sugar molecules as direct building blocks, such as polysaccharides and glycosylated molecules, remain unsolved. Within this lecture, the design, construction and optimization of a microbial platform that is suitable for supporting efficient production of polysaccharides and glycosylated molecules are presented. As an example, biosynthesis of trehalose, a disaccharide of glucose, is used to demonstrate the applicability of this platform.

Engineering Clostridium Thermocellum to Produce Ethanol from Cellulose at High Yield and Titer

Olson, D.,
Hon, S.,
Lanahan, A.,
Zheng, T.,
Lo, J.,
Zhou, J.,
Tian, L., LISBP, Toulouse University
Rydzak, T.,
Guss, A.,
Lynd, L. R., Dartmouth College

Ethanol is the biofuel produced in the largest amount worldwide. *Clostridium thermocellum* is a promising candidate for ethanol production from cellulose due to its native ability to consume cellulose, and thus engineering efforts have focused on methods for improving ethanol production. To date, we have created over 40 different strains that employ a variety of engineering strategies, however we have only been able to achieve an ethanol yield of 75% of the theoretical maximum and 25 g/l titer. To allow for further improvements, we suspected that we might need an exogenous ethanol production pathway. *Thermoanaerobacterium saccharolyticum* has been engineered to produce ethanol at 90% of the theoretical maximum yield and 70 g/l titer. We identified the genes responsible for the ethanol production pathway in this organism and transferred the pathway to *C. thermocellum.*

Enhanced Production of 2-Fucosyllactose in Engineered Escherichia coli.

Jung, S. M., Seoul National University

2'-Fucosyllactose (2-FL) is one of most abundant functional oligosaccharides in human milk, which is involved in many biological functions for human health. To date, most microbial systems for 2-FL production have been limited to use E. coli JM strains since they cannot metabolize lactose. In this study, E. coli BL21star(DE3) was engineered through deletion of the whole endogenous lactose operon and introduction of the modified lactose operon containing lacZ□M15 from E. coli K-12. Expression of genes for guanosine 5'-diphosphate (GDP)-l-fucose biosynthetic enzymes and heterologous α-1,2-fucosyltransferase (FucT2) from Helicobacter pylori allowed the engineered E. coli BL21star(DE3) to produce 2-FL with 3-times enhanced yield than the non-engineered E. coli BL21star(DE3). In addition, the titer and yield of 2-FL were further improved by adding the three aspartate molecules at the N-terminal of FucT2. Overall, 6.4 g/L 2-FL with the yield of 0.225 g 2-FL/g lactose was obtained in fed-batch fermentation of the engineered E. coli BL21star(DE3) expressing GDP-l-fucose biosynthetic enzymes and three aspartate tagged FucT2.

Enhanced Xylitol Production in Engineered Saccharomyces Cerevisiae By Dual Utilization of NADPH and NADH Cofactors

Jo, J. H.,
Kim, H. J.,

Xylitol, a natural sweetener, can be produced by hydrogenation of xylose in hemicelluloses. In microbial processes, utilization of only NADPH cofactor limited xylitol biosynthesis not to reach an industrial level. To overcome this drawback, *Saccharomyces cerevisiae* D452-2 was engineered to express two types of a xylose reductase(XR) with either NADPH-dependence or NADH-preference. Engineered *S. cerevisiae* DWM expressing both the XRs exhibited 10% higher xylitol productivity than yeast strains expressing each XR in batch culture. Furthermore, the coexpression of S. cerevisiae *ZWF1* and *ACS1* genes in the DWM-ZWF1-ACS1 strain increased intracellular concentrations of NADPH and NADH, and improved maximum xylitol productivity by 25%. Finally, the optimized fed-batch fermentation of *S. cerevisiae* DWM-ZWF1-ACS1 resulted in 196.2 g/L xylitol concentration, 4.27 g/L-h productivity and almost the theoretical yield. Expression of the two types of XR utilizing both NADPH and NADH is a promising strategy to meet the industrial demands for microbial xylitol production.

Experimental Evolultion of Escherichia coli Under Various Chemical Compounds Using High-Throughput Automated Culture System

Horinouchi, T., Quantitative Biology Center, RIKEN
Suzuki, S., Quantitative Biology Center, RIKEN
Shimizu, H., Osaka University
Furusawa, C., Osaka University

Biological systems can adapt to various environmental conditions by changing their phenotype and genotype. Experimental evolution is a powerful tool to trace phenotypic and genotypic changes in adaptive evolution. By integrating experimental evolution approach and recent omics technologies including high-throughput sequencing, we can obtain genome-wide information on the adaptation dynamics to environmental changes, which facilitate better understandings of the nature of adaptive evolution and provide possible strategies for bioengineering. However, to maintain a large number of long-term cultures is often hard at work for experimental biologists, which limits the number of cultures to be maintained simultaneously. To overcome the limitation of manual culture experiments, we developed a fully-automated culture system which can maintain massive parallel experimental evolutions by a rapid serial transfers [1]. Using this system, we performed experimental evolutions of *Escherichia coli* MDS42 strain under various stressors including acid, alkali, saline, and surfactant agent. First, we obtained stress tolerant strains which can grow faster than parent strain in each stress. Next, we performed transcriptome analysis and genome re-sequence analysis of these tolerant strains. We found that the overall gene expression changes during experimental evolution were similar among independently evolved tolerant strains for same stressors. We also found that several mutations were commonly fixed in same genes among independently evolved tolerant strains for same stressors. We evaluated the effects of commonly fixed mutations to introduce mutations in the genome of the parent strain. From these results, we can discuss about stress adaptation mechanism and breeding strategies for useful microorganisms.

[1] T. Horinouchi *et al.*, **J. Lab. Autom.**, 19(5), pp. 478-482, 2014.

Targeted Gene Deletion Using DNA-Free RNA-Guided Cas9 Nuclease Accelerates Adaptation of CHO Cells to Suspension Culture

Lee, N., KAIST
Yoo, C., KAIST
Shin, J., KAIST
Cho, S., KAIST
Cho, B. K., KAIST

Chinese hamster ovary (CHO) cells are the preferred host for the production of a wide array of biopharmaceuticals. Thus, efficient and rational CHO cell line engineering methods have been in high demand to improve quality and productivity. Here, we provide a novel genome engineering platform for increasing desirable phenotypes of CHO cells based upon the integrative protocol of high-throughput RNA sequencing and DNA-free RNA-guided Cas9 (CRISPR associated protein9) nuclease based genome editing. For commercial production of therapeutic proteins, CHO cells have been adapted for suspension culture in serum-free media, which is highly beneficial in productivity and economical aspects. To engineer CHO cells for rapid adaptation to suspension culture, we exploited strand-specific RNA-seq to identify genes differentially expressed according to their adaptation trajectory in serum-free media. More than 180 million sequencing reads were generated and mapped to the currently available 109,152 scaffolds of the CHO-K1 genome. We identified significantly downregulated genes according to the adaptation trajectory and then verified their effects using the genome editing method. Growth-based screening and targeted amplicon sequencing revealed that the functional deletions of *Igfbp4* and *AqpI* gene accelerate suspension adaptation of CHO-K1 cells. The availability of this strand-specific transcriptome sequencing and DNA-free RNA-guided Cas9 nuclease mediated genome editing facilitates the rational design of the CHO cell genome for efficient production of high quality therapeutic proteins.

Metabolic Flux Analysis in Insulin-Stimulated Adipocytes Form Temporal Patterns of Intracellular Metabolome Data

Ohno, S., University of Tokyo
Krycer, J. R., University of Sydney
Yugi, K., University of Tokyo
Quek, L. E., University of Sydney
Hirayama, A., Keio University
Soga, T., Keio University
Kuroda, S., University of Tokyo
James, D. E., University of Sydney

Background

The hormone insulin is secreted from beta-cells in the pancreas, and plays a great role for maintaining blood glucose homeostasis. In adipocytes, insulin activates glucose uptake and utilization, and lipid biosynthesis. Abnormalities of blood insulin level are widely known to be related to type-II diabetes mellitus, thus deep understanding of adipocyte metabolism by insulin stimulation has been desired. However, how metabolic fluxes change in adipocytes by insulin is poorly understood. In this study, we estimated metabolic fluxes in insulin-stimulated differentiated 3T3-L1 adipocytes from intracellular metabolome data.

Materials and Methods

3T3-L1 adipocyte cell lines were cultured, and the media were swapped with those including U-^{13}C glucose and insulin. Cell samples were obtained at 1-60 min after insulin stimulation (100 nM), and intracellular metabolite concentrations and their mass isotopomer abundances were measured by CE-MS. Control experiments without insulin addition were also examined. Measured experimental metabolome data seemed to consist of two phases, the early transient and the late sustained phase, after insulin stimulation. Then we estimated time average fluxes within each phase by a modified metabolic flux analysis approach, considering temporal patterns of metabolite concentrations and mass isotopomer fractions. The optimal time dividing two experimental phases were also estimated.

Results

^{13}C -tracer experiments for adipocyte cell lines and the modified metabolic flux analysis approach exhibited different characteristics in the estimated metabolic fluxes between the early and the late experimental phase, and between with or without insulin stimulation. Regarding experiments with insulin, large absolute fluxes from glucose to triacylglycerol through upstream of glycolysis were estimated in the early phase, which decreased by more than 80% in the late phase. In contrast, less than 30% of absolute fluxes in the early phase decreased in glycogen biosynthesis pathway in the late phase. Regarding experiments without insulin stimulation, overall absolute fluxes were estimated to be much smaller than those without insulin, but a flux ratio at the branching point of glucose-6-phosphate and triose-phosphate were similar between with or without insulin stimulation.

In summary, in adipocytes, insulin stimulation lead to a large increase in fluxes of glucose uptake, upstream of glycolysis, and triacylglycerol biosynthesis pathway, whereas increased fluxes in glycogen biosynthesis pathway were relatively small. In future, we will develop more sophisticated method to estimate dynamics of metabolic fluxes. Moreover, we will integrate the fluxome, metabolome, and phosphoproteome data which we've already obtained, for comprehensive understanding of insulin action on adipocyte cells.

Transcriptional and Translational Landscape of the Model Antibiotic Producer Streptomyces Coelicolor

Jeong, Y., KAIST
Bucca, G., Surrey
Cho, S., KAIST
Kim, S. C., Korea Advanced Institute of Science and Technology
Smith, C. P., Surrey
Cho, B. K., KAIST

Individual *Streptomyces* species have the genetic potential to produce a diverse array of natural products of considerable commercial, medical and veterinary importance. However, the genetic information flow of their unique high G+C genomes by transcription and translation processes remains largely unexplored. In order to harness their full biosynthetic potential, it will be important to develop a detailed understanding of the regulatory networks that orchestrate their diverse metabolism. Here, we reveal extensive translational control of the secondary metabolic genes of the model streptomycete, *Streptomyces coelicolor* A3(2) through the genome-scale integration of transcriptome and translatome data. Our systematic study determined 3,570 transcription start sites and identified a high proportion (~21%) of leaderless mRNAs and 230 small RNAs; this enabled deduction of promoter architecture on a genome-scale. The comprehensive translational landscape was determined by using massively parallel sequencing of ribosome-protected mRNA fragments (Ribo-seq). Interestingly, our study reveals that the translation efficiency of secondary metabolic genes was negatively correlated with transcription and that several key antibiotic regulatory genes were translationally-induced at transition growth phase. These findings could lead to the design of new approaches to antibiotic discovery and development.

Regulation of Arginine Metabolism during Soy Sauce Fermentation Process By Metabolic Engineering of Moromi Microbes

Fang, F.,
Zhang, J., Jiangnan University
Chen, J., Jiangnan University
Du, G., Jiangnan University

The industrial fermentation for fermented foods or condiments production is a process usually using a complex microflora as the producer. Lactic acid bacteria (LAB) play important roles in food mature processing and aroma components production, whereas the utilization of arginine through the arginine deiminase (ADI) pathway by LAB may cause the accumulation of citrulline, which is the precursor of ethyl carbamate, the group 2A carcinogen.

LAB utilize arginine as the supplement for obtaining energy, and ammonia released with arginine metabolism through the ADI pathway also contributes to their resistance to acidic stress. Generally the end product for ADI pathway is ornithine and not much of citrulline is produced. However, the conversion of citrulline to ornithine may not be sufficient when LAB strains are exposed to stress (i.e. osmotic stress). This is the main reason that citrulline accumulated in the moromi during soy sauce fermentation. Given the fact that genetic modification of microorganisms for food production is not acceptable, also the natural microbiota of the moromi can hardly be constantly genetic modified, metabolic engineering of the ADI pathway is done by strain interference. A *Bacillus* strain isolated from the moromi was selected for regulation of arginine metabolism during soy sauce fermentation. We successfully increased the efficacy of citrulline decomposition in the moromi with the employment of this *Bacillus* strain. The result demonstrated a good example in regulation of a substance metabolism without genetic engineering a strain, which might give a better knowledge for engineering microbial metabolism pathways in a complex microflora.

Biosynthesis of L-Arginine Using Metabolically Engineered Corynebacterium Glutamicum

Park, S. Y., Korea Advanced Institute of Science and Technology (KAIST)
Park, S. H., Korea Advanced Institute of Science and Technology (KAIST)
Kim, H. U., Korea Advanced Institute of Science and Technology (KAIST)
Kim, T. Y., Korea Advanced Institute of Science and Technology (KAIST)
Park, J. S., Daesang corporation research center
Kim, S. S., Daesang corporation research center
Lee, S. Y., Korea Advanced Institute of Science and Technology (KAIST)

For overproduction of L-arginine, *Corynebacterium glutamicum* was genetically engineered. As a strategy for enhancing cell tolerance to L-arginine, random mutagenesis was firstly conducted. After that, the genes *argR* and *farR* encoding the proteins which repress arginine expressing operon were inactivated. Then, *pgi* gene encoding phosphoglucose isomerase was downregulated by point mutation and the *opcA, pgl, tal, tat* and *zwf* genes in an opreon were overexpressed for the stronger PPP flux. For blocking the export of L-glutamate, which can be further conversed toward L-arginine, *Ncgl1221* gene encoding L-glutamate exporter protein was inactivated. The next step was enhancing expression rate of the *argF* and *carAB* genes for optimized conversion of L-ornithine to L-citrulline. The last step was overexpressing of *argGH* operon to resolve the bottleneck of arginine synthesis. The final strain went through the large scale fed-batch fermentation in 1,500 L bioreactor and produced 81 g/L of L-arginine. The strategies used here for overproduction of arginine will provide useful information in metabolic engineering of Corynebacteria strains. [This work was supported by the Technology Development Program to Solve Climate Changes on Systems Metabolic Engineering for Biorefineries from the Ministry of Science, ICT and Future Planning (MSIP) through the National Research Foundation (NRF) of Korea (NRF-2012M1A2A2026556 and NRF-2012M1A2A2026557).]

Development of a Gene Knockout System in E. coli Using Integration-Helper Plasmid

Lee, J. E., Chemical & Biomolecular Engineering, Korea Advanced Institute of Science and Technology (KAIST), Daejeon, South Korea
Song, C. W., Korea Advanced Institute of Science and Technology (KAIST)
Lee, S. Y., Korea Advanced Institute of Science and Technology (KAIST)

In this study, more rapid and efficient engineering method using integration helper plasmid in *E. coli* was developed. The integration helper plasmid, pCW611, has two recombinases which are expressed in reverse direction by two independent inducible systems. By using this system, required time and effort can be significantly reduced because the iterative transformation of the helper plasmid and curing steps are not required. We could delete one target gene in 3 days by using pCW611. To verify the usefulness of this gene manipulation system, the deletion experiments were performed for knocking out four target genes individually (*adhE*, *sfcA*, *frdABCD*, and *ackA*) and two genes simultaneously for two cases (*adhE-aspA* and *sfcA-aspA*). Also, fumaric acid producing *E. coli* strain was developed by deleting four target genes (*fumB*, *iclR*, *fumA*, and *fumC*) in 10 days as a proof-of-concept study.

(This work was supported by the Advanced Biomass Research and Development Center of Korea (NRF-2010-0029799) through the Global Frontier Research Program of the Ministry of Science, ICT and Future Planning (MSIP) through the National Research Foundation (NRF). Development of systems metabolic engineering platform technologies for biorefineries; NRF-2012M1A2A2026556 and NRF-2012M1A2A2026557) funded by the Ministry of Education, Science and Technology).

Metabolic Engineering of Corynebacterium Glutamicum for Making Cell Factory for Ornithine Biosynthesis

Lee, J. E., Chemical & Biomolecular Engineering, Korea Advanced Institute of Science and Technology (KAIST), Daejeon, South Korea
Kim, S. Y., Chemical & Biomolecular Engineering, Korea Advanced Institute of Science and Technology (KAIST), Daejeon, South Korea
Lee, J., Korea Advanced Institute of Science and Technology (KAIST)
Lee, S. Y., Korea Advanced Institute of Science and Technology (KAIST)

L-Ornithine is non-essential amino acid which has versatile applications. Here, we metabolically engineered *Corynebacterium glutamicum* ATCC 13032 strain for overproduction of L-ornithine. Firstly, the *proB* and *argF* genes were knocked out to block the conversion of L-ornithine to citrulline and block the competitive flux, respectively. Next, the *argF* gene encoding the protein which repress the L-arginine operon was knocked out to increase the ornithine production. The engineered strain produced 230 mg/L of L-ornithine from glucose as carbon source in flask cultivation. This strain was further enhanced by overexpression of the *argCJBD* genes from arginine overproducer *C. glutamicum* ATCC 21831 strain using expression plasmid resulting 7.19 g/L of L-ornithine production. To make enough NADPH pool, the carbon flux was changed to the pentose phosphate pathway. For redirection of flux, the *pgi* and *zwf* genes were overexpressed by changing the start codons and replacing the native promoter of *tkt* operon with strong *sod* promoter. The final strain YW06 (pSY223) went through the fed-batch cultivation and produced 51.5 g/L of L-ornithine after 40 h with the productivity of 1.29 g/L/h. This results represents the possibility of producing L-ornithine efficiently with *C. glutamicum* strains. [This work was supported by the Technology Development Program to Solve Climate Changes on Systems Metabolic Engineering for Biorefineries from the Ministry of Science, ICT and Future Planning (MSIP) through the National Research Foundation (NRF) of Korea (NRF-2012M1A2A2026556 and NRF-2012M1A2A2026557).]

Further Improvement of Bacillus subtilis cell factory Producing scyllo-Inositol, a Promising Therapeutic Agent for Alzheimer's Disease

Tanaka, K., Kobe University
Natsume, A.,
Takenaka, S., Kobe University
Yoshida, K. I.,

scyllo-Inositol (SI) is one of the inositol stereoisomers, rare in the nature, and expected as a promising disease-modifying therapeutic agent for Alzheimer's disease. On the other hand, *myo*-inositol (MI) is another inositol stereoisomer most abundant in nature and thus supplied cheap from agricultural wastes including rice bran.

By using *Bacillus subtilis* strains with genetic modifications in inositol metabolism, we have developed the bioconversion processes from MI into SI. Previously, in a modified strain inactive in inositol catabolism, the constitutive and simultaneous overexpression of IolG and IolW, which are the two inositol dehydrogenases responsible for the conversion, enabled us to demonstrate an efficient conversion with a rate up to 10 g/ L/ 48 h.

In order to devise further elevation in the conversion efficiency, we attempted to improve the substrate uptake by overexpressing *iolT* for the major MI transporter. In addition, *E. coli pntAB* encoding the membrane-bound transhydrogenase was introduced aiming at enhanced supply of NADPH required for the rate-limiting IolW reaction. These two additional modifications successfully elevated the conversion efficiency with an improved rate up to 18 g/ L/ 48h.

13c-Metabolic Flux Analysis for Mevalonate-Producing Strain of Escherichia coli

Wada, K., Osaka University
Toya, Y., Osaka University
Matsuda, F., Osaka University
Shimizu, H., Osaka University

Introduction

Mevalonate (MVA) is used to produce various useful products such as drugs, cosmetics and food additives, therefore the biological production of MVA is highly desired. An MVA-producing strain of *Escherichia coli* was constructed by introducing *mvaES* genes from *Enterococcys faecalis*. These enzymes catalyze the condensation of acetyl-CoA and reduction of them using NADPH. The distribution of flux at acetyl-CoA node and the flux of NADPH generation to drive the MVA biosynthesis pathway may effect on the MVA productivity. In the present study, the effect of MVA production on the central carbon metabolism was investigated by ^{13}C-metabolic flux analysis. This analysis is a method used to estimate the intracellular reaction rate of pathways based on ^{13}C-labelling experiment.

Materials and methods

The control and engineered strains were cultured in synthetic medium with [1-^{13}C] glucose as sole carbon source. The ^{13}C-experiment of proteinogenic amino acids at the exponential growth phase was measured by GC-MS. Metabolic fluxes were estimated by minimizing the residual sum of squares between the experimentally measured and model predicted ^{13}C-enrichment. The 95% confidence intervals of each flux were calculated by a grid search method. These flux estimations were performed by in-house software OpenMebius. The relationship between MVA production rate and each flux variability (flux solution space) was calculated on Matlab using CobraToolbox.

Results and discussion

The engineered strain produced 1.84 mmol/gDCW/h yielding 22% (C-mol/C-mol) of MVA from glucose in the aerobic exponential growth phase. Acetate production rate in the engineered strain was ten times lower than the control strain and MVA was produced by only the engineered strain. The response to MVA production in central metabolic pathway was discussed the differences of flux distribution between control and engineered strains. Although 45% and 43% of acetyl-CoA was diverted to acetate production and to the tricarboxylic acid cycle in the control strain, whereas the ratios were decreased to 4% and 27% in the engineered strain. It seems that acetyl-CoA was diverted to MVA synthesis pathway in the engineered strain due to overexpression of enzymes of the pathway. NADPH production is important for MVA synthesis as it provides the reducing power for MVA synthesis. Although the oxidative pentose-phosphate pathway is considered as the NADPH generating pathway in *E. coli*, the flux was not significantly different between the control and engineered strains. The production/consumption balance of NADPH suggested that most NADPH for MVA synthesis was obtained from the transhydrogenase reaction in the engineered strain. The flux solution space predicted that repression of citrate synthase is effective for enhancing MVA production.

Increased Acetyl-CoA Derived Production in Yeast through the Use of an Optimized Heterologous Phosphoketolase Pathway

Bergman, A., Chalmers University of Technology
Chen, Y.,
Siewers, V., Chalmers University of Technology
Nielsen, J., Chalmers University of Technology

Petroleum is a declining resource and easily accessible reserves are assumed to be depleted in a foreseeable future, which is why it is important to develop fuels derived from renewable resources. One alternative is to develop and use cell factories, which can directly convert plant-based material into fuels, of which a successful example is ethanol production by yeast. However, since ethanol cannot cover the full range of applications of liquid fuels today, it is also important to put effort in developing renewable energy dense fuel compounds, such as alkanes or fatty acid ethyl esters.

This work investigates overproduction of fatty acids - the precursor for energy dense biofuels - in *Saccharomyces cerevisiae* through utilization of an alternative route for cytosolic acetyl-CoA production in yeast. The aim is to find efficient enzymes that catalyze the formation of acetyl-CoA directly from intermediates in the glycolysis and pentose phosphate pathway (PPP), which potentially could also provide additional NADPH required as a cofactor for fatty acid biosynthesis. Therefore, an *in vitro* evaluation of several heterologous phosphoketolases and phosphotransacetylases was conducted, and catalytically efficient enzymes were selected for *in vivo* confirmation. The most effective candidates displayed beneficial effects on growth rate when expressed in a pyruvate decarboxylase negative strain, likely due to the increased capacity of the strain to produce the essential cytosolic acetyl-CoA. Furthermore, selected candidates will be expressed in a strain with an acetyl-CoA carboxylase modified to have decreased posttranslational inactivation - thus yielding more of the fatty acid precursor malonyl-CoA - and the produced fatty acids will be quantified. In summary, the optimized heterologous route is capable of redirecting carbon flux towards C2-synthesis in yeast, and constitutes a promising tool to improve formation of acetyl-CoA derived products.

Identification and Functional Characterisation of Novel Aspergillus Niger and Trichoderma Reesei Sugar Transporters
Sloothaak, J.,

Identification and functional characterisation of novel *Aspergillus niger* and *Trichoderma reesei* sugar transporters

J Sloothaak[1], JAT Tamayo-Ramos[1], DI Odoni[1], T Laothanachareon[1,2], VAP Martins dos Santos[1], PJ Schaap[1]

Abstract

Microbial cell factories can be used to replace the petrochemical production of chemicals and fuels. The use of second generation feedstocks as substrate avoids competition with the demand for food and feed. Effective transport of the constituent sugars into the cell factories is considered a bottleneck, so molecular engineering designs must consider sugar porters. The filamentous fungi *Aspergillus niger* and *Trichoderma reesei* are able to utilise a large range of sugars found in plant waste biomass, and are among the most important workhorses of today's industrial biotechnology. Both fungi have over 200 predicted sugar porter proteins, but the characteristics of these remain largely unknown. We aim to identify and characterise the most important transporter proteins for the uptake of the constituent sugars in second generation feedstocks.

We used a combination of *in silico* and *in vivo* approaches to select likely transporter candidates. Hidden Markov models, trained with previously characterised transporters, were used to predict transporter function in the *A. niger* and *T. reesei* proteomes *in silico*. Quantitative membrane proteomics, performed on *Aspergillus niger* growing on various carbon sources, was used to obtain abundance profiles of putative sugar transporters *in vivo*. Likely transporter candidates were selected and expressed in a suitable *S. cerevisiae* host to study the function *ex vivo*. Plate and liquid growth profiles were obtained and uptake characteristics were determined using radioactively labeled sugars.

Several novel *Aspergillus niger* and *Trichoderma reesei* sugar porters of the main constituents of second generation feedstocks were characterized. These transporters can be applied in the design of improved and novel cell factories for the sustainable production of biochemicals and biofuels.

Affiliations

[1] Laboratory of Systems and Synthetic Biology, Wageningen University, Stippeneng 4, 6708WE, Wageningen, The Netherlands

[2] Enzyme Technology Laboratory and Integrative Biorefinery Laboratory, National Center for Genetic Engineering and Biotechnology (BIOTEC), National Science and Technology Development Agency, Thailand Science Park, 113 Pahonyothin Road, Pathumthani 12120, Thailand

Sugar Uptake and Kinetics in Aspergillus niger

Lameiras, F., Technical university of Delft
Heijnen, J. J., Technical university of Delft
van Gulik, W., Technical university of Delft

Aspergillus niger predominantly grows on dead decaying plant material which mostly consists of polysaccharides like hemicellulose and pectin. These structural heteropolysacharides represent a significant part of agriculture waste that is currently insufficiently used.

Having the capacity to degrade plant cell wall wastes, this filamentous ascomycete can hydrolyse sugar polymers due to high levels of excreted enzymes, and further ferment both hexose as well as pentose monomer substrates.

Besides glucose, sugar monomers in hemicellulose can include xylose (in higher amounts), mannose, rhamnose, arabinose, and small amounts of galacturonic acid. Pectins on the other hand are rich in galacturonic acid. The mentioned substrates were selected for kinetic and uptake studies.

Preliminary experiments consisted of batch cultivations on both single carbon sources and an equimolar mixture of these. All carbon sources have been characterized in terms of biomass yields and maximum growth rates. Interestingly, in combination, some of the carbon sources were consumed simultaneously and some sequentially.

Previously, a well-defined chemostat protocol and tools for quantitative metabolomics were developed and optimized for *A. niger* (Lameiras, Heijnen, & van Gulik, 2015). With these tools, an accelerostat experiment was performed on a mixture of six sugars for characterization of the transport kinetics. It was found that the uptake of glucose, xylose and mannose could be described with Michaelis Menten kinetics, while the uptake of arabinose, galacturonic acid and rhamnose appeared to be inhibited by the presence of other sugars.

References:

Lameiras, F., Heijnen, J. J., & van Gulik, W. M. (2015). Development of tools for quantitative intracellular metabolomics of Aspergillus niger chemostat cultures. *Metabolomics*, *11*(5), 1253–1264. doi:10.1007/s11306-015-0781-z

Electrochemistry of Oral Plaque Microbes: A New Insight into the Oral Peridontis Disease

Naradasu, D. Jr., The university of tokyo, Japan

Name : NARADASU DIVYA
Nationality : INDIAN
Theme : Electron transportation study in oral cavity/ plaque

Nowadays number of people are suffering from oral diseases caused by microbes which usually grow on our teeth resulted in formation of plaque on the bone or its dissolution.

The bone or teeth dissolution by microbial flora is stimulated by protons accumulation in the plaque biofilm. The acidification is rapid and the supragingival pH can reach around 4 within several minutes, which has been considered due to the amassing of lactic acid with pK_a of 3.9 (Fig. 1). However, recent microbial community analysis revealed that there are 1200 Ð 70000 kinds of microbial strains in the plaque biofilm, indicating the microbial community is much more complex than previously thought and thus lactate should be rapidly consumed as electron donor. Thus the mechanism how microbial flora make acidic condition at the mineral surface remained ambiguous.

We recently found that one of the microbes in the plaque biofilm is capable of performing Extracellular electron transport (EET) during their anaerobic respiration process consuming lactate as a sole electron donor. Extracellular electron transport is the mechanism in which microbes couple their metabolism with electron transport process via the outer membrane-bound cytochrome c-type cytochromes across the outer membrane from inside of cell.

Thus, EET by microbes adhered to the surface of gum or bone possibly triggers proton accumulation in the biofilm by consuming lactic acid, acetic acid and formic acid that mutans bacteria produce.

Biofilm stay in the depth of bone, which makes the treatment of plaque difficult, and the microorganism found in plaque is known to be predominant in formation of thick nanowires and secretes extracellular polysaccharides to sustain the biofilm from previous studies.

With the help of these studies, we may find out why protons are being accumulated while electrons are being transferred and how microbes are changing their redox environment to acidic condition and still surviving under acidic stress. Thus, a way to suppress, remove the plaque, and kill the microbes by altering their nature like changing or manipulating the redox environments can be found.

Sequential (multiple) Gene Knockout in Clostridium Acetobutylicum Using Mobile Group II Intron

Cho, S. J., Chemical & Biomolecular Engineering, Korea Advanced Institute of Science and Technology (KAIST), Daejeon, South Korea
Jang, Y. S., Korea Advanced Institute of Science and Technology (KAIST)
Lee, J., Korea Advanced Institute of Science and Technology (KAIST)
Lee, S. Y., Korea Advanced Institute of Science and Technology (KAIST)

Clostridium acetobutylicum is one of the promising microorganisms for the efficient production of renewable chemicals and biofuels. However, due to the lack of efficient genetic manipulation tools, strain improvement has been rather slow. Fortunately, mobile group II intron was successfully applied to gene knockout of *Clostridium* species, such as *Clostridium perfringens* (Chen group) and *Clostridium acetobutylicum* (Minton group and Yang group) in 2005 and 2007, respectively, for the first time. Nevertheless, construction of a multiple gene knockout mutant was not easy, due to the limit of resources for genetic engineering, such as selection marker and temperature-sensitive origin of replication. Since the knockout system based on the mobile group II intron was constructed in a replicable plasmid, curing of the plasmid was required prior to the disruption of the next gene. However, the curing process is not easy, since the replication of the knockout vector is rather stable. We developed a multiple gene-knockout system that does not require marker pop-out process by using a mobile group II intron *Ll.ltrB*and two different antibiotics markers, erythromycin and thiamphenicol resistance genes. By using this strategy, a quintuple knockout mutant has been developed, recently. We will discuss an improved sequential (multiple) gene knockout strategy. [This work was supported by the Technology Development Program to Solve Climate Changes on Systems Metabolic Engineering for Biorefineries from the Ministry of Science, ICT and Future Planning (MSIP) through the National Research Foundation (NRF) of Korea (NRF-2012M1A2A2026556 and NRF-2012M1A2A2026557); by the C1 Gas Refinery Project funded by the MSIP through the NRF of Korea (2015M3D3A1A01064918); and the Advanced Biomass R&D Center of Korea (2011-0028386) through the Global Frontier Research Program of the MSIP.]

Synthesis of a Fitc Conjugated Anti-gp120 Aptamer for the Detection of HIV Infected Cells By Flow Cytometry

Alexandre, K. B., Council for Scientific and Industrial Research
Mufhandu, H. T.,

Aptamers are nucleic acid molecules generated by systematic evolution of ligands by exponential Enrichment or SELEX. These molecules potentially have an important diagnostic and analytical application given their ability to bind different targets with high affinity and specificity. Currently, antibodies are the most commonly used molecules for these purposes in applications such as flow cytometry and ELISA. However, antibodies do also have various limitations. For example, they are often expensive to generate as they require the use of animals; and some target molecules may have limited immunogenicity. It is in this context that aptamers can be used as substitutes for antibodies since they are generated by a relatively inexpensive chemical synthesis. They can also be designed to bind any target.

The viral protein gp120 plays an important role in HIV-1 life cycle as it promotes the virus entry into target cells. Once a productive infection is established infected cells can be distinguished from uninfected ones by their expression of gp120 on their cellular membrane.

Thus, the aim of this study is to synthesize a FITC-conjugated anti-gp120 aptamer for the detection of HIV-1 infected cells by flow cytometry.

We isolated a gp120 binding RNA aptamer and showed that it inhibit HIV-1 infectivity. We showed that this aptamer tightly binds a consensus HIV-1 subtype C gp120. We mapped its binding sites on gp120 and found that it interacts with the coreceptor binding site on the viral glycoprotein, at the base of the V3 loop, and in the bridging sheet within the conserved V1/V2 stem-loop. We now plan to conjugate this oligonucleotide to fluorescein (FITC) by standard solid phase chemistry using cyanoethyl phosphophoramidites. The ability of the FITC-conjugated aptamer to specifically bind gp120 will be tested by flow cytometry using the cell lines 293T expressing gp120 on their surfaces. Parent 293T cells, not expressing gp120, will be used as controls.

The final outcome of this project is a FITC-conjugated aptamer capable of detecting HIV-1 infected cells by flow cytometry.

Development of Tools for Precise Genome Engineering and Biosynthetic Pathway Construction in Lactic Acid Bacteria

Dudnik, A., Technical University of Denmark
Forster, J., Technical University of Denmark

Strains of Lactic Acid Bacteria (LAB) have a broad range of applications in food industry, including manufacturing of cheese, sausages, and other fermented products. Due to their robustness and stress tolerance, LAB are also being explored as potential candidates for production of fine chemicals within a frame of several projects, including BacHBerry. The latter is focused on identification and production of novel high-value plant-borne polyphenolic compounds, such as flavonoids using bacterial cell factories. Many LAB strains are easily genetically accessible, as there exist efficient transformation protocols, as well as expression vectors, and classical tools for genome modification. Major drawbacks of currently available genome modification strategies are that they are time consuming, require several rounds of selection, and in most cases only a single locus can be targeted at a time. Recent studies were able to overcome these issues by using CRISPR/Cas (Clustered, Regularly Interspaced, Short Palindromic Repeats – CRISPR-associated proteins) and recombineering (recombination-mediated genetic engineering). The main goal of this project is to assemble a toolbox for rapid engineering of *L. lactis*, a well-studied species of LAB, and construction of efficient production strains. First we plan to introduce and optimize a system for genome engineering of *L. lactis* in order to have a tool for rapid site-directed mutagenesis, as well as insertion and deletion of genetic elements. Moreover, we also aim at establishing a set of compatible vectors that would allow simultaneous and tuneable co-expression of complex biosynthetic pathways. Obtained materials would be further used for construction and optimization of platform strains for production of polyphenolic compounds.

Production of Isopropanol-Butanol-Ethanol As a Fuel By Metabolic Engineering of Clostridial Species

Cho, S. J., Chemical & Biomolecular Engineering, Korea Advanced Institute of Science and Technology (KAIST), Daejeon, South Korea
Lee, J., Korea Advanced Institute of Science and Technology (KAIST)
Jang, Y. S., Korea Advanced Institute of Science and Technology (KAIST)
Lee, S. Y., Korea Advanced Institute of Science and Technology (KAIST)

1-Butanol is an attractive biofuel that can replace gasoline, but its low octane rate might lead to abnormal combustion. In the present study, we developed engineered *Clostridium acetobutylicum* strains that can produce high amount of isopropanol-butanol-ethanol (IBE) mixture. In addition to introduction of the adhB-593 gene encoding NADPH-dependent primary/secondary alcohol dehydrogenase, co-overexpression of the hydGB-593 gene encoding putative ferredoxin:NADP+ reductase improved IBE titer further. The final strain was employed as a host for a large-scale fermentation to examine its possibility for industrial IBE production. Even though the genomes of many clostridial strains harbor the orthologous genes to hydGB-593, *C. acetobutylicum* does not harbor this gene. Our results suggest that *C. acetobutylicum* might have a different mechanism of NADPH regeneration from those in other clostridial species, and the introduction of hydGB-593 might enhance the NADPH regeneration in *C. acetobutylicum*. [This work was supported by the Technology Development Program to Solve Climate Changes on Systems Metabolic Engineering for Biorefineries from the Ministry of Science, ICT and Future Planning (MSIP) through the National Research Foundation (NRF) of Korea (NRF-2012M1A2A2026556 and NRF-2012M1A2A2026557) and by the C1 Gas Refinery Project funded by the MSIP through the NRF of Korea (2015M3D3A1A01064918).]

How to Characterize and Exploit Intracellular Transport Processes By Metabolic Engineering

Steiger, M. G., University Leiden
Punt, P. J., TNO Microbiology and Systems Biology
Ram, A. F. J.,
Mattanovich, D., Austrian Centre of Industrial Biotechnology (ACIB GmbH)
Sauer, M., Austrian Centre of Industrial Biotechnology (ACIB GmbH)

The *in vitro* characterization of transmembrane carrier proteins is challenging, because it normally requires the overexpression and reconstitution into liposomal systems. Here, we present a new strategy making use of an *in vivo* metabolic engineering approach to describe the substrate specificity of a mitochondrial carrier namely MttA of *Aspergillus terreus.* This mitochondrial carrier protein is involved in the biosynthesis of itaconic acid. We analyzed the transport specificity of MttA making use of different metabolically engineered *Aspergillus niger* strains. It was found that MttA preferentially transports *cis*-aconitic acid over citric acid and does not transport itaconic acid. The expression of MttA in selected *A. niger* strains results in secretion of aconitic acid (up to 9 g/L). Furthermore, the mitochondrial localization of this protein is confirmed using fluorescence microscopy. MttA can be used in further strain engineering strategies to transport *cis*-aconitic acid from the mitochondria to the cytosol for the production of itaconic acid or related metabolites.

Thus, metabolic engineering can be used for both the *in vivo* characterization of transport protein functions like MttA and to make use of this protein by creating aconitic acid producing strains.

Bridging Adaptive Laboratory Evolution and Metabolic Engineering for Construction of an Improved Glycerol Utilizing Saccharomyces Cerevisiae Strain

Strucko, T., Technical University of Denmark
Zirngibl, K., EMBL
Jouhten, P., EMBL Heidelberg
Mohamed, E. T., DTU, Center for Biosustainability
Feist, A., Technical University of Denmark
Patil, K. R., EMBL
Forster, J., Technical University of Denmark

With increasing interest in biosustainable technologies, the demand for converting available non-saccharide carbon sources most efficiently is increasing. Highly abundant crude glycerol, a major waste residue in biodiesel production, has attracted attention as an alternative carbon source for microbial fermentation processes. The most commonly known microbial cell factory, the yeast *Saccharomyces cerevisiae*, has been extensively applied for the production of a wide range of scientifically and industrially relevant products using saccharides as carbon source. However, popular wild-type laboratory yeast strains, commonly applied in metabolic engineering studies, cannot utilize glycerol as a sole carbon source. Moreover, previous attempts of rational metabolic engineering approaches did not result in significantly improved glycerol utilization in *S. cerevisiae.*

In this work, an adaptive laboratory evolution approach to obtain *S. cerevisiae* strains with an improved ability to grow on glycerol was applied. A broad array of evolved strains, which exhibited a significant increase in the specific growth rate and a much shortened lag phase, were isolated. The best performing strains were further analyzed by using whole genome re-sequencing and classical genetics in order to identify causative mutations. A set of mutations in genes previously not linked to glycerol utilization was identified, which allowed successful re-engineering of the wild-type CEN.PK113-7D. In conclusion, the knowledge acquired in this study may be further applied for rational *S. cerevisiae* strain improvement for using glycerol as a carbon source in industrial biotechnology processes.

This work is a part of the *DeYeastLibrary* consortium financed by ERA-IB

DeYeastLibrary - Designer yeast strain library optimized for metabolic engineering applications http://www.era-ib.net/deyeast-library

Transcriptional Regulation during Lipid Accumulation in Yarrowia Lipolytica

Kerkhoven, E., Chalmers University of Technology
Pomraning, K., Pacific Northwest National Laboratory
Baker, S. E., Pacific Northwest National Laboratory
Nielsen, J., Chalmers University of Technology

Oleaginous yeasts such as *Yarrowia lipolytica* are capable of accumulating lipids up to 70% of their biomass, predominantly in the form of triacylglycerols (TAGs), and this has fuelled interest in exploiting these fungi for the production of biodiesel. To optimise yields, we are studying the metabolic fluxes and their regulation in *Y. lipolytica*in different growth conditions on a genomic scale.

A multi-factorial chemostat experiment was designed, including two growth conditions (nitrogen and carbon limitation) and two strains (a wild-type and a DGA1 overexpressor). These cultures were sampled for various omics technologies (RNAseq, proteomics, metabolomics, lipidomics) and integrative data analysis was performed with this comprehensive dataset and the aid of a genome-scale model of *Y. lipolytica* metabolism.

Lipid accumulation is substantially increased upon nitrogen limitation and DGA1 overexpression, however, both transcriptomics and proteomics indicate the lack of regulation of expression levels of lipid biosynthetic enzymes. The genes that were differentially expressed showed a strong correlation between transcript and protein levels, indicating that many of these are under transcriptional regulation.

The strongly correlated genes were enriched for involvement in amino acid biosynthesis. More precisely, specifically leucine biosynthesis was transcriptionally downregulated upon high lipid accumulation in the DGA1 strain during lipid accumulation. This was further supported by metabolomics data, which indicated the intermediate of leucine biosynthesis 2-isopropylmalate as overflow metabolite during nitrogen limitation in the WT strain, while absent at high lipid accumulation during nitrogen limitation in the DGA1 strain.

While the lipid biosynthetic pathway is not transcriptionally regulated at high lipid accumulation, leucine biosynthesis is specifically down-regulation, having the effect that carbon flux is redirected from amino acid biosynthesis towards lipid accumulation. This suggests new and unexpected targets for further metabolic engineering.

Comprehensive Analysis of Glucose and Xylose Metabolism in Escherichia coli Under Aerobic and Anaerobic Conditions By 13c-Metabolic Flux Analysis

Gonzalez, J., University of Delaware
Antoniewicz, M. R., University of Delaware

Glucose and xylose are the two most abundant sugars derived from the breakdown of lignocellulosic biomass. In this work, we have applied parallel labeling experiments and integrated ^{13}C metabolic flux analysis (^{13}C-MFA) to elucidate metabolism of glucose and xylose in *E. coli* under aerobic and anaerobic conditions. While aerobic glucose metabolism is well understood, until now, there have been only a handful of studies focusing on anaerobic glucose metabolism in *E. coli*, and no ^{13}C-MFA studies on xylose metabolism have been published. As such, this work fills a gap in current understanding regarding metabolism of these two important sugars. Specifically, in this study we performed parallel labeling experiments with [1,2-^{13}C]glucose, [1,6-^{13}C]glucose, [1,2-^{13}C]xylose and [5-^{13}C]xylose, the four isotopic tracers that were determined to be optimal for ^{13}C-MFA. These same tracers were used to perform ^{13}C-MFA of an *E. coli* strain that can co-utilize xylose and glucose, creating a complete picture of the metabolism of these sugars. Additionally, isotopic labeling experiments were performed with [U-^{13}C]glucose and [U-^{13}C]xylose to measure, for the first time, turnover of macromolecules including proteins, RNA and lipids. A common assumption in ^{13}C-MFA is that turnover of macromolecules can be neglected. Here, we demonstrate that this assumption is largely valid for *E. coli* grown under aerobic conditions. However, under anaerobic growth conditions, significant turnover of lipids was detected for both glucose and xylose. Additionally, there was significant turnover of RNA when cells were grown anaerobically on xylose. Taken together, this study provides valuable new information on metabolic conversions of two important sugars by *E. coli* that can be used for future metabolic engineering efforts.

13c Metabolic Flux Analysis of the Green Microalga Chlorella vulgaris Under Autotrophic, Mixotrophic, and Heterotrophic Conditions

McConnell, B. O., University of New Hampshire
Antoniewicz, M. R., University of Delaware

One of the most pressing grand challenges facing humanity is to sustainably support a human population that is predicted to increase to 9 billion people by mid-century. Meeting this challenge will require the development of new and innovative renewable production systems. One approach to address the problems of dwindling reserves of non-renewable resources and rising levels of carbon dioxide (CO_2) in the atmosphere is to capture concentrated CO_2 emerging from waste treatment facilities, petroleum refineries, coal and other energy sources and convert the CO_2 into useful products, including dietary supplements, medicinals, and biofuels. Microalgae species represent a potential renewable source of such products due to their ability to use light to fix CO_2 into biomolecules.

Chlorella is a genus of unicellular, non-motile, photosynthetic, eukaryotic microalgae which can survive in fresh and salt water. In regards to microalgal biodiesel production economics, non-polar lipid content is the most important factor and one of the best lipid producing *Chlorella* species is *C. vulgaris*. *Chlorella vulgaris* UTEX 395 is a model green microalga with great biodiesel production potential and a sequenced genome. However, due to inefficiencies in carbon metabolism, critical technical bottlenecks still remain to economic production of desirable products. Therefore, it is important to obtain a quantitative understanding of *Chlorella vulgaris* UTEX 395 metabolism using ^{13}C Metabolic Flux Analysis (^{13}C-MFA).

In this work, we performed labeling switch experiments and took advantage of the fact that *C. vulgaris* can grow autotrophically, mixotrophically, and heterotrophically and can easily change between these modes of growth. First, cells were grown heterotrophically on fully ^{13}C labeled glucose. Next, the fully ^{13}C labeled cells were transferred into fresh media. Autotrophic, mixotrophic, and heterotrophic unlabeling conditions were used. No significant turnover of proteinogenic amino acids was observed during CO_2-enriched autotrophic unlabeling or heterotrophic unlabeling. However, during autotrophic unlabeling with air there was significant turnover of proteinogenic amino acids related to photorespiration (alanine, glycine, serine, aspartate, and glutamate) consistent with low concentration of CO_2 near RuBisCO.

Next, metabolic fluxes were quantified using ^{13}C-MFA. In these experiments, we also measured net production and consumption of CO_2 and O_2 in real time using an on-line process mass spectrometer to further improve flux resolution. For heterotrophic and mixotrophic growth, labeled glucose tracers were used. For photosynthetic growth, a combination of labeled and unlabeled carbon dioxide and amino acids were used as substrates. This is in contrast to the current ^{13}C-MFA methods for photosynthetic growth which use high concentrations of ^{13}C labeled bicarbonate. With our approach, we were able to quantify significant photorespiratory fluxes which are not captured using other methods. The metabolic flux maps enabled us to identify inefficiencies in central carbon metabolism under the different growth conditions and suggest metabolic engineering targets to improve *Chorella vulgaris*'s metabolic efficiency.

13c-Metabolic Flux Analysis of the Pentose Phosphate Pathway Using GC-MS Analysis of RNA and Glycogen

Au, J., University of Delaware
Long, C. P., University of Delaware
Antoniewicz, M. R., University of Delaware

The pentose phosphate pathway (PPP) plays a key role in cellular metabolism, generating precursors for amino acid and nucleotide biosynthesis, as well as cofactors in the form of NADPH. Additionally, the PPP has been the target of metabolic engineering strategies for the production of various industrially relevant compounds. ^{13}C- Metabolic flux analysis (^{13}C-MFA) has proven to be a powerful tool in gaining a quantitative understanding of metabolic pathways. However, there are inherent difficulties in measuring fluxes in the PPP due to the reversibility of the sugar phosphate interconversions and the relative paucity of measured fragments derived from upper central carbon metabolism in typical GC-MS analysis of proteinogenic amino acids.

In this work, we present a method to incorporate glycogen and RNA ^{13}C-labeling to gain significant new observability in the PPP for flux analysis. These measurements provide direct information on glucose 6-phosphate (G6P) and ribose 5-phosphate (R5P), respectively. A significant advantage over existing methods is that relatively little biomass is required for these measurements. Two fragments are obtained for each sugar, providing additional positional labeling information. These methods were applied for the analysis of both *E. coli* and CHO cells, using parallel labeling experiments that have been specifically optimized for this purpose. In each system, ^{13}C-MFA was able to resolve net and exchange PPP fluxes using only glycogen and RNA data. This approach reduces the amount of biomass and sample analysis required by providing direct labeling information of PPP intermediates. Overall, this method provides a rigorous technique for measuring PPP fluxes. More generally, these measurements can also be added to compliment commonly used amino acid labeling information for precise ^{13}C-MFA.

Elucidating Physiology of Complex Microbial Systems through Co-Culture 13c-Metabolic Flux Analysis

Gebreselassie, N. A., University of Delaware
Antoniewicz, M. R., University of Delaware

Microbial communities carry out efficient biological transformations that are the result of multiple complementary metabolic systems working together. In bioprocess applications, co-culture systems have unique advantages over mono-culture systems in optimizing substrate utilization and product yield, i.e. co-culture systems allow for the independent optimization and assembly of various complementary metabolic capabilities into functional systems, where the diversity of metabolic pathways and ability of microorganisms to exchange metabolites dramatically increases the space of possible metabolic conversions.

Co-culture systems are now increasingly applied in metabolic engineering to optimize utilization of mixed substrates, for production of mixed products, and distribution of complex metabolic pathways among species. Among the many omics tools used to gain insight into the physiology of these systems, fluxomics is the most direct and relevant method to study the *in vivo* metabolic state of co-cultures. In the past, ^{13}C-metabolic flux analysis (^{13}C-MFA) of co-culture systems required physical separation of proteins or cells to resolve individual population fluxes. In this work, we demonstrate for the first time that it is possible to determine metabolic flux distributions in multiple species simultaneously without the need for physical separation of cells or proteins, or overexpression of species-specific products. Instead, metabolic fluxes for each species in a co-culture are estimated directly from isotopic labeling of total biomass obtained using conventional mass spectrometry approaches such as GC-MS. In addition to determining metabolic fluxes, the approach quantifies the relative population size of each species in a mixed culture and interspecies metabolic exchange, thus enabling detailed studies of species dynamics and interactions in co-cultures.

We will demonstrate application of our novel co-culture ^{13}C-MFA approach on the following three diverse and industrially relevant co-culture systems:

i. A co-culture system in which each species consumes the same substrate but metabolizes the substrate through different pathways, specifically a co-culture of *E. coli* knockouts *Δpgi* and *Δzwf*.
ii. A co-culture system in which each species consumes the same substrate but produces a different product, specifically a co-culture of wild-type *E. coli* and *S. cerevisiae.*
iii. A co-culture system in which one species consumes a particular substrate and produces a by-product that serves as the substrate for the second species, specifically a glucose consuming *E. coli* and a glucose non-consuming *E. coli* strain (*ΔptsI*) that uses acetate produced by the first strain as its carbon source.

Through the use of co-culture ^{13}C-MFA, we are able to elucidate the metabolic state of the various co-culture systems as well as investigate species dynamics and interspecies metabolite exchange. Moreover, an improved metabolic network model of *S. cerevisiae* for ^{13}C-MFA was developed through parallel tracer experiments. In summary, our work provides a powerful novel approach to study co-culture systems that will greatly facilitate rational design of co-culture systems in metabolic engineering.

Decoding the Genome-Wide Regulatory Landscape of Escherichia coli in Response to Various Environmental Stresses

Seo, S., Seoul National University
Palsson, B. O., University of California, San Diego

Understanding the genome-wide regulatory network in response to environmental stress is prerequisite to rationally design microbes surviving against stresses such as metal iron depletion, oxidative stress, low pH stress, and osmotic stress. However, the full regulatory potential of microbes remains undefined. Here, we comprehensively reconstructed transcriptional regulatory networks of several transcription factors related to iron depletion, oxidative stress, low pH stress, and osmotic stress in *Escherichia coli* K-12 MG1655 by using genome-wide measurements. Integrative data analysis revealed that each transcription factor exhibits complex role in many fundamental cellular processes to coordinately respond to iron starvation stress, low pH stress, oxidative stress, and osmotic stress as well. These types of new knowledges can be further utilized to find proper engineering targets to improve cellular tolerance towards theses stresses. Also, we believe that the incorporation of these types of comprehensive operon structures that account for cellular regulation along with regulatory network into current computational model would make it possible to mechanistically model and predict the complex regulatory interactions and thus allow us to more accurately compute complex phenotypes and design genomes for synthetic cells in the future.

Applied Metabolic Engineering of Saccharomyces Cerevisiae for the Manufacturing of Diverse Valuable Classes of Molecules

Salerno, G., Evolva SA

Nature, and notably so foods, contains a treasure trove of small molecule ingredients that can improve health, wellness and nutrition. But most of these ingredients have "issues": the plant or animal that makes the ingredient is too rare, too hard to grow or does not make enough of it. Hence, the ingredient is not available at the right quality or price that allows most of us to benefit from it in a sustainable manner.

Evolva is leader in metabolic engineering of *Saccharomyces cerevisiae* yeast for the production by fermentation of a diverse array of small molecule ingredients. Evolva has introduced a number such molecules to the market, such as Vanillin, Resveratrol, Valencene and Nootkatone and is on the way to introduce even more, such as Stevia sweeteners and Saffron ingredients. , All of these molecules are manufactured by fermentation from abundant inexpensive raw materials and delivered to the market via a reliable cost-effective production at consistent quality.

Of particular interest to Evolva are the terpenes and one class of these is constituted by the naturally ubiquitous sesquiterpenes (C15). These are known to work a in a wide range of high value applications, including flavors and fragrances, cosmetic products, biofuels, pharmaceutical products and insect control. In spite of proven efficacy there has, however, been relatively little commercial development of sesquiterpene ingredients, mostly due to their high production cost. We will present a number of Evolva ingredient projects in the terpene space, with particular emphasis on a range of sesquiterpenes.

evolva

Applied metabolic engineering of *Saccharomyces cerevisiae* for the manufacturing of diverse valuable classes of molecules

Ângela Carvalho, Klavs Riishede Hansen, Esben Halkjær Hansen, Jean Amick, Giovanni Salerno, Jørgen Hansen. Evolva Biotech A/S, Lersø Parkallé 42-44, DK-2100 Copenhagen Ø – DK

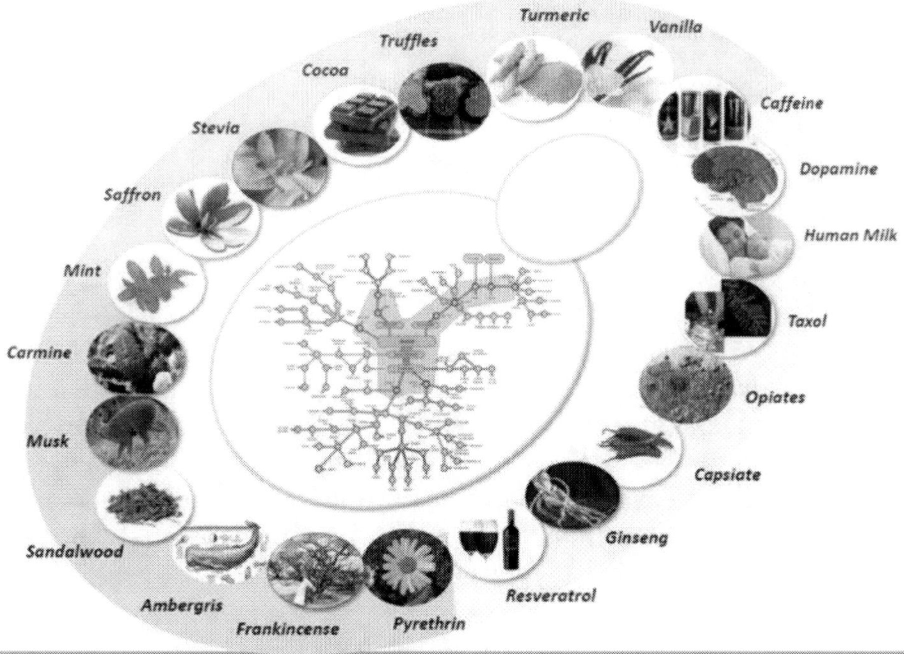

Evolva Confidential Material

Slide 1

Bioconversion of Methane By the Aerobic Methanotrophic Bacterium Methylomicrobium Buryatense

Lidstrom, M. E., University of Washington
Puri, A., University of Washington
Chu, F., University of Washington
Gilman, A., University of Washington
Fu, Y., University of Washington
Hendershott, M., University of Washington
Beck, D., University of Washington
Kalyuzhnaya, M. G., San Diego State University
Yan, X.,

Biological conversion of biogas-derived methane to liquid products is a promising strategy for the production of fuels and chemicals from an inexpensive renewable resource. The aerobic methane-oxidizing bacterium *Methylomicrobium buryatense* is a robustly growing, alkaliphilic organism with a naturally high lipid content that was originally isolated on natural gas from a soda lake in Russia. We have developed a set of genetic tools to carry out metabolic engineering in this methanotroph, and applied these plus metabolic models and systems analysis to further increase the lipid content of *M. buryatense.* Our results demonstrate that this bacterium can potentially be used as a platform for generating fuels and chemicals from methane

Redirecting Metabolic Pathway By UTR Engineering to Improve L-Tyrosine Production in Escherichia coli

Min, B. E., POSTECH
Jung, G. Y., POSTECH

L-tyrosine is a commercially important compound in the food, pharmaceutical, chemical, and cosmetic industries. Microbial production of L-tyrosine have led to higher production yield of L-tyrosine, but it still remain to be achieve for optimizing the pathway such as translation-level expression control and carbon flux rebalancing around phosphoenolpyruvate (PEP) node. Here, we redirected metabolic pathway by altering gene expression levels to improve L-tyrosine production in *Escherichia coli.* To optimize the L-tyrosine biosynthetic pathway, a synthetic constitutive promoter and a synthetic 5′-untranslated region (5′-UTR) were introduced for each gene of interest to allow for control at both transcription and translation levels. Carbon flux rebalancing was achieved by controlling the expression level of PEP synthetase using UTR Designer. The L-tyrosine productivity of the engineered *E. coli* strain was increased through pathway optimization resulting in 3.0 g/L of L-tyrosine titer, 0.0354 g L-tyrosine/h/g DCW of productivity, and 0.102 g L-tyrosine/g glucose yield. Thus, this work demonstrates that metabolic pathway redirecting by UTR engineering is an effective strategy for improvement of L-tyrosine production.

Oxygenation Cascade Analysis in Conversion of N-Octane Catalyzed By Cytochrome P450 CYP102A3 Mutants at P331 Site

Wang, S.,
Liu, Z., Institute of Chemical and Engineering Sciences

To improve the hydroxylation efficiency of cytochrome P450 CYP102A3 to substrate *n*-octane, a previously reported triple mutant F88V/S188Q/A330V with altered regioselectivity was selected and site-saturation mutagenesis was performed at its P331 site which is adjacent to one of the active sites A330. Using whole-cell biotransformations to analyze the created mutants, four better mutants P331A, P331T, P331S and P331V were achieved and exhibited significantly improved conversion rate towards octane, which are 33%, 33%, 22% and 28%, respectively. Whereas the activity of P331R was greatly reduced and P331K gave almost zero conversion of *n*-octane. Besides the main product octanols, different octanones and 1,7-octanediol were also detected for some of the mutants. The above results demonstrated that the P331 site of CYP102A3 also plays an important role on the octane oxidation and CYP102A3 is a functionally flexible biocatalyst that can be optimised for a variety of industrial applications.

Adaptive Laboratory Evolution of Product-Tolerant Hosts for Enhanced Production of Biobased Chemicals

Lennen, R. M., University of Wisconsin-Madison
Mohamed, E. T. T.,
Malla, S.,
Chekina, K., Technical University of Denmark
Sommer, M.,
Feist, A., Technical University of Denmark
Nielsen, A. T.,
Herrgård, M. J., Technical University of Denmark

Product toxicity is frequently encountered as a major barrier to achieving economically viable cell factory processes. Even relatively non-toxic products can impose significant cellular stresses at economically-relevant titers, often in excess of 100 g/L, that are required for bulk biobased chemicals such as polymer precursors and fuels. To address this issue, we have utilized a robotic platform to rapidly evolve parallel populations of *E. coli* K-12 MG1655 for enhanced growth in the presence of toxic concentrations of 11 chemicals representing diverse functional classes (including dicarboxylic acids, diamines, monocarboxylic acids, diols, and aromatic acids) that have significant interest as polymer precursors, chemical intermediates, and biofuel precursors. Whole genome resequencing was performed for over 220 isolates from 85 independently evolved populations. Based on these results, causative loss-of-function mutations were determined by screening knockout libraries and generation of combinatorial knockouts, while causative coding mutations were determined in several strains by conjugation-based genome shuffling. Cross-compound screening of all evolved isolates has also revealed specific sets of mutations that confer tolerance toward multiple classes of compounds. We have inferred numerous novel mechanisms of tolerance to different compound classes through the use of these strategies.

There is a risk that cells that have evolved tolerance to high concentrations of an exogenously added chemical can exhibit alterations in metabolic flux, cell wall properties, and other cellular processes that render them to be poorer endogenous producers of that chemical. Possible solutions to this problem are screening introduced production pathways in many isolates from distinct evolutionary lineages, or reconstructing only minimal sets of a variety of causal mutations for tolerance in production strains. An example case will be presented where significantly improved endogenous production of isobutyrate was achieved using evolved strains as hosts.

Heterologous Gene Expression in Saccharomyces Cerevisiae for Higher Isobutanol Production

Morita, K., Osaka University
Matsuda, F., RIKEN CSRS
Ishii, J., Kobe University
Kondo, A., Graduate School of Engineering, Kobe University
Shimizu, H., Osaka University

□Introduction□ The bio-production of isobutanol by *Escherichia coli* and *Corynebacterium glutamicum* has been attained by introducing the Ehrlich pathway (Atsumi *et.al.*, 2008, Smith *et.al.*, 2010). *Saccharomyces cerevisiae* is also one of the most promising hosts for isobutanol production since its stress tolerance against harsh conditions during fermentation is useful for a cost-effective biorefinery process. Although Ehrlich pathway was introduced into *S. cerevisiae*, the isobutanol production by the engineered strains were still insubstantial, indicating that further engineering was required considering intercellular cofactor balances and compartmentation of the metabolic pathways. In this study, metabolically engineered *S. cerevisiae*strains for isobutanol production were constructed by the flux balance analysis (FBA)-based metabolic design and the expression of heterologous genes.

□Methods□ The FBA was performed using the backbone model of the carbon central metabolism of *S. cerevisiae* (iBKSce50) by MATLAB R2013a with glpk solver. The glucose and oxygen up take rates were set to 10 and 0.1 mmol $gDW^{-1}h^{-1}$, respectively, in all analyses. Engineered yeast strains were constructed by introducing the plasmid vectors harboring the multi-copy type 2μ *ori* sequence and target genes into the yeast laboratory strain YPH499 using lithium-acetate methods. The constructed strains were cultivated in 5 mL of synthetic dextrose medium under semi-anaerobic conditions. Isobutanol titers at 48 h after the start of cultivation was determined by using gas chromatography.

□Result□**Introduction of bacterial pathways:** Isobutanol is synthesized from the intermediate of valine biosynthesis, 2-ketoisovalate, through decarboxylation by 2-ketoacid decarboxylase and reduction by alcohol dehydrogenase. The isobutanol yield of *S. cerevisiae* strain over-expressing *kivd* gene from *Lactococcus lactis*, *S. cerevisiae ADH6* and *ILV2* were 0.78±0.17% mol/mol-glucose (YKP003 strain). The FBA-based metabolic simulation indicated that cofactor (NADPH) imbalance could be improved by additional introductions of phosphoenolpyruvate carboxylase (PPC). Based on the findings, YKP006 strain was constructed by introducing PPC coding gene from *Synechocystis*sp. PCC 6803. The fermentation test indicated that isobutanol yield was 1.13±0.02% mol/mol-glucose and 1.4 times higher than that of control strain.

Pathway construction in cytosol: The precursor for isobutanol synthesis, 2-ketoisovalate, was produced from pyruvate via the valine biosynthesis pathway localized in mitochondria. For more improvement of isobutanol production in *S. cerevisiae*, a valine biosynthetic pathway has been constructed in the cytosol by expression of three *S. cerevisiae* genes, *ILV2c*, *ILV5c*, and *ILV3c,* lacking the mitochondria transit sequences

(Brat *et.al.*, 2012). For further activation of cytosolic pathway, genes for valine biosynthesis were cloned from various organisms and introduced into *S. cerevisiae*. Among the gene tested, overexpressions of ALS gene from *Lactobacillus plantarum*, KARI gene from *Sulfolobus tokodaii* (YKI024 strain: 1.10±0.04% mol/mol-glucose) and the DHAD gene from *Zymomonas mobilis* (YKI015 strain: 1.27±0.10% mol/mol-glucose) increased isobutanol yield two folds compared to the control strain (0.52±0.08% mol/mol-glucose), respectively.

Effect of Nutrient Starvation on 3-Hydroxypropionic Acid Production By Recombinant Escherichia coli at Stationary Phase

Tokuyama, K., Osaka University
Toya, Y., Osaka University
Matsuda, F., RIKEN CSRS
Shimizu, H., Osaka University

1. Background

Escherichia coli is able to catabolize the extracellular glucose even under a starvation of essential nutrients. It has been reported that the limiting nutrients such as nitrogen, sulfur, magnesium, or phosphorus distinctively affect the carbon central metabolism and the biochemical productivity of *E. coli* at the stationary growth phase. In this study, the effects of nutritional starvations on the fermentation and metabolic profiles were characterized for a metabolically engineered *E. coli* strain producing 3-hydroxypropionic acid (3HP).

2. Materials & methods

E. coli strain MG1655(DE3) was transformed with pETM6 derived vectors. The *E. coli* cells grown in M9 minimal medium containing glucose as a sole carbon source were inoculated in nitrogen, sulfur, magnesium, or phosphorus free M9 minimal medium, and cultivated under aerobic condition at 37°C. Concentrations of glucose, 3HP and byproducts in the culture broth were determined by an HPLC. Intracellular metabolites were extracted from the cells with chloroform-water-methanol and quantified using LC-MS/MS and GC-MS.

3. Results

3HP-producing strain was constructed by overexpression of maloyl-CoA reductase encoded by *mcr* gene derived from *Chloroflexus aurantiacus* in *E. coli* MG1655(DE3). In order to evaluate the effect of single nutrient starvations on 3HP production during the stationary phase, the 3HP-producing strain was cultured in M9 medium lacking nitrogen, sulfur, magnesium, or phosphorus, respectively. The comparison of the fermentation profiles showed that larger 3HP yields were observed under the sulfur (6.0% C-mol 3HP/ C-mol glucose) and magnesium starvation (5.8%) conditions than the starvations of nitrogen (0.2%) and phosphorus (undetectable). The byproduction of acetate was occurred in the all conditions. The specific glucose consumption rate in magnesium starvation (2.31 mmol/gCDW/h) was faster than that in nitrogen, phosphorus or sulfur starvation (0.33, 1.11 or 0.78 mmol/gCDW/h, respectively).

The metabolic profiling analysis of the intermediates of the central carbon metabolism revealed that the concentration of malonyl-CoA, a precursor for 3HP synthesis, was larger in magnesium or sulfur starvation than that in nitrogen or phosphorus starvation. The results indicated that the accumulation of malonyl-CoA enhanced the 3HP production in sulfur and magnesium starvation. The high ATP concentration in magnesium or sulfur starvation would contribute the malonyl-CoA accumulation since acetyl-CoA carboxylase requires ATP for a conversion of acetyl-CoA to malonyl-CoA. The active glucose uptake in magnesium starvation could be explained by an accumulation of phosphoenolpyruvate, the substrate of phosphotransferase system (PTS) for glucose transport. On the other hand, the significant accumulation of α-ketoglutarate in nitrogen starvation (4.1 mM) suggested that the slow glucose uptake was derived from the allostric inhibition of PTS by α-ketoglutarate (Ki = 1.3 mM). These results suggest that the choice of lacking nutrients affected the fermentation and metabolic profiles in the 3HP-producing strain during the stationary phase.

Engineering Escherichia coli for Production of Glycolic Acid from D-Xylose through Dahms Pathway

Cabulong, R. B., Myongji University
Valdehuesa, K. N. G., Myongji University
Ramos, K. R. M., Myongji University
Maza, P. A. M. A., Myongji University
Pangan, J. O., Myongji University
Bañares, A. B., Myongji University
Nisola, G. M., Myongji University
Lee, S. P., Myongji University
Lee, W. K., Myongji University
Chung, W. J., Myongji University

Glycolic acid (GA) is the simplest form of α-hydroxy acid and it is extensively used in cosmetics and dermatology for its exfoliating and moisturizing properties. Its corresponding polymeric form, polyglycolic acid (PGA), has excellent gas barrier properties which makes it a good packaging material. In addition, the use of PGA in PET bottles and medical sutures has also led to increasing demand for GA. Recent reports on the biosynthetic production of ethylene glycol from D-xylose have identified GA as a by-product during fermentation. On the other hand, studies that were aimed at GA production in recombinant microorganisms have employed the glyoxylate shunt. In this study, GA production was implemented by recruiting the Dahms pathway in *E. coli*. D-xylose is oxidized to D-xylonic acid by a heterologous xylose dehydrogenase (Xdh). D-xylonic acid is the converted to glycolaldehyde and pyruvate through the inherent D-xylonic acid metabolism in *E. coli*. Pyruvate is further assimilated into the central carbon metabolism, while glycolaldehyde is either reduced to ethylene glycol or oxidized to GA. The metabolic engineering strategies carried out to produce mainly GA include (1) optimization of the expression levels of enzymes catalyzing the conversion of D-xylose to glycolaldehyde, (2) recruitment of alternative dehydratase, (3) blocking ethylene glycol formation by deletion of relevant aldehyde reductases, (4) blocking GA oxidation reaction by disruption of the *glcD* gene which codes for glycolate oxidase, and (5) overexpression of glyoxylate shunt enzymes glyoxylate reductase (*ycdW* gene), isocitratelyase (*aceA* gene), and isocitrate dehydrogenase kinase/ phosphatease (*aceK* gene) to channel the TCA cycle towards GA formation. *This work was supported by Basic Science Research Program through the National Research Foundation of Korea funded by the Ministry of Education (No. 2009-0093816), and Korea Research Fellowship Program through the NRF funded by the Ministry of Science, ICT and Future Planning (No. 2015H1D3A1062172).*

A Novel Platform for Secretory Production of Recombinant Proteins in Saccharomyces Cerevisiae

Bae, J. H., KRIBB
Sung, B. H., KRIBB
Kim, H. J.,
Park, S. H.,
Lim, K. M.,
Kim, M. J., KRIBB
Lee, C. R., University of Science and Technology
Sohn, J. H., KRIBB

In spite of the higher preference of *Saccharomyces cerevisiae* as an expression host for the secretory production of recombinant proteins, only a limited number of secretion signals are available. To expand the repertoire of effective secretion signals in *S. cerevisiae*, we developed a target protein-specific translational fusion partner (TFP) screening system that involves recruitment of an optimal secretion signal and fusion partner from genome. Around 4 x 10^3 TFP library was constructed from a genomic and truncated cDNA library by using the invertase-based signal sequence trap technique. The efficiency of the system was demonstrated using two rarely secreted proteins, human interleukin (hIL)-2 and hIL-32. Optimal TFPs for secretion of hIL-2 and hIL-32 were easily selected, yielding secretion of these proteins up to hundreds of mg/L. Moreover, numerous uncovered yeast secretion signals and fusion partners were identified, leading to efficient secretion of various recombinant proteins. Selected TFPs were found to be useful for the hypersecretion of other recombinant proteins at yields of up to several g/L. This screening technique could provide new methods for the production of various types of difficult-to-express proteins

Agar Metabolism of the New Marine Bacterium Cellulophaga Lytica W5C and Its Potential Use in Marine Biomass Conversions

Ramos, K. R. M., Myongji University
Valdehuesa, K. N. G., Myongji University
Cabulong, R. B., Myongji University
Nisola, G. M., Myongji University
Maza, P. A. M. A., Myongji University
Pangan, J. O., Myongji University
Bañares, A. B., Myongji University
Lee, S. P., Myongji University
Hong, S. K.,
Lee, W. K., Myongji University
Chung, W. J., Myongji University

Marine macroalgae, considered as third generation biomass, are a promising feedstock source for the production of various biofuels and fine chemicals. Red macroalgae, in particular, contains agarose as one of its main components. This polysaccharide accounts for up to 52 dry weight percent in *Gelidium amansii*. Agarose is made up of alternating units of D-galactose and 3,6-anhydro-L-galactose (L-AHG). Because of the high carbohydrate content of red macroalgae, specifically the fermentable D-galactose, it is a viable resource of sugars. The products of agarose hydrolysis are also known to have medical and pharmaceutical effects. Due to the potential of red macroalgae as biomass resource, studies on organisms capable of degrading agar or agarose have steadily increased in the past years. Moreover, researchers have also begun to elucidate the metabolic pathway for the rare sugar, L-AHG. In this study, an agar-degrading bacterium designated as W5C, was isolated from marine red macroalgae, *Palmaria* sp., collected from the seashore of Yeosu, South Korea. Preliminary characterization identified the bacterium as Gram-negative, catalase- and gelatinase-positive, and oxidase- and amylase-negative. Based on 16S rRNA sequencing and phylogenetic analysis, strain W5C belonged to the *Cellulophaga* genus and is highly related to *Cellulophaga lytica*. Full genome sequencing of the isolate and homology searches identified four putative agarase-encoding genes. The genes encoding for the agarolytic enzymes were cloned in *Escherichia coli* for characterization. Additionally, the strain possesses genes involved in the utilization of L-AHG. The respective enzymes were cloned and characterized as well. Elucidation of the entire agarose metabolic pathway in the *Cellulophaga* isolate may be applied to marine biomass utilization for the synthesis of biofuels and fine chemicals.
This work was supported by Basic Science Research Program through the National Research Foundation of Korea funded by the Ministry of Education (No. 2009-0093816), and Korea Research Fellowship Program through the NRF funded by the Ministry of Science, ICT and Future Planning (No. 2015H1D3A1062172).

Development of Platform Yeast Strain Capable of Direct Fermentation of Raw Biomass to Ethanol

Kahar, P., Kobe University
Itomi, A., Kobe University
Thontowi, A., Indonesia Institute of Science (LIPI)
Otsuka, H., Kobe University
Kihara, C., Kobe University
Lee, J., Kobe University
Betha Juanssil Fero, A., Indonesia Institute of Science (LIPI)
Djohan, A. C.,
Sunarya, Y.,
Ishii, J., Kobe University
Ogino, C., Graduate School of Engineering, Kobe University
Prasetya, B.,
Kondo, A., Graduate School of Engineering, Kobe University

Lignocellulosic biomass, particularly from agricultural and forestry sectors, represents a rather unused low-cost source for production of biofuel (alkanes and fatty alcohols), commodity chemicals (waxes), fine chemicals (sesquiterpenes, bioplastics) and proteins (enzymes, pharmaceuticals). To convert them into such products, the presence of the robust microbial platform is required. However, significant challenges exist for the efficient cost-effective conversion of biomass to such materials. One technical barrier for biomass conversion is the stress condition caused by toxic compounds released from the chemical pretreatment process or the enzymatic digestion of biomass in which remarkably interfere with microbial growth and subsequent fermentation. We believe they are many broad yeast candidates available in nature, which can move beyond difficulties to out boundary so that the good platform for fermentation of lignocellulosic biomass could be built without too much genetic modification. However, the selection of appropriate candidates is not easy due to the unique adaptation possessed by many yeasts upon the environmental stress. The intensive screening should be conducted using natural resources, such as soil, leaf, fermented foods. However, to screen across such a wide range of resources requires arduous, cost- and time-consuming processes as well as knowledges. In this study, we have developed rapid method for screening the candidate yeasts from huge number of yeast strains deposited in NITE Biological Resource Center (NBRC) and also in Indonesia Culture Collection (InaCC), which capable of direct fermentation of lignocellulosic sugar lysate into ethanol. By screening of more than 3000 yeast strains, we found some potential candidates for rapid fermentation of sugars, from the lignocellulosic sugar lysate and also from the synthetic medium containing high concentrated inhibitory chemical cocktails. Most of yeasts screened as candidate strains are belong to *Saccharomyces cerevisiae* with specific phenotypes such as high-temperature adaptive, flocculation and rapid invasive growth.

Alternative Inverse Metabolic Engineering Based on Transient Acclimation of Yeast Improves Acid-Containing Xylose Fermentation and Tolerance to Formic and Acetic Acids

Hasunuma, T.,
Kondo, A., Graduate School of Engineering, Kobe University

Improving the production of ethanol from xylose is an important goal in metabolic engineering of *Saccharomyces cerevisiae*. Furthermore, *S. cerevisiae* must produce ethanol in the presence of weak acids (formate and acetate) generated during pretreatment of lignocellulosic biomass. In this study, weak acid-containing xylose fermentation was significantly improved using cells that were transiently acclimated to the weak acids during pre-cultivation. Transcriptome analyses showed levels of transcripts for transcriptional/translational machinery–related genes (*RTC3* and *ANB1*) were enhanced by formate and acetate acclimation. Recombinant yeast strains overexpressing *RTC3* and *ANB1* demonstrated improved ethanol production from xylose in the presence of the weak acids, along with improved tolerance to the acids. Novel metabolic engineering strategy based on the combination of transient acclimation and system-wide analysis was developed, which can develop stress-tolerant strains in a short period of time, although conventional evolutionary engineering approach has required long periods of time to isolate inhibitor-adapted strains.

Influence of Key Genes in Arginine Metabolism on Stress Tolerance in Lactococcus Lactis NZ9000

Zhang, J.,
Zhang, M., Jiangnan University
Du, G., Jiangnan University
Chen, J., Jiangnan University

The aim of the present study was to find key factors of arginine metabolism on the stress tolerance. Arginine metabolism was switched by argininosuccinate synthase (ASS) or argininosuccinate lyase (ASL) overexpression in *L. lactis* NZ9000 to enhance the stress tolerance. The recombinant strains exhibited higher growth performance, viability and fermentation performance compared with the control strain under environmental stresses. Analysis of the physiological data showed that the recombinant cells exhibited higher intracellular pH, intracellular NH_4^+ and ATP content, and H^+-ATPase activity under acid stress, and the content of amino acid in arginine deiminase (ADI) pathway was significantly higher than the control strain. Further transcriptional analysis showed that the expression of the genes related to aspartate synthesis and arginine catabolism were up-regulated. These results suggest that up-regulating arginine metabolism flux can improve the multiple-stress tolerance. As arginine synthesis pathway widely exists in various kinds of microorganisms, results presented in this study provide new idea to enhance stress tolerance of tons of microorganisms especially industrial strains.

Modulating Arginine Biosynthesis in E. coli nissle to Consume Ammonia in a Mouse Model of Hyperammonemia

Kotula, J. W., Synlogic, Inc.
Tucker, A., Synlogic, Inc.
Falb, D., Synlogic, Inc.
Isabella, V. M., Synlogic, Inc.
Silva, A., Synlogic, Inc.
Millet, Y., Synlogic, Inc.
Anderson, C., Synlogic, Inc.
Miller, P., Synlogic, Inc.

Urea cycle disorders (UCD) are a collection of genetic conditions in which the patient is unable to convert nitrogen to urea, resulting in the toxic accumulation of ammonia systemically (hyperammonemia). Up to 70% of the excess blood ammonia originates from the gastrointestinal tract, making a gut microorganism-based therapeutic a promising platform to absorb the excess ammonia.

Towards this goal, we have engineered strains of the probiotic bacterium *Escherichia coli* Nissle to consume ammonia by upregulating expression of the arginine (Arg) biosynthetic genes. Here we compare two strategies for optimizing expression of the Arg pathway in Nissle: deletion of the ArgR transcriptional repressor that regulates arginine biosynthesis (Nissle Δ*argR*) and mutation of ArgR binding sites in each *arg* locus (Nissle ΔArgBox). We demonstrate that both strategies yield increased arginine biosynthesis relative to wildtype Nissle, with the Nissle Δ*argR* stain producing the highest levels of arginine. We used metabolite profiling and qPCR to characterize the arginine biosynthetic pathway performance in the Nissle Δ*argR* and Nissle DArgBox strains. The metabolic profile of the Nissle ΔArgBox strain revealed several important differences when compared with the Nissle Δ*argR* strain. Overexpression of ArgG (argininosuccinate synthase) increases arginine yield in the Nissle DArgBox, but arginine production was still below that of the Nissle Δ*argR* strain. We have performed qRTPCR and RNAseq experiments to identify other potential genetic modifications that may improve arginine production in the Nissle ΔArgBox strain.

Concurrent with Nissle ΔArgBox strain optimization, we assessed the performance of the Nissle Δ*argR* strain as a therapy in a genetic mouse model of UCD (spf-ASH). The Nissle Δ*argR* strain effectively reduced blood ammonia levels in these mice, validating the Arg biosynthesis approach for reducing blood ammonia levels from the gastrointestinal tract.

Fitness Selection By Genome Wide Regulator Engineering and Gene Translation Regulation

Wang, Z., Tianjin University
Choudhury, A., University of Colorado, Boulder
Garst, A., University of Colorado Boulder
Gill, D. R. T., University of Colorado

Regulation of gene transcription and translation is fundamentally important for microbes to maintain normal cellular processes and respond to environmental changes. Modification of gene transcription (Alper et al., 2006; Alper & Stephanopoulos, 2007) and translation (Haft et al., 2014) was employed to adapt to novel environments through few mutations with large effects. The most typical case study, global transcription machinery engineering (gTME), has successfully been demonstrated by engineering sigma factors to improve three distinct phenotypes (ethanol tolerance, metabolite overproduction, and multiple phenotypes)(Alper & Stephanopoulos, 2007). However, limited regulators or landscape were rationally designed for kinds of environmental changes. Herein, global regulators, regulators regulating over teen genes and significant genes involved in transcription and translation were selected for modification. The potential interesting sites, including DNA binding sites, active sites and dimerization sites, were collected and designed. Saturated mutagenesis was then performed by CRISPR EnAbled Trackable genome Engineering (CREATE) for the selected interesting sites (Andrew Garst, submitted.). The library was subject to multiple environmental stresses like short chain alcohol, low pH, high temperature, salts, furfural and antibiotics. From these selections, multiple novel beneficial mutations were identified with enhanced tolerance to these stresses. These mutations might enable better understanding of different tolerance mechanisms.

Reference:

Alper, H., Moxley, J., Nevoigt, E., Fink, G.R., Stephanopoulos, G. 2006. Engineering yeast transcription machinery for improved ethanol tolerance and production. *Science*, **314**(5805), 1565-8.

Alper, H., Stephanopoulos, G. 2007. Global transcription machinery engineering: a new approach for improving cellular phenotype. *Metab Eng*, **9**(3), 258-67.

Andrew Garst, M.B., Gur Pines, Sean Lynch, Andrea Halweg-Edwards, Zhiwen Wang, Ryan T. Gill. submitted. Trackable genome scale engineering with single nucleotide resolution.

Haft, R.J., Keating, D.H., Schwaegler, T., Schwalbach, M.S., Vinokur, J., Tremaine, M., Peters, J.M., Kotlajich, M.V., Pohlmann, E.L., Ong, I.M., Grass, J.A., Kiley, P.J., Landick, R. 2014. Correcting direct effects of ethanol on translation and transcription machinery confers ethanol tolerance in bacteria. *Proc Natl Acad Sci U S A*, **111**(25), E2576-85.

Construction of an Efficient Biosynthesis Pathway in Sacchromyces Cerevisiae for Astaxanthin Production

Ma, T., School of Pharmaceutical Sciences
Li, X., Key Laboratory of Combinatorial Biosynthesis and Drug Discovery (Wuhan University), Ministry of Education
Ye, Z., Key Laboratory of Combinatorial Biosynthesis and Drug Discovery (Wuhan University), Ministry of Education
Shi, B., Key Laboratory of Combinatorial Biosynthesis and Drug Discovery (Wuhan University), Ministry of Education
Xiao, E., Key Laboratory of Combinatorial Biosynthesis and Drug Discovery (Wuhan University), Ministry of Education
Zhu, F., Key Laboratory of Combinatorial Biosynthesis and Drug Discovery (Wuhan University), Ministry of Education
Deng, Z., Wuhan University
Liu, T., E-mail: liutg@whu.edu.cn

Construction of An Efficient Biosynthesis Pathway in *Sacchromyces cerevisiae* for Astaxanthin Production

Tian Ma[1], Xiaowei Li[1], Zilin Ye[1], Bin Shi[1], Erting Xiao[1], Fayin Zhu[1], Zixin Deng[1], Tiangang Liu[1,2]

[1]Key Laboratory of Combinatorial Biosynthesis and Drug Discovery (Wuhan University), Ministry of Education, Wuhan University School of Pharmaceutical Sciences, Wuhan, PR China

[2]Hubei Engineering Laboratory for Synthetic Microbiology, Wuhan Institute of Biotechnology, Wuhan, PR China

Correspondence: Prof. Tiangang Liu, Key Laboratory of Combinatorial Biosynthesis and Drug Discovery (Wuhan University), Ministry of Education, Wuhan University School of Pharmaceutical Sciences, Donghu Road 185, 430071 Wuhan, PR China.

E-mail: liutg@whu.edu.cn

Keywords:

Astaxanthin, Metabolic engineering, Carotenoids, *Sacchromyces cerevisiae*

Astaxanthin as a highly valued keto-carotenoid has kinds of beneficial effects on human health, including cancer prevention, immune response enhancement, cardiovascular disease prevention, biological antioxidant protection and so on. Now, it has numerous applications, such as nutritional supplements, pharmaceuticals, cosmetics, food colorants, and animal feed additives, and the worldwide market of astaxanthin is expected to increase dramatically. But in nature, the content of astaxanthin produced by microbes or plants is really low. The market demand for biosynthetic astaxanthin and the efficiency of astaxanthin biosynthesis drive the development in metabolic engineering of native astaxanthin producers and heterologous hosts. Due to the biosafety and robustness in fermentation industry, *Sacchromyces cerevisiae* has become an important platform for heterologous biosynthesis of carotenoids. In this study, a series of galactose induced GAL promoters were tested in *S. cerevisiae* and chosen for astaxanthin biosynthesis pathway construction. Then heterologous genes from different sources, responsible for the biosynthesis of beta-carotene and astaxanthin, were introduced into *S. cerevisiae* to produce astaxanthin. Afterwards, the MVA module was balanced by monitoring the metabolic intermediates and adjusting the expression level of targeted protein. Finally, the upstream module and cofactor were balanced for further improved the production level. Combination of those metabolic engineering strategies with fermentation engineering, the engineered strain showed good performance in astaxanthin production.

Metabolic Enginnering of Yeast for Production of Medicinal Triterpenoids Maslinic Acid and Corosolic Acid

Dai, Z., Tianjin Institute of Industrial Biotechnology, Chinese Academy of Sciences
Zhang, X., Chinese Academy of Sciences

Maslinic acid and corosolic acid produced by the medicinal plants have diverse biological activities, including antibacterial, anti-human immunodeficiency virus (HIV) and/or antitumor activities. Here, we identified the synthetic pathways of maslinic acid and corosolic acid, and produced these triterpenoids from sugars by metabolically engineered yeasts. Firstly, 99 candidate P450 genes were obtained by transcriptome sequencing of methyl jasmonate-induced *Crataegus pinnatifida* cells, which were then screened in *Saccharomyces cerevisiae* platform strain producing oleanolic acid. A cytochrome P450 gene MAA45, which could catalyze the C-2a oxidation of oleanolic acid, was identified by LC-MS and NMR analysis. This gene was also evaluated for its potential oxidizing activities in *S. cerevisiae* platform strain producing ursolic acid, and was shown to have the ability to generate corosolic acid. MAA45 was characterized as a multifunctional enzyme with oleanane- and ursane-type triterpenoids 2a-oxidase activities. In addition, *S. cerevisiae* cell factory BY-T30-OA was engineered to produce sufficient amounts of metabolic precursors (oleanolic acid and ursolic acid), and the MAA45 gene was integrated into this platform strain to improve the production of targeted triterpenoids. These findings will be useful in the heterologous production of pharmacologically and industrially important triterpenoids, including maslinic acid and corosolic acid.

Improving the Availability of Oxygen to Achieve High Cell Density in Escherichia coli and Halomonas TD01

Ouyang, P., Tsinghua University

Technology of high cell density fermentation is required for industrial production. However, the key factor limiting the cell high density is the shortage of oxygen at fermentation anaphase. It is urgent to develop a variety of oxygen utilization technology in order to increase the cell density of aerobic microorganisms in fermentation.

So far, there are mainly two strategies to increase cell density. One is the optimization of the oxygen utilization conditions in fermentation, while the other one is the update of microbial strains, which hold a better oxygen utilization efficiency by molecular modification. Expression of hemoglobin was a common choice. However, obstacles still existed against the inter-hemoglobin contacts. Therefore, it is a hotspot to increase the contact efficiency between hemoglobin and oxygen instead.

In the periplasmic space of microorganism, the oxygen could be captured by mature hemoglobin efficiently, which could reduce the transport loss of oxygen and increase the utilization efficiency. It is a possible way to increase the cell density. In our study, the cell density was significantly increased by 150% via using the twin-arginine translocase (Tat) pathway to export active *Vitreoscilla* hemoglobin to the periplasm in *Escherichia coli* and *Halomonas* TD01.

Metabolic Engineering of Corynebacterium Glutamicum for the Utilization of Biomass-Derived Carbon and Energy Sources

Radek, A., Forschungszentrum Jülich GmbH
Witthoff, S., Forschungszentrum Jülich GmbH
Krumbach, K., Forschungszentrum Jülich
Noack, S., Forschungszentrum Jülich
Marienhagen, J., Forschungszentrum Juelich

Methanol is a pure and inexpensive raw material, which is mainly produced from fossil-fuel-based synthesis gas. Over the past years however, new approaches were developed for its production from renewable carbon sources. Methanol is an important carbon feedstock in the chemical industry, but has found only limited application in biotechnological production processes. This can be mostly attributed to the inability of most microbial platform organisms to utilize methanol as carbon and energy source. With the aim to turn methanol into a suitable feedstock for microbial production processes, we engineered the industrially important, but non-methylotrophic bacterium Corynebacterium glutamicum towards the utilization of methanol as auxiliary carbon source in a sugar-based medium [1,2].

Additionally, biomass-derived D-xylose represents an economically interesting substrate for the sustainable microbial production of value-added compounds. All currently described *C. glutamicum* strains utilize D-xylose via the commonly known isomerase pathway that leads to a significant carbon loss in the form of CO_2, in particular, when aiming for the synthesis of α-ketoglutarate and its derivatives (e.g. l-glutamate). Driven by the motivation to engineer a more carbon-efficient *C. glutamicum* strain, we functionally integrated the Weimberg pathway from *Caulobacter crescentus* in *C. glutamicum* [3]. This five-step pathway enabled a recombinant *C. glutamicum* strain to utilize D-xylose in D-xylose/D-glucose mixtures and thus represents a promising starting point for the engineering of efficient production strains, exhibiting only minimal carbon loss on D-xylose containing substrates.

[1] Witthoff S., Mühlroth A., Marienhagen J., Bott M. (2013) C1 metabolism in *Corynebacterium glutamicum*: an endogenous pathway for oxidation of methanol to carbon dioxide. *Appl. Environ. Microbiol.* 79(22): 6974–6983.

[2] Witthoff S., Schmitz K., Niedenführ S., Nöh K., Noack S., Bott M., Marienhagen J. (2015). Metabolic engineering of *Corynebacterium glutamicum* for the metabolization of methanol. *Appl. Environ. Microbiol.* 81(6): 2215–2225.

[3] Radek A., Krumbach K., Gätgens J., Wendisch V. F., Wiechert W., Bott M., Noack S., Marienhagen J. (2014). Engineering of *Corynebacterium glutamicum* for minimized carbon loss during utilization of D-xylose containing substrates. *J. Biotechnol.* 192: 156–160.

Development of Klebsiella Oxytoca Genome-Scale Metabolic Model and in silico Analyses for 2,3-Butanediol Production

Park, J. M., GS Caltex
Rathnasingh, C., GS Caltex
Song, H., GS Caltex
Seung, D., GS Caltex

Klebsiella oxytoca is known as one of the most promising 2,3-butanediol (2,3-BD) producers and its whole genome sequences have been reported recently. In order to assess its metabolic characteristics and to optimize its metabolic pathways for 2,3-BD production with a system-wide scope, we constructed a genome-scale metabolic model of *K. oxytoca* (KoxGSC145) based on genome annotation information, metabolic pathway databases, and various experimental data. The model contains 1,457 stoichiometric reactions and 1,099 metabolites. The model was thoroughly refined and verified by comparing *in silico* results with experimental data based on constraints-based flux analyses. Then, we applied the model to predict its physiological responses of *K. oxytoca* according to environmental and genetic perturbations and also to design metabolic engineering strategies to improve its metabolic performance for 2,3-BD production. Firstly, *in silico* analysis, which tested the effect of augmenting the metabolic flux pool of 2,3-BD precursors, elucidated that increasing the pyruvate pool is primarily important for 2,3-BD synthesis. Secondly, we performed *in silico* single gene knockout simulation for 2,3-BD overproduction, and investigated the changes of the *in silico* flux solution space of a *ldhA* gene knockout mutant in comparison with that of the wild-type strain. Finally, the KoxGSC1457 model was used to optimize the oxygen levels during fermentation for 2,3-BD production. The KoxGSC1457 model successfully investigated metabolic characteristics of *K. oxytoca* at systems level. [This work was supported by the Industrial Strategic Technology Development Program (No. 10050407) funded by the Ministry of Trade, Industry & Energy (MOTIE, Korea).]

Metabolic Engineering of Klebsiella Pneumoniae Based on in silico Analyses for the Enhanced Production of 2,3-Butanediol

Rathnasingh, C., GS Caltex
Park, J. M., GS Caltex
Kim, D. K., GS Caltex
Song, H., GS Caltex

The production of 2,3-butanediol (2,3-BD) was improved in *Klebsiella pneumoniae* by deleting the pathway genes responsible for the formation of lactic acid, ethanol, and acetic acid. Despite the fact that the cell growth and 2,3-BD production rates of the *K. pneumoniae ΔldhA ΔadhE Δpta-ackA* strain were lower than those of the wild-type strain, the mutant produced 2,3-BD at high titer and yield in the batch fermentation. Under the fed-batch fermentation, the mutant produced 91 g/L of 2,3-BD with a yield and productivity of 0.45 g per g glucose and 1.62 g/L/h, respectively. The metabolic characteristics of the mutants were consistent with the results of *in silico* simulation. *K. pneumoniae* knockout mutants developed with the support of *in silico* investigation could produce enhanced 2,3-BD with high titer, yield, and productivity. [This work was supported by the Industrial Strategic Technology Development Program (No. 10050407) funded by the Ministry of Trade, Industry & Energy (MOTIE, Korea).]

Electrospun Nanofibrous Rhpdgf and Collagen Loaded Scaffolds to Promote the Healing of Diabetic Wounds

Liu, S. J.,

This work developed nanofibrous growth factor-loaded collagen/poly-D-L-lactide-glycolide (PLGA) scaffolds that enabled the sustained release of recombinant human platelet-derived growth factor (rhPDGF) to treat wounds that are associated with diabetes. PLGA, rhPDGF, and collagen were firstly dissolved in 1,1,1,3,3,3-hexafluoro-2-propanol, from which nanofibrous scaffolds were fabricated by electrospinning. An elution method and an enzyme-linked immunosorbent assay kit were utilized to determine the *in vivo* and *in vitro* release rates of growth factors from the scaffolds. The nanofibrous scaffolds released high and effective concentrations of rhPDGF for more than three weeks. The nanofibrous rhPDGF-loaded collagen/PLGA scaffolds were more hydrophilic than collagen/PLGA or virgin PLGA scaffolds and were able to contain more water. Furthermore, the rhPDGF-loaded collagen/PLGA scaffolds considerably promoted the healing of diabetic wounds. The scaffolds caused more re-epithelialization and contained more collagen I in diabetic rats than in the control rats, as determined from the expressed *matrix metalloproteinase 9* in the hair canals in developing hair follicles. The experimental results herein suggest that the nanofibrous rhPDGF-loaded collagen/PLGA scaffolds enhanced epithelialization in the treatment of diabetic wounds and were very effective accelerators of the healing of such wounds in the early stages of that healing.

Optimal Pathway Rebalancing for Efficient Conversion of Glycerol to 3-Hydroxypropionic Acid in Escherichia coli

Kwak, D. H., POSTECH
Lim, H. G., POSTECH
Noh, M. H., POSTECH
Park, S., Pusan National University
Jung, G. Y., POSTECH

3-Hydroxypropionic acid (3-HP), which can be converted into many chemicals such as acrylic acid, acrylamide, and propiolactone, is an important platform chemical. It can be biologically produced from glycerol by two consecutive enzymatic reactions, dehydration of glycerol to 3-hydroxypropionaldehyde (3-HPA) and oxidation of 3-HPA. The pathway has been proved efficient, but imbalance between the rates of the two enzymatic reactions often results in the accumulation of the toxic 3-HPA, which severely reduces cell viability and 3-HP production. In this study, we used UTR engineering to maximally increase the activities of glycerol dehydratase (GDHt) and aldehyde dehydrogenase (ALDH) for the high conversion of glycerol to 3-HP. Thereafter, the activity of GDHt was precisely controlled to avoid the accumulation of 3-HPA by varying expression of *dhaB1*, a gene encoding a main subunit of GDHt. The optimally balanced *E. coli* HGL_DBK4 showed a substantially enhanced 3-HP titer (40.51 g/L) and productivity (1.35 g/L/h) compared with the parental strain. Especially, the yield on glycerol, 0.97 g 3-HP/g glycerol, in a fed-batch experiment, was the highest ever reported. Our study demonstrates that metabolic balancing through UTR engineering is an effective strategy for the development of efficient microorganisms for bio-refinery production.

Metabolic Engineering of Klebsiella Oxytoca Based on in silico Genome-Scale Analyses to Enhance 2,3-Butanediol Production

Park, J. M., GS Caltex
Rathnasingh, C., GS Caltex
Kim, D. K., GS Caltex
Song, H., GS Caltex
Seung, D., GS Caltex

2,3-butanediol (2,3-BD) has applicability for the manufacture of 1,3-butadiene and methyl ethyl ketone, moistening and softening agents, and cosmetics. *Klebsiella oxytoca* is a promising microorganism to produce 2,3-BD. In order to enhance 2,3-BD production, we engineered genetically *K. oxytoca* strain based on *in silico* genome-scale analyses. The *in silico* gene knockout simulation of *K. oxytoca* was carried out for 2,3-BD overproduction by using genome-scale metabolic model of *K. oxytoca*, KoxGSC1457, which is composed of 1,457 reactions and 1,099 metabolites. The double knockouts of *ldhA* and *pflB* genes were targeted with the criteria of minimization of byproducts formation and maximization of 2,3-BD production. The constructed *K. oxytoca ΔldhA ΔpflB* strain showed higher 2,3-BD yields and higher final concentrations than those obtained from the wild-type and *ΔldhA* strains. Optimizing the condition of fed-batch fermentation, we achieved the titer of >110 g/L 2,3-BD, with a productivity of 2.1 g/L/h and yield of 0.45 g/g glucose. [This work was supported by the Industrial Strategic Technology Development Program (No. 10050407) funded by the Ministry of Trade, Industry & Energy (MOTIE, Korea).]

Metabolic Engineering of Klebsiella Oxytoca Strain to Enhance (R,R)-2,3-Butanediol Production

Park, J. M., GS Caltex
Rathnasingh, C., GS Caltex
Song, H., GS Caltex
Seung, D., GS Caltex

2,3-butanediol exists in three types of stereoisomoers, (*R,R*)-BD, (*S,S*)-BD, and *meso*-BD, and each stereoisomer has different physiochemical properties. *meso* and *(R,R)*-2,3-BD have similar applicability in chemical industry sectors, such as 1,3-butadiene and methyl ethyl ketone. However, their applications are strongly limited to some special sectors by their different physiochemical properties. *meso*-2,3-BD is applicable to a component of cosmetics as an moistening agent. Only *(R,R)*-2,3-BD has an applicability in agricultural industry, which is to activate plants' own defense systems. Natural *Klebsiella oxytoca* mainly produces *meso*-2,3-BD (> 90% of total 2,3-BD). In order to expand the potential applications of *K. oxytoca* strain, we engineered the metabolism of *K. oxytoca* strain. We introduced an exogenous gene encoding (*R,R*)-2,3-BD dehydrogenase from *Paenibacillus polymyxa* into the site of *budC* gene in *K. oxytoca* strain and increased the expression of of (*R,R*)-2,3-BD dehydrogenase further. The final strain could produce up to 106.7 g/L of (*R,R*)-2,3-BD [*meso*-2,3-BD, 9.3 g/L] in fed-batch fermentation, with a productivity of 3.1 g/L/h and a yield of 0.40 g/g. [This work was supported by the Industrial Strategic Technology Development Program (No. 10050407) funded by the Ministry of Trade, Industry & Energy (MOTIE, Korea).]

Development of Crispr Based Genome Edition and Pathway Optimization Techniques for Metabolic Engineering Applications

Bi, C., Chinese Academy of Sciences
Li, Q., Chinese Academy of Sciences
Zhang, X., Chinese Academy of Sciences
Zhao, D., Chinese Academy of Sciences

Microbial genome edition is to modify chromosome in way of deletion, insertion and replacement, which is one of the most important technique in Synthetic Biology and Metabolic Engineering research. The emergence of CRISPR/Cas9 technique inspires various genomic edition methods. In this research, we perused the goal of development of the easiest method for *Escherichia coli* genome edition and technique for multiplex optimization of native metabolic pathway by chromosomal editing. For this purpose, we designed modular plasmid assembly strategy, compared effect of different length of homologous region for recombination, and tested different sets of recombinases. The genome editing technique we developed only requires the work of one plasmid construction and one transformation of practice to edit a genomic locus within 3 days with minimal lab work. By further development, multiple loci were able to be edited simultaneously. Base on this, regulation sequences of several genes in *E. coli* MEP pathway were disturbed for various expression levels. Thus a metabolic pathway library with different expression of each gene were obtained. And strains with highest production of β-carotene which represented optimized MEP pathway could be selected. In this article, we described the process of method development.

Genomic Analysis for the Mechanism of Leucine Overproduction in a Corynebacterium Glutamicum

Tonouchi, N., Ajinomoto Co. Inc.
Koseki, C., Ajinomoto Co. Inc.
Nakanishi, K., Ajinomoto Co. Inc.

Amino acid industry has a long history for over 50 years. At present, many kinds of amino acid are produced by fermentation. In the early days of amino acid production, the producer strains were bred by repeated conventional mutagenesis and selection. However, due to the limited technology available at the time, little was studied for the genetic analysis of the strain to reveal the mechanism of overproduction.

The wild type strain has a regulation system to prevent the overproduction of each amino acid; feedback inhibition and repression at each key-enzyme step in the bio-synthetic pathway. Therefore, to achieve the overproduction, mutation in the genes of coding the key enzyme to release of the feedback inhibition and repression are essential. The most important key-enzymes for leucine production are α-isopropyl malate synthase (IPMS) and acetohydroxylate synthase (AHAS), which are coded by *leuA* and *ilvBN* genes, respectively.

Corynebacterium glutamicum strain NB-12 is a leucine producer that is bred by conventional mutagenesis. Whole genome sequence was determined by SOLiD system and compared with the wild type strain. A single mutation was observed in the coding regions of both the *leuA* (G564D) and *ilvBN* (G235D) genes. These findings suggest that the mutations lead to the release of feedback inhibition: in fact, the introduction of the two mutant genes into wild type *E.coli* is resulted in leucine accumulation. However, concerning feedback repression, no mutation was observed in the promoter region of either gene. Instead, NB-12 is known to be an Ile auxotroph, and a mutation was observed in the *ilvA* gene encoding threonine dehydratase (A95T). These results suggest that the addition of limited amount of isoleucine leads to the release of feedback repression of the genes encoding these key enzymes. In the present study, specific mutations and mechanism of leucine overproduction will be discussed.

Engineering Saccharomyces Cerevisiae for Prostratin Production
Wong, J.,

The need for effective anti-HIV drugs is urgent. Over 34 million people worldwide are infected with HIV, many of whom live in countries where medicine is inaccessible due to cost. Prostratin, a natural product derived from *Homalanthus nutans,* a small rainforest tree grown in Samoa, is being actively developed as an adjuvant therapy to clear latent viral reservoirs, the primary obstacle to eradication of HIV. Currently, prostratin is produced via a semisynthetic route. However, such processes are inefficient and costly; the development of a microbial production system promises to significantly reduce the cost of prostratin. The biosynthetic pathway of prostratin begins with the non-mevalonate pathway, after which prenyltransferases and a diterpene synthase produce casbene, a major precursor to prostratin. The biosynthetic steps in the prostratin pathway after casbene are unknown, although several groups have postulated that P450s catalyze the ring closures seen in prostratin. A *Saccharomyces cerevisiae* strain was constructed with an upregulated isoprenoid pathway for production of the diterpene casbene at titers of up to 500mg/L, a 16-fold increase from the highest previously reported titers. Using coexpression analysis and comparative genomics on the Euphorbiaceae species *Homalanthus nutans*, *Jatropha curcas*, *Euphorbia resinifera*, and *Euphorbia fischeriana*, we found P450 oxidative enzymes that modify casbene. The resulting P450s were integrated into the casbene overproducing yeast strain using CRISPR-Cas genome engineering, resulting in the production of oxidized casbanes, which are important intermediates in the prostratin pathway. This result is a significant step towards the development of an industrial microbial system for the production of prostratin.

Metabolic Engineering of Corynebacterium Glutamicum for Cysteine Fermentation

Hirasawa, T., Advanced Low Carbon Technology Research and Development Program, Japan Science and Technology Agency (JST, ALCA)
Kondoh, M., Tokyo Institute of Technology
Kishino, M., Tokyo Institute of Technology

Cysteine is one of the sulfur-containing amino acids and has been utilized for production of pharmaceuticals and cosmetics. Cysteine is mainly produced by extracting from hairs, but awareness to consider the influence of the production process to the safety and environment has been increasing. Therefore, it has been expected that the methods to produce cysteine should be changed to fermentation by microorganisms. We have performed metabolic engineering studies on cysteine fermentation by *Corynebacterium glutamicum*, which has been utilized for fermentative production of amino acids such as glutamic acid and lysine.

In cysteine biosynthesis pathways, serine O-acetyltransferase (SAT) coded by *cysE* gene is subjected to feedback inhibition by cysteine. For creating cysteine-producing *C. glutamicum* strain, we introduced the mutant *cysE* gene encoding feedback-resistant SAT and the constructed recombinant strain could indeed produce cysteine. Moreover, in order to increase the productivity of cysteine by *C. glutamicum*, the metabolic engineering strategies were derived by *in silico* metabolic simulation using a genome-scale metabolic model of *C. glutamicum* (Shinfuku et al. 2009 Microb. Cell Fact. 8:43). Simulation results suggest that the increase in the metabolic fluxes of the pentose phosphate pathway for enhancement of NADPH supply to sulfate reduction reactions and the increase in acetyl-CoA supply for SAT reaction would be effective to enhance cysteine production by *C. glutamicum*. In addition, we have examined the effects of overexpression of the genes encoding the possible cysteine exporters on cysteine production by *C. glutamicum*.

This work is financially supported by the Ministry of Agriculture, Forestry and Fisheries, Japan.

Overproduction of w-Hydroxypalmitic Acid from Glucose in Escherichia coli

Kim, J., Seoul National University
Sung, C., Seoul National University
Kim, M., Seoul National University
Jung, E., Seoul National University
Kim, B. G., Seoul National University

Fatty acid and its dervatives are good starting materials for various functional chemicals. Among the compounds, hydroxyfatty acids, especially ω-hydroxyfatty acids, are becoming highly significant ingredients of ceramides, lubricants, therapeutics, and plastic monomers in cosmetic, food, and chemical industry. Here, we have produced ω-hydroxypalmitic acid from glucose in *E. coli* BW25113 by introducing biosysthesis pathway of ω-hydroxypalmitic acid and applying strategies for overproduction of fatty acid and cofactor optimization. To introduce biosynthesis pathway of ω-hydroxypalmitic acid, CYP153A from *Marinobacter aquaeolei* was expressed with *camA/camB* from *Pseudomonas putida*. β-hydroxyacyl-ACP dehydratase (*fabZ*) and acyl-ACP thioesterase from *Lactobacillus reuteri* were expressed and acyl-CoA synthase (*fadD*) was deleted in order to overproduce long-chain fatty acids and block the degradation of fatty acid, respectively. The engineered *E. coli* without CYP153A produced 1.01 g /L of free fatty acid (C16 503.0mg/L, C18 508.4mg/L) which were 1.8-fold higher than that of wild type strain. Transaldolase A (*talA*), which connects pentose phosphate pathway and hexose diphosphate pathway, was deleted to increase intracellular NADPH concentration, which is required in biosynthesis of fatty acid. The strategy slightly increased the production free fatty acid to 1.08 g/L (C16 503.0mg/L, C18 508.4mg/L). CYP153A and CamA/CamB were applied into the engineered *E. coli* showing similar amount of production of ω-hydroxypalmitic acid (211.7 mg/L). Increase in precursor concentration did not improved increase of ω-hydroxypalmitic acid which indicates the hydroxylation step was the rate-limiting step. To maximize production of ω-hydroxypalmitic acid, NADH was overproduced by replacing alcohol dehydrogenase (*adh*)in *E. coli* chromosome with NAD^+-dependent formate dehydrogenase (*fdh*) from *Candida boidinii*. Such strategy showed 2.8-fold increase in production of ω-hydroxypalmitic acid (610.8mg/L). Quantification of NADH and NAD+ showed that expression of *fdh* and deletion of *adh* increase the ratio of $NADH/NAD^+$. Throughout this study, the ω-hydroxypalmitic acid overproduction was achieved by increasing production of free fatty acid and enhancing activity of P450 by increasing NADH pool. Further increment would be achieved by engineering P450 by increasing activity or electron transfer efficiency to release the rate-limiting step.

Engineering for the Production of Renewable Dicarboxylic Acids from Fatty Acid Based Feedstocks

Hutagalung, A.,

Verdezyne is an industrial biotechnology company that uses engineered yeast to convert fatty acids derived from non-food sources to several different dicarboxylic acids, or diacids. These diacids are commonly used in a diverse range of industrial and consumer products. An example is the twelve-carbon-long dodecanedioic acid, which is found in products from toothbrushes and cosmetics to lubricants, adhesives and car parts. Another example is adipic acid, which is mostly used as a monomer for the production of nylon. Both of these diacids are produced from petroleum-based inputs. Verdezyne has developed technologies with the intent of commercializing processes that replace important petroleum-based chemical intermediates with renewable and sustainable alternatives.

The production of diacids from long chain fatty acids presents numerous challenges from metabolic engineering through fermentation to purification. Some of the challenges at the level of metabolic engineering include the characterization of enzyme substrate specificity and the appropriate regulation of enzyme expression and intracellular metabolite trafficking. In order to exercise greater control over the length of the product(s) formed by our engineered strains, we have also performed some targeted protein engineering of enzymes in the two key pathways that we have manipulated, namely omega- and beta-oxidation, which occur in the cytoplasm and the peroxisome, respectively. The omega-oxidation pathway oxidizes the terminal carbon of a fatty acid to produce a diacid, while beta-oxidation shortens the diacid to produce one acetyl-CoA and a diacid that is shorter by two carbons. This presentation will focus on some of the genetic changes and protein engineering that have allowed for the production of diacids of different chain lengths from long chain fatty acids as the starting feedstock.

A Cas9-Based Toolkit to Program Expression Context in Saccharomyces Cerevisiae

d'Espaux, L., Lawrence Berkeley National Lab
Reider Apel, A., Lawrence Berkeley National Lab
Wehrs, M., Lawrence Berkeley National Lab
Sachs, D., Lawrence Berkeley National Lab
Garber, M., Lawrence Berkeley National Lab
Mukhopadhyay, A.,
Keasling, J. D., Joint Bioenergy Institute

From bread, beer and wine to synthetic biology, *Saccharomyces cerevisiae* has long been our go-to eukaryotic host. More recently, CRISPR/Cas9-directed homologous recombination has allowed for rapid and efficient chromosomal integration of heterologous genes and pathways. What remains relatively unexplored is how best to choose gene expression parts—integration locus, promoter, etc.—for a gene of interest. Here we develop a toolkit of well-characterized parts to program gene expression. Our toolkit consists of (1) two dozen genomic integration loci and (2) four dozen native promoters characterized at different growth phases and in different media; (3) modular protein tags to control protein expression level, solubility, and sub-cellular localization; and (4) a web tool to use these data to design donor DNA cassettes. These integration cassettes can be generated in 1-2 rounds of PCR and integrated with high efficiency using pre-cloned Cas9-gRNA plasmids targeting our integration sites. We demonstrate using our toolkit to program various expression contexts for genes of interest as a way to develop and optimize engineered strains.

Extremely Thermophilic Organisms for Efficient Glucose and Xylose Fermentations

Antoniewicz, M. R., University of Delaware
Cordova, L., University of Delaware
Lu, J., University of Delaware
Cipolla, R., University of Delaware

In this work, we describe two new extremely thermophilic organisms that can efficiently ferment glucose and xylose at high temperatures. The first organism is new *Geobacillus* strain LC300 with extremely fast growth rates on glucose and xylose at 72 °C, 2.15 h^{-1} on glucose and 1.52 h^{-1} on xylose (doubling time less than 30 minutes). The corresponding specific glucose and xylose utilization rates are 5.55 g/g/h and 5.24 g/g/h, respectively. As such, *Geobacillus* LC300 grows 3-times faster than *E. coli* on glucose and xylose, and has a specific xylose utilization rate that is 3-times higher than the best metabolically engineered organism to date. To gain more insight into the metabolism of *Geobacillus* LC300 its genome was sequenced using PacBio's RS II single-molecule real-time (SMRT) sequencing platform and detailed ^{13}C-metabolic flux analysis (^{13}C-MFA) was performed using parallel labeling experiments. Additionally, to further demonstrate the biotechnological potential of this organism, *Geobacillus* LC300 was grown to high cell-densities in fed-batch culture, where cells maintained high xylose utilization rate under low dissolved oxygen concentrations. All of these characteristics make *Geobacillus* LC300 an attractive host for future metabolic engineering and biotechnology applications.

The second strain we describe is an evolved *Thermus thermophilus* strain LC113 that efficiently co-utilizes glucose and xylose, the two most abundant sugars in lignocellulosic biomass, at high temperatures and without any carbon catabolite repression. To generate the strain, wild-type *T. thermophilus* HB8 was first evolved on glucose to improve its growth characteristics. The cells were then further evolved on xylose. The resulting strain, designated *T. thermophilus* LC113, was characterized in detail via growth studies and ^{13}C-MFA with [1,6-^{13}C]glucose, [5-^{13}C]xylose, and [1,6-^{13}C]glucose + [5-^{13}C]xylose as isotopic tracers. Compared to the starting strain, the evolved strain had an increased growth rate, increased biomass yield, increased tolerance to high temperatures up to 90 °C, and gained the ability to grow on xylose. At the optimal growth temperature of 81 °C the maximum growth rate of *T. thermophilus* LC113 on glucose and xylose was 0.44 and 0.46 h^{-1}, respectively. In medium containing both glucose and xylose the strain efficiently co-utilized the two sugars. ^{13}C-MFA results provided insights into the flexible metabolism of *T. thermophilus* LC113 that allows efficient glucose and xylose co-utilization. ^{13}C-MFA revealed that metabolic fluxes in the upper part of metabolism, i.e. upper glycolysis and non-oxidative pentose phosphate pathway, adjusted flexibly to varying sugar availability. On the other hand, fluxes in the lower part of metabolism including TCA cycle remained remarkably constant. This is the first time that ^{13}C-MFA has been applied to elucidate co-utilization of glucose and xylose in any organism. Therefore, this study also serves as an important benchmark for future investigations of glucose and xylose co-utilization in other strains.

Engineering of Escherichia coli for Krebs Cycle-Dependent Production of Malic Acid

Auriol, C., Toulouse White Biotechnology
Debora, T., Université de Toulouse; INSA, UPS, INP; LISBP
Baylac, A., Toulouse White Biotechnology Center
François, J. M., Toulouse White Biotechnology
Walther, T.,

The four-carbon dicarboxylic acid malate is a Krebs cycle intermediate and a compound of considerable economic interest. It is mainly used as an acidulant in the food and beverage industry, or as a precursor for specialty chemicals.

Since in *Escherichia coli*, malate is mainly produced during anaerobic fermentation that is redox neutral, cofactor supply may become limiting in biosyntheses that employ additional reduction steps for the conversion of malate into other target molecules. To construct a platform strain in which malate production is increased, and in which sufficient NAD(P)H and ATP cofactors are produced for additional bioconversion steps, we developed a strain that produces malate via the Krebs cycle and glyoxylate shunt. Following this strategy, the ATP and NAD(P)H output of the malate-producing pathway can be increased at the expense of a decreased maximum theoretical yield which drops to 1.33 mol/mol.

As acetate is the major metabolic by-product during aerobic growth on glucose, we engineered *E. coli* by inactivating both malate dehydrogenases *(Dmdh Dmqo)* and malic enzymes (*DsfcA DmaeB*), and the major acetate-producing pathway (*DackA-pta*). Significant malate secretion was found to depend on the expression of a malate-insensitive Ppc$_{K620S}$ mutant. In baffled flask cultures with glucose, the resulting strain MG1655 *Dmdh Dmqo DsfcA DmaeB DackA-pta* pACT3-*ppc*$_{K620S}$produced malate, fumarate, acetate and pyruvate at molar yields of 0.48, 0.11, 0.33 and 0.28, respectively. In controlled bioreactor cultures this strain accumulated 25 g/l malate with a volumetric productivity of 0.53 g/l/h at a yield of 0.5 mol/mol.

Acetate production could not be totally prevented by *DackA-pta* deletion and strains deprived of the ATP/acetate-producing pathway showed retarded growth. Therefore, we chose to engineer alternative metabolic targets to increase malate production without deleting acetate pathway. Finally, we inactivated the transcriptional repressors, IclR and ArcA, and additionally overexpressed the NADH-insensitive citrate synthase mutant GltA$_{R164L}$ to increase carbon flux into the TCA cycle and consequently minimized the carbon overflow through the acetate pathway. In baffled flask cultures with glucose as the carbon source, the optimized strain MG1655 *Dmdh Dmqo DsfcA DmaeB DiclR DarcA* pACT3-*ppc*$_{K620S}$ –*gltA*$_{R164L}$ produced malate, fumarate, acetate and pyruvate at molar yields of 0.8, 0.11, 0.33 and 0.28, respectively.

Metabolic Engineering of Gluconobacter Oxydans Wsh-003 for the Direct Production of 2-Keto-L-Gulonic Acid from D-Sorbitol

Zhou, J., Jiangnan University

2-Keto-L-gulonic acid (2-KLG), the direct precursor of vitamin C, is currently produced by a two-step fermentation route from D-sorbitol. However, this route involves three bacteria (*Gluconobacter oxydans*, *Bacillus megaterium* and *Ketogulonicigenium vulgare*), making the mix-culture system redundant and complicated. Thus, replacement of the conventional two-step fermentation process with a one-step process could be revolutionary in vitamin C industry. Based on the genomic analysis of *Gluconobacter oxydans* and *Ketogulonicigenium vulgare*, a group of dehydrogenases, which use pyrroloquinoline quinine (PQQ) as the cofactor, responsible for the stepwise conversion of D-sorbitol to L-sorbose, L-sorbosone, and finally 2-KLG, were identified. In order to achieve the direct production of 2-keto-L-gulonic acid from D-sorbitol, systematic metabolic engineering were performed on an industrial strain *G. oxydans* WSH-003. Coordination expression of these dehydrogenases in an industrial strain *G. oxydans* WSH-003 achieved the one-step production of 2-KLG. Based on a CRISPR-Cas9/Red system established in the strain, the efficiency of 2-KLG production were further significantly enhanced by integrative expression of dehydrogenases and PQQ biosynthesis genes with a series of constitutive promoters with gradient strength, and knockout of a series identified non-essential membrane proteins.

Study of Non-Alcoholic Fatty Liver Disease Using a Novel Enrichment Analysis That Integrates Transcriptomics and Metabolite Concentrations Data

Pandey, V., EPFL
Hatzimanikatis, V., Swiss Federal Institute of Technology of Lausanne (EPFL)

Non-alcoholic fatty liver disease (NAFLD) includes a spectrum of liver damages ranging from steatosis

to steatohepatitis (NASH), and rarely, progression to liver cirrhosis. Molecular interactions of NAFLD are poorly understood and its analysis requires a detailed understanding of the underlying metabolic and regulatory processes on the molecular level. Thus, to understand the molecular interactions of NAFLD, we developed a novel method we call minimal network enrichment analysis (MiNEA), and we performed MiNEA using constraints inferred from metabolite concentrations and transcriptomics data from a NASH mouse model system.

Furthermore, we compare whether the molecular interactions of NAFLD differ between mouse and human, we also performed the MiNEA analysis using transcriptomics data from humans with NASH. In both human and mouse oxidative stress pathways are found to be upregulated through superoxide anion synthesis in NASH phenotype. Additionally, in mouse, synthesis of ceramide is identified as upregulated in steatosis but downregulated in NASH. Glutamine synthesis is upregulated in mouse steatosis, while it is downregulated in human NASH.

Expression of Varied GFPs in Saccharomyces cerevisiae: Comparison of Codon-Optimized and non-Codon-Optimized GFPs

Matsuno, T., Kobe University
Kaishima, M.,
Ishii, J., Kobe University
Kondo, A., Graduate School of Engineering, Kobe University

Green fluorescent protein (GFP), which was originally isolated from jellyfish, is a widely used tool in biological research, and homologs from other organisms are available. However, researchers must determine which GFP is the most suitable for a specific host. Here, we expressed GFPs from several sources in the yeast *Saccharomyces cerevisiae*, which represents an ideal eukaryotic model. Five non-codon-optimized GFPs (EGFP, AcGFP1, TagGFP2, mUkG1 and ZsGreen) and seven codon-optimized GFPs (yEGFP, yAcGFP1, yTagGFP2, ymUkG1, yZsGreen, ymWasabi and ymNeonGreen) were used to compare the protein expression levels and the green fluorescence intensities. The commercially available GFPs that have been optimized for mammalian codon usage (EGFP, AcGFP1 and TagGFP2) unexpectedly exhibited extremely low expression levels and low fluorescence intensities in *S. cerevisiae*. The mean fluorescence intensities of these codon-optimized GFPs (yEGFP, yAcGFP1 and yTagGFP2) were markedly 22-, 67- and 101-fold higher than those of corresponding non-codon-optimized GFPs. The codon-optimization of other GFPs also increased both the expression levels and the fluorescence intensity of the cells. Among the codon-optimized GFPs tested, monomeric ymUkG1 and tetrameric yZsGreen exhibited the highest fluorescence intensities in *S. cerevisiae*. Furthermore, we applied the codon-optimized ymUkG1 to the GÎ³ recruitment system, which enables to detect protein-protein interactions and to screen for protein variants presenting desirable affinities. The expression of ymUkG1 as a fusion-tagged protein successfully improved the reporting ability of the GÎ³ recruitment system.

Keywords:
GFP, codon-optimization, *Saccharomyces cerevisiae*

Overproduction of Taxadiene in Engineered Escherichia coli

Bian, G.,
Liu, T., Wuhan Institute of Biotechnology

Taxol (Paclitaxel) is one of the blockbuster anticancer drugs in the clinic with excellent activity against a range of cancers. Recent developments in metabolic engineering and synthetic biology have allowed overproduction of terpenoid based on microbial fermentation rather than the time consuming and low yield plant-based production, resulting in remarkable breakthroughs in the production of complex natural products such as taxadiene, the precursor of taxol. Here, we constructed a rationally designed recombinant *E. coli* strain that could be engineered for high-level production of taxadiene. Base on the instruction of in vitro reconstitution of mevalonate (MVA) pathway, the production of two universal 5-carbon isoprenoid precursors, isopentenyl pyrophosphate (IPP) and dimethylallyl pyrophosphate (DMAPP) were optimized, together with the heterologous taxadiene forming pathway carrying geranyl diphosphate synthase (GGPPS) and taxadiene synthase (TS) of different origin, taxadiene production in *E. coli* was significantly increased at the shake-flask scale. This work could be regarded as a novel platform for engineering of other terpenoid synthesis from mevalonate pathway. Successful overproduction of taxadiene has encouraged the hope that the terpenoid metabolic pathway can be rationally engineered for enhanced productivity with limited engineering procedure.

Pushing Product to Its Limitation: Systems Metabolic Engineering of Corynebacterium Glutamicum for Glutamate Overproduction

Chen, Z., Tsinghua University
Wu, Y.,

Glutamate is a very important amino acid with an annual production of over 2 million tons. Although large efforts have been done to improve its production, the current yield of glutamate by industrial Corynebacterium glutamicum strains is still far from the therotical maximum. In this work, we use genome-scale model to guide the systems metabolic engineering of *C.glutamicum* for high yield production of glutamate. Main strategies include : 1) increase the precursor availability by enhancement of anaplerotic fluxes and reduction of fluxes towards downstrean TCA pathway; 2) reduction of CO2 emission by modication the fluxes towards pentose phosphate pathway and pyruvate oxidation; 3) introduction of phosphoketolase for acetyl-coA generation; 4) engineering the energy metabolism; 5) engineering glutamate secrete pathway. By combing with media and process optimization, the final engineered strain can produce more than 150 g/L glutamte with the yield ~ 1.0 mol/mol glucose.

Introducing Extra NADPH Consumption Ability Significantly Increases the Photosynthetic Efficiency of Cyanobacteria

Zhou, J., Institute of Microbiology, Chinese Academy of Sciences
Li, Y., Institute of Microbiology, Chinese Academy of Sciences

Increasing photosynthetic efficiency is crucial to increasing biomass production to meet the growing demands for food and energy. Theoretical arithmetic analysis indicated that 2.57 ATP/2 NADPH are generated in light reactions, whereas 3 ATP/2 NADPH are required for CO_2 fixation in dark reactions. This suggests that the light reactions and dark reactions are imperfectly coupled due to shortage of ATP supply, or accumulation of NADPH. Here we hypothesized that solely increasing NADPH consumption might increase the coupling of light reactions and dark reactions, thereby increasing the photosynthetic efficiency. To test this hypothesis, a synthetic NADPH consumption pathway was constructed and introduced into the cyanobacterium *Synechocystis* sp. PCC 6803. The resulting extra NADPH-consuming mutant grew much faster and achieved a higher biomass concentration. Further analyses of the photosynthetic oxygen evolution rate, chlorophyll fluorescence and P700 indicated that the activities of both photosystem II and photosystem I and the light saturation point of the NADPH-consuming mutant all significantly increased, conceiving that the ATP production was increased. Thus, we demonstrated that stimulating NADPH consumption is a promising strategy to improve photosynthetic efficiency and to enable utilization of high-intensity lights.

Shotgun Screening for Investigating the Effective Factors for Producing Heparosan

Yamazaki, S., Ajinomoto Co., Inc.
Ono, T., Ajinomoto Co., Inc.
Mihara, Y., Ajinomoto Co., Inc.
Tonouchi, N., Ajinomoto Co. Inc.
Mori, K. I., Ajinomoto Co., Inc.

Glycosaminoglycans, linear mucopolysaccharides, such as hyaluronate, chondroitin sulfate, and heparosan, have become a focus of research as useful compounds for producing cosmetics and pharmaceuticals. Heparosan produced by *Escherichia coli* K5 strain, a pathogenic bacteria, as a capsular polysaccharide has been reported to be converted into a sulfated polysaccharide with anti-inflammatory activity. The biosynthesis and efflux system of heparosan in *E. coli* have been elucidated; however, there is little information about factors that effectively increase heparosan production. In the present study, we developed a heparosan producer strain from non-pathogenic bacteria, and isolated novel factors for increasing heparosan production.

Heparosan production was studied by introducing the capsule biosynthetic genes of the *E. coli* K5 strain (called Region 2) into *E. coli* B strain. However, heparosan accumulation in the B strain with introduced region 2 was lower than that in the wild-type *E. coli* K5 strain. Enhanced expression of the *rfaH* gene, an activator of heparosan efflux, effectively increased heparosan production. However, little is known about other factors. Therefore, the purpose of the present study was to investigate possible genes for an increased production of heparosan using shotgun cloning.

The study was designed as follows. *E. coli* BL21 (DE3), with introduced Region 2 and enhanced *rfaH* gene expression, was used as the recipient strain. The fragments of *E. coli* K5 genomic DNA (~4 kb) were randomly ligated into the pSTV28 vector and transformed into the recipient strain. We performed screening for novel factors increasing heparosan production using a shotgun cloning approach with the entire genome of the *E. coli* K5 strain. Transformants with increased heparosan production were selected, and the introduced genes were evaluated.

Using this screening approach, 30 genes involved in the carbohydrate metabolism, efflux pump, and stress tolerance were identified from the library of 1,698 clones. Possible functions of the genes and mechanisms of the increase of heparosan production will be discussed.

Producing Industrial Enzyme from Cellulosic Biomass Towards Engineering Xylose Utilizing Ability into Protein Expression Systems

Zhu, T.,
Li, Y., Institute of Microbiology, Chinese Academy of Sciences

Growing energy crisis and environmental pressures led to renewed interest in cellulosic biomass as a renewable feedstock for the production of fuels, chemicals and bulk enzymes, which demands xylose utilization from microbial cell factories. However, many good industrial enzyme producers cannot effectively metabolize xylose.

In this study, using *Pichia pastoris*, one of the most widely used expression systems as an example, we demonstrated successful engineering of xylose metabolizing ability into protein expression systems. A heterologous XI (xylose isomerase) pathway was introduced into *P. pastoris* GS115 by overexpressing the *Orpinomyces* spp. XI or/and the endogenous XK (xylulokinase) gene, and evolutionary engineering strategies were also applied. Results showed that the XI pathway could be functionally expressed in *P. pastoris*. After 50 generation of sequential batch cultivation, a set of domesticated recombinant *P. pastoris* strains with different performance metrics on xylose were obtained. One evolved strain showed the highest xylose assimilation ability, whose cell yield on xylose can even be comparable to that on glucose or glycerol. This strain also showed significantly increased β-mannanase production when cultured on xylose medium. Furthermore, transcription analysis of xylose pathway genes suggested that overexpression of XI and XK might be the key factors affecting effective xylose assimilation.

This study is the first work demonstrating the construction of efficient xylose utilizing *P. pastoris* strains, thus providing a basis for using cellulosic biomass for bulk industrial enzyme production.

Investigating Metabolite Profile and Relative Flux Ratio Under Growth Phase of Ralstonia Eutropha H16

Im, D.,
Oh, M., Korea University

To achieve high yield of biochemical synthesis, understanding overall metabolism, such as the concentrations of metabolites or fluxes among metabolites of microbe is important. However, even with intensively studied microbe strains, it is difficult to estimate *in-vivo* concentration of metabolites or determine active pathways in specific phase due to low stability of metabolites. Advanced analysis instruments, e.g. NMR, GC/MS, LC/TOFMS with proper and strain-optimized sampling methodology have enabled us to determine intracellular metabolite pool sizes and active metabolic pathways. To compare metabolic differences according to carbon sources, *Ralstonia eutropha* H16, a natural PHB producer, was cultivated under 2 different carbon sources and intracellular metabolite samples were analyzed by LC/TOFMS. To obtain reliable *in-vivo* metabolite concentration data, we compared cell harvesting methods and compared cell extraction methods. Results showed that rapid filtration provided more consistent values of metabolite concentrations. On the other hand, *R. eutropha* H16 underwent serious metabolite leakage during methanol quenching possibly by cell membrane damage. This damage resulted in detecting smaller amount of intracellular metabolites and relatively inconsistent metabolite concentrations among samples. Between the two extraction methods, boiling ethanol method showed a little higher-efficiency. We can speculate that chloroform/methanol/water extraction needs more modification to improve the efficiency. The optimal quenching and extraction methods were successfully applied to monitoring metabolite profiling of *R. eutropha* H16 grown with fructose or acetate. To determine relative flux ratio among ED, EMP and PP pathway of *R. eutropha* H16 under exponential growth phase, 1-13C labeled fructose was fed and proteins were harvest at 3 different time points to confirm stable flux ratio. Relative flux ratio was calculated by analyzing ion fragments of proteiongenic amino acids using GC/MS.

Engineering Membrane Phospholipids Increases E. coli Robustness and Production of Biorenewables

Tan, Z., Iowa State University
Yoon, J. M.,
Chen, Y., Iowa State University
Lombardo, J., University of Missouri
Nielsen, D. R., Arizona State University
Shanks, J. V., Iowa State University
Jarboe, L., Iowa State University

Production of fuels and chemicals by using renewable material is limited by toxicity from end-products, many of which generally cause membrane damage to host strains. Herein, two distinct membrane properties, i.e. membrane rigidity and membrane integrity, were improved to construct a robust membrane, by which increases *E. coli* robustness and production of biorenewables.

Firstly, novel trans unsaturated fatty acids (TUFA) were produced and incorporated into the membrane of *E. coli* MG1655 by expression of cis-trans isomerase (Cti) from *Pseudomonas aeruginosa*. While the engineered strain was found to have no increase in membrane integrity, a significant increase in membrane rigidity and decrease in membrane fluidity was observed. As a result, tolerance to exogenously added octanoic acid and octanoic acid production were both increased relative to the wild-type strain. This membrane engineering strategy to improve octanoic acid tolerance was found to require fine-tuning of TUFA abundance. Besides improving tolerance and production of carboxylic acids, TUFA production also enabled increased tolerance in *E. coli* to other bio-products, e.g. alcohols, organic acids, aromatic compounds, a variety of adverse industrial conditions, e.g. low pH, high temperature, and also elevated production of styrene, another versatile bio-chemical product.

Secondly, since a phospholipid molecule generally consists of a hydrophilic phosphate head and two hydrophobic fatty acid tails and most previous engineering studies attempted to engineer membrane fatty acid tails, little was known about effect of phospholipids head engineering. To this end, we proposed the idea that engineering phospholipid head to booster the robustness of microbes. Specifically, key enzymes candidates which are responsible for different phospholipids biosynthesis were chosen and engineered in *E. coli* MG1655. Results showed that increase of phospholipid A significantly increased membrane integrity and membrane electrochemical potential. In the case of fatty acids, this membrane engineering was also found to increase both tolerance to exogenously added fatty acids by 29.1% at most and fatty acids production by 42% relative to the wild-type strain. There was a positive relationship between phospholipid A content and tolerance increase to fatty acids. Besides increasing tolerance and titer of carboxylic acids, phospholipid A production also enabled increased tolerance of *E. coli* to other important bio-products, e.g. ethanol, hexanol, toluene, styrene. Moreover, tolerance to several important inhibitors in hydrolysate of lignocellulose, e.g. furfural, hydroxymethylfurfural (HMF), acetate and a variety of adverse industrial conditions, e.g. low pH, high temperature, were also elevated by phospholipid A production.

Both of the two membrane engineering strategies permitted enhanced growth due to alleviating bio-product toxicity, demonstrating the general effectiveness of this membrane engineering strategy towards improving strain robustness.

Metabolic Engineering of Clostridium Pasteurianum

Brost, R., Institute of Bioprocess and Biosystems Engineering
Zeng, A. P., Hamburg University of Technology

Bioproduction of fuels and chemicals has become a research field of increasing interest, as rising CO_2 levels in the atmosphere and growing awareness of its consequences indicate the need for more sustainable alternatives compared to the petroleum-based industries. *Clostridium pasteurianum* is a gram-positive, spore-forming anaerobe which is of great industrial interest as it can produce not only high-energy biofuels such as butanol and ethanol but also 1,3-propanediol which is a precursor for the production of various polymers. *C. pasteurianum* has the ability to utilize glycerol, a waste byproduct from biodiesel and bioethanol production, as a substrate for solvent production [1]. Furthermore, *C. pasteurianum* has recently been shown to be potentially electroactive [2], which makes it a good candidate for development of electricity-driven bioprocesses.

Whereas metabolic engineering has already been applied to various clostridium stains, efficient genetic tools are still highly demanded for *C. pasteurianum*. Genetic manipulation of *C. pasteurianum* is expected to overcome the low productivity of current strains and to further increase its bioindustrial potential. We have developed genetic tools for metabolic engineering of *C. pasteurianum* in order to create a platform microorganism that can be adapted to different substrates and produce increased amounts of desired products by making use of its electroactivity. To this end *E. coli-C. pasteurianum* shuttle vectors were created for homologous and heterologous overexpression in *C. pasteurianum*. Transformation of *C. pasteurianum* was facilitated by *in vivo* methylation of foreign DNA to protect shuttle vectors from clostridial restriction endonucleases. Strategies to further increase transformation efficiency and to enable efficient gene knockouts in *C. pasteurianum* are currently unter development.

References

[1] W. Sabra, C. Groeger, P.N. Sharma, and A.-P. Zeng (2014) Appl. Microbial and Biotechnol. 98:4267–4276.

[2] O. Choi, T. Kim, H. M. Woo, and Y. Um (2014) Scientific reports, 4.

Controlling Bacterial Transcription with an Archaeal Transcription Regulator

Liu, H.,
Bervoets, I., Free University of Brussels
Peeters, E.,

During the bottom-up engineering of synthetic genetic circuits, it is essential to minimize cross-talk with the host metabolism and to attain orthogonality. In this context, it is a promising approach to leverage gene regulatory parts from across the borders of the bacterial and archaeal domains of life. Archaea harbour a simplified version of the eukaryotic RNA polymerase II machinery, with a 13-subunit RNA polymerase and additional basal transcription factors TATA binding protein and transcription factor B. In contrast, bacterial and archaeal regulatory transcription factors are similar and typically harbour two domains of which one helix-turn-helix DNA-binding domain. In this work, we aim to build an orthogonal genetic circuit by introducing an archaeal transcription factor, the Lrp-type beta-alanine-responsive activator BarR from *Sulfolobus acidocaldarius*, in a bacterial system.

Two compatible plasmids were introduced into *E. coli*, one harbouring a *barR* gene under control of an IPTG-inducible promoter and one harbouring promoter/operator sequences fused to a fluorescent reporter gene. Upon combining the natural archaeal *barR* promoter/operator region to the reporter gene, expression was observed indicating that despite the fundamental difference between the respective transcription machineries, the bacterial RNA polymerase recognizes and initiates transcription from an archaeal promoter region. Moreover, upon inducing BarR expression, a strong repression was observed, presumably due to steric promoter occlusion. Next, we performed similar experiments using plasmid constructs in which different versions of the BarR operator were fused to a bacterial promoter and the reporter gene. This revealed that fusion constructs harbouring a single operator site displayed a slight BarR-dependent transcriptional activation, but only in the stationary growth phase. This activation was not observed anymore when cells were grown in the presence of beta-alanine. Altogether, this work demonstrates that prokaryotic transcription factors have a flexible nature and are capable of interacting with both bacterial and archaeal transcription machineries, in a positive or negative manner.

Engineering Outer Membrane Proteins to Increase E. coli Membrane Integrity and Production of Fatty Acids

Tan, Z., Iowa State University
Black, W.,
Jarboe, L., Iowa State University

Constructing microbial biocatalysts producing biorenewables at economically viable yields and titers is often hampered by toxicity of products. For production of fatty acids, membrane damage is deemed as the principal mechanism of toxicity, particularly in regards to membrane integrity. Previous engineering studies have attempted to engineer membrane fatty acids of phospholipids ("membrane wall"), with the goal of increasing *E. coli* fatty acid tolerance and production, have had mixed results. Herein, we used a novel approach to reconstruct the *E. coli* membrane by engineering membrane proteins ("membrane gates"). Two outer membrane proteins, FX and OP were chosen for engineering. Results showed that, OP mutant not only increased short chain fatty acids (SCFA) production, but also increased long chain fatty acids (LCFA) production. However, although FX and FY were considered as partners in transport of fatty acids, their engineering effects on FA production are totally different. Knockout of FY increased FA production, while FX mutant decreased LCFA production significantly by 34% ($P<0.01$). Instead, elevated expression level of FX contributed to LCFA production. Further membrane characterization results showed that, either disruption of OP or FX overexpression increased membrane integrity remarkably while disruption of FX decreased integrity during FA production process. There was a positive relationship between FX abundance and LCFA titer within a certain range. Besides its individual effect, the two engineering manipulations, disruption of OP and FX overexpression have a synergistic role in further increasing fatty acids production. These engineering expands the understanding of membrane proteins in strengthening membrane properties and strain robustness.

Omics Analyses for Hyper Glucose-Tolerance Using Acetic Acid Bacteria

Hadano, H., BOST Kindai Univ.
Takebe, S., BOST Kindai Univ.
Matsushita, K., Yamaguchi University
Azuma, Y., BOST Kindai Univ.

Acetic acid bacteria (AAB), forming a divergent phylogenetic group within alpha-proteobacteria, are characterized by their abilities to transform alcohols and sugars into the corresponding organic acids. The capabilities have been historically used for vinegar fermentation from ethanol. Whereas, natural habitats of AAB are widespread niches accumulating sugars, e.g. fruits, flowers and soils near fruit plants. Among AAB, two phylogenetically closed species, *Tanticharoenia sakaeratensis* and *Asaia bogorensis*, show a distinctive ability to grow in media containing 30% glucose. Interestingly, both of AAB could not multiply in media containing even 2% of salt, which gives lower osmotic pressure than glucose 30%. It indicates that osmotic tolerance of the AAB is not for general substances but specific to sugars, and that the AAB must include mechanisms for the hyper glucose tolerance. To clarify mechanisms for the hyper glucose-tolerance, comparative omics analyses, including genome DNA sequencing, transcriptome and proteome analyses, were performed. The results illustrated that the two AAB altered expressions of similar genes and systems under different conditions with low and high glucose concentrations. In both AAB gene expressions for glycolysis and pentose phosphate pathways decreased, and ones for antioxidant enzymes (such as superoxide dismutase (SOD) and peroxidase) and cytochrome o ubiquinol oxidase increased. On the contrary, the omics analyses indicated a variety uniqueness of each bacterium. For instance, under high glucose conditions, *T. sakaeratensis* repressed a main energy metabolism including NADH dehydrogenase (complex I) and induced gene expression of pyruvate decarboxylase, whereas *A. bogorensis* promoted expressions of more stress responsible genes involved in anti-oxidation, such as DNA starvation/stationary phase protection protein (Dps) and osmotically inducible peroxiredoxin OsmC. Because the complex I is a main superoxide producer, *T. sakaeratensis* needs less antioxidant enzymes than *A. bogorensis*. We herein propose that enzymes managing oxidative stresses are engaged in overcoming the high glucose stress, and the acquisitions of the glucose tolerance systems in the two AAB were evolutionally independent.

This work was financially supported by the Advanced Low Carbon Technology Research and Development Program (ALCA), and a Grant-in-Aid for Scientific Research from the Ministry of Education, Culture, Sports, Science and Technology of Japan (KAKEN-HI: 22510222).

Transcription Activator-like Effectors: Adding Another Layer to the Onion of Bacterial Gene Regulation

Politz, M. C., University of Wisconsin-Madison
Copeland, M. F., University of Wisconsin-Madison
Johnson, C., University of Wisconsin-Madison
Markley, A. L.,
Pfleger, B. F., University of Wisconsin Madison

Over the past decade, novel *trans*-acting regulatory elements that can be reprogrammed to repress any desired gene of interest have been discovered. The flexibility offered by these *trans*-regulators has the potential to revolutionize the fields of synthetic biology and metabolic engineering by allowing researchers to create layers of synthetic regulation overtop of endogenous regulatory mechanisms. One such *trans*-regulatory element is the class of proteins known as Transcription Activator-Like Effectors (TALEs). These proteins, which were originally identified as virulence factors in plant pathogenic bacteria, have gained traction in biotechnology applications due to their customizable DNA binding specificity. Here, we demonstrate that a TALE designed to bind the *lac* operator is able to inhibit transcription initiation in *E. coli*. We have also developed a TALE induction system mediated by protein degradation. By expressing a site-specific protease any TALE of interest constructed using our novel scaffold can be degraded, thereby promoting target gene expression. Understanding the attributes of TALEs as well as other customizable *trans*-acting regulatory systems, such as CRISPR/Cas and sRNAs, will enable researchers to optimize any genetic system of interest.

Constitutive Expression of CO-Dehydrogenase and CO-Hydrogenase in Citrobacter Amalonaticus Y19

Ainala, S. K., Pusan National University
Seol, E., School of Chemical and Biomolecular Engineering
Park, K. S.,
Park, S., Pusan National University

Citrobacter amalonaticus Y19 has a great potential as a biocatalyst for production of hydrogen (H_2) from oxidation of carbon monoxide (CO), via so-called the water gas shift reaction (WGSR). It has a high H_2 production activity and can grow heterotrophically on various sugars. However, transcription of the gene clusters for CO-dependent H_2 production, involving CO-dehydrogenase (CODH), CO-dependent hydrogenase (CO-*hyd*) and their maturation factors (CO-*hyp*), strictly required the presence of CO to activate the transcriptional regulatory protein CooA and glucose-deficient condition to avoid carbon catabolite repression (CCR). To express CO-dependent H_2 production activity constitutively while growing Y19 on glucose, the native CO-inducible promoters were replaced by strong and engineered constitutive promoters. All promoter-replaced strains (Y19-PRs) showed CO-dependent H_2 production activity while growing on glucose in CO-free condition. The best engineered strain Y19-PR1 showed a similar CO-dependent H_2 production activity as wild type Y19. Compared to wild type Y19, transcription of the CODH gene cluster in Y19-PR1 increased by 1.5-fold but that of the CO-*hyd* operon genes remained at a similar level. The Y19-PR1 should be considered a better biocatalyst than its wild type counterpart due to its constitutive expression of the CO-dependent H_2 production activity during the growth on glucose.

Discover Variations, Building Blocks or Pathways in Your Research More Effectively with Integrated Omics Technologies

Fu, A., Wuhan University

Discover variations,building blocks or pathways in your research more effectively with in tegrated omics technologies

As we launch our Metabolic Engineering research, probably, the first step of our efforts is to pick out useful genetic elements or functional pathways in some specific organisms or available databases. Thanks to the advancing sequencing technology, especially to the si ngle molecular technologies which is characterized by the super long sequencing reads an d simultaneously detection of epigenomic information, nowadays, these mining works in Wuhan institute of biotechnology (WIB) have been accelerated substantially. The public service platform of WIB consist of six major processional platforms of DNA sequencing , Mass spectrometry, high throughput screening.etc. With the integration of those platform s, we have already provided our partners hundreds of complete genomes of microbes, fun gal and parasite genomes with full-length assembled chromosomes, viral genomes from single long reads, at least 10X fold i mproved assembles of plants,animals and human genomes,highest, along with services of isoforms and new genes detection with full-length RNA sequencing, subsequent metabonomics analysis, high throughput protein exp ression and identification. Moreover,In addition to its experimental expertise, WIB was al so renowned for its bioinformatics talent —
a trait that continues with several groups at the WIB. One of our strengths is that we go b eyond generating efficient, high quality omics data.

Manipulation of Methyl-Directed Mismatch Repair System to Engineer the Robustness of Clostridium Tyrobutyricum

Zhu, L. Jr., Nanjing Tech University
Wu, Q. Jr., Nanjing Tech University
Jiang, L. Sr., Nanjing Tech University
Huang, H., Nanjing Tech University

High robustness performance of industrial microbes is crucial for successful bioprocesses. Microbial production of toxic products and tolerance to biomass hydrolysates requires robust cell growth and metabolism under tough conditions. To improve the aero- and acid tolerance of the organic-acid-producing *Clostridium tyrobutylicum*, a novel adaptive evolutionary strategy based on stress-induced mutagenesis (SIM) was developed using non-dividing cells. First of all, the *mutS/L* operon essential for methyl-directed mismatch repair (MMR) activity was inactivated from the genome of *C. tyrobutylicum* to generate hypermutable cells with over 300-fold increased mutation rates. Secondly, a periodic SIM based adaptive evolution procedure, which synchronized the mutagenesis and the selection process in a single plate-incubation step, was developed using the *mutL*-deleted mutant. *C. tyrobutylicum* mutants tolerant to high concentrations of butyric acid (100 g/L), NaCl (90 g/L), and high temperature (55 °C) were obtained. Combined transcriptome and phenome analysis of *C. tyrobutylicum* with deficient MMR activity was carried out to better understand the global effects on the regulatory networks. Our analysis showed that 95 (2.1%) of all *C. tyrobutylicum* genes were induced and 178 (4.0%) genes were repressed, including those for trehalose biosynthesis, nucleotides biosynthesis, carbon metabolism, amino acid utilization, except for acid resistance. Also regulated were the master regulators (ArcA, EvgA, H-NS and RpoS) and gene/operon-specific transcription factors (GadX, GadW, AppY, YdeO, KdgR). These results indicate that stress-induced adaptive evolution in non-dividing cells is an effective approach that can improve microbial tolerance against various stresses and generate robust microbial strains suitable for production of chemicals.

Oligo- and dsDNA-Mediated Genome Editing Using a Teta Dual Selection System and Its Application in Escherichia coli

Kim, K., Ulsan National Institute of Science and Technology
Ryu, Y. S., UNIST
Chandran, S. P., Ulsan National Institute of Science and Technology (UNIST)
Lee, S. K., Ulsan National Institute of Science and Technology

Escherichia coli has been used widely as a cell factory, which can make various value-added materials by engineering the genome or/and introducing the heterologous genes related to the target material-producing biosynthetic pathway of interests. Thus, precise and seamless genome modification method to engineer targeted genome is needed to construct potent biosynthetic system. Herein, we report a promising method in *E. coli* that relies on the insertion of an optimized *tetA* dual selection cassette followed by replacement of the same cassette with short, single-stranded DNA (oligos) or long, double–stranded DNA and isolation of recombinants by the optimized selection process on $NiCl_2$. This method could be rapidly and successfully used for genome engineering, including deletions, insertions, replacements, and point mutations.

By using our method, we tested the strength of various promoters by expressing *gfp* (Green Fluorescent Protein) under them on the chromosome using the TetA dual selection system. Previously, metabolic engineering and synthetic biology in precedent researches usually introduce target genes on plasmid due to its ability to control expression level according to their various copy number from low to high and easy introduction to desired strains. There are some limitations about its maintenance such as instability and robustness. The copy number variation of plasmid can affect consistent expression of the genes related to specific metabolic pathway, which can lead to variation of yield of target materials among the same strains. Besides, the plasmid is easy to insert into the desired strain by transformation, but also easy to escape from the strain according to condition like the origin of replication, at the same time. Thus, we have tried to combine the genome editing method and promoter engineering for constant and high level gene expression system on genome. We tested various inducible promoters, constitutive promoters and generated new promoters for genome-level gene expression. Moreover, increment and optimization of gene expression and construction of biosynthetic system on genome aim to overcome the disadvantages of plasmid-based metabolic engineering and promise stable expression on genome. In conclusion, this method can be applied to achieve scarless, proficient, and targeted genome editing for strain construction, metabolic engineering, and synthetic biology.

A Rapid and Robust Gene Knock-in/Knock-out System in Geobacillus Thermoglucosidasius Based on the Pyre marker.

Sheng, L., University of Nottingham

Dr L. Sheng, Prof N. Minton, University of Nottingham, UK.

The genus *Geobacillus* has increasingly attracted interests as thermophilic bioprocessing platforms, due to desirable properties such as catabolic versatility [1,2]. To fully utilize their potentials, both product yield and specificity needs to be improved via metabolic engineering, requiring gene knock-in/knock-out tools. Here, adopting the ACE strategy by heap et al [3] we developed a set of rapid strategies for gene knock-in/knock-out in *Geobacillus thermoglucosidasius*. The product of *pyrE* gene catalyzes an essential step in uracil biosynthesis but also metabolizes 5-fluororotic acid (5-FOA) to toxic metabolites, hence its disruption gives both positive and negative selectivity. A *ΔpyrE strain,* involving truncating an internal part of the gene leaving the first 300 bp and stop codon intact was obtained. For gene knock-in, by using a vector with homology arms that repaired the disrupted *pyrE* while simultaneous integrates DNA cargo*, the* trans-2-enoyl-CoA reductase gene from *T. caldaria* was successfully integrated at the *pyrE* locus, confirmed by both PCR and Western. The positive selectivity based on growth on uracil free medium resulted in 100% correct integrant with a turnaround time of just 4 days. For gene knock-out, the *ΔpyrE* phenotype was firstly complemented by using heterologous *pyrE* amplified from *G. Kaustophilus and G. thermoleovorans* fused to and co-expressed from the antibiotic selection marker. By co-utilizing uracil auxotrophy and the temperature sensitivity of the complementation vector we developed, single-crossover integrant was obtained at 100% efficiency when grown above 60 °C in the absence of uracil, avoiding the need for screening. Double crossover integrant, either corresponding to wild type or desired deletion mutants, were then selected based on the counter-selectivity of 5-FOA resistance. Turnaround time was only 6 days with high efficiency that screening of only 8-12 colonies was adequate. Lastly, the truncated *pyrE* was repaired back to wild type following the same procedure for gene knock-in. To exemplify the rapidity and robustness of the technique, the TMO production strain [2], consisting of an *ldh* deletion, *pfl* deletion and *pdh* up-regulation via promoter replacement, were recreated in only 30 days. Although the intermediate strains (*Δldh, Δldhpdhup)* showed the expected fermentation profile, the final strain didn't. While repeated construction of the strain led to no avail, Genome scale analysis revealed potential SNPs in both TM242 and our strain in contributing to the phenotype observed.

In summary, we established a full set of recombination based genetic tools, with very high efficiency and robustness, in the host *G. thermoglucosidasius*, offering new scopes for metabolic engineering in this organism.

1. Suzuki, H., Murakami, A. & Yoshida, K. Counterselection system for Geobacillus kaustophilus HTA426 through disruption of pyrF and pyrR. *Appl. Environ. Microbiol.* **78,**7376–83 (2012).

2. Cripps, R. E. *et al.* Metabolic engineering of Geobacillus thermoglucosidasius for

high yield ethanol production. *Metab. Eng.* **11,**398–408 (2009).

3. Heap, J. T. *et al.* Integration of DNA into bacterial chromosomes from plasmids without a counter-selection marker. *Nucleic Acids Res.* **40,** e59 (2012).

Integrating Transcriptomic and Metabolomic Data with Thermodynamically Consistent Metabolic Models

Hernandez Gardiol, D. F., EPFL
Pandey, V., EPFL
Hatzimanikatis, V., Swiss Federal Institute of Technology of Lausanne (EPFL)

Constraint based models are a powerful mathematical framework for integrating experimental data from different sources. In this study we demonstrate the added value in combining various levels of data. The first is metabolomics and thermodynamics via Thermodynamics-Based Flux Analysis (TFA) constraints. TFA constraints guarantee the thermodynamic consistency between measured metabolomics levels and flux directionalities. We further introduce transcriptomics-based constraints (derived from IMAT), that allow us to integrate transcriptomics together with TFA constraints for the first time. We apply the methodology in an E. coli genome scale model using available metabolomics and transcriptomics data.

Despite integrating several levels of data, the problem is underdetermined and there was a multitude of alternative solutions that can explain the observed\integrated data. Oftentimes results are only reported for a single solution that is chosen arbitrarily, and alternative possible solutions are ignored. In the studies here we also calculated alternate optimal solutions and we showed that alternative solutions can differ drastically different from one another and they should always be taken into account in order to have unbiased predictions.

We also found that transcriptomics-based predictions differ with and without thermodynamics. Without thermodynamics-based constraints, we found that the transcriptomics-based constraints predict thermodynamically infeasible fluxes at optimality. When we forced consistency with both transcriptomics and metabolomics data sets simultaaneously, we produced more precise and accurate predictions of metabolic fluxes, according to C13 measurements. These results further emphasize that all models should be thermodynamically consistent.

Quantifying Variation within the Bacterial Species E. coli

Monk, J. M., University of California, San Diego
Koza, A., Technical University of Denmark
Campodonico, M. A., Technical University of Denmark
Machado, D.,
Seoane, J., The Novo Nordisk Foundation Center for Biosustainability
Palsson, B. O., University of California, San Diego
Herrgård, M. J., Technical University of Denmark
Feist, A. M., University of California, San Diego

Escherichia coli strains are widely used in academic and applied research as well as in biotechnology for production of various compounds. Despite its status as a model organism, strain-specific differences and their underlying contributing factors are still not well characterized. These differences have a major impact on cell physiology and for the applied purposes of synthetic biology, metabolic engineering, and process design. In this study, strain-specific differences are quantified in seven widely-used applied biotechnology strains of *E. coli* (BL21, C, Crooks, DH5a, K-12 MG1655, K-12 W3110, and W) using genomics, phenomics, transcriptomics and genome-scale modelling to guide the choice of strain for a given product. Metabolic physiology and gene expression varied widely with downstream implications for productivity, product yield, and titre. Further, these differences can be linked to differential regulatory structure. Analysing high flux reactions and the expression levels of their encoding genes revealed a quantitative link between these sets and show that often, these sets are correlated with strain-specific caveats. Integrated modelling also revealed that certain strains are better suited to produce a given compound or express a desired construct considering native expression states of pathways that enable high-production phenotypes. The result of this study is a resource comparing strains in an important model species and a general strategy for choosing a host strain or chassis selection for applied biotechnology.

Simultaneous Production of Prodigiosin and Serrawettins As the Main Bioactive Compounds By Serratia surfactantfaciens Sp. Nov

Su, C.,
Sun, Y.,
Li, Z.,

A Gram-negative, red-pigmented bacterial strain YD25^T was isolated from rhizosphere soils under continuously planting burley tobacco in Fujian Province, China. The bacterium was characterized by a polyphasic approach, including 16S rRNA gene sequence analysis, multilocus sequence analysis (MLSA), ortho average nucleotide identity (OrthoANI) calculations, genome to genome distance calculator (GGDC) analysis, fatty acid methyl esters analysis, determination of DNA G+C content and an analysis of phenotypic features. On the basis of 16S rRNA gene sequence similarity, strain YD25^T was shown to be related to *Serratia nematodiphila* DSM 21420^T (99.53 %) and *S. marcescens* LMG 2792^T (99.47 %), respectively. However, the results of MLSA, OrthoANI calculations, GGDC analysis as well as physiological and biochemical tests allowed genotypic and phenotypic differentiation of strain YD25^T from recognized *Serratia* species. On the basis of polyphasic characterization data obtained in the present study, the strain YD25^T as a new species for which the name *Serratia surfactantensis* sp. nov., is proposed, with YD25^T (=CCTCC M2013617^T) being designated as the type strain.

By liquid chromatography, mass spectrometry and nuclear magnetic resonance, the main antibacterial compounds simultaneously produced by *S. surfactantensis* YD25^T were identified to be prodigiosin and serrawettin W2. In addition, *S. surfactantensis* YD25^T produces another six homologous of serrawettin W2. So far, *S. surfactantensis* YD25^T is the first one which produces prodigiosin and a mixture of serrawettin W2 analogues simultaneously. According to the antibacterial tests, serrawettin W2 shows potent antimicrobial activity against the tested Gram-positive and Gram-negative bacteria, including nine clinical isolates *Staphylococcus aureus*strains. Meanwhile, serrawettin W2 inhibits the growth of Hela cells and shows less cytotoxicity to nonmalignant cell lines.

Analysis of the draft genome sequence of *S. surfactantensis* YD25^T resulted in the identification of a putative cyclic lipopeptide (CLP) antibiotics serrawettin W2 biosynthesis gene cluster. The biosynthesis gene cluster is a total of 28 kb interrupted DNA sequence containing polyketide synthases (PKS) gene and non-ribosomal peptide synthetases (NRPS) gene. The PKS gene is consisted of a ketosynthase (KS), a keto reductase (KR) and an enoyl reductase (ER) domain. In the putative NRPS gene, there are a total of five modules and each module consists of condensation (C), adenylation (A) and thiolation (T) domains and the five modules harbors one thioesterification (TE) domain. The genome-based identification and structure prediction of serrawettin W2 provides the proof of the chemical analyses in the *S. surfactantensis* YD25^T.

A Crispr/Cas9 Mediated Genome Engineering Examples for Heterologous Carotenoid Production in Saccharomyces Cerevisiae

Roubos, J. A., DSM Biotechnology Center
Verwaal, R., DSM Biotechnology Center
Vonk, B., DSM
Buiting-Wiessenhaan, N., DSM

One of the key drivers for successful metabolic engineering is the efficiency by which genomes can be edited. CRISPR/Cas9 systems have quickly become an important new method for precision genome engineering of yeast and many other species. For example, the introduction of designed double-stranded breaks in *S.Â cerevisiae* allows for rather efficient â??marker-freeâ?? recovery of cells which have been modified by donor-DNA mediated homologous recombination. Although, various strategies have been published for the use of CRISPR/Cas9 systems in yeast, there is room for further improvement. Key elements of a CRISPR/Cas9 system are the expression of Cas9, the design and expression of the sgRNA(s), and the design and introduction of donor DNA fragments. We present various novel strategies for the introduction and use of Cas9, sgRNA and donor DNA to efficiently introduce mutations and heterologous pathways as well as the deletion of genomic DNA fragments. The introduction of a carotenoid pathway with three genes from *Xanthophyllomyces dendrorhous* in *Saccharomyces cerevisiae* is our testbed. Both singleplex and multiplex approaches will be demonstrated. The resulting carotenoid producing strains are characterized using proteomics and direct product measurements.

Development of Software Tools for Effective Analysis and Application of Fungal Secondary Metabolism Gene Clusters

Machida, M., Technology Research Association of Highly Efficient Gene Design (TRAHED)
Umemura, M., Technology Research Association of Highly Efficient Gene Design (TRAHED)
Takeda, I., Tokyo University of Agriculture and Technology
Asai, K.,
Ohyama, A., In silico Biology, Inc.

Genome sequencing of filamentous fungi revealed secondary metabolism (SM) genes much more than expected before the sequencing by searching sequence similarity to so-called □gCore Genes□h such as PKSs and NRPSs. Furthermore, discovery of SM gene clusters that could not be predicted only from sequence similarity such as the kojic acid biosynthesis gene cluster strongly suggests existence of far more SM genes[1)] in the genomes. In order to analyze fungal SM genes more comprehensively, we have developed software tools for the motif-independent prediction of SM biosynthesis gene clusters, MIDDAS-M[2)] and MIPS-CG[3)]. Recently, the gene cluster producing a circular peptide compound, ustiloxin B, but without NRPS was successfully determined by MIDDAS-M and successive experimental analyses. This was the first discovery of RiPS (Ribosomal Peptide Synthesis) pathway gene clusters from filamentous fungi[4)]. Differently from MIDDAS-M, which requires transcriptome information, MIPS-CG is able to predict SM gene clusters only from genome sequences without knowledge of SM gene sequences including the □gCore Genes□h. By comparing ten fungal genome sequences, MIPS-CG detected 21 SM gene clusters from the 24 gene clusters that have been identified from the ten species to date.

Once SM gene clusters are predicted, their function should be experimentally validated by referring to the known SM gene clusters. In order to facilitate the analysis, we have developed a software tool, □gin silico Molecular Cloning Genome Design Suite□h (IMCDS). IMCDS allows visual analysis and comparison between predicted and known SM gene clusters under multiple representation capability including gene expression and nucleotide sequence views. It includes automatic multi-color visualization of genes based on their function and designing primers for the experimental analyses such as preparation of a DNA fragment to disrupt a target gene or to replace its promoter for over-expression. IMCDS also includes multiple tools for wide-variety of sequence analyses including gene prediction, similarity search, translation of nucleotide to amino acid sequences, prediction of secondary structure, in silico plasmid construction and so on. Our software tools are thought to dramatically facilitate analysis and application of SM gene clusters

1) Terabayashi, Y, et al., *Fungal Genet Biol*, **47**, 953-61 (2010).

2) Umemura, M. et al., *PLoS One*, **8**, e84028 (2013).

3) Takeda, I. et al., *DNA Res*, **21**, 447-57 (2014).

4) Umemura, M. et al., *Fungal Genet Biol*, **68**, 23-30 (2014).

Novel Propeptides for Enhanced Secretion from Lactococcus Lactis

Ow, D. S. W., Bioprocessing Technology Institute, A*STAR
Lim, P. Y., Bioprocessing Technology Institute, A*STAR
Tan, L. L., Molecular Engineering Laboratory, A*STAR
Wong, F. T., Molecular Engineering Laboratory, A*STAR

Lactic acid bacteria (LAB) are a "generally regarded as safe" microorganisms which are traditionally used in food fermentation and as chassis for heterologous protein production. In recent years, they have also emerged as delivery tools for protein therapeutics. However, these applications are heavily depended on the secretion efficiencies and capabilities of LAB. Secretion of proteins will not only simplify the purification process but also oftentimes increase total protein production. Besides utilizing an efficient signal peptide, importance of a suitable propeptide for soluble, secreted heterologous protein production was also emphasized in previous studies (1, 2). To expand the propeptide library for protein secretion in LAB, we identified three propeptide candidates from *in silico* mining of LAB genomes. Here, we will examine and compare these propeptides for enhanced protein secretion in *Lactococcus lactis*. These will also be compared to a known and highly productive propeptide "LEISSTCDA" (1). Secretion efficiencies and yields of these propeptides will be characterized and examined for a variety of model proteins. From the success of these propeptides to enhance secretion, we predict that this collection of propeptides will be invaluable to future heterologous protein secretion optimisation in LAB.

(1) Le Loir Y, Gruss A, Ehrlich SD, Langella P. A Nine-Residue Synthetic Propeptide Enhances Secretion Efficiency of Heterologous Proteins in*Lactococcus lactis. Journal of Bacteriology*. 1998;180(7):1895-1903.

(2) Le Loir Y, Nouaille S, Commissaire J, Brétigny L, Gruss A, Langella P. Signal Peptide and Propeptide Optimization for Heterologous Protein Secretion in *Lactococcus lactis. Applied and Environmental Microbiology*. 2001;67(9):4119-4127

Engineering Production of Cholesterol Drugs Simvastatin and Lovastatin in Saccharomyces Cerevisiae

Bond, C. M., University of California Los Angeles
Tang, Y., University of California, Los Angeles

Polyketide-derived natural products are a rich source of bioactive molecules, and fungal polyketides have been an important source of pharmaceuticals. These molecules are challenging to directly synthesize due to their complexity, so they are usually produced industrially through fermentation. This can be limiting due to hosts that are difficult to engineer for production of more potent analogs or increased titers. As there is great interest in producing natural product analogs, a more tractable host is desired. The yeast *Saccharomyces cerevisiae* is a well-studied organism that has long been in use industrially. An engineered drop-in yeast host for the high production of fungal polyketides and their analogs would shift the industrial paradigm of native host polyketide production and enable greater ease in the characterization of new polyketide clusters.

The lovastatin pathway from filamentous fungi *Aspergillus terreus* was used for the initial development of such a yeast host because of the importance lovastatin and its analogs play as cholesterol medications. Episomal and genomic expression of the seven genes of the lovastatin pathway from *Aspergillus terreus*, or engineered versions of enzymes, has led to the production of lovastatin and blockbuster cholesterol-lowering drug simvastatin in *S. cerevisiae*. Over 35 mg/L of the first major precursor, DMLA, has been produced de novo in yeast using the promiscuity of enzymes from the lovastatin pathway and the analogous compactin pathway from *Penicillium citrinum*. The strains and strategies developed in the process of optimizing lovastatin and simvastatin production in yeast will introduce a cheaper method of synthesizing the blockbuster drug simvastatin and provide tools for improved production and engineering of other fungal polyketide pathways, smoothing the pipeline from the characterization of new natural product clusters to the industrial-scale production of the polyketides they encode.

Improve the Protein Synthesis through Engineering the Endoplasmic Reticulum Proliferation Pathway in Pichia Pastoris

Zhao, J. Sr.,
Tian, C. Sr.,

Abstract

Pichia pastoris□one of the most widely used expression system, has been developed as production platform for thousands of proteins successfully, but the productivities of most target proteins were still low. Using the metabolic engineering for the modification of the *Pichia pastoris* host cell to improve the heterologous protein expression has been becoming a newly research hotspot. Endoplasmic reticulum is the major place for synthesis and folding of the nascent proteins, and the proliferation of endoplasmic reticulum has been thought as one of the necessity to enhance the heterologous protein production. However, rare study has focused on engineering this important organelle for enhancing the protein production in *P. pastoris*. Previously, the gene encoding DGKpp involved in membrane regeneration and phospholipid metabolism showed significantly upregulated expression levels in the engineered *Pichia* strains with the capability of high heterologous protein production and it is a new hint to convince us to do the ER proliferation engineering for heterologous protein production. In this study, combining the bioinformatics, metabolic engineering and modern biological techniques, the molecular function of DGKpp on the endoplasmic reticulum proliferation will be explored. Finally, two industrial enzymes will be selected to test their heterologous expression in the ER-proliferated *Pichia* chassis.

Keywrods Metabolic engineering, *Pichia pastoris*, Endoplasmic reticulum proliferation, phospholipid metabolism

Boosting Bioprocess Performance By Yeast Extracts: It´s All in the Mix

Bekers, K. M.,
van de Werf, M. J.,

Yeast extract is widely used in industry to boost cell growth and enhance productivity, titer and product yields in fermentation processes. Opposed to chemically defined media, it has the benefits of providing not only ready to use complete building block for cells, but it also contains a range of cofactors, vitamins and growth factors. However, being such a complex mixture of nutrients, generally little is known on which components in the yeast extract are responsible for its beneficial results. Moreover, it is generally not known to biotechnologists that yeast extract can be highly different in composition. As a result, process optimization by adding yeast extract is generally only reviewed on a very basic level, and only a few different yeast extracts are tested.

To review the impact yeast extract can have on fermentation processes, a study was conducted in which the effect of three highly different yeast extract (combination)s on the performance of a range of micro-organisms was evaluated. In particular, (combinations of) Ohly® KAF/KAT (a yeast extract high in free amino acids), Ohly® Flav-R-Ound (a yeast extract containing relatively high nucleotide concentrations) and Ohly® Flav-R-Tide (a yeast extract high in peptides) were tested in this study. They were reviewed with a 10-point experimental design to find an optimal blend for a range of industrial relevant organisms. In addition, the same blend optimization was reviewed for one organism at different conditions.

The results show that there is clearly not a ´golden standard´ of a yeast extract that works optimally with all micro-organisms, or even for the same organism cultivated under different conditions. With that, it also shows a potential for process optimization, if the right combination of yeast extracts is supplied. The experimental approach of testing ten blends as applied in this research shows to be a simple pragmatic approach to improve industrial production processes without the need for indepth knowledge about the exact composition of the yeast extract.

A Small-Scale Multibioreactor System for Fast and Scalable Bioprocess Optimization of Microbial Cell Factories for Small Molecule Production

Klein, T., Technical University of Denmark
Schneider, K., Technical University of Denmark
Borodina, I., Technical University of Denmark
Kildegaard, K. R., Technical University of Denmark
Lennen, R. M., University of Wisconsin-Madison
Herrgård, M. J., Technical University of Denmark
Luo, H., Technical University of Denmark

The fast advance of recombinant DNA technologies allows for rapid development of novel microbial cell factories with native and artificial pathways for small molecule production. Due to the complexity of biological systems, strain design often ends in a variety of potential production strains that need to be evaluated and characterized. Additionally, the physiology of the production strain often has a tremendous impact on strain performance, which calls for cultivation conditions comparable to the final production process. To address these issues, various small-scale multibioreactor systems have emerged in the last decade. The main requirements for such systems are (i) flexibility with regard to substrates and forced nutrient limitations, control of growth rates, cultivation conditions like pH or temperature and (ii) transferability of these results to larger fermentation scales. Physiological characteristics like yields and rates are independent of the cultivation scale and thus well transferable into production scale bioreactors. Continuous cultivation is the method of choice for thorough and precise physiological characterization of microbial cell factories. Here we present a system of 24 parallel bioreactors with 7 mL working volume for continuous cultivation with online monitoring of dissolved oxygen. Stirring allows for high oxygen transfer rates with kLa values up to 120 h^{-1}. We show two case studies where this system was applied for (i) finding optimal cultivation conditions (feeding rate, pH and temperature) for production of the human molecule melatonin with engineered *E. coli* and (ii) investigating the impact of different nutrient limitations (C, N, P) on the physiology and productivity of 3-HP producing *S. cerevisiae*. In both cases, results from the small-scale system runs were confirmed in 1L fed-batch fermentations to demonstrate their scalability and show the applicability of the small-scale cultivation system for fast characterization and optimization of microbial bioprocesses.

Beyond Amino Acids - Systems Metabolic Engineering of Corynebacterium Glutamicum for Materials and Health Care Products

Becker, J., Saarland University
Kind, S.,
Schäfer, R.,
Rohles, C.,
Giesselmann, G.,
Zelder, O.,
Schröder, H.,
Wittmann, C., Saarland University

For now almost six decades, *Corynebacterium glutamicum* has established as major industrial producer for l-amino acids, especially l-lysine [1, 2]. Systems biology and systems-level engineering substantially broadened the product spectrum, and leveraged the integration of *C. glutamicum* into the rising era of biorefinery application. Today, its product spectrum has the potential to serve consumer markets for health, nutrition, textiles, housing, energy, and agriculture [2]. In this work, we developed genetically defined producer strains of *C. glutamicum* for aspartate-derived products being applicable as materials and health-care products.

The production of diaminopentane – a valuable monomer building block for polyamide production – was realized on basis of the lysine-hyperproducer *C. glutamicum* LYS-12 [1]. Extending the terminal pathway by optimized expression of lysine decarboxylase [3], combined with elimination of by-product formation [4] and transport engineering [5] bred the streamlined *C. glutamicum* DAP-16 strain that produced diaminopentane with a high yield of 0.5 mol mol^{-1}, a final titer of 88 g L^{-1}, and a productivity of 2.2 g L^{-1} h^{-1} during fed-batch fermentation [6]. Diaminopentane was then purified to 99% purity by solvent extraction and distillation. Subsequent polymerization with sebacic acid from castor oil produced the 100 % bio-based polyamide PA5.10 exhibiting excellent material properties for industrial application [6]. Production of the chemical chaperon and cosmetic ingredient ectoine recruited a heterologous pathway from *Pseudomonas stutzeri*. Genome-based integration into one of the two lysine-biosynthetic branches enabled ectoine formation with simultaneous down-regulation of carbon flux towards lysine. Full elimination of lysine secretion and implementation of a fed-batch strategy elevated production up to a molar yield of 30 % whereby one of the highest reported ectoine productivities of 6.7 g L^{-1} d^{-1} was achieved [7]. Beyond, basic producers for metabolically related products are currently constructed as proof-of-concepts, awaiting further engineering towards attractive titers and yields.

[1] Becker J, Zelder O, Haefner S, Schroeder H, Wittmann C: From zero to hero – design-based systems metabolic engineering of *Corynebacterium glutamicum* for L-lysine production. ME 2011, 13:159.

[2] Becker J, Wittmann C. 2015. Advanced Biotechnology: Metabolically Engineered Cells for the Bio-Based Production of Chemicals and Fuels, Materials, and Health-Care

Products. AngewandteChemie 54:3328.

[3] Kind S, Jeong WK, Schröder H, Wittmann C. 2010. Systems-wide metabolic pathway engineering in *Corynebacterium glutamicum* for bio-based production of diaminopentane. ME 12(4):341

[4] Kind S, Jeong WK, Schröder H, Zelder O, Wittmann C. 2010. Identification and elimination of the competing N-acetyldiaminopentane pathway for improved production of diaminopentane by *Corynebacterium glutamicum*. AEM 76(15):5175.

[5] Kind S, Kreye S, Wittmann C. 2011. Metabolic engineering of cellular transport for overproduction of the platform chemical 1,5-diaminopentane in *Corynebacterium glutamicum*. ME 13(5):617

[6] Kind S, Neubauer S, Becker J, Yamamoto M, Völkert M, Abendroth GV, Zelder O, Wittmann C. 2014. From zero to hero - Production of bio-based nylon from renewable resources using engineered *Corynebacterium glutamicum*. ME 25:113

[7] Becker J, Schäfer R, Kohlstedt M, Harder BJ, Borchert NS, Stöveken N, Bremer E, Wittmann C. 2013. Systems metabolic engineering of *Corynebacterium glutamicum* for production of the chemical chaperone ectoine. MCF 12:110

Regulation of the Pentose Phosphate Pathway in Saccharomyces Cerevisiae

Kim, S. R., Kyungpook National University
Seol, J.,
Lee, Y.,
Jeong, D.,
Wemmer, D. E.,
Kim, K. H., Korea University Graduate School
Jin, Y. S., University of Illinois at Urbana-Champaign

Saccharomyces cerevisiae is one of the most preferred industrial hosts due to stress tolerance and the ease of genetic manipulation. Numerous efforts have been conducted to extend its substrate range, and a number of studies reported engineered strains fermenting xylose efficiently. However, several questions remain to be solved, and one of them is how the native regulatory systems of engineered *S. cerevisiae* respond to xylose and its intermediates. In a prior study, we reported that the deletion of *PHO13*, encoding alkaline phosphatase, enhanced xylose fermentation significantly. Later, we identified some evidence that Pho13 is involved in transcriptional and metabolic regulations of the pentose phosphate pathway. In this presentation, transcriptomic and metabolomic data will be presented, which led to the discovery of a novel regulatory mechanism of the pentose phosphate pathway.

Rewiring Methanol-free AOX1 Promoter for Recombinant Proteins Expression in Pichia pastoris

Cai, M.,
Shi, L., State Key Laboratory of Bioreactor Engineering, East China University of Science and Technology, Shanghai, China
Wang, X., East China University of Science and Technology
Wang, J., East China University of Science and Technology

The alcohol oxidase 1 (*AOX1*) promoter (P_{AOX1}) of *Pichia pastoris* is frequently used to express recombinant proteins as it leads to high levels of expression in response to methanol. Large amounts of methanol, however, can pose a considerable safety risk in industrial bioprocess. To develop a methanol-free approach, we identified and investigated the roles of three transcriptional suppressors and one activator of P_{AOX1}. Their transcription regulation mechanisms on P_{AOX1} were first investigated. Then, deletion of the three repressors – Mig1, Mig2, and Nrg1 – and overexpression of the activator – Mit1 – lead to P_{AOX1} expression in response to glycerol. A novel *P. pastoris* strain was generated in which Mit1 was overexpressed under the control of P_{AOX1} in the triple deletion mutant Δ*mig1*Δ*mig2*Δ*nrg1*. When cultured in glycerol, GFP regulated by P_{AOX1} achieved higher levels of expression to the wide-type strain induced with methanol. The cellular processes of the triple deletion mutant overexpressing Mit1 were studied systematically by RNA-Seq. This novel strain represents a safe and efficient alternative to the traditional methanol-based expression systems.

Simultaneous Utilization of Glucose and Xylose for Efficient Succinate Production by Engineered Corynebacterium glutamicum

Mao, Y., Tianjin University
Chen, T., Tiamjin University
Wang, Z., Tianjin University

Pentose utilization is considered as an important segment for lignocellulosic biorefinery due to its abundance in hemicellulosic wastes. In this study, *Corynebacterium glutamicum* was engineered for efficient succinate production. The *xylA* and *xylB* genes from *Xanthomonas campestris* and the endogenous *tkt* and *tal* genes were co-overexpressed, which enabled the strain to produce succinate with a high yield of 0.98 g g^{-1} D-xylose. Expression of *araE* from *Bacillus subtilis* further increased succinate production rate by 11%. Finally, the engineered *C. glutamicum* CGL31 strain could simultaneously and completely consume glucose and xylose and produce 109.5 g L^{-1} succinate with a productivity of 6.84 g h^{-1} L^{-1} and a yield of 0.81 g g total sugars $^{-1}$, which suggests it has great potential to efficiently produce succinate from sugar mixtures derived from lignocellulose.

Keywords: *Corynebacterium glutamicum*, xylose, succinate, transketolase, transaldolase

Construction, Model-based Analysis and Characterization of a Synthetic Promoter Library for Fine-tuned Gene Expression in Bacillus subtilis

Liu, D., Tiamjin University
Meng, J., Tianjin University
Wang, Z., Tianjin University
Zhao, X., Tianjin University

Bacillus subtilis is the best characterized Gram-positive bacteria, possesses unique advantages as a host for producing microbial enzymes and industrially important biochemicals. To extend its productivity, *Bacillus subtilis* needs to undergo further engineering, including the development of applicable promoter libraries. We constructed constitutive promoter library based on -10 -35 consensus sequences recognized by sigA, *Bacillus subtilis* -16 consensus and an AT-rich region upstream. Using green fluorescent protein as a reporter, a total of 5000 promoters were randomly picked and the GFP expression level was investigated. The strength of the library covers a range of 200 fold. Firstly, 225 potential promoters of various strengths were isolated and sequenced for promoter analysis. A model describing the promoter strength was built and validated. Secondly, the library was employed to precisely regulate adenylosuccinate (sAMP) synthase, which is encoded by *purA*, one of the most important enzymes involved in the *de novo* biosynthesis of purine metabolites. We assess the impact of sAMP levels on inosine production. Finally, we developed a strategy to improve promoter strength by constructing the promoter clusters consisted of multiple promoters in tandem stemmed from our constructed library. Application of the promoter clusters in secretory production of endoxylanase proved that it was efficient. This is the first report of synthetic promoters constructed in *Bacillus subtilis*, and our screening strategy together with the use of synthetic promoters of various strengths will contribute to the future engineering of *Bacillus subtilis*.

Enhanced Chiral D-(-)-2,3-Butanediol Production by Engineered Corynebacterium glutamicum from Glucose and Xylose Mixtures

Fu, J., Tianjin University
Mao, Y., Tianjin University
Tao, R., Tianjin University
Hou, G., Tianjin University
Wang, Z., Tianjin University
Chen, T., Tiamjin University
Zhao, X., Tianjin University

The platform chemical 2,3-butanediol has a wide range of applications, and especially can be used as an alternative for biofuel production. Noticeably, chiral pure D-(-)-2,3-butanediol with spatial configuration has unique applications i.e. antifreeze and chiral intermediates. Herein, the non-pathogenic *Corynebacterium glutamicum* (could not naturally produce 2,3-butanediol) was engineered to produce chiral D-(-)-2,3-butanediol from glucose, xylose and arabinose mixtures. Initially, with the purpose of D-(-)-2,3-butanediol production in *C. glutamicum*, its native *butA* gene (coding meso-2,3-butanediol dehydrogenase) was deleted, and an artificial gene cluster, encoding the 2,3-butanediol biosynthetic pathway from *Bacillus subtilis* was introduced. The resulted mutant could produce 20.78 g/L D-(-)-butanediol from 50 g/L glucose in flask experiment in minimum medium with a yield of 0.43 g/ g glucose, and 78 g/L D-(-)-butanediol within 50h in the 5-L fermentor. Subsequently, to synchronously assimilate glucose and xylose as substrate, the xyloseisomerase and xylulokinase coding genes *xylA* and *xylB* from *Xanthomonas campestris* were introduced into *C. glutamicum*. In shake-flask fermentations, 51 g/L D-(-)-2,3-butanediol with a yield of 0.35 g/ g mixed sugar was obtained by the optimum strain CGF32, which offered a novel strategy on engineering microbe as cell factory for the production of high-valued chemicals from renewable resource.

Keywords
Metabolic engineering; D-(-)-2,3-butanediol; *Corynebacterium glutamicum*; xylose.

Synthetic Metabolic Solutions for Enhanced Agricultural Production with Cyanobacteria

Malatinszky, D.,
Valton, M.,
Rowland, J.,
Steuer, R., Imperial College London
Jones, P. R., Imperial College London

The global population continues to increase and food self-sustainability is still an issue for many. Synthetic biology can contribute towards enhancing our food self-sustainability and simultaneously minimizing the impact of the agricultural industry on planetary resources. Two such strategies can reduce our reliance on fossil fuels and enhance productivity. In the first case we are constructing synthetic metabolic solutions that lower the resource demand (in this case N fertilizer), and in the second case the aim is to directly enhance the photosynthetic yield by minimizing the negative impact of photorespiration. In both cases, cyanobacteria acts as the provider, with the intention to either supply nutrients to crop plants, or directly provide feed and food. Taking a systems metabolic engineering approach, our first goal was to enhance our understanding of a model nitrogen fixing cyanobacteria, by constructing a highly curated two-cell model of *Anabaena sp.* PCC 7120. Optimum solutions for metabolic exchange between heterocysts and vegetative cells were found to exceed the exchange hypothesized from wet-lab experiments. Strains with modified N-metabolism were thereafter used to construct model synthetic consortia and evaluated under varying conditions. The N-metabolism of an *amt* mutant of *Anabaena* 7120 responded differently to other model N organisms (e.g. *Rhodobacter capsulatus*) in terms of NH_3-excretion, but displayed a similar conditional capability to rescue non N-fixing strains as reported previously for *Azotobacter vinelandii*. In order to implement entirely synthetic pathways that allow for enhanced photosynthetic yield, we are developing more robust engineering methods and studying the conditional importance of the three different photorespiration routes in *Synechocystis spp.* PCC 6803.

Development of Metabolite Profiling and Metabolomics Tools for Pseudomonas Taiwanensis

Wordofa, G., Technical University of Denmark

Strain development for the production of fuels and chemicals has traditionally relied on measurements of metabolite products titers to assess the performance of engineered strains. Today, improved metabolomics and targeted metabolite profiling methods make it possible to collect a broader range of measurements of the metabolic state of an organism. Despite of the challenges applying global metabolomics and targeted metabolite profiling methods, it is clear that their application provides significant benefits in both pathway discovery and synthetic biology. However, global metabolomics and metabolite profiling of strains is a challenging task due to the chemical and structural diversity of metabolites from polar to non-polar molecules. As a result of this diversity, methods used for quenching of metabolism and metabolite extraction significantly influence the number and intensity of metabolites detected by universal detection technologies.

In recent years, different sophisticated protocols including quenching with cold solvent mixtures and fast filtration have been developed for the quantification of intracellular metabolites of different strains especially yeast and bacteria. The direct application of these methods and other techniques to *P. taiwanensis* VLB120 strain, which is the only known solvent-tolerant *Pseudomonas* strain, is not well studied. Therefore, I am conducting a comprehensive examination of the metabolite profile of *Pseudomonas* VLB120 strains using different analytical techniques including gas and liquid chromatogram – mass spectrometry.

Auto-Regulation of Ethanol and Post-Squalene Biosynthesis in Engineered Saccharomyces Cerevisiae for Squalene Overproduction

Wang, Y., Tsinghua University
Rasool, A., Beijing Institute of Technology
Li, C., Beijing Institute of Technology

Many metabolic engineering strategies are devised to redirect the metabolic flux in engineered pathway. However, some metabolic pathways are auto-regulated leading to the diversion of precursors flux into cognate pathways. In this study, the effects of squalene overproduction on ethanol and post-squalene biosynthesis in *Saccharomyces cerevisiae* were investigated. Genes related to pre-squalene biosynthesis in *Saccharomyces cerevisiae* include *HMG1* encoding for 3-hydroxy-3-methyl glutaryl (HMG) CoA reductase, *IDI1* encoding for isopentenyl diphosphate isomerase-1, *ERG20* encoding for farnesyl diphosphate synthase and *ERG9* encoding for squalene synthase. These four genes were overexpressed in *Saccharomyces cerevisiae* for squalene overproduction and evaluation of its effects on ethanol synthesis as well as post-squalene biosynthesis pathways. Squalene was produced 10 folds (34 mg/L, without terbinafine) and 35.024 folds (119.081 mg/L, with terbinafine) higher in the so constructed *Saccharomyces cerevisiae* M1EG than the wild type strain. Meanwhile, ethanol production was reduced 2.067 folds (60 mg/L) in the engineered *Saccharomyces cerevisiae* compared with wild type strain. The expression of genes involved in ethanol metabolic pathway were also decreased in different extent when squalene was overproduced with the highest decrease up to 493 folds. Ergosterol is a product in post-squalene pathway. In engineered *S. cerevisiae* M1EG, ergosterol production was also simultaneously enhanced by 1.184 folds (10.692 mg/L) with the overproduction of squalene. The same phenomenon appeared in another engineered *Saccharomyces cerevisiae* FOH-2 overexpressing genes involved in terpenoid backbone biosynthesis pathway for squalene overproduction. Compared with the wild type strain, squalene synthesis was enhanced in the engineered *S. cerevisiae* FOH-2. On the meanwhile, ethanol production was reduced by 3.618 folds with the decreased expression of ethanol synthesis genes. Besides, ergosterol production was also enhanced along with squalene overproduction.

Our results indicate that ethanol production and post-squalene biosynthesis pathway were auto-regulated to lower the metabolic burden of overexpressed pre-squalene biosynthesis pathway in *Saccharomyces cerevisiae*. The study contributes to the research of metabolic flux regulation and is useful for further metabolic pathway control as well as the improvement of products synthesis.

Keywords: squalene, pre-squalene, post-squalene, ethanol production, *Saccharomyces cerevisiae*

Hydrogen Production, Adaptive Evolution and Transcriptomeanalysis of Metabolic Engineered Thermoanaerobacterium sp. strain

Zhu, M., South China University of Technology
Yang, X., South China University of Technology
Li, S., South China University of Technology

Hydrogen is regarded as an attractive future energy carrier for its high energy content andÂ zero CO2 emission. The strictly anaerobic, gram-positive bacterium, *ThermoanaerobacteriumÂ aotearoense* SCUT27 is a promising strain for hydrogen production because it has the capabilityÂ of utilizing glucose and xylose almost simultaneously. However, a metabolic pathway analysisÂ pointed that lactate occupied most of the liquid metabolites and consumed a large amount ofÂ NADH. In order to promote hydrogen yield, the lactate metabolic pathway was blocked to redirectÂ the carbon flux and total cellular NADH distribution. After gene modification, an increase of 2
and 2.5 folds for H2 yield and production rate were achieved, accompanied by 2.4 folds forÂ ethanol yield. The maximum H2 yields using the *Î?ldh* mutant were 2.71, 1.45 and 2.28 mol H2Â mol-1 sugar under glucose, xylose and glucose/xylose mixture tests, respectively. Moreover, weÂ tried to use agro-industrial by-products as substrate to yield hydrogen. After an optimization ofÂ acid hydrolysis by Response Surface Methodology (RSM), sugarcane bagasse was utilized by *T.Â aotearoense* SCUT27*Î?ldh*, to result a best hydrogen yield of 1.86 mol H2/mol total sugar and aÂ hydrogen production rate (HPR) of 0.52 L/L Â· h in a 5-L fermentor.

More work was performed on *T. aotearoense* SCUT27*Î?ldh*. In order to reduce the lag timeÂ when the strain grew in the medium containing high concentration sugar (>80 g/L) and enhanceÂ hydrogen productivity, it was adaptively evolved to yield the mutant strain, SCUT27*Î?ldh* G3#4.Â Compared with SCUT27*Î?ldh*, G3#4 decreased the lag time from 82 h to 12 h (from inoculation toÂ OD600 = 0.5) in the medium with 120 g/L carbon source. In a 5-L fermentor containing 2 LÂ medium, the maximum hydrogen production and HPR were 3.09 L and 0.28 L/LÂ·h. TranscriptomeÂ data showed that the UDP-Glucose was accumulated in SCUT27*Î?ldh* G3#4, which is the
precursor of trehalose, and the latter one acted as a compatible solute to help cell survive fromÂ high permeability environment.

Keywords:
Thermoanaerobacterium aotearoense / Metabolic engineering / Hydrogen production / AdaptiveÂ evolution

References

1. Li, S., Lai, C., F., Cai, Y., H., *et al., Bioresour. Technol.* 2010, 101, 8718-8724
2. Yang, X., Lai, Z., Lai, C., *et al., Biotechnol. Biofuels* 2013, 6:124
3. Lai, Z., Zhu, M., Yang, X., *et al., Biotechnol. Biofuels* 2014, 7:119
4. Yang, X., Zhu, M., Huang, X., *et al., Bioresour. Technol.* 2015, 198, 47-54
5. Tikam C., Lisa S., Paolo G. *International Journal of Food Microbiology* 185

(2014) 140â??157

Microbial Sulfur Metabolism and Cysteine Fermentation

Takumi, K., Ajinomoto Co., Inc
Yamazaki, S., Ajinomoto Co., Inc.
Takei, K.,
Nonaka, G., Ajinomoto Co., Inc.

Cysteine (cystine) is an important amino acid in pharmaceutical, food, and cosmetic industry. However, cysteine is toxic to cells even at low concentrations; thus, its intracellular levels are stringently controlled through several modes of regulation, which hinders the achievement of cysteine fermentative production. These regulatory modes include the feedback inhibition of key metabolic enzymes, degradation, efflux pumps, and the transcriptional regulation of biosynthetic genes. Accordingly, shedding light on how to manipulate these complex regulations are crucial for optimizing metabolic flux toward cysteine over production. Searching and engineering high capacity efflux pump of cysteine is also crucial for both Down-regulating cysteine intracellular level to avoid toxic effects, and secreting it into the fermentation media to facilitate further isolation process. Recently, we have succeeded in establishing the technology for fermentative production of cysteine and launched a commercial basis production. In this presentation we discuss the microbial sulfur metabolism and metabolic engineering of the cysteine-producing strains.

Production of Manoyl Oxide, a Precursor to the Medical Active Compound Forskolin, in Synechocystis PCC 6803

Englund, E., Uppsala University
Andersen-Ranberg, J., University of Copenhagen
Hamberger, B., University of Copenhagen
Lindberg, P.,

Terpenoids, also known as isoprenoids, are a large and diverse family of natural compounds including sterols, carotenoids and hormones, many of which are commercially interesting. Forskolin is a complex C_{20} diterpenoid that is produced in the root cork cells of the plant *Coleus forskohlii*, and has been shown to have pharmacological activities. Because of its complex molecular structure, chemical synthesis of forskolin is not economically attractive, and extraction and purification from *C. forskohlii* plants is laborious and resource intensive.

We have engineered the unicellular cyanobacterium *Synechocystis* PCC 6803 to produce the forskolin precursor 13R-manoyl oxide (13R-MO) by expressing the two *C. forskohlii* enzymes CfTPS2 and CfTPS3. These two nonnative genes were integrated into the genome at three different locations with two different promoters, with the top producing strains yielding production titers of 0.25 mg/L/g DCW 13R-MO. To further increase the yield, 13R-MO producing strains were engineered by introduction of selected enzymes from the *C. forskohlii* non-mevalonate pathway (MEP), which resulted in improving the titer to 0.46 mg/L/g DCW in the highest producing strain. By cultivating the strains at different light intensities, large differences in product accumulation were observed, attributed to changes in promoter activities. Finally, we quantified carotenoid and chlorophyll content in our different strains since those pigments are made or partially made from the same terpenoid precursor as 13R-MO and the results suggests that heterologous expression of 13R-MO influences pigment composition. This work is of interest for future engineering of terpenoid products in cyanobacteria and provides insight into promoter selection.

Overexpressing Photosynthetic Carbon Flux Control Enzymes in Synechocystis PCC 6803

Liang, F., Uppsala University
Lindblad, P., Uppsala University

It is environmentally sustainable to convert inorganic carbon directly into desired compounds with photoautotrophic organisms, like higher plants, algae, and cyanobacteria. The Calvin cycle is the primary carbon fixation pathway in photoautotrophic organisms. In higher plants, Ribulose-1,5-biphosphate carboxylation (RuBP) carboxylation and RuBP regeneration together control the carbon flux in this process based on experimental data and dynamic mimic data. Rubilose-1,5-carboxylase/oxygenase (RuBisCO) is the enzyme catalyzing RuBP carboxylation. For RuBP regeneration, Seduheptoluse-1,7-biphosphatase (SBPase), aldolase (FBA) and transketolase (TK) are the three main enzymes controlling photosynthetic rate. Data from cyanobacteria is still missing. We succeeded overexpressing native RuBisCO, SBPase (fructose-1,6-/sedoheptoluse-1,7-biphosphatase, FBP/SBPase), FBA and TK separately in *Synechocystis* PCC 6803. The engineering strains showed different phenotypes compared to control strain under 100 Î¼mol photons m^{-2} s^{-1} light conditions in shaking flasks. Growth rate, maximal *in vivo* oxygen evolution rate and dry weight increased in RuBisCO, FBP/SBPase, FBA overexpressed strains. TK overexpressed strain showed chlorotic phenotype and increased dry weight. These results indicate these four enzymes are key enzymes in Calvin cycle in *Synechocystis* PCC 6803 and subsequently are potential targets to increase photosynthetic rates and yields.

Metabolic and Cell Surface Engineering to Produce Isobutanol from Lignocellulosic Biomass in Yeast Saccharomyces cerevisiae

Ishii, J., Kobe University
Matsuda, F., Osaka University
Ishikado, S., Kobe University
Ida, K.,
Kondo, A., Graduate School of Engineering, Kobe University

The production of higher alcohols including C3-C5 alcohols in engineered bacteria has received significant attention, because they can be used as fuels, fuel additives, or commodity chemicals. The budding yeast, *Saccharomyces cerevisiae*, has considerable potential as a producer of higher alcohols because of its capacity to naturally fabricate fusel alcohols, in addition to its robustness and tolerance to low pH. However, because its natural productivity is not significant, we considered a strategy of genetic engineering to increase production of higher alcohols. For example, the branched-chain higher alcohol isobutanol, which is involved in valine biosynthesis, offers higher octane values than their straight-chain counterparts with equivalent carbon numbers and can be used with current infrastructure in addition to possessing almost the same capabilities (energy density, hygroscopicity and octane number) as gasoline. However, its natural productivity in yeast is not significant. In this study, we show the engineering strategies of metabolic pathway and cell surface display to improve the isobutanol production in *S. cerevisiae*. To relieve the cofactor imbalance, we tested activation of the transhydrogenase-like shunt comprising pyruvate carboxylase (Pyc2), malate dehydrogenase (Mdh2) and malic enzyme (Mae1), the bypass for converting NADH to NADPH. As another approach, we tested the multiple gene knockout, which is related to the pyruvate metabolism, to improve the isobutanol production. Finally, we displayed the cellulose enzymes (endoglucanase and Î²-glucosidase) on the yeast cell surface and successfully produced isobutanol from cellobiose and Î²-glucan.

This work was supported by Special Coordination Funds for Promoting Science and Technology, Creation of Innovation Centers for Advanced Interdisciplinary Research Areas (Innovative Bioproduction Kobe; iBioK), MEXT, Japan, and in part by Industrial Technology Research Grant Program in 2011 from New Energy and Industrial Technology Development Organization (NEDO) of Japan.

AUTHOR INDEX

AUTHOR INDEX

AUTHOR INDEX

AUTHOR INDEX

AUTHOR INDEX

AUTHOR INDEX

AUTHOR INDEX

AUTHOR INDEX

AUTHOR INDEX

AUTHOR INDEX

AUTHOR INDEX